INTRODUCTION TO SEMICONDUCTOR MATERIALS AND DEVICES

INTRODUCTION TO SEMICONDUCTOR MATERIALS AND DEVICES

M. S. TYAGI
Department of Electrical Engineering
Indian Institute of Technology, Kanpur
India

JOHN WILEY & SONS
New York   Chichester   Brisbane   Toronto   Singapore

Copyright © 1991, by John Wiley & Sons, Inc.

All rights reserved. Published simultaneously in Canada.

Reproduction or translation of any part of
this work beyond that permitted by Sections
107 and 108 of the 1976 United States Copyright
Act without the permission of the copyright
owner is unlawful. Requests for permission
or further information should be addressed to
the Permissions Department, John Wiley & Sons.

Library of Congress Cataloging in Publication Data:

Tyagi, M. S. (Man S.), 1934-
   Introduction to semiconductor materials and devices/M. S. Tyagi.
   Includes bibliographical references.
   ISBN 0-471-60560-3
   1. Semiconductors.   I. Title.
TK7871.85.T93   1991
621.381'52—dc20

89-24852
CIP

Printed in the United States of America

10 9 8 7 6 5 4 3 2 1

To
PROFESSOR HEINZ BENEKING
who introduced me to semiconductors

PREFACE

This book provides an introduction to the physics of semiconductor materials and devices. It is intended primarily as a textbook in a two-semester undergraduate course for students in electrical engineering, applied physics, and materials science. However, parts of the book can also be used for a first-level graduate course on semiconductor devices. In addition, the book can serve as a reference book for scientists and engineers who require a thorough and up-to-date understanding of device physics and the basics of semiconductor technology.

The text is organized into five parts. The first two chapters provide basic material on atomic physics, crystal structure, and energy bands which is needed for understanding semiconductors. These chapters may be skipped over by the student who has already taken courses in atomic and solid state physics. The second part, Chapters 3 through 5, describes the basic properties of semiconductors. Chapter 3 develops the theory of semiconductors by taking the elemental semiconductor silicon as an example. However, a thorough discussion of III-V and II-VI semiconductors is also provided. Chapter 4 presents current transport from an elementary viewpoint, and Chapter 5 is devoted to nonequilibrium carrier transport in semiconductors.

The third part, Chapters 6 through 10, considers the physics of junctions and interfaces. We start with the p-n junction which is the basic building block for a large number of semiconductor devices. The static characteristic of the p-n junction diode is covered in detail in Chapter 7, and Chapter 8 focuses on electrical breakdown in a reverse-biased diode. Chapter 9 is of special importance, for it covers the various aspects of the dynamic behavior of junction diodes. Chapter 10 is concerned with majority carrier devices such as tunnel, Schottky barrier, and heterojunction diodes.

The fourth part, Chapters 11 through 18, provides a fairly comprehensive and up-to-date discussion of semiconductor devices. Chapter 11 is devoted to two terminal devices that are used at microwave frequencies. Optoelectronic devices—solar cells, photodetectors, light-emitting diodes, and semiconductor lasers—form the subject matter of Chapter 12. The next two chapters discuss the various aspects of bipolar junction transistors. Chapters 15 and 16 deal with junction and surface field-effect devices, respectively. Chapter 17 presents a unified view of circuit models for bipolar and field-effect transistors. Power semiconductor devices, namely, power rectifiers and thyristors, are described in Chapter 18.

The last part of the book comprises the remaining two chapters. Chapter 19 deals with processing technology from crystal growth to the lithographic process of pattern transfer along with a brief discussion of bipolar and MOS integrated devices. Chapter 20 provides a brief account of measurements of basic semiconductor parameters.

In each chapter mathematical equations are used as an aid to illustrate the concepts developed, but they have been kept to a minimum. All the chapters contain simple illustrations as an aid to the reader. Numerical examples have been worked out to suggest the orders of magnitude of the quantities involved. A large number of problems are presented at the end of each chapter; they are an integral part of the development of the topics. Some of the problems are simple, whereas others require an in-depth understanding of the subject matter.* References have been given where necessary, and a list of additional reading material is suggested at the end of each chapter.

The entire book can be covered in two semesters. Chapters 1 through 10 and Chapter 20 form a full one-semester course on the physics of semiconductors, junctions, and interfaces. The remaining chapters on various devices may be taken up in the second semester. The text is organized according to a modular scheme. Except for Chapters 3 through 7, most of the others are fairly independent and self-contained. Thus, the instructor will have considerable freedom to orient the course to the requirements of the ongoing projects in his or her department.

I would like to express my deep sense of gratitude to the reviewers, Professors R. S. Muller, University of California, Berkeley; E. D. Smith, the University of Toledo; Cary Y. Yang, Santa Clara University; and Major Edward S. Kolesar, Jr., Air Force Institute of Technology, Wright Patterson AFB, OH. Their numerous suggestions and critical remarks have contributed immensely to the manuscript. I am also grateful to Major Kolesar for going through the manuscript very carefully and helping me to improve it in many ways.

I am indebted to several of my colleagues and to both past and present students who have helped me in ways too numerous to mention. I am particularly grateful to Professors S. C. Agarwal, K. K. Sharma, R. K. Gupta, and Jitendra Kumar; to Drs. V. Singh, and A. P. Shukla for many helpful discussions; and to Dr. S. C. Sen for encouragement. I would also like to thank Dr. S. N. Gupta, Central Electronics Engineering Research Institute, Pilani, Dr. A. K. Gupta, REC, Kurukshetra, and Dr. K. M. K. Srivatsa and my students M. R. N. Murthy, V. Raghavan, D. S. Rao, and M. S. K. Reddy for reading parts of the manuscript and making suggestions for improvement.

I wish to thank R. Pandey for his untiring efforts in typing the various versions of the manuscript and to J. N. Tripathy, Quasim Hussain, S. P. Singh, and Shiuli Gupta for their assistance in preparing the artwork. My sincere appreciation is extended to A. S. Mandal who spent long hours with me in checking the typed manuscript and preparing the problem solutions. My thanks are also due to Professor S. Sampath, ex-director, Indian Institute of Technology, Kanpur, for his continued interest and encouragement and to the Quality Improvement Program of the Institute for providing financial assistance to prepare the final manuscript.

I also extend my appreciation to Christina Kamra of John Wiley and Sons for her keen interest in the project. Finally, I am grateful to my family members for

---

*A solutions manual is available to teachers/instructors on request from the publisher.

their support, especially to my son Paritosh and my daughter Prachi who were left alone while their father was busy with the writing of this book. I am not expressing my thankfulness to my wife Chandra because, according to Hindu tradition, wife and husband are a single entity.

I.I.T. Kanpur
November 1988

Man S. Tyagi

CONTENTS

PREFACE vii

PART 1 BASIC PHYSICS 1

CHAPTER 1 Review of Atomic Structure and Statistical Mechanics 3

Introduction 3
1.1 Early Ideas on Atomic Structure 3
1.2 Wave Particle Duality 6
1.3 Quantum Mechanics 6
1.4 The Schrödinger Wave Equation 9
1.5 Some Examples of Solutions of the Schrödinger Wave Equation 11
1.6 The Electronic Structure of Atoms and the Periodic Table of Elements 16
1.7 Statistical Mechanics 19

CHAPTER 2 Crystalline Solids and Energy Bands 27

Introduction 27
2.1 The Bonding of Atoms 27
2.2 The Crystalline State 29
2.3 Crystal Defects 36
2.4 Lattice Vibrations and Phonons 39
2.5 Energy Bands in Solids 44
2.6 Metals and Insulators 54

PART 2 FUNDAMENTALS OF SEMICONDUCTORS — 59

CHAPTER 3 Semiconductor Materials and Their Properties — 61

Introduction — 61

3.1 Semiconducting Materials — 61

3.2 Elemental and Compound Semiconductors — 62

3.3 The Valence Bond Model of the Semiconductor — 66

3.4 The Energy Band Model — 73

3.5 Equilibrium Concentrations of Electrons and Holes Inside the Energy Bands — 78

3.6 The Fermi Level and Energy Distribution of Carriers Inside the Bands — 87

3.7 The Temperature Dependence of Carrier Concentrations in an Extrinsic Semiconductor — 90

3.8 Heavily Doped Semiconductors — 92

CHAPTER 4 Carrier Transport in Semiconductors — 102

Introduction — 102

4.1 The Drift of Carriers in an Electric Field — 102

4.2 Variation of Mobility with Temperature and Doping Level — 108

4.3 Conductivity — 110

4.4 Impurity Band Conduction — 114

4.5 The Hall Effect — 115

4.6 Nonlinear Conductivity — 118

4.7 Carrier Flow by Diffusion — 121

4.8 Einstein Relations — 122

4.9 Constancy of the Fermi Level Across a Junction — 124

CHAPTER 5 Excess Carriers in Semiconductors — 129

Introduction — 129

5.1 Injection of Excess Carriers — 130

5.2 Recombination of Excess Carriers — 132

5.3 Mechanisms of Recombination Processes — 134

5.4 Origin of Recombination Centers — 144

5.5 Excess Carriers and Quasi-Fermi Levels — 147

5.6 Basic Equations for Semiconductor Device Operations — 148

5.7 Solution of Carrier Transport Equations—An Illustration — 152

PART 3 JUNCTIONS AND INTERFACES   159

CHAPTER 6 p-n Junctions   161
Introduction   161
6.1 Description of p-n Junction Action   161
6.2 The Abrupt Junction   166
6.3 Example of an Abrupt p-n Junction   174
6.4 The Linearly Graded Junction   177
6.5 The Diffused Junction   182

CHAPTER 7 Static I-V Characteristics of p-n Junction Diodes   187
Introduction   187
7.1 The Ideal Diode Model   187
7.2 Real Diodes   196
7.3 Temperature Dependence of the I-V Characteristic   201
7.4 High-Level Injection Effects   203
7.5 Example of a p-n Junction Diode   205

CHAPTER 8 Electrical Breakdown in p-n Junctions   213
Introduction   213
8.1 Phenomenological Description of Breakdown Mechanisms   214
8.2 Theoretical Treatment of Internal Field Emission   216
8.3 Zener Breakdown in p-n Junctions   221
8.4 Secondary Multiplication in Semiconductors   224
8.5 Avalanche Breakdown in p-n Junctions   225
8.6 Effect of Junction Curvature and Crystal Imperfections on the Breakdown Voltage   231
8.7 Distinction Between the Zener and Avalanche Breakdown   234
8.8 Applications of Breakdown Diodes   234

CHAPTER 9 Dynamic Behavior of p-n Junction Diodes   238
Introduction   238
9.1 Small-Signal ac Impedance of a Junction Diode   239
9.2 The Charge Control Equation of a Junction Diode   246
9.3 Switching Transients in Junction Diodes   248
9.4 High-Speed Switching Diodes   258

CHAPTER 10 Majority Carrier Diodes — 263

Introduction — 263
10.1 The Tunnel Diode — 263
10.2 The Backward Diode — 270
10.3 The Schottky Barrier Diode — 270
10.4 Ohmic Contacts — 289
10.5 Heterojunctions — 291

PART 4 SEMICONDUCTOR DEVICES — 299

CHAPTER 11 Microwave Diodes — 301

Introduction — 301
11.1 The Varactor Diode — 301
11.2 The p-i-n Diode — 306
11.3 The IMPATT Diode — 311
11.4 The TRAPATT Diode — 320
11.5 The BARITT Diode — 323
11.6 Transferred-Electron Devices — 327

CHAPTER 12 Optoelectronic Devices — 337

Introduction — 337
12.1 The Solar Cell — 337
12.2 Photodetectors — 351
12.3 Light Emitting Diodes — 357
12.4 Semiconductor Lasers — 363

CHAPTER 13 Bipolar Junction Transistors I: Fundamentals — 378

Introduction — 378
13.1 Principle of Operation — 378
13.2 Fabrication Methods and Doping Profiles — 383
13.3 Analysis of the Ideal Diffusion Transistor — 385
13.4 Real Transistors — 393
13.5 Static I-V Characteristics in the Normal Active Region — 397
13.6 Charge Control Equations — 405

CHAPTER 14 Bipolar Junction Transistors II: Devices — 412

Introduction — 412
14.1 The Diffusion Transistor at High Frequencies — 412
14.2 The Drift Transistor — 414
14.3 High-Frequency Performance — 420
14.4 High-Frequency and Microwave Transistors — 425
14.5 Power Transistors — 427
14.6 The Switching Transistor — 435

CHAPTER 15 Junction and Metal-Semiconductor Field-Effect Transistors — 443

Introduction — 443
15.1 Principle of Operation — 444
15.2 Static I-V Characteristics of the Idealized Model — 448
15.3 JFET Structures — 451
15.4 Basic Types of MESFETs — 455
15.5 Models for I-V Characteristics of Short-Channel MESFETs — 456
15.6 High-Frequency Performance — 461
15.7 MESFET Structures — 464

CHAPTER 16 MOS Transistors and Charge-Coupled Devices — 469

Introduction — 469
16.1 Semiconductor Surfaces — 470
16.2 C-V Characteristics of the MOS Capacitor — 474
16.3 The Si-SiO$_2$ System — 483
16.4 Basic Structures and the Operating Principle of MOSFET — 484
16.5 Current-Voltage Characteristics — 487
16.6 Transistor Ratings and Frequency Limitations — 497
16.7 Short-Channel Effects — 498
16.8 MOSFET Structures — 503
16.9 Charge-Coupled Devices — 505

CHAPTER 17 Circuit Models for Transistors — 516

Introduction — 516
17.1 Two-Port Network Description of a Bipolar Transistor — 517

17.2	Models for Bipolar Transistors	518
17.3	Circuit Models for JFETs and MESFETs	528
17.4	Models for MOS Transistors	533

CHAPTER 18 Power Rectifiers and Thyristors — 538

	Introduction	538
18.1	Power Rectifiers	538
18.2	Thyristors	543
18.3	Some Special Thyristor Structures	554
18.4	Bidirectional Thyristors	557
18.5	Field-Controlled Thyristor	559

PART 5 SEMICONDUCTOR TECHNOLOGY AND MEASUREMENTS — 563

CHAPTER 19 Technology of Semiconductor Devices and Integrated Circuits — 565

	Introduction	565
19.1	Crystal Growth and Wafer Preparation	566
19.2	Methods of p-n Junction Formation	570
19.3	Growth and Deposition of Dielectric Layers	583
19.4	The Planar Technology	588
19.5	Masking and Lithography	590
19.6	Pattern Definition	592
19.7	Metal Deposition Techniques	593
19.8	General Remarks on Integrated Devices	595
19.9	Bipolar Integration	596
19.10	MOS Integration	605

CHAPTER 20 Semiconductor Measurements — 613

	Introduction	613
20.1	Conductivity Type	613
20.2	Resistivity	614
20.3	Hall Effect Measurements	618
20.4	Drift Mobility	622
20.5	Minority Carrier Lifetime	627
20.6	Diffusion Length	632

APPENDIXES

A	List of Symbols	637
B	Physical Constants	643
C	International System of Units	644
D	Unit Prefixes	645
E	Properties of Some Important Semiconductors at 300 K	646
F	Important Properties of SiO_2 and Si_3N_4 at 300 K	647
G	Electrical Noise	648
	Answers to Selected Problems	649
	Index	657

PART 1
BASIC PHYSICS

1
Review of Atomic Structure and Statistical Mechanics

INTRODUCTION

This book focuses on semiconductor devices. These devices operate by controlling the flow of electrons in a semiconductor which is a crystalline solid. The physical, chemical, and electrical properties of a solid depend on the arrangement of atoms in the substance. This arrangement, in turn, is determined by the arrangement of electrons in an atom of the substance. Hence, it is necessary to understand the electronic structure of the atom. The transport of charge through a conductor or a semiconductor also depends on the properties of the electron. The main thing we need to know is how the conduction electrons in a solid are distributed in energy. This information is provided by the distribution functions of statistical mechanics.

In this introductory chapter, we present a brief account of the electronic structure of the atom and statistical mechanics. Starting from the early ideas of atomic theory, we review the quantum mechanical concepts necessary for understanding atomic structure and electronic conduction in solids. The chapter closes with a brief description of statistical mechanics.

1.1 EARLY IDEAS ON ATOMIC STRUCTURE

The concept of the atom as a building block of matter dates back to the early Greek philosophers in 400 B.C., but detailed knowledge of atomic structure is quite new. The first experimental evidence of the existence of atoms came from Dalton's law of combining weights in 1811. The electrical nature of an atom was established later in the nineteenth century when Michael Faraday (1791–1867) discovered the laws of electrolysis.

Toward the end of the nineteenth century, in 1897, the electron was discovered by J. J. Thomson (1856–1940) and was shown to be a universal constituent of matter. Subsequently, it was demonstrated that an atom is made up of a positively charged nucleus surrounded by a sea of electrons that have a negative charge and occupy most of the volume of the atom. Niels Bohr (1885–1962) attempted to explain the stability of this so-called nuclear atom. He assumed that the electrons around the nucleus revolve in circular orbits, and he presented the following three postulates:

1. An electron inside the atom exists in only one of a set of discrete energy levels.
2. The only energy levels in which the electrons can exist are those for which the angular momentum of the electron is $nh/2\pi$, where n is an integer. Each of these levels is known as a stationary state.
3. As long as the electron remains in one of its allowed stationary states, it can neither gain nor lose energy. Absorption or emission occurs only when it makes the transition from one energy level to another. If E_1 represents the energy of the initial state and E_2 that of the final state, then

$$E_2 - E_1 = nh\nu \tag{1.1}$$

where h is Planck's constant and ν is frequency.

The lightest atom is that of hydrogen which contains only one electron. The allowed values of electron energy for this atom can be easily deduced from the above basic assumptions. Let r_n denote the radius of the nth orbit. Equating the centrifugal force with the force of electrostatic attraction between the electron and the proton gives

$$m_o \omega^2 r_n = \frac{q^2}{4\pi \varepsilon_o r_n^2} \tag{1.2}$$

where m_o is the mass of the electron, ω is its angular frequency, q is the fundamental unit of charge, and ε_o is the permittivity of free space. From postulate 2 the condition for quantization of angular momentum can be written as

$$m_o \omega r_n^2 = \frac{nh}{2\pi} \tag{1.3}$$

Eliminating ω between Eqs. (1.2) and (1.3), we obtain

$$r_n = \varepsilon_o \frac{n^2 h^2}{q^2 \pi m_o} \tag{1.4}$$

The total energy E_n of the electron in the nth orbit is given by

$$E_n = \text{potential energy} + \text{kinetic energy}$$
$$= \frac{-q^2}{4\pi \varepsilon_o r_n} + \frac{m_o \omega^2 r_n^2}{2} = \frac{-q^2}{8\pi \varepsilon_o r_n}$$

where we have made use of Eq. (1.2). Substituting the value of r_n from Eq. (1.4), we can write E_n as

$$E_n = \frac{-q^4 m_o}{8\varepsilon_o^2 n^2 h^2} \tag{1.5}$$

EARLY IDEAS ON ATOMIC STRUCTURE

When numerical values of the various quantities (see Appendix B) are substituted in Eqs. (1.4) and (1.5) we obtain

$$r_n = 0.53 n^2 \text{ Å}$$

and

$$E_n = -\frac{13.6}{n^2} \text{ eV} \tag{1.6}$$

Figure 1.1 shows electron energy levels for various values of n. All the observed hydrogen spectral lines are obtained by the difference in the energy of these levels, and the possible transitions are shown by arrows. Equation (1.6) shows that as n is increased the energy levels become crowded together and ultimately assume an almost continuous distribution. The spectral lines appear in several groups labeled the Lyman, Balmer, and Paschen series after their early investigators.

The state for $n = 1$ has the energy -13.6 eV and is called the ground state of the hydrogen atom. The negative sign means that it is the binding energy. To remove the electron from the ground state to infinity requires an energy of 13.6 eV, which is called the *ionization energy* of the hydrogen atom. The radius of the ground state orbit is 0.53 Å, so that the diameter of the spherical model of the hydrogen atom is approximately 1×10^{-8} cm. Both the ionization energy and the size of the atom are in good agreement with the experimental findings.

Attempts to apply Bohr's theory to helium and to more complex atoms failed because the assumption of circular orbits did not yield the correct values of energy in these atoms. In addition, the theory could not explain why some transitions between electronic states in complex atoms are forbidden. As will be shown in the following pages, these difficulties are overcome when quantum mechanics is used in making the calculations.

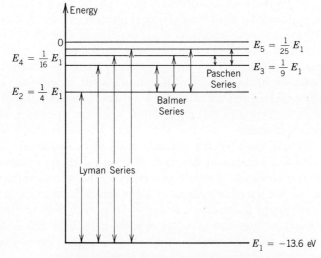

FIGURE 1.1 Energy levels of the hydrogen atom.

1.2 WAVE PARTICLE DUALITY

It is a known fact that electromagnetic radiation shows the dual nature of a wave and a particle. The wave theory of light is quite satisfactory for long waves used in radio communication. However, inferences drawn from the interaction of matter with short wavelength radiation in visible and ultraviolet regions suggest that light consists of a stream of particles, each having an energy hν. This particle is known as a *photon*. A photon always travels with the velocity of light and will not exist at velocities lower than that of light. For this reason a photon has no rest mass. Like photons, electrons exhibit a dual nature. Besides the electron, heavier particles such as protons, neutrons, neutral atoms, or molecules also show the wavelike behavior.

In 1924, Louis de Broglie (1892–) suggested that a particle of momentum p shows wavelike behavior with the wavelength λ given by

$$\lambda = \frac{h}{p} \tag{1.7}$$

Since $h = 6.626 \times 10^{-34}$ J-Sec, it is clear that the wavelike behavior is exhibited only by particles like electrons and protons that have a small value of p because of their small mass. Heavier objects of macroscopic dimensions have a vanishingly small de Broglie wavelength, and their motion can be described by the laws of classical mechanics. A few disturbing questions now arise. Is light a wave phenomenon or a bundle of particles like photons? Is an electron a particle or a wave, or is it both? These are perplexing questions and will remain so as long as we adhere to the classical explanation of physical phenomena. Efforts to avoid these difficulties have led to a unified theory in which the concepts of wave and particle have lost their fundamental distinction. This new theory is known as *quantum* or *wave mechanics*.

1.3 QUANTUM MECHANICS

1.3.1 Introductory Ideas

There are three main differences between the classical and the quantum mechanical approach to describing physical phenomena. First, classical mechanics deals with macroscopic objects, and our experience with these objects can be described in infinite detail. For example, these objects can move with any speed, have any energy, and can occupy any position in space. In contrast, quantum mechanics abandons this hypothesis of infinite detailed experience. The new theory selects only a certain class of all possible observations and provides mathematical rules for the selection of this class. These rules are known as the postulates of quantum mechanics.

Second, the elements of observation and calculation are identical in classical mechanics but not in quantum mechanics. Every observation in classical mechanics provides a number, and every analysis deals with numbers. For example, we can measure the volume of a gas directly, and we can also calculate it from knowledge of temperature and of gas pressure. Both the measurement and calculations give numbers in terms of some basic unit that is directly measurable. In contrast, quantum mechanics allows greater freedom of abstract notations, which may not have immediate physical meaning.

The last point of difference between the two theories has to do with how sharply each can predict data. In classical mechanics, data can be determined as precisely as one chooses, and no restrictions are placed on the accuracy of either the measurement or the calculations. In quantum mechanics, the sharp prediction of observation is prevented by Heisenberg's uncertainty principle which states that for any pair of variables that are canonically conjugate to each other, both variables cannot be measured simultaneously to any degree of precision. Two pairs of canonically conjugate variables of interest to us are the momentum and position and energy E and time t. If Δx and Δp represent the uncertainties in position x and momentum p of a particle, respectively, then the uncertainty principle states that

$$\Delta x \cdot \Delta p \simeq h/2\pi \tag{1.8}$$

Thus, if we want to locate the position of the particle precisely by making $\Delta x = 0$, then nothing will be known about its momentum because Δp becomes infinite. Similarly, if ΔE and Δt are the uncertainties in the measurement of E and t, respectively, then

$$\Delta E \cdot \Delta t \simeq h/2\pi \tag{1.9}$$

Note that the uncertainty in the simultaneous measurement of a pair of canonically conjugate variables in Eqs. (1.8) and (1.9) does not arise because of errors associated with the measuring instruments. This uncertainty is inherent in the structure of quantum mechanics and is a fundamental property of matter.

1.3.2 Basic Postulates

The basic principles of quantum mechanics were developed during the late 1920's based on two different but related approaches. Erwin Schrödinger (1887–1961) developed a scheme of mechanics known as *wave mechanics*. This scheme is based on the physical ideas of Planck's quantum theory and de Broglie's ideas of the wave nature of matter. Werner Heisenberg (1901–1976) formulated an alternative approach in terms of matrix algebra called *matrix mechanics*. Because the Schrödinger wave mechanics is simpler, we will confine our discussion to it.

We will first introduce the notion of state of a body. Any object of physical inquiry is termed a *physical system,* and a collection of physical systems is called an *ensemble*. Properties like position, momentum, and energy associated with a system are called the *dynamical variables* or the *observables*. In classical mechanics the state of the system is defined in terms of these dynamical variables. For example, the state of a system with a number of particles at any time is defined by designating the position and momentum coordinates of all the particles at that time. In quantum mechanics, the state of the system is defined by a state function Ψ that contains all the information we can obtain about the system. Quantum mechanics enables us to determine Ψ functions for different situations and also provides us with rules for extracting useful information from a given Ψ. The most straightforward way to approach quantum mechanics is to consider the rules defined in several postulates which are assumed to be true. In the statement of postulates, we will consider only one coordinate x and the time t. Other coordinates can be added when needed. The following choice of postulates is in no way unique, and a different set of postulates can be used [1].

Postulate I: There exists a state function $\Psi(x, t)$ which contains all the measurable information about each particle of a physical system.

In order to construct a bridge between the states and the measurement, we need to develop the notion of an operator. Any mathematical operation (i.e., addition, multiplication, differentiation) can be represented by a symbol, which is called an *operator*. For example, $\partial/\partial x$ is called a *differential operator*.

Postulate II: Every dynamical variable has a corresponding operator. This operator is used on the state function to obtain measurable information about the system.

The operators corresponding to the various dynamical variables are as follows:

Dynamical Variable	Quantum Mechanical Operator	
Position x	$\hat{x} \equiv x$	
Momentum p_x	$\hat{p}_x \equiv \dfrac{\hbar}{j} \dfrac{\partial}{\partial x}$	(1.10)
Total energy E	$\hat{E} \equiv -\dfrac{\hbar}{j} \dfrac{\partial}{\partial t}$	
Potential energy $V(x)$	$\hat{V} \equiv V(x)$	

Here $\hbar = h/2\pi$, and the "hat" used on the quantity signifies the operator corresponding to that quantity. If an operator $\hat{y}$ acting on Ψ produces a linear homogeneous equation of the form

$$\hat{y}\Psi = y\Psi \tag{1.11}$$

where y is a real number, then y is called the eigen value of the operator and a function that satisfies Eq. (1.11) is called an *eigen function*. The only possible values which a measurement of dynamical variable can yield are the eigen values given by Eq. (1.11).

Postulate III: (a) The state function $\Psi(x, t)$ and its space derivative $\partial \Psi / \partial x$ must be continuous, finite, and single-valued for all values of x. (b) The function Ψ must be normalized.

Condition (a) contains a requirement that all physical systems should meet. For example, no state function describing a physical system in the real world can be infinite.

The normalization conditions (b) requires that

$$\int_{-\infty}^{\infty} \Psi \Psi^* dx = 1 \tag{1.12}$$

where Ψ^* is the complex conjugate of Ψ. In a more general three-dimensional case, Eq. (1.12) can be written as

$$\iiint \Psi \Psi^* d\tau = 1 \tag{1.13}$$

where $d\tau$ represents an element of volume about the point where Ψ and Ψ^* are determined. It is obvious that $\Psi \Psi^*$ is a positive real number equal to $|\Psi|^2$.

Postulate IV: The average value $\langle y \rangle$ of any dynamical variable y, corresponding to state function Ψ, is given by

$$\langle y \rangle = \int_{-\infty}^{\infty} \Psi^* \hat{y} \Psi \, dx \qquad (1.14)$$

where Ψ is assumed to be normalized. This postulate forms the necessary link between states and the experience because Eq. (1.14) enables us to calculate observable quantities that we can compare with measurements. Note that $\langle y \rangle$ in Eq. (1.14) is the expected value of many observations; for this reason it is called the *expectation value*.

Postulate V: The operator for any measurement, for which an operator is not postulated, is determined by expressing the measured quantities in terms of the four basic variables of Eq. (1.10) and by substituting the operators for these variables.

The state functions describing a particle (or a system of particles) are generally in the form of a wave. Therefore, they are also called *wave functions*, and the field of quantum mechanics is also known as *wave mechanics*. With regard to the physical significance of Ψ, assume that Ψ applies to a single particle such as an electron. Then $\Psi \Psi^* d\tau$ is interpreted as the statistical probability that the particle is found in a volume element $d\tau$ at any instant of time. Thus, $\Psi \Psi^*$ represents the probability density, and Eq. (1.13) states that the probability of finding the particle within "all space" is unity.

1.4 THE SCHRÖDINGER WAVE EQUATION

Let us consider a particle with potential energy $V(x, y, z)$, momentum **p** and mass m. The total energy E (also known as classical Hamiltonian) of the particle can be written as

$$E = \frac{p^2}{2m} + V(x, y, z) \qquad (1.15)$$

where $p^2 = p_x^2 + p_y^2 + p_z^2$. Using the operator for p_x from Eq. (1.10), we find that the operator corresponding to p_x^2 is $-\hbar^2(\partial^2/\partial x^2)$ and that corresponding to $p^2/2m$ is $(-\hbar^2/2m)\nabla^2$. Replacing the dynamical variables in Eq. (1.15) by their corresponding operators and operating them on Ψ lead to the following differential equation:

$$-\frac{\hbar^2}{2m} \nabla^2 \Psi + V(x, y, z) \Psi = -\frac{\hbar}{j} \frac{\partial \Psi}{\partial t} \qquad (1.16)$$

where

$$\nabla^2 = \left(\frac{\partial^2}{\partial x^2} + \frac{\partial^2}{\partial y^2} + \frac{\partial^2}{\partial z^2} \right)$$

Equation (1.16) is known as the *Schrödinger wave equation*. This equation is analogous to the wave equation of classical physics and represents the behavior shown by a material particle (see Prob. 1.9). The potential energy term in Eq. (1.16) forms the necessary link between the quantum mechanics and the real world. The differences between the state functions for different situations lie in the variations of the environment of the particle. These differences can be expressed conveniently in terms of potential energy whose gradient at various points yields the forces acting on the particle. Hence, the potential energy contains all environmental factors that could influence the state of a particle.

We now consider conservative systems in which the total energy of the system remains constant. This is the case for an isolated system that has no interaction with the ambient. Typical examples are the motion of a particle in a static field of force and the motion of an electron in a hydrogen atom. In systems of this type, the potential energy V may be regarded as independent of time, and using the method of separation of variables we can express Ψ as

$$\Psi(x, y, z, t) = \psi(x, y, z)\phi(t) \tag{1.17}$$

Substituting for $\nabla^2 \Psi$ from Eq. (1.17) into Eq. (1.16), we obtain

$$-\frac{\hbar^2}{2m} \frac{\nabla^2 \psi}{\psi} + V(x, y, z) = -\frac{\hbar}{j} \frac{1}{\phi}\left(\frac{d\phi}{dt}\right) \tag{1.18}$$

Here the right-hand side is a function of time alone, and the left-hand side is a function of space coordinates alone. Thus, each side must be equal to some constant, say, E. This results in two simple differential equations

$$\nabla^2 \psi + \frac{2m}{\hbar^2}(E - V)\psi = 0 \tag{1.19}$$

and

$$-\frac{\hbar}{j\phi}\left(\frac{d\phi}{dt}\right) = E \tag{1.20}$$

where E is the total energy of the particle. Equation (1.19) is the time-independent Schrödinger equation since ψ is the time-independent part of the wave function. Upon integration, Eq. (1.20) yields

$$\phi(t) = A \exp\left(-\frac{jEt}{\hbar}\right) \tag{1.21}$$

and the wave equation can be written as

$$\Psi(x, y, z, t) = A\psi(x, y, z) \exp\left(-\frac{jEt}{\hbar}\right) \tag{1.22}$$

where A is a constant of integration. It is often more convenient to solve the time-independent equation first, and the time dependence, when needed, can be expressed by using Eq. (1.22).

For exponential time dependence as given by Eq. (1.22), we have

$$\Psi\Psi^* = \psi\psi^* \tag{1.23}$$

This means that the probability density is time independent. Such wave functions are said to be stationary, and the states represented by these functions are called *stationary states*. In many situations the time-dependent equation Eq. (1.20) has a number of solutions, each with a discrete value of E corresponding to different states of the system. In such cases the general solution is obtained by the linear superposition of these functions, and the wave function is given by

$$\Psi = \sum_{n=1}^{n} A_n \psi_n(x, y, z) \exp(-jE_n t/\hbar) \tag{1.24}$$

where A_n are constants and ψ_n is the solution of Eq. (1.19) corresponding to $E = E_n$.

1.5 SOME EXAMPLES OF SOLUTIONS OF THE SCHRÖDINGER WAVE EQUATION

1.5.1 A Free Particle

A particle moving in a region of constant potential energy is not subjected to any force and is called a *free particle*. Consider a free particle moving along the x axis. Since the potential energy is arbitrary, we can take $V = 0$, and Eq. (1.19) can be written as

$$\frac{d^2\psi}{dx^2} + k^2\psi = 0 \tag{1.25}$$

where

$$k = \frac{\sqrt{2mE}}{\hbar} = \frac{p}{\hbar} = \frac{2\pi}{\lambda} \tag{1.26}$$

Equation (1.25) has the solution

$$\psi = A_1 \exp(\pm jkx)$$

and

$$\Psi(x, t) = A_1 \exp[j(\pm kx - \omega t)] \tag{1.27}$$

where

$$\omega = \frac{E}{\hbar} = \frac{\hbar k^2}{2m} \tag{1.28}$$

It can be seen that the wave function given by Eq. (1.27) represents a stationary state and is an eigen function of the momentum and energy operators (see Prob. 1.11). Since $\Psi\Psi^* = \psi\psi^* = AA^*$ is independent of x and t, the probability of finding the particle is the same everywhere and the wave function conveys no knowledge about the precise location of the particle.

1.5.2 The Harmonic Oscillator

A particle bound to its equilibrium position by a force that is directly proportional to the displacement from that position executes simple harmonic motion and is called a *harmonic oscillator*. The small-amplitude motion of a pendulum is a typical example. We are interested in this problem mainly because these ideas will be used in the study of mechanical vibrations of lattice atoms in crystals.

Consider a particle of mass m executing simple harmonic motion along the x axis and being acted on by restoring force $F = -dV/dx = -\gamma_p x$. The potential energy of the particle is $V(x) = \gamma_p x^2/2$ and its total energy E is given by

$$E = \frac{\gamma_p x^2}{2} + \frac{p^2}{2m} \tag{1.29}$$

The classical equation of motion of the particle is obtained from Newton's second law

$$\frac{d^2x}{dt^2} + \omega_o^2 x = 0 \tag{1.30}$$

where $\omega_o = \sqrt{\gamma_p/m} = 2\pi\nu_o$ and ν_o denotes the frequency of oscillation. Equation (1.30) has the solution

$$x = A_o \sin(\omega_o t + \theta_o) \tag{1.31}$$

where A_o and θ_o are constants of integration. The classical motion of the particle is shown in Fig. 1.2a where $V(x)$ is plotted as a function of x. At $x = 0$, $V(x) = 0$ and the kinetic energy has its maximum value. At $x = \pm x_o$ the kinetic energy becomes zero, and the particle reverses its direction. These points are called the *classical turning points*. At these points $E = V(x_o) = \gamma_p x_o^2/2$ and thus,

$$x_o = \pm\sqrt{2E/\gamma_p} \tag{1.32}$$

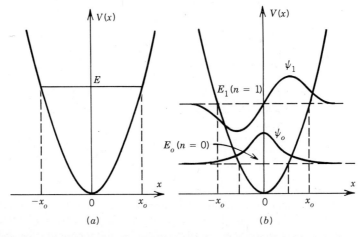

FIGURE 1.2 Diagrams illustrating the classical and the quantum mechanical motions of the harmonic oscillator: (a) classical motion and (b) the energy levels and wave functions for the two lowest energy states calculated using quantum mechanics.

In classical solutions there is no limit on E; it can even be zero, causing the amplitudes of oscillations to vanish. However, the situation is very different for quantum mechanical solutions. When we substitute for $V(x)$, the time-independent Schrödinger equation (1.19) becomes

$$\frac{d^2\psi}{dx^2} + \frac{2m}{\hbar^2}\left(E - \frac{\gamma_p x^2}{2}\right)\psi = 0 \tag{1.33}$$

The solution of this equation shows that the energy of the oscillator is given by [2]

$$E = \left(n + \frac{1}{2}\right)h\nu_o = \left(n + \frac{1}{2}\right)h\nu_o \tag{1.34}$$

where the quantum number $n = 0, 1, 2, 3,$ and so on. The following features of the solution are noteworthy:

1. The energy E of the system is quantized and can change only in units of $h\nu_o$. This is in contrast to the classical solution where E is a continuous variable.
2. The lowest value of E is not zero but is $h\nu_o/2$. This energy is known as zero-point energy. Figure 1.2b shows the wave functions for the two lowest energy states. It is seen that ψ has a finite value beyond the classical turning points. Thus, there is a finite probability of finding the particle in the classically forbidden region.

1.5.3 Particle in a Square Potential Well

The motion of a particle in a potential well of finite height assumes importance because of its similarity with the electron motion in crystalline solids. Figure 1.3a shows a one-dimensional potential well of width $2W$. The potential energy is zero for $-W < x < W$ and is V_o for all other values of x. It is assumed

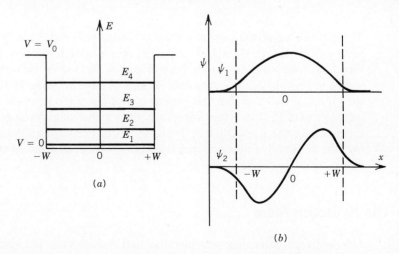

FIGURE 1.3 (a) Energy levels, and (b) the wave functions corresponding to the lowest two energy states of a square potential well of height $V_O = (h^2/2mW^2)$. After H. A. Watson (ed), Microwave Semiconductor Devices and their Circuit Applications, p. 30, McGraw-Hill, New York, 1969.

that the energy E of the particle is less than V_o, so that classically the particle cannot get out of the well. Inside the well the Schrödinger equation becomes

$$\frac{d^2\psi}{dx^2} + k_1^2\psi = 0 \qquad -W < x < W \tag{1.35}$$

with

$$k_1 = \left(\frac{2mE}{\hbar^2}\right)^{1/2}$$

In regions outside the well we can write

$$\frac{d^2\psi}{dx^2} - k_2^2\psi = 0 \quad \text{for } -W > x > W \tag{1.36}$$

where

$$k_2 = \left[\frac{2m(V_o - E)}{\hbar^2}\right]^{1/2}$$

The solution of Eq. (1.35) is obtained in terms of sine and cosine functions, while that of Eq. (1.36) is a decaying exponential. Requiring that both ψ and $d\psi/dx$ have to be continuous at $x = \pm W$, we obtain

$$k_1 W \tan k_1 W = k_2 W$$
$$k_1 W \cot k_1 W = -k_2 W \tag{1.37}$$

and from the definitions of k_1 and k_2 we have

$$(k_1 W)^2 + (k_2 W)^2 = \frac{2mV_o W^2}{\hbar_2} \tag{1.38}$$

These three equations can be solved graphically. For the case where the right-hand side of Eq. (1.38) is $4\pi^2$, there are four stationary energy levels shown in Fig. 1.3a. The wave functions ψ_1 and ψ_2 for the lowest two energy levels are shown in Figure 1.3b. It is seen that ψ_1 and ψ_2 have exponential-like tails outside the potential well and that the penetration of the wave function into this region increases with E. If V_o is now increased, the number of allowed energy levels also increases. In the limit when V_o becomes infinite, the particle cannot penetrate into the classically forbidden region and the number of allowed energy levels becomes infinite (see Sec. 2.5.2).

1.5.4 The Hydrogen Atom

The hydrogen atom has one electron and one proton. Because the proton is 1837 times heavier than the electron, we regard the proton mass to be infinite and we assume that the motion of the system is that of an electron moving in a Coulomb field of a fixed nucleus. Since the problem is spherically symmetrical, the spherical coordinate system is used in the calculations. The potential energy of the electron at a distance r from the proton is $V(r) = -q^2/4\pi\varepsilon_o r$. Substituting this value in Eq. (1.19), we obtain

$$\nabla^2\psi + \frac{2m_o}{\hbar^2}\left[E + \frac{q^2}{4\pi\varepsilon_o r}\right]\psi(r,\theta,\phi) = 0 \qquad (1.39)$$

This equation can be solved by using the method of separation of variables and writing $\psi(r,\theta,\phi) = R(r)\Theta(\theta)\Phi(\phi)$. The solution is given in all the standard texts on quantum mechanics (see reference 2 at the end of this chapter).

The electron wave functions are characterized by three quantum numbers: n, l, and m_l, each associated with one degree of freedom. Here n is called the *principal* or the *radial quantum number* and can have any positive integral value. The energy of an electron corresponding to any quantum state is given by Eq. (1.5). The quantum number l is associated with the orbital motion of the electron and is known as the *orbital quantum number*. For a given n it can have values $l = 0, 1, 2 \ldots (n-1)$. The quantum number m_l specifies the z component of the angular momentum and is called the *magnetic quantum number*. This name arises because in the presence of an externally applied magnetic field, say, along the z axis, an orbital with a given value of n and l can reorient itself in $(2l + 1)$ different ways, each of which corresponds to a different value of the electron energy. Thus, m_l can have any one of the $(2l + 1)$ integral values ranging from $-l$ to l including zero.

The ground state wave function along with the radial position probability $|\psi|^2 \cdot r^2$ is shown in Fig. 1.4. Also shown is the potential energy of the electron and the three lowest energy levels. It is seen that the electron wave function is spread over a region of space. However, the position probability density has its maximum at a distance that is equal to the first Bohr orbit.

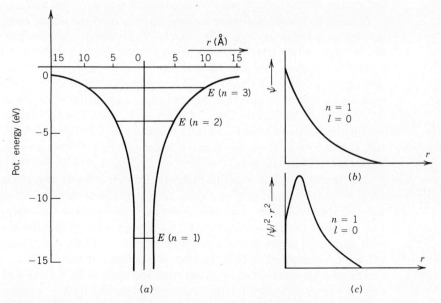

FIGURE 1.4 Results from the solution of the Schrödinger equation for the electron in the hydrogen atom: (*a*) the potential energy curve and the first three energy levels; (*b*) the wave function and (*c*) the radial position probability for the state with $n = 1$ and $l = 0$.

The orbitals corresponding to $n = 1, 2, 3$, and 4 have been named the K, L, M, and N orbitals, respectively. In addition, the energy levels for $l = 0, 1, 2, 3$ are referred to as s, p, d, and f. The state of an electron in the atom is designated first by writing the principal quantum number and then by writing the letter corresponding to the l value. Thus, an energy level for which $n = 2$ and $l = 0$ is called the $2s$ level and one for $n = 4$ and $l = 3$ is called the $4f$ level.

1.6 THE ELECTRONIC STRUCTURE OF ATOMS AND THE PERIODIC TABLE OF ELEMENTS

1.6.1 The Electronic Structure of Atoms

In a multielectron atom, the motion of any electron is influenced not only by the positive charge at the nucleus, but also by the presence of other electrons. Therefore, an accurate solution of the Schrödinger equation is not possible. An approximate method, the central field model, is often used to obtain wave functions and energy levels. In this method, it is assumed that the electrons form a spherically symmetrical cloud of charge around the nucleus, so that the field resulting from this charge distribution is radial and the potential energy of any individual electron can be expressed as a function of r only. The solution of the Schrödinger equation for this problem shows that the wave functions of a single electron are again characterized by three quantum numbers n, l, and m_l, all of which have a similar set of values as discussed above. The total wave function of the system may now be expressed as a product of one electron wave function. Because of the interaction between the electrons, the radial wave function in a multielectron atom has a different form than that in a hydrogen atom. Thus, states having the same value of n but different values of l no longer have the same energy because the energy also depends on l. It is therefore customary to characterize the energy levels of electrons by the quantum numbers n, l, and m_l.

It has been observed that in a multielectron atom the actual number of available states is twice as large as predicted by the above combinations of n, l, and m_l. To explain this phenomenon it is postulated that the electron not only orbits around the nucleus, but also spins around its own axis. Thus, it can exist in two spin states corresponding to spin in two unique directions, say, clockwise and counterclockwise. A quantum number m_s is associated with the electron spin which can take only two values $+1/2$ or $-1/2$ depending on whether the spin is parallel or antiparallel to the orbital momentum. With spin included, the quantum state of an electron in an atom is characterized by a set of four quantum numbers: n, l, m_l, and m_s.

The Pauli exclusion principle states that in a system no two electrons can occupy the same quantum state. Thus, in a multielectron atom no two electrons have an identical set of quantum numbers. This limits the number of electrons that can go to the s, p, d, and f levels. It becomes obvious that the maximum possible number of electrons for a given l is $2(2l + 1)$. Thus, the s, p, d, and f levels can have a maximum of 2, 6, 10, and 14 electrons, respectively.

The average distance of electrons from the nucleus for a given n is appreciably different from its values for the states characterized by $n + 1$ and $n - 1$. Thus, we can talk of "shells" of electrons around the nucleus, each shell corresponding to a given value of n. The electrons that have the same value of n but different values of l are said to form subshells. The innermost shell corresponding to

$n = 1$ can occupy 2 electrons. Similarly, a maximum of 8 electrons can occupy the $n = 2$ shell and 18 electrons the $n = 3$ shell. When the maximum permissible number of electrons is present for a given n, the electrons are said to form a closed shell. Similarly, one can talk of a closed subshell. The average charge distribution resulting from a closed shell or subshell is spherically symmetrical.

Consider an atom having Z protons and thus a nuclear charge Zq. To make the atom neutral we need to add Z electrons to it. The Pauli exclusion principle allows only 2 electrons for the $n = 1$ shell. The lowest energy state that the third electron can have is the $2s$ shell for which $n = 2$ and $l = 0$. Two electrons can go to this shell. Similarly, 6 electrons can go to the next higher energy $2p$ shell and so on. The process will continue until the Z lowest levels are filled with electrons.

1.6.2 Periodic Table of Elements

The electrons in the outermost shell of an atom are called *valence electrons* and play an important role in nature. They determine the chemical behavior of an atom, and they are responsible for the bonding of atoms in the formation of molecules and solids. In addition, these electrons determine the electrical conduction in solids. It is observed that atoms with the same number of valence electrons exhibit similar chemical behavior although they have different atomic numbers. This similarity in chemical behavior has led to the classification of elements in the form of the periodic table.

Table 1.1 shows elements of similar chemical properties in the same column, known as a *group*, and the elements with consecutive values of the atomic number Z in the same row. The electronic configuration in the outermost shells is also shown. The table has seven periods. Each period starts with an element that has one valence electron and ends with an element that has a completely filled valence shell and is chemically inert.

The periodic table can be understood with the help of the four quantum numbers and the Pauli exclusion principle. The first period begins with hydrogen and ends with He, with the filling of the $1s$ shell. Thus, He has the electronic configuration $1s^2$. In the next atom Li, the third electron goes to the next higher energy $2s$ subshell. Thus, Li has only one valence electron and is placed in the first group below hydrogen. As one moves to the right, one electron is added with each increase in Z and the added electron goes to the next available higher energy state. Thus, the $2s$ subshell gets filled with Be, and the $2p$ subshell starts filling with the next element boron. The period ends with the filling of the $n = 2$ shell and the element is inert gas Ne. The next element Na has one valence electron in the $3s$ state and thus has properties similar to Li. The third period terminates with the filling of the $3p$ subshell, and Ar is again an inert gas. The next electron in the element with $Z = 19$ should normally be expected to go to the $3d$ subshell since one expects this to be the higher energy level next to the $3p$ level. However, it has been shown that in many electron atoms, the energy is lowest for the state for which $(n + l)$ is lower. Thus, the 19th electron in K goes to the $4s$ subshell since $(n + l) = 4$ for this shell, while it is 5 for the $3d$ subshell. Similarly, the 20th electron in Ca also occupies the $4s$ subshell. When two subshells have the same value of $(n + l)$, the state with lower n has the lower energy. Thus, in scandium the 21st electron goes to the $3d$ subshell. The rest of the table can be constructed along similar lines.

The valency of an element is determined by the number of valence electrons. In chemical reactions atoms tend to lose or gain enough electrons so as to ac-

TABLE 1.1
The Periodic Table of Elements

I A	II A		III B	IV B	V B	VI B	VII B		Transition Elements		I B	II B	III A	IV A	V A	VI A	VII A	VIII
1 1.008 H $1s$																		2 4.0 He $1s^2$
3 6.94 Li $2s$	4 9.01 Be $2s^2$												5 10.81 B $2s^2 2p$	6 12.01 C $2s^2 2p^2$	7 14.01 N $2s^2 2p^3$	8 16.00 O $2s^2 2p^4$	9 19.00 F $2s^2 2p^5$	10 20.18 Ne $2s^2 2p^6$
11 22.99 Na $2p^6 3s$	12 24.31 Mg $2p^6 3s^2$												13 26.98 Al $3s^2 3p$	14 28.09 Si $3s^2 3p^2$	15 30.97 P $3s^2 3p^3$	16 32.06 S $3s^2 3p^4$	17 35.45 Cl $3s^2 3p^5$	18 39.95 Ar $3s^2 3p^6$
19 39.10 K $3p^6 4s$	20 40.08 Ca $3p^6 4s^2$		21 44.96 Sc $3d 4s^2$	22 47.90 Ti $3d^2 4s^2$	23 50.94 V $3d^3 4s^2$	24 52.00 Cr $3d^5 4s$	25 54.94 Mn $3d^5 4s^2$	26 55.85 Fe $3d^6 4s^2$	27 58.93 Co $3d^7 4s^2$	28 58.71 Ni $3d^8 4s^2$	29 63.54 Cu $3d^{10} 4s$	30 65.37 Zn $3d^{10} 4s^2$	31 69.72 Ga $4s^2 4p$	32 72.59 Ge $4s^2 4p^2$	33 74.92 As $4s^2 4p^3$	34 78.96 Se $4s^2 4p^4$	35 79.91 Br $4s^2 4p^5$	36 83.80 Kr $4s^2 4p^6$
37 85.47 Rb $4p^6 5s$	38 87.62 Sr $4p^6 5s^2$		39 88.91 Y $4d 5s^2$	40 91.22 Zr $4d^2 5s^2$	41 92.91 Nb $4d^4 5s$	42 95.94 Mo $4d^5 5s$	43 99 Tc $4d^6 5s$	44 101.1 Ru $4d^7 5s$	45 102.91 Rh $4d^8 5s$	46 106.4 Pd $4d^{10}$	47 107.87 Ag $4d^{10} 5s$	48 112.40 Cd $4d^{10} 5s^2$	49 114.82 In $5s^2 5p$	50 118.69 Sn $5s^2 5p^2$	51 121.75 Sb $5s^2 5p^3$	52 127.60 Te $5s^2 5p^4$	53 126.90 I $5s^2 5p^5$	54 131.30 Xe $5s^2 5p^6$
55 132.91 Cs $5p^6 6s$	56 137.34 Ba $5p^6 6s^2$		57 138.91 La $5p^6 5d 6s^2$	72 178.49 Hf $5d^2 6s^2$	73 180.95 Ta $5d^3 6s^2$	74 183.85 W $5d^4 6s^2$	75 186.21 Re $5d^5 6s^2$	76 190.2 Os $5d^6 6s^2$	77 192.22 Ir $5d^9$	78 195.08 Pt $5d^9 6s$	79 196.97 Au $5d^{10} 6s$	80 200.59 Hg $5d^{10} 6s^2$	81 204.38 Tl $6s^2 6p$	82 207.2 Pb $6s^2 6p^2$	83 208.98 Bi $6s^2 6p^3$	84 (209) Po $6s^2 6p^4$	85 (210) At $6s^2 6p^5$	86 (222) Rn $6s^2 6p^6$
87 (223) Fr $6p^6 7s$	88 226.02 Ra $6p^6 7s^2$		89 227.03 Ac $6d 7s^2$															

RARE EARTHS

58 140.12 Ce	59 140.91 Pr	60 144.24 Nd	61 (145) Pm	62 150.36 Sm	63 151.96 Eu	64 157.25 Gd	65 158.92 Tb	66 162.50 Dy	67 164.93 Ho	68 167.26 Er	69 168.93 Tm	70 173.04 Yb	71 174.97 Lu
90 232.04 Th	91 231.04 Pa	92 238.04 U	93 237.05 Np	94 (244) Pu	95 (243) Am	96 (247) Cm	97 (247) Bk	98 (251) Cf	99 (252) Es	100 (258) Fm	101 (258) Md	102 (259) No	103 (260) Lw

quire a stable structure like that of the inert gases requiring 8 electrons in the outermost orbit. Elements of group I, II, and III lose one, two, and three electrons, respectively, thus acquiring an inert gas-like structure that results in positively charged ions. These elements are called *electropositive*. Elements of the VII, VI, and V groups usually gain one, two, or three electrons, respectively, to become negative ions and are called *electronegative*. Group IV elements fall between the electronegative and electropositive elements.

1.7 STATISTICAL MECHANICS

The objective of statistical mechanics is to treat the behavior of a very large number of identical systems in a probabilistic fashion without going into the details of each and every individual component of the ensemble. The results obtained using this approach predict the average behavior of the system on the basis of the most probable values of the properties under consideration. The most important characteristic of statistical mechanics is a distribution function. For a given system containing N particles at an equilibrium temperature T, such a function gives us the average number of particles that will show a specified dynamical behavior. A brief account of three distribution functions that govern the distribution of particles among the various available energy states is presented below.

1.7.1 The Maxwell-Boltzmann Distribution

The Maxwell-Boltzmann (MB) distribution is applicable to classical particles such as gas molecules. These particles are distinguishable, and any number of them can occupy a given state. A set of identical particles are said to be distinguishable if they do not interact with each other except for occasional random collisions. This means that their wave functions do not overlap. This condition is met when the density of particles is so low that the average separation between them is large compared to the de Broglie wavelength of the particle. For example, at very low temperatures helium atoms are indistinguishable and have strong interactions among them. However, at higher temperatures their average separation becomes large enough to treat them as distinguishable.

Consider a system of N distinguishable particles with constant total energy E. Let this energy be distributed in n energy levels having energies $E_1, E_2, E_3, E_i, \ldots E_n$. Let $N_1, N_2, N_3, N_i, \ldots N_n$ be the number of particles in the corresponding levels. Of all the chance distributions of N particles among n levels, some will occur with high probability and others with low probability. The distribution having the maximum probability of occurrence is the one that can be realized in a maximum number of statistically independent ways. This problem is analogous to that of distributing N balls into n boxes of different sizes. Let g_i be the ratio of the size of the ith box to the size of all the n boxes. The probability P of the distribution that N_1 balls be in box 1, N_2 be in box 2, and so on, is given by

$$P = \frac{N!}{\prod_{i=1}^{n} N_i!} \prod_{i=1}^{n} g_i^{N_i} \qquad (1.40)$$

where the Π notation is used to indicate an extended product. We assume that N is very large, so that using Stirling's approximation, $\ln N! = (N \ln N - N)$, the above relation can be written as

$$\ln P = N \ln N - \sum_{i=1}^{n} N_i \ln N_i + \sum_{i=1}^{n} N_i \ln g_i \qquad (1.41)$$

In applying Eq. (1.41) to an ideal gas of noninteracting particles, we note that the total number of molecules N is given by

$$\sum_i N_i = N \qquad (1.42)$$

and the total energy E of the system is given by

$$\sum_i E_i N_i = E \qquad (1.43)$$

To determine the most probable distribution, we maximize $\ln P$ with the conditions (1.42) and (1.43). For thermal equilibrium at a temperature T, it can be shown that the probability of occupation of the jth energy level is given by [2]

$$f(E_j) = \frac{N_j}{N} = \frac{g_j \exp(-E_j/kT)}{\sum_j g_j \exp(-E_j/kT)} \qquad (1.44)$$

For the case when all g_j's are equal, Eq. (1.44) becomes

$$f(E_j) = \frac{\exp(-E_j/kT)}{\sum_j \exp(-E_j/kT)} \qquad (1.45)$$

which expresses the Maxwell-Boltzmann distribution. The above relation has been derived for quantized energy levels. For a system with a continuous distribution of energy, the denominator in Eq. (1.45) can be written as $1/B(T)$, and the probability $f(E)$ that a state at energy E is occupied is given by

$$f(E) = B(T) \exp(-E/kT) \qquad (1.46)$$

As an illustration, let us consider an ideal gas at temperature T. We assume that there is no variation in the potential energy, so that it can be taken as zero and we can write

$$E = \frac{m}{2}(v_x^2 + v_y^2 + v_z^2) \qquad (1.47)$$

where each molecule has been assumed to have mass m and v_x, v_y, and v_z are the components of the velocity of a molecule. We now determine the distribution function of molecules in an element $dv_x dv_y dv_z$ around the point (v_x, v_y, v_z). From Eqs. (1.46) and (1.47) we obtain

$$f(v_x, v_y, v_z) dv_x dv_y dv_z = B \exp\left[-\frac{m(v_x^2 + v_y^2 + v_z^2)}{2kT}\right] dv_x dv_y dv_z \qquad (1.48)$$

To evaluate B, we normalize the distribution so that

$$B \int_o^\infty \exp\left(-\frac{mv_x^2}{2kT}\right)dv_x \int_o^\infty \exp\left(-\frac{mv_y^2}{2kT}\right)dv_y \int_o^\infty \exp\left(-\frac{mv_z^2}{2kT}\right)dv_z = 1$$

which gives

$$B = \left(\frac{m}{2\pi kT}\right)^{3/2} \qquad (1.49)$$

Substituting this value of B in Eq. (1.48), we get the Maxwellian distribution of velocities. The distribution function for speed v is obtained by noting that the element $dv_x dv_y dv_z$ in spherical coordinates is given by $4\pi v^2 dv$. Thus, from Eqs. (1.48) and (1.49) we get

$$f(v)dv = \left(\frac{m}{2\pi kT}\right)^{3/2} 4\pi v^2 \exp\left(-\frac{mv^2}{2kT}\right)dv \qquad (1.50)$$

This distribution is plotted in Fig. 1.5 as a function of the normalized velocity $v/\sqrt{2kT/m}$. The function is zero for $v = 0$, and it also goes to zero for infinite speed. In between a maximum occurs at $v_m = \sqrt{2kT/m}$. From Eq. (1.50) it can be shown (see Prob. 1.14) that the average velocity $\bar{v} = \sqrt{8kT/m}$ and the average energy is $3kT/2$. The velocity corresponding to the average energy is the root mean square velocity $v_{\rm rms} = \sqrt{3kT/m}$. It can also be seen from Fig. 1.5 that the majority of the particles have their velocities lying in the range 0 to $\sqrt{6kT/m}$, which corresponds to thermal energies ranging from 0 to $3kT$.

1.7.2 The Fermi-Dirac Distribution

In the MB distribution, the particles were considered to be distinguishable and any number of them could occupy a given energy state. However, in the Fermi-Dirac (FD) statistics, which is applicable to electrons in solids, the particles are indistinguishable and the Pauli exclusion principle requires that one quantum

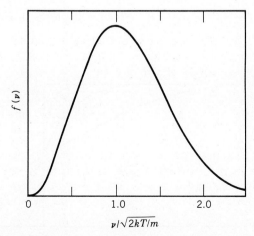

FIGURE 1.5 Maxwell's distribution of velocities giving the fraction $f(v)$ of gas molecules that have velocity between v and $(v + dv)$ at temperature T.

state be occupied only by one particle. Now let us consider a system with n energy levels. Let the ith level have a degeneracy g_i, which means that it has g_i quantum states. The number of independent ways of realizing a distribution of N_i particles in g_i states at the ith level is $g_i!/N_i!(g_i - N_i)!$, and the total number of ways of realizing a distribution of $N_1, N_2 \ldots N_n$ indistinguishable particles in n energy levels is the product of the factors of the above form over all the levels. Thus

$$P = \prod_{i=1}^{n} \frac{g_i!}{N_i!(gi - Ni)!} \tag{1.51}$$

We now have to maximize P with the condition that the total number of particles in the system and the total system energy remain constant. This leads to the following expression [2] for $f(E_i)$:

$$f(E_i) = \frac{N_i}{g_i} = \frac{1}{1 + \exp\left(\dfrac{E_i - E_F}{kT}\right)} \tag{1.52}$$

The parameter E_F is called the *Fermi energy* or the *Fermi level*. Note that the function $f(E_i)$ can have a maximum value of unity corresponding to an energy at which every state is occupied by a particle. The minimum value of $f(E_i)$ is zero, which means that none of the energy levels is occupied. When the energy levels are crowded together into a continuum, the subscript i is dropped and Eq. (1.52) is written as

$$f(E) = \frac{1}{1 + \exp\left(\dfrac{E - E_F}{kT}\right)} \tag{1.53}$$

The FD distribution function is plotted in Fig. 1.6 for several values of T. Note that at $T = 0°K$ the distribution is characterized by the step function

$$f(E) = 1 \quad \text{for } E < E_F$$

and

$$f(E) = 0 \quad \text{for } E > E_F$$

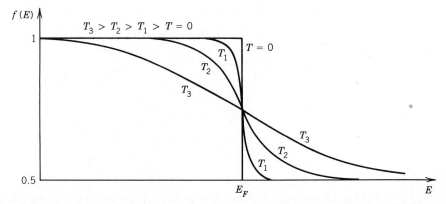

FIGURE 1.6 The Fermi distribution function as a function of energy for several temperatures.

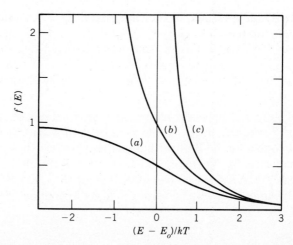

FIGURE 1.7 Distribution functions for (a) Fermi-Dirac, (b) Maxwell-Boltzmann, and (c) Bose-Einstein statistics as a function of $(E - E_O)/kT$. (From J.C. Slater, *Introduction to Chemical Physics*, p. 84, McGraw-Hill, New York, 1938. Reprinted by permission of author.)

Thus, all the states below E_F are filled while all those above E_F are completely empty. As the temperature is increased above $T = 0$, some of the states below E_F become empty and the particles that leave these states fill the states above E_F. Thus, the edges of the step are rounded, and the function changes rapidly from nearly unity to nearly zero over an energy range of a few kT around E_F. Moreover, the probability of occupation of the energy level at E_F for any finite value of T is one-half. It is also obvious that when $(E - E_F)$ is large compared to kT, the unity term in Eq. (1.53) can be neglected and

$$f(E) \simeq \exp(E_F/kT)\exp(-E/kT) = B(T)\exp(-E/kT) \qquad (1.54)$$

Thus, the FD distribution becomes equivalent to the MB distribution. Physically, this means that for $(E - E_F) \gg kT$, the probability of occupation of the quantum states is so small that the Pauli exclusion principle is of no consequence.

1.7.3 The Bose-Einstein Distribution

Particles like electrons are indistinguishable and obey the Pauli exclusion principle. However, some particles like photons and phonons (quanta of lattice vibrations), are indistinguishable but do not obey the Pauli exclusion principle. The distribution function of indistinguishable particles that do not obey the exclusion principle can be expressed as [2]

$$f(E) = \frac{1}{\exp[(E - E_B)/kT] - 1} \qquad (1.55)$$

This relation is called the *Bose-Einstein (BE) distribution law.*

In the case of particles like He atoms, which obey BE statistics, the parameter E_B is obtained from the condition that the total number of particles in the system be constant. However, for particles like photons and phonons which have no rest mass, this condition is not valid because photons and phonons can be destroyed and recreated. For such particles $E_B = 0$, and the BE statistics takes the simple form

$$f(E) = \frac{1}{\exp(E/kT) - 1} \tag{1.56}$$

Figure 1.7 shows the above three distributions plotted as a function of $(E - E_o)/kT$ where E_o is E_F for the FD distribution and E_B for the BE distribution. It is seen that for the BE distribution the curve becomes asymptotically infinite as E approaches E_B. Therefore, it follows that all the allowed states must have energy greater than E_B. It is also evident that for $(E - E_o) > 2kT$, both the FD and BE distributions converge toward the MB distribution.

REFERENCES

1. R. L. LONGINI, *Introductory Quantum Mechanics for the Solid State,* Wiley Interscience, New York, 1970.
2. J. P. MCKELVEY, *Solid State and Semiconductor Physics,* Harper and Row, New York, 1966.

ADDITIONAL READING LIST

1. M. BORN. *Atomic Physics,* Hafner Publishing Company, New York, 1962.
2. P. A. LINDSAY, *Introduction to Quantum Mechanics for Electrical Engineers,* McGraw-Hill, London, 1967.
3. J. C. SLATER, *Introduction to Chemical Physics,* McGraw-Hill, New York, 1963.
4. E. SCHRÖDINGER, *Statistical Thermodynamics,* Cambridge University Press, London, 1960.
5. F. REIF, *Fundamentals of Statistical and Thermal Physics,* McGraw-Hill, New York, 1965.
6. F. F.Y. WANG, *Introduction to Solid State Electronics,* North-Holland, New York, 1980.

PROBLEMS

1.1 Calculate the energy of a photon with wavelength $\lambda = 5000$ Å and $\lambda = 5$ Å. Express your answers both in electron volt and in joule units.

1.2 Using the Bohr formula for electron energy in a hydrogen atom, derive an expression for the wavelength of spectral lines due to transitions from excited states to the ground state. Also calculate the wavelengths of the first four lines.

1.3 Calculate the velocity of an electron in the ground state of the hydrogen atom and compare it with the velocity of light in a vacuum. Assuming that the relativistic effects become important when the velocity of the electron becomes comparable to the velocity of light, determine whether an excited state exists where the relativistic treatment of the hydrogen atom becomes necessary.

1.4 Calculate the de Broglie wavelength (a) of an object of mass 1 gm and (b) of a neutron with kinetic energy of $3kT/2$ at $T = 300$ K. Is it necessary to consider the wave properties of the matter in these cases?

1.5 Consider a pendulum bob having a mass of 50 gm and moving with a velocity of 3 m/sec. Suppose that the momentum p is not known more accurately than $\Delta p = 10^{-6} p$. What limitations does quantum mechanics impose on the simultaneous measurements of the position?

1.6 An oscillator has a resonant frequency of 10 MHz. What is the energy of the quantum of radiation associated with this oscillator? How many quanta of energy is 10^{-6} joule?

1.7 Estimate the size of the hydrogen atom by assuming that the uncertainty in the momentum of the electron can be calculated by setting Δp_x equal to the value of momentum corresponding to the kinetic energy in the ground state and that the total energy of the system is to be maximized. The resulting uncertainty in position Δx is a crude measure of the radius of the hydrogen atom.

1.8 Using the results of Eq. (1.10), determine the quantum mechanical operator associated with (a) velocity, (b) acceleration, and (c) force.

1.9 The equation

$$\nabla^2 \Psi = \frac{1}{v^2} \frac{\partial^2 \Psi}{\partial t^2},$$

where Ψ represents the wave amplitude and v the wave velocity, is known as a wave equation in classical physics. Assuming that the time dependence of Ψ is given by $\Psi(x, t) = \psi(x)\exp(j\omega t)$ and using the de Broglie relation for the matter waves, show that the wave equation is identical to the time-independent Schrödinger equation.

1.10 If Ψ_1 and Ψ_2 are two solutions of the Schrödinger equation, show that $\Psi = a_1\Psi_1 + a_2\Psi_2$ is also a solution. This shows the superposition property of the wave functions.

1.11 Show that the free particle wave function of Eq. (1.27) is an eigen function of the momentum and the energy operators.

1.12 Consider an electron in a one-dimensional square potential well of width 3 Å and infinitely high sides: Take $V = 0$ inside the well.
(a) Calculate the lowest three energy levels and the velocity of the electron in the ground state.
(b) Calculate the normalized wave functions of the electron for the lowest three energy levels.
(c) Calculate the expectation values of x, the momentum p_x, and the energy E.

1.13 Calculate the energy and the wavelength of the photon in the transition $2p \rightarrow 1s$ for the hydrogen atom. Explain how these quantities will be affected in an Li atom.

1.14 From the Maxwellian distribution given in Eq. (1.50) show that

$$f(E)\,dE = 2\left[\frac{E}{\pi(kT)^3}\right]^{1/2} \exp(-E/kT)\,dE$$

where $E = mv^2/2$ is the kinetic energy of the gas molecule. This expression gives the fraction of molecules that have their energy between E and $(E + dE)$. From this expression show that the average energy of a molecule is $3kT/2$.

1.15 Show that the probability that a state ΔE above E_F is occupied equals the probability that a state ΔE below E_F is empty.

2
CRYSTALLINE SOLIDS AND ENERGY BANDS

INTRODUCTION

In Chapter 1 we saw that the valence electrons in an atom determine its chemical behavior and influence the type of bond formed between the atoms that results in the formation of molecules and solids. The strength with which the atoms are bound together determines the various phases of the substance. For example, when the atoms of a substance are separated by a large distance, they form a gas. When the spacing is decreased to make the average distance between the atoms comparable to the size of the atom, the substance behaves as a liquid. Finally, when the atoms are brought so close together that their outer electron orbits overlap, a very strong interaction between the atoms results, which binds them together to form a solid phase.

This chapter discusses the origin of forces that hold the atoms together and describes the various types of bonds between the atoms. A brief account of the crystal structure and crystal defects necessary for understanding the properties of semiconductors is then presented. Finally, we will discuss the problem of allowed energy levels of electrons in a crystalline solid, a subject that is of prime interest in determining the electrical conduction in metals and semiconductors.

2.1 THE BONDING OF ATOMS

A solid is formed by the chemical bonding of a large number of atoms together. In this process the total energy of the system is reduced. The energy of bonding is measured by the energy required to separate 1 gram atom or 1 mole of a solid into individual atoms (or molecules) and is known as *cohesive energy*. This energy is different from *binding energy* which denotes the strength of a bond and is the negative of the energy required to break the bond [1]. For example, the binding energy of an Si—Si bond is −1.84 eV, and the binding energy of an electron in the ground state of hydrogen atom is −13.6 eV.

When two atoms are brought together to form a molecule, two kinds of forces become important, attractive and repulsive. As the separation is decreased, the attractive forces come into play first. However, with a gradual decrease in the spacing between the two atoms, a point is reached where the outer electron orbits of the atoms start overlapping, and the repulsive forces become important. The chemical bond between the atoms is formed at the equilibrium spacing where the two forces balance each other and the total potential energy reaches its minimum. Depending on the nature of the interaction between the atoms, there are four different types of chemical bondings: ionic, covalent, metallic, and molecular.

2.1.1 The Ionic Bond

The ionic bond is formed predominantly between the electronegative and electropositive elements. The electropositive atom loses its valence electrons and becomes a positively charged ion. The lost electrons are gained by the electronegative element which becomes negatively charged. After this transfer of electrons, both ions have 8 electrons in their valence orbits and are held together by the electrostatic force of attraction between negative and positive charges. A typical example is the NaCl molecule. Here the Na atom becomes positively charged by losing an electron to Cl which becomes negatively charged. The two ions then attract each other and are held together by simple electrostatic force.

The ionic bond is fairly strong, and the ionic substances are usually hard and have high melting and boiling points. Since the valence electrons are rather tightly bound to their respective ions, the movement of these electrons under an applied electric field is not possible and most ionic substances are insulators at room temperature. At higher temperature, the ions themselves become mobile, giving rise to ionic conduction.

2.1.2 The Covalent Bond

The covalent bond is formed by the sharing of electrons between the bonded atoms. This sharing results from the overlap of the bonding orbitals, and the shared electrons can come either from one or from all the atoms that take part in the bonding. The covalent bond is a bond between the atoms of the same polarity; hence, it is also known as a *homopolar bond*.

The covalent bonds of crystals formed by the group IV elements C, Ge, and Si are of most interest to us. Let us consider the diamond crystal formed by the bonding of carbon atoms. The ground state electronic configuration of carbon is $1s^2 2s^2 2p^2$. When carbon atoms are brought together, one of the two $2s$ electrons gets excited to the $2p$ subshell, giving four unpaired electrons in the $2s^1 2p^3$ hybrid state. These unpaired electrons then make covalent bonds with the four nearest atoms.

Unlike the ionic bond, the covalent bond is highly directional, since the electron pair bond is formed along a line joining the two atoms forming the bond. The bond is very strong, and the materials formed have high melting and boiling points. In addition, the covalently bonded substances are relatively poor conductors of electricity at normal temperatures.

2.1.3 The Metallic Bond

The metallic bond is formed between electropositive elements. The valence orbit of all metal atoms is either an s subshell or a p subshell with two electrons. The size of the outer shells in these atoms is rather large, and the valence electrons are not as tightly bound to the nucleus as in nonmetals. Binding in metallic crystals occurs because, when the atoms are brought together, it is possible for the valence electrons to be near the positive ions (the region of low potential energy) and, at the same time, to have wave functions existing throughout the crystal. Thus, a metallic bond has two characteristic features. First, the bond remains unsaturated, and a large number of atoms can be held together by the mutual sharing of electrons. Second, the density of electrons between the atoms is considerably lower than that allowed by the Pauli exclusion principle. As a result, the electrons are able to move freely throughout the metal without experiencing significant change in their energy. These "free" electrons account for all the important properties of the metals.

2.1.4 The Molecular Bond

This type of bonding occurs in inert gases as well as in some organic molecules and arises solely from the dipolar forces between the bonded species. Consider two inert gas atoms at a separation larger than the radii of the valence shells of the atoms. The center of symmetry of the electron and nuclear charge is momentarily displaced, and the neutral atom at any instant of time has a fluctuating dipole moment whose average value is zero. This fluctuating dipole moment induces an instantaneous dipole moment in the other atom. The interaction between the reference and the induced dipole moment is attractive and serves to bind the atoms together. The dipolar forces are quite weak, and the substances exhibiting this type of bonding are characterized by low melting and boiling points, and are poor conductors of electricity.

2.2 THE CRYSTALLINE STATE

2.2.1 The Structure of Solids

Solid substances can generally be divided into two categories: amorphous and crystalline. Figure 2.1a shows a two-dimensional representation of an amorphous solid. Here the atoms are arranged in an irregular and random manner like the molecules in a liquid. The atoms have only short-range order, which means that a definite arrangement of atoms in the substance extends only over a distance of a few atomic spacings. Glass is a typical example of an amorphous solid. A crystalline solid has a regular periodic arrangement of atoms. If this regular arrangement of atoms extends over the whole sample, the material is said to be a single crystal (Fig. 2.1b). A polycrystalline material consists of groups or clusters of single crystals of various orientations that are joined together (Fig. 2.1c). The line separating the crystal sections of two different orientations is known as a *grain boundary*.

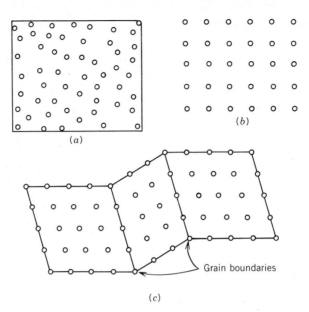

FIGURE 2.1 Two-dimensional representations of atoms in different types of solids: (a) amorphous, (b) single crystal, and (c) polycrystalline materials.

2.2.2 Crystal Structure

(a) UNIT CELL AND BRAVAIS LATTICES

A crystal is a three-dimensional structure composed of atoms arranged on a lattice. A lattice is a periodic arrangement of points in space and is defined by three fundamental translation vectors **a**, **b**, **c** such that the arrangement of atoms looks the same at any point **r** as it looks from any other point **r**′ given by

$$\mathbf{r}' = \mathbf{r} + (n_1 \mathbf{a} + n_2 \mathbf{b} + n_3 \mathbf{c}) \equiv \mathbf{r} + \mathbf{T} \tag{2.1}$$

where n_1, n_2, and n_3 are integers and the vector $\mathbf{T} = n_1 \mathbf{a} + n_2 \mathbf{b} + n_3 \mathbf{c}$ is called the *lattice translation vector*. A crystal structure is formed by associating a basis of atoms identically to each of the lattice points. A unit cell is the smallest structure that will generate the entire crystal from simple translations in three directions.

Figure 2.2 depicts a two-dimensional crystal structure formed by attaching an atom to every lattice point. Several possible ways to select a unit cell are shown in Fig. 2.2b. Note that each of the choices 1, 2, and 3 has only one lattice point per unit cell and defines a primitive unit cell. Choice 4 has two lattice points per unit cell and represents a nonprimitive cell. It is customary to choose the dimensions of the unit cell in such a way that the length of the edges of the cell is minimized and there is an atom at each corner. This would eliminate choices like 2 and 3.

In three dimensions, there are only 14 ways of arranging the lattice points in space which satisfy Eq. (2.1). These 14 lattices are called *Bravais lattices* and have been grouped into seven crystal systems [2] that are characterized by the lengths a, b, c and their angles α, β, and γ as defined in the unit cell of Fig. 2.3. A space lattice is called *simple* if the unit cell in Fig. 2.3 has lattice points only at the corners. It is termed body centered if there is also a lattice point at the center

THE CRYSTALLINE STATE

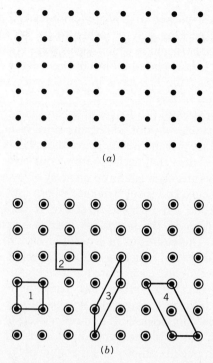

FIGURE 2.2 A two-dimensional crystal showing (*a*) a two-dimensional lattice and (*b*) the crystal structure formed by associating one atom to each lattice point. Four possible choices of unit cell are also shown.

of the cube, and it is named face centered when there are lattice points at the corners and at the center of all the faces of the unit cell.

(b) ATOMIC PACKING IN CRYSTALS

A crystal structure is formed by the packing of atoms together. We can assume the atoms to be hard spherical balls. The simplest crystal structure is that of a simple cubic lattice whose unit cell has one atom at each of the eight corners. However, crystals found in nature do not favor a simple cubic structure, but

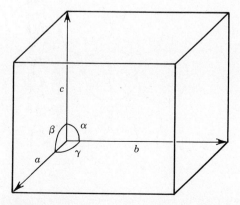

FIGURE 2.3 Crystal axes and angles in a unit cell.

rather a close-packed or a nearly close-packed structure. As shown in Fig. 2.4, the closest packing may be achieved in the following manner. A layer A of spheres is constructed in which each sphere is surrounded by six spheres. A similar layer B is now placed on top of the A layer as shown in Fig. 2.4a. Note that each sphere of the B layer is in contact with six spheres of this layer and with three spheres of layer A.

If a close-packed third layer of spheres is now placed above the B layer, there are two possible ways of adding the third layer. In Fig. 2.4a the spheres of the third layer are placed directly above the spheres of A layer. The fourth layer may now be placed so that each sphere of this layer is placed above a sphere of the B layer. This arrangement forms an array of layers $ABAB$.... and results in a hexagonal close-packed (hcp) structure whose unit cell is shown in Fig. 2.5a.

The second close-packed structure is obtained by putting the spheres of the third layer in the interstices formed by the three spheres of the B layer (Fig. 2.4b). This results in an array of layers $ABCABC$..... and leads to the face-centered cubic (fcc) structure whose unit cell is shown in Fig. 2.5b. This cell has eight corner atoms, each shared among eight cells, and six face atoms, each of which is shared between two cells. Thus, the total number of atoms per unit cell is 4. The distance between the nearest neighbors is $a/\sqrt{2}$ where a denotes the length of the cube edge. This unit cell obviously is not a primitive cell. The primitive cell of this lattice is shown by the dashed lines in Fig. 2.5b. However, the nonprimitive cell is simpler and is commonly used to describe this structure. It is

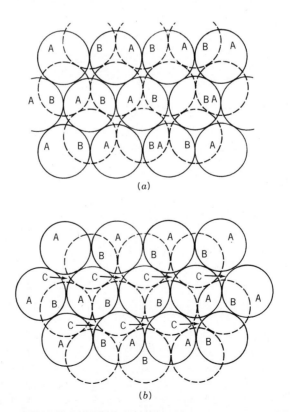

FIGURE 2.4 Close-packed arrangement of spheres in (a) an hcp structure and (b) an fcc structure. The arrangement of layers in (a) is $ABABA$... and in (b) it is $ABC\ ABC$....

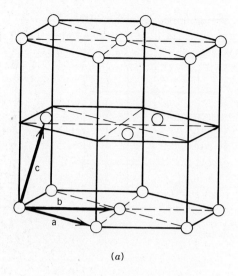

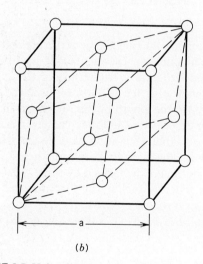

FIGURE 2.5 Unit cells of (a) hcp and (b) fcc structures.

evident that in a close-packed arrangement each atom has 12 nearest neighbors, 6 in the same layer and 3 each in the layers immediately above and below this layer. This number of nearest neighbors in a crystal arrangement is known as the *coordination number*. No crystal structure with a coordination number greater than 12 is possible.

(c) SOME SIMPLE CRYSTAL STRUCTURES

The simplest crystal structures are those of NaCl and CsCl, shown in Fig. 2.6a and b, respectively. In the NaCl structure all the Na^+ ions are situated at the lattice points of one fcc sublattice, while all the Cl^- ions are situated on the lattice points of another sublattice. The coordination number for this structure is 6. The CsCl structure is body centered with Cl^- ions at the body center position and the Cs^+ ions at the corner of the cube (or vice versa). The coordination number of this structure is 8.

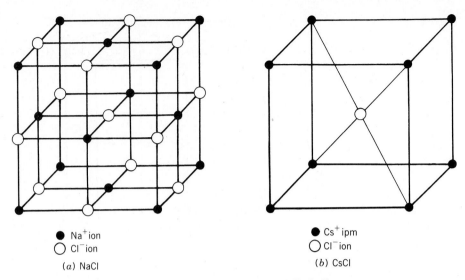

- ● Na⁺ ion
- ○ Cl⁻ ion

(a) NaCl

- ● Cs⁺ ipm
- ○ Cl⁻ ion

(b) CsCl

FIGURE 2.6 Crystal lattice of (a) NaCl and (b) CsCl structures.

The diamond and zinc blende structures are of interest to us because the majority of semiconductor crystals belong to this class. A nonprimitive unit cell of the diamond structure is shown in Fig. 2.7. The structure can be visualized as two interpenetrating fcc sublattices displaced from one another along the cube diagonal by 1/4 the length of that diagonal. In terms of close-packing, the structure can be described as follows. The layer A' of atoms is placed above the layer A of atoms, so that each atom of A' touches an atom of layer A and rests vertically above it. The layer B of atoms is then stacked in such a way that each atom of A' rests in the hollow formed by three atoms of the B layer. A further B' layer of atoms is added in such a way that atoms of layer B' are placed vertically over those of layer B in exactly the same manner as atoms of layer A' were placed over the atoms of A layer. Next, a layer C is added such that each atom of B' is in the

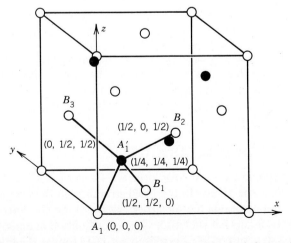

FIGURE 2.7 A nonprimitive unit cell of the diamond lattice. The shaded atoms belong to the second sublattice which is displaced from the first along the major cube diagonal by one-fourth the length of the diagonal.

hollow formed by three atoms of layer C. Finally, a C' layer of atoms is added to have atoms of this layer resting vertically above the atoms of C layer. The resulting structure has two sublattices ABC and $A'B'C'$, and the stacking follows the sequence $AA'BB'CC'AA'\ldots$. The coordination number of this structure is 4.

The unit cell in Fig. 2.7 has 8 atoms at the corners, 6 face-centered atoms, and 4 atoms in the body of the cube. Thus, there are 8 atoms per unit cell. The position of various atoms can be conveniently specified in submultiples of the cube edge a. If we take the position of the corner atom A_1 as the origin, the coordinates of the three face-centered atoms B_1, B_2, and B_3, are as shown. The coordinates of the nearest atom A'_1 of the second sublattice are $(1/4, 1/4, 1/4)$. If A_1 belongs to the A layer, then A'_1 belongs to the A' layer. Note that each of the atoms A_1, B_1, B_2, and B_3 is situated at a distance of $\sqrt{3}a/4$ from A'_1, and this is the nearest neighbor distance in the diamond structure. The diamond structure is relatively empty because only 34 percent of the unit cell volume is occupied by atoms, compared to 74 percent in close-packed fcc and hcp structures (see Prob. 2.5).

The zinc blende (ZnS) structure is similar to the diamond structure except that Zn atoms are placed on one fcc sublattice and S atoms on the other. Thus, each Zn atom is surrounded by four S atoms and vice versa. Elements Si and Ge crystallize in diamond lattice, and the compound semiconductors GaAs and GaP have the zinc blende structures.

2.2.3 Crystal Planes and Directions

In a crystalline solid it is often necessary to define a direction or a crystal face. Miller notation is used for defining planes and directions in a crystal. In this notation we specify three indices that are related to the intercepts on a set of coordinate axes. In a cubic crystal, the coordinate axes x, y, z are taken parallel to the edges of the unit cell and a lattice point is taken as the origin. If a represents one side of the unit cube, then any plane in the cell will cut the coordinate axes x, y, z at points that are integral multiples of a. Let these points be $x = s_1 a$, $y = s_2 a$, and $z = s_3 a$. The plane is now defined by the following procedure: (1) Determine the value of intercepts s_1, s_2, and s_3 on the three axes; (2) take the reciprocals of these numbers and clear the fractions by multiplying each of them by their lowest common denominator. The set of integral numbers (hkl) obtained after this procedure denotes the Miller indices of the plane. As an example, consider the plane of Fig. 2.8 which intersects the x axis at a, the y axis at $2a$, and the z axis at $3a$. The intercepts on the three axes are 1, 2, 3. The Miller indices of this plane are obtained by taking the reciprocals of these numbers, that is, 1, 1/2, 1/3, and by clearing the fractions by multiplying each of the reciprocals by 6. The indices of the desired plane therefore are (632).

If a plane is parallel to an axis, its intercept on that axis is infinite and the resulting Miller index is zero. When the intercept on an axis is negative, a bar is placed over the corresponding index. Thus, $(1\bar{2}4)$ is a plane that has a negative intercept on the y axis. It should be clear that all parallel planes have the same Miller indices. The three most important planes (100), (110), and (111) of a cubic crystal cell are shown in Fig. 2.9. In diamond-type crystals the (111), $(\bar{1}11)$, $(1\bar{1}1)$, and $(11\bar{1})$ planes are equivalent in the sense that the crystal has the same symmetry about these planes. A family of equivalent planes like this is denoted by curly brackets as $\{111\}$.

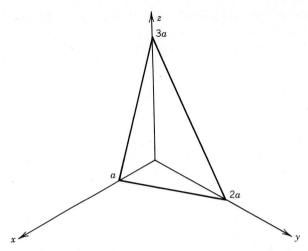

FIGURE 2.8 Miller indices of a plane in a cubic crystal. The plane has the indices (632).

A direction in the crystal is defined by noting the coordinates of the nearest lattice point on the line and expressing them as multiples of the basic unit cell vectors **a**, **b**, and **c**. Thus, the direction index of a vector $\mathbf{r} = u\mathbf{a} + v\mathbf{b} + w\mathbf{c}$ drawn from the origin is written with square brackets as $[uvw]$. In a cubic crystal, the separation d between two adjacent parallel planes with Miller indices (hkl) is given by

$$d = \frac{a}{\sqrt{(h^2 + k^2 + l^2)}} \tag{2.2}$$

2.3 CRYSTAL DEFECTS

Real crystals depart from perfect periodicity first because of the presence of lattice defects, and second because of thermal vibrations of lattice atoms around their equilibrium positions. Crystal defects may be classified as point defects,

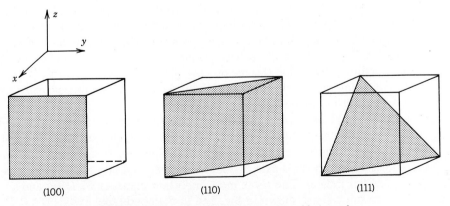

FIGURE 2.9 Three important planes of a cubic crystal.

The indices u and v should not be confused with u and v used for the velocity of sound and velocity of electron in the next section.

CRYSTAL DEFECTS　　　　　　　　　　　　　　　　　　　　　　　　　　　　　　　**37**

line defects (or dislocations), and gross defects like twinning. A study of these defects is important because they influence the mechanical, chemical, and electrical properties of the material.

2.3.1 Point Defects

A point defect is localized about a point in the crystal. Two elementary point defects are vacancy and interstitial. A vacancy is created when an atom moves out of its regular site, and an interstitial is formed when an atom gets located in one of the interstitial voids in the crystal. If an atom migrates from its regular site to the surface of the crystal and settles there at a lattice site, the defect is known as the *Schottky defect*. On the other hand, when an atom creates a vacancy interstitial pair by moving from its regular site into an interstitial position, the defect is known as the *Frenkel defect*. A two-dimensional representation of Schottky and Frenkel defects is shown in Fig. 2.10.

At temperatures above 0°K thermal agitation of crystal atoms causes some of these atoms to leave their lattice sites, which creates a certain concentration of Schottky and Frenkel defects. The concentration of a particular type of defect depends on the energy of formation of that defect. In the diamond-type structure, the energy required for the formation of a Frenkel defect is smaller than that required for a Schottky defect. In substances whose crystals are grown at high temperatures, thermal agitation causes a large number of vacancies, and when the crystal is cooled to room temperature, many of them get frozen in the lattice.

Point defects involving chemical impurities at substitutional or interstitial sites can also occur depending on the size of the impurity atom. For example, Au in the Si lattice occupies about 90 percent substitutional sites, whereas impurities like Zn and Cu occupy predominantly interstitial sites.

2.3.2 Dislocations

A dislocation is a line defect that results when a part of a crystal slips relative to another part over an atomic plane within the crystal. This happens when the crystal is subjected to stresses beyond its elastic limit.

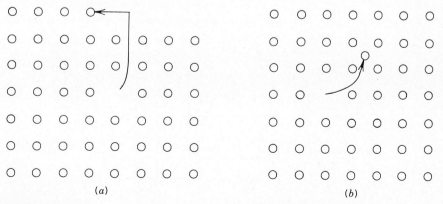

FIGURE 2.10 Schematic representation of (*a*) Schottky and (*b*) Frenkel defects in a two-dimensional crystal.

Consider the situation shown in Fig. 2.11 where stress is applied to a crystal such that the lower half of the crystal moves to the right with respect to the upper half. The dislocation line marks the boundary of the region that has slipped with respect to the rest of the crystal. In Fig. 2.11, this line (AD) is at right angles to the slip direction. This type of dislocation is called an *edge dislocation*. Here, the effect of the dislocation is to introduce an extra plane of atoms (ABCD) into the crystal. Another form of dislocation, known as a *screw dislocation,* is shown in Fig. 2.12. Here the direction of slip is parallel to the dislocation line (AF). The edge and screw dislocations may be regarded as different aspects of the same internal slip phenomenon. In the diamond-type structure, slipping that leads to dislocation formation always occurs between (111) parallel planes which are joined only by one bond per atom, such as the bond between the atoms A-A′ in Fig. 2.7.

A crystal defect closely related to dislocation is a stacking fault. Referring to the diamond lattice, we see that the normal stacking sequence along the (111) direction is AA′ BB′ CC′ AA′ ..... When this sequence is violated, a stacking fault results. Thus, the sequence AA′ BB′ AA′ BB′ CC′ involves a stacking fault at the plane B′A.

2.3.3 Twinning and Inclusions

Twinning is a gross defect that occurs when one part of the crystal forms the mirror image of the other, with the two parts remaining in intimate contact over the bounding surface (Fig. 2.13). The bounding plane of reflection is known as the *twin plane.* Twinning may occur during the crystal growth and may also be produced by mechanical shear of successive planes of atoms. A twinned material has a very high dislocation content.

Sometimes small particles of metals, dielectrics, or refractory materials get incorporated into the crystal during its growth. These particles form a precipitate near the dislocation sites or near vacancies. This type of defect is called an *inclusion*. For example, Si crystals may contain precipitates of carbon, SiO_2, and many heavy metals.

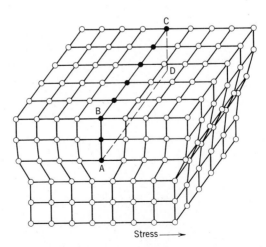

FIGURE 2.11 Schematic diagram of a crystal containing edge dislocation. The dislocation line AD is normal to the slip direction.

CRYSTAL DEFECTS

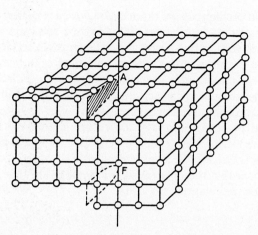

FIGURE 2.12 Geometry of a screw-type dislocation. The dislocation line is parallel to the slip direction.

Crystal defects have considerable influence on material properties such as resistivity and dielectric strength. Dislocations and other related defects act as recombination centers for mobile carriers in semiconductors and thus reduce the excess carrier lifetime (see Chapter 5). Dislocations also give rise to enhanced diffusion effects in their vicinity during the fabrication of devices and result in the segregation of metallic impurities around them.

2.4 LATTICE VIBRATIONS AND PHONONS

2.4.1 Normal Modes of Vibrations

In a crystal the lattice atoms do not remain stationary but execute oscillations around their mean equilibrium positions because of thermal agitation. These

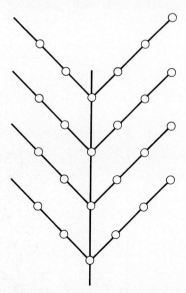

FIGURE 2.13 Two-dimensional schematic representation of a twinned crystal.

thermal oscillations are often called *lattice vibrations*. However, a lattice is a geometrical configuration of points in space and cannot vibrate. Only the atoms associated with lattice points have the freedom to vibrate.

One way to treat the thermal vibrations of lattice atoms is to regard the entire crystal as an assembly of identical harmonic oscillators. However, this viewpoint does not provide a correct description of thermal vibrations. Instead, the crystal should be considered as a single entity with its normal modes of vibrations. The normal modes of a crystal lattice are similar to the modes of vibrations of a stretched string whose ends are clamped on a rigid support. The three lowest frequency waves for the transverse vibrations of the string are shown in Fig. 2.14. It is evident that the only possible vibrations are those for which the wavelength λ satisfies the relation

$$L = \frac{n\lambda}{2} \tag{2.3}$$

where L is the length of the string and n has only positive integral values. The mode corresponding to $n = 1$ is the first normal mode, that corresponding to $n = 2$ is the second normal mode, and so on. The frequency v of the wave is related to λ by the relation $u = v\lambda$ where u is the wave velocity.

The normal modes of a crystal lattice can be described in the same way by assuming that the atoms at the ends have zero displacement. Inside the crystal, the atoms can have displacement either along the direction or perpendicular to the direction of wave propagation. Thus, a crystal has transverse and longitudinal modes of vibrations. For both modes of vibrations, the crystal appears as a continuous structure to long wavelength waves. However, when the wavelength becomes comparable to the interatomic spacing, the crystal has to be treated as a structure composed of discrete atoms bound together by lattice forces.

2.4.2 Wave Motion in a One-dimensional Lattice

Consider a one-dimensional infinite chain of identical atoms each of mass m and spaced at a distance a from each other (Fig. 2.15). We assume that the restoring

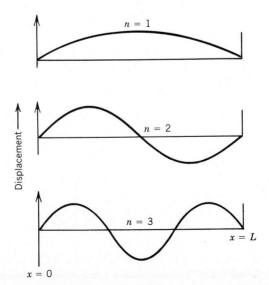

FIGURE 2.14 The three lowest normal modes of a string clamped rigidly at the two ends.

LATTICE VIBRATIONS AND PHONONS

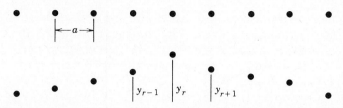

FIGURE 2.15 Wave motion of a one-dimensional lattice consisting of equally spaced identical atoms.

forces acting on any atom result only from the forces exerted by the nearest neighbors. Let y_r, y_{r-1}, and y_{r+1} represent the displacements of the rth, $(r-1)$th and $(r+1)$th atoms, respectively. If these displacements are small compared to the spacing between the atoms, then the equation of motion for the rth particle can be written as

$$m\frac{d^2y_r}{dt^2} = \gamma_p(y_{r+1} + y_{r-1} - 2y_r) \tag{2.4}$$

where γ_p is the force constant. Since the disturbance is propagated as a wave, we assume a solution of the form

$$y_r = A\exp[-j(\omega t - \beta x)] \tag{2.5}$$

where $x = ra$, ω is the angular frequency, and β is the propagation constant. Substituting for y_{r-1}, y_r, and y_{r+1} from Eq. (2.5) into Eq. (2.4) and solving for ω yields

$$\omega = \sqrt{\frac{4\gamma_p}{m}}\left|\sin\frac{\beta a}{2}\right| \tag{2.6}$$

The absolute sign in Eq (2.6) is used because we regard ω as positive. Figure 2.16 shows a plot of ω as a function of β. Such a plot is called a *dispersion curve*. The phase velocity v_{ph} of the wave is given by $v_{ph} = \omega/\beta$, and the group velocity v_g is given by $v_g = d\omega/d\beta$. It is seen from Fig. 2.16 that for values of β near the origin ω varies linearly with β, and $v_{ph} = v_g = a\sqrt{\gamma_p/m} = u$ is constant independent of β. Since $\omega = 2\pi v$ and $v_{ph} = v\lambda$, we have $\beta = \omega/v\lambda = 2\pi/\lambda$. As β increases, the relation between ω and β becomes nonlinear and both v_{ph} and v_g start decreasing with ω and frequency reaches its maximum value at $\beta = \pi/a$. This makes $\lambda = 2a$, which corresponds to the Bragg reflection of the elastic wave from the successive atoms. Thus, at this value of β, standing waves are established and no energy is propagated in the crystal.

Equation (2.6), shows that ω is a periodic function of β with a period $2\pi/a$. On an ω versus β plot, there will be many values of β corresponding to the same values of ω. However, all possible values of ω can be obtained by considering the values of β in the fundamental interval between $-\pi/a$ and π/a. For a chain of N atoms only $(N-2)$ atoms are free to vibrate; hence, there are $(N-2)$ allowed values of β, each corresponding to one normal mode. When N is very large, the number of normal modes can be regarded as N, and each mode represents a degree of freedom. In a three-dimensional crystal with N atoms, there will be N longitudinal and $2N$ transverse modes giving $3N$ degrees of freedom.

The longitudinal modes of vibration represent compressional waves with atoms vibrating along the direction of wave propagation. On the other hand, the transverse modes are shear waves, and in these modes the atoms vibrate perpendicular

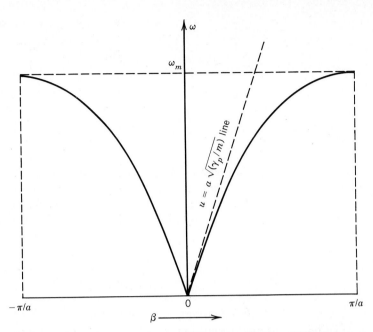

FIGURE 2.16 Dispersion curve for a one-dimensional lattice of identical atoms.

to the direction of wave propagation. Transverse modes cannot be propagated in liquid and gas media where the shear modulus is zero.

Next, we consider the one-dimensional chain of atoms shown in Fig. 2.17. Here all the even-numbered atoms have mass M and the odd-numbered atoms mass m. Figure 2.18 shows the dispersion relation for this diatomic lattice [2]. The plot has two branches. The lower branch has $\omega = 0$ at $\beta = 0$ and reaches its maximum value $\omega_m = \sqrt{2(\gamma_p/M)}$ at $\beta = \pi/2a$. This branch describes the usual acoustical waves in solids, and the vibrational modes associated with this branch are called *acoustical modes*. The upper branch has its maximum frequency $\omega_m = \sqrt{2\gamma_p(M + m)/Mm}$ at $\beta = 0$, and ω decreases with β reaching its minimum value at $\beta = \pi/2a$. The modes of this branch are called *optical modes* because the frequency range of these modes falls within the infrared region.

The physical characteristics of optical and acoustical modes are shown in Fig 2.19. In the case of acoustical modes, the two different kinds of atoms vibrate back and forth in the direction (Fig. 2.19a), whereas they move in opposite directions in the case of optical modes (Fig. 2.19b).

For a crystal like NaCl, where the Na and Cl atoms have different masses, the acoustical and optical branches are separated from each other (Fig. 2.18). However, for crystals with a diamond-type structure, the atoms of the two sublattices are of the same kind, and the acoustical and optical branches should meet at $\beta = \pi/2a$. This result is true only for a one-dimensional chain of atoms. In a three-dimensional crystal we have acoustical and optical modes in longitudinal

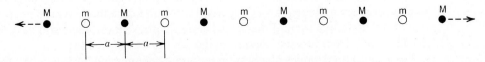

FIGURE 2.17 A linear diatomic array of atoms.

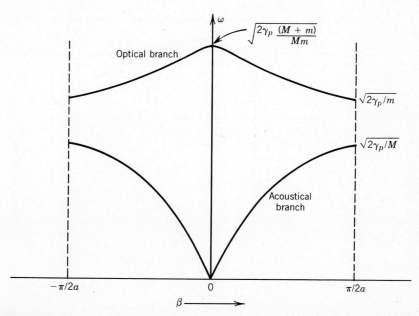

FIGURE 2.18 Dispersion curves for the diatomic linear array of atoms showing acoustical and optical branches.

and transverse directions. Since each of the acoustical and optical modes is identical in the two transverse directions, we have four different kinds of vibrations, namely, transverse optical (TO), longitudinal optical (LO), transverse acoustical (TA), and longitudinal acoustical (LA). Figure 2.20 shows the measured dispersion relation for the transverse and longitudinal modes in Si in the [100] direction.

2.4.3 Phonons

For small vibrational amplitudes each of the optical and acoustical modes can be treated as a simple harmonic oscillator with energy $E = (n + 1/2) h\upsilon$ where n is a positive integer. The quantum of energy $h\upsilon$ is called a *phonon*. Both the photon and the phonon obey the BE statistics. However, conceptually they are different. The phonon is a quantum of mechanical energy of lattice vibrations, whereas the

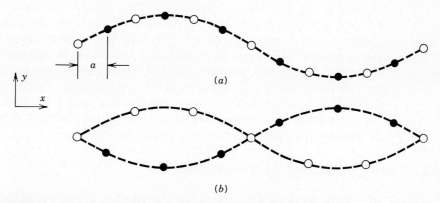

FIGURE 2.19 Motion of atoms in (*a*) transverse acoustical and (*b*) transverse optical modes in a diatomic linear array of atoms.

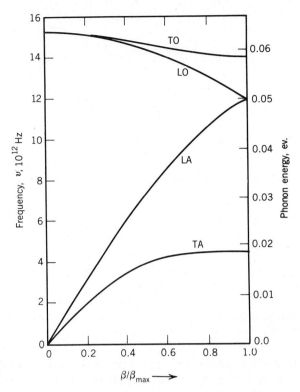

FIGURE 2.20 Measured dispersion curves of lattice vibrations in [100] silicon. (From B. N. Brockhouse, Lattice vibrations in silicon and germanium, *Phys. Rev. Letters*, Vol. 2, p. 256. Copyright © 1959. Reprinted by permission of the American Physical Society, New York.)

photon is a quantum of electromagnetic energy. The momentum associated with a photon of energy $h\nu$ is $h\nu/c$, whereas for a phonon it is $h\nu/u$. Since the velocity of sound u in a solid medium is small compared to the velocity of light c, a phonon has a large momentum while the momentum of a photon is very small.

2.5 ENERGY BANDS IN SOLIDS

The energy band picture can be developed either by the atomistic (i.e., tight binding) or the one-electron approach. In the atomistic approach, electrons are assumed to be tightly bound to individual atoms. As atoms are brought together to form a crystal, interaction between the neighboring atoms causes the electron energy levels of individual atoms to spread into bands of energies. In the one-electron approximation, we study the behavior of a single electron in the potential field established by the lattice atom cores and modified by the presence of all the other free electrons. The various permissible energy levels obtained for this electron represent the allowed energy levels of all the electrons. A brief account of these two viewpoints is presented in this section.

2.5.1 The Tight Binding Approximation

In order to understand the formation of energy bands from tight binding approximation, let us consider the two tuned circuits shown* in Fig 2.21. When these

---

*L used here for inductance should not be confused with length L.

ENERGY BANDS IN SOLIDS

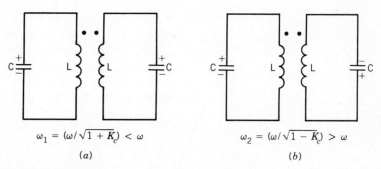

FIGURE 2.21 Modes of oscillation of two coupled tuned circuits: Two oscillators vibrate, (a) in phase, and (b) out of phase.

circuits are separated by a large distance, there is no interaction between them and each circuit has a frequency of oscillation, $\omega = 1/\sqrt{LC}$. As the two oscillators are brought close together, energy is transferred from one circuit to the other because of the presence of mutual inductance, and the original frequency ω splits into two frequencies $\omega_1 = \omega/\sqrt{1 + K_c}$ and $\omega_2 = \omega/\sqrt{1 - K_c}$. Here the coupling coefficient K_c is the ratio of mutual inductance between the two circuits to the self-inductance L. The frequency ω_1 corresponds to the symmetrical mode of oscillation in which the voltages on the two capacitors are in phase, but for the modes represented by ω_2, the two oscillators vibrate in opposite phase. The frequency difference $(\omega_2 - \omega_1)$ will increase with increasing value of K_c. Now consider that N identical resonant circuits, each having a resonant frequency ω, are brought together to form a set of N-coupled oscillators. The mutual coupling between the circuits will cause the N natural frequencies to split into a band of frequencies starting from some lowest frequency ω_l to some highest frequency ω_h. For a sufficiently large value of N, both ω_i and ω_h, as well as the band width $(\omega_h - \omega_l)$, become independent of N and the band can be treated to form a continuous distribution of frequencies.

The above considerations also apply to electronic states in atoms. We can assume that each electron energy level is an oscillator similar to the resonant circuit. Figure 2.22 shows the energy levels of electrons in a sodium atom. The solid dots denote the number of electrons in each level, and $E = 0$ represents the energy of an electron outside the atom. The energy band development of a sodium crystal is shown in Fig. 2.23. We consider a crystal containing N atoms that are arranged in a periodic array. When the atoms are far apart, there is no interaction between them, and the allowed states of the system are just the states of a single atom repeated N times in space. As the interatomic spacing is decreased, the valence orbitals of the atoms begin to overlap and the energy of the $3s$ subshell spreads into a band. Similarly, the empty $3p$ subshell also spreads into a band. A further reduction in the interatomic spacing causes broadening of the bands formed by the $3s$ and $3p$ energy levels. In addition, the $2s$ and $2p$ levels spread into bands. At equilibrium separation r_{eq} the bands formed by the $3s$ and $3p$ states have overlapped. The overlapped band has $8N$ states but only N electrons, so that only a small fraction of the available states in the band are occupied by electrons. If the number of atoms is large, the spacing between the energy levels will be small, and the band can be treated as having a continuous distribution of energy states.

Figure 2.24 shows the potential energy profile for Na atoms at equilibrium spacing r_{eq} along a line. It is seen that outside the crystal, to the right, the potential energy rises toward $E = 0$ asymptotically. However, because of mutual interac-

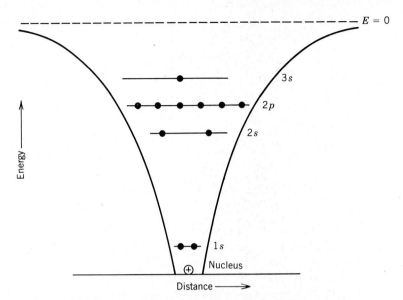

FIGURE 2.22 Electronic energy levels in a sodium atom.

tion between the neighboring atoms, the energy between them lies considerably below $E = 0$. Note that the electrons belonging to each atom in the 2s and 2p bands are still separated from neighboring atoms by potential humps and are not able to move from one atom to another. The band formed by the 3s electrons is only half filled, and the electrons in this band can move freely from one atom to another because their energies lie above the potential humps. In a three-dimensional sodium crystal, the bands formed by 3s and 3p electrons overlap,

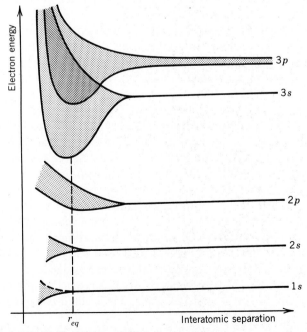

FIGURE 2.23 Development of energy bands in a sodium crystal. After J. C. Slater, Introduction to Chemical Physics, p. 494, McGraw-Hill, New York, 1938.

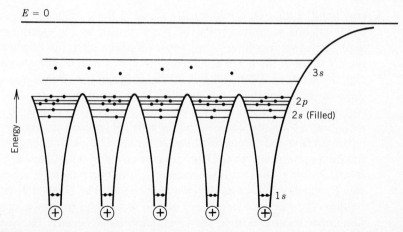

FIGURE 2.24 Potential profile and electron energy levels along an array of atoms in a sodium crystal.

and some of the electrons in the 3s band are contained in the energy levels of the 3p band.

Not all substances have energy bands like those of sodium. The development of energy bands in a diamond crystal is shown in Fig. 2.25. A carbon (C) atom has 6 electrons with the $1s^2\,2s^2\,2p^2$ configuration. Consider an assembly of N equidistant C atoms. When the interatomic spacing is large, the energy levels of electrons belonging to various subshells are isolated levels. As the interatomic spacing is decreased, the $6N$ states of the $2p$ subshell start spreading into an en-

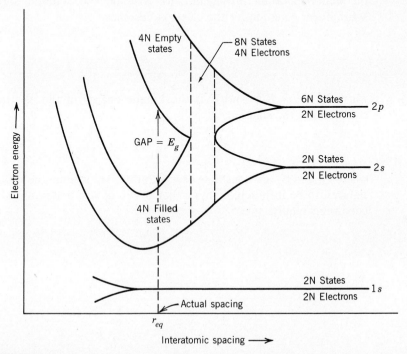

FIGURE 2.25 Development of energy bands in a diamond crystal. (From R. B. Adler, A. C. Smith, and R. L. Longini, *Introduction to Semiconductor Physics*, SEEC, vol. 1, p. 78. Copyright © 1964. Reprinted by permission of John Wiley & Sons, Inc., New York.)

ergy band. Similarly, the lower $2N$ states in the $2s$ subshell spread into another band which is separated from the band formed by the electrons in the $2p$ subshell. At yet smaller interatomic separation, the $6N$ states of the upper band and the $2N$ states of the lower band merge to give a wider energy band with $8N$ states and $4N$ electrons. With still further reduction in the interatomic spacing, the $2N$ states in the $2s$ subshell and the filled $2N$ states in the $2p$ subshell come together and spread into a band of $4N$ states, all of which are occupied by electrons. The remaining $4N$ empty states in the $2p$ subshell spread into another band which is separated from the completely filled lower band by a forbidden energy region. This situation corresponds to the formation of covalent bonds beween the neighboring atoms. At the equilibrium separation r_{eq}, the two bands are separated by an energy E_g which is known as the *band gap energy*. The completely filled lower band contains the valence electrons of all the atoms and is known as a *valence band*. The upper band is termed the *conduction band* because the excitation of electrons from the valence band into this band is responsible for electrical conduction.

2.5.2 Energy-Momentum Relation for Electrons in Solids

The one-electron approximation leads to a deeper insight and provides quantitative information about the electron energies and momenta. However, this approach is valid only for valence electrons that are loosely bound to the atoms.

(a) THE WAVE MECHANICS OF FREE ELECTRONS

Let us consider an electron in a one-dimensional potential well of length L with vertical sides similar to those shown in Fig. 1.3a. Since the potential energy in the well is constant throughout, we write $V = 0$, and the time-independent Schrödinger equation for the electron becomes

$$\frac{d^2\psi}{dx^2} + k^2\psi = 0 \tag{2.7}$$

where $k^2 = 2m_o E/\hbar^2$ and E is the kinetic energy of the electron. The above equation has the solution

$$\psi(x) = C_1 \sin kx + C_2 \cos kx \tag{2.8}$$

where C_1 and C_2 are constants of integration. We assume the walls of the potential well to be infinitely high, so that $\psi = 0$ at $x = 0$ and also at $x = L$. These conditions require that

$$k = \frac{n\pi}{L} \tag{2.9}$$

and

$$E = \frac{p^2}{2m_o} = \frac{\hbar^2}{2m_o}k^2 \tag{2.10}$$

This relation shows that $p = \hbar k$. Here p is the electron momentum and k is known as the wave number. When dealing with electron motions in solids, it is

ENERGY BANDS IN SOLIDS

customary to plot E as a function of k. The relation describes a parabola and is shown in Fig 2.26a. From this plot we obtain

$$v = \frac{1}{\hbar}\frac{dE}{dk} \quad (2.11)$$

and

$$m_o = \hbar^2 \bigg/ \left(\frac{d^2E}{dk^2}\right)$$

where v is the electron velocity. Equations (2.9) and (2.10) show that for a given energy E, there are two values of k, namely, $k = n\pi/L$ and $k = -n\pi/L$. Substituting either of these values in Eq. (2.10) yields

$$E_n = \frac{n^2\pi^2\hbar^2}{2m_oL^2} \quad (2.12)$$

where $n = 1, 2, 3\ldots\ldots$. Each value of n corresponds to the one-electron energy level in the crystal. In any laboratory-size solid, L is very large compared to atomic dimensions, and Eq. (2.12) may be assumed to represent a continuous distribution of energies.

(b) ELECTRON MOTION IN A PERIODIC POTENTIAL [3]

In a real crystal, electrons do not move in a constant potential but in the periodic potential established by the ion cores (Fig. 2.24). Let us consider the one-dimensional motion of an electron in a periodic potential with a period a such that

$$V(x) = V(x + a) \quad (2.13)$$

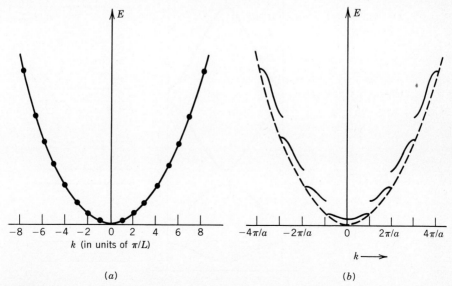

FIGURE 2.26 E-k diagrams for electrons in a one-dimensional solid. Electrons (a) in a constant potential and (b) in a periodic potential with period a.

The Schrödinger wave equation then admits the solutions of the form

$$\psi_k(x) = u_k(x)\exp(\pm jkx) \tag{2.14}$$

where $k = 2\pi/\lambda$ and $u_k(x)$ is a periodic function of x with period a. Solutions 2.14 are known as Bloch functions.

The effect of periodic potential is to distort the E versus k curve from its parabolic shape and to introduce discontinuities for certain values of k. These features are shown in Fig. 2.26b. It is seen that the curve has discontinuities occurring at values of k given by

$$k = \pm \frac{n\pi}{a} \tag{2.15}$$

where n has integral values. Since $k = 2\pi/\lambda$, the above relation means that $2a = n\lambda$, which is the condition for Bragg reflection at normal incidence. Thus, when Eq. (2.15) is satisfied, the electron wave cannot propagate in the crystal.

The values of k given by Eq. (2.15) mark the boundaries of the various energy zones. Evidently, k may vary from $-\infty$ to ∞. However, since k is periodic with a period of $2\pi/a$, it is sufficient to consider k values only in the interval from $-\pi/a$ to π/a and to express all energies in terms of k values contained in this interval. Such a reduced representation is shown in Fig 2.27. From these plots it is clear that as E increases, the width of allowed bands increases, but the forbidden gap between the adjacent bands is reduced.

In the above discussion we have assumed a linear array of infinite atoms. For an array of N atoms spaced a distance a apart, we have $L = aN$ and the separation between the two adjacent k values is π/L. Since there are $2\pi/a$ values of k within a band, each band has $2N$ electron states.

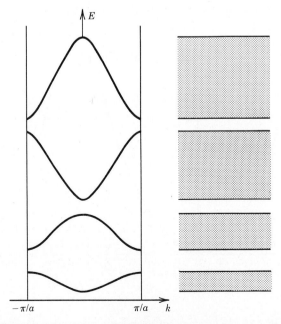

FIGURE 2.27 The E-k plot of Fig. 2.26b in a reduced representation.

ENERGY BANDS IN SOLIDS 51

2.5.3 Crystal Momentum and Effective Mass

The Bloch functions Eq. (2.14) represent waves in which an electron moves unhindered through the crystal with a velocity $v = d\omega/dk$. To discuss the motion of an electron, it is convenient to introduce the quantity $P = \hbar k$, called the *crystal momentum*. It is then possible to describe the motion of an electron in terms of P because its rate of change equals the externally applied force on the electron. Since $E = \hbar\omega$, we have

$$v = \frac{d\omega}{dk} = \frac{1}{\hbar}\frac{dE}{dk} = \frac{dE}{dP} \qquad (2.16)$$

which is equivalent to the classical relation Eq. (2.11) for the free electron. Let an electric field $\mathcal{E}_o$ produce a force $F = -q\mathcal{E}_o$ on the electron. The electron will gain energy from the field at a rate

$$\frac{dE}{dt} = -q\mathcal{E}_o \cdot v = F\left(\frac{d\omega}{dk}\right) = \frac{F}{\hbar}\left(\frac{dE}{dk}\right) \qquad (2.17)$$

Also

$$\frac{dE}{dt} = \frac{dE}{dk} \cdot \frac{dk}{dt}$$

From these two relations we obtain

$$F = \hbar\frac{dk}{dt} = \frac{dP}{dt} \qquad (2.18)$$

which is Newton's second law. Thus, the crystal momentum P is analogous to the electron momentum p for a free electron. However, it should be noted that P is not the actual momentum of the electron. This is because in addition to the external force F, an electron in a crystal is also acted on by internal lattice forces that are quite complex.

It is now easy to show that the acceleration dv/dt of the electron can be written as (see Prob. 2.13)

$$\frac{dv}{dt} = \frac{F}{m^*} \qquad (2.19)$$

where

$$m^* = \frac{\hbar^2}{d^2E/dk^2} \qquad (2.19a)$$

is known as the effective mass. The effective mass m^* is different from the actual gravitational mass of the electron because of the existence of the lattice forces. If it were possible to determine the resultant F_1 of the complex lattice forces, we could write

$$F_{\text{total}} = F_1 + F = m_o\frac{dv}{dt} \qquad (2.20)$$

Since it is not possible to determine F_1, we simply write $F = m^*(dv/dt)$. Obviously, m_o and m^* are not the same. Note that the actual behavior of the electron in a crystal can be described only by quantum mechanics. In writing the classical relation Eq. (2.19), the major quantum mechanical features of the electronic motion are taken care of by writing m^* for the actual mass m_o.

We will now attempt to understand the motion of electrons in a crystal using the concept of effective mass. Consider a single electron in an otherwise empty band shown in Fig 2.28a. The plots of v and m^* are also shown. Note that m^* is a constant near the band edges but varies with E in the middle of the band. Let the electron start from the lowest energy state at the bottom of the band and be accelerated to the right by an external field $\mathscr{E}_o$. Near $k = 0$, the velocity v increases linearly with k and the electron behaves as a free particle of mass m^*. When the electron has gained sufficient momentum from the field and k becomes large, the interaction of the electron wave with the lattice potential begins to interfere with its motion. As a consequence, the electron acceleration goes down and m^* starts increasing. The velocity v reaches its maximum value at the inflection point P where m^* becomes infinite. Beyond this point v begins to decrease, and the electron experiences a strong retarding force that tends to reduce its velocity to zero. In this region, m^* becomes negative. When the electron reaches the band edge (point Q) it suffers Bragg reflection which changes the sign of its momentum. Thus, the electron reappears at the edge R where v is zero and m^* is negative. From this point onward the electron begins to accelerate against the applied force and v becomes negative. At the inflection point S the velocity against the field starts decreasing and m^* becomes positive. Eventually, the

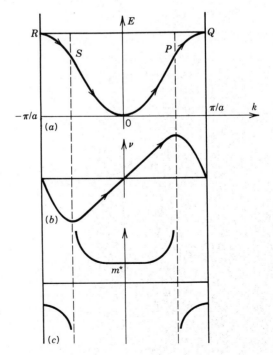

FIGURE 2.28 Plots of (a) electron energy, (b) electron velocity and (c) electron effective mass as a function k in a band.

electron is brought to rest at $k = 0$ where it is ready to repeat the above cycle once again.

From the above discussion, it is clear that under the influence of an applied field, an electron in a perfect crystal oscillates back and forth in k space, absorbing energy from the field for half the period and giving it back in the remaining half. In real crystals, this cycle is interrupted by loss of energy owing to scattering collisions resulting from the interaction of electrons with crystal imperfections. This process is illustrated in the partially filled band shown in Fig 2.29. An electric field that causes the electron to move to the right may cause an electron A near the top of the distribution to jump over to states B, C, and so on. Once the electron has reached state C, one of two things may happen. The electron may be accelerated to state D at the top of the band from where it suffers Bragg reflection; or, alternatively, it may return from C to its original state A after losing the gained momentum and energy to the lattice in a scattering collision. The energy loss causes the joule heating of the crystal. Note that application of the external force F causes more electrons in the positive half of the k space compared to the negative half. This new distribution is characterized by a net velocity in the direction of F and results in an electric current. In a completely filled band, there are no higher energy states which the electrons can occupy, and the only collision mechanism is the Bragg reflection. The entire electron population in the filled band executes this cycle of behavior without contributing to any current.

It should be clear now that the electron population near the top of a band behaves like particles of negative mass and negative charge. If a few vacant states are created near the top in an otherwise filled band, the electrons in these states will be accelerated against the direction of the applied force. This motion can be described more easily in terms of vacant states, each state being characterized by a positive charge and a positive mass. Such a state is called a *hole*.

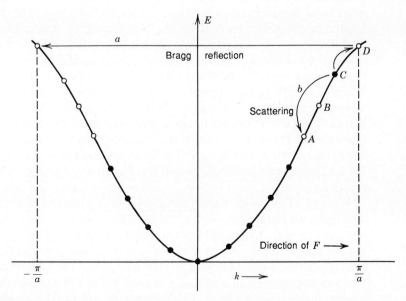

FIGURE 2.29 Two modes of interaction of an electron near the top of the band with the lattice; arrow *a* shows a Bragg reflection and *b* shows lattice scattering.

2.5.4 Energy Bands in Three-dimensional Crystals

In a three-dimensional crystal, both **P** and **k** are vector quantities and the Schrödinger wave equation has the solutions of the form

$$\psi_\mathbf{k}(\mathbf{r}) = u_\mathbf{k}(\mathbf{r})\exp(j\mathbf{k} \cdot \mathbf{r}) \tag{2.21}$$

where the function $u_\mathbf{k}(\mathbf{r})$ has the periodicity of the lattice, **r** is the spatial coordinate vector, and **k** has three components k_x, k_y, and k_z. Like the case of a one-dimensional lattice, the energy levels of electrons in a three-dimensional periodic potential fall into allowed bands separated by forbidden gaps.

The E versus **k** diagram in a three-dimensional crystal is rather complex. Since E is now a function of three variables, a pictorial representation of the $E - \mathbf{k}$ relation is difficult. Therefore, it is customary to draw several diagrams of E as a function of **k** along the principal directions of the crystal.

The motion of electrons in a three-dimensional crystal is similar to the motion in a one-dimensional crystal. The effective mass m^* now becomes a tensor with nine components given by

$$\frac{1}{m^*_{ij}} = \frac{1}{\hbar^2}\left(\frac{\partial^2 E}{\partial k_i \partial k_j}\right) \tag{2.22}$$

with i and j assuming the values x, y, and z. For coordinate axes along the symmetry directions of the crystal, the off-diagonal terms of the tensor become zero and we obtain three values of m^* along the three principal axes.

2.6 METALS AND INSULATORS

In the last section we saw that no electric current can arise from a completely filled band. Similarly, no current can result from an empty band that has no electrons. Any electronic conductivity in a crystal must, therefore, arise from the motion of electrons in bands that are only partially filled. This observation provides the basis for distinguishing between metals and insulators.

In a metal, the top band occupying the valence electrons is only partially filled. When an electric field is applied, electrons in this band can increase their energy by moving to higher levels, and this explains the high conductivity of metals. The partial filling of the top band can be accomplished by one of the following two possibilities. The number of valence electrons may not be sufficient to fill the band (Fig. 2.24), or a completely filled valence band may overlap in energy with the next empty allowed band. The second case is encountered more frequently in practice.

Figure 2.30 shows the energy band diagram of a metal with overlapping bands. The figure also depicts the energy of electrons along an array of atoms which terminates at $x = 0$. To the right of this point, the potential energy asymptotically rises to $E = 0$. At $E = 0$, the electron is in a vacuum just outside the metal and has no kinetic energy. This level is, therefore, called the *vacuum level*. The occupancy of energy states in a band is governed by the FD statistics and is characterized by the Fermi level (E_F). The energy required to take an electron from E_F to the vacuum level is the *work function* of the metal. Thus, the work function ($q\phi_m$) is given by $q\phi_m = [E(=0) - E_F]$.

METALS AND INSULATORS

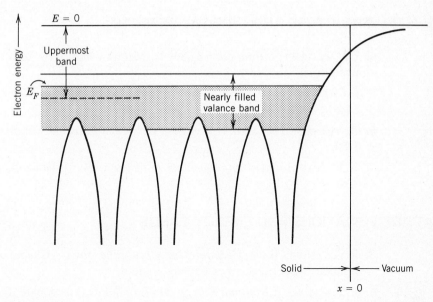

FIGURE 2.30 Energy band diagram of a metal showing overlapping bands and potential rise at the surface.

The number of valence electrons in an insulator is just sufficient to fill the valence band. Above the valence band there is an empty conduction band (see Fig. 2.25) which is separated from the valence band by a forbidden energy region of several electron volts. At 0°K the valence band is completely filled, and the conduction band is completely empty. Thus, there is no electron current, and the conductivity is zero. At higher temperatures, a few electrons are excited from the valence band to the conduction band, and a finite but negligibly small conductivity is observed. Between metals and insulators lies an important class of materials known as semiconductors. These materials will be considered in the next chapter.

REFERENCES

1. H. F. Wolf, *Semiconductors,* Wiley Interscience, New York, 1971.
2. J. P. McKelvey, *Solid State and Semiconductor Physics,* Harper and Row, New York, 1966, Chapter 3.
3. C. Kittel, *Introduction to Solid State Physics,* John Wiley, New York, 1976.

ADDITIONAL READING LIST

ATOMIC BONDING

1. L. Pauling, *The Nature of the Chemical Bond,* Cornell University Press, Ithaca, N.Y., 1960.
2. R. L. Sproull, *Modern Physics,* John Wiley, New York, 1963, Chapter 8.
3. J. C. Slater, *Introduction to Chemical Physics,* McGraw-Hill, New York, 1963, Chapter 22.

CRYSTAL STRUCTURE AND CRYSTAL DEFECTS

4. L.V. AZAROFF, *Introduction to Solids,* McGraw-Hill, New York, 1960, Chapters 2 to 6.
5. F.C. PHILLIPS, *An Introduction to Crystallography,* John Wiley, New York, 3d ed., 1963.
6. A. HOLDEN, *The Nature of Solids,* Columbia University Press, New York, 1965.
7. H.G. VAN BUEREN, *Imperfections in Crystals,* North-Holland, Amsterdam, 1961.

LATTICE VIBRATIONS AND ENERGY BANDS

8. R.A. SMITH, *Wave Mechanics of Crystalline Solids,* Chapman and Hall, London, 1969, especially Chapters 4 and 8.
9. L. PINCHERLE, *Electronic Energy Bands in Solids,* Macdonald, London, 1971.
10. F.F.Y. WANG, *Introduction to Solid State Electronics,* North-Holland, New York, 1980, Chapters 6, 7, 8, and 11.

PROBLEMS

2.1 Prove that the force between two spherically symmetrical charge distributions separated by a distance r is the same as the force between two point charges separated by the same distance.

2.2 Explain why metals are able to form alloys when mixed in arbitrary proportions.

2.3 Determine the number of lattice points per unit cell in simple, body-centered, and face-centered cubic lattices.

2.4 Find the nearest neighbor distance in the following lattices: (a) a body-centered cubic, (b) a face-centered cubic, and (c) a diamond lattice.

2.5 The packing efficiency of a lattice is defined as the ratio of the maximum volume that can be filled by hard spheres placed at lattice points to the total volume. Show that the packing efficiency is 52 percent for a simple cubic lattice, 74 percent for an fcc lattice, and 34 percent for a diamond lattice.

2.6 The lattice constant of a diamond crystal is 3.56 Å. Calculate the number of atoms per cm^2 in the diamond crystal in the (100) and (111) planes.

2.7 Find the number of equivalent (100), (110), and (111) planes in a cubic crystal.

2.8 Prove that in a cubic crystal the direction $[hkl]$ is perpendicular to the (hkl) plane.

2.9 Consider a cubic crystal with lattice constant a. Prove that the first (hkl) plane not passing through the origin intersects the $x, y,$ and z axes at a/h, a/k, and a/l, respectively.

2.10 Consider a one-dimensional lattice with five particles. The particles at the ends cannot vibrate. The lattice executes longitudinal vibrations. Show that all the possible motions of the particles can be described by the first three normal modes. Sketch the waveforms of these modes.

2.11 The optical phonon energy in Si at $\beta = 0$ is 0.063 eV. Using BE statistics for phonons, determine the average energy per phonon for optical and low energy acoustical modes at 300 K.

2.12 A one-dimensional crystal has a lattice constant $a = 2.5$ Å. Determine the free electron momentum and the energy at which the first Bragg reflection occurs.

2.13 Show that the acceleration of an electron in a one-dimensional crystal with periodic potential is given by Eq. (2.19) with m^* given by Eq. (2.19a).

PART 2
FUNDAMENTALS OF SEMICONDUCTORS

3
SEMICONDUCTOR MATERIALS AND THEIR PROPERTIES

INTRODUCTION

Semiconductors are important because they have found applications in almost all branches of industry and areas of daily life. Moreover, the study of semiconductors provides an opportunity to apply and test much of the theory of solids and to obtain information on transport properties which could not be obtained from the study of metals alone. There are two reasons why observation and interpretation of electronic processes is easier in semiconductors than in metals and other crystals. First, semiconductor crystals can be grown with purity far in excess of what has been achieved in the case of metals and insulators. Second, the mobile carrier concentrations in a semiconductor are low, so that the carriers can be treated as distinguishable, noninteracting particles like molecules of an ideal gas. As discussed in Sec. 1.7, for noninteracting particles the Pauli exclusion principle is of little consequence. Thus, instead of the Fermi-Dirac statistics we can use the much simpler Maxwell-Boltzmann (MB) statistics. It then becomes possible to obtain analytical solutions for many problems that cannot be solved accurately using the FD statistics. The condition for which the MB statistics can be used in a semiconductor is stated in Sec. 3.5.2.

This chapter presents a detailed discussion of semiconductors. The different types of semiconductors are described, and the theory of these materials is developed by taking elemental semiconductors as an example.

3.1 SEMICONDUCTING MATERIALS

In the previous chapter we saw that substances in nature can be classified as metals and insulators. Metals are good conductors of electricity, whereas insulators have very high values of resistivity. The resistivity scale of various substances in nature is shown in Fig. 3.1. On this scale semiconductors are materials that have resisitivity ranging from about 10^{-3} to 10^8 Ω $-$ cm. This classification, however, is quite arbitrary because it does not say in what respect semiconductors are different from metals and insulators.

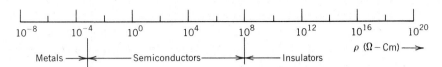

FIGURE 3.1 Resistivity scale of materials found in nature.

A more satisfactory definition of semiconductors is provided by the band theory of solids. As shown in Chapter 2, in metals the bonding of atoms is such that free electrons become available after the bond formation. However, in an insulator all the valence electrons are used to form the bonds, and the resulting valence band is separated from the conduction band by an energy gap. Consequently, mobile carriers in insulators can be produced only when energy is supplied from an external source (such as thermal radiation) to raise electrons from the valence band to the conduction band. Semiconductors have an energy band structure similar to that of insulators. However, the energy gap is smaller (generally less than 2.5 eV), so that a significant number of electrons can be thermally excited from the valence band to the conduction band at room temperature.

3.1.1 Classification of Semiconductors [1]

Semiconductors can be classified in several ways. One classification may be according to the nature of the current carriers. Thus, we have ionic and electronic semiconductors. In ionic semiconductors, the conduction takes place through the movement of ions and is accompanied by mass transport. In an electronic semiconductor, current is carried by electrons and no mass transport is involved.

Semiconductors may also be classified as elemental and compound semiconductors. In an elemental semiconductor, all the constituent atoms are of the same kind, whereas a compound semiconductor is composed of atoms of two or more different elements.

Yet another way of classifying semiconductors is according to their structure. Accordingly, a semiconductor may be amorphous, polycrystalline, or single crystal. The nature of the crystalline structure has a significant influence on the properties of the semiconductor. Amorphous semiconductors have a short-range order and can be considered to be quasi-periodic. They display poor characteristics in terms of carrier mobility and lifetime. In the majority of the semiconductor devices, single-crystal materials are used because of their superior electrical characteristics. Polycrystalline semiconductors show electrical behavior similar to that of single crystals, but their conductivity is significantly lower because of the presence of grain boundaries that obstruct the flow of mobile carriers. Therefore, they have found only limited use. In this book we will be concerned mainly with single-crystal electronic semiconductors.

3.2 ELEMENTAL AND COMPOUND SEMICONDUCTORS [1]

3.2.1 Elemental Semiconductors

Two important semiconductors are germanium and silicon, which belong to group IV of the periodic table. All the group IV elements have two electrons in their outermost p subshell and crystallize in a diamond-type structure in which the neighboring atoms are bound by homopolar cohesive forces. These forces determine the two important properties of these materials, namely, the melting point and the energy gap. Some important characteristics of group IV semiconductors are given in Table 3.1.

The cohesive forces are weakened with an increase in atomic number, and this is reflected by an increase in the size of the atom and a decrease in the melting point and the band gap. Thus, there is a gradual transition from strongly insulating to essentially metallic behavior. The diamond, for example, is an insulator, whereas near room temperature α-Sn behaves almost like a metal.

3.2.2 Compound Semiconductors

In recent years a number of compound semiconductors have found applications in various devices. The most important among them are the intermetallic III-V and II-VI compounds. Some IV-VI compounds also exhibit semiconducting properties, but most of them have a small band gap that limits their use to infrared detectors and lasers.

III-V Compound Semiconductors These semiconductors are compounds formed from the elements in group IIIA (all metals) with those in group VA. Ignoring the compounds of thallium and bismuth, we present the remaining sixteen:

Group III A ↓	Group V A →			
	N	P	As	Sb
B	BN	BP	BAs	BSb
Al	AlN	AlP	AlAs	AlSb
Ga	GaN	**GaP**	**GaAs**	**GaSb**
In	InN	**InP**	**InAs**	**InSb**

Not all these compounds are potential semiconductors. Boron and nitrogen compounds, for example, have been included only for the sake of completeness. Aluminum compounds are not very stable and usually disintegrate with time. The six most important semiconductors are printed in bold type.

Most of the III-V semiconductors have the zinc blende crystal structure. Eight valence electrons are shared between a pair of nearest atoms, so that, on an aver-

TABLE 3.1
Some Properites of Group IV Elements

Element	Atomic Number	Band Gap E_g at 300 K (eV)	Melting Point (°C)	Covalent Radius (Å)
Carbon (C)	6	5.30	3800	0.77
Silicon (Si)	14	1.12	1417	1.17
Germanium (Ge)	32	0.67	937	1.22
Tin (α-Sn)	50	0.08	232	1.40

age, each atom has four valence electrons. This suggests that the bonding has a covalent character, and to a first approximation, the cohesion between the atoms is homopolar. Therefore, we would expect the properties of these compounds to be similar to those of corresponding group IV elements. However, since the elements of group III are more electropositive, and those of group V are more electronegative than the group IV elements, the bonding in III-V compounds has a partial ionic character as well. Therefore, the cohesive force between atoms represents the cohesive force of the covalent bonding plus an additional term because of ionic contribution. As a result, the cohesive force and the strength with which the valence electrons are bound to the atoms are higher for these crystals than those for the corresponding group IV semiconductors. This is evidenced by the fact that each of these compounds has a higher melting point and a larger band gap compared to group IV elements with the same lattice spacing, atomic number, and density. As an example, consider InSb and α-Sn. These substances have the same lattice constant and the same average atomic number and atomic weight. On the basis of purely covalent bonding, the two should exhibit the same properties. However, because of the partial ionic character of bonds in InSb, it has a higher melting point (525°C) and a larger band gap ($E_g = 0.17$ eV at 300 K).

Because of their predominantly covalent character, the III-V semiconductors form a link between group IV elements and II-VI compounds. Therefore, most of their characteristics can be deduced from those of neighboring materials. To apply these considerations, Table 3.2 is drawn by assigning isoelectronic III-V compounds to their group IV elements. Here, the average total number of electrons (core plus valence electrons) per atom in a row is nearly the same, and this is what is meant by the term *isoelectronic*. It is seen that certain material properties change gradually as one moves from one end to the other in horizontal and vertical directions. These characteristics were predicted by H. J. Welker [2] even before many of the compounds were obtained in the laboratory. They can be summarized as follows [1]:

Vertical Direction The cohesive energy and the energy gap E_g decrease with increasing atomic number. This is because in an atom with a large number of electrons, the core electrons repel each other with a greater force. This results in weak cohesion of atoms and a loose binding of valence electrons. Decreasing binding energy results in wider allowed bands and narrower band gaps.

TABLE 3.2
Variation in the Properties of Isoelectronic Group IV and III-V Compounds

Group IV Semiconductors			III-V Compounds			
C				BN			
SiC			BP		AlN		
Si		BAs		AlP		GaN	
SiGe	BSb		AlAs		GaP		InN
Ge		AlSb		GaAs		InP	
GeSn			GaSb		InAs		
α-Sn				InSb			

Cohesive energy ↑→

Band gap ↑→

Horizontal Direction   Both the cohesive energy and the energy gap increase in horizontal transition from elemental to compound semiconductors. This happens because of increased cohesion due to the ionic contribution.

The technical importance of III-V semiconductors is that they provide a wider choice of band gap than do elemental semiconductors. Some of these compounds have energy gaps corresponding to the visible region of spectrum and have found applications in light-emitting diodes. Others, like InSb and InAs, are used in galvanomagnetic devices.

A major problem with compound semiconductors is that their preparation in single-crystal form is more difficult. The bond formation process usually encourages a uniform growth, and we would, therefore, expect the stoichiometry of the compound to be established during crystal growth. However, because of the different vapor pressures of the constituents near the melting point, one component may vaporize more rapidly, causing an excess of the other type of atoms. These excess atoms may either get trapped interstitially in the lattice or may precipitate out to form a second phase. Either of these possibilities destroys the stoichiometry.

The III-V compounds have a general formula $A^{III}B^V$. However, several of these compounds have identical structures and similar lattice constants. Such compounds exhibit complete solid solubility in each other and structures of the form $A_x^{III}B_{1-x}^{III}C^V$ are possible, with x ranging continuously from 0 to 1. The properties of such compounds vary smoothly with composition. Thus, it is possible to grow crystals of a compound like $Al_xGa_{1-x}As$. Similarly, the growth of crystals having two different group V elements like GaP_yAs_{1-y} is also possible.

II-VI Compound Semiconductors   These semiconductors are formed by combining the elements in groups IIB and VIA in the periodic table. A list of all the II-VI compounds is given below:

Group II B ↓	Group VI A →			
	O	S	Se	Te
Zn	ZnO	**ZnS**	**ZnSe**	**ZnTe**
Cd	CdO	**CdS**	**CdSe**	**CdTe**
Hg	HgO	HgS	HgSe	HgTe

Among the oxides, ZnO has found use in both electrophotography as a photosensitive material and in paints and varnishes. Little information is available about the semiconducting properties of CdO and HgO. The remaining three group VI elements are called *chalcogenides*. Among the chalcogenides of mercury both HgSe and HgTe are semimetals, whereas HgS is a semiconductor. The remaining six compounds printed in bold letters form a group of potential II-VI semiconductors.

The crystal structure of these compounds is rather complex. CdS and CdSe crystallize in the hcp wurtzite structure, and ZnTe and CdTe in the zinc blende structure, whereas ZnS and ZnSe can exist in both these forms. The bonding in these compounds is a mixture of covalent and ionic types. This is because the average number of electrons per atom is still 4, but group VI atoms are considerably more electronegative than group II atoms, and this introduces ionicity. The ionic character varies considerably over the whole range and increases as the atomic size decreases. Thus, while ZnS is predominantly ionic, the bonding is nearly covalent in HgTe. The ionic character has the effect of binding the valence electrons rather tightly to the lattice atoms. Thus, the band gaps of these compounds are larger than those of the covalent semiconductors of comparable atomic weights, although their hardness does not match that of covalent semiconductors. Because of their large band gaps, these materials are considered potential sources of electroluminescent devices.

Like III-V compounds, ternary compounds of the form $A^{II}B_x^{VI}C_{1-x}^{VI}$ can also be obtained from many of these semiconductors. Thus, it has been possible to grow crystals of ZnS_xSe_{1-x}, with x ranging from 0 to 1. Here again the properties of the resulting compound vary gradually with the fraction x.

Narrow Gap Semiconductors Another class of semiconductors that have received considerable attention in recent years because of their applications as infrared detectors and radiation sources are IV-VI narrow gap semiconductors. Four of these compounds, namely, PbS, PbSe, PbTe, and SnTe, crystallize in the simple cubic NaCl crystal structure. The bonding in these compounds is mainly ionic with some covalent contribution.

With the exception of SnTe, all of these compounds have a small band gap and low effective mass of the carriers. In IV-VI compounds and their alloys, single crystals of somewhat inferior quality can be easily prepared. These crystals have a number of defects such as voids, inclusions, dislocations, and point defects. High-quality crystals are difficult to prepare.

Other Semiconductors Several other compounds have been found to have semiconducting properties. These include a number of simple metallic oxides (such as MgO and Cu_2O), a large number of transition metal oxides (TiO_2, V_2O_3, Fe_2O_3, and NiO), and several other oxides having the general formula AB_2O_4, where A is a second group element and B is a transition metal. Typical examples of this class are $MgFe_2O_4$ and $ZnFe_2O_4$. Some of these compounds have found use in devices such as thermistors, but many others are only of academic interest.

3.3 THE VALENCE BOND MODEL OF THE SEMICONDUCTOR

After presenting a brief account of different types of semiconductors, our next endeavor will be to study the basic principles of semiconductor theory. In doing so, we will be concerned mainly with the elemental semiconductors. There are two reasons for making this choice. First, elemental semiconductors have the simplest structure and their properties are easy to understand. A second and more compelling reason is that these materials are widely used in semiconductor devices. During the early days of transistor development, Ge was the dominant material used in making diodes and transistors. However, most of the present-day devices employ Si. Next comes GaAs which has found applications in a number of devices where it has established its superiority over Si. However, GaAs technology is more involved, and the material is much more expensive than Si. There are three main reasons why Si has become the device material of choice:

1. Si is the most readily available material and is found in its oxides and silicates which form about 25 percent of the earth's crust.
2. Si has a wider band gap than Ge. Thus, Si devices can be operated at temperatures up to 200°C compared to Ge devices which cannot be operated above 80°C.

3. A layer of stable native oxide (SiO$_2$) can be grown thermally on the surface of Si. This oxide layer can be used for selective masking of the silicon surface. The grown SiO$_2$ also makes the surface of the device passive and can be used for protecting the device against the effects of the ambient. No other semiconductor is known to form a stable oxide.

Two approaches can be employed to study semiconductors: the valence bond model and the energy band model. The valence bond model emphasizes events in space and time, whereas the energy band model does the same in energy and momentum. Each of these models gives some information that the other cannot, and in some respects, they complement each other.

3.3.1 Intrinsic Semiconductors

As we saw in the previous chapter, Ge and Si crystallize in a diamond lattice, with each atom bound to its four nearest neighbors by covalent bonds. These neighbors are all equidistant from the central atom and lie at the four corners of a tetrahedron (Fig. 3.2). The distance between the nearest neighbors is $\sqrt{3}a/4$ where a is the side of a unit cell (Fig. 2.7). Each bond has two electrons with opposite spins, so that the central atom appears to have eight electrons in its valence orbit which it shares with its four nearest neighbors. A large crystal of silicon can be generated by repeating this unit cell in space.

Let us consider a pure crystal of Si containing no impurities and no lattice defects. A two-dimensional representation of such a crystal is shown in Fig. 3.3. Here the circles represent the atomic core with four positive charges, and the dots represent the valence electrons. At 0°K all the electrons are tightly bound in the bonds, and the crystal is a perfect insulator. At higher temperatures lattice atoms vibrate around their equilibrium positions. These vibrations occasionally impart enough energy to some of the bound electrons, enabling them to leave the bond and move freely inside the crystal. Fig. 3.3a illustrates the creation of a free electron by breaking a covalent bond at position A. The vacant site created by the broken bond has a net positive charge left on the atomic core. This va-

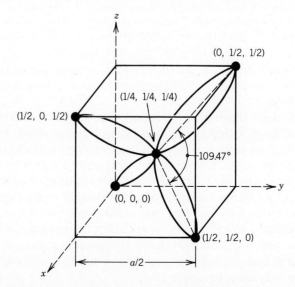

FIGURE 3.2 Tetrahedron configuration of Si crystal showing the four nearest neighbors.

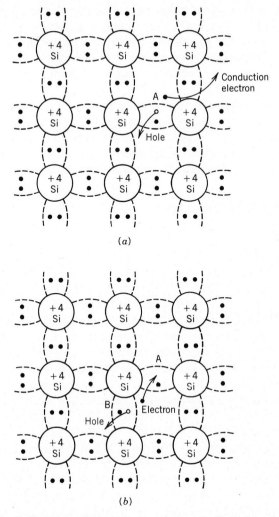

FIGURE 3.3 A two-dimensional schematic representation of silicon crystal showing the atom cores and valence electrons. (*a*) Thermal generation of an electron–hole pair and (*b*) motion of a hole from location *A* to *B*. (From S. M. Sze, *Semiconductor Devices: Physics and Technology*, p. 9. Copyright © 1985. Reprinted by permission of John Wiley & Sons, Inc., New York.)

cancy behaves like a particle of positive charge and mass equal to that of an electron and is called a *hole*. Like a free electron, a hole can also move freely through the crystal.

It might seem quite disturbing that a vacant site, short of an electron, can move through the crystal. One way to visualize the movement of a hole is shown in Fig. 3.3*b*. Here, as a result of thermal motion, an electron in a covalent bond at *B* adjacent to the hole jumps into the original vacant site *A*, thus transferring the location of the hole to the site from where the electron came. In this way, the motion of a hole can be regarded as the transfer of ionization from one atom to another by the jumping of the bound electron from a covalent bond to a neighboring vacant site. This process does not affect the motion of a free electron. Thus, electrons and holes can move freely through the crystal independent of each other. This picture of a hole has a number of serious limitations which will become clear from the following discussion.

Section 3.8.1 will show that at room temperature the wave function of a hole (or an electron) extends over a region of space that encompasses several thousand lattice atoms. Thus, the size of a hole is orders of magnitude larger than the interatomic spacing. This being the case, any notion of a hole movement being created by the jumping of a bound electron between neighboring bonds is meaningless.

Another difficulty is encountered when one tries to explain the Hall effect (see Sec. 4.5). If the motion of a hole can be represented by the motion of an electron in the opposite direction, then the Hall voltage will have the same negative sign regardless of whether electrons or holes are moving. However, the measured Hall voltage is positive when the current is carried by holes.

Thus, it is clear that the mental picture of a hole provided by the valence bond model, though helpful in visualizing how a hole is created, is not useful in understanding the motion of a hole as a particle. This should not be surprising inasmuch as neither a hole nor an electron is a classical particle.

In the above process, the electron–hole pairs were produced by the thermal vibrations of lattice atoms. This process of pair generation is called the *thermal generation process*. In addition to this process, there is a reverse process of *recombination* in which a freely moving electron loses its ionization energy by reentering into a broken covalent bond. In this way an electron–hole pair disappears. When a semiconductor crystal is in equilibrium at a given temperature, a certain number of bonds always remain broken. Any additional breaking of bonds above their thermal equilibrium value is exactly balanced by the reverse process of recombination.

The semiconductor under discussion will have as many vacant sites as the number of free electrons. Such a semiconductor is called an *intrinsic semiconductor*, and the holes and electrons created in this way are called *intrinsic charge carriers*. The total number of holes and electrons in a piece of semiconductor is rarely of any interest. What is usually important is the number of carriers per unit volume, known as *carrier density* or *carrier concentration*, and it is expressed as the number of carriers per cm^3. Thus, for an intrinsic semiconductor

$$p = n = n_i \tag{3.1}$$

where p and n represent the hole and electron concentrations, respectively, and the subscript i is used to denote that the semiconductor is intrinsic. The intrinsic concentration n_i varies with the temperature and is different for different semiconductors. The values of n_i for Si, Ge, and GaAs at 300 K are given in Chapter 4 in Table 4.2.

3.3.2 Extrinsic Semiconductors

Apart from thermal generation, charge carriers in semiconductors can also be created by adding certain types of impurity atoms. For silicon and germanium, the impurities of most interest are the elements of groups III and V shown in the periodic table. These impurities are intentionally added after purifying the crystal. The process of adding impurities to a semiconductor is called *doping*, and the impurity-added semiconductors are known as *doped semiconductors*. By suitable doping, a crystal can be made to have unequal values of electron and hole concentrations. The resulting semiconductor is said to be *extrinsic*. Almost all the semiconductor devices employ extrinsic semiconductors.

(a) N–TYPE SEMICONDUCTORS

Let us assume that a very small amount of a group V element (As, P, Sb) is added to an otherwise pure crystal of silicon. Upon incorporation into the crystal, these impurities occupy lattice sites that are normally occupied by Si atoms. Therefore, these dopants are called *substitutional impurities*. Fig. 3.4 shows the situation when an As atom replaces a Si atom in the lattice. An As atom has five valence electrons. Four of these make covalent bonds with the Si atoms, but the fifth electron cannot be accommodated in the bonding arrangement and is bound to the impurity atom only by weak electrostatic forces. At 0°K this electron will move around the impurity atom just like an electron in a hydrogen atom. Therefore, we can use the Bohr theory of the hydrogen atom to calculate the radius of its ground state orbit (r_o) and the ionization energy (E_I). However, two changes must be made. First, the effective mass of the electron m_e^* has to be used in place of the free electron mass m_o. Second, the free space permittivity (ε_o) has to be replaced by the permittivity of the semiconductors (ε_s). Thus, we can write

$$r_o = 0.53 \left(\frac{\varepsilon_s}{\varepsilon_o}\right)\left(\frac{m_o}{m_e^*}\right) \text{ Å} \qquad (3.2a)$$

and

$$E_I = 13.6 \left(\frac{m_e^*}{m_o}\right)\left(\frac{\varepsilon_o}{\varepsilon_s}\right)^2 eV \qquad (3.2b)$$

where 0.53 Å is the ground state Bohr radius and 13.6 eV is the ionization energy of the electron in the hydrogen atom. For germanium, $m_e^* \approx 0.5 m_o$ and $\varepsilon_s = 16\varepsilon_o$. Substituting these values in Eqs. (3.2a) and (3.2b) yields a ground state radius of approximately 16 Å and an ionization energy of 0.026 eV. For silicon, the corresponding values are 6.3 Å and 0.1 eV, respectively.

From the above discussion, it is clear that a very small energy is required to ionize the extra electron attached to the impurity atom. The magnitude of thermal energy kT/q at 290 K is 0.025 eV, which is comparable to the electron ioniza-

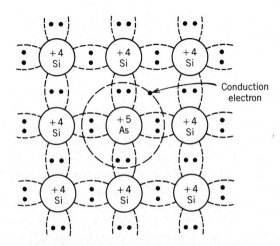

FIGURE 3.4 A donor atom in a silicon crystal.

tion energy. Thus, at ordinary temperatures the thermal vibrations of the lattice atoms will impart enough energy to the electron to shake it free, whereupon it becomes a conduction electron. The impurity atom left behind becomes a positive ion. This ion is immobile under an applied electric field because it is bound to neighboring atoms by the usual covalent bonds. When a freely moving electron gets attracted to it owing to electrostatic attraction, the electron will not stay long because the thermal vibrations of the lattice will again shake it free. Since the impurity atom donates one free electron to the crystal, it is called a *donor atom* or simply a *donor*.

Now consider a large number of donor atoms randomly distributed in the semiconductor. At a sufficiently high temperature, each of these atoms donates one conduction electron to the crystal. Therefore, the electron concentration in the semiconductor rises above n_i and becomes higher than the hole concentration. Such a semiconductor is known as an *n-type semiconductor*. In an *n*-type semiconductor, the electrons are in the majority and are termed the *majority carriers*, whereas the holes, which are in the minority, are the *minority carriers*.

A rise in the electron concentration above n_i depresses the hole concentration below n_i. This happens because of the electron–hole pair recombination process. The rate of recombination R is proportional to the electron and hole concentrations (n and p), and can be written as $R = \alpha_r pn$, where α_r is a constant of proportionality. For an intrinsic semiconductor in thermal equilibrium $p = n = n_i$ and $R = \alpha_r n_i^2$. Since this rate must be balanced by the thermal generation rate G_{th}, we have

$$R = G_{th} = \alpha_r n_i^2 \tag{3.3}$$

At a given temperature, G_{th} is found to be independent of doping of the semiconductor provided that the dopant concentration remains small (i.e., less than 1 dopant atom for 10^6 atoms). Now consider a crystal with the equilibrium concentrations n_o and p_o, where the subscript o is used to denote thermal equilibrium. For this crystal $R = \alpha_r p_o n_o = G_{th}$. Thus, if n_o is increased above n_i by doping the semiconductor, p_o must be decreased to keep $p_o n_o = n_i^2$. The physical process is as follows: The donor atoms contribute additional electrons to the crystal. Some of these electrons fall back into the broken bonds causing a decrease in the hole concentration below n_i.

(b) P–TYPE SEMICONDUCTORS

An atom of group III elements (B, Al, In, Ga) also occupies a substitutional position in the Si (or Ge) lattice but has only three valence electrons. These electrons complete three of the covalent bonds, leaving one bond vacant which accepts an electron from a neighboring atom and creates a hole in the crystal as shown in Fig. 3.5a for an Al atom in the Si lattice. When this happens, a negative charge is left on the Al atom. The hole is attracted toward this negative charge and moves around it in a Bohr-like orbit as shown in Fig. 3.5b. At 0°K the hole remains bound to the impurity atom but breaks away from it at higher temperatures, whereupon it wanders freely in the crystal. The binding energy of the hole can be calculated from Eq. (3.2b) after replacing m_e^* by the hole effective mass m_h^*. The calculated value is in reasonably good agreement with the experimentally determined values in Ge, but only an order of magnitude agreement is seen in the case of Si.

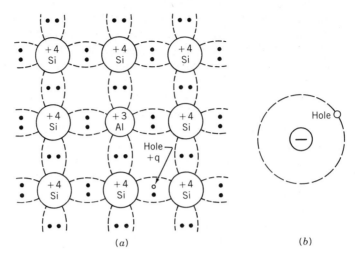

FIGURE 3.5 (a) An acceptor atom in silicon crystal and (b) hole motion around the negatively charged acceptor atom.

An impurity atom that contributes a hole is called an *acceptor atom* because it accepts a bound electron from the covalent bond. When a number of acceptor atoms are added to the semiconductor, most of these atoms ionize at room temperature, contributing one hole per impurity atom. The resulting crystal has an excess of holes compared to the electrons and is called a *p-type semiconductor*. Like the donor atom, the ionized acceptor atom remains immobile in an electric field.

In a semiconductor crystal that is doped with both donors and acceptors, the extra electrons attached to the donor atoms fall into the incomplete bonds of the acceptor atoms so that neither electrons nor holes will be produced. This process is known as *compensation,* and the semiconductor having both donors and acceptors is said to be compensated. If the number of donors exceeds the number of acceptors, the semiconductor becomes the *n*-type, and vice versa for a *p*-type semiconductor. When the two types of impurities are equal in number, the material is said to be fully compensated. The electron–hole concentrations in such a semiconductor are equal, but many of the properties (e.g., resistivity) are different from those of the intrinsic crystal.

(c) DOPING OF COMPOUND SEMICONDUCTORS

Compound semiconductors have more than one kind of atom. Therefore, the role of an impurity in these semiconductors depends largely on which kind of atom is replaced by the impurity atom. In III-V semiconductors, group II elements replace the trivalent atoms and act as acceptors. The most important donors are atoms from group VI which replace the pentavalent atoms. Atoms of group IV are called *amphoteric impurities.* They act as donors when they are built into the sites occupied by group III elements; they act as acceptors when they replace the pentavalent atoms.

In some compound semiconductors like GaP, it has been observed that even isovalent atoms (those having the same number of valence electrons) act as donors and acceptors. For example, nitrogen (N) is an isovalent acceptor in GaP. The electronegativity of an N atom is greater than that of the P atom. When the N atom replaces P in GaP, it becomes negatively charged by attracting an extra electron to itself. The electrostatic field of this negative charge can loosely bind a hole that becomes free only at high temperatures. Similarly, a much less electronegative isovalent atom, like Bi, substituted for P in GaP acts as a donor. In a III-V semiconductor (such as GaAs) the presence of a Ga atom at an As site (or vice versa) would create an electrically active defect. However, in covalently bonded compounds no more than one atom out of 10^9 atoms is situated on the wrong sublattice, and the effect of this type of defect is not very significant.

In II-VI semiconductors, group III elements replace the group II elements and act as donors. Elements of group VII (Cl, Br, and I) also act as donors by replacing group VI elements. Similarly, a p-type semiconductor can be obtained by doping either with group I or with group V elements. Semiconductors that can be made both p- and n-type are said to be *ambipolar*. Elemental semiconductors (Ge and Si) and most of the III-V compounds are ambipolar. II-VI semiconductors that have a larger ionic contribution in bonding tend to be *unipolar*. Some of them are found only n-type semiconductors while others are only the p-type. The mechanism that gives rise to this behavior is known as *self-compensation*.

To understand self-compensation, let us assume that ZnTe is to be doped with iodine (I) to make it n-type. For an I atom to replace a Te atom, the semiconductor temperature has to be raised, and this creates zinc vacancies by evaporation of Zn atoms. Each Zn vacancy acts as a double acceptor by accepting two free electrons. Let $\Delta E(Zn)$ be the energy required to create a Zn vacancy, and let $\Delta E(I)$ be the energy released when a donor electron drops into the acceptor level created by the Zn vacancy. Since the binding energy of donors and acceptors is small, $\Delta E(I) \simeq E_g$. Now suppose that a Zn vacancy is created in ZnTe doped with I. If the two donor electrons (from two I atoms) can drop into the Zn vacancy, an energy of $2E_g$ is regained in the process. If $2E_g > \Delta E(Zn)$, the energy of the system is lowered by the creation of Zn vacancies, and the trapping of the donor electrons by these vacancies will frustrate all efforts to dope ZnTe with I atoms. Thus, ZnTe always remains p-type. This is because the Zn atom is smaller than the Te atom, and its binding energy is lower than that of a Te atom. Just the opposite is true for CdS. Here the small-sized S atoms have smaller binding energies than the larger Cd ions. Thus, CdS always has an excess of sulphur vacancies, which act as donors, and is always n-type. The same is true for CdSe and ZnSe. However, in CdTe the energy E_g is small, and self-compensation is not complete, so CdTe is ambipolar.

In narrow gap semiconductors PbS, PbSe, and PbTe, doping is easily affected by deviation from stoichiometric composition. Excess metal atoms act as donors, whereas excess chalcogenide atoms act as acceptors. Impurity doping is possible with Tl as an acceptor and Bi as a donor.

3.4 THE ENERGY BAND MODEL

3.4.1 The Energy Band Structure of Semiconductors

In a three-dimensional crystal **k** has three components—k_x, k_y, k_z—and a two-dimensional representation of the E versus **k** relation is difficult. Because of crystal symmetry, it becomes more meaningful to express E as a function of **k** along the principal directions of the crystal. Fig. 3.6 shows the E-**k** plots for two semiconductors along the [100] and [111] directions. In Fig. 3.6a it is seen that the lowest conduction band minimum and the valence band maximum are situated at the same value of **k**. Semiconductors with this type of band structure are called

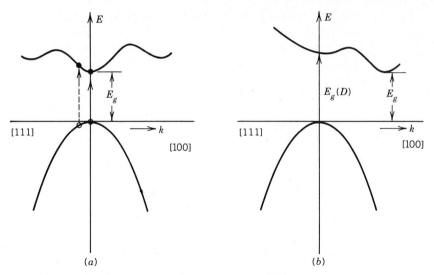

FIGURE 3.6 E-**k** plots for (a) direct gap and (b) indirect gap semiconductors.

direct gap semiconductors. The significance of this term is that a photon of energy $h\nu = E_g$ can excite an electron from the top of the valence band directly into a state at the bottom of the conduction band, as shown by the vertical line in Fig. 3.6a. If the photon energy $h\nu > E_g$, the excess energy is shared between the electron and the hole, and both have the same value of **k** (dashed line). Fig. 3.6b is the E-**k** diagram for an indirect gap semiconductor. Here the valence band maximum and the lowest conduction band minimum are situated at different values of **k**. In such semiconductors, the direct transition of the electron from the valence band to the conduction band by a photon of energy E_g is not possible. This happens because a photon has very small momentum, whereas in a nonvertical transition over E_g, an electron suffers a large change in momentum. However, a nonvertical transition can occur indirectly by the cooperation of a lattice phonon which can supply the required momentum. In this case, the minimum photon energy required to produce an electron–hole pair will be larger than E_g. Vertical transitions are also possible for $h\nu > E_g(D)$, where $E_g(D)$ is the direct gap energy. Elemental semiconductors Ge and Si have an indirect band gap, as do GaP and all Al compounds. Other III-V and most II-VI semiconductors have direct band gap.

Figure 3.7 shows the E-**k** diagram of a substance in which the conduction and valence bands overlap slightly. Because of this overlap, some states in the conduction band get filled leaving behind an equal number of empty states in the valence band, thus, making available a finite number of electron–hole pairs, even at 0°K. Such substances are called *semimetals*. Typical examples are Bi, As, Sb, and some of their alloys.

3.4.2 Energy Band Representation of Semiconductors

(a) INTRINSIC SEMICONDUCTOR

The energy band scheme in Fig. 3.6 is too detailed. One generally represents the electron energy in the valence and conduction bands separated by a band gap as shown in Fig. 3.8 for an intrinsic semiconductor. Here E_V represents the valence band maximum, and E_C represents the conduction band minimum. No distinc-

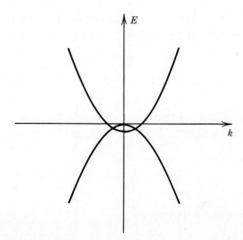

FIGURE 3.7 Energy band diagram for a semimetal.

tion is made between the direct and indirect gap semiconductors. The vertical distance in Fig. 3.8 represents electron energy that increases upwards, and therefore, the electrostatic potential increases downwards. The horizontal coordinate in the diagram has no significance since it is a point diagram that is valid at each point in the interior of the semiconductor (but not at the surface). We recall that the energy band picture of a solid is based on the coupling of identical atoms that are arranged in a periodic array. Thus, the energy band diagram is a characteristic of the material, provided that the volume of the sample contains a sufficiently large number of unit cells—a requirement that is met in all laboratory-size samples.

Let us now consider the intrinsic semiconductor shown in Fig. 3.8. At 0°K the valence band is completely filled, the conduction band has no electrons, and the material is an insulator. At a finite temperature, thermal vibrations of lattice atoms produce a certain concentration n_i of electrons in the conduction band and an equal number of empty states in the valence band. In the presence of an externally applied field, electrons will conduct in the conduction band, and the vacant sites (i.e., the holes) will conduct in the valence band.

How an empty state in a valence band acts as a hole can be seen as follows: Suppose there are $4N$ states in the valence band and the current arising from

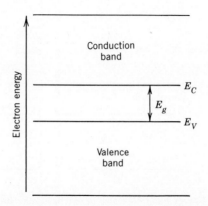

FIGURE 3.8 A two-dimensional energy band diagram of an intrinsic semiconductor.

each of these states is $q\mathbf{v}_i$ where $\mathbf{v}_i$ is the velocity of electron in the ith energy state. Thus, for the completely filled band, the current density $\mathbf{J}$ can be written as

$$\mathbf{J} = -\frac{q}{V}\sum_{i=1}^{4N}\mathbf{v}_i = 0 \tag{3.4}$$

where V is the volume. Now assume that some jth state near the top of the valence band is empty. The current density $\mathbf{J}_p$ arising from the band is then given by

$$\mathbf{J}_p = -\frac{q}{V}\sum_{i\neq j}\mathbf{v}_i \neq 0 \tag{3.5}$$

and this can be written as

$$\mathbf{J}_p = -\frac{q}{V}\left[\sum_{i=1}^{4N}\mathbf{v}_i - \mathbf{v}_j\right] = 0 + \frac{q}{V}\mathbf{v}_j \tag{3.6}$$

Thus, in the absence of the jth electron, the current density is given by $q\mathbf{v}_j$.

(b) EXTRINSIC SEMICONDUCTORS

The energy band diagram in Fig. 3.8 can be easily modified to include the effect of the electron energy levels of donor and acceptor atoms. We have seen that the fifth electron of a donor atom is loosely bound and can be set free by supplying an energy E_I which is only about tens of milli electron volt. Thus, the electron attached to a donor atom must have an energy level E_d below the conduction band edge E_C such that $E_C - E_d = E_I$. This level is shown in Fig. 3.9a. Obviously, E_d lies in the otherwise forbidden region. It is customary to show the donor level by a dashed line because donor atoms are randomly distributed in the crystal and do not form a periodic structure like the host atoms. It is to be noted that the level represented by E_d corresponds to the ground state of the electron attached to the donor atom. There could be many excited states between E_d and E_C, but these states are of little interest to us because at ordinary temperatures almost all the donors are ionized and their electrons are set free into the conduction band. The energy band diagram for a p-type semiconductor is shown in Fig. 3.9b. The acceptor energy level E_a lies in the forbidden gap close to the valence band edge E_V.

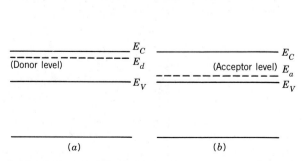

FIGURE 3.9 Energy band diagrams of extrinsic semiconductor showing (a) the donor level and (b) the acceptor level.

THE ENERGY BAND MODEL

It is now easy to see what happens in a semiconductor that is doped with both donors and acceptors. In such a semiconductor, electrons from the level E_d instead of going to the conduction band will drop into the acceptor states at E_a; each such jump eliminates one electron–hole pair that would have been there, provided this jump could be prevented. If there are more donors than acceptors, then all the states at E_a will be filled by the electrons from the donor levels and the leftover electrons of the donor atoms will be excited to the conduction band to make the semiconductor n-type. If there is an excess of acceptors, then the semiconductor will become p-type.

3.4.3 The Energy Levels of Impurities in Si and GaAs

Our discussion thus far has been confined to the donor and acceptor atoms that give rise to energy levels close to band edges. These levels are called *shallow levels*. Several impurities give rise to energy levels deep in the band gap; they are known as *deep impurities*.

Figure 3.10 depicts the energy levels of the commonly known impurities in Si and GaAs. The ionization energies for important impurity atoms are also given. Note that many of these impurities give rise to more than one energy level.

3.4.4 The Energy-Momentum Relation in Valence and Conduction Bands

From the energy band diagram of a semiconductor, it is clear that an electron is not available in the conduction band unless its total energy E exceeds E_C. An electron with energy $E = E_C$ is said to have zero kinetic energy. If the energy E

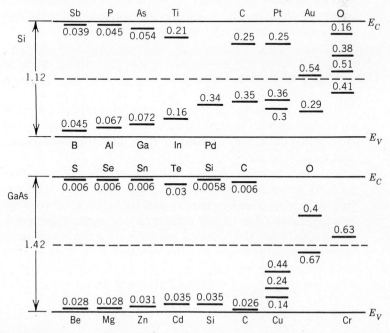

FIGURE 3.10 Energy levels of impurities in Si and GaAs. (From S. M. Sze, *Semiconductor Devices: Physics and Technology*, p. 23. Copyright © 1985. Reprinted by permission of John Wiley & Sons, Inc., New York.)

is greater than E_C, the difference $(E - E_C)$ will appear as the kinetic energy of the electron in the conduction band. Assuming the electron to be a classical particle with effective mass m_e^* and momentum **p**, we can write energy E as

$$E = E_C + \frac{\mathbf{p}^2}{2m_e^*} = E_C + \frac{1}{2}m_e^*\mathbf{v}^2 \qquad (3.7a)$$

Similarly, a hole at E_V has a zero kinetic energy, and its energy E in the valence band can be expressed as

$$E = E_V - \frac{\mathbf{p}^2}{2m_h^*} \qquad (3.7b)$$

These relations tell us how the energy of electrons and holes in the conduction and valence bands depends on their momenta.

3.5 EQUILIBRIUM CONCENTRATIONS OF ELECTRONS AND HOLES INSIDE THE ENERGY BANDS

Now that the general features of energy bands in semiconductors have been described, the important problem to investigate is how the electrons and holes are distributed among the various available states inside the bands when thermal equilibrium prevails. Note that there are no allowed states in the forbidden gap. To calculate the electron and hole concentrations in the conduction and valence bands, we need to know how many energy states are available in a given energy interval and how these states are occupied by the electrons. The product of the density of states and the probability of occupancy of these states give the number of electrons in the energy interval of interest.

3.5.1 The Density of States Function

A function that gives the number of available electron states per unit volume in an energy interval between E and $E + dE$ is designated by $N(E)dE$, where $N(E)$ is known as the *density of states function* and represents the number of states per unit volume per eV.

To derive an expression for $N(E)$, we need to consider a crystal of finite volume. We assume that the constant energy surfaces are spherical so that the effective mass m^* is a scalar quantity and the relation between energy and momentum is parabolic. Let us now consider a free carrier (electron or hole) of mass m^* in the solid. Since the potential energy is arbitrary, we assume $V = 0$. The time-independent Schrödinger equation can then be written as

$$\nabla^2 \psi + \mathbf{k}^2 \psi = 0 \qquad (3.8a)$$

with

$$\mathbf{k}^2 = \frac{2m^*}{\hbar^2}E \qquad (3.8b)$$

CONCENTRATIONS OF ELECTRONS AND HOLES INSIDE THE ENERGY BANDS

Consider a volume in **k** space with components k_x, k_y, and k_z as shown in Fig. 3.11. Since $\mathbf{k}^2 = k_x^2 + k_y^2 + k_z^2$, Eq. (3.8b) can be written as

$$E = \frac{\hbar^2}{2m^*}(k_x^2 + k_y^2 + k_z^2) \tag{3.9}$$

Using the method of separation of variables, we can write

$$\psi(x,y,z) = \psi(x)\psi(y)\psi(z) \tag{3.10}$$

Substituting for $\psi(x,y,z)$ from Eq. (3.10) and for **k** from Eq. (3.9) into Eq. (3.8a), we can show that $\psi(x)$ has the solution

$$\psi(x) = B_1 \exp(jk_x x) + B_2 \exp(-jk_x x) \tag{3.11}$$

Since a laboratory-size crystal is large compared to the dimensions of a unit cell, we assume $\psi(x)$ to be periodic with periodicity L, so that*

$$\psi(x + L) = \psi(x) \tag{3.12}$$

With this condition, Eq. (3.11) will be satisfied only when

$$k_x = \frac{2\pi}{L} n_x \tag{3.13}$$

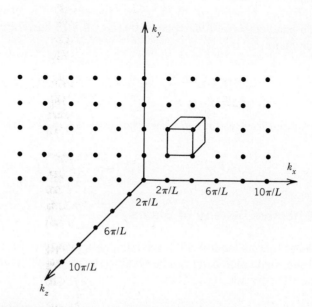

FIGURE 3.11 The k-space for a free carrier of mass m^* with a unit cell.

*Equation (3.12) is known as periodic boundary condition.

where n_x is a positive integer. Similar solutions can be found for $\psi(y)$ and $\psi(z)$. The vector **k** thus has the components

$$k_x = \frac{2\pi}{L}n_x, \quad k_y = \frac{2\pi}{L}n_y, \quad k_z = \frac{2\pi}{L}n_z \qquad (3.14)$$

where n_y and n_z are also positive integers. From Fig. 3.11 the volume of a unit cell in **k** space is $(2\pi/L)^3 = 8\pi^3/L^3$. Therefore, the volume of a unit cell in **k** space for a crystal of unit volume is $8\pi^3$. For a crystal of unit volume, the volume of a spherical shell in **k** space having radius between k and $(k + dk)$ is given by

$$\frac{4\pi}{3}[(k + dk)^3 - k^3] \simeq 4\pi k^2 dk$$

Thus, the number of unit cells in this volume is $k^2 dk/2\pi^2$. The density of available electron states is twice as large as this because each cell can have two electrons with different spins. Hence, the number of electron states per unit volume of the crystal within an infinitesimal range dk of k is given by

$$N(k)dk = \frac{k^2}{\pi^2} dk \qquad (3.15)$$

Using Eq. (3.8b), we can write this as

$$N(E)dE = \frac{4\pi}{h^3}(2m^*)^{3/2}E^{1/2}dE \qquad (3.16)$$

This equation gives the density of states for a particle of mass m^* with energy between E and $E + dE$. This result can be applied to electrons in the conduction band after replacing m^* by m_e^*. Since the energy E is measured from the conduction band edge E_C, E is replaced by $(E - E_C)$. Thus, for electrons in the conduction band

$$N(E)dE = \frac{4\pi}{h^3}(2m_e^*)^{3/2}(E - E_C)^{1/2}dE, \quad \text{for } E > E_C \qquad (3.17a)$$

A similar expression can be written for holes in the valence band:

$$N(E)dE = \frac{4\pi}{h^3}(2m_h^*)^{3/2}(E_V - E)^{1/2}dE, \quad \text{for } E < E_V \qquad (3.17b)$$

3.5.2 The Effective Density of States

Under thermal equilibrium, the number of electrons dn_o in an energy interval dE of the conduction band can be obtained by multiplying the function $N(E)dE$ with the FD function $f(E)$. Thus,

$$dn_o = N(E)f(E)dE$$
$$= \frac{4\pi}{h^3}(2m_e^*)^{3/2}(E - E_C)^{1/2}\left[1 + \exp\left(\frac{E - E_F}{kT}\right)\right]^{-1}dE \qquad (3.18)$$

The density of electrons in the conduction band can be obtained by integrating Eq. (3.18) from E_C to some value E_{top} at the top of the conduction band. The integral, however, becomes complicated. For ease of calculation, we limit the range of application of Eq. (3.18) by assuming that $(E_C - E_F) > 3kT$. With this limitation, the FD distribution function can be approximated by the Boltzmann distribution:

$$f(E) = \left[1 + \exp\left(\frac{E - E_F}{kT}\right)\right]^{-1} \simeq \exp\left[-\frac{(E - E_F)}{kT}\right]$$

The integral of Eq. (3.18) can now be written as

$$n_o = \frac{4\pi}{h^3}(2m_e^*)^{3/2} \int_{E_C}^{E_{\text{top}}} (E - E_C)^{1/2} \exp\left[-\frac{(E - E_F)}{kT}\right] dE \quad (3.19)$$

In a typical semiconductor, the width of the conduction band is several eV, but most of the electrons are confined in the vicinity of E_C so that the upper limit of integration can be replaced by infinity. The resulting integral can be easily evaluated after substituting

$$\frac{E - E_C}{kT} = y.$$

Thus

$$n_o = \frac{4\pi}{h^3}(2m_e^* kT)^{3/2} \exp\left(-\frac{E_C - E_F}{kT}\right) \int_0^\infty y^{1/2} \exp(-y)\, dy$$

$$= N_C \exp\left[-\frac{E_C - E_F}{kT}\right] \quad (3.20)$$

where

$$N_C = 2\left(\frac{2\pi m_e^* kT}{h^2}\right)^{3/2} \quad (3.21)$$

N_C is called the *effective density of states in the conduction band*. It represents the number of states required to be placed at the energy E_C, which, after multiplying with the probability of occupation of this level, gives the number of electrons in the conduction band. Note that in reality the density of states at E_C is zero and Eq. (3.20) is just an artifice to obtain a simple expression for n_o.

To compute the equilibrium hole concentration p_o, we note that the distribution function $f_h(E)$ of a hole is that of a state not occupied by an electron. Thus

$$f_h(E) = 1 - f(E) = 1 - \left[1 + \exp\left(\frac{E - E_F}{kT}\right)\right]^{-1} = \frac{1}{1 + \exp\left(\frac{E_F - E}{kT}\right)} \quad (3.22)$$

A derivation similar to that implemented for the electrons leads to (see Prob. 3.12)

$$p_o = N_V \exp\left(-\frac{E_F - E_V}{kT}\right) \quad (3.23)$$

$$N_V = 2\left(\frac{2\pi m_h^* kT}{h^2}\right)^{3/2} \qquad (3.24)$$

where N_V represents the *effective density of states in the valence band*.

Equations (3.21) and (3.24) show that both N_C and N_V depend on temperature. Since the effective masses of electrons and holes are not the same, N_C and N_V for a semiconductor are different (see Table 4.2).

3.5.3 The Equilibrium Electron–Hole Product

Since the expressions for N_C and N_V are independent of the doping of the semiconductor, relations (3.20) and (3.23) are valid for any semiconductor provided that the Fermi level E_F lies in the forbidden gap and is at least $3kT$ away from the band edges. These relations are also valid for an intrinsic semiconductor. If E_i denotes the energy of the Fermi level in an intrinsic semiconductor, then Eqs. (3.20) and (3.23) can be written as

$$n_i = N_C \exp\left(-\frac{E_C - E_i}{kT}\right) \qquad (3.25)$$

and

$$p_i = n_i = N_V \exp\left(-\frac{E_i - E_V}{kT}\right) \qquad (3.26)$$

Combining Eq. (3.25) with (3.20) and Eq. (3.26) with (3.23) leads to

$$n_o = n_i \exp\left(\frac{E_F - E_i}{kT}\right) \qquad (3.27)$$

and

$$p_o = n_i \exp\left(\frac{E_i - E_F}{kT}\right) \qquad (3.28)$$

These relations show that $p_o n_o = n_i^2$. It is clear from Eq. (3.27) that for an n-type semiconductor ($n_o > n_i$) the Fermi level E_F lies above E_i and hence is close to the conduction band edge E_C. Similarly, for a p-type semiconductor the Fermi level lies close to the valence band edge E_V. Multiplying Eq. (3.20) by Eq. (3.23) yields

$$p_o n_o = n_i^2 = K_1^2 T^3 \exp(-E_g/kT) \qquad (3.29)$$

where

$$K_1^2 = 32\left(\frac{\pi^2 k^2 m_e^* m_h^*}{h^4}\right)^{3/2}$$

The effective masses m_e^* and m_h^* which enter into the expressions for N_C and N_V are called *density of states effective masses*. Equation (3.29) is an important relation since it shows that, for any semiconductor in thermal equilibrium, the product of electron and hole concentrations is independent of doping.

We would like to emphasize that Eqs. (3.19) through (3.29) are valid only for semiconductors in which the mobile carrier concentration is small. Consequently, the FD distribution function can be replaced by the Maxwell-Boltzmann distribution. Physically, this means that the number of mobile carriers in the bands is so small that only a very small fraction of available states is occupied. This is known as the *diluteness condition* because the electrons (solute) form a dilute solution in the available energy states (solvent). A dilute electron gas is said to be nondegenerate, and the resulting semiconductor is called a *nondegenerate semiconductor*.

3.5.4 The Temperature Dependence of Intrinsic Carrier Concentration

From Eq. (3.29) the intrinsic concentration n_i can be written as

$$n_i = K_1 T^{3/2} \exp(-E_g/2kT) \tag{3.30}$$

In Ge and Si, the band gap E_g decreases monotonically with temperature and can be expressed as

$$E_g(T) = E_{go} - b_1 T \tag{3.31}$$

where E_{go} is the extrapolated value of the band gap at 0°K and b_1 is the rate of decrease of the energy gap with temperature. The linear form of Eq. (3.31) is fairly accurate from 100 K to 400 K, but is grossly inaccurate at low temperatures. Substituting Eq. (3.31) into Eq. (3.30) leads to the equation

$$n_i = K_2 T^{3/2} \exp(-E_{go}/2kT) \tag{3.32}$$

where

$$K_2 = K_1 \exp(b_1/2k)$$

By substituting the numerical values of m_e^*, m_h^*, b_1, and E_{go}, n_i as a function of temperature can be determined. However, because there are many difficulties in the independent measurements of these quantities, such an effort does not lead to results in close agreement with the experimentally determined values. From measurements, it is found that for temperatures above 50 K [4]

$$n_i(T) = 1.76 \times 10^{16} T^{3/2} \exp\left(-\frac{4550}{T}\right) \text{ cm}^{-3} \quad \text{for Ge} \tag{3.33a}$$

and

$$n_i(T) = 3.88 \times 10^{16} T^{3/2} \exp\left(-\frac{7000}{T}\right) \text{ cm}^{-3} \quad \text{for Si} \tag{3.33b}$$

3.5.5 Ionization of Impurities

We can calculate the values of the carrier concentration n_o and p_o from Eqs. (3.20) and (3.23) if the Fermi energy E_F is known. In a semiconductor, E_F is determined from the charge neutrality condition which requires a knowledge of ionization of donor and acceptor impurities. Thus, it is necessary to examine the subtle feature of the occupation statistics of these impurities. In this discussion, we will assume a nondegenerate semiconductor. Let us first consider the case of donor levels. When a donor atom (e.g., As) is ionized, it loses an electron to the conduction band and becomes positively charged. This situation can be represented by

$$As^{\circ} - e \Longleftrightarrow As^{+}$$

If N_d^+ represents the concentration of ionized donors and N_d^o that of the neutral donors, the total donor concentration N_d is given by $N_d = N_d^o + N_d^+$. One might conclude that the probability of occupation of the donor level E_d by an electron can be obtained simply by substituting $E = E_d$ in the FD distribution function. However, this is not correct because of the spin degeneracy of the donor levels. When a donor atom is ionized, there are two possible quantum states corresponding to each of the two allowed spins. The electron can occupy any one of these states with the condition that as soon as one of them is occupied, the occupancy of the other is precluded. Taking this into account, the occupation probability of level E_d is given by

$$f(E_d) = \frac{N_d^o}{N_d} = \left[1 + \frac{1}{2}\exp\left(\frac{E_d - E_F}{kT}\right)\right]^{-1} \qquad (3.34)$$

Hence, the concentration of ionized donors is

$$N_d^+ = (N_d - N_d^o) = N_d\left[1 + 2\exp\left(\frac{E_F - E_d}{kT}\right)\right]^{-1} \qquad (3.35)$$

A donor state is said to be filled when it is occupied by an electron, and it is said to be empty when it is ionized. In contrast, an acceptor state is empty when it is neutral and is filled when ionized. The transition process at an acceptor atom (e.g., Al) is

$$Al^{\circ} + e \Longleftrightarrow Al^{-}$$

The probability of occupation of the acceptor level at E_a, if spin degeneracy is taken into consideration, is given by

$$f(E_a) = \frac{N_a^-}{N_a} = \left[1 + 2\exp\left(\frac{E_a - E_F}{kT}\right)\right]^{-1} \qquad (3.36)$$

where N_a^- is the concentration of ionized acceptors and N_a is the total acceptor concentration. The concentration of neutral acceptors is $N_a^o = N_a - N_a^-$. Because of two-fold degeneracy of the valence band in Si and Ge, the factor 2 in Eq. (3.36) should be replaced by 4.

In an n-type semiconductor, the ratio of the concentration of electrons bound to donor atoms (N_d^o) to the total electron concentration ($n_o + N_d^o$) can be obtained by using Eq. (3.20) and by neglecting the unity term in Eq. (3.34). That is

$$\frac{N_d^o}{n_o + N_d^o} \approx \left[1 + \frac{N_C}{2N_d}\exp\left(-\frac{E_C - E_d}{kT}\right)\right]^{-1} \qquad (3.37)$$

This fraction will remain small if the second term on the right-hand side in Eq. (3.37) is very large compared to unity, which means

$$N_d \ll \frac{N_C}{2} \exp\left(-\frac{E_C - E_d}{kT}\right) \qquad (3.38)$$

Near 300 K, $(E_C - E_d)$ is on the order of kT in both Ge and Si; thus, the exponential factor in Eq. (3.38) is on the order of unity and the above condition reduces to $N_d \ll N_C/2$. Since $N_C = 2.8 \times 10^{19}$ cm^{-3} at 300 K for Si, this condition is satisfied for all values of N_d which are orders of magnitude lower than this number. This means that the fraction of unionized donors under this condition is very small, and the donors are almost completely ionized. The same is also true for acceptors.

To clarify these ideas, let us consider a silicon sample at 300 K doped with 2×10^{16} donors cm^{-3}. Assuming that all of them are ionized, we can write $N_d^+ = N_d \simeq n_o = 2 \times 10^{16}$ and $N_d^o \simeq 0$. Substituting the value of N_C and n_o in Eq. (3.20) we obtain

$$\frac{N_C}{n_o} = \exp[(E_C - E_F)/kT] \simeq 10^3$$

or

$$(E_C - E_F) = 6.9kT$$

Now

$$(E_d - E_F) = (E_C - E_F) - (E_C - E_d) \simeq 6kT, \quad \text{since } (E_C - E_d) \simeq kT.$$

Substituting this value in Eq. (3.34) gives

$$f(E_d) = \frac{N_d^o}{N_d} = [1 + \frac{1}{2}\exp(6)]^{-1} \simeq 2\exp(-6) \simeq \frac{1}{200}$$

Thus, out of 200 donor atoms, only one atom has its electron attached to the level E_d. Therefore, the donors are more than 99 percent ionized. Since there are $N_C = 2.8 \times 10^{19}$ states at energy E_C, out of which only 2×10^{16} are occupied, the diluteness condition is also satisfied.

3.5.6 Equilibrium Electron–Hole Concentrations

Let us consider a homogeneous piece of nondegenerate semiconductor to which N_d donors cm^{-3} and N_a acceptors cm^{-3} have been added. Since the semiconductor as a whole remains space-charge neutral

$$N_d^+ + p_o - N_a^- - n_o = 0$$

or

$$n_o - p_o = N_d^+ - N_a^- \qquad (3.39)$$

The left-hand side of this relation represents the mobile charge and the right-hand side the immobile charge. The value of N_d^+ can be obtained from Eq. (3.35) provided E_F and T are known, and the value of N_a^- can be similarly determined using Eq. (3.36), so that the difference $(n_o - p_o)$ is known. At temperatures above 100 K, practically all the donors and acceptors are ionized and Eq. (3.39) can be rewritten as

$$n_o - p_o = N_d - N_a \tag{3.40}$$

From Eq. (3.29)

$$p_o n_o = n_i^2$$

From the identity $(p_o + n_o)^2 = (p_o - n_o)^2 + 4 p_o n_o$, we obtain

$$(p_o + n_o) = \pm[(N_d - N_a)^2 + 4n_i^2]^{1/2} \tag{3.41}$$

In this equation, only the positive sign has physical significance because $(p_o + n_o)$ is always positive. From Eqs. (3.40) and (3.41) we obtain

$$n_o = \frac{1}{2}[(N_d - N_a) + \sqrt{(N_d - N_a)^2 + 4n_i^2}]$$

$$p_o = \frac{1}{2}[(N_a - N_d) + \sqrt{(N_d - N_a)^2 + 4n_i^2}] \tag{3.42}$$

These relations show that if the net impurity concentration $N_d - N_a = 0$, then $p_o = n_o = n_i$ and the semiconductor is intrinsic. If $(N_d - N_a)$ is positive, $n_o > p_o$ and the semiconductor is n-type. On the other hand, when $N_a > N_d$ the semiconductor is p-type.

For an n-type semiconductor having $(N_d - N_a)^2 \gg 4n_i^2$,

$$n_o \simeq (N_d - N_a) = N_d \quad \text{if} \quad N_a = 0$$

and

$$p_o = n_i^2/N_d \tag{3.43}$$

Similarly, for a p-type semiconductor with $(N_a - N_d)^2 \gg 4n_i^2$,

$$p_o \simeq (N_a - N_d) = N_a \quad \text{if} \quad N_d = 0$$

and

$$n_o = n_i^2/N_a \tag{3.44}$$

Semiconductors that satisfy Eqs. (3.43) and (3.44) are considered to be strongly extrinsic. As the temperature of an extrinsic semiconductor is gradually raised, n_i increases with temperature and eventually a point is reached where the semiconductor starts behaving like an intrinsic semiconductor. We will assume the condition $|N_d - N_a| = 5n_i$ as the onset point of intrinsic behavior.

As an example, consider a Ge sample having $n_o = 2 \times 10^{13}$ cm^{-3} and $p_o = 3.5 \times 10^6$ cm^{-3} at 200 K where all the dopants are ionized. Let us calculate the values of n_o and p_o at 300 K.

$$\text{At } 200\,\text{K}, (N_d - N_a) = (n_o - p_o) \simeq 2 \times 10^{13} \text{ cm}^{-3}$$

This is the net dopant concentration in the sample and will remain the same at 300 K. The value of n_i at 300 K as obtained from Table 4.2 is 2.4×10^{13} cm^{-3}. Thus, we can write

$$n_o - p_o = 2 \times 10^{13}$$

and

$$n_o p_o = (2.4 \times 10^{13})^2 = 5.76 \times 10^{26}$$

Solving these equations yields $n_o = 3.6 \times 10^{13}$ cm^{-3} and $p_o = 1.6 \times 10^{13}$ cm^{-3}.

3.6 THE FERMI LEVEL AND ENERGY DISTRIBUTION OF CARRIERS INSIDE THE BANDS [3]

3.6.1 Calculation of the Fermi Level

The position of the Fermi level in a semiconductor is determined from the charge neutrality condition

$$p_o - n_o + N_d^+ - N_a^- = 0$$

For a nondegenerate semiconductor, n_o is given by Eq. (3.20) and p_o by Eq. (3.23). Substituting for N_d^+ from Eq. (3.35) and for N_a^- from (3.36), we can write the above relation as

$$N_V \exp\left(-\frac{E_F - E_V}{kT}\right) - N_C \exp\left(-\frac{E_C - E_F}{kT}\right) + N_d\left[1 + 2\exp\left(\frac{E_F - E_d}{kT}\right)\right]^{-1}$$
$$- N_a\left[1 + 2\exp\left(\frac{E_a - E_F}{kT}\right)\right]^{-1} = 0 \quad (3.45)$$

This equation can be solved for E_F provided that all the other quantities are known. We will now apply Eq. (3.45) for some simple cases.

(a) AN INTRINSIC SEMICONDUCTOR

In the case of an intrinsic semiconductor, $(N_d^+ - N_a^-) = 0$, and denoting E_F by E_i, we can write Eq. (3.45) as

$$\exp\left(\frac{E_C + E_V - 2E_i}{kT}\right) = \frac{N_C}{N_V} = \left(\frac{m_e^*}{m_h^*}\right)^{3/2}$$

Since $E_C = E_V + E_g$, the above relation gives

$$E_i = E_V + \frac{E_g}{2} - \frac{3}{4}kT \ln\left(\frac{m_e^*}{m_h^*}\right) \qquad (3.46)$$

It can be seen that if $m_e^* = m_h^*$, the Fermi level lies exactly at the midpoint in the band gap. If $m_e^* > m_h^*$, the third term in Eq. (3.46) is negative and E_i is shifted from the center of the band gap toward the valence band. On the other hand, for $m_h^* > m_e^*$ the intrinsic Fermi level shifts toward the conduction band. For both Si and Ge, and for many other semiconductors, the deviation caused by the third term in Eq. (3.46) is quite small, and E_i is generally taken to be at the center of the band gap. However, for a semiconductor such as InSb, $m_h^* \simeq 20 m_e^*$ and E_i is shifted well toward the conduction band at 300 K.

(b) A STRONGLY EXTRINSIC SEMICONDUCTOR

Let us consider a semiconductor doped with N_d donors cm^{-3}. Assuming that all the donor atoms are ionized and p_o is negligible compared to n_o, we can write Eq. (3.45) as

$$N_C \exp\left(-\frac{E_C - E_F}{kT}\right) = N_d$$

Substituting for N_C from Eq. (3.25) and solving for E_F yields

$$E_F = E_i + kT \ln\left(\frac{N_d}{n_i}\right) \qquad (3.47a)$$

When the semiconductor is not strongly extrinsic, N_d will be different from n_o. In such a case, E_F can be directly obtained from Eq. (3.27).

$$E_F = E_i + kT \ln(n_o/n_i) \qquad (3.47b)$$

Similarly, for a p-type sample doped with N_a acceptors cm^{-3}

$$E_F = E_i - kT \ln(p_o/n_i) \qquad (3.48)$$

As an example, we calculate the position of E_F in a Si sample at 300 K doped with 3×10^{13} acceptors cm^{-3} (all of which are ionized).
Since $(p_o - n_o) = N_a = 3 \times 10^{13}$ cm^{-3} and $n_i = 1.5 \times 10^{10}$ cm^{-3} we have $p_o = 3 \times 10^{13}$ cm^{-3} and $n_o = 7.5 \times 10^6$ cm^{-3}. From Eq. (3.48) we have

$$E_F = E_i - 25.86 \times 10^{-3} \ln \frac{3 \times 10^{13}}{1.5 \times 10^{10}} \text{ eV}$$

$$= E_i - 0.196 \text{ eV}$$

Thus, the Fermi level E_F in the sample lies 0.196 eV below the center of the band gap.

3.6.2 Distribution of Electrons and Holes Inside the Bands

In Sec. 3.5.2 we obtained the equilibrium electron and hole concentrations in terms of the effective densities of states N_C and N_V and the Fermi level E_F. In that treatment, N_C was interpreted as the density of states at the conduction band edge E_C, and the electron concentration was obtained by multiplying this density with the probability of occupation at E_C. Physically, such an interpretation is untenable since both the density of states and the probability of occupation vary with the electron energy inside the band. Therefore, it is desirable to know the actual distribution of electrons in the conduction band and the distribution of holes in the valence band. The electron concentration in a small interval dE around E in the conduction band is given by Eq. (3.18). By varying E and choosing dE to be small, we can plot $n(E)\,dE$ as a function of E. To illustrate the procedure, let us consider silicon at 300 K doped with 1.5×10^{16} donors cm^{-3}. The Fermi level is then approximately $7kT$ below E_C, and the semiconductor is nondegenerate. The density of states in the conduction and valence bands is given by Eq. (3.17a) and (3.17b), respectively, and is plotted in Fig 3.12a. The probability of occupation $f(E)$ is plotted in Fig. 3.12b. When $f(E)$ is multiplied with the density of states $N(E)$ in the conduction band, we obtain the electron concentration $n(E)$.

The distribution function of holes in the valence band is shown by the dashed lines in Fig. 3.12b. The resulting electron and hole concentrations are plotted in Fig. 3.12c as a function of energy. Most of the electrons in the conduction band are confined within a narrow region (about $3kT$) from the band edge E_C, which is a typical feature of classical particles. Similarly, the majority of holes are confined within a region of about $3kT$ below the valence band edge E_V. Accordingly, since all the electrons (and also holes) are confined within a narrow energy range

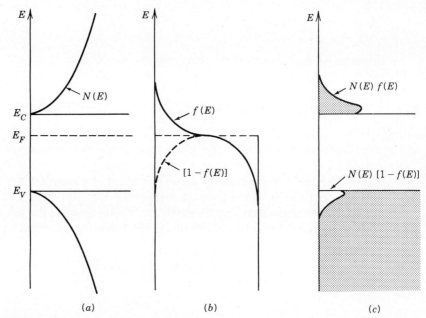

FIGURE 3.12 Energy distribution of electrons and holes in an n-type semiconductor at room temperature. (a) Density of states, (b) distribution functions of electrons and holes, and (c) electron and hole concentrations inside the bands.

3.7 THE TEMPERATURE DEPENDENCE OF CARRIER CONCENTRATIONS IN AN EXTRINSIC SEMICONDUCTOR [3]

Until now, we have assumed that all the donors and acceptors are ionized. While this is a good approximation above 100 K, it is not valid at lower temperatures. Therefore, we must now consider the important case when the temperature is so low that the impurity atoms will not be fully ionized. To be specific, let us consider a semiconductor doped only with donors. The charge neutrality condition can then be written as

$$N_d^+ + p_o = n_o \qquad (3.49)$$

where N_d^+ is given by Eq. (3.35). At low temperatures, E_F may lie above E_d, but the condition $(E_C - E_F) > 3kT$ can still be satisfied if N_d is small enough to keep $n_o \ll N_C$ at all temperatures. (Note that $kT < 1 \times 10^{-3}$ eV at $T = 10$ K). Under these conditions n_o and p_o are given by Eqs. (3.20) and (3.23), respectively, and Eq. (3.49) can be written as

$$N_d \left[1 + 2 \exp\left(\frac{E_F - E_d}{kT}\right) \right]^{-1} + N_V \exp\left[-\frac{E_F - E_V}{kT}\right] = n_o \qquad (3.50)$$

We also have

$$n_o = N_C \exp\left[-\frac{E_C - E_F}{kT}\right]$$

Eliminating E_F between these two relations, we obtain

$$N_d N_C \left[N_C + 2n_o \exp\left(\frac{E_C - E_d}{kT}\right) \right]^{-1} + \frac{N_C N_V}{n_o} \exp(-E_g/kT) = n_o \qquad (3.51)$$

For silicon doped with 1.2×10^{16} phosphorus atoms cm^{-3}, this equation has been solved for n_o as a function of temperature assuming $m_e^* = m_h^* = m_o$ and $(E_C - E_d) = 45 \times 10^{-3}$ eV. Figure 3.13 shows n_o plotted as a function of $1/T$ from 33 K to 600 K. Depending on the temperature, several approximations of Eq. (3.51) are possible. For temperatures up to 300 K, the second term on the left-hand side in Eq. (3.51) can be neglected. This leads to the following quadratic equation for n_o

$$2n_o^2 + n_o N_C' - N_d N_C' = 0 \qquad (3.52a)$$

where

$$N_C' = N_C \exp\left[-\frac{E_C - E_d}{kT}\right] \qquad (3.52b)$$

THE TEMPERATURE DEPENDENCE OF CARRIER CONCENTRATIONS

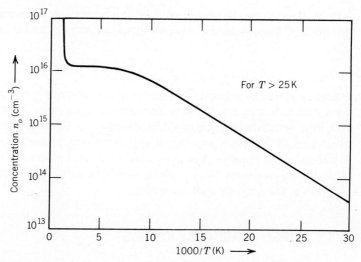

FIGURE 3.13 Electron concentration as a function of $1/T$ for a Si sample doped with 1.2×10^{16} phosphorus atoms cm^{-3}.

From Eq. (3.52a) we obtain

$$n_o = \frac{-N'_C \pm \sqrt{N'^2_C + 8N_d N'_C}}{4}$$

Since n_o is positive even when $N_d = 0$, only the positive sign before the square root leads to an admissible solution and

$$n_o = \frac{N'_C}{4}\left[\left(1 + \frac{8N_d}{N'_C}\right)^{1/2} - 1\right] \qquad (3.53)$$

At very low temperatures, the exponent in Eq. (3.52b) is large, and N'_C is quite small so that

$$n_o \simeq \left(\frac{N_d N'_C}{2}\right)^{1/2} = \left(\frac{N_d N_C}{2}\right)^{1/2} \exp\left[-\frac{E_C - E_d}{2kT}\right] \qquad (3.54)$$

Since the pre-exponential factor in Eq. (3.54) is only weakly dependent on temperature, n_o varies as $\exp[-(E_C - E_d)/2kT]$ and a plot of $\ln n_o$ versus $1/T$ is a straight line with a slope of $(E_C - E_d)/2k$. This behavior is seen in Fig. 3.13 for temperatures below 50 K. The range of temperature in which Eq. (3.54) is valid is known as the *impurity ionization range*. As T increases further, N'_C also increases, and a situation is reached where the second term in Eq. (3.53) becomes small compared to unity. At these temperatures

$$n_o \simeq \frac{N'_C}{4}\left[\frac{4N_d}{N'_C}\right] = N_d \qquad (3.55)$$

which means that all the donor impurities are completely ionized. The temperature range in which Eq. (3.55) is valid, is known as the *exhaustion* or the *saturation range* since n_o in this range saturates to a value N_d. Finally, at very high temperatures, the first two terms in Eq. (3.51) can be neglected in comparison to the third term, the semiconductor becomes intrinsic, and n_o can be expressed as

$$n_o = n_i = \sqrt{N_C N_V} \exp(-E_g/2kT) \qquad (3.56)$$

Thus, a plot of n_o versus $1/T$ can be used to determine the impurity ionization energy according to Eq. (3.54) at low temperatures, and E_g according to Eq. (3.56) at high temperatures. Figure 3.14 shows [4] a semilogarithmic plot of n_o as a function of $1/T$ for a number of experimentally measured germanium samples with different dopings. The general features of the curve in Fig. 3.13 are evident in all samples except No. IV which is very heavily doped. Similar considerations apply for the ionization of acceptors.

3.8 HEAVILY DOPED SEMICONDUCTORS

In the theory of semiconductors presented so far, we have considered only lightly doped materials in which the following assumptions about the density of quantum states have been made: (1) The density of states in the conduction and the valence bands arises only because of host atoms and is proportional to the kinetic energy of carriers. (2) The density of states because of dopant atoms is a delta function that is situated in the band gap. (3) These densities of states remain independent

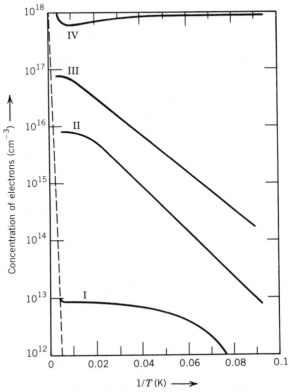

FIGURE 3.14 Electron concentrations as a function of $1/T$ for a number of arsenic doped germanium samples. (From R. B. Adler et al, reference 4, p. 47. Copyright © 1964. Reprinted by permission of John Wiley & Sons, Inc., New York.)

of the concentration of impurity atoms. None of these assumptions is valid at heavy doping. When the concentration of dopant impurities becomes large, the majority carriers in the crystal become so numerous that their presence alters the lattice periodic potential, and several changes occur in the energy band structure of the semiconductor.

In order to describe the energy band structure of a heavily doped semiconductor, three additional effects have to be considered. First, we must consider many body effects to account for the energies of interaction among the mobile carriers and between carriers and ionized impurities. Second, the randomness of the dopant impurities causes fluctuations in the local electrostatic potential, which results in the formation of band tails. Finally, we have to use the FD distribution instead of the MB distribution because, at heavy doping, the majority carrier population becomes degenerate. These effects will be considered below in some detail assuming a heavily doped crystal of an n-type semiconductor.

3.8.1 Many Body Effects

In a heavily doped n-type semiconductor, many body effects involve ionized donor–electron interaction, electron–electron interaction, and electron–hole interaction. These interactions affect the density of quantum states in the conduction and valence bands. These effects have been reviewed by a number of investigators [5,6].

(a) DONOR–ELECTRON INTERACTION

We have seen that in a lightly doped n-type semiconductor, the donor impurity forms an energy level E_d below the conduction band that is a delta function. The coulomb screening of donor ions by the majority carrier electrons reduces the impurity ionization energy $E_I = (E_C - E_d)$. The reduction in the donor ionization energy causes the donor level to move toward the conduction band edge E_C as the donor concentration is gradually increased and ultimately merges into the conduction band. Measurements of the ionization energy of arsenic in germanium have shown that E_I can be expressed by the following empirical relation:

$$E_I = E_{IO}[1 - (N_d/N_{\text{crit}})^{1/3}] \tag{3.57}$$

where E_{IO} is the ionization energy in a lightly doped crystal and $N_{\text{crit}} = 2 \times 10^{17}$ cm^{-3} is the critical donor concentration at which the ionization energy will vanish. From the known value of $E_{IO} = 0.0127$ eV for arsenic, and using Bohr's model for an electron attached to the donor atom, it can be shown (see Prob. 3.23) that E_I drops to zero when the average spacing between the arsenic atoms is about 3 r_o where r_o is the ground state Bohr radius of the unbonded electron attached to the arsenic atom. Note that the electron donor interaction causes only a shift in the donor level toward E_C, but no change occurs in the position of band edges E_C and E_V so that the energy gap E_g remains unchanged.

(b) ELECTRON–ELECTRON INTERACTION

Electron–electron interaction results in a downward shift in the conduction band edge E_C. This shift is caused by electron exchange energy which evolves from the Pauli exclusion principle. When the electron concentration in the semiconductor becomes sufficiently large, their wave functions begin to overlap. Consequently, the Pauli exclusion principle becomes operative, and the electrons spread in their momenta in such a way that the overlapping of the individual electron wave functions is avoided.

Let us now estimate the electron concentration at which the overlapping of individual electron wave functions starts. If we assume conduction electrons as particles of mass m_e^* in thermal equilibrium at a temperature T, then the energy distribution is Maxwellian, and a large majority of electrons are contained in the energy range from 0 to $3kT$ (see Fig. 1.5), with an average energy of $3kT/2$. Since the momentum $\mathbf{p}$ is related to the kinetic energy $(E - E_C)$ by the relation $|\mathbf{p}| \simeq \sqrt{2m_e^*(E - E_C)}$, the corresponding spread in the momentum is from 0 to $\sqrt{6m_e^*kT}$, and momentum corresponding to the average energy is $\sqrt{3m_e^*kT}$. Thus, the uncertainty in the momentum will be $\Delta p = \sqrt{3m_e^*kT}$, and the uncertainty in position Δr is obtained from Heisenberg's uncertainty relation giving

$$\Delta r \simeq \frac{h}{\sqrt{3m_e^*kT}}$$

Substituting $m_e^* = m_o$ and $T = 300$ K in the above relation gives $\Delta r = 58.8$ Å. For a uniform spacing of particles of Δr in each direction, there will be one particle in every cube of volume $(\Delta r)^3$, and the number of particles at which the electron wave functions begin to overlap is $1/(\Delta r)^3 = 4.9 \times 10^{18}$ cm^{-3} which is somewhat lower than the effective density of states N_C in the conduction band.

(c) ELECTRON–HOLE INTERACTION

The majority carriers in a semiconductor (i.e., electrons in an n-type semiconductor) screen the minority carriers in the same way as they screen the ionized impurity atoms. In an n-type semiconductor, the electron–hole coulombic interaction reduces the hole potential energy and causes the valence band edge E_V to move up toward the conduction band. This upward shift of E_V has been termed the hole correlation energy [5]. The hole–hole interaction energy in a heavily doped n-type semiconductor remains negligible because of the very low concentration of holes.

From the above discussion, we conclude that the effect of many body interactions on the band structure of a heavily doped semiconductor is to produce a downward shift in E_C and an upward shift in E_V. This situation is shown in Fig 3.15. Furthermore, the electron-ionized donor interaction causes a shift in E_d, but have no effect on E_C and E_V.

3.8.2 The Effect of Randomness in Impurity Distribution

The donor and acceptor atoms in a semiconductor have a random distribution. This randomness of ionized impurities causes fluctuations in the local potential. The effect of these fluctuations remains negligible in a lightly doped semiconductor, but in a heavily doped semiconductor they produce a spatially dependent distortion of the quantum density of states. The statistical average of the density of states over the entire crystal, which defines the macroscopic properties of the semiconductor, then shows a tailing of the valence and conduction bands into the energy gap [7]. As a consequence, the bands are no longer parabolic near the edges.

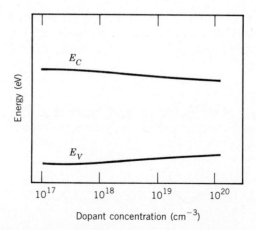

FIGURE 3.15 Energy band diagram showing the shift of the band edges E_C and E_V in a heavily doped semiconductor.

A high density of randomly distributed dopant impurities further complicates the band structure by creating a band of impurity states. Thus, the discrete donor level in the n-type semiconductor will spread into a band at heavy doping. However, this impurity band formation has little effect on the band gap in Ge and Si, because before any significant spread occurs in the donor level, it moves upward and merges with the conduction band. Once the donor level E_d has reached the conduction band, its spread does not give rise to localized states in the band gap.

Figure 3.16 shows the variation in the local electron–hole potential energies in a heavily doped n-type semiconductor under thermal equilibrium. Since the Fermi level is constant throughout, variations in the local donor concentration cause variations in the conduction band edge. For example, in a local region where the donor concentration $N_d(x)$ is higher than its average value, the conduction band edge $E_C(x)$ comes close to the Fermi level. Since the rigid band gap $E_g(x)$ is not affected by the random distribution of impurity atoms, $E_V(x)$ also moves parallel to $E_C(x)$. Similarly, in a region where $N_d(x)$ is lower than its average value, $E_C(x)$ moves away from the Fermi level and E_V comes closer to it. Thus, it is clear that $E_g(x)$ remains constant throughout the semiconductor. Note that here $E_g(x)$ is lower than its value E_g in the case of an intrinsic or lightly doped semiconductor. It has been reduced below its value E_g because of the many body interactions. This band gap is measured in the optical absorption experiment (see Sec. 5.1) and defines the optical band gap E_g^{opt}. The difference $\Delta E_g^{opt} = (E_g - E_g^{opt})$ is called the *optical band gap reduction*. From Eq. (3.29) it is clear that a decrease in the band gap causes an increase in the $p_o n_o$ product.

Figure 3.16 shows that the variations in E_C and E_V result in tail states in the band gap. These tail states cause a further increase in the $p_o n_o$ product above its value n_i^2. On the macroscopic level, these changes reflect an additional reduction in the band gap. Thus, the band gap E_g' in Fig. 3.16 is lower than the optical band gap E_g^{opt}. The $p_o n_o$ product can now be obtained by substituting E_g' for E_g in Eq. (3.29). That is

$$p_o n_o = K_1^2 T^3 \exp(-E_g'/kT)$$

The actual $p_o n_o$ product, however, will be lower than predicted by the above relation because of majority carrier degeneracy.

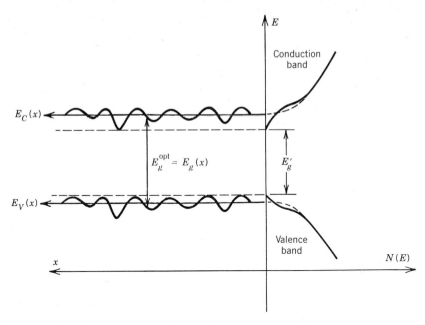

FIGURE 3.16 Spatial variation in the band edges E_C and E_V due to random distribution of impurities in a semiconductor at heavy doping. (From D. S. Lee and J. G.. Fossum, reference 5, p. 630. Copyright © 1983 IEEE. Reprinted by permission.)

3.8.3 The Effect of Carrier Degeneracy

To this point, our discussion has been based on Maxwell-Boltzmann (MB) statistics. This has allowed us to express the thermal equilibrium electron and hole concentrations in terms of N_C and N_V using Eqs. (3.20) and (3.23), respectively. Consequently, we obtained the relation $p_o n_o = n_i^2$. However, it is to be recalled that the MB statistics are valid only for a nondegenerate semiconductor in which the Fermi level lies in the band gap at least $3kT$ away from the band edges. In a heavily doped n-type semiconductor, the Fermi level does not remain in the band gap but moves into the conduction band. A semiconductor in which the Fermi level lies inside either the conduction or the valence band is called a *degenerate semiconductor*. Such a semiconductor has a large concentration of majority carriers and behaves like a metal. In the case of a degenerate semiconductor, we have to use FD statistics instead of MB statistics. Use of the FD statistics reduces the $p_o n_o$ product below n_i^2. However, the band gap narrowing causes an increase in this product. Thus, to incorporate the effect of band gap reduction and the majority carrier degeneracy in a single expression, we can define an effective electrical band gap E_g^{elect} and write

$$p_o n_o = n_{ie}^2 = K_1^2 T^3 \exp(-E_g^{\text{elect}}/kT) \tag{3.58}$$

In view of Eq. (3.29), this relation can be rewritten as

$$n_{ie}^2 = n_i^2 \exp(\Delta E_g^{\text{elect}}/kT) \tag{3.59}$$

where $\Delta E_g^{\text{elect}} = (E_g - E_g^{\text{elect}})$ is the electrical band gap narrowing and n_{ie} is the effective intrinsic concentration in the heavily doped semiconductor. The elec-

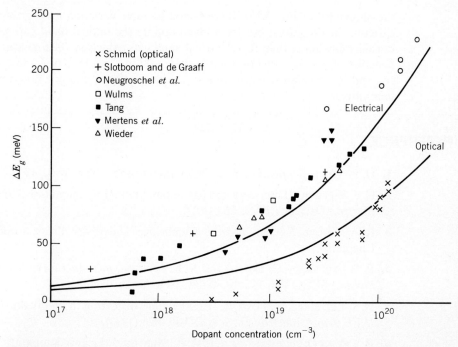

FIGURE 3.17 Electrical and optical band gap reductions in heavily doped silicon as a function of dopant impurity concentration. The solid curves are based on theoretical calculations; the points are experimental data. (From R. J. Overstraeten and R. P. Mertens, reference 6, p. 1079. Copyright © 1987. Reprinted by permission of Pergamon Press, Oxford.)

trical band gap E_g^{elect} has no other significance except that it permits us to express the measured $p_o n_o$ product by Eq. (3.58). Equation (3.59) is an important relation and is used to describe minority carrier transport in devices that employ heavily doped regions of semiconductors.

Experimentally determined values of the optical and the electrical band gap narrowing in heavily doped silicon as a function of dopant impurity concentration

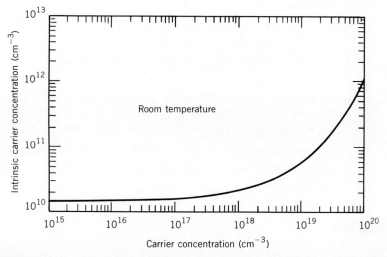

FIGURE 3.18 Average value of the effective intrinsic carrier concentration n_{ie} as a function of dopant concentration in heavily doped silicon. (From B. J. Baliga, reference 8, p. 28. Copyright © 1987. Reprinted by permission of John Wiley & Sons, Inc., New York.)

are shown [6] in Fig. 3.17. There is considerable variation among the results of different investigators, but for a given doping the optical band gap reduction is considerably lower than the electrical band gap reduction. The measured n_{ie} as a function of dopant concentration is shown [8] in Fig 3.18. The band gap narrowing and n_{ie} are almost insensitive to the dopant type and are functions of concentrations only.

REFERENCES

1. H. F. WOLF, *Semiconductors,* Wiley-Interscience, New York, 1971.
2. H. J. WELKER, "Discovery and development of III-V compounds," *IEEE Trans. Electron Devices ED-23,* 664 (1976).
3. R. A. SMITH, *Semiconductors,* Cambridge University Press, London, 1978, Chapter 4.
4. R. B. ADLER, A. C. SMITH, and R. L. LONGINI, *Introduction to Semiconductor Physics,* SEEC Vol. 1, John Wiley, New York, 1964.
5. D. S. LEE and J. G. FOSSUM, "Energy-band distortion in highly doped silicon," *IEEE Trans. Electron Devices ED-30,* 626 (1983).
6. R. J. VAN OVERSTRAETEN and R. P. MERTENS, "Heavy doping effects in silicon," *Solid-State Electron.* 30, 1077 (1987).
7. E. O. KANE, "Band tails in semiconductors," *Solid-State Electron.* 28, 3 (1985).
8. B. J. BALIGA, *Modern Power Devices,* John Wiley, New York, 1987.

ADDITIONAL READING LIST

1. O. MADELUNG, *Physics of III-V Compounds,* John Wiley, New York, 1964.
2. B. RAY, *II-VI Compounds,* Pergamon Press, Oxford, 1969.
3. D. R. LOVETT, *Semimetals and Narrow Bandgap Semiconductors,* Pion Limited, London, 1977.
4. R. E. HUMMEL, *Electronic Properties of Materials — An Introduction for Engineers,* Springer, New York, 1985.
5. K. SEEGER, *Semiconductor Physics: An Introduction,* 3d ed., Springer, Berlin, 1985.
6. E. MOOSER, "Bonds and bands in semiconductors," Chapter 1 in *Crystalline Semiconducting Materials and Devices,* P. N. BUTCHER, N. H. MARCH, and M. P. TOSI (eds.), Plenum Press, New York, 1986.
7. S. T. PANTELIDES, A. SELLONI, and R. CAR, "Energy-gap reduction in heavily doped silicon: causes and consequences," *Solid-State Electron.* 28, 17–24 (1985).

PROBLEMS

3.1 The melting points of ionic crystals are generally lower and their band gaps much larger than those of covalently bonded semiconductors. Explain these observations in terms of bonding of atoms.

PROBLEMS

3.2 The band gaps and melting points of III-V semiconductors are higher than those of isoelectronic elemental group IV semiconductors. However, in the case of II-VI semiconductors, the band gap is higher, but the corresponding melting point is lower. Explain these observations in terms of atomic cohesion and crystal bonding.

3.3 Calculate the number of atoms per cm^3 of the following: (a) Si, (b) Ge, and (c) GaAs.

3.4 The length of a side of unit cube of GaAs is 5.65 Å. Draw a diagram of a unit cube of GaAs in which all the corner atoms are Ga atoms. Taking one of the corner atoms as the origin, write down the coordinates of the four nearest and the next nearest arsenic atoms. Also write down the coordinates of the nearest gallium atoms and their distance from the origin.

3.5 The density of GaP at 300 K is 4.13 gm-cm^{-3}, and its molecular weight is 100.7. Calculate its lattice constant. How will this be affected if the crystal is cooled down to liquid nitrogen temperature?

3.6 The Bohr model-like solution for the ionization energy of group III and Group V impurities in Si and Ge leads to values independent of the size of the impurity atoms. The actual value of this energy is found to decrease with the increasing atomic number of the impurity and is only in qualitative agreement with the calculated value. Explain these discrepancies.

3.7 (a) Draw the energy band diagram of a 2 cm long n-type compensated Si sample doped with boron and phosphorus. The sample has Fermi-level $4kT$ below the conduction band edge E_C. Taking the potential at the center of the gap to be zero, write down the electrostatic potential at E_C, E_d, E_F and E_a. Assume $T = 300$ K and use Fig. 3.10 for the values of E_d and E_a.
(b) Draw the energy band diagram when a potential of 2 V is applied across the sample.

3.8 A photon of monochromatic light of wavelength 5000Å is absorbed in GaAs and excites an electron from the valence band into the conduction band. Using the electron and hole effective masses given in Table 4.2, calculate the kinetic energy of electron and hole.

3.9 Consider the direct recombination of electron–hole pairs in GaAs at 300 K. What is the maximum wavelength of the emitted radiation? What is the wavelength corresponding to the average energy of transition? See Table 4.2 for the band gap of GaAs.

3.10 The energy gap of Si and Ge decreases monotonically with temperature. The principal reason for this decrease is the thermal expansion of the lattice. Explain how the sign of the temperature dependence can be inferred from the plot of Fig. 2.25.

3.11 A 1 cm^3 of a metal has a density of states distribution $N(E)dE = 6.82 \times 10^{21} \sqrt{E} dE$, where E is measured from the bottom of the band. The number of electrons in a small energy interval between $E_1 = 4.6$ eV and $E_2 = 4.601$ eV is 2.07×10^{15}. Determine (a) the fraction of the states between E_1 and E_2 occupied by electrons and the position of the Fermi level, and (b) the number of electrons between E_1 and E_2 at 0 °K.

3.12 Show that for a nondegenerate semiconductor the hole concentration in the valence band is given by Eq. (3.23).

3.13 A Si sample is doped with 10^{15} phosphorus atoms cm^{-3}. At 300 K calculate the electron and hole concentrations and the position of the Fermi level. Assume that the distribution of states in the conduction band is given by $N(E)dE = 8 \times 10^{20} \sqrt{E}\, dE$ cm^{-3}. Calculate the number of electrons between the energy interval $1.9\, kT$ to $2.1 kT$ above the band edge E_C.

3.14 Show that the average kinetic energy of a hole in the valence band for a nondegenerate semiconductor is $3kT/2$.

3.15 Show that the effective density of states N_C represents the density of states in a strip $1.2kT$ wide near the edge of the conduction band.

3.16 In a semiconductor sample, donor and acceptor levels are 0.3 eV apart from each other. If 80 percent of the acceptors are ionized at 300 K, evaluate the fraction of ionized donors. If the donor level is $2kT$ below the conduction band edge, determine the position of the Fermi level.

3.17 A Ge sample is uniformly doped with 5×10^{16} atoms cm^{-3} of In. Assume all the atoms are ionized and take $n_i = 2 \times 10^{13}$ cm^{-3} at $T = 290$ K. (a) Calculate the electron and hole concentrations in the sample. (b) Assuming that the intrinsic concentration in Ge increases by 6 percent per °K rise in temperature, estimate the temperature at which the sample becomes intrinsic.

3.18 Figure P3.18 shows the intrinsic concentration n_i plotted as a function of $1/T$ for Si and Ge. Determine the band gap E_{go} of these materials from the plots.

3.19 A piece of Si is doped with 1.2×10^{16} atoms cm^{-3} of boron. Assuming that all these atoms are ionized at 300 K, determine the position of the Fermi level and plot the electron and hole concentrations as a function of energy in the respective bands.

3.20 A uniformly doped silicon sample has a donor concentration of 5×10^{15} cm^{-3} and an acceptor concentration of 1.1×10^{16} cm^{-3}. (a) Calculate n_o and p_o at 300 K and at 150 K. (b) Determine the temperature at which the sample becomes intrinsic. (c) If this sample is further doped with 4×10^{15} donors cm^{-3}, calculate the new electron and hole concentrations at 300 K.

3.21 Calculate the displacement of the intrinsic Fermi level E_i from the center of the band gap in the case of (a) Si, (b) GaAs, (c) InSb, and (d) InAs. Use the data from Table 4.2 and Appendix E.

3.22 A silicon sample is doped with 1.2×10^{15} atoms cm^{-3} of Al for which $(E_a - E_V) = 0.057$ eV. (a) Show that at $0°K$, $E_F = (E_V + E_a)/2$. (b) Calculate the position of Fermi level at 50 K, 100 K, 200 K, 500 K, and 800 K.

3.23 The impurity ionization energy E_I of As is given by Eq. (3.57) where $N_{\text{crit}} = 2 \times 10^{17}$/cm^3. Show that E_I becomes zero when the average distance between the neighboring As atoms is $3\, r_o$, where r_o is the ground state radius of the donor electron attached to the As atom. Assume $E_{I0} = 0.013$ eV.

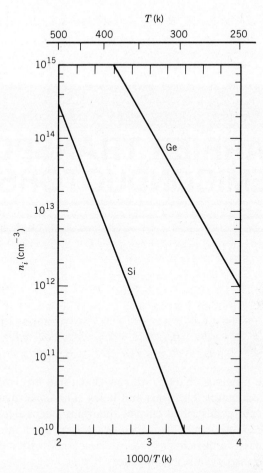

FIGURE P3.18 Intrinsic carrier concentration plotted as a function of $1/T$ for silicon and germanium. From B. G. Streetman, *Solid State Electronic Devices* 2nd ed. © 1980, p. 78. Reprinted by permission of Prentice Hall, Inc., Englewood Cliffs, New Jersey.)

3.24 Phosphorus in Si has a donor level 0.045 eV below E_C. Using Eq. (3.2b), calculate the effective mass of electron and the ground state Bohr radius r_o. It is observed that the ionization energy of phosphorus becomes zero at a dopant concentration of 3×10^{18} cm^{-3}. Calculate the average separation between the impurity atoms at this doping and compare it with r_o.

4 CARRIER TRANSPORT IN SEMICONDUCTORS

INTRODUCTION

In the previous chapter, we saw that there are two types of charge carriers in semiconductors: electrons and holes. The exact dynamic behavior of these carriers can be described only by quantum mechanics. However, their response to applied fields can be described by classical mechanics using the concept of effective mass. In this chapter, we are concerned with a detailed study of electron and hole transport in semiconductors.

There are two mechanisms by which current can flow in semiconductors: drift and diffusion. The drift current results from the movement of electrons and holes under the action of an applied electric field and is similar to that in a metal. The diffusive motion of carriers is caused by gradients in carrier concentrations. Concentration gradients can be established by the graded doping of a semiconductor, as well as by the injection of carriers from a suitable source. Besides drift and diffusion, the other basic process is recombination. Taken together with the carrier generation, these processes form the physical basis for all semiconductor devices.

4.1 THE DRIFT OF CARRIERS IN AN ELECTRIC FIELD

4.1.1 Relaxation Time and Mobility

In Chapter 2 we saw that an electron can move unhindered through a perfectly periodic crystal except that it suffers Bragg reflection at the end of the zone. Thus, its motion in response to an externally applied small electric field $\mathscr{E}$ can be described by the relation

$$\mathbf{F} = -q\mathscr{E} = m_e \frac{d\mathbf{v}}{dt} \qquad (4.1)$$

THE DRIFT OF CARRIERS IN AN ELECTRIC FIELD

where **v** is the electron velocity and m_e is the conductivity effective mass that is different from the density of state effective mass m_e^*.

In real crystals, the situation is different because of the presence of imperfections. Crystal imperfections cause a local change in the periodic potential and set up an electric field. When an electron passes through the neighborhood of an imperfection, it interacts with this field, and this interaction changes the direction of the electron's motion. Thus, the electron is scattered by imperfections. In each scattering event, the electron is said to collide with the imperfection, and each collision causes a change in the electron momentum.

Let us now consider a semiconductor crystal at temperature T. The electrons and holes in the crystal are in random motion because of their thermal energy. Disregarding the nature of the scattering centers for the moment, we can visualize the motion of an electron as shown in Fig. 4.1a. The trajectory consists of a series of straight lines between collisions. The free times between collisions are denoted by t_1, t_2, t_3, and so on, and the average of these times is the mean free time between collisions $\bar{\tau}$. Each scattering event causes an energy exchange between the electron and the lattice. In some collisions, the electron loses energy to the lattice, while in others it gains energy from the lattice atoms. Since no energy is being supplied from outside, the net rate of energy exchange is zero. Because the motion of the electrons, as well as the collisions, are random, each electron effectively returns to its original position after a large number of collisions. Thus, the average velocity of the electron is zero. However, its random thermal velocity v_{th} is the rms velocity and is given by $v_{th} = \sqrt{3kT/m_e}$. At room temperature, v_{th} is on the order of 10^7 cm sec^{-1}.

When an electric field is applied to the crystal, the electrons are accelerated by the field during their motion between the collisions. As a result, the free paths become curved, like the trajectory of a projectile moving under gravity (Fig. 4.1b). It is now seen that after a number of collisions, the electron has traversed a distance l along the direction of the applied force. This motion is known as the *drift motion*. During its free flight between two collisions, the electron gains kinetic energy from the electric field and transfers this energy to the lattice

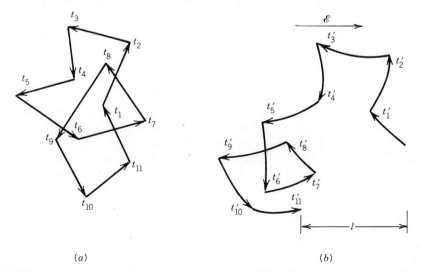

FIGURE 4.1 Path of a free electron in a crystal (a) under thermal equilibrium and (b) under the influence of an electric field.

upon collision at the end of its flight. This energy is dissipated as joule heating in the crystal.

Thus, it is clear that in the presence of an electric field, an electron acquires a drift velocity. The resultant velocity, which is the vector sum of the drift velocity and the random thermal velocity, is then higher than v_{th}. Since the average distance between two collisions (i.e., the mean free path) is not altered by the field, the effect of increase in the electron velocity is to make the collisions more frequent. An increase in the electric field causes an increase in the resultant velocity. Consequently, the electrons suffer more collisions per unit time, and more energy is lost to the lattice. The net result is that electrons and holes in a crystal cannot be accelerated continuously to any extent but, rather, gain an average drift velocity. The collisions act as a resistive force that increases with increasing velocity. This situation is similar to that encountered in the motion of a projectile through a viscous medium. Thus, a real crystal is analogous to a viscous medium in which Eq. (4.1) is modified as

$$\mathbf{F} = m_e \frac{d\mathbf{v}}{dt} + \gamma_v(E)\mathbf{v}(E) \tag{4.2}$$

where $\gamma_v(E)$ is a constant of proportionality. Note that γ_v and the resultant velocity $\mathbf{v}$ are functions of the electron energy E.

At a low electric field such that the drift velocity $v_d(E)$ is small compared with v_{th}, the latter remains unaffected by the field and Eq. (4.2) can be written as

$$\mathbf{F} = m_e \frac{d\mathbf{v}_d(E)}{dt} + m_e \frac{\mathbf{v}_d(E)}{\tau_c(E)} \tag{4.3}$$

where $\tau_c(E) = m_e/\gamma_v(E)$ has the significance of a relaxation time. This interpretation can be seen by examining what happens when the force $\mathbf{F}$ is abruptly removed after attaining its steady state. The above equation then becomes

$$\frac{d\mathbf{v}_d(E)}{dt} + \frac{\mathbf{v}_d(E)}{\tau_c(E)} = 0 \tag{4.4}$$

which has the solution

$$\mathbf{v}_d(E,t) = \mathbf{v}_d(E,0) \exp[-t/\tau_c(E)] \tag{4.5}$$

where $\mathbf{v}_d(E,0)$ is the drift velocity at time $t = 0$ when the force is removed. As seen in Eq. (4.5), the drift velocity decays with the characteristic time $\tau_c(E)$ as the system relaxes to its equilibrium state.

The situation represented by Eq. (4.3) is a transient state that remains only for time periods on the order of 10^{-12} sec. The transients, therefore, will die quickly, giving rise to a steady state situation in which

$$\mathbf{v}_d(E) = \frac{\mathbf{F}}{m_e}\tau_c(E) = \frac{-q\mathscr{E}\tau_c(E)}{m_e} \tag{4.6}$$

In general, both $\mathbf{v}_d(E)$ and $\tau_c(E)$ are functions of energy, and when dealing with a large number of electrons, these quantities have to be averaged out over the entire population. We assume that $\tau_c(E)$ is nearly the same for all the electrons, so that a suitable average is possible for both $\mathbf{v}_d(E)$ and $\tau_c(E)$. Hence, the average drift velocity $\mathbf{v}_d$ for electrons can be written as

$$\mathbf{v}_d = \langle \mathbf{v}_d(E) \rangle = -\mu_n \mathscr{E} \tag{4.7}$$

where

$$\mu_n = \frac{q\langle \tau_c(E) \rangle}{m_e} = \frac{q\tau_c}{m_e} \tag{4.8a}$$

is called the *mobility of an electron*. Similarly, we define *hole mobility* μ_p as

$$\mu_p = \frac{q\langle \tau_c(E) \rangle}{m_h} \tag{4.8b}$$

where m_h is the conductivity effective mass of a hole. Mobility represents the drift velocity per unit field and is always defined to be a positive quantity. Thus, for electrons

$$\mathbf{v}_d = (-\mathbf{v}_n) = \mu_n \mathscr{E} \tag{4.9a}$$

and for holes

$$\mathbf{v}_d = \mathbf{v}_p = \mu_p \mathscr{E} \tag{4.9b}$$

Mobility is commonly expressed in $cm^2\ V^{-1}\ sec^{-1}$. The relaxation time τ_c does not have any other significance except that predicted by Eq. (4.5). However, if the scattering is isotropic, which means that the collisions occur completely at random and that all directions of deflection of the carrier after collision are equally probable, then τ_c is the same as the mean free time. As an example, consider intrinsic Ge at 300 K. From Table 4.2 we have $\mu_n = 3900\ cm^2\ V^{-1}\ sec^{-1}$ and $m_e = 0.12\ m_o$. Thus, substituting $\mu_n = 0.39\ m^2\ V^{-1}\ sec^{-1}$ and $m_o = 9.1 \times 10^{-31}$ kg, we obtain

$$\tau_c = \frac{\mu_n m_e}{q} = \frac{0.39 \times 0.12 \times 9.1 \times 10^{-31}}{1.6 \times 10^{-19}} = 2.66 \times 10^{-13}\ sec$$

for electrons. Similarly for holes, $\mu_p = 0.19\ m^2\ V^{-1}\ sec^{-1}$ and $m_h = 0.23\ m_o$ which gives $\tau_c = 2.48 \times 10^{-13}$ sec.

4.1.2 Carrier-Scattering Mechanisms

Measurements have shown that mobilities vary with temperature as well as with the impurity concentration in a semiconductor. To understand these variations, knowledge of the various scattering mechanisms in a semiconductor is necessary. There are four main sources of scattering for the mobile carriers [1]. They are: (a) lattice vibrations, (b) ionized impurities, (c) neutral impurities, and (d) electrons and holes. Together these mechanisms determine the mean free time, and, thereby, the mobility.

(a) SCATTERING BY LATTICE VIBRATIONS

In Chapter 2 we saw that the atoms of a periodic lattice execute vibrations about their equilibrium positions. In a purely covalent crystal like Ge and Si, electrons and holes are predominantly scattered by the longitudinal lattice vibrations that produce a series of compressions and dilations. At any instant, the lattice atoms are squeezed together in some regions and are pulled apart in others, and this disturbs the periodicity of the lattice. These disturbances cause local changes in the dielectric constant of the crystal and alter the periodic potential in the vicinity of the compressions and dilations. A local change in potential sets up an electric field that scatters the moving carriers.

An increase in temperature increases the amplitude of lattice vibrations and excites normal modes of higher energies. Therefore, the lattice scattering of carriers will be more effective at higher temperatures. This produces a relaxation time (τ_L) which decreases with increasing temperature. The resulting lattice scattering limited mobility is given by

$$\mu_L = B_1 T^{-a} \tag{4.10}$$

where B_1 is a constant of proportionality and a is found to vary from 1.6 to 2.8. From purely theoretical considerations involving longitudinal acoustical modes, a is expected to be 3/2. Departures from the simple $T^{-3/2}$ law are attributed to a variety of causes such as nonisotropic effective masses, scattering caused by transverse acoustic phonons, and scattering by optical phonons.

(b) IONIZED IMPURITY SCATTERING

Ionized impurity scattering is essentially coulomb scattering. The charge associated with an impurity ion causes a change in the lattice potential near its location and thus produces an electric field that deflects the moving carrier. Figure 4.2 shows the manner in which an electron is scattered by a positively charged donor ion. Had the ion not been present, the electron would have gone along the path shown by the dashed line. The path is bent because of the presence of the positive charge. At low temperatures the thermal speed is small, and a carrier spends more time in the neighborhood of the impurity, increasing the effectiveness of scattering. As the temperature increases, scattering becomes less effective. Consequently, the relaxation time for impurity scattering (τ_I) increases

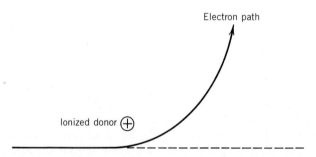

FIGURE 4.2 Coulomb scattering of a conduction electron by a positively charged donor ion.

with temperature, and the ionized impurity scattering limited mobility (μ_I) is given by the relation

$$\mu_I = B_2 T^{3/2} \qquad (4.11)$$

where B_2 is a constant of proportionality.

(c) NEUTRAL IMPURITY SCATTERING

The size of a neutral impurity atom is different from that of the host lattice atom, and this difference causes a strain in the lattice. This strain produces a bump in the potential near the impurity atom, and the electric field produced scatters the carrier when it gets close enough to the impurity atom. Neutral impurity scattering in semiconductors is only of secondary importance because the concentration of neutral impurities in a semiconductor is negligibly small compared to that of ionized donors or acceptors.

(d) CARRIER-CARRIER SCATTERING

In a semiconductor, holes and electrons may deflect each other during their movement through the crystal. The hole–electron interaction is not important in an extrinsic semiconductor for the majority carriers because the minority carriers are in short supply. The collisions between similar types of carriers are unimportant for the current transport because the measured current is proportional to the sum of momenta of all the carriers. If one carrier collides with the other, there is an exchange of momentum between the two, but the total momentum of the group as a whole remains unaltered.

Thus, the dominant modes of carrier scattering which determine the overall carrier mobility are lattice and ionized impurity scatterings. When a number of different mechanisms contribute to the carrier scattering simultaneously, each of these can be treated as an independent event. Therefore, the total probability of scattering is the product of the individual probabilities. Thus, the resultant mean free time is obtained by reciprocal addition of the mean free times of the different mechanisms (see Prob. 4.6). The same is true for the relaxation time. Applying this principle to the case in which only the ionized impurity and lattice scatterings are important enables us to write

$$\frac{1}{\tau_c} = \frac{1}{\tau_I} + \frac{1}{\tau_L} \qquad (4.12)$$

In view of Eq. (4.12), the effective mobility μ can be expressed as

$$\frac{1}{\mu} = \frac{1}{\mu_I} + \frac{1}{\mu_L} \qquad (4.13)$$

As an example, consider a silicon sample with $n_o = 10^{15}$ cm^{-3} at 300 K. Let an electric field $\mathscr{E} = \mathscr{E}_o \exp(j\omega t)$ be applied to the sample. We will determine the frequency at which the current density (J) lags the electric field by 60°.

From Table 4.2, we get $m_e = 0.26\, m_o$, and taking $\mu_n = 0.128\, m^2\, V^{-1}\, sec^{-1}$ we obtain $\tau_c = 1.9 \times 10^{-13}$ sec. Since the hole concentration in the sample is negligibly small, the current density $J = qn_o v_d$. Substituting this value in Eq. (4.3) yields:

$$\frac{dJ}{dt} + \frac{J}{\tau_c} = \frac{-q^2 n_o}{m_e} \mathscr{E}_o \exp(j\omega t) \equiv C_1 \exp(j\omega t)$$

This equation has the steady-state solution of the form

$$J = A \exp[j(\omega t - \theta)]$$

Substituting this value in the above equation and separating the real and imaginary parts, we obtain

$$A \cos\theta = \frac{\tau_c}{1 + \omega^2 \tau_c^2} C_1 \quad \text{and} \quad A \sin\theta = \frac{\omega \tau_c^2}{1 + \omega^2 \tau_c^2} C_1$$

From which $\tan\theta = \omega\tau_c$. Thus, at $\theta = 60°$, $\omega\tau_c = \sqrt{3}$, so that $\omega = 2\pi f = \sqrt{3}/\tau_c$ or $f = 1.45 \times 10^{12}$ Hz.

4.2 VARIATION OF MOBILITY WITH TEMPERATURE AND DOPING LEVEL

4.2.1 The Temperature Dependence of Mobility

The mobility of electrons in n-type Ge as a function of temperature is plotted in Fig. 4.3 for three samples having different impurity concentrations as indicated [1]. For the lightly doped sample No. 1, the ionized impurity scattering is

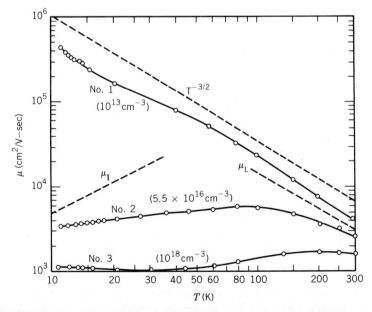

FIGURE 4.3 Electron mobility in n-type germanium as a function of temperature for various doping levels. (From R. B. Adler et al, reference 1, p. 33. Copyright © 1964. Reprinted by permission of John Wiley & Sons, Inc., New York.) These plots and those of Fig. 4.8 are from the data of E. M. Conwell, Proc. IRE 40, 1327 (1952).

negligible, so that μ_I in Eq. (4.13) is large compared to μ_L, and the mobility is limited by the lattice scattering at all temperatures. For sample No. 2, which has a higher dopant concentration, the scattering of electrons by ionized impurities is more effective at low temperatures, and therefore, mobility increases with temperature. However, at higher temperatures, the lattice scattering dominates and the mobility starts decreasing. In the intermediate temperature range, the mobility exhibits a broad maximum. Finally, for sample No. 3, the impurity concentration is so large that the ionized impurity scattering is dominant in the entire range of temperature.

Figure 4.4 shows the hole mobility in p-type silicon as a function of temperature for two samples with different dopings. The trend as observed above for the n-type Ge sample No. 2 is also evident from these plots.

4.2.2 Variation of Mobility with Impurity Concentration

As the dopant concentration in a semiconductor increases, the carrier mobility at a given temperature decreases because of the increased scattering by the impurity ions. Figure 4.5 shows the mobility versus impurity concentration plots for holes and electrons in Ge and Si at room temperature. The mobility reaches a maximum value at low impurity concentrations corresponding to the lattice scattering limitation. The electron–hole mobility curves in silicon can be represented by the following empirical relation [2]

$$\mu = \mu_{min} + \frac{\mu_{max} - \mu_{min}}{1 + (N^+/N_{ref})^\alpha} \tag{4.14}$$

where N^+ represents the total concentration of ionized impurities and the other parameters have the following values.

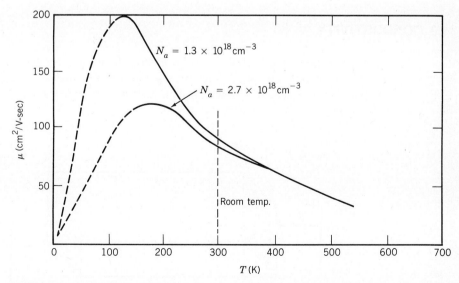

FIGURE 4.4 Hole mobility in p-type silicon as a function of temperature. (From A. S. Grove, *Physics and Technology of Semiconductor Devices*, p. 110. Copyright © 1967. Reprinted by permission of John Wiley & Sons, Inc., New York.) These plots are based on the work of J. L. Pearson and J. Bardeen, Phys. Rev. 75, 865 (1949).

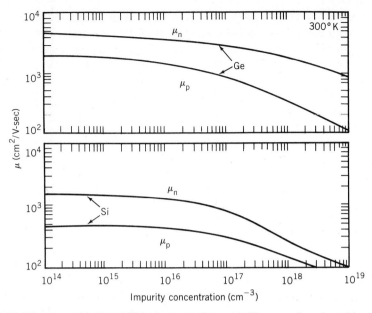

FIGURE 4.5 Electron and hole mobilities in germanium and silicon as a function of dopant impurity concentrations. (From S. M. Sze, reference 5, p. 29. Copyright © 1981. Reprinted by permission of John Wiley & Sons, Inc., New York.)

	μ_{max} (cm^2/V-sec)	μ_{min} (cm^2/V-sec)	N_{ref} (cm^{-3})	α
Electrons	1360	92	1.3×10^{17}	0.91
Holes	495	47.7	6.3×10^{16}	0.76

4.3 CONDUCTIVITY

Let us consider a rectangular bar of a homogeneous semiconductor of length l and cross-sectional area A as shown in Fig. 4.6. Let a voltage V be applied across the bar so that a current I flows through it. If n represents the electron concen-

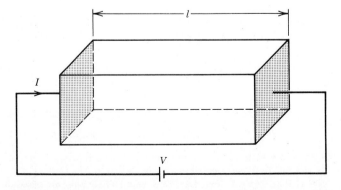

FIGURE 4.6 A rectangular bar of a semiconductor with applied voltage.

tration and p the hole concentration in the semiconductor, the electron and hole current densities $\mathbf{J}_n$ and $\mathbf{J}_p$ are given by

$$\mathbf{J}_n = qn\mu_n \mathscr{E}$$

and

$$\mathbf{J}_p = qp\mu_p \mathscr{E}$$

The total current density $\mathbf{J}$ through the bar is

$$\mathbf{J} = \mathbf{J}_n + \mathbf{J}_p = q(n\mu_n + p\mu_p)\mathscr{E} \tag{4.15}$$

The resistance R of the bar can be written as

$$R = \frac{V}{I} = \rho\frac{l}{A} = \frac{l}{\sigma A} \tag{4.16}$$

where ρ represents the resistivity of the semiconductor and its reciprocal σ represents its conductivity. From Eq. (4.16) we obtain

$$\frac{I}{A} = |J| = \sigma \frac{V}{l}$$

For a uniformly doped bar, V/l denotes the electric field. Thus, in general, we can write

$$\mathbf{J} = \sigma \mathscr{E} \tag{4.17}$$

Comparing Eqs. (4.15) and (4.17), we obtain

$$\sigma = q(\mu_n n + \mu_p p) = \sigma_n + \sigma_p \tag{4.18}$$

where σ_n and σ_p represent the electron and hole components of the conductivity, respectively. For an intrinsic semiconductor $n = p = n_i$, and the intrinsic conductivity σ_i is given by

$$\sigma_i = q(\mu_n + \mu_p)n_i \tag{4.19}$$

Since the mobilities μ_n and μ_p are not widely different from each other, σ_p and σ_n in Eq. (4.19) are comparable. For an extrinsic semiconductor with equilibrium concentrations n_o and p_o, Eq. (4.18) becomes

$$\sigma = q(\mu_n n_o + \mu_p p_o) \tag{4.20}$$

In a strongly extrinsic semiconductor, the contribution of minority carriers to the total conductivity is negligible. As we have just seen, the carrier mobilities depend on the total concentration of ionized impurities ($N_d^+ + N_a^-$). However, n_o and p_o depend on the difference between the donor and acceptor concentrations. Thus, in general, the conductivity (or resistivity) in Eq. (4.20) must be calculated using the values of n_o and p_o and the mobility data of Fig. 4.5. For uncompensated materials, only one type of impurity is present, and the resistivity is a simple function of concentration of that type of impurity. Figure 4.7 shows the room temperature resistivity as a function of impurity concentration for *n*- and *p*-type Ge, Si, and GaAs containing only one type of impurity.

In semiconductors both the carrier concentration and mobility are temperature sensitive. Thus, the temperature dependence of conductivity is determined by the temperature dependence of both these quantities. Let us first consider an intrinsic semiconductor. Substituting the value of n_i from Eq. (3.32) into Eq. (4.19) gives

$$\sigma_i = q(\mu_n + \mu_p) K_2 T^{3/2} \exp(-E_{go}/2kT)$$

which, after substituting the temperature dependence of μ_p and μ_n, can be expressed as

$$\sigma_i = \text{Const} \cdot T^{a_o} \exp(-E_{go}/2kT) \tag{4.21}$$

where a_o has a value of about -1. In Eq. (4.21), most of the variation of σ_i is contained in the exponential factor, so that a plot of $\ln \sigma_i$ as a function of $1/T$ gives a straight line whose slope can be used to determine E_{go}.

The situation is different for extrinsic semiconductors. Semilog plots of σ as a function of $1/T$ for three *n*-type Ge samples are shown in Fig. 4.8. Let us first

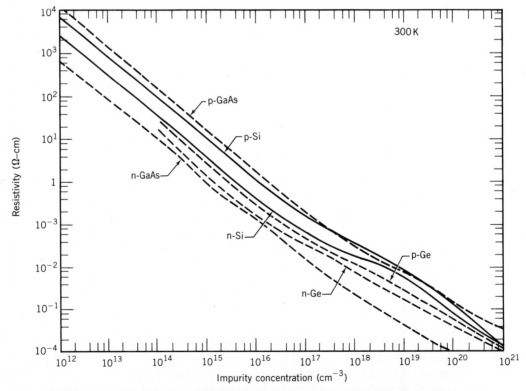

FIGURE 4.7 Resistivity as a function of dopant impurity concentration for uncompensated germanium, silicon, and gallium arsenide at 300 K. (From S. M. Sze, reference 5, pp. 32–33. Copyright © 1981. Reprinted by permission of John Wiley & Sons, Inc., New York.)

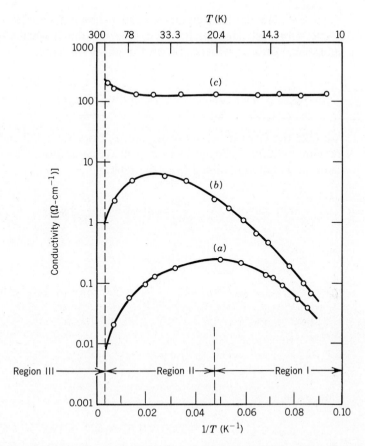

FIGURE 4.8 Logarithmic plots of conductivity as a function of $1/T$ for three n-type germanium samples. In sample (a) region I corresponds to the impurity ionization range, region II represents saturation, and region III is the intrinsic range. (From R. B. Adler et al, reference 1, p. 51. Copyright © 1964. Reprinted by permission of John Wiley & Sons, Inc., New York.)

consider the curve for sample (a) which, though strongly extrinsic, is not heavily doped. At low temperatures (region I) the conductivity increases rapidly with increasing temperature because of an increase in the electron concentration n_o. In this region, the rate of intrinsic carrier generation is so small that p_o is negligible and $\sigma = q\mu_n n_o$. Although μ_n also varies with temperature, this change is negligibly small compared to that of n_o (see Fig. 3.13). At temperatures around 30 K (region II), all the donors are ionized and the intrinsic carrier generation is still negligible, so that n_o cannot increase further. Therefore, the conductivity in this region decreases with temperature because of the decrease in mobility which is dominated by lattice scattering. Finally, at still higher temperatures (region III) the electron and hole concentrations become large and the semiconductor becomes intrinsic. The curve for sample (b) also exhibits a similar trend.

Next, we consider the temperature dependence of σ for sample (c). This sample is nearly degenerate, and the ionization energy for donors is zero. Thus, all the donors are ionized even at the lowest temperature, and there is no change in n_o except at high temperatures. Because of the presence of a large number of donor ions, the mobility in this sample is dominated by ionized impurity scattering and varies slowly with increasing temperature.

As an example, let us calculate the maximum possible resistivity of Ge at 300 K. Obviously, the resistivity will be maximum when the conductivity is a minimum. Since $p_o = n_i^2/n_o$, from Eq. (4.20) we have

$$\sigma = q\left(\mu_n n_o + \mu_p \frac{n_i^2}{n_o}\right)$$

We take the intrinsic values of μ_p and μ_n from Table 4.2 and assume that these remain constant independent of dopant concentration. The condition for minimum in σ is then obtained from the relation

$$\frac{d\sigma}{dn_o} = q\left(\mu_n - \mu_p \frac{n_i^2}{n_o^2}\right) = 0$$

which gives

$$n_o = \sqrt{\frac{\mu_p}{\mu_n}} n_i = \sqrt{\frac{1900}{3900}} \times 2.4 \times 10^{13} = 1.67 \times 10^{13} \text{ cm}^{-3}$$

and

$$p_o = n_i^2/n_o = 3.45 \times 10^{13} \text{ cm}^{-3}$$

This gives

$$\rho_{\max} = [1.6 \times 10^{-19}(3900 \times 1.67 \times 10^{13} + 1900 \times 3.45 \times 10^{13})]^{-1}$$
$$= 47.8 \text{ }\Omega\text{-cm}.$$

which is slightly higher than the intrinsic resistivity (45 Ω-cm).

4.4 IMPURITY BAND CONDUCTION

In Sec. 3.8 we saw that the ionization energy of shallow donors and acceptors decreases with the increasing concentration of dopants and becomes zero when the impurity concentration reaches a critical value. If such a semiconductor is cooled to temperatures near 0°K, the conductivity does not drop to zero as in the case of a lightly doped semiconductor, but has a finite value. This happens because of impurity band conduction.

When the donor concentration reaches the critical value, it is possible for an electron (or a hole in a p-type semiconductor) to move from the ground state of one donor to that of the neighboring ionized donors by the process of quantum mechanical tunneling discussed in Chapter 8. In an applied electric field, the migration of carriers in one direction becomes energetically favorable compared to others, and the impurity conduction now results with a small activation energy by the "hopping" of the carrier from one impurity site to another. For nondegenerate materials the distance between neighboring impurity atoms is large and the hopping of carriers becomes highly improbable.

4.5 THE HALL EFFECT

4.5.1 Phenomenological Description

Thus far, we have discussed only the motion of carriers in an applied electric field. The motion of electrons and holes in the presence of electric and magnetic fields is important and gives rise to a number of galvanomagnetic effects. The most important of these effects is the Hall effect. An elementary treatment of the Hall effect is presented in this section.

Consider a rectangular bar of a uniformly doped, strongly n-type semiconductor of length l, width W, and thickness d. Let an electric field $\mathcal{E}_x = V_x/l$ be applied along the x direction and a magnetic field B_z along the z direction as shown in Fig. 4.9. In the presence of the electric field, electrons move from right to left with a velocity v_{nx} given by

$$(-v_{nx}) = \mu_n \mathcal{E}_x$$

When a particle of charge q moves in a magnetic field, the field exerts a force $q(\mathbf{v} \times \mathbf{B})$ on the particle which causes it to move in a direction perpendicular to both the electric and magnetic fields. In this case, the particle is an electron of charge $-q$. Therefore, it will be deflected toward the front face ($-y$ direction) of the bar. When a steady current is flowing along the x direction, electrons move to the front side of the bar in response to the magnetic field and a positive charge of ionized donors is left on the back face. Thus, an electric field $\mathcal{E}_y$ is developed between the two faces that opposes the motion of electrons toward the front face. A Hall voltage V_H can be measured between the two faces, with the front face negative with respect to the back face. If the carriers are holes instead of electrons, it can easily be seen that the sign of V_H is reversed.

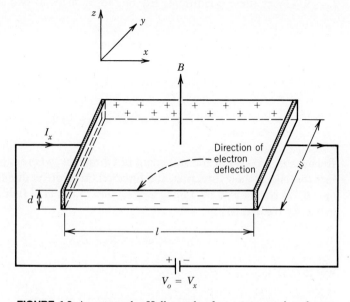

FIGURE 4.9 A rectangular Hall sample of an n-type semiconductor.

4.5.2 Mathematics of the Hall Effect

For developing a simple expression for the Hall voltage, we assume that all the conduction electrons have the same drift velocity $\mathbf{v}_n$ and the same value of τ_c. The resultant force acting on any electron then is given by the Lorentz force equation

$$\mathbf{F} = -q(\mathscr{E} + \mathbf{v}_n \times \mathbf{B}) \qquad (4.22)$$

From Eq. (4.3) $\mathbf{F}$ is given by

$$\mathbf{F} = m_e \frac{d\mathbf{v}_n}{dt} + m_e \frac{\mathbf{v}_n}{\tau_c} \qquad (4.23)$$

In the steady state, the first term in Eq. (4.23) is zero, and combining the above two relations yields

$$\mathscr{E} + \mathbf{v}_n \times \mathbf{B} = -\frac{\mathbf{v}_n}{\mu_n} \qquad (4.24a)$$

where

$$\mu_n = \frac{q\tau_c}{m_e}$$

Using the relation $\mathbf{J} = -qn_o\mathbf{v}_n$ for electrons, we can write Eq. (4.24a) as

$$\mathscr{E} = \frac{\mathbf{J}}{q\mu_n n_o} + \frac{\mathbf{J}}{qn_o} \times \mathbf{B} \qquad (4.24b)$$

This vector equation can be simplified further noting that $B_x = B_y = 0$ and $J_y = J_z = 0$. Under these conditions, Eq. (4.24b) reduces to two scalar equations

$$\mathscr{E}_x = \frac{J_x}{\sigma} \qquad (4.25a)$$

and

$$\mathscr{E}_y = -\frac{J_x B_z}{qn_o} = -\mu_n B_z \mathscr{E}_x \qquad (4.25b)$$

Relation (4.25a) is just the statement of Ohm's law, whereas (4.25b) expresses the fact that along the y direction, the force on an electron due to the magnetic field $(q\mu_n B_z \mathscr{E}_x)$ is balanced by a force $(-q\mathscr{E}_y)$ due to the Hall field. Equation (4.25b) is generally written as

$$\mathscr{E}_y = R_H J_x B_z \qquad (4.26)$$

where

$$R_H = -\frac{1}{qn_o}$$

is called the *Hall constant*. A similar analysis can be carried out for a p-type semiconductor, and it can be shown that

$$\mathcal{E}_y = R_H J_x B_z = \mu_p B_z \mathcal{E}_x \tag{4.27}$$

with

$$R_H = \frac{1}{qp_o}$$

where p_o is the equilibrium concentration of holes in the sample. It is thus seen that R_H has a negative sign for electrons and a positive sign for holes. Knowing the dimensions of the bar, we can determine the values of R_H and the mobility μ_n (or μ_p). If W denotes the width of the bar along the y direction, and V_H is the measured Hall voltage, then $|\mathcal{E}_y| = V_H/W$. From Eq. (4.26) we obtain

$$R_H = \frac{V_H d}{I_x B_z} \tag{4.28}$$

where $I_x = J_x W d$ is the current along the x direction. The carrier mobility μ_H can be determined from the knowledge of R_H and the resistivity of the bar, or directly from Eq. (4.27). Thus, writing μ_H for μ_p in Eq. (4.27), or for μ_n in Eq. (4.25b), we get

$$\mathcal{E}_y = \frac{V_H}{W} = \mu_H B_z \mathcal{E}_x$$

which gives

$$\mu_H = \frac{l}{W}\left(\frac{V_H}{V_x B_z}\right) \tag{4.29}$$

where $|\mathcal{E}_x| = V_x/l$ and V_x is the applied voltage. The measurement of the Hall voltage and the dimensions of the bar thus enable us to calculate the sign and the concentration of the current carriers as well as the carrier mobility.

The mobility determined by the Hall voltage measurement is called the *Hall mobility* and is higher from the conductivity mobility defined by Eqs. (4.8a) and (4.8b) by a factor of $3\pi/8$. This factor arises from the quantum mechanical averaging process which has been neglected in the above treatment (see Chapter 20).

Interpreting the Hall effect for semiconductors containing comparable concentrations of holes and electrons is more involved, but the situation can be analyzed in the same manner as was done above. In such cases it can be shown [3] that (see Prob. 4.15)

$$\mathcal{E}_y = \left(\frac{\mu_p^2 p_o - \mu_n^2 n_o}{\mu_p p_o + \mu_n n_o}\right)\mathcal{E}_x B_z \tag{4.30}$$

and

$$R_H = \frac{\mu_p^2 p_o - \mu_n^2 n_o}{q(\mu_p p_o + \mu_n n_o)^2} \tag{4.31}$$

These equations reduce to those obtained for the *n*-type semiconductor when $p_o = 0$ and to those for the *p*-type semiconductor when $n_o = 0$.

Let us calculate the values of R_H and μ_H for a sample with $l = 2$ cm, $W = d = 0.4$ cm, $B_z = 5 \times 10^3$ Gauss, $V_x = 1.5$ V, $I_x = 7.5$ mA, and $V_H = +6$ mV in Fig. 4.9.

We have 10^4 Gauss = 1 weber m^{-2} = 10^{-4} weber cm^{-2}. Thus, from Eq. (4.28) we obtain

$$R_H = \frac{6 \times 10^{-3} \times 0.4}{7.5 \times 10^{-3} \times 5 \times 10^{-5}} = 6.4 \times 10^3 \text{ cm}^3 \text{ coulomb}^{-1}.$$

Since R_H is positive, the material is *p*-type and $p_o = 9.76 \times 10^{14}$ cm^{-3}. Now from Eq. (4.29)

$$\mu_H = \frac{2 \times 6 \times 10^{-3}}{0.4 \times 1.5 \times 5 \times 10^{-5}} = 400 \text{ cm}^2 \text{ V}^{-1} \text{ sec}^{-1}$$

4.6 NONLINEAR CONDUCTIVITY

Our discussion of conductivity until now was based on Eq. (4.9) that, in general, can be written as

$$\mathbf{v}_d = \mu \mathcal{E} \tag{4.32a}$$

This relation shows that the drift velocity is directly proportional to electric field $\mathcal{E}$ and the mobility μ is a constant independent of $\mathcal{E}$. When this assumption is valid, the conductivity is said to be linear. The meaning of this statement becomes clear when we rewrite Eq. (4.32a) after multiplying both sides with the concentration and the charge of carriers. Assuming a strongly *n*-type semiconductor, we observe that Eq. (4.32a) becomes

$$\mathbf{J} = \mathbf{J}_n = \sigma \mathcal{E} \tag{4.32b}$$

where σ is independent of $\mathcal{E}$ since μ is assumed to be constant. This linear relation between $\mathbf{J}$ and $\mathcal{E}$ (or between $\mathbf{v}_d$ and $\mathcal{E}$) is a statement of Ohm's law.

As was mentioned earlier, Eq. (4.32b) is valid only for low electric fields such that the average drift velocity of carriers is small compared with their random thermal velocity v_{th}. When the electric field becomes large enough to make v_d comparable to v_{th}, Ohm's law is no longer valid and the conductivity becomes nonlinear. The departure from Ohm's law at high electric fields can be more readily observed in semiconductors than in metals. The explanation for this is as follows: In metals the electron gas is degenerate, and the average kinetic energy of an electron is on the order of the Fermi energy which is several electron volts and corresponds to a large value of thermal velocity v_{th}. In order to observe the deviation from Ohm's law, and to make v_d comparable to v_{th}, the energy gained by an electron from the applied field must be comparable to its thermal energy. Obtaining this energy in metals requires fantastically large fields at which the metal will melt and evaporate because of joule heating. On the other hand, electron population in a semiconductor is nondegenerate, and the average thermal energy of an electron is only 3kT/2. Hence, a change on the order of kT in the electron energy will cause a perceptible change in its average energy. Changes of this magnitude can be easily achieved by the application of an electric field. It is for these reasons that the nonlinear conductivity could not be observed by early investigators who were mostly concerned with metals. The first observation of nonlinear conductivity was made [4] on germanium in 1951. Since then, the phenomenon has been observed in all the known semiconductors of importance.

To get an idea about how nonlinear conductivity arises, let us consider a piece of a semiconductor. As explained earlier, in the absence of an external field, the electrons and holes remain in thermal equilibrium with the lattice atoms and their net energy exchange with the lattice is zero. Thus, the average kinetic energy of a lattice atom is the same as that of a free carrier, and both have the same temperature T.

If an electric field is now applied to the crystal, the carriers gain energy from the field. They try to dissipate this excess energy to the lattice by emitting more phonons than they were emitting under thermal equilibrium. This mode of energy transfer, however, is fairly inefficient, and the carriers are not able to transfer all their excess energy to the lattice as quickly as they gain it from the field. Therefore, in the presence of an electric field a carrier has more kinetic energy than it had under thermal equilibrium, and since the kinetic energy is a measure of temperature, the carrier temperature T_e rises above the lattice temperature T. Thus, the electrons and holes are heated by the field, and they become warm in the presence of a small electric field. As the temperature T_e rises gradually above T, the carriers start emitting more phonons and a steady state is reached when the rate of energy loss due to phonon emission is balanced by the rate of energy gain from the electric field. For a carrier with average drift velocity $\mathbf{v}_d$ and charge q, the power gained from the field is $q\mathbf{v}_d \cdot \mathscr{E}$, and if there are n_o such carriers in a unit volume, the power gained from the field is

$$\text{Power gained/volume} = qn_o\mathbf{v}_d \cdot \mathscr{E} = \mathbf{J} \cdot \mathscr{E}$$

The carriers will reach a steady state at some temperature T_e when all this power is dissipated to the lattice in the form of the emission of phonons. It might be argued that in the steady state the lattice is receiving the same power from the carriers as the carriers are receiving from the field. Hence, the lattice temperature should also rise and should ultimately approach the carrier temperature T_e. However, this does not happen, for two reasons. First, the heat capacity of the lattice is enormously large compared to that of the electrons, and second, the crystal can lose heat by conduction, convection, and radiation. But, when the product $\mathbf{J} \cdot \mathscr{E}$ becomes large, the crystal is also heated up.

As long as T_e is only slightly above T, Ohm's law is obeyed and the velocity $\mathbf{v}_d$ is proportional to the electric field, making μ a constant. At such values of the electric field, the electrons and holes are only "warm." As the field is increased gradually, T_e rises continuously and, at some value of the field, T_e becomes significantly higher than T. The carriers now interact more strongly with the lattice, and their average drift velocity falls below the linear projection of its low field value. Consequently, the mobility starts decreasing with field, and the current-voltage characteristic of the semiconductor shows considerable deviation from Ohm's law.

At relatively lower values of the field ($\approx 10^3$ V/cm), the average energy gained by a carrier in a mean free path is small, and the carrier loses this energy to the lattice in the form of low-energy longitudinal acoustical phonons. However, when the field is increased to a value such that the average energy of the carrier is on the order of the optical phonon energy, the dominant mode of scattering is the emission of optical phonons. At an electric field of about 10^4 V/cm, both in Ge and Si the energy loss to the lattice is mainly in the form of optical phonons. When this happens, the average drift velocity of the carriers approaches a limiting value independent of the electric field. This scattering limited-saturation velocity v_s is given by [5]

$$v_s = \left[\frac{8E_r \tanh(E_r/2kT)}{3\pi m_e}\right]^{1/2} \quad (4.33)$$

where E_r is the optical phonon energy at zero momentum and m_e is the electron-effective mass. A similar relation can be written for holes.

For an electric field higher than that required for drift velocity saturation, the temperature T_e increases rapidly with field and the carriers become "hot." Hot carriers have found a number of applications (see Chapter 8). Figure 4.10 shows the drift velocity for holes and electrons plotted as a function of the electric field for Ge and Si. The situation for GaAs is more complicated (see Chapter 11).

For silicon, the results of Fig. 4.10 can be represented by the following expressions [2]:

$$|v_n| = v_s \frac{(\mathscr{E}/\mathscr{E}_c)}{[1 + (\mathscr{E}/\mathscr{E}_c)^2]^{1/2}} \quad \text{for electrons} \quad (4.34a)$$

and

$$|v_p| = v_s \left[\frac{\mathscr{E}/\mathscr{E}_c}{1 + \mathscr{E}/\mathscr{E}_c}\right] \quad \text{for holes} \quad (4.34b)$$

The values of v_s and the critical field $\mathscr{E}_c$ are given in Table 4.1.

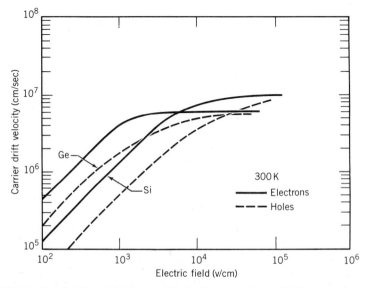

FIGURE 4.10 Drift velocities of electrons and holes in germanium and silicon as a function of the electric field. (From S. M. Sze, reference 5, p. 46. Copyright © 1981. Reprinted by permission of John Wiley & Sons, Inc., New York.)

TABLE 4.1
Saturated Carrier Velocity (v_s) and Critical Electric Field ($\mathscr{E}_c$) for Electrons and Holes in Silicon

	v_s (cm/Sec)	$\mathscr{E}_c$ (Volt/cm)
Electrons	1.1×10^7	8×10^3
Holes	9.5×10^6	1.95×10^4

4.7 CARRIER FLOW BY DIFFUSION

Diffusion is a manifestation of the random thermal motion of particles and shows up as a net current of particles from regions where they are heavily concentrated to neighboring regions of lower concentration. From the theory of diffusion in gases, we know that when two volumes of a gas with different densities are brought together, the molecules move from the high-density to the low-density region. Thus, it is clear that the diffusion flow is proportional to the concentration gradient of particles under consideration, and occurs in the direction of the negative gradient.

Let us consider the diffusive motion of particles whose concentration is decreasing with distance x, as shown in Fig. 4.11. This plot, for example, may be assumed to represent the hole concentration along a bar of a semiconductor. If we consider a surface normal to the x axis at x_o, it is clear that, because of thermal motion, the particles will be crossing the plane at x_o from the left as well as from the right. Since the concentration to the left of x_o is higher than that to the right, more particles from the left will be crossing the plane at x_o than from the right. This results in a net flow of particles from the left to the right. The particle flux across the surface at x_o is defined as the net number of particles crossing the surface per unit area per unit time. Since the net flow in Fig. 4.11 is along the positive x direction while the concentration gradient is along the negative x direction, the particle flux $\mathbf{F}_o$ is given by

$$\mathbf{F}_o = -D \frac{dC_B}{dx}\bigg|_{x=x_o}$$

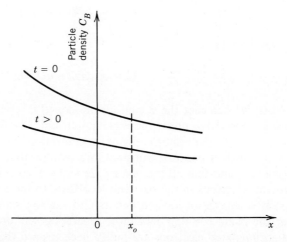

FIGURE 4.11 Schematic representation of diffusion motion of particles.

where C_B represents the particle concentration and dC_B/dx is the concentration gradient. The constant of proportionality D is known as the diffusion coefficient (or diffusion constant). In three dimensions, the above relation can be written as

$$\mathbf{F}_o = -D\nabla C_B$$

If the particles are holes, the diffusion current density $\mathbf{J}_{pD}$ is defined by

$$\mathbf{J}_{pD} = -qD_p\nabla p \qquad (4.35a)$$

where D_p and ∇p represent the diffusion coefficient for holes and the hole concentration gradient, respectively. The corresponding relation for electrons is given by

$$\mathbf{J}_{nD} = -(-q)D_n\nabla n = qD_n\nabla n \qquad (4.35b)$$

Diffusion currents are important in semiconductors but not in metals. The reason why is that semiconductors have two types of carriers: electrons and holes. A local increase in the concentration of one of these carriers is accompanied by an increase in the other, and the electrical neutrality is maintained even in the presence of a concentration gradient. Metals have only one type of carrier, and the self-annihilating large fields set up in the metals resist any effort to create a concentration gradient.

4.8 EINSTEIN RELATIONS

Although the drift and diffusion processes appear to be different, a close relationship exists between the mobilities and diffusion coefficients. This is because both D and μ are determined by the thermal motion and scattering of the free carriers. Before establishing this relationship, let us determine the expressions for hole and electron currents in the presence of both drift and diffusion. The current density in a semiconductor in which the concentration gradient and electric fields are simultaneously present is obtained by adding the drift and diffusion current components. Thus, for holes

$$\mathbf{J}_p = q\mu_p p\mathscr{E} - qD_p\nabla p \qquad (4.36a)$$

and for electrons

$$\mathbf{J}_n = q\mu_n n\mathscr{E} + qD_n\nabla n \qquad (4.36b)$$

Next, we consider the nonuniformly doped n-type bar of the semiconductor shown in Fig. 4.12a under thermal equilibrium. The bar is intrinsic to the left of $x = 0$, while to the right the donor concentration increases gradually up to $x = l$ beyond which it becomes constant. We assume that the semiconductor is nondegenerate, and that all the donors are ionized. As a result of the concentration gradient, electrons in the bar tend to diffuse to the left of $x = l$, leaving behind a positive charge of ionized donors and causing some electrons to accumulate near the plane $x = 0$. This charge separation sets up an electric field in which the accumulated electrons are pulled back toward the plane at $x = l$.

EINSTEIN RELATIONS

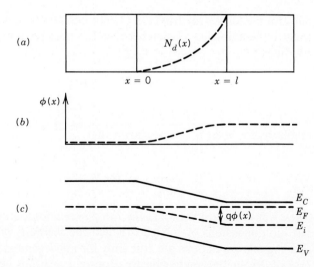

FIGURE 4.12 A bar of n-type semiconductor with graded impurity distribution under thermal equilibrium; (a) the dopant distribution, (b) potential variation, and (c) the energy band diagram.

The equilibrium potential distribution across the bar is shown in Fig. 4.12b, and its energy band diagram is shown in Fig. 4.12c. In drawing the energy band diagram, we note that in thermal equilibrium the Fermi level E_F must be the same throughout (see Sec. 4.9). Note that E_F coincides with E_i at $x = 0$ where $n_o = n_i$, and E_i gradually moves away from E_F as n_o rises above n_i. Since the band structure of the semiconductor is not changed by doping, the band edges E_C and E_V undergo the same change as the intrinsic Fermi level E_i.

In thermal equilibrium, electrons tend to diffuse down the concentration gradient, causing a diffusion current from right to left. The presence of an electric field tends to cause a drift current of electrons in the opposite direction. Since no net current flows through the semiconductor, the two currents must add up to zero. As the concentration gradient and the field exist only along the x axis, the current flow is one-dimensional and we obtain

$$J_n = qD_n \frac{dn_o}{dx} + q\mu_n n_o \mathscr{E} = 0 \tag{4.37}$$

Note that in Eq. (4.37) $\mathscr{E}$ is negative. For a nondegenerate semiconductor

$$n_o(x) = n_i \exp\left(\frac{E_F - E_i(x)}{kT}\right) \tag{4.38a}$$

We assume that this relation is valid at all points in the semiconductor and that the electron concentration is not influenced by the small charge imbalance that establishes the electric field. This is a valid assumption (see Sec. 5.6). The electrostatic potential $\phi(x)$ at any location x in the bar is defined by the relation $E_i(x) = -q\phi(x)$. We can, therefore, write $E_F - E_i(x) = E_i(0) - E_i(x) = -q[\phi(0) - \phi(x)]$. The potential $\phi(x)$ is assumed to be zero at the location $x = 0$ where the bar is intrinsic. Thus, $\phi(0) = 0$, and Eq. (4.38a) can be written as

$$n_o(x) = n_i \exp\left(\frac{q\phi(x)}{kT}\right) \tag{4.38b}$$

Substituting for dn_o/dx from Eq. (4.38b) into (4.37), and noting that $\mathscr{E} = -d\phi/dx$ leads to the following relation between the electron mobility μ_n and the diffusion coefficient D_n

$$D_n = \frac{kT}{q}\mu_n \qquad (4.39a)$$

Similarly, for holes it can be shown that

$$D_p = \frac{kT}{q}\mu_p \qquad (4.39b)$$

These equations are known as the *Einstein relations,* and the constant kT/q which has the dimensions of voltage is called the *thermal voltage.* It should be noted that the above relations are true only for nondegenerate semiconductors. In the case of degenerate semiconductors Einstein relations become quite complex [6]. Einstein relations will be found extremely useful in our discussion of current flow problems in a number of semiconductor devices.

4.9 CONSTANCY OF THE FERMI LEVEL ACROSS A JUNCTION

In the energy band diagram in Fig. 4.12c, we had assumed that in thermal equilibrium the Fermi level is the same throughout. In this section, we will justify this assumption. We will show that when two materials that are in physical contact are in thermal equilibrium, the Fermi level is constant through the junction. For the sake of simplicity, let us consider two metals 1 and 2 in contact with each other. Let E_{F1} and E_{F2} denote the positions of the Fermi levels in metals 1 and 2, respectively, and let E_1 and E_2 be any two energy levels in the conduction bands. Let us consider the energy band diagram when $E_1 = E_2$ (Fig. 4.13). In this situation, E_{F1} is higher than E_{F2}. Since the probability of occupation of a quantum state by an electron is 1/2 at the Fermi level, a larger fraction of available states above E_1 is occupied by electrons in metal 1 than that in metal 2. This means that the average energy of an electron is higher in metal 1 than in metal 2. Consequently, electrons will flow from metal 1 to metal 2 until the energy level at which the probability of occupation is 1/2 is the same throughout the system. This situation occurs when the two Fermi levels are aligned, which implies that $E_{F1} = E_{F2}$.

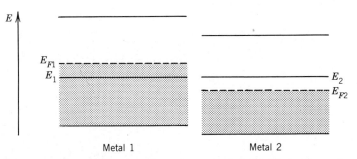

FIGURE 4.13 Energy band diagram of two metals in contact before attaining thermal equilibrium.

There is an instructive way to show that in thermal equilibrium E_{F1} should be equal to E_{F2}. Referring to Fig. 4.13, let us consider a small energy interval ΔE around an absolute value of energy E that is identical in the two metals. We assume that both metals have partially filled electron states in the energy interval ΔE. Therefore, electrons are able to move horizontally from metal 1 to metal 2 without the expenditure of any energy. Let $N_1(E)$ and $f_1(E)$ represent the quantum density of states per unit energy and the FD distribution function, respectively, in metal 1, and let $N_2(E)$ and $f_2(E)$ be the corresponding quantities in metal 2. The electron current that crosses the junction from metal 1 to metal 2, in the energy interval ΔE, is proportional to the number of electrons $[N_1(E)f_1(E)\Delta E]$ in metal 1 and the number of vacant states $[N_2(E)\{1 - f_2(E)\}\Delta E]$ in metal 2. Thus, we can write

$$I_{12} \propto [N_1(E)f_1(E)\Delta E][N_2(E)\{1 - f_2(E)\}\Delta E] \qquad (4.40a)$$

Similarly, the reverse electron current from metal 2 to metal 1 is given by

$$I_{21} \propto [N_2(E)f_2(E)\Delta E][N_1(E)\{1 - f_1(E)\}\Delta E] \qquad (4.40b)$$

The principle of detailed balance states that, for any system in thermal equilibrium, the rate of a physical process in a given energy interval must be balanced by its inverse process occurring in the same energy range. Thus, thermal equilibrium requires that $I_{12} = I_{21}$, and from Eqs. (4.40a) and (4.40b) we get $f_1(E) = f_2(E)$ or

$$\left[1 + \exp\left(\frac{E - E_{F1}}{kT}\right)\right] = \left[1 + \exp\left(\frac{E - E_{F2}}{kT}\right)\right] \qquad (4.41)$$

This equation can be satisfied only when $E_{F1} = E_{F2}$.

In the above example, we have considered the contact between two metals. However, the result is general and is valid for any two materials in physical contact which are in thermal equilibrium.

REFERENCES

1. R. B. ADLER, A. C. SMITH and R. L. LONGINI, *Introduction to Semiconductor Physics,* SEEC Vol. 1, John Wiley, New York, 1964.

2. D. M. CAUGHEY and R. E. THOMAS, "Carrier mobilities in silicon empirically related to doping and field," *Proc. IEEE* 55, 2192 (1967). Also G. BACCARANI and P. OSTOJA, *Solid-State Electron.* 18, 579 (1975).

3. R. A. SMITH, *Semiconductors,* Cambridge University Press, London, 1978.

4. E. J. RYDER and W. SHOCKLEY, "Mobilities of electrons in high electric fields," *Phys. Rev.* 81, 139 (1951).

5. S. M. SZE, *Physics of Semiconductor Devices,* John Wiley, New York, 1981, Chapter 1.

6. F. A. LINDHOLM and R. W. AYERS, "Generalized Einstein relation for degenerate semiconductors," *Proc. IEEE* 56, 371 (1968). Also H. KROEMER, *IEEE Trans. Electron Devices* ED-25, 850 (1978).

ADDITIONAL READING LIST

1. J. L. MOLL, *Physics of Semiconductors*, McGraw-Hill, New York, 1964.
2. B. R. NAG, *Theory of Electrical Transport in Semiconductors*, Pergamon Press, Oxford, 1972.
3. H. F. WOLF, *Semiconductors*, Wiley Interscience, New York, 1971.
4. J. A. PALS, "Basic properties in semiconductor devices," Chapter 13 in *Crystalline Semiconducting Materials and Devices*, P. N. BUTCHER, N. H. MARCH, and M. P. TOSI (eds.), Plenum Press, New York, 1986.
5. F. F.Y. WANG, *Introduction to Solid State Electronics*, North-Holland, New York, 1980, Chapters 4 and 12.

PROBLEMS

4.1 Argue why the concept of mobility is meaningless for an electron moving in a vacuum.

4.2 Explain why the carrier mobility in group II-VI semiconductors is lower than that in group III-V and IV semiconductors.

4.3 A GaAs sample is doped so that the electron and hole components of current are equal in an applied electric field. Calculate the equilibrium electron–hole concentrations, the net doping, and the sample resistivity at 300 K.

4.4 A 2-cm long bar of *n*-type Ge has a cross-sectional area of 0.1 cm^2 and a resistivity of 10 Ω-cm. A 10 V battery is connected across the bar. (a) How long does it take for an electron to drift across the bar? (b) How much energy does an electron deliver to the lattice during its transit through the bar?

4.5 A sample of Ge has an intrinsic resistivity of 60 Ω-cm. Calculate the value of current density in an applied field of 10 mV cm^{-1} if there are 10^{13} donors cm^{-3} and 4×10^{12} acceptors cm^{-3}. Take $\mu_n = 4200$ cm^2 V^{-1} sec^{-1} and $\mu_p = 2000$ cm^2 V^{-1} sec^{-1}. Assume that all the impurities are ionized.

4.6 Show that the probability that an electron with a mean free time $\bar{\tau}$ in a semiconductor remains unscattered for a time t is proportional to $\exp(-t/\bar{\tau})$. Consider a semiconductor with n different scattering processes independent of each other. Show that the resultant mean free time $\bar{\tau}$ is given by $1/\bar{\tau} = 1/\tau_1 + 1/\tau_2 \ldots + 1/\tau_n$.

4.7 Determine the number of free electrons per atom in a substance whose density is 8.2 gm cm^{-3}, atomic weight 55, electron mobility 430 cm^2 V^{-1} sec^{-1}, and resistivity 1.65×10^{-7} Ω-cm.

4.8 Assuming all the impurities to be ionized, determine how many grams of boron should be added to 1 kg of Ge to obtain a resistivity of 0.2 Ω-cm. Take hole mobility $\mu_p = 1250$ cm^2 V^{-1} sec^{-1}.

4.9 Assuming that μ_p and μ_n are independent of dopant concentrations, calculate the maximum possible values of resistivity in Si and GaAs at 300 K. Compare these values with the intrinsic resistivity of these semiconductors.

4.10 Estimate the temperature dependence of intrinsic resistivity of Si at 300 K. Express your results in terms of percentage change per °C rise in temperature.

4.11 A 300-Ω resistor is connected in series with a piece of intrinsic Ge. Near 300 K the value of the resistor changes by 1 percent $°K^{-1}$. Calculate what should be the resistance of the Ge sample to ensure that the resistance of the combination remains invariant when the temperature is changed by a few °C around 300 K. Ignore the temperature dependence of μ_n and μ_p.

4.12 The resistivity ρ_o of a Ge sample is measured at 300 K. The sample is then remelted and doped with 4.4×10^{16} arsenic atoms cm^{-3}. A new crystal is grown that has a resistivity of 0.1 Ω-cm and is n-type. Determine the type and concentration of dopant atoms in the original sample and value of ρ_o. Assume $\mu_n = 2\mu_p = 3000$ cm^2 V^{-1} sec^{-1} where necessary.

4.13 A sample of intrinsic semiconductor has a resistance of 10 Ω at 364 K and 100 Ω at 333 K. Assuming that this change is caused entirely by the temperature variation in n_i, calculate the band gap E_{go} of the semiconductor.

4.14 A bar of n-type Si with a cross-sectional area of 1 mm $\times$ 1 mm and a length of 1 cm is connected to a 2 V battery and is carrying a current of 2 mA at 300 K. (a) Calculate the thermal equilibrium electron and hole concentrations in the bar. (b) Calculate the dopant concentration assuming that only donors with energy level 0.148 eV away from the conduction band are present. (c) Determine the temperature at which the bar will become intrinsic.

4.15 Show that the Hall field and the Hall constant for a semiconductor having comparable electron and hole concentrations are given by Eqs. (4.30) and (4.31), respectively. Determine the type and concentration of dopant for which the Hall coefficient will become zero in (a) Ge, (b) Si, and (c) GaAs. Assume $T = 300$ K.

4.16 A Ge sample doped with 2.56×10^{13} Ga atoms cm^{-3} is cut with a rectangular cross-section of 3 mm $\times$ 3 mm and length of 12 mm. A magnetic field of 6K Gauss is applied perpendicular to the 3 mm $\times$ 12 mm faces, and a current of 10 mA is flowing along the length of the sample in the negative x direction (Fig. 4.9). Calculate the magnitude and sign of the Hall voltage. How will this voltage change when the above sample is replaced by an intrinsic Ge sample of the same dimensions? Assume $T = 300$ K.

4.17 Assuming that the volume of an n-type semiconductor bar with graded doping remains space-charge neutral, determine the impurity distribution $N_d(x)$ and the electric field in the bar. Where are the charges that produce the field located? Assume the problem to be one-dimensional.

4.18 The hole diffusion coefficient in a Si sample is 12 cm^2 sec^{-1}. Determine the mean free path and the carrier drift velocity in a field of 200 V cm^{-1}. Also calculate the energy gained by the carrier in a mean free path. Assume $T = 300$ K.

TABLE 4.2
Properties of Ge, Si and GaAs at 300 K

Property	Ge	Si	GaAs
Atomic/molecular weight	72.6	28.09	144.63
Density (g cm^{-3})	5.33	2.33	5.32
Dielectric constant	16.0	11.9	13.1
Effective density of states			
Conduction band, N_C (cm^{-3})	1.04×10^{19}	2.8×10^{19}	4.7×10^{17}
Valence band N_V (cm^{-3})	6.0×10^{18}	1.02×10^{19}	7.0×10^{18}
Electron affinity (eV)	4.01	4.05	4.07
Energy gap, E_g (eV)	0.67	1.12	1.43
Intrinsic carrier concentration, n_i (cm^{-3})	2.4×10^{13}	1.5×10^{10}	1.79×10^{6}
Lattice constant (Å)	5.65	5.43	5.65
Effective mass			
Density of states m_e^*/m_o	0.55	1.18	0.068
m_h^*/m_o	0.3	0.81	0.56
Conductivity m_e/m_o	0.12	0.26	0.09
m_h/m_o	0.23	0.38	
Melting point (°C)	937	1415	1238
Intrinsic mobility			
Electron (cm^2 V^{-1} sec^{-1})	3900	1350	8500
Hole (cm^2 V^{-1} sec^{-1})	1900	480	400

5
EXCESS CARRIERS IN SEMICONDUCTORS

INTRODUCTION

The term *excess carriers* is used for electrons and holes that are in excess of their thermal equilibrium values. Excess carriers can be created in a semiconductor by a variety of processes such as optical excitation, electron bombardment, or injection from a contact. A study of the behavior of excess carriers is important because most of the semiconductor devices operate under nonequilibrium conditions in which the electron and hole concentrations are significantly different from their thermal equilibrium values.

In the case of metals, excess carriers can be introduced only by producing an electrical charge on the specimen, and the field resulting from the mutual repulsion of these charges forces the charge to reside on the surface. Thus, no change in the bulk properties is observed. In the case of semiconductors, because of the existence of two types of carriers, an excess carrier density can be created in the bulk without introducing any significant space charge. Hence, a large concentration of excess carriers can be maintained in a semiconductor under nonequilibrium conditions, and this alters the conduction properties of the specimen.

When excess electrons are introduced in a metal, the process of lattice collisions restores equilibrium. In semiconductor crystals, the processes that restore equilibrium are the diffusion and drift of carriers into or out of the region or the recombination of carriers inside the region. We discussed the drift and diffusion of carriers in the last chapter. In this chapter we will discuss the carrier recombination in detail. Starting with the introduction of excess carriers, we will first treat the kinetics and then the mechanisms of excess carrier recombination. This discussion is followed by a brief account of the origin of recombination centers. Finally, we will discuss the nonequilibrium transport of excess carriers and formulate a set of equations to solve the carrier flow problems in semiconductors.

5.1 INJECTION OF EXCESS CARRIERS

5.1.1 Injection Mechanism

As an illustration of the injection of excess carriers, we consider the process of optical absorption in semiconductors. Figure 5.1 shows a schematic arrangement for studying the photoelectric effects in a semiconductor crystal. A monochromatic beam of light of energy $h\nu$ is allowed to fall on the semiconductor of thickness d. A photodetector is placed behind the crystal, which measures the intensity of the transmitted beam. If $I(0)$ is the intensity of the incident beam, then the intensity $I(d)$ of the transmitted beam is given by

$$I(d) = I(0) \exp(-\alpha d) \tag{5.1}$$

The coefficient α is called the *absorption coefficient* and has a unit of cm^{-1} if d is measured in centimeters. The typical variation of α with photon energy in a semiconductor is shown in Fig. 5.2. It is seen that when $h\nu$ is small compared to E_g, the coefficient α is negligible, and most of the light is transmitted through the crystal, making it transparent to radiation at these frequencies. The small absorption observed in this region is caused by free electrons that are excited from lower energy states to higher energy states in a given band. This free carrier absorption obviously does not create excess electron–hole pairs. As the photon energy increases, α begins to rise rapidly when $h\nu$ approaches E_g. A photon with energy $h\nu \geq E_g$ is absorbed in the semiconductor because it has enough energy to break the covalent bond and create an electron–hole pair. For these frequencies, the crystal becomes opaque to the incident radiation. If a photon has energy considerably in excess of E_g, then an electron excited from the valence band to the conduction band will have energy in excess of the average thermal energy. This situation is shown in Fig. 5.3. The excited electron now loses energy to the lattice during scattering collisions until its velocity reaches the thermal velocity at that temperature. The process of creating excess carriers by shining light is known as *photogeneration*. Photogenerated carriers cause an increase in the conductivity of the semiconductor. A plot of conductivity as a function of the photon energy will have a shape similar to the curve of Fig. 5.2.

Introduction of excess electron-hole pairs in a semiconductor is also known as *injection of excess carriers*. When excess carriers are injected in a semicon-

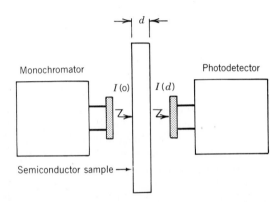

FIGURE 5.1 Schematic diagram for measuring photoabsorption in a semiconductor.

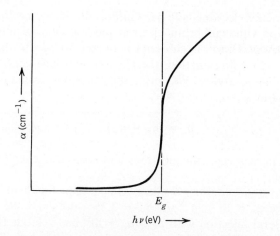

FIGURE 5.2 The optical absorption coefficient as a function of photon energy in a semiconductor. (From B. G. Streetman, *Solid State Electronic Devices*, 2nd ed., © 1980, p. 96. Reprinted by permission of Prentice Hall, Inc., Englewood Cliffs, New Jersey.)

ductor, the *pn* product exceeds n_i^2. An opposite situation is also possible in which $pn < n_i^2$. This happens when the electron–hole concentrations are below their thermal equilibrium values. We then talk of extraction of carriers from the semiconductor.

5.1.2 Low- and High-Level Injections

The carrier injection level is determined by the concentration of excess carriers. When the excess carrier concentration is small compared to the equilibrium majority carrier concentration, it is said to be a low-level injection. If this concentration becomes comparable to or exceeds the equilibrium majority carrier concentration, the injection is called a high-level injection. For example, consider a homogeneously doped bar of semiconductor irradiated with photons of energy $h\nu > E_g$. We assume the bar is thin enough to allow light to penetrate the sample

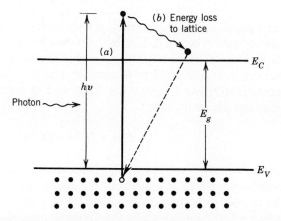

FIGURE 5.3 The optical absorption of a photon with energy $h\nu > E_g$. (*a*) An electron–hole pair is created, and (*b*) the excited electron loses its energy to the lattice phonons and attains thermal equilibrium. (From B. G. Streetman, *Solid State Electronic Devices*, 2nd ed., © 1980, p. 94. Reprinted by permission of Prentice Hall, Inc., Englewood Cliffs, New Jersey.)

and produce electron–hole pairs uniformly. To be specific, let the semiconductor be n-type with equilibrium electron and hole concentrations given by n_{no} and p_{no}, respectively. Here the subscript n is used to signify that the semiconductor is n-type. After illumination, the electron and hole concentrations change to the new values n_n and p_n. We define the excess electron and hole concentrations by the relations

$$n_e = (n_n - n_{no}), \qquad p_e = (p_n - p_{no}) \tag{5.2}$$

In the present case, the electrons and holes are generated in pairs, and we have $n_e = p_e$.

Let us now consider a silicon sample doped with 10^{15} donors cm^{-3}. At 300 K, we have $n_{no} = 10^{15}$ cm^{-3} and $p_{no} = 2.25 \times 10^5$ cm^{-3}. We first consider the case of a low-level injection with $p_e = n_e = 10^{12}$ cm^{-3}. From Eq. (5.2) we obtain $p_n = p_{no} + p_e = 10^{12}$ cm^{-3} and $n_n = n_{no} + n_e \approx 10^{15}$ cm^{-3}. Thus, the low-level injection is essentially a minority carrier injection because the concentration of majority carriers remains practically unchanged. As an example of high-level injection, we take $p_e = n_e = 10^{17}$ cm^{-3}. Thus, $n_n = 1.01 \times 10^{17}$ cm^{-3} and $p_n = 1 \times 10^{17}$ cm^{-3}. The majority carrier concentration now has also changed, and both n_n and p_n have become nearly equal. Although high-level injection is encountered in a number of semiconductor devices, to keep the mathematics simple our treatment in this chapter will be limited mainly to low-level injection.

5.2 RECOMBINATION OF EXCESS CARRIERS

As we have already seen, in thermal equilibrium the rate of thermal generation G_{th} is balanced by the rate of recombination R. When the semiconductor is illuminated to produce electron–hole pairs by photogeneration at a rate of G_L pairs per unit volume per second, the concentration of carriers increases above its equilibrium value. With the increase in concentration, the recombination rate also increases and a steady state is reached when

$$G_{th} + G_L - R = 0 \tag{5.3}$$

Let us now assume that excess carriers have been generated by light and that the steady state is reached. Let the light now be suddenly switched off. As soon as the light is switched off, the excess carriers decay by recombination and the carrier concentration tends to revert to its equilibrium value. As long as the total concentrations n and p are not too large, we can assume that the rate of pair recombination obeys the law

$$R(n, p, T) = \alpha_r(T) pn \tag{5.4}$$

In dealing with the recombination of excess carriers, it is not the absolute rate of recombination which is important, but the *net rate* of recombination in excess of the thermal generation rate defined by $U = R - G_{th}$. Thus, when the light is switched off, the rate at which the excess carriers decay at a given temperature is given by

$$U = R - G_{th} = \alpha_r[pn - n_i^2] \tag{5.5}$$

Obviously, this is the rate at which the electrons and holes in the semiconductor disappear, so that this relation can also be written as

$$-\frac{dp}{dt} = -\frac{dn}{dt} = \alpha_r[pn - n_i^2] \tag{5.6}$$

which is a general relation describing the balance between generation and recombination processes. As long as the *pn* product exceeds n_i^2, the right-hand side of Eq. (5.6) is positive and both dp/dt and dn/dt decay with time, indicating the recombination of excess carriers. When the *pn* product is decreased below n_i^2 by extracting carriers, the time derivatives in Eq. (5.6) become positive and there is a net generation of carriers.

5.2.1 Excess Carrier Lifetime

Let us consider the recombination of excess carriers just after the light is switched off. Besides the recombination in the bulk, electron–hole pairs also recombine at the surface. However, for the sake of simplicity, we ignore surface recombination and assume the electron–hole concentrations to be constant throughout the sample. Writing $p = (p_o + p_e)$ and $n = (n_o + n_e)$ we can write Eq. (5.5) as

$$U = \alpha_r[(p_o + p_e)(n_o + n_e) - n_i^2]$$
$$= \alpha_r[p_o n_e + n_o p_e + p_e n_e] \tag{5.7}$$

This is a nonlinear equation in p_e and n_e and is difficult to solve. The equation can be simplified further when the $p_e n_e$ term is negligibly small compared to the first two terms, which is the case for low-level injection. To be specific, let us consider a strongly *n*-type semiconductor doped uniformly with N_d donors cm^{-3}. Assuming that all the donor atoms are ionized, we have $n_o = n_{no} = N_d$ and $p_o = p_{no} = n_i^2/N_d$. Since $p_e = n_e$, ignoring the $p_e n_e$ term in Eq. (5.7) yields

$$U = -\frac{dp_n}{dt} = \alpha_r(n_{no} + p_{no})p_e \tag{5.8a}$$

Note that $\alpha_r(n_{no} + p_{no})$ has the dimension of reciprocal time and can be written as

$$\tau_p = \frac{1}{\alpha_r(n_{no} + p_{no})} \simeq \frac{1}{\alpha_r N_d} \tag{5.8b}$$

The constant τ_p is known as the minority carrier (hole in this case) lifetime and is a function of temperature because α_r is temperature dependent. From Eqs. (5.8a) and (5.8b) we have

$$\frac{dp_n}{dt} = -U = -\frac{p_e}{\tau_p} \tag{5.9}$$

which has the solution

$$p_e(t) = p_e(0) \exp(-t/\tau_p) \tag{5.10a}$$

or

$$p_n(t) = p_{no} + [p_n(0) - p_{no}] \exp(-t/\tau_p) \quad (5.10b)$$

where $p_e(0) = [p_n(0) - p_{no}]$ is the excess carrier concentration at $t = 0$ when the light is switched off. It can be shown (see Prob. 5.3) that τ_p is the average time that a hole spends before recombining with an electron. The lifetime τ_p is independent of injected carrier concentration as long as the low-level injection condition is maintained.

A similar analysis can be used to study recombination of excess electrons in a p-type semiconductor to yield

$$n_e(t) = n_e(0) \exp(-t/\tau_n) \quad (5.11)$$

where $\tau_n = 1/\alpha_r N_a$ is the excess electron lifetime. It should be noted that both Eq. (5.10) and Eq. (5.11) describe the decay of excess minority carriers. Since the electrons and holes recombine in pairs, the excess majority carriers decay at the same rate as the excess minority carriers. However, since the minority carriers are in short supply, they set the pace for recombination. Majority carriers follow the minority carriers, which ultimately leads to the characterization of the recombination process by a minority carrier lifetime.

5.3 MECHANISMS OF RECOMBINATION PROCESSES [1, 2]

Electron–hole pairs can recombine in a semiconductor in two ways. First, an electron can drop directly from the conduction band into an unoccupied state in the valence band. This is known as *direct band-to-band recombination*. Second, an electron initially makes a transition to an energy level lying deep in the band gap, and it subsequently captures a hole from the valence band. This is known as *indirect recombination*. In the process of electron–hole pair recombination, an energy equal to the difference between electron and hole energy is released. This energy can be emitted as a photon, in which case the recombination is said to be radiative. Alternatively, the energy can be dissipated to the lattice in the form of phonons. A third possibility is that the energy can be imparted as a kinetic energy to a third mobile carrier. This process is called the *Auger process*. Note that both the phonon and Auger recombinations are nonradiative. These different processes will now be considered.

5.3.1 Direct Band-to-Band Recombination

In the case of direct band-to-band recombination, an electron in the conduction band falls into a vacant state in the valence band, thus neutralizing an electron–hole pair. This process is illustrated schematically in Fig. 5.4. Here an electron loses energy on the order of the band gap. In a direct gap semiconductor, the energy is emitted as a photon and the recombination is said to be radiative. In an indirect gap semiconductor, band-to-band transitions involve a large change in electron momentum, and the momentum conservation condition requires either emission or absorption of a phonon. Consequently, direct band-to-band recombination has a very low probability in indirect gap semiconductors.

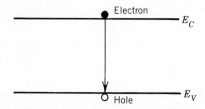

FIGURE 5.4 Direct band-to-band radiative recombination of an electron–hole pair.

All the available information about the band-to-band radiative recombination has been obtained from the optical absorption coefficient $\alpha(v)$ of the semiconductor. In a wide gap semiconductor, the free carrier absorption remains negligibly small, and $\alpha(v)$ characterizes the production of electron–hole pairs by the incident radiation. The rate G_r at which electron–hole pairs are generated radiatively under thermal equilibrium can be determined by multiplying $\alpha(v)$ with the incident photon flux $P(v)\,dv$ and integrating the product over all energies. Thus

$$G_r = \int_0^\infty c'\alpha(v)P(v)\,dv \tag{5.12}$$

where c' is the velocity of light in the semiconductor and $P(v)\,dv$ is given by Planck's radiation law. The integral in Eq. (5.12) is determined graphically from the known plots of $P(v)$ and $\alpha(v)$ as a function of the photon energy [3].

The radiative recombination rate R_r can be written as $R_r = \alpha_r pn$. In thermal equilibrium, this rate must be balanced by the thermal generation rate G_r, so that

$$G_r = \alpha_r p_o n_o = \alpha_r n_i^2$$

The excess carrier radiative recombination rate $U_r = (R_r - G_r)$ is given by

$$U_r = \alpha_r(pn - n_i^2) = \frac{G_r}{n_i^2}(pn - n_i^2) \tag{5.13a}$$

which, after substituting $p = p_o + p_e$ and $n = n_o + p_e$, becomes

$$U_r = \frac{G_r}{n_i^2}[n_o + p_o + p_e]p_e \tag{5.13b}$$

The radiative lifetime τ_r is now obtained as

$$\tau_r = \frac{n_i^2}{G_r(n_o + p_o + p_e)} \tag{5.14}$$

For a strongly n-type semiconductor having a donor concentration N_d, we have $n_o \simeq N_d, p_o \simeq 0$ and the low-level injection lifetime is obtained by neglecting p_e in Eq. (5.14), giving

$$\tau_r \simeq \frac{n_i^2}{G_r N_d} \tag{5.15a}$$

At high-level injection, p_e is large compared to $(p_o + n_o)$ and

$$\tau_r = \frac{n_i^2}{G_r p_e} \tag{5.15b}$$

The recombination coefficient $\alpha_r = (G_r/n_i^2)$ for some important semiconductors is given in Table 5.1.

TABLE 5.1
Values of the Recombination Coefficient (G_r/n_i^2) at 300 K

Semiconductor	Ge	Si	GaP	GaAs
G_r/n_i^2 (cm^3 sec^{-1})	5.3×10^{-14}	1.8×10^{-15}	5.4×10^{-14}	7.2×10^{-10}

Note: Data in Table 5.1 is taken with permission from M. H. Pilkuhn, Light Emitting Diodes, in Handbook on Semiconductors, T. S. Moss (ed), Vol. 4, Device Physics edited by C. Hilsum, North-Holland Amsterdam, 1981, p. 545.

The indirect gap semiconductors Ge, Si, and GaP have a very low value of the recombination coefficient, and the radiative lifetime in these semiconductors is several orders of magnitude higher than the measured lifetime. For example, in Si doped with 10^{15} donors cm^{-3} the lifetime calculated from Eq. (5.15a) is 0.56 sec, whereas the measured value may range from a fraction of a millisecond to a few microseconds. Thus, the radiative recombination remains insignificant in these semiconductors and lifetime is controlled by indirect recombination via localized states in the band gap.

Note that both G_r and n_i^2 increase with temperature. However, G_r is somewhat more sensitive to temperature variations than n_i^2. Consequently, the radiative lifetime in Eq. (5.15a) is a slowly decreasing function of temperature.

5.3.2 Indirect Recombination via Deep Energy Levels in the Band Gap

Impurity atoms other than donors and acceptors and some types of crystal defects in a semiconductor, introduce localized energy levels deep in the band gap away from the band edges. These levels act as stepping stones for electrons between the conduction and valence band, making a substantial enhancement in the recombination process. Depending on its location in the band gap, a deep level may act as an electron or a hole trap or a recombination center. An electron trap has a high probability of capturing a conduction electron and setting it free after some time. Similarly, a hole trap has a high probability of capturing a hole that is subsequently released into the valence band. At a recombination center the probabilities of electron and hole captures are nearly equal. Thus, an electron capture is followed by a hole capture, and this results in the elimination of an electron–hole pair. The potential energy of the pair is lowered in two stages. Part of the energy is released when the electron makes a transition from a state in the conduction band to the deep level center, and the rest is released when the trapped electron recombines with a hole. In general, in both steps the energy is dissipated in the form of phonons, and the recombination is nonradiative.

The mechanism of indirect recombination through the deep level centers has been investigated by Hall, Shockley, and Read [4]. Many recombination centers have more than one energy level, but in most cases only one level dominates the recombination. For this reason, we will keep our discussion confined to single-level recombination. The four steps that occur in the recombination of an electron–hole pair through a deep-level center are illustrated in Fig. 5.5. The arrows in this figure indicate the transition of the electron in the process. Process (a) depicts the capture of an electron from the conduction band by the center located at energy E_t. The inverse process of emission of the electron from the center into the conduction band is process (b). Process (c) represents the capture of a hole from the valence band by the center. Here the center emits the captured electron into the valence band, which is equivalent to the transfer of a hole from the valence band to the center. Finally, in process (d), the center captures an electron from the valence band leaving behind a hole in the band. This is equivalent to the center's emission of a hole into the valence band. It should now be clear that if the level is to act as an efficient recombination center, the electron capture process (a) must be followed by the hole capture process (c) and both these processes should have nearly the same probability. If process (a) is followed by the electron emission process (b), then the center acts as an electron trap. Similarly, when process (c) is immediately followed by process (d), the center acts as a hole trap.

Now consider a semiconductor with a density of recombination centers N_t located at energy E_t. The probability that the center is occupied by an electron is given by

$$f(E_t) = \frac{1}{1 + \exp\left(\frac{E_t - E_F}{kT}\right)} \equiv f \qquad (5.16)$$

We will next calculate the rates of the individual processes. The rate of the electron capture process R_a should be proportional to the concentration of free electrons n in the conduction band and should also be proportional to the concentration $N_t(1 - f)$ of the unoccupied centers. Therefore

$$R_a = \sigma_{cn} v_{th} n N_t (1 - f) \qquad (5.17)$$

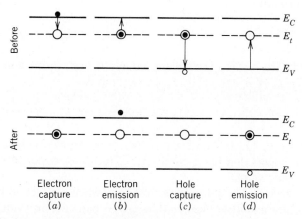

FIGURE 5.5 Schematic representation of electron–hole pair recombination through the deep-level recombination center at thermal equilibrium.

where v_{th} represents the thermal velocity of the electron and σ_{cn} is the capture cross-section. An electron with velocity v_{th} must, on the average, come within a cross-sectional area σ_{cn} of a trap to be captured, and thus sweeps out an effective volume $\sigma_{cn} v_{th}$ per second.

The electron emission rate R_b should be proportional to the concentration $N_t f$ of the occupied centers. It should also be proportional to the emission probability e_n that an electron makes a transition from an occupied center into the conduction band. Thus

$$R_b = e_n N_t f \tag{5.18}$$

The emission probability e_n depends on the density of empty states in the conduction band and the location of energy E_t with respect to the conduction band edge E_C.

The rate of the hole capture process R_c should be proportional to the concentration of the occupied centers $N_t f$ and the hole concentration

$$R_c = \sigma_{cp} v_{th} p N_t f \tag{5.19}$$

where σ_{cp} represents the capture cross-section of holes. Finally, the rate of the hole emission process R_d can be written as

$$R_d = e_p N_t (1 - f) \tag{5.20}$$

where e_p is the emission probability of a hole into the valence band.

The emission probabilities e_n and e_p are determined from the condition that, under thermal equilibrium, the rate of electron capture must be balanced by the rate of electron emission, and a similar balance must exist between hole capture and hole emission. These statements are justified in view of the principle of *detailed balance,* which states that for any system in thermal equilibrium the rate of a physical process and its reverse must balance each other. Thus, equating $R_b = R_a$ and writing

$$n = n_o = n_i \exp[(E_F - E_i)/kT]$$

we obtain

$$e_n = \sigma_{cn} v_{th} n_i \exp[(E_t - E_i)/kT] \tag{5.21}$$

Similarly, e_p is obtained by equating Eqs. (5.19) and (5.20) after substituting $p = p_o = n_i \exp[(E_i - E_F)/kT]$. That is

$$e_p = \sigma_{cp} v_{th} n_i \exp[(E_i - E_t)/kT] \tag{5.22}$$

Equation (5.21) shows that e_n increases exponentially as E_t rises above E_i toward the conduction band. Similarly, e_p increases exponentially as E_t approaches the valence band.

We will now consider the situation when the semiconductor is illuminated uniformly, and a photogeneration rate G_L pairs per unit volume per sec is maintained. Since photogeneration increases the concentration of electrons in the conduction band, the electron capture rate R_a increases above its equilibrium value. In steady state the rate at which the electrons enter into the conduction band (i.e., $G_L + R_b$) must be equal to the rate R_a at which the electrons leave the band. Thus

$$G_L + R_b - R_a = 0 \tag{5.23}$$

Similarly, the detailed balance for holes in the valence band leads to

$$G_L + R_d - R_c = 0 \tag{5.24}$$

From these relations we obtain

$$U = R_a - R_b = R_c - R_d \tag{5.25}$$

In the presence of photogeneration, the probability of occupancy of the center by an electron will become dependent on the injection level G_L. The new value of f can be obtained by substituting the values of R_a, R_b, R_c, and R_d in Eq. (5.25) and solving the resulting expression for f. This gives

$$f = \frac{\sigma_{cn} n + \sigma_{cp} p_1}{\sigma_{cn}(n + n_1) + \sigma_{cp}(p + p_1)} \tag{5.26}$$

Finally, substituting this value of f in the rates of the individual processes, we get

$$U = (R_a - R_b) = \frac{\sigma_{cp}\sigma_{cn} v_{th} N_t (pn - n_i^2)}{\sigma_{cn}(n + n_1) + \sigma_{cp}(p + p_1)} \tag{5.27}$$

where

$$n_1 = n_i \exp[(E_t - E_i)/kT]$$
$$p_1 = n_i \exp[(E_i - E_t)/kT] \tag{5.28}$$

Note that n_1 and p_1 represent the equilibrium electron and hole concentrations that would result when the Fermi level lies at the trap level E_t. After making the substitutions

$$\tau_{po} = \frac{1}{\sigma_{cp} v_{th} N_t} \quad \text{and} \quad \tau_{no} = \frac{1}{\sigma_{cn} v_{th} N_t} \tag{5.29}$$

Eq. (5.27) can be rewritten as

$$U = \frac{(pn - n_i^2)}{\tau_{po}(n + n_1) + \tau_{no}(p + p_1)} \tag{5.30}$$

We note that the driving force for the recombination is the term $(pn - n_i^2)$. To see the effect of the level E_t on the rate of recombination, we assume $\tau_{po} = \tau_{no} = \tau_o$ for simplicity. Substituting for n_1 and p_1, we can write Eq. (5.30) as

$$U = \frac{(pn - n_i^2)}{\tau_o \left[n + p + 2n_i \cosh\left(\frac{E_t - E_i}{kT}\right) \right]} \tag{5.31}$$

For given values of p and n, the net rate of recombination U will be maximum when the third term in the denominator approaches its minimum value. It is clear that this term will have its minimum value of $2n_i$ when E_t lies at the intrinsic level E_i. If E_t moves away from E_i toward the conduction band, then n_1 increases. In such a case, the emission probability e_n of the electron increases, and the trapped electron, instead of recombining with a hole, is emitted back into the conduction band and the center behaves as an electron trap rather than a recombination center. On the other hand, if E_t moves away from E_i toward the valence band, then e_p increases and the center becomes a hole trap. Thus, a recombination center is most effective when its energy is in the middle of the band gap.

(a) LOW-LEVEL INJECTION LIFETIME

For an n-type semiconductor, the excess carrier lifetime is defined by Eq. (5.9), which can be written as

$$\tau_p = \frac{p_n - p_{no}}{U} = \frac{p_e}{U}$$

Substituting for U from Eq. (5.30), we obtain

$$\tau_p = p_e \left[\frac{\tau_{po}(n + n_1) + \tau_{no}(p + p_1)}{(pn - n_i^2)} \right] \tag{5.32}$$

Making the substitutions $n = n_{no} + n_e$, $p = p_{no} + p_e$, and remembering that $n_e = p_e$ and n_{no} is large compared to p_{no} and p_e, we can reduce the above relation to

$$\tau_p = \frac{\tau_{po}(n_{no} + n_1) + \tau_{no}(p_e + p_1)}{n_{no}} \tag{5.33}$$

Now consider the case of a strongly extrinsic n-type semiconductor for which the recombination centers are situated at the energy level E_i. For this case, $n_1 = p_1 = n_i \ll n_{no}$, and the above relation simplifies to

$$\tau_p \simeq \tau_{po} = \frac{1}{\sigma_{cp} v_{th} N_t} \tag{5.34}$$

Physically, this means that the lifetime is determined by the capture cross-section for holes. This is understandable because the Fermi level is well above E_t, and prior to injection, all the recombination centers are occupied by electrons. Thus, it is the hole capture that limits the rate of recombination. Hence, τ_p is independent of electron concentration.

For a strongly p-type semiconductor with level E_t lying close to the center of band gap, it can similarly be shown that

$$\tau_n \simeq \tau_{no} = \frac{1}{\sigma_{cn} v_{th} N_t} \tag{5.35}$$

In silicon and germanium the capture cross-section for holes is larger than that for electrons, and this implies that $\tau_{no} > \tau_{po}$.

As the semiconductor becomes less extrinsic, the Fermi level gets closer to the recombination level and fewer of the centers are occupied by the majority carriers. Therefore, their efficiency for capturing minority carriers decreases. Consequently, the minority carrier lifetime has a tendency to increase with increasing resistivity of the semiconductor.

(b) HIGH-LEVEL INJECTION LIFETIME

At high-level injection, the excess carrier concentration is no longer negligible compared to the equilibrium majority concentration, and the lifetime τ_p in Eq. (5.32) becomes

$$\tau_p = \frac{\tau_{po}(n_{no} + p_e + n_1) + \tau_{no}(p_e + p_1)}{(n_{no} + p_e)} \quad (5.36)$$

Thus, τ_p increases above its low-level injection value as p_e becomes comparable to n_{no}. In the limit when $p_e \gg n_{no}$, relation (5.36) reduces to

$$\tau_p(\text{high}) \simeq \tau_{po} + \tau_{no} \quad (5.37)$$

which is also independent of the doping level.

The parameters τ_{po} and τ_{no} increase with temperature because of a sharp decrease in the respective capture cross-sections with increasing temperature. Thus, the Shockley-Read-Hall (SRH) lifetime has a strong tendency to increase with rising temperature.

5.3.3 Auger Recombination [2, 3]

Auger recombination is a three-particle process in which either two electrons and one hole, or two holes and one electron, are involved. This type of recombination is possible for both direct band-to-band and indirect recombination processes involving traps. A brief account of these processes is presented below.

(a) BAND-TO-BAND AUGER RECOMBINATION

The band-to-band Auger recombination for a direct gap semiconductor with parabolic valence and conduction bands is illustrated in Fig. 5.6. The two electrons and one hole (eeh) process is shown in Fig. 5.6a. Here electron 1 in the conduction band makes a transition to an empty state 1' in the valence band. The energy of the electron–hole pair is then transferred to the nearby electron 2, which rises high in the conduction band to the state 2'. The excited electron now loses its kinetic energy as lattice phonons and returns to thermal equilibrium as depicted in Fig. 5.3. It is evident from Fig. 5.6a that the transition of an electron from the conduction band minimum to a vacant state at the valence band maximum is not possible because it is incompatible with the conservation of energy and momentum of the system. The condition for the conservation of momentum requires an activation energy on the order of the band gap energy E_g. Auger recombination involving two holes and one electron is shown in Fig. 5.6b. Here the recombining electron 1 imparts its energy to a hole at 2', which transfers to state 2 deep in the valence band.

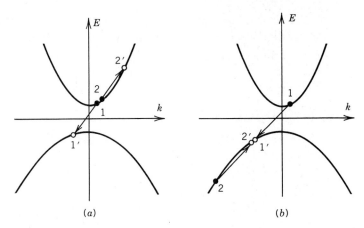

FIGURE 5.6 Band-to-band Auger recombination in a direct gap semiconductor with parabolic valence and conduction bands; (a) the two electrons and one hole (eeh) process and (b) the one electron and two holes (ehh) process.

From the reaction kinetic viewpoint, the Auger recombination is a third-order process, and the recombination rate for the eeh-process can be written as $C_n n^2 p$, where C_n is the Auger coefficient. Similarly for the ehh-process, the rate of recombination is $C_p p^2 n$, and the total rate of Auger recombination R_A is given by

$$R_A = C_n n^2 p + C_p p^2 n \tag{5.38}$$

where C_p is the Auger coefficient for the ehh process. In thermal equilibrium, the rate of recombination must be balanced by the Auger generation rate G_A giving

$$G_A = C_n n_o^2 p_o + C_p p_o^2 n_o = (C_n n_o + C_p p_o) n_i^2 \tag{5.39}$$

and the excess carrier recombination rate $U_A = (R_A - G_A)$ can be written as

$$U_A = C_n(n^2 p - n_i^2 n_o) + C_p(p^2 n - n_i^2 p_o) \tag{5.40}$$

where n and p are given by Eq. (5.2). Under low-level injection, the following expression for the Auger lifetime $\tau_A = p_e/U_A$ is obtained from Eq. (5.40):

$$\tau_A = [C_n(2n_i^2 + n_o^2) + C_p(2n_i^2 + p_o^2)]^{-1}$$

It is evident from this equation that

$$\tau_A = \frac{1}{C_n n_o^2} = \frac{1}{C_n N_d^2} \tag{5.41a}$$

for a strongly n-type semiconductor and

$$\tau_A = \frac{1}{C_p N_a^2} \tag{5.41b}$$

MECHANISMS OF RECOMBINATION PROCESSES [1,2]

for a strongly p-type semiconductor. In the case of high-level injection, such that $p_e \gg n_o$ and $p_e \gg p_o$, it can be seen from Eq. (5.40) that

$$\tau_A = \frac{1}{(C_n + C_p)p_e^2} \tag{5.42}$$

Experimentally determined values of C_n and C_p vary over a wide range [2], but in Ge as well as in Si, both C_p and C_n are lower than 3×10^{-31} cm^6 sec^{-1}. Thus, band-to-band Auger recombination in these semiconductors becomes important only when the mobile carrier concentration is higher than about 10^{18} cm^{-3}. This situation can occur either in a heavily doped semiconductor or with a very high-level injection.

Both C_n and C_p increase with temperature because of an increase in the Auger generation rate G_A. Consequently, the Auger lifetime decreases with temperature.

(b) TRAP-AIDED AUGER RECOMBINATION

Trap-aided Auger processes may become significant at high carrier concentrations and are less understood than the band-to-band Auger process. The two relevant trap Auger effects are shown in Fig. 5.7. In process (a) a trapped electron makes a transition to the valence band by giving its energy to an electron in the conduction band. The rate of this process can be witten as $T_a p n n_t$, where n_t represents the concentration of trapped electrons and T_a is the trap Auger coefficient. This process can become important only in a very heavily doped n-type semiconductor with a high concentration of traps below the Fermi level. Process (b) shows recombination involving two holes and one electron. Here the electron makes a transition to an unoccupied trap level in the band gap and gives its energy to a hole in the valence band. The rate of this process can be expressed as $T_b p n p_t$, where p_t is the concentration of empty traps and T_b is the respective trap Auger coefficient. This process may acquire importance in heavily doped p-type semiconductors with a large density of traps located above the Fermi level.

In an n-type semiconductor, the bulk lifetime τ_p' is obtained from the relation

$$\frac{1}{\tau_p'} = \frac{1}{\tau_p} + \frac{1}{\tau_r} + \frac{1}{\tau_A} \tag{5.43}$$

where τ_r is the radiative recombination lifetime, τ_A is the Auger lifetime, and τ_p is the SRH lifetime. A similar relation can be written for the bulk lifetime τ_n' in

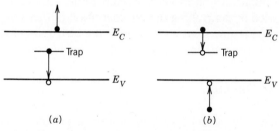

FIGURE 5.7 Schematic illustrations of two possible trap-aided Auger processes in a semiconductor.

a p-type semiconductor. In the case of Si and Ge, the last two terms in Eq. (5.43) are negligibly small unless the carrier concentration is very high, so we can write $\tau_p' = \tau_p$.

As an example, let us consider an n-type Si sample with $N_d = 10^{17}$ cm^{-3} and $N_t = 2 \times 10^{13}$ cm^{-3}. It is given that at 300 K the hole capture cross-section is 10^{-14} cm^{-2}, $v_{th} = 1.15 \times 10^7$ cm sec^{-1}, and $C_n = 2 \times 10^{-31}$ cm^6 sec^{-1}. We will now estimate the bulk lifetime τ_p' under low-level injection conditions.

Substituting for τ_p, τ_r, and τ_A from Eqs. (5.34), (5.15a), and (5.41a), respectively, and using the value of (G_r/n_i^2) from Table 5.1 we obtain

$$\frac{1}{\tau_p'} = \sigma_{cp} v_{th} N_t + \frac{G_r}{n_i^2} N_d + C_n N_d^2$$

$$= (10^{-14} \times 1.15 \times 10^7 \times 2 \times 10^{13} + 1.8 \times 10^{-15}$$

$$\times 10^{17} + 2 \times 10^{-31} \times 10^{34}) \text{ sec}^{-1}$$

$$= 2.3 \times 10^6 + 1.8 \times 10^2 + 2 \times 10^3 \simeq 2.3 \times 10^6 \text{ sec}^{-1}$$

which yields a bulk lifetime $\tau_p' = \tau_p = 4.3 \times 10^{-7}$ sec.

5.4 ORIGIN OF RECOMBINATION CENTERS

The physical origin of recombination centers in a semiconductor is related to the presence of impurities, imperfections caused by radiation damage, and unsaturated bonds left near the surface. A brief account of these origins is presented in this section.

5.4.1 Impurities

As mentioned earlier, the impurities that give rise to energy levels near the middle of the band gap act as efficient recombination centers. Typical examples are gold and platinum in silicon and copper in germanium.

Properties of gold in silicon have been extensively investigated. A gold atom in silicon introduces two energy levels — an acceptor level 0.54 eV below E_C and a donor level 0.29 eV above E_V. The acceptor level near the center of the gap is found to be dominant in controlling the carrier recombination. In the case of platinum, the level 0.36 eV above the valence band dominates the recombination process. Lifetime control by gold and platinum doping requires the diffusion of these impurities into silicon at temperatures in excess of 800°C. The lifetime can be varied by controlling the concentration of these impurities. For platinum doping, the lifetime is very sensitive to the resistivity of the silicon. Thus, control of lifetime using platinum is difficult, and gold doping is commonly used.

5.4.2 Radiation Damage

High-energy particles such as neutrons, protons, electrons, and γ-rays can displace atoms in a semiconductor lattice from their normal positions, creating vacancies and interstitials. These vacancies and interstitials can form more complex lattice defects that behave like deep impurities. The minority carrier lifetime in a semiconductor irradiated with high-energy particles decreases with an increase in the radiation dose.

Electron irradiation as a means of excess carrier lifetime control has received considerable attention because it can be performed after device fabrication. This technology is an attractive alternative to gold and platinum diffusion, for it provides a clean and simple process with greater control and uniformity than can be achieved by the impurity diffusion process. The main disadvantage of irradiation is that the radiation-induced recombination centers tend to anneal out after a period of time, even when the device is not subjected to high temperatures.

Besides irradiation, recombination centers are also introduced during processes such as diffusion and oxidation which are used to fabricate devices. High surface concentration diffused layers and mechanical damage on the surface produce a strained surface layer that diffuses rapidly into the semiconductor producing vacancies and vacancy complexes. These defects may act as recombination centers.

5.4.3 Surface States

At the surface of a semiconductor, the lattice is abruptly terminated, resulting in unsaturated bonds for the surface atoms. This rather drastic irregularity introduces a large density of localized energy levels in the forbidden gap. These levels are known as *surface states*. In contrast to an electron in the bulk which belongs to the entire crystal, a surface electron is confined only to an atomic layer. Surface states that have energy levels near the middle of the band gap act as efficient recombination centers.

The density of surface states on clean surfaces is on the order of 10^{15} cm^{-2}, which is about the same as the number of unsaturated bonds. Actual surfaces that are exposed to air are always covered with an oxide layer. Because of the presence of this oxide layer, some of the unsaturated bonds get saturated, which causes a reduction in the density of surface states. The typical density of these states on the oxide-covered surfaces of silicon and germanium ranges from 10^{11} to 10^{12} states per cm^2. Thermally oxidized silicon surfaces can have densities that are orders of magnitude smaller than these values.

Carrier recombination at the surface is characterized by a surface recombination velocity, which is defined as the ratio of the excess carrier flux reaching the surface to the concentration of excess carriers at the surface. In order to determine the surface recombination velocity in terms of the physical parameters of the surface, let us consider the n-type semiconductor shown in Fig. 5.8. In this sample, a region of thickness x_o near the surface has an increased concentration of recombination centers. Thus, we expect that, when the sample is illuminated with light, the net carrier recombination rate U will be enhanced in this region. As a result of this situation, the surface concentration $p_n(0)$ will be lower than that in the bulk. Because of this concentration gradient, electrons and holes from the bulk will diffuse to the surface and establish a net flux of excess carriers. Confining our attention to the minority carriers (holes), we define the surface recombination velocity S by the relation

$$-D_p \frac{\partial p_n}{\partial x}\bigg|_{x=0} = S(p_n(0) - p_{no}) \qquad (5.44)$$

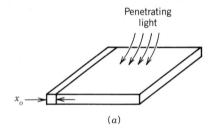

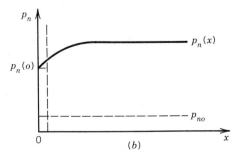

FIGURE 5.8 Excess carrier recombination near a surface having a high density of recombination centers. (*a*) Illuminated semiconductor and (*b*) excess minority carrier distribution.

For low-level injection, the carrier recombination rate is given by Eq. (5.9), which after using Eq. (5.34) can be written as

$$U = \frac{p_n - p_{no}}{\tau_p} = \sigma_{cp} v_{th} N_t (p_n - p_{no})$$

If N_{to} represents the concentration of centers in the region of thickness x_o, then the total number of centers in this layer is $N_{to}x_o = N_{st}$, and the total number of carriers recombining per unit time in the surface layer is given by

$$U_s = \sigma_{cp} v_{th} N_{st} (p_n(0) - p_{no}) \tag{5.45}$$

In steady state, this rate must be balanced by the flux given by Eq. (5.44). Thus, S can be expressed as

$$S = \sigma_{cp} v_{th} N_{st} \tag{5.46}$$

where N_{st} represents the number of centers per unit area of the surface. For a perfectly reflecting surface, S should be zero, whereas for a perfectly absorbing surface, it should be infinite, which implies that no excess carrier concentration can be maintained at a perfectly absorbing surface. In many semiconductors, the rate of recombination at the surface is much larger than in the interior. In such cases, the effective lifetime is obtained by the reciprocal addition of the two lifetimes (see Sec. 20.5).

Surface recombination velocity is a sensitive function of the conditions at the surface. For sand-blasted surfaces, the typical values of S may be as high as 10^5 cm/sec, whereas for clean etched surfaces, this value may be as low as 10 to 100 cm/sec.

5.5 EXCESS CARRIERS AND QUASI-FERMI LEVELS

The concept of the Fermi level used in Eqs. (3.27) and (3.28) is meaningful only under thermal equilibrium conditions when no excess carriers are present. However, under nonequilibrium conditions, it is often convenient to write the steady-state electron and hole concentrations in terms of the quasi-Fermi levels E_{Fn} and E_{Fp} defined by the relations

$$n = n_i \exp\left(\frac{E_{Fn} - E_i}{kT}\right) \tag{5.47a}$$

and

$$p = n_i \exp\left(\frac{E_i - E_{Fp}}{kT}\right) \tag{5.47b}$$

Since n and p are different from n_o and p_o, it should be obvious that E_{Fn} and E_{Fp} represent two separate energy levels. In fact, by multiplying the above two relations, we obtain

$$pn = n_i^2 \exp\left(\frac{E_{Fn} - E_{Fp}}{kT}\right) \tag{5.48}$$

This relation shows that as long as the pn product is different from n_i^2, E_{Fn} and E_{Fp} will be displaced from each other.

The concept of quasi-Fermi levels is important in semiconductors because their gradients determine the current. For example, consider the case where electron current is flowing owing to both drift and diffusion processes. The current density $J_n(x)$ can then be written as

$$J_n(x) = q\mu_n n(x)\mathcal{E}(x) + qD_n\frac{dn}{dx}$$

Substituting the value of $n(x)$ from Eq. (5.47a) and expressing the field $\mathcal{E}(x) = (dE_i/dx)/q$, it can be shown (see Prob. 5.15) that

$$J_n(x) = \mu_n n(x)\left(\frac{dE_{Fn}}{dx}\right) = \sigma_n(x)\frac{1}{q}\left(\frac{dE_{Fn}}{dx}\right) \tag{5.49a}$$

and similarly for holes

$$J_p(x) = \mu_p p(x)\left(\frac{dE_{Fp}}{dx}\right) = \sigma_p(x)\frac{1}{q}\left(\frac{dE_{Fp}}{dx}\right) \tag{5.49b}$$

These equations are equivalent to the modified forms of Ohm's law where the gradients of the quasi-Fermi potentials $(1/q)E_{Fn}$ and $(1/q)E_{Fp}$ determine the total field. If the two quasi-Fermi levels are constant throughout the semiconductor, these gradients are zero and current cannot flow.

5.6 BASIC EQUATIONS FOR SEMICONDUCTOR DEVICE-OPERATION [5]

All semiconductor devices operate under nonequilibrium conditions in which currents of electrons and holes flow through the semiconductor. The carrier currents at any point in the device can be calculated from the knowledge of the electric field and the carrier concentration gradient at that point. In general, we may not know these quantities, and two additional relations are needed to solve the carrier flow problems. These relations are the Gauss's law and the continuity equation. Gauss's law relates the electric field to the charge density in the semiconductor and thus enables us to determine the electric field for a given charge distribution. The continuity equation expresses the principle of particle conservation by relating the net rate of change of carriers in a volume to the processes that introduce and remove these carriers. Gauss's law, the continuity equations for electrons and holes, along with the current flow equations, form a set of equations for analysis of semiconductor devices.

5.6.1 Gauss's Law

Gauss's law states that the total normal electrical flux coming out of a closed surface equals the charge enclosed by the surface. If ρ denotes the volume charge density and $\mathbf{D}$ the electrical flux density, then Gauss's law can be expressed as

$$\oint_s \mathbf{D} \cdot d\mathbf{s} = \int_{\text{Vol}} \rho \cdot dv$$

When the surface integral on the left-hand side is converted into a volume integral using the Divergence theorem, the above equation becomes

$$\int_{\text{Vol}} \nabla \cdot \mathbf{D} \, dV = \int_{\text{Vol}} \rho \cdot dV \tag{5.50}$$

Since this relation is true for any arbitrary volume, the integrands on the two sides of Eq. (5.50) must be equal. Thus

$$\nabla \cdot \mathbf{D} = \rho \tag{5.51}$$

which is one of Maxwell's equations for an electromagnetic field. In semiconductor devices, we are interested in the electric field $\mathscr{E}$ rather than the electrical flux density $\mathbf{D}$. For an isotropic semiconductor of permittivity ε_s, these two quantities are related by

$$\mathbf{D} = \varepsilon_s \mathscr{E} \tag{5.52}$$

and thus

$$\nabla \cdot \mathscr{E} = \frac{\rho}{\varepsilon_s} \tag{5.53}$$

Noting that $\mathscr{E} = -\nabla \phi$, we obtain

$$\nabla^2 \phi = -\rho/\varepsilon_s \tag{5.54}$$

Relation (5.54) is known as the *Poisson* equation. When ρ as a function of space coordinates is specified, Eq. (5.54) can be solved to obtain the electric field $\mathscr{E}$ and the potential ϕ as a function of x, y, and z.

5.6.2 The Continuity Equations

When excess carriers are introduced in a region of a semiconductor, they disappear not only by recombination, but also by flowing out of the volume. Similarly, the number of carriers inside the volume can be increased by photogeneration as well as by the flux of carriers into the volume.

To investigate this problem further, let us consider a unit volume of an n-type semiconductor. We will focus our attention on one type of carrier, say, holes. The rate at which the hole concentration changes in the volume can be written as

$$\frac{\partial p}{\partial t} = G_{th} + G_L - R(n,p) - \frac{1}{q} \nabla \cdot \mathbf{J}_p \tag{5.55a}$$

Here G_{th} represents the rate of thermal generation of holes, and G_L denotes the rate of generation from other sources. The third term on the right-hand side is the rate of recombination, and the last term is the divergence term that we have assumed to be negative denoting a net flux of holes entering the volume. A similar equation can be written for electrons

$$\frac{\partial n}{\partial t} = G_{th} + G_L - R(n,p) + \frac{1}{q} \nabla \cdot \mathbf{J}_n \tag{5.55b}$$

The last term on the right-hand side in Eq. (5.55b) has a positive sign because of the negative charge of the electron. When a simple concept of lifetime holds, we can write $(R - G_{th}) = U = p_e/\tau_p$ for holes. In addition, since $p_e = (p - p_o)$, we can write Eq. (5.55a) as

$$\frac{\partial p_e}{\partial t} = G_L - U - \frac{1}{q} \nabla \cdot \mathbf{J}_p \tag{5.56a}$$

Similarly, the rate of change of electron concentration is

$$\frac{\partial n_e}{\partial t} = G_L - U + \frac{1}{q} \nabla \cdot \mathbf{J}_n \tag{5.56b}$$

5.6.3 Current Flow Equations

In order to arrive at explicit solutions of Eqs. (5.56a) and (5.56b), it is necessary to express $\mathbf{J}_p$ and $\mathbf{J}_n$ in terms of the electron and hole concentrations. This can be done by writing the current density equations

$$\mathbf{J}_p = q\mu_p p \mathscr{E} - qD_p \nabla p$$

and

$$\mathbf{J}_n = q\mu_n n \mathscr{E} + qD_n \nabla n \tag{5.57}$$

These equations describe the static and dynamic behavior of mobile carriers in semiconductors under the influence of an electric field and concentration gradients.

5.6.4 Carrier Transport Equations

The transport equations for holes and electrons in a semiconductor are obtained after substituting the values of $\mathbf{J}_p$ and $\mathbf{J}_n$ from Eq. (5.57) into Eqs. (5.56a) and (5.56b). The solution of the resulting equations under appropriate boundary conditions describes the electron and hole distributions as a function of space coordinates and time and thus, furnishes a complete description of carrier transport under nonequilibrium situations.

A general solution of the transport equations is rather difficult, and several approximations have to be made. In most cases we will be concerned only with one-dimensional current flow in which both $\mathscr{E}$ and the concentration gradients are along the x direction. We can then write

$$\nabla p = \frac{\partial p}{\partial x} \quad \text{and} \quad \nabla n = \frac{\partial n}{\partial x}$$

and substituting for $\mathbf{J}_p$ and $\mathbf{J}_n$ from Eq. (5.57) into (5.56a) and (5.56b), we obtain

$$\frac{\partial p_e}{\partial t} = G_L - U - \mu_p \mathscr{E} \frac{\partial p}{\partial x} - \mu_p p \frac{\partial \mathscr{E}}{\partial x} + D_p \frac{\partial^2 p}{\partial x^2} \tag{5.58a}$$

$$\frac{\partial n_e}{\partial t} = G_L - U + \mu_n \mathscr{E} \frac{\partial n}{\partial x} + \mu_n n \frac{\partial \mathscr{E}}{\partial x} + D_n \frac{\partial^2 n}{\partial x^2} \tag{5.58b}$$

Here the electric field $\mathscr{E}$ consists of the externally applied field and the internal field that arises owing to the imbalance between the excess electron and hole concentrations. In the above equations, there are three unknowns p, n and $\mathscr{E}$, and to obtain a solution for these quantities, a third relation is needed. This relation is obtained by writing the Poisson equation Eq. (5.53) as

$$\frac{d\mathscr{E}}{dx} = \frac{q}{\varepsilon_s}[N_d - N_a + p_o - n_o + p_e - n_e] \tag{5.59}$$

Analytical solutions of Eqs. (5.58a), (5.58b), and (5.59) are not possible unless some approximations are made. For simplicity we consider a homogeneously doped semiconductor. The sum of the first four terms on the right-hand side of Eq. (5.59) is then zero and

$$\frac{d\mathscr{E}}{dx} = \frac{q}{\varepsilon_s}[p_e - n_e] \tag{5.60}$$

Another approximation that is needed in obtaining an analytical solution is that at any time the excess electron and hole concentrations are equal at all points in the semiconductor. Clearly, this assumption is not quite correct because if this were the case, no internal field would exist and the electrons and holes would diffuse with the same diffusion constant. This cannot be the case since we know that D_p and D_n are different. However, the internal fields that are produced in practical situations are such that there is only a small imbalance in the electron–hole concentrations, and charge neutrality is approximately valid. As an example, let us consider n-type silicon with $n_o = 10^{15}$ cm^{-3} and $p_e = 10^{12}$ cm^{-3}. If the electron concentration n_e departs 1 percent from p_e, then from Eq. (5.60), this results in $d\mathscr{E}/dx \simeq 10^3$ V cm^{-1}. Internal fields of this magnitude are rarely encountered in a homogeneous semiconductor. Thus, it is clear that in practical situations, the excess electron and hole concentrations are nearly equal. This condition is known as the *quasi-neutrality condition*. Substituting $n_e = p_e$ and eliminating $d\mathscr{E}/dx$ in Eq. (5.58a) and (5.58b) and noting that $\partial p/\partial x = \partial p_e/\partial x$, we obtain the following equation:

$$\frac{\partial p_e}{\partial t} = G_L - \frac{p_e}{\tau_p} - \mu_a \mathscr{E} \frac{\partial p_e}{\partial x} + D_a \frac{\partial^2 p_e}{\partial x^2} \qquad (5.61)$$

where

$$D_a = \frac{(n+p)D_p D_n}{nD_n + pD_p} \quad \text{and} \quad \mu_a = \frac{(n-p)\mu_n \mu_p}{n\mu_n + p\mu_p} \qquad (5.62)$$

D_a and μ_a are known as the ambipolar diffusion constant and ambipolar mobility, respectively, and Eq. (5.61) is the so-called ambipolar transport equation. Both D_a and μ_a depend on the excess carrier concentration p_e, and it is impossible to solve Eq. (5.61) in a closed form for any arbitrary value of p_e. Two special circumstances in which Eq. (5.61) can be solved are discussed in the following subsections.

(a) LOW-LEVEL INJECTION IN A STRONGLY EXTRINSIC SEMICONDUCTOR

In the case of low-level injection in a strongly n- or p-type semiconductor, the ambipolar diffusion constant and mobility reduce to their respective values for the minority carriers as can be seen from Eq. (5.62). Thus, for low-level injection in a strongly n-type semiconductor, $n \gg p$, and we obtain $D_a = D_p$ and $\mu_a = \mu_p$. We have already seen that under low-level injection conditions, the excess carrier lifetime is the minority carrier lifetime. Thus, at low-level injection, the transport and recombination parameters are those of minority carriers, and the behavior of majority carriers is the same as that of minority carriers.

(b) VERY HIGH-LEVEL INJECTION

When the concentration of excess carriers is large compared to the equilibrium majority carrier concentration, $p \approx n$ and Eq. (5.62) reduces to

$$D_a = \frac{2D_n D_p}{D_n + D_p}, \quad \text{and} \quad \mu_a = \frac{(n_o - p_o)\mu_n \mu_p}{n(\mu_n + \mu_p)} \simeq 0 \qquad (5.63)$$

These relations are also true for low-level injection in an intrinsic semiconductor. Under these conditions, the field term in Eq. (5.61) becomes zero and the transport is dominated by diffusive motion with the diffusion constant given by Eq. (5.63). The fact that the field term vanishes at high-level injection greatly simplifies the analysis.

5.7 SOLUTION OF CARRIER TRANSPORT EQUATIONS — AN ILLUSTRATION

As an example of the solution to the carrier transport equations, we consider the uniformly doped, long n-type semiconductor bar shown in Fig. 5.9a. The face of the bar near $x = 0$ is assumed to be strongly illuminated with ionizing radiation that produces electron–hole pairs at a uniform rate of G_L pairs per unit volume per sec in the thin layer of thickness x_o, and no excess pairs are produced in the rest of the bar. Thus, the surface generation rate of the carrier is $G_L x_o$ pairs per unit area per sec. The excess carrier concentrations established in the layer x_o cause diffusion of the carriers inward into the body of the semiconductor. We assume that no carrier recombination occurs at the surfaces so that the excess carrier concentration across any plane perpendicular to the x axis is constant. In a steady state, the surface generation rate must be balanced by the flux of electrons and holes to the right at $x = 0$, which means that $J_p(0) = -J_n(0) = qG_L x_o$. Since the electrons and holes recombine in pairs, no net current can flow. Thus, at any plane to the right of $x = 0$, we must have $J_p(x) + J_n(x) = 0$. This leads to

$$\mu_n n_n \mathscr{E} + D_n \frac{dn_e}{dx} + \mu_p p_n \mathscr{E} - D_p \frac{dp_e}{dx} = 0 \tag{5.64}$$

The assumption of quasi-neutrality requires that $p_e = n_e$ and $dp_e/dx = dn_e/dx$. Making these substitutions and solving Eq. (5.64) for $\mathscr{E}$ yields

$$\mathscr{E} = -\frac{(D_n - D_p)}{(\mu_n n_n + \mu_p p_n)} \left(\frac{dp_e}{dx} \right) \tag{5.65}$$

The ratio of the drift to the diffusion current density for holes is given by

$$\frac{J_{pf}}{J_{pD}} = -\frac{\mu_p p_n \mathscr{E}}{D_p \left(\dfrac{dp_e}{dx} \right)}$$

where the subscript f is used for the drift field. Substituting the value of $\mathscr{E}$ from Eq. (5.65) and making use of the Einstein relation for holes lead to

$$\frac{J_{pf}}{J_{pD}} = \frac{(\mu_n - \mu_p) p_n}{\mu_n n_n + \mu_p p_n} = \frac{(D_n - D_p) p_n}{D_n n_n + D_p p_n} \tag{5.66}$$

For low-level injection, $n_n \gg p_n$, and it is seen that J_{pf} is negligible compared to J_{pD}. Similarly for electrons, the ratio J_{nf}/J_{nD} at low-level injection can be shown to be given by

$$\frac{J_{nf}}{J_{nD}} = \left(1 - \frac{\mu_p}{\mu_n} \right) = \left(1 - \frac{D_p}{D_n} \right) \tag{5.67a}$$

which obviously is not negligible as long as D_p is different from D_n. In fact, from Eq. (5.67a) it is seen that

$$J_{nf} = J_{nD}\left(1 - \frac{D_p}{D_n} \right) = J_{nD} - J_{pD} \tag{5.67b}$$

Thus, it is clear that under low-level injection, the minority carrier current (holes in this case) is dominated by diffusion and is not appreciably influenced by the field. The majority carrier current, on the other hand, is strongly influenced by the electric field. These results can be easily understood by referring to the sketches in Fig. 5.9. In the illuminated bar of Fig. 5.9a, the excess holes and electrons generated near the surface tend to diffuse away from the plane $x = 0$. Since the diffusion constant for electrons is higher than that for holes, the electrons diffuse away faster than the holes, which raises the hole concentration near $x = 0$ above the electron concentration. The resulting distribution of excess carriers is as shown in Fig. 5.9b. The space-charge and electric field distributions are shown in Fig. 5.9c and Fig. 5.9d, respectively. Since the quasi-neutrality condition is approximately satisfied at each point, the field established by this charge imbalance is small. In this field, the holes experience a force along the field direction and a drift current of holes flows along the positive direction of x. The electrons drift in the opposite direction, but the conventional drift current of electrons flows in the same direction as the hole drift current. This current obviously opposes the electron diffusion current. Under low-level injection, the concentration of holes is small compared to that of electrons, and the drift component of the hole current

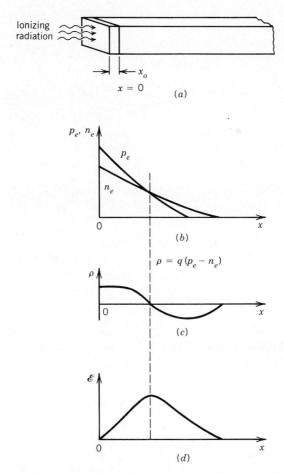

FIGURE 5.9 Steady state injection from the surface of a bar. (a) Ionizing radiation falling on the surface, (b) excess electron and hole distributions, (c) space charge, and (d) electric field set up by the charge imbalance.

is negligible compared to the diffusion component. The drift current of electrons, however, is still appreciable because it is proportional to the product of the small field and a very large concentration of majority carriers.

Because the hole current flows mainly by diffusion, the field term in Eq. 5.58a can be neglected, and since $G_L = 0$ in the region to the right of $x = 0$, the steady state continuity equation, after substituting $U = p_e/\tau_p$, reduces to

$$\frac{d^2 p_e}{dx^2} - \frac{p_e}{L_p^2} = 0 \tag{5.68}$$

where $L_p = \sqrt{D_p \tau_p}$ is called the diffusion length of holes. We no longer need the partial derivative since $\partial p_e/\partial t = 0$ in the steady state. The above equation has the solution

$$p_e(x) = A_1 \exp(-x/L_p) + A_2 \exp(x/L_p)$$

where A_1 and A_2 are constants of integration. The hole concentration $p_e(x)$ must reduce to zero as x approaches infinity. Therefore, we have $A_2 = 0$ and

$$p_e(x) = p_e(0) \exp(-x/L_p) \tag{5.69}$$

where $A_1 = p_e(0)$ represents the excess hole concentration at $x = 0$. The hole current at $x = 0$ is given by

$$J_p(0) = -qD_p \left(\frac{dp_e}{dx}\right)_{x=0} = \frac{qD_p p_e(0)}{L_p}$$

In the steady state, the hole flux $J_p(0)/q$ must equal the hole generation rate $G_L x_o$. Thus

$$p_e(0) = \frac{G_L x_o L_p}{D_p} \tag{5.70}$$

It is evident from Eq. (5.69) that the excess hole concentration decreases exponentially with x. In fact, Eq. (5.69) has the same form as Eq. (5.10a), and it can be shown that L_p is the average distance a hole diffuses before recombining.

The neglect of minority carrier drift current is not justified at high-level injection when the majority and the minority carrier concentrations become equal. Substituting $p_n = n_n$ in Eq. (5.66), we obtain

$$J_{pf} = \frac{D_n - D_p}{D_n + D_p} J_{pD} \tag{5.71}$$

and the total current density J_p becomes

$$J_p = J_{pf} + J_{pD} = -q \frac{2D_p D_n}{D_n + D_p} \left(\frac{dp_n}{dx}\right) = -qD_a \left(\frac{dp_n}{dx}\right) \tag{5.72}$$

Since $n_n = p_n$, the same expression is also obtained for J_n. Thus, under high-level injection, not only the majority but also the minority carrier current has drift and diffusion components. This current, however, can be expressed only as a diffusion current by employing the ambipolar diffusion constant, and the same is true for the majority carrier current. The physical significance of this result is that, under high-level injection, sufficient internal field is created by the charge imbalance to change the diffusion constants D_n and D_p to a single value D_a. The excess carrier concentration does not drift under these circumstances. Although electrons and holes do drift into and out of the distribution, the distribution itself remains stationary.

REFERENCES

1. R. A. SMITH, *Semiconductors,* Cambridge University Press, London, 1978.
2. M. S. TYAGI and R. VAN OVERSTRAETEN, "Minority carrier recombination in heavily doped silicon," *Solid-State Electron.* 26, 577 (1983).
3. J. S. BLAKEMORE, *Semiconductor Statistics,* Pergamon Press, Oxford, 1962.
4. R. N. HALL, "Electron-hole recombination in germanium," *Phys. Rev.* 87, 387 (1952); W. SHOCKLEY and W.T. READ, "Statistics of the recombination of holes and electrons," *Phys. Rev.* 87, 835 (1952).
5. J. P. MCKELVEY, *Solid State and Semiconductor Physics,* Harper and Row, New York, 1966.

ADDITIONAL READING LIST

1. W. SHOCKLEY, *Electrons and Holes in Semiconductors,* Van Nostrand, Princeton, N.J., 1950.
2. B. G. STREETMAN, *Solid State Electronic Devices,* Prentice-Hall, Englewood Cliffs, N.J., 1979, Chapter 4.
3. D. K. SCHRODER, "The concept of generation and recombination lifetimes in semiconductors," *IEEE Trans. Electron Devices* ED-29, 1336 (1982).
4. T. S. MOSS (ed.), *Handbook on Semiconductors,* Vol. 2, Optical Properties of Solids, North-Holland, Amsterdam, 1980.
5. P.T. LANDSBERG, "The band-band Auger effects in semiconductors," *Solid-State Electron.* 30, 1107–1115 (1987).
6. F. F.Y. WANG, *Introduction to Solid State Electronics,* North-Holland, New York, 1980, Chapter 13.

PROBLEMS

5.1 A 0.5-μm thick sample of GaAs at 300 K has an area of 1 cm^2. The sample is uniformly illuminated with monochromatic light of $h\upsilon = 2.1$ eV. The absorption coefficient α at this wave length is 4×10^4 cm^{-1}. The power incident on the sample is 12 mW. (a) Calculate the power absorbed by the sample; (b) determine how much power is dissipated by the excess electrons to the lattice before recombining; and (c) determine the number of photons per sec falling on the sample and the number of photons per sec emitted from the electron–hole pair recombination.

5.2 Let the sample in Prob. 5.1 be n-type with $n_o = 10^{16}$ cm^{-3}. (a) Assuming that each of the absorbed photons produce one electron–hole pair in the sample, calculate the excess electron and hole concentrations in the steady-state. (b) Calculate the photoconductivity of the sample and determine whether it is the case of low- or high-level injection. Use the data from Table 4.2 for GaAs.

5.3 Using the result of Eq. (5.10a), show that the average time a hole spends before recombining with an electron is τ_p.

5.4 A homogeneous semiconductor bar is illuminated uniformly by a penetrating light that generates electron–hole pairs at a constant rate G_L cm^{-3} sec^{-1}. Assuming low-level injection, (a) calculate the excess carrier concentration as a function of time if the light is switched on at $t = 0$; and (b) determine the steady state values of electron and hole concentrations and show that the photo conductivity $\Delta\sigma$ of the sample is given by $\Delta\sigma = q(\mu_n + \mu_p)G_L\tau_p$.

5.5 A p-type Ge sample with a resistivity of 40 Ω-cm at 300 K is uniformly illuminated with light that generates 10^{13} excess electron–hole pairs per cm^3 per sec. In the steady state, calculate the change in resistivity of the sample caused by the light. If the light is switched off at $t = 0$, calculate the time required for the excess conductivity to drop to 10 percent of its value at $t = 0$. Assume $\tau_n = 10^{-6}$ sec.

5.6 A Si sample is doped with 10^{15} donors cm^{-3}. Calculate the excess electron and hole concentrations required to increase the sample conductivity by 15 percent. What carrier generation rate is required to maintain these excess concentrations? Assume $\mu_p = 0.3 \mu_n$, $\tau_p = 10^{-6}$ sec and $T = 300$ K.

5.7 A sample of n-type Si has a dark resistivity of 1 KΩ-cm at 300 K. The sample is illuminated uniformly to generate 10^{21} electron–hole pairs per cm^3 per sec. The hole lifetime in the sample is 1 μ sec. Calculate the sample resistivity and the percentage change in the conductivity after illumination owing to majority and minority carriers.

5.8 An n-type bar of GaAs has a length l and an area of cross-section A. The bar is illuminated with light that generates G_L electron–hole pairs cm^{-3} sec^{-1} uniformly. Assuming the hole mobility to be negligible in comparison to electron mobility, show that the steady state photocurrent in the bar can be written as $I_{ph} = qAlG_L\tau_p/\tau_t$, where τ_t is the average transit time of electron through the bar.

5.9 For a semiconductor with indirect recombination characterized by a single trap level at E_t and $\tau_{po} = \tau_{no}$, show that under low-level injection the maximum possible lifetime occurs when E_F lies at E_i, and that this maximum is given by

$$\tau_p = \tau_{po}\left[1 + \cosh\left(\frac{E_t - E_i}{kT}\right)\right]$$

5.10 Gold in n-type Si can be represented by a single acceptor level located 0.54 eV below the conduction band. For this impurity the capture rate constants are $\sigma_{cn}v_{th} = 1.65 \times 10^{-9}$ cm^3 sec^{-1} and $\sigma_{cp}v_{th} = 1.15 \times 10^{-7}$ cm^3 sec^{-1}. Calculate the low- and high-level injection lifetimes at 300 K for a Si sample doped with 5×10^{14} donor atoms cm^{-3} and 3.8×10^{13} atoms cm^{-3} of gold.

5.11 Consider an infinitely long, square but thin bar of an n-type semiconductor illuminated by light that creates electron–hole pairs uniformly at a rate G_L pairs cm^{-3} sec^{-1}. The surface at $x = 0$ has the surface recombination velocity S cm sec^{-1}. Set up the continuity equation for holes and determine their concentration $p_n(x)$ as a function x. Determine $p_n(x)$ when S approaches infinity.

5.12 Excess carriers are injected in a region of a uniformly doped n-type semiconductor with $n_o = 10^{14}$ cm^{-3}. The excess carrier concentration is maintained at 2×10^{18} cm^{-3} throughout the region. (a) Calculate the Auger recombination lifetime assuming $C_n = 2.7 \times 10^{-31}$ cm^6 sec^{-1} and $C_p = 1.1 \times 10^{-31}$ cm^6 sec^{-1}. (b) If the measured lifetime in the above sample is 4×10^{-7} sec, determine the lifetime in the absence of Auger recombination.

5.13 A Si sample is doped uniformly with 10^{15} donors cm^{-3} and has $\tau_p = 1$ μsec. (a) Determine the photogeneration rate that will produce 2×10^{13} excess pairs cm^{-3} in the steady state. (b) Calculate the conductivity of the sample and the position of the electron and hole quasi-Fermi levels in the steady state at 300 K.

5.14 Show that the nonequilibrium pn product in a semiconductor with band gap E_g is the same as the equilibrium product $p_o n_o$ with band gap $E_g - (E_{Fn} - E_{Fp})$.

5.15 Prove relations (5.49a) and (5.49b).

5.16 A Si sample has $n_o = 10^{15}$ cm^{-3} at 300 K. Assume $D_p = 12.5$ cm^2 sec^{-1} and $D_n = 3D_p$. Sketch the ambipolar diffusion coefficient for injected carrier concentrations from 10^{13} cm^{-3} to 10^{17} cm^{-3}.

PART 3
JUNCTIONS AND INTERFACES

6
p-n JUNCTIONS

INTRODUCTION

A p-n junction is the metallurgical boundary between the n- and p-regions of a semiconductor crystal. A junction cannot be formed by simply pressing the pieces of *p*- and *n*-type semiconductors together. Such an arrangement, in fact, will produce discontinuities in the regular crystal lattice and will stop the flow of electrons and holes across the interface. The junction can be formed by starting with a block of, for example, an n-type semiconductor and converting a part of it to the *p*-type by one of several methods described in Sec. 19.2. The p-n junction is the basic building block of a large number of devices including integrated circuits. It is, therefore, essential that its basic properties and principle of operation be thoroughly understood.

In this chapter, we first provide a qualitative description of the p-n junction action. Next, the transition region in an abrupt p-n junction is discussed in detail by examining the electric field and potential distributions under static conditions. Finally, we consider the properties of linearly graded and other types of p-n junctions encountered in semiconductor devices.

6.1 DESCRIPTION OF p-n JUNCTION ACTION

6.1.1 Junction in Thermal Equilibrium

Figure 6.1 shows the schematic diagram of a p-n junction. The two-dimensional representation shown is sufficient in most cases since the junction area is usually much larger than the depths of the n- and p-regions; thus, the current flow is essentially one-dimensional. To visualize how equilibrium is reached, we begin by considering the initially separated p- and n-regions. Let the two now be brought into intimate contact. Since the hole concentration is higher on the p-side than on the n-side, holes diffuse from the p- to the n-region and recombine with electrons. As the holes move out of the p-region, they leave behind uncompensated

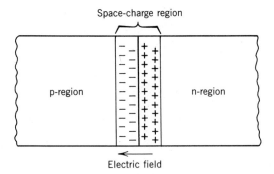

FIGURE 6.1 Schematic diagram of a p-n junction under thermal equilibrium.

negatively charged acceptor ions. Similarly, some positively charged donor ions near the junction are left uncompensated as electrons diffuse from the n-side to the p-side. Thus, a double layer of charges is formed near the junction having negative space charge on the p-side and a positive space charge on the n-side. Since this space-charge region is depleted of mobile carriers, it is also called the *depletion region*. The n- and p-regions on the two sides of the depletion region contain no net charge, and each of these will be referred to as the neutral region. The space-charge region creates an electric field that is directed from the positive charge of donor ions toward the negative charge of acceptor ions. This built-in field presents a barrier to the flow of majority carriers but aids the flow of minority carriers. Thus, holes tend to diffuse over the barrier constituting a hole diffusion current from the p- to the n-side. However, holes near the depletion region edge on the n-side tend to drift with the field, and this flow constitutes a drift current of holes from the n- to the p-side. Likewise, electrons tend to diffuse from the n- to the p-side and have a tendency to drift in the opposite direction. At thermal equilibrium the magnitude of the built-in field is such that the diffusion of electrons and holes is exactly balanced by their drift tendencies and no net current flows across the junction. The electric field in the space-charge region causes a potential difference between the p-and the n-neutral regions. This potential difference is called the *built-in voltage*.

6.1.2 Application of Bias

Application of an external voltage to the p-n junction is called *biasing*. Biasing disturbs the balance between the diffusion and drift currents of electrons and holes. As a consequence, a net current flows across the junction. Since the space-charge region is depleted of mobile carriers, its resistance is much higher than that of the neutral n- and p-regions, and practically all the bias voltage appears across this high-resistivity region.

Figure 6.2a shows the space-charge layer and the neutral regions in a p-n junction at zero bias. When a positive voltage is applied to the p-side, the junction is said to be forward biased (Fig. 6.2b). The positive charge on the p-contact pushes holes toward the junction. These holes move to the edge of the depletion region and neutralize some of the negatively charged acceptor ions. Similarly, electrons flow from the neutral n-region into the depletion region and neutralize a part of the positive charge of ionized donors. This decrease in the space charge reduces the depletion region width and lowers the potential barrier that was op-

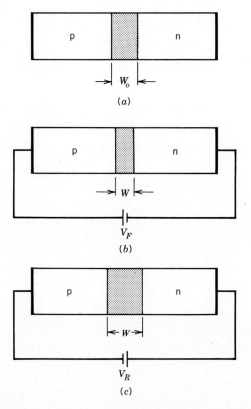

FIGURE 6.2 Space-charge layer width in a p-n junction under: (a) zero bias, (b) forward bias, and (c) reverse bias.

posing the diffusion of electrons and holes at thermal equilibrium. This situation allows more electrons and holes to diffuse over the barrier, while the drift current of minority carriers in the opposite direction remains practically unchanged. Thus, a net current flows through the junction which becomes appreciable even for small values of forward bias.

The junction is said to be reverse biased when the p-region is made negative with respect to the n-region (Fig. 6.2c). The majority carriers are now pulled away from the edges of the depletion region. Therefore, the depletion region widens, and the potential barrier to the flow of majority carriers is increased. The increase in the potential barrier reduces the diffusion current of majority carriers, but the drift current of minority carriers, which opposes the diffusion current, remains almost unaffected. Because the minority carrier concentration on both sides of the junction is low, only a small number of minority carriers can reach the boundaries of the depletion region. As a consequence, the reverse current in a p-n junction is small and tends to become independent of the applied bias.

6.1.3 Energy Band Diagrams

Energy band diagrams provide good insight into the behavior of the p-n junction. In the band diagrams shown in Fig 6.3, it has been assumed that both sides of the junction are uniformly doped and that the transition between the n- and p-regions is abrupt. Fig. 6.3a shows the initially separated n- and p-type semi-

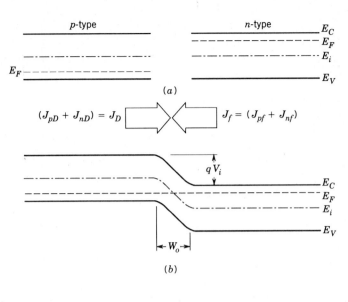
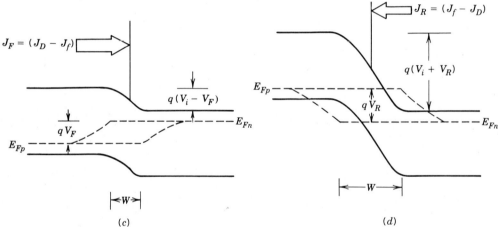

FIGURE 6.3 Energy band diagrams of a p-n junction: (a) separated p- and n-type semiconductors, (b) junction under thermal equilibruim, (c) forward bias, and (d) reverse bias.

conductors. When the two are brought into contact, majority carriers flow across the junction down the concentration gradients until the Fermi levels on the two sides are aligned (see Sec. 4.9). This situation is shown in Fig. 6.3b. Far from the junction, the electron and hole concentrations on the two sides remain undisturbed, so that the positions of the conduction and valence band edges, (E_C and E_V) relative to the Fermi level E_F, remain unchanged. In the junction depletion region, both E_C and E_V on the p-side must approach their respective values on the n-side gradually, and this results in the bending of the energy bands. Since the electrostatic potential in the diagrams in Fig. 6.3 increases downwards, the n-region is at a higher potential than the p-region. The bending in E_C and E_V corresponds to an electric field that is directed from the n- to the p-region and whose magnitude at any point is given by the slope of the band edges. The change in E_C (or E_V) which occurs over the junction depletion region is qV_i where V_i is the built-in voltage.

The diffusion and drift tendencies for electrons can be expressed in terms of the corresponding current densities J_{nD} and J_{nf}. Likewise, J_{pD} and J_{pf} express the diffusion and drift tendencies for holes. Since there is no net current across the junction in thermal equilibrium, we must have

$$J_n + J_p = 0 \tag{6.1}$$

where

$$J_n = J_{nD} + J_{nf}$$

and

$$J_p = J_{pD} + J_{pf} \tag{6.2}$$

For Eq. (6.1) to be satisfied, it is necessary that J_n and J_p be zero separately. If this were not so, then one would have $J_n = -J_p$, which means that both electrons and holes move across the junction in the same direction. Obviously, this is not possible. Hence, Eq. (6.1) requires that

$$J_n = J_p = 0 \tag{6.3}$$

The balance between the drift and diffusion processes is shown in Fig. 6.3b.

Application of a bias voltage (V_a) across the junction modifies the equilibrium energy band diagram by changing the barrier potential and altering the curvature of the bands. Figure 6.3c shows the energy band diagram for a forward bias $V_a = V_F$ in the steady state situation. The junction potential is decreased from V_i to $(V_i - V_F)$, and a net current flows because of the increased diffusion of electrons and holes across the junction. For usual values of V_F, the forward current density J_F remains lower than the equilibrium current densities J_D and J_f shown in Fig. 6.3b. However, if V_F is made to approach V_i, the potential barrier will disappear and the structure will behave as a single piece of semiconductor carrying a large current that will be limited only by the overall resistance of the circuit.

The energy band diagram for a reverse voltage $V_a = -V_R$ is shown in Fig. 6.3d. The barrier potential is now increased to $(V_i + V_R)$, and only a small drift current density J_R of minority carriers flows across the junction.

Figures 6.3c, d also show the quasi-Fermi levels E_{Fn} and E_{Fp}. As discussed in Sec. 5.5, the gradients of E_{Fn} and E_{Fp} determine the electron and hole currents in a semiconductor. Thus, the forms of quasi-Fermi energy levels are related to the majority and minority carrier currents. Note that E_{Fn} in the n-region is displaced from E_{Fp} in the neutral p-region by qV_a, where $V_a = V_F$ for forward bias and $V_a = -V_R$ for reverse bias. Under forward bias (Fig. 6.3c), E_{Fn} and E_{Fp} are different not only in the junction region, but also in the neutral n- and p-regions near the edges of the depletion layer. The splitting of the quasi-Fermi levels in the neutral regions indicates the presence of excess carriers in these regions and is intimately related to the minority carrier currents. In the reverse biased junction in Fig. 6.3d, the splitting of the quasi-Fermi levels in the neutral n- and p-regions indicates the extraction of minority carriers from these regions. Note that in Figs. 6.3c and d, E_{Fn} and E_{Fp} remain constant in the depletion region.

6.2 THE ABRUPT JUNCTION

The most commonly used methods of forming p-n junctions are alloying, diffusion, and epitaxial growth. These methods are described in Sec. 19.2. In general, the transition from the *p*-type to the *n*-type region may follow a complex distribution of donor and acceptor impurities. However, for ease of mathematical analysis, the junction is usually approximated by either an abrupt or a linearly graded junction.

The abrupt junction is a good approximation for junctions fabricated by alloying, epitaxial growth, and for diffused junctions in some situations. In practice, the change from the *n*- to the *p*-type may occur over 30 to 40 atomic spacings, but for making theoretical calculations the transition is assumed to be abrupt at the metallurgical boundary (Fig. 6.4a).

6.2.1 Calculation of the Built-in Voltage

The built-in voltage across a p-n junction in thermal equilibrium is the contact potential difference that develops when two materials with different work functions are joined together. The difference between the work functions of the p- and the n-semiconductors is equal to the difference between the Fermi levels when they are separated. For nondegenerate semiconductors, the Fermi levels in the n- and the p-regions are obtained from Eqs. (3.47) and (3.48), and the built-in voltage can be determined from their difference (see Prob. 6.2). Here we will follow an alternative approach based on the balance between drift and diffusion currents.

Figure 6.4a shows the impurity profile of the abrupt junction. The exact impurity distribution (dashed lines) is approximated by the step distribution. The mobile carrier concentrations under thermal equilibrium are shown in Fig. 6.4b. We assume all the impurity atoms to be ionized. The electron and hole concen-

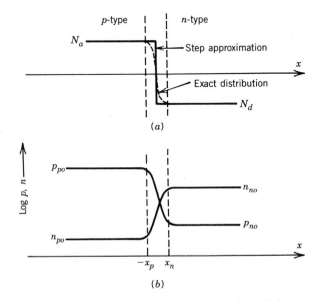

FIGURE 6.4 Impurity distribution and thermal equilibrium electron–hole concentrations in an abrupt p-n junction.

THE ABRUPT JUNCTION

trations in the neutral n-region will be $n_{no} = N_d$ and $p_{no} = n_i^2/n_{no}$, and in the neutral p-region $p_{po} = N_a$ and $n_{po} = n_i^2/p_{po}$. In the depletion region, the hole concentration decreases smoothly from p_{po} on the p-side to p_{no} on the n-side. Similarly, the electron concentration changes from n_{no} on the n-side to n_{po} on the p-side. These sharp gradients in electron and hole concentrations result in diffusion and drift currents across the junction as explained earlier. We have already seen in Eq. (6.3) that in thermal equilibrium, the hole and electron current densities are separately zero. Thus, confining our attention to holes, we write

$$J_p = q\mu_p p(x)\mathcal{E}(x) - qD_p \frac{dp}{dx} = 0 \tag{6.4}$$

Using the Einstein relation for holes and solving for $\mathcal{E}(x)$, we obtain

$$\mathcal{E}(x) = \frac{kT}{q} \cdot \frac{1}{p(x)} \cdot \frac{dp}{dx} \tag{6.5}$$

The built-in voltage V_i can now be obtained by integrating Eq. (6.5) from $x = -x_p$ to $x = x_n$,

$$V_i = -\int_{-x_p}^{x_n} \mathcal{E}(x)\,dx = \frac{kT}{q} \ln \frac{p_{po}}{p_{no}} \tag{6.6}$$

Since $p_{po} = N_a$ and $p_{no} = n_i^2/N_d$, we can write

$$V_i = \frac{kT}{q} \ln\left(\frac{N_d N_a}{n_i^2}\right) \tag{6.7}$$

The positive sign of V_i shows that the n-side is at a higher potential than the p-side. Thus, the built-in voltage depends on the donor and acceptor concentrations on the two sides of the junction. Returning to Eq. (6.6), we see that

$$p_{no} = p_{po} \exp\left(-\frac{V_i}{V_T}\right) \tag{6.8}$$

where $V_T = kT/q$ is the thermal voltage. If we start with $J_n = 0$, we get

$$n_{po} = n_{no} \exp\left(-\frac{V_i}{V_T}\right) \tag{6.9}$$

Note that Eqs. (6.8) and (6.9) are obtained by using the Einstein relation for the particles of a Boltzmann gas. These equations, therefore, are called the *Boltzmann relations*. Since the drift and diffusion tendencies for holes and electrons balance independently at each and every point in the depletion region, the hole concentration $p(x)$ at any location x is related to p_{po} by the relation

$$p(x) = p_{po} \exp\left(\frac{-V(x)}{V_T}\right) \tag{6.10}$$

where $V(x)$ is the potential difference between the point x and the edge of the space charge layer $-x_p$. A similar relation can be obtained for the electron concentration $n(x)$ (see Prob. 6.3).

6.2.2 The Electric Field and Potential Distributions

The electric field and potential distributions in a p-n junction are obtained from the solution of the one-dimensional Poisson equation. To simplify the calculations, the entire semiconductor is divided into three distinct zones; the n- and the p-type neutral regions and the space-charge region. The neutral regions are assumed to be field free.

Let us now consider a p-n junction across which a voltage $(V_i - V_a)$ exists. The applied voltage V_a may be either negative or positive. The relationship between the charge distribution and the electrostatic potential ϕ is given by the Poisson equation

$$\frac{d^2\phi}{dx^2} = -\rho/\varepsilon_s \tag{6.11}$$

where ε_s is the permittivity of the semiconductor and ρ is the space-charge density. Assuming all the donors and acceptors to be ionized, $\rho = q(N_d - N_a + p - n)$. The electron and hole concentrations (n and p) vary with the electrostatic potential ϕ, and an exact solution of Eq. (6.11) becomes complicated. An additional simplifying assumption made in the analysis is the *depletion approximation*. Accordingly, the space-charge region is regarded as completely free of mobile carriers, so that $n = p = 0$ and Eq. (6.11) takes the form

$$\frac{d^2\phi}{dx^2} = -\frac{q}{\varepsilon_s}(N_d - N_a) \tag{6.12}$$

Referring to Fig. 6.5a, we take the origin at the metallurgical junction. The space-charge layer widths x_n and x_p can now be calculated separately. In the n-region, Eq. (6.12) becomes

$$\frac{d^2\phi}{dx^2} = -\frac{q}{\varepsilon_s}N_d \quad \text{for } 0 < x \leq x_n$$

$$= 0 \quad \text{for } x > x_n \tag{6.13}$$

Integrating Eq. (6.13) with respect to x and using the condition that the field vanishes at x_n, we write $d\phi/dx = 0$ at $x = x_n$ to obtain

$$\mathcal{E}(x) = -\frac{d\phi}{dx} = \mathcal{E}_m\left(1 - \frac{x}{x_n}\right) \tag{6.14}$$

where the maximum field $\mathcal{E}_m$ occurs at $x = 0$ and is given by

$$\mathcal{E}_m = -\frac{qN_d}{\varepsilon_s}x_n \tag{6.15}$$

THE ABRUPT JUNCTION

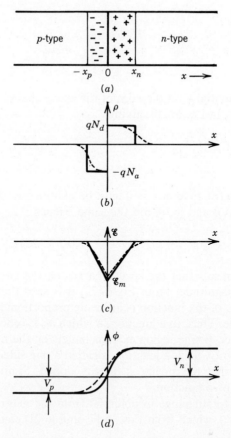

FIGURE 6.5 Properties of an abrupt p-n junction: (a) depletion region, (b) space charge, (c) electric field, and (d) potential distributions. Dashed lines show the results obtained by considering a gradual transition between the neutral and depleted regions, whereas the solid lines are the results of depletion approximation obtained using Eqs. (6.14), (6.18), (6.16), and (6.20).

The expression for $\mathscr{E}(x)$ can be integrated once again to yield

$$\phi(x) = \frac{qN_d}{\varepsilon_s}\left(x_n x - \frac{x^2}{2}\right) \tag{6.16}$$

In deriving this relation, we have used the boundary condition $\phi = 0$ at $x = 0$. The potential V_n at the neutral edge x_n can be obtained from Eq. (6.16) as

$$V_n = \frac{qN_d}{2\varepsilon_s}x_n^2 \tag{6.17}$$

Similar calculations made on the p-side of the junction lead to the following expressions:

$$\mathscr{E}(x) = \mathscr{E}_m\left(1 + \frac{x}{x_p}\right) \tag{6.18}$$

$$\mathscr{E}_m = -\frac{qN_a}{\varepsilon_s}|x_p| \tag{6.19}$$

and

$$\phi(x) = \frac{qN_a}{\varepsilon_s}\left(x_p x + \frac{x^2}{2}\right) \tag{6.20}$$

The potential V_p at the edge of the space-charge layer is obtained by substituting $x = -x_p$ in Eq. (6.20), giving

$$V_p = -\frac{qN_a}{2\varepsilon_s}x_p^2 \tag{6.21}$$

The electric field at $x = 0$ must be continuous, so that the values of $\mathscr{E}_m$ given by Eqs. (6.15) and (6.19) are the same. Hence

$$N_a|x_p| = N_d|x_n| \tag{6.22}$$

which means that the space charges on the two sides of the junction have the same magnitude. From Eq. (6.22) it is seen that the depletion region width on any side of the junction is inversely proportional to the dopant concentration on that side. Thus, in a junction in which N_a is orders of magnitude higher than N_d, the space-charge region extends largely on the n-side, making x_p negligible compared to x_n. Such a junction is called a one-sided abrupt p$^+$-n junction where p$^+$ means that the p-side is heavily doped. Similarly, we can have a one-sided abrupt n$^+$-p junction.

The total change in potential from the neutral p- to n-side is the difference $(V_n - V_p)$, which from Eqs. (6.17) and (6.21) can be written as

$$V_t = (V_n - V_p) = \frac{q}{2\varepsilon_s}(N_d x_n^2 + N_a x_p^2) \tag{6.23a}$$

$$= -\frac{\mathscr{E}_m}{2}[|x_n| + |x_p|] \tag{6.23b}$$

It is seen from Eq. (6.23a) that, in an asymmetrical junction (N_a and N_d are different), the major portion of the potential drop occurs on the low doped side where the space-charge layer is thicker. Equations (6.22) and (6.23a) can be solved for x_n and x_p. The total width of the space-charge layer $W = |x_n| + |x_p|$ can now be determined, and the following relations are obtained (see Prob. 6.6):

$$W = W^*\sqrt{(V_i - V_a)} \tag{6.24a}$$

with

$$W^* = \sqrt{\frac{2\varepsilon_s}{q}\left(\frac{1}{N_d} + \frac{1}{N_a}\right)} \tag{6.24b}$$

and

$$|x_p| = \left(\frac{N_d}{N_d + N_a}\right)W \tag{6.25}$$

$$|x_n| = \left(\frac{N_a}{N_d + N_a}\right)W \tag{6.26}$$

Note that V_a is positive for forward bias and negative for reverse bias. The characteristic constant W^* in Eq. (6.24b) represents the width of the space-charge layer when the total voltage $(V_i - V_a)$ is 1 V and is known as the *width constant*.

The space-charge density, electric field, and potential distributions in a typical asymmetric p-n junction are shown in Fig. 6.5. The dashed curves indicate the form of distributions obtained from exact calculations, whereas the solid curves are the results of the depletion approximation.

6.2.3 Validity of the Depletion Approximation [1]

Thus far, we have assumed that there are no mobile carriers in the space-charge region and that the transition from the neutral region to the depletion region is abrupt. These assumptions are not quite correct because the mobile carrier concentrations decrease exponentially with voltage in the space-charge layer, changing from its majority value at one edge to its minority value at the other edge (Fig. 6.4b). Thus, it is clear that the electron–hole concentrations may be negligible compared to the donor and acceptor concentrations near the center of the space-charge layer, but it is not so near the edges x_n and x_p. Let us assume that the depletion approximation is valid from the point where the majority carrier concentration becomes 1 percent of its value in the neutral region. This requires a potential change of $4.6\,V_T$, which is about 0.12 V at 300 K. Consequently, the depletion approximation is valid only when the total potential drop across each side is large compared to this value. This condition is certainly satisfied when the junction is reverse biased by 1 V or more. However, the depletion approximation will not be valid for a forward-biased junction when $(V_i - V_a)$ is comparable to $4.6\,V_T$.

A peculiar situation is encountered in a one-sided abrupt junction. Here, the potential change across the space-charge layer on the heavily doped side may remain small compared to $4.6\,V_T$, even when a voltage of several volts exists across the junction. As a result, the depletion approximation is valid on the lightly doped side, but not on the heavily doped side. Consequently, the total width of the space-charge layer calculated using the depletion approximation will be fairly accurate, but the width on the heavily doped side will be quite small compared to its actual value (see Sec. 6.3).

6.2.4 Depletion Layer Capacitance

As we have seen above, the space-charge region in a p-n junction consists of a dipole layer of charges produced by donor and acceptor ions. In the depletion approximation, the magnitude of the space charge per unit area Q_d on either side of the junction can be written as

$$Q_d = qN_a|x_p| = qN_d|x_n| \tag{6.27}$$

Since x_n and x_p are functions of the junction voltage, Q_d varies with the voltage. A change in the bias voltage causes readjustment of x_n and x_p to their appropriate new values. This change is accomplished by the movement of majority carriers in or out of the space-charge layer. The situation is similar to the charging and discharging of a parallel plate capacitor. However, the junction space-charge layer capacitance differs from an ordinary parallel plate capacitor in one important respect. In the parallel plate capacitor, the charges reside on the two plates, and the capacitance is independent of voltage. In the depletion layer, the charges are distributed in a region of space whose thickness increases nonlinearly with the voltage. Therefore, the charge Q_d also varies with the voltage in a nonlinear manner. Because of its nonlinear behavior, the junction capacitance C_j can be defined for only a small change dV_a in the junction voltage

$$C_j = A \frac{dQ_d}{dV_a} \tag{6.28}$$

Substituting for Q_d from Eq. (6.27) and using Eq. (6.25), we obtain

$$C_j = qA \left(\frac{N_d N_a}{N_d + N_a} \right) \frac{dW}{dV_a} \tag{6.29}$$

Finally, substituting for dW/dV_a from Eq. (6.24a) into Eq. (6.29) leads to

$$C_j = A \left[\frac{q \varepsilon_s}{2(V_i - V_a)} N_I \right]^{1/2} \tag{6.30}$$

where $N_I = (N_d N_a)/(N_d + N_a)$ represents the effective dopant concentration. There are two useful ways in which Eq. (6.30) can be rewritten:

$$C_j = \frac{C_o}{\left(1 - \frac{V_a}{V_i}\right)^{1/2}} \tag{6.31}$$

where C_o is the zero-bias capacitance. In addition, substituting for $(V_i - V_a)$ from Eq. (6.24) into Eq. (6.30) yields

$$C_j = \frac{\varepsilon_s A}{W} \tag{6.32}$$

This relation shows that the junction capacitance can be regarded as the capacitance of a parallel plate capacitor with separation between the plates as the depletion region width W.

The relation in Eq. (6.32) further shows that for a given bias, the capacitance C_j can be uniquely defined only for small changes in the junction voltage that cause a negligible change in W. This can be understood with the help of Fig. 6.6. Here the junction has a depletion region width W at the bias voltage V_a. When the voltage is changed between $V_a + \Delta V_a$ and $V_a - \Delta V_a$, the depletion region width varies between $W - \Delta W$ and $W + \Delta W$. As long as ΔW is small compared to W, the capacitance C_j is given by Eq. (6.32). If ΔV_a is large enough to make ΔW comparable to W, then Eq. (6.32) does not give the proper value of C_j.

From Eq. (6.30) it is seen that C_j varies inversely with the square root of $(V_i - V_a)$. Since this relation is based on the depletion approximation, it is valid

THE ABRUPT JUNCTION

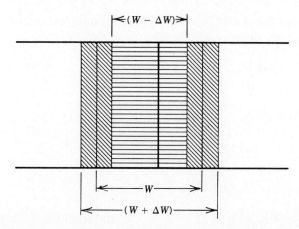

FIGURE 6.6 Change in the depletion region width caused by an increment ΔV_a in the junction bias.

under reverse bias. It may also be valid under a small forward bias if V_i is of the order of 1 V.

Measurement of C_j as a function of reverse bias can be used to determine the built-in voltage V_i and the dopant concentration near the junction. According to Eq. (6.30), a plot of $1/C_j^2$ as a function of V_a is a straight line whose intercept on the voltage axis gives V_i, and the slope can be used to determine the effective dopant concentration N_I.

As an example, consider the $1/C_j^2$ versus V_a plot of a silicon p^+-n-n^+ junction diode shown in Fig. 6.7. The diode has a junction area of 10^{-3} cm^2. The n-region is grown on the n^+-substrate and has a thickness W_1. Let us determine V_i, the dopant concentrations on the two sides of the p^+-n junction, and the value of W_1.

Equation (6.30) can be written as

$$\frac{1}{C_j^2} = \frac{2(V_i - V_a)}{A^2 q \varepsilon_s N_I}$$

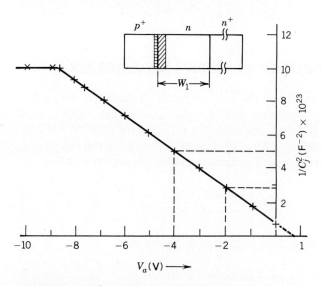

FIGURE 6.7 $1/C_j^2$ versus V_a plot for an abrupt p^+-n junction.

From the plot of Fig. 6.7, we find the intercept on the voltage axis ($V_i = 0.68$ V) and the slope

$$\frac{d(1/C_j^2)}{dV_a} = -\frac{(5.0 - 2.9) \times 10^{23}}{-4 + 2} = \frac{2.1 \times 10^{23}}{2} (F^{-2}V^{-1}) = \frac{2}{A^2 q \varepsilon_s N_I}$$

which gives

$$N_I = \frac{4}{2.1 \times 10^{23} \times 10^{-6} \times 1.6 \times 10^{-19} \times 8.85 \times 10^{-14} \times 11.9}$$

$$= 1.13 \times 10^{14} \text{ cm}^{-3}$$

At 300 K, $kT/q = 25.86 \times 10^{-3}$ V, and from Eq. (6.7), we have

$$0.68 = 25.86 \times 10^{-3} \ln\left(\frac{N_d N_a}{2.25 \times 10^{20}}\right)$$

or

$$N_a N_d = 5.91 \times 10^{31} \text{ cm}^{-6}$$

Also

$$N_I = \frac{N_d N_a}{N_d + N_a} = 1.3 \times 10^{14} \text{ cm}^{-3}$$

From these equations, we obtain $N_d \approx 1.3 \times 10^{14}$ cm^{-3} and $N_a = 4.55 \times 10^{17}$ cm^{-3}. Figure 6.7 further shows that $1/C_j^2 (= 10 \times 10^{23})$ becomes constant and independent of the bias for reverse voltage in excess of 8.7 V. This happens because the depletion region has reached the end of the n-zone. At this point $C_j = 10^{-12}$ F, and from Eq. (6.32) we get

$$W = W_1 = \frac{\varepsilon_s A}{C_j} = \frac{8.85 \times 10^{-14} \times 11.9 \times 10^{-3}}{10^{-12}} = 1.05 \times 10^{-3} \text{ cm}.$$

6.3 EXAMPLE OF AN ABRUPT p-n JUNCTION

To clarify the various points discussed in Sec. 6.2, let us consider a Si abrupt p-n junction at 300 K with $N_a = 10^{18}$ cm^{-3} and $N_d = 10^{15}$ cm^{-3}. Taking $n_i = 1.5 \times 10^{10}$ cm^{-3} and using Eq. (6.7), we obtain $V_i = 0.752$ V. For Si $\varepsilon_s = 8.85 \times 10^{-14} \times 11.9 = 1.05 \times 10^{-12}$ F/cm. Substituting the values of ε_s, N_d, and N_a in Eq. (6.24), we obtain the zero-bias depletion region width as

$$W_o = \left[\frac{2 \times 1.05 \times 10^{-12} \times 0.752}{1.6 \times 10^{-19}}(10^{-15} + 10^{-18})\right]^{1/2} = 9.93 \times 10^{-5} \text{ cm}.$$

From Eqs. (6.25) and (6.26) we get

$$|x_n| = \frac{9.93 \times 10^{-5} \times 10^{18}}{10^{18} + 10^{15}} = 9.92 \times 10^{-5} \text{ cm} \quad \text{and} \quad |x_p| = 9.92 \times 10^{-8} \text{ cm}$$

EXAMPLE OF AN ABRUPT p-n JUNCTION

The maximum field in the depletion region is obtained from Eq. (6.23b):

$$\mathcal{E}_m = -\frac{2V_i}{W_o} = -\frac{1.54 \times 10^5}{9.93} = -1.51 \times 10^4 \text{ V/cm}$$

From Eqs. (6.17) and (6.21), the voltages V_n and V_p on the n- and p-sides of the junction are obtained as $V_n = 0.75$ V and $V_p = -0.75 \times 10^{-3}$ V. These results show that most of the space-charge layer extends on the n-side of the junction and practically all the potential appears on this side. The zero-bias capacitance obtained from Eq. (6.32) is 1.057×10^{-8} F cm^{-2}.

We will now estimate the equilibrium diffusion and drift components of the hole and electron currents. The variation of electron and hole concentrations in the space-charge layer is plotted in Fig 6.8a. In the neutral n-region $p_{no} = n_i^2/N_d = 2.25 \times 10^5$ cm^{-3}. The hole concentration changes from p_{po} to p_{no} over the

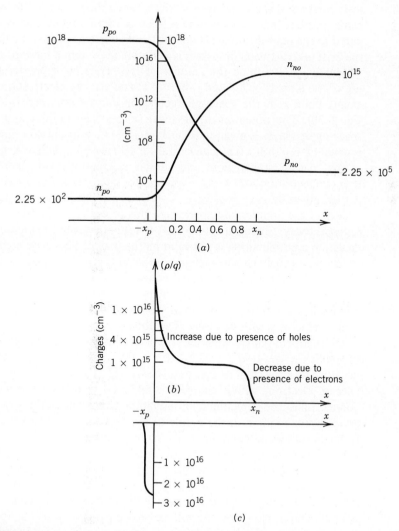

FIGURE 6.8 Plots of (a) mobile carrier concentrations and (b) space-charge distributions for the abrupt Si p-n junction.

space-charge layer width W_o. Hence, the average concentration gradient across the space-charge region is

$$\frac{dp}{dx} = -\frac{p_{po} - p_{no}}{W_o} = -\frac{10^{18}}{9.93 \times 10^{-5}} \approx -1 \times 10^{22} \text{ cm}^{-4}$$

If we assume an effective diffusion constant $D_p = 6.5$ cm^2 sec^{-1} in the space-charge layer, the hole diffusion current density J_{pD} is

$$J_{pD} = -qD_p\frac{dp}{dx} = 1.6 \times 10^{-19} \times 6.5 \times 1 \times 10^{22} = 1.04 \times 10^4 \text{ A cm}^{-2}$$

For a junction area of 10^{-2} cm^2 this corresponds to a current of 104 A across the junction. This is a very large current. Similarly, taking $D_n = 23$ cm^2 sec^{-1}, we obtain a diffusion current density $J_{nD} = 37$ A cm^{-2}, which is small compared to J_{pD}. The diffusion current densities J_{pD} and J_{nD} are balanced by equal and opposite drift currents of the respective carriers. When the junction is forward biased, the actual current that flows is only a fraction of an ampere and is negligible compared to the equilibrium drift and diffusion currents. Thus, the balance between the drift and diffusion processes is maintained even in a forward-biased junction.

Finally, we examine the validity of the depletion approximation. We make use of the already calculated values of x_n and x_p. The electrostatic potential $\phi(x)$ at any point x in the space charge region can be obtained from Eq. (6.16) or Eq. (6.20). The potential difference $V(x)$ between point x and edge x_p of the space-charge layer is given by $V(x) = \phi(x) - V_p$, and the hole concentration $p(x)$ is given by Eq. (6.10). The concentration $p(x)$ was obtained at a number of points in the space-charge region, and similar calculations were made for the electron concentration $n(x) = n_{no} \exp[-q(V_n - \phi(x))/kT]$. The space-charge density is then given by $\rho(x) = q(N_d - N_a + p - n)$. The value of $\rho(x)$ on the p- and n-sides of the junction obtained using this procedure is plotted in Figs. 6.8b and c, respectively. Since $V_p (= 0.75$ mV) is negligibly small compared to $4.6V_T$, the depletion approximation is not valid on the p-side. This is obvious from Fig. 6.8c which shows that the space-charge layer on the p-side is not depleted of mobile carriers. Even at the metallurgical boundary, charge density has a value of 2.8×10^{16} cm^{-3}, which is small compared to the acceptor concentration N_a. Hence, the width $|x_p| = 9.92 \times 10^{-8}$ cm in the above calculations is grossly in error. The actual value of x_p will be larger than this.

On the n-side of the junction it is seen that the space-charge layer is depleted of mobile carriers from $x = 0.2x_n$ to $x = 0.8x_n$. Near the edge x_n the space-charge density decreases from its constant value of 10^{15} charges cm^{-3} because of the presence of electrons whose charge is subtracted from the positive charge of the donors. Thus, the depletion approximation, though not strictly valid, represents a good approximation on the n-side of the junction, and the value of x_n obtained from the above calculations will not be significantly different from its actual value. The same is also true for the total space-charge layer width W_o.

It should now be clear that when the junction is reverse biased by tens of volts, the width x_n will increase and a larger fraction of it will be depleted of mobile carriers. This will improve the accuracy of the calculations made using the depletion approximation. However, the voltage on the p-side will still be too small to allow the calculations of x_p by this method. For a forward-biased junction, the fraction of x_n which is depleted of mobile carriers will decrease, and the validity of the depletion approximation will become questionable even on the n-side.

6.4 THE LINEARLY GRADED JUNCTION

6.4.1 Condition for Rectification [2]

The linearly graded junction is a good approximation for the junctions formed by deep diffusion. Figure 6.9 shows the impurity distribution in a linearly graded junction. The dopant concentration on both sides of the junction varies linearly with the distance from the junction and can be written as

$$(N_d - N_a) = Kx \tag{6.33}$$

where x represents the distance away from the junction and K is the grading constant. It is not possible to obtain rectifying action for all values of K. To study its effect on the junction properties, let us write the one-dimensional Poisson equation:

$$\frac{d^2\phi}{dx^2} = -\frac{q}{\varepsilon_s}[N_d - N_a + p - n] \tag{6.34}$$

Assuming that both sides of the junction are nondegenerate, we can write

$$n = n_i \exp(\phi/V_T) \quad \text{and} \quad p = n_i \exp(-\phi/V_T) \tag{6.35}$$

After substituting these values of n and p in Eq. (6.34), we get

$$\frac{d^2\phi}{dx^2} = \frac{2qn_i}{\varepsilon_s}\left[\sinh\frac{\phi}{V_T} - \frac{Kx}{2n_i}\right] \tag{6.36}$$

By defining a dimensionless potential function U as

$$U = \phi/V_T \tag{6.37}$$

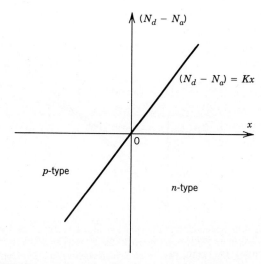

FIGURE 6.9 Impurity distribution in a linearly graded p-n junction.

and substituting

$$\frac{Kx}{2n_i} = Z$$

the above relation is transformed to

$$\frac{d^2U}{dZ^2} = \left(\frac{2n_i}{KL_{Di}}\right)^2 [\sinh U - Z] \tag{6.38}$$

where

$$L_{Di} = (\varepsilon_s V_T / 2\dot{q} n_i)^{1/2} \tag{6.39}$$

is the intrinsic Debye length in the semiconductor. Solution of Eq. (6.38) is plotted in Fig. 6.10 for various values of $B = (KL_{Di}/2n_i)$ for Si. It is observed that for small values of B, the space charge extends throughout the n- and p-regions, and the junction effects are absent. When B becomes large compared to unity, an identifiable space-charge region is formed around the junction. But the assumption of a depletion region bounded by regions of space-charge neutrality on either side is not valid until B becomes larger than 10^4. It is to be noted that the neutral regions to the left and to the right of the space-charge layer also contain some charge because of the impurity gradient. Therefore, these regions are not neutral but rather quasi-neutral. The results of the calculation of the space-charge layer width, electric field, and potential distributions presented below are accurate as long as the electric field in the quasi-neutral regions is negligibly small compared to the average field in the space-charge layer.

6.4.2 The Built-in Voltage

The built-in voltage of a linearly graded junction can be obtained in a similar manner as for the abrupt junction. Since the impurity gradient has the same value

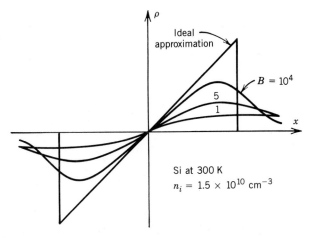

FIGURE 6.10 Space-charge distribution in a linearly graded p-n junction for various values of grading constant. (Based on the work of S. P. Morgan and F. M. Smits, Bell. Syst. Tech J. 39, 1573, (1960). From J. L. Moll, *Physics of Semiconductors*, p. 123. Copyright © 1964. Reprinted by permission of McGraw-Hill, Inc., New York.)

on both sides of the junction, the space-charge region widens equally on the two sides. Thus, if W_o represents the space-charge layer width of the unbiased junction, then

$$x_n = \frac{W_o}{2} \quad \text{and} \quad x_p = -\frac{W_o}{2}$$

Followng Eq. (6.7), the built-in voltage can be approximately written as

$$V_i = V_T \ln\left[\frac{N_d(W_o/2)N_a(-W_o/2)}{n_i^2}\right]$$

From Eq. (6.33) we obtain

$$N_d(W_o/2) = N_a\left(-\frac{W_o}{2}\right) = \frac{KW_o}{2}$$

and thus

$$V_i = V_T \ln\left[\frac{K^2 W_o^2}{4n_i^2}\right] \tag{6.40}$$

In Eq. (6.40), there are two unknowns, V_i and W_o. Another relation between these two quantities is obtained from the solution of the Poisson equation (see Eq. 6.45b). Combining these two relations, we can obtain V_i as a function of K. In Fig. 6.11a, V_i is plotted as a function of K for Si, Ge, and GaAs p-n junctions at 300 K.

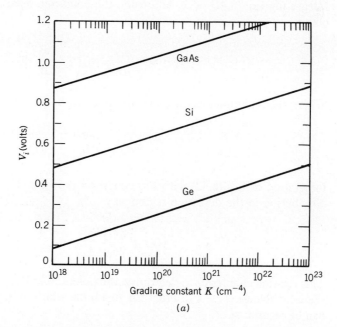

(a)

FIGURE 6.11 (a) Built-in voltage for germanium, silicon, and gallium arsenide linearly graded junctions as a function of grading constant K at 300 K. (From S. M. Sze, *Physics of Semiconductor Devices*, 1e, p. 94. Copyright © 1969. Reprinted by permission of John Wiley & Sons, Inc., New York.)

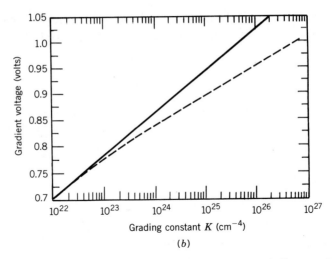

FIGURE 6.11 (*b*) Gradient voltage as a function of K for heavily doped silicon p-n junctions. The dashed line shows the effect of band gap narrowing. (From K.W. Teng and S. S. Li, reference 3, p. 284. Copyright © 1985. Reprinted by permission of Pergamon Press, Oxford.)

6.4.3 The Electric Field and Potential Distributions

Let us consider a junction across which a voltage $(V_i - V_a)$ exists. Using the depletion approximation, we can write Eq. (6.34) as

$$\frac{d^2\phi}{dx^2} = -\frac{q}{\varepsilon_s}Kx, \qquad -\frac{W}{2} \leq x \leq \frac{W}{2} \tag{6.41}$$

Integrating Eq. (6.41) once, and using the condition $d\phi/dx = 0$ at $x = W/2$, we obtain

$$\mathscr{E}(x) = -\frac{d\phi}{dx} = \mathscr{E}_m\left(1 - \frac{4x^2}{W^2}\right) \tag{6.42}$$

Where the maximum electric field $\mathscr{E}_m$ is given by

$$\mathscr{E}_m = -\frac{qKW^2}{8\varepsilon_s} \tag{6.43}$$

Integration of Eq. (6.42) with the condition $\phi = 0$ at $x = 0$ gives the potential distribution in the depletion region as

$$\phi(x) = \frac{qK}{2\varepsilon_s}\left(\frac{W^2}{4}x - \frac{x^3}{3}\right) \tag{6.44}$$

This relation enables us to determine the potential at $W/2$ and $-W/2$, and hence, the total voltage $(V_i - V_a)$. From Eq. (6.44) the width of the space-charge layer W can be written as

$$W = \left[\frac{12\varepsilon_s}{qK}(V_i - V_a)\right]^{1/3} \tag{6.45a}$$

THE LINEARLY GRADED JUNCTION

A relation between the zero-bias width W_o and the built-in voltage is obtained from Eq. (6.45a). Thus

$$W_o = \left[\frac{12\varepsilon_s}{qK}V_i\right]^{1/3} \quad (6.45b)$$

The electric field and potential, as given by Eqs. (6.42) and (6.44), are plotted in Fig. 6.12b and c, respectively. The space change distribution is shown in Fig. 6.12a.

6.4.4 The Junction Capacitance

The junction capacitance can be obtained in a manner similar to that of the abrupt junction. The magnitude of the depletion region charge (Q_d) on either side of the junction is given by

$$Q_d = \frac{q}{2} \cdot \left(\frac{W}{2}\right) \cdot \left(\frac{KW}{2}\right) = \frac{q}{8}KW^2 \quad (6.46)$$

From this relation, C_j is obtained as

$$C_j = A\frac{dQ_d}{dV_a} = \frac{qA}{4}KW\left(\frac{dW}{dV_a}\right)$$

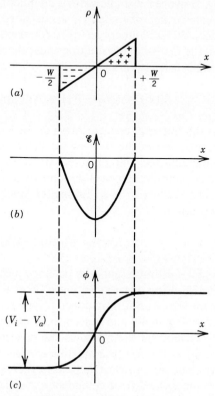

FIGURE 6.12 Plots of (a) space-charge, (b) electric field, and (c) potential distributions in a linearly graded junction.

When the values of W and (dW/dV_a) are substituted from Eq. (6.45a), the result is

$$C_j = A\left[\frac{qK\varepsilon_s^2}{12(V_i - V_a)}\right]^{1/3} = \frac{\varepsilon_s A}{W} \qquad (6.47)$$

The depletion region capacitance can also be obtained from an expression identical to Eq. (6.47) but replacing V_i by the gradient voltage, which is given by

$$\frac{2}{3}V_T \ln\left[\frac{K^2\varepsilon_s V_T}{8qn_i^3}\right]$$

The gradient voltage as a function of K for Si p-n junction (at 300 K) is plotted in Fig. 6.11b. The solid line is the result of conventional calculations, and the dashed line shows the effect of band gap narrowing resulting from the heavy doping [3].

6.5 THE DIFFUSED JUNCTION [4]

The diffusion process for making a p-n junction is described in Sec. 19.2. The dopant distribution is of the complementary error function (erfc) type for a diffusion made from an infinite source, and is a Gaussian type in the case of diffusion performed using a finite source. Diffused junctions, in general, cannot be treated analytically. However, a shallow diffused junction into a uniformly doped semiconductor can be approximated by a one-sided abrupt junction. On the other hand, for deep diffused junctions (diffusion depth in excess of about 5 μm), both the erfc and the Gaussian distributions near the junction can be approximated by an exponential distribution of the form

$$N(x') = (N_d - N_a) = N_o \exp(-x'/\lambda) - N_B \qquad (6.48)$$

where N_o is the impurity concentration at the surface, N_B is the background impurity concentration in the starting sample, x' is the distance from the surface into the semiconductor, and λ is the characteristic length associated with the diffusion. The constant N_o can be eliminated by using the condition $N(x_j) = 0$ at the metallurgical boundary $x' = x_j$. Taking the junction point as the origin, Eq. (6.48) can be rewritten as

$$N(x) = N_B[\exp(-x/\lambda) - 1] \qquad (6.49)$$

where $x = (x' - x_j)$ represents the distance measured from the junction.

Figure 6.13a shows the impurity profile of a p-n junction made by diffusing n-type dopant into a uniformly doped p-type semiconductor. The distribution is represented by Eq. (6.49) and is seen to be essentially exponential on the n-side, but almost constant on the p-side of the junction. For small values of x/λ, such that $\exp(-x/\lambda) \approx 1 - x/\lambda$, the junction can be treated as linearly graded. As we have seen in Sec. 4.8, an electric field in the exponentially doped quasi-neutral n-region balances the diffusion tendency of electrons. However, there is no field in the homogeneously doped p-region. The electric field for the entire structure is shown in Fig. 6.13b, and the space-charge density obtained by differentiating

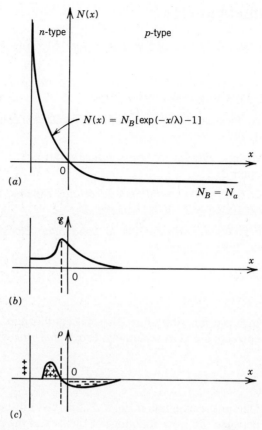

FIGURE 6.13 The diffused junction: (a) dopant concentration, (b) electric field, and (c) space-charge distributions.

the field is shown in Fig. 6.13c. Since the field is constant in the quasi-neutral n-region, there is no space charge in this region but rather a sheet of positive charge on the surface. The remainder of the positive charge is confined to the depletion region on the n-side. These two charges balance the negative charge contained in the depletion region on the p-side of the junction.

REFERENCES

1. P. E. GRAY, D. DEWITT, A. R. BOOTHROYD, and J. F. GIBBONS, *Physical Electronics and Circuit Models of Transistors,* SEEC Vol. 2, John Wiley, New York, 1964.

2. S. P. MORGAN and F. M. SMITS, "Potential distribution and capacitance of a graded p-n junction," *Bell System Tech. J.* 39, 1573 (1960).

3. K.W. TENG and S. S. LI, "Theoretical calculations of Debye length, built-in potential and depletion layer width versus dopant density in a heavily doped p-n junction diode," *Solid-State Electron.* 28, 277 (1985).

4. S. K. GHANDHI, *Semiconductor Power Devices,* John Wiley, New York, 1977, Chapter 2.

ADDITIONAL READING LIST

1. J. L. MOLL, *Physics of Semiconductors,* McGraw-Hill, New York, 1964, Chapter 7.
2. R. S. MULLER and T. I. KAMINS, *Device Electronics for Integrated Circuits,* 2d ed., John Wiley, New York, 1986.
3. D. A. FRASER, *The Physics of Semiconductor Devices,* Clarendon Press, Oxford, 1977.
4. S. M. SZE, *Physics of Semiconductor Devices,* John Wiley, New York, 1981, Chapter 2.
5. B. G. STREETMAN, *Solid State Electronic Devices,* 2d ed. Prentice-Hall, Englewood Cliffs, N.J., 1979, Chapter 5.
6. F. F.Y. WANG, *Introduction to Solid State Electronics,* North-Holland, New York, 1980, Chapter 14.

PROBLEMS

6.1 In a p-n junction under thermal equilibrium, both the electron and hole currents are zero separately. Prove that this is possible only when

$$\frac{D_p}{\mu_p} = \frac{D_n}{\mu_n} = \frac{kT}{q}$$

6.2 Obtain the expression Eq. (6.7) for the built-in voltage from the difference between the work functions of the *n*- and the *p*-type semiconductors forming the junction.

6.3 Show that the electron–hole product is constant equal to n_i^2 throughout the space-charge region of a p-n junction in thermal equilibrium. How will this product change when the junction is biased? Find out its value for a Ge p-n junction at (a) $V_a = 0.2$ V and (b) $V_a = -1$ V. Take $T = 300$ K.

6.4 For an abrupt p-n junction show that

$$\frac{dV_i}{dT} = \frac{V_i - V_g}{T} + \frac{dV_g}{dT} - \frac{3k}{q}$$

where V_g is the voltage corresponding to the band gap of the semiconductor. For a silicon p-n junction with $N_d = 10^{15}$ cm^{-3} and $N_a = 10^{18}$ cm^{-3}, calculate dV_i/dT at 300 K and V_i at 305 K. Assume $dV_g/dT = -2.75 \times 10^{-4}$ V °K^{-1}.

6.5 A Ge p-n junction has a donor concentration of $10^{-8} N$ on the n-side and an acceptor concentration of $2 \times 10^{-6} N$ on the p-side where N denotes the number of Ge atoms cm^{-3}. Calculate the built-in voltage at 290 K and determine the temperature at which V_i decreases by 4 percent. Assume $E_g = 0.67$ eV and $dE_g/dT = 0$.

6.6 Using the results of Eqs. (6.22) and (6.23a) obtain Eqs. (6.24), (6.25) and (6.26).

6.7 A Si abrupt p-n junction has $N_d = 10^{15}$ cm^{-3} on the n-side and $N_a = 4 \times 10^{18}$ cm^{-3} on the p-side. At 300 K calculate (a) the built-in voltage, (b) the zero-bias depletion region width, and (c) the maximum electric field in the depletion region at zero bias. Repeat your calculations for (b) and (c) for a reverse bias of 3 V.

6.8 Show that the maximum electric field in a p-n junction is twice the average field. A Si p-n junction has $N_a = 10^{15}$ cm^{-3} on the p-side and $N_d = 2 \times 10^{16}$ cm^{-3} on the n-side. Using the depletion approximation, determine the values of x_n and x_p and the maximum electric field $\mathscr{E}_m$ at a reverse bias of 10 V. Assume $T = 300$ K.

6.9 The donor and acceptor concentrations on the n- and p-sides of a Si abrupt p-n junction are equal to 10^{16} cm^{-3}. The whole semiconductor is illuminated uniformly such that the hole concentration in the neutral n-region rises to 10^{13} cm^{-3}. No current is allowed to flow. What will be the reading of a voltmeter whose positive terminal is connected to the p-side at 290 K?

6.10 A single crystal Si sample with $N_d = 3 \times 10^{14}$ cm^{-3} is alloyed to another Si sample with $N_d = 7.5 \times 10^{18}$ cm^{-3} to form an abrupt junction. (a) Calculate the built-in voltage of the junction at 300 K and plot (approximately) the space-charge distribution. (b) Draw the energy band diagram of the junction and find out if it can be used as a rectifier.

6.11 An abrupt Si p-n junction at 300 K has an area of 1 mm^2. The measured junction capacitance C_j as a function of the bias is given by the relation

$$\frac{1}{C_j^2} = \frac{5}{11} \times 10^6 (3 - 5V_a)$$

where C_j is expressed in μF and V_a in volt. (a) Determine the built-in voltage and the depletion region width at zero bias. (b) Calculate the dopant concentration on the two sides of the junction.

6.12 A Si abrupt p-n junction has $N_a = 3 \times 10^{18}$ cm^{-3} on the p-side and an area of 1.6×10^{-3} cm^2. The junction capacitance is 18 pF at a reverse bias of 3.2 V and 12 pF at 8.2 V. Calculate the built-in voltage and the donor concentration on the n-side.

6.13 The depletion region capacitance of a GaAs p$^+$–n junction diode is measured as a function of the reverse bias, and the following data are obtained

Reverse bias (V)	0	0.5	1.0	3.0	5.0
$C_j(pF)$	19.9	17.3	15.6	11.6	9.8

Check if it is an abrupt junction diode.

6.14 Show that the maximum electric field in the depletion region of a linearly graded p-n junction is 3/2 times of the average field.

6.15 A Si diode has a linearly graded p-n junction with a grading constant of 2×10^{21} cm^{-4} and a junction area of 0.5 mm^2. (a) Using depletion approximation, calculate the maximum electric field in the space-charge layer and the junction capacitance at reverse biases of 3 V and 25 V. (b) If a tank circuit is made using the above diode with a 2 mH inductor, calculate the resonant frequencies at the two bias voltages in (a). Assume $T = 300$ K.

6.16 The dopant distribution in a silicon sample varies as

$$N(x') = [N_{ao} \exp(-x'/\lambda_1) - N_{do} \exp(-x'/\lambda_2)]$$

where $N_{ao} = 10^{18}$ cm^{-3}, $\lambda_1 = 10^{-4}$ cm, and $\lambda_2 = 2 \times 10^{-4}$ cm. (a) If a p-n junction is desired at 1 μm from the surface $x' = 0$, determine the value of N_{do} and the donor concentration at the junction. (b) Assume that the dopant distribution is linearly graded near the junction. Determine the grading constant K and the maximum electric field in the junction depletion region at zero bias.

6.17 Show that the depletion region capacitance of a p-n junction for any arbitrary dopings on the two sides can always be expressed by Eq. (6.32).

7
STATIC I-V CHARACTERISTICS OF p-n JUNCTION DIODES

INTRODUCTION

In Chapter 6 we saw that when bias is applied to a p-n junction, the thermal equilibrium balance between the drift and diffusion processes is disturbed and a net current flows. This current increases exponentially with the forward bias but attains a constant small value for a sufficiently large reverse bias. A structure having a p-n junction with ohmic contacts at the ends of the neutral n- and p-regions is called a *p-n junction diode*.

In this chapter we discuss the current-voltage characteristics of a p-n junction diode under static conditions, which means that the carrier distributions do not change with time. We first present a model based on several simplifying assumptions. This model is referred to as the ideal diode. Next, we compare the current-voltage (I-V) characteristic of real p-n junction diodes with that of the ideal diode and examine the validity of our assumptions. Subsequently, the temperature dependence of the I-V characteristics and modifications introduced at high-level injection are considered. Finally, an example of a p-n junction diode is provided to illustrate the various points.

7.1 THE IDEAL DIODE MODEL

When a voltage is applied to a p-n junction diode, the electron and hole distributions in the space-charge and neutral regions are changed from their equilibrium values. The diode current is intimately related to these distributions. To be specific, let us consider the forward-biased diode (Fig. 7.1a). The hole and electron distributions in various regions of the diode are shown in Fig. 7.1b. The applied bias causes the majority carriers to diffuse across the junction. As a result,

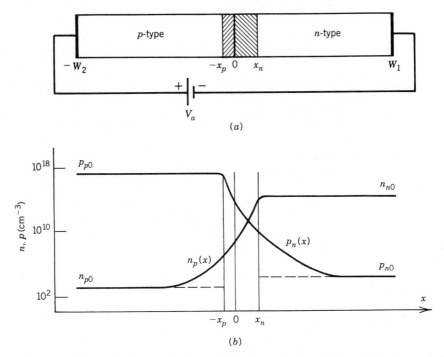

FIGURE 7.1 (*a*) Schematic diagram of a forward-biased p-n junction diode and (*b*) electron-hole distributions in the various regions.

the concentration of minority carriers in the neutral n- and p-regions is increased. To maintain the space-charge neutrality in these regions, the majority carrier concentrations also increase by the same amount. These changes, however, are not visible in the plots of Fig. 7.1b because, under the level of injection considered here, the changes in the majority concentrations are negligibly small compared to their thermal equilibrium values.

It is not necessary to specify the exact form of the electron–hole distributions at this stage. However, if we confine our attention to holes in the n-region, it becomes clear that an increase in the hole concentration at the edge x_n causes a flux of holes toward the contact at W_1. As these holes move away from x_n, they recombine with electrons and their concentration decreases gradually, reaching an equilibrium value at a few diffusion lengths away from x_n. Accordingly, in the neutral *n*-region the hole current has its maximum value at x_n and decreases continuously to become zero at large distances. Similarly, in the neutral p-region, the electron current has its maximum value at $-x_p$ and decreases away from it. Since the electron-hole currents are functions of position, a reference point is needed at which they can be easily evaluated. The edges of the space-charge layer x_n and $-x_p$ are two such reference points. The hole and electron currents evaluated at these points are added together to obtain the total diode current.

7.1.1 Simplifying Assumptions

In the ideal diode model, an abrupt junction is assumed between the uniformly doped n- and p-regions. In addition, the following simplifying assumptions are made:

1. The current flow across the diode is one-dimensional.

THE IDEAL DIODE MODEL

2. All of the applied voltage appears across the space-charge region, and the neutral regions are field free. Thus, the minority carrier currents in the neutral n- and p-regions can be described as diffusion currents only.
3. There is no net generation or recombination of carriers in the junction depletion region, which means that the electron and hole currents remain constant throughout this region.
4. Low-level injection prevails on both sides of the junction. This condition places a restriction on the current that can be drawn through the ideal diode and is not valid at high currents.
5. The Boltzmann relation for electrons and holes is valid throughout the depletion region. As we have seen in Sec. 6.2, the Boltzmann relation is a direct consequence of the balance between the drift and diffusion processes in the depletion region. Since the current drawn through the junction is quite small compared to the equilibrium diffusion or drift current, this balance is approximately maintained.

7.1.2 Solution of the Continuity Equation

Let us consider a voltage V_a applied to the diode in Fig. 7.1a and confine our attention to holes in the n-region to the right of x_n. Since there is no external source of carrier generation, the continuity equation for holes in this region is obtained by putting $G_L = 0$ in Eq. (5.56a):

$$\frac{\partial p_e}{\partial t} = -\frac{p_e}{\tau_p} - \frac{1}{q} \nabla \cdot \mathbf{J}_p \tag{7.1}$$

where p_e represents the excess hole concentration. According to assumption (2), we neglect the field in the n-region and write

$$J_p = -q\, D_p \left(\frac{\partial p_e}{\partial x}\right) \tag{7.2}$$

Moreover, for one-dimensional current flow $\nabla \cdot \mathbf{J}_p = \partial J_p / \partial x$. Since $\partial p_e / \partial t = 0$ in the steady state, the hole transport equation in the n-region becomes

$$\frac{d^2 p_e}{dx^2} - \frac{p_e}{L_p^2} = 0 \tag{7.3}$$

which is identical to Eq. (5.68). The general solution of Eq. (7.3) is

$$p_e(x) = A_1\, exp(-x/L_p) + A_2\, exp(x/L_p) \tag{7.4}$$

The constants A_1 and A_2 are determined from the boundary conditions which the hole distribution satisfies at x_n and W_1. To determine the boundary condition at x_n we use assumption (5). When voltage V_a is applied to the junction, the hole concentration at $-x_p$ is related to that at x_n by the Boltzmann relation

$$p_p(-x_p) = p_n(x_n)\, \exp[(V_i - V_a)/V_T] \tag{7.5}$$

From Eq. (6.8) we have

$$p_{po} = p_{no}\, \exp(V_i/V_T)$$

Combining this with Eq. (7.5) gives

$$\frac{p_p(-x_p)}{p_{po}} = \frac{p_n(x_n)}{p_{no}} \exp(-V_a/V_T) \tag{7.6}$$

At low-level injection, the hole concentration $p_p(-x_p)$ is not changed significantly from p_{po} and Eq. (7.6) becomes

$$p_n(x_n) = p_{no} \exp(V_a/V_T) \tag{7.7}$$

The excess hole concentration $p_e(x_n) = [p_n(x_n) - p_{no}]$ is thus given by

$$p_e(x_n) = p_{no}[\exp(V_a/V_T) - 1] \tag{7.8}$$

The boundary condition at W_1 is dependent on the surface condition and is characterized by the surface recombination velocity S defined by Eq. (5.44), which can be rewritten as

$$p_e(W_1) = \frac{J_p(W_1)}{qS} \tag{7.9}$$

At an ohmic contact, the value of S is very large so that it can be treated as a perfectly absorbing surface with $S = \infty$. Thus

$$p_e(W_1) = 0 \tag{7.10}$$

The constants A_1 and A_2 are obtained by substituting the boundary conditions (7.8) and (7.10) into (7.4). The excess hole concentration $p_e(x)$ can then be written as (see Prob. 7.1).

$$p_e(x) = p_{no}[\exp(V_a/V_T) - 1]\frac{\sinh\left(\frac{W_1 - x}{L_p}\right)}{\sinh\left(\frac{W_1 - x_n}{L_p}\right)} \tag{7.11}$$

The hole current density $J_p(x)$ is now obtained from Eq. (7.2) as

$$J_p(x) = \frac{qD_p}{L_p} p_{no}[\exp(V_a/V_T) - 1]\frac{\cosh\left(\frac{W_1 - x}{L_p}\right)}{\sinh\left(\frac{W_1 - x_n}{L_p}\right)} \tag{7.12}$$

This current has its maximum value at x_n given by

$$J_p(x_n) = \frac{qD_p}{L_p} p_{no} \coth\left(\frac{W_1 - x_n}{L_p}\right)[\exp(V_a/V_T) - 1] \tag{7.13}$$

The minority carrier curent $J_n(-x_p)$ resulting from the injection of electrons into the p-region can be found by a similar analysis:

$$J_n(-x_p) = \frac{qD_n}{L_n} n_{po} \coth\left(\frac{|W_2 - x_p|}{L_n}\right)[\exp(V_a/V_T) - 1] \tag{7.14}$$

where D_n and L_n represent the diffusion coefficient and the diffusion length of electrons in the p-region.

7.1.3 The Ideal Diode Equation

We stated in assumption (3) above that the electron and hole currents remain constant in the junction depletion region. Accordingly,

$$J_n(-x_p) = J_n(x_n)$$

and the total current I at x_n is given by

$$I = A[J_p(x_n) + J_n(x_n)] \qquad (7.15)$$

where A represents the junction area. Substituting for $J_p(x_n)$ and $J_n(x_n)$ from Eqs. (7.13) and (7.14) into Eq. (7.15) and writing $p_{no} = n_i^2/N_d$ and $n_{po} = n_i^2/N_a$, we obtain

$$I = qAn_i^2\left[\frac{D_p}{L_p N_d}\coth\left(\frac{W_n}{L_p}\right) + \frac{D_n}{L_n N_a}\coth\left(\frac{W_p}{L_n}\right)\right]\left[\exp\left(\frac{V_a}{V_T}\right) - 1\right] \qquad (7.16)$$

where

$$W_n = |W_1 - x_n| \quad \text{and} \quad W_p = |W_2 - x_p| \qquad (7.16a)$$

Note that W_n and W_p represent the widths of the neutral n- and p-regions, respectively. We will now consider two limiting cases of interest depending on the values of W_n and W_p.

(a) LONG-BASE DIODE

When the neutral region widths W_n and W_p are very long, so that $W_n > 6L_p$ and $W_p > 6L_n$, both coth terms in Eq. (7.16) tend to unity and

$$I = I_s[\exp(V_a/V_T) - 1] \qquad (7.17)$$

where

$$I_s = qAn_i^2\left(\frac{D_p}{N_d L_p} + \frac{D_n}{N_a L_n}\right) \qquad (7.17a)$$

is called the *reverse saturation current*. This name is given because for a reverse voltage in excess of about $3V_T$, the current saturates at $-I_s$. Equation (7.17) is the so-called *ideal diode equation* [1].

(b) SHORT-BASE DIODE

In a short-base diode, the neutral region widths W_n and W_p are much shorter than the diffusion lengths L_p and L_n. Consequently, there is little recombination in the neutral-base regions, and most of the injected carriers recombine at the ohmic contacts. Current-voltage relation for a short-base diode can be directly obtained from Eq. (7.16). For small values of x, $\coth x \simeq 1/x$, and Eq. (7.16) becomes

$$I = qAn_i^2 \left(\frac{D_p}{N_d W_n} + \frac{D_n}{N_a W_p} \right) [\exp(V_a/V_T) - 1] \qquad (7.18)$$

Note that this relation has the same form as Eq. (7.17) except that the diffusion lengths L_p and L_n have been replaced by W_n and W_p. Since W_n and W_p are much smaller than the respective diffusion lengths, the pre-exponential factor in Eq. (7.18) is larger than I_s given by Eq. (7.17a).

The pre-exponential term in Eq. (7.18) varies with the applied bias since both W_n and W_p change with the junction voltage. To see the effect of the bias voltage V_a on the diode current, let us consider a p^+-n junction diode. Neglecting the $D_n/N_a W_p$ term, we obtain

$$I = \frac{qAn_i^2 D_p}{N_d[W_1 - W^*\sqrt{(V_i - V_a)}]} [\exp(V_a/V_T) - 1] \qquad (7.19)$$

where we have used the relations $W_n = (W_1 - x_n)$ and $x_n = W^*\sqrt{(V_i - V_a)}$. For a forward bias $V_a = V_F$, and x_n decreases with V_a causing a decrease in the pre-exponential term of Eq. (7.19). However, in the case of a reverse-biased diode $V_a = -V_R$, and Eq. (7.19) becomes

$$I = -\frac{qAn_i^2 D_p}{N_d[W_1 - W^*\sqrt{(V_i + V_R)}]} \qquad (7.19a)$$

As V_R is increased gradually, $W^*\sqrt{(V_i + V_R)}$ increases and a point may be reached where this term becomes equal to W_1 and the reverse current attains a very high value. This phenomenon is called *punch-through* because the depletion region punches through the neutral n-region. When punch-through occurs, the junction voltage loses its control over the diode current. This current is now limited by the space-charge effects in the depletion region.

Short-base diodes are sometimes used in special applications where high switching speeds are required. The majority of general-purpose diodes fall between a short- and long-base diode. The ideal diode equation, however, assumes a long-base diode; most of our further discussion in this chapter will be confined to it.

(c) THE IDEAL DIODE I-V CHARACTERISTICS

The ideal diode current voltage characteristic is described by Eqs. (7.17) and (7.17a). It is seen that in the forward direction, the current I rises exponentially with the voltage, whereas in the reverse direction, it saturates to a value I_s. These features are evident from Fig. 7.2.

Our discussion until now has been confined to an abrupt p-n junction. In a linearly graded junction, the calculation of diode current is complicated by the fact that, because of concentration gradients, an electric field is also present in the neutral n- and p-regions. This field is in a direction so as to retard the flow of injected minority carriers. However, it does not have any marked influence on the form of the I-V characteristic. When the grading constant K is sufficiently

THE IDEAL DIODE MODEL

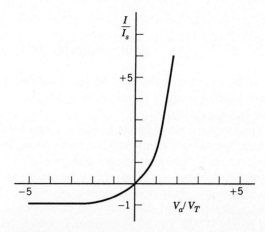

FIGURE 7.2 Ideal diode current-voltage characteristic.

large to produce a well-defined space-charge region at the junction, the Boltzmann boundary condition Eq. (7.7) is still satisfied. Thus, we obtain the ideal diode equation for the linearly graded and other types of junctions independent of the impurity distribution [2].

7.1.4 Minority and Majority Carrier Currents

The steady state hole distribution $p_e(x)$ in the neutral n-region of a p-n junction diode is given by Eq. (7.11). In the limit when W_1 approaches infinity, $p_e(x)$ can be written as

$$p_e(x) = p_{no}[\exp(V_a/V_T) - 1] \exp\left[-\frac{(x - x_n)}{L_p}\right] \quad (7.20)$$

which represents the excess hole distribution in an ideal diode. Note that Eq. (7.20) can be obtained directly from Eq. (7.4) after imposing the boundary condition $p_e(x) = 0$ as x tends to infinity. This makes $A_2 = 0$ (see Eq. (5.69)). To simplify Eq. (7.20) further, we transfer the origin to x_n and make the substitutions $(x - x_n) = y$ and $p_{no}[\exp(V_a/V_T) - 1] = p_e(0)$. We then obtain

$$p_e(y) = p_e(0)\exp(-y/L_p) \quad (7.21)$$

which shows that the excess hole concentration decays exponentially with distance y. Since the total concentration is $p_n(y) = p_e(y) + p_{no}$, the above relation becomes

$$p_n(y) = p_e(0)\exp(-y/L_p) + p_{no} \quad (7.22)$$

A similar expression can be written for the electron concentration $n_p(y')$ in the neutral p-region:

$$n_p(y') = n_e(0)\exp(-y'/L_n) + n_{po} \quad (7.23)$$

where

$$n_e(0) = n_{po}[\exp(V_a/V_T) - 1]$$

and y' is the distance measured from the depletion region edge x_p. These relations will now be used to study the electron and hole distributions under forward- and reverse-bias conditions.

When a forward bias is applied, holes are injected from the p- into the n-region, and their concentration at $y = 0$ increases above its thermal equilibrium value. The injected holes produce a space-charge and to maintain charge neutrality in the n-region electrons are drawn from the ohmic contact at W_1 (Fig. 7.1a). As the excess holes move away from the depletion region edge x_n, some of them recombine with electrons and their concentration decays exponentially with y as predicted by Eq. (7.21). The excess majority carriers (i.e., electrons) are required to maintain the space-charge neutrality and have nearly the same distribution as holes. In the steady state, the loss of holes and electrons owing to recombination is compensated for by the injection of holes from the p-side and a flux of electrons from the ohmic contact in the opposite direction. Similar considerations can be used to obtain the carrier distributions in the neutral p-region. The steady state electron and hole distributions on the two sides of the junction are shown in Fig. 7.3a. The carrier concentrations are plotted on a linear scale, and a break

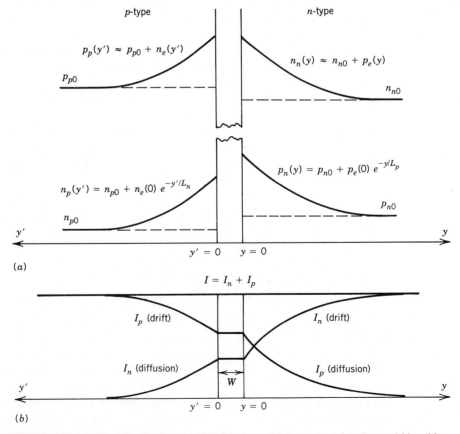

FIGURE 7.3 (a) Carrier distributions and (b) electron and hole currents in a forward-biased long-base p-n junction diode.

in the concentrations on each side is shown to represent the widely different minority and majority carrier concentrations on the same plot.

The electron and hole currents are shown in Fig. 7.3b. These currents are intimately related to the carrier distributions in Fig. 7.3a. Confining our attention to the n-side of the junction, we note that the hole current in the neutral n-region flows mainly by diffusion, since the electric field here has been assumed to be negligibly small. Thus, we can write

$$I_p(y) = -qAD_p\left(\frac{dp_e}{dy}\right) = \frac{qAD_p p_e(0)}{L_p}\exp(-y/L_p)$$

It is evident from this relation that the hole current also decays exponentially as we move away from the junction edge $y = 0$. This is because the hole concentration gradient decreases as carriers are lost by recombination. The loss of hole current is, however, compensated by the electron current and the total current remains constant throughout the diode. The electron current I_n in the n-region is now obtained by subtracting I_p from the total current I. Far away from the junction, all the excess holes disappear by recombination so that I_p becomes zero and the current is carried entirely by electrons. In moving away from W_1 toward the junction, I_n decreases gradually, and at the edge $y = 0$, only that part of the electron current remains, which is injected into the p-region.

The flow of electrons from the ohmic contact at W_1 toward the junction is caused by a small electric field in the neutral n-region. It is clear from the plots of Fig. 7.3a that the concentration gradients of electrons and holes are nearly the same in this region. Thus, the excess electrons diffuse away from the junction, causing a diffusion current of electrons that flows opposite to the hole current. Because of their higher diffusion coefficient, electrons diffuse away from the edge faster than holes. This creates a small deviation from the space-charge neutrality and produces an electric field as discussed in Sec. 5.7. In this field, electrons drift from the contact at W_1 toward the junction. During their motion, electrons lose energy to the lattice by scattering collisions. This causes a voltage drop in this region, which we have neglected in the above analysis. The electron current I_n in Fig. 7.3b is obtained by subtracting the electron diffusion current from the drift current (see Sec. 7.5 for details). Holes also drift in the field produced by the charge imbalance, but, because of their small concentration in the n-region, the drift current of holes remains negligible compared to the diffusion current. The electron and hole currents on the p-side can be obtained from similar considerations. It should be clear now that in a forward-biased diode, the n- and p-regions on the two sides of the depletion layer do not remain completely space-charge free but are quasi-neutral.

Figure 7.4 shows the steady-state electron hole distributions when the junction is sufficiently reverse biased. It is evident from Eq. (7.7) that for a reverse voltage in excess of 0.1 V, the hole concentration at the edge x_n (i.e., $y = 0$) is decreased to a value that is practically zero. Thus, the holes in the neutral n-region diffuse toward the junction causing a diffusion current of holes. Similarly, on the p-side of the junction the electron concentration approaches zero at the edge $y' = 0$, and electrons from the neutral p-region diffuse toward the junction. Note that the majority carriers on both sides undergo the same variations as the minority carriers. The electron and hole currents can be obtained from these distributions in the same way as for the forward-biased diode.

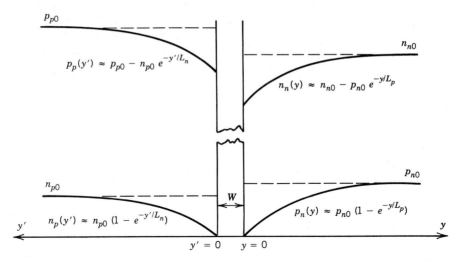

FIGURE 7.4 Electron and hole distributions in a long-base p-n junction diode at a reverse bias in excess of $5V_T$.

The carrier distributions in Fig. 7.4 show that the average distance over which the minority carriers diffuse toward the junction is only one diffusion length. Thus, the reverse saturation current in a long-base diode is caused by the collection of holes generated within a distance L_p from the junction on the n-side, and by the collection of electrons generated within a distance L_n from the depletion region edge on the p-side (see Prob. 7.5). Since a reverse bias of about 0.1 V reduces the minority carrier concentrations at the edges of the depletion region to negligibly small values, a further increase in the reverse voltage has little effect on these concentrations. Hence, the reverse current saturates and becomes independent of bias voltage.

7.2 REAL DIODES

The ideal diode equation is not quite accurate in describing the I-V characteristics of real diodes, and several deviations from the ideal behavior are observed over a significant range of useful biases. The departures are attributable to the following effects: (a) The generation and recombination of electron–hole pairs in the depletion region, (b) voltage drops associated with the electric field in the neutral n- and p-regions, (c) the current arising from leakage across the surface of the junction, and (d) the onset of high-level injection. The first three effects will be considered in this section, whereas the high-level injection is discussed in Sec. 7.4.

7.2.1 Carrier Generation-Recombination in the Junction Depletion Region

Like neutral n- and p-regions, the depletion region contains generation-recombination centers. Under a forward bias, the injected electrons and holes move through this region, and some of them are lost by recombination. This gives rise to a recombination current that is to be added to the current given by the idealized model. Under the reverse-bias condition, carriers are pushed out of this region, and the electron and hole concentrations fall far below their equilibrium values. Thus, there is a net generation of carriers, giving rise to a generation current.

To obtain mathematical expressions for the generation-recombination current in the depletion region, we make use of the Shockley-Read-Hall theory discussed in Sec. 5.3.2. We have seen that when a voltage V_a is applied across the junction, the *pn*-product in the depletion region is given by

$$pn = n_i^2 \exp(V_a/V_T) \tag{7.24}$$

The excess carrier recombination rate U in the depletion region of a biased p-n junction can be obtained after substituting for the *pn* product from Eq. (7.24) into Eq. (5.30), giving

$$U = \frac{n_i^2[\exp(V_a/V_T) - 1]}{\tau_{po}(n + n_1) + \tau_{no}(p + p_1)} \tag{7.25}$$

For a reverse-biased junction, V_a is negative and U represents a net rate of generation of electron–hole pairs. If the reverse voltage is higher than 0.1 V, both p and n become negligible and U can be written as

$$U = -\frac{n_i^2}{\tau_{po}n_1 + \tau_{no}p_1} \tag{7.26}$$

This expression can be further simplified by assuming that $\tau_{po} = \tau_{no} = \tau_o$, and that the trap level E_t is situated at the intrinsic Fermi level E_i so that $n_1 = p_1 = n_i$. For this case Eq. (7.26) reduces to

$$U = -\frac{n_i}{2\tau_o} \tag{7.27}$$

where τ_o is the effective lifetime of carriers in the depletion region. The generation current I_{gen} is obtained by integrating U over the entire width of the depletion region. Thus

$$I_{\text{gen}} = qA \int_{-x_p}^{x_n} U\, dx = -\frac{qAn_i W}{2\tau_o} \tag{7.28}$$

where A and W represent the junction area and the depletion region width, respectively. The negative sign indicates that the current is in the reverse direction.

Under a forward bias, the integral in Eq. (7.28) is not easy to evaluate, and a number of simplifying assumptions are made to deduce the dependence of current on the applied voltage. We again assume $\tau_{no} = \tau_{po} = \tau_o$ and $n_1 = p_1 = n_i$ in Eq. (7.25), and write

$$U = \frac{n_i^2[\exp(V_a/V_T) - 1]}{\tau_o(n + p + 2n_i)} \tag{7.29}$$

It is evident from this relation that for a given bias, U reaches its maximum value at a location in the depletion region where the sum $(p + n)$ is minimized. Since the *pn*-product is given by Eq. (7.24), it can be easily seen that $(p + n)$ is minimized when

$$p = n = n_i \exp(V_a/2V_T)$$

Substituting this condition in Eq. (7.29) leads to

$$U_{\max} = \frac{n_i[\exp(V_a/V_T) - 1]}{2\tau_o[\exp(V_a/2V_T) + 1]} \quad (7.30)$$

Assuming that $U = U_{\max}$ over the whole of the depletion region, the recombination current $I_{\text{rec}} = qAWU_{\max}$ can be written as

$$I_{\text{rec}} = \frac{qAn_iW}{2\tau_o}\left[\exp\left(\frac{V_a}{2V_T}\right) - 1\right] \quad (7.31)$$

Note that the term outside the bracket in Eq. (7.31) is the same as the generation current in Eq. (7.28). One can thus combine the two relations and write

$$I_{rg} = I_{Ro}[\exp(V_a/2V_T) - 1] \quad (7.32)$$

where

$$I_{Ro} = \frac{qAn_iW}{2\tau_o} \quad (7.32a)$$

The current I_{rg} in Eq. (7.32) represents the generation current $I_{\text{gen}} = -I_{Ro}$ in a reverse-biased diode and a recombination current in a forward-biased diode.

7.2.2 The Electric Field in the Quasi-Neutral *n*- and *p*-Regions

In the derivation of the ideal diode equation, the neutral n- and p-regions were assumed to be field free. This assumption is not quite correct as we have seen that a small electric field exists in both these regions, and it is responsible for the majority carrier drift current. For low-level injection, this electric field is proportional to the total diode current. Consequently, the voltage drops across the neutral regions are linearly dependent on the current. Thus, it is possible to represent each of these regions by a resistance. The total voltage drop across the neutral regions and the ohmic contacts on the two sides of the junction can be represented by IR_s, where I is the current and R_s is the diode series resistance. The voltage V_j which appears across the junction is the difference between the applied voltage V_a and the IR_s drop so that

$$V_j = V_a - IR_s \quad (7.33)$$

Replacing V_a by V_j in the ideal diode equation, we obtain

$$I = I_s\left[\exp\left(\frac{V_a - IR_s}{V_T}\right) - 1\right] \quad (7.34)$$

REAL DIODES

The IR_s term remains negligible when I is small but becomes important at high currents.

7.2.3 The I-V Characteristics of Real Diodes

The current-voltage characteristics of real p-n junction diodes can be explained by considering the above effects.

(a) REVERSE CURRENT

The total reverse current through a p-n junction diode is obtained by adding the depletion region generation current to the diffusion current. Assuming a long-base diode and a reverse bias in excess of $3kT/q$, the total current can be written as

$$I = -(I_s + I_{Ro}) \tag{7.35}$$

where I_s is given by Eq. (7.17a) and I_{Ro} by Eq. (7.32a). Note that I_s is independent of bias, but I_{Ro} increases with the applied voltage because of the increase in the depletion region width W. Consequently, the diode current fails to saturate.

The temperature dependence of both I_s and I_{Ro} is contained in n_i, which increases rapidly with temperature. Ge has a large value of n_i at 300 K. Consequently, in Ge p-n junction diodes, I_s dominates at room temperature and above, and the reverse current obeys the ideal diode equation. However, in Si and GaAs junctions, the depletion region generation current I_{Ro} may become several orders of magnitude higher than I_s at room temperature because of the low values of n_i. Since I_s rises as n_i^2, whereas I_{Ro} rises as n_i, at sufficiently high temperatures, I_s will become higher than I_{Ro}. This tendency is clearly seen from the reverse I-V characteristics of the Si diodes shown [3] in Fig. 7.5. Here I_s dominates above 175° C, whereas I_{Ro} is dominant at lower temperatures. It should also be realized that when a Ge diode is cooled to low temperatures, so that n_i is sufficiently reduced, the current I_{Ro} will become larger than I_s, and the ideal diode equation will no longer be valid.

At room temperature, the magnitude of the reverse current in many Si diodes is found to be far in excess of that predicted by the carrier generation in the junction depletion region. Surface leakage is the main cause of this excess current. Surface leakage occurs because the entire reverse voltage is developed across the depletion region, and this produces high electric field at the surface. The junction surface is usually contaminated with ionic impurities that get deposited during the numerous chemical treatments used in device fabrication. These ions can move under the action of the applied field. Positive ions may also invert the surface on the p-side of the junction to n-type, thus establishing a direct conducting channel over the surface [3]. A surface leakage current flows through this channel which rises with the voltage in some unpredictable manner.

(b) FORWARD CURRENT

For a forward-biased diode, the total current is obtained by adding the depletion region recombination current to the ideal diode current. When the bias voltage is higher than 0.1 V, the current can be written as

$$I = I_s \exp(V_F/V_T) + I_{Ro} \exp(V_F/2V_T) \tag{7.36}$$

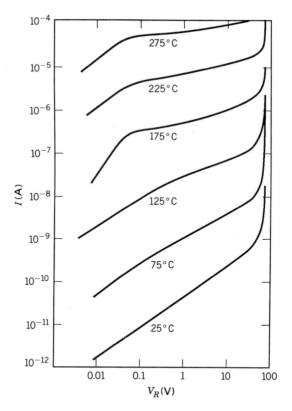

FIGURE 7.5 Measured reverse I-V characteristics of a silicon p-n junction diode at different temperatures. (From A. S. Grove, reference 3, p. 178. Copyright © 1967. Reprinted by permission of John Wiley & Sons, Inc., New York.)

For a Ge diode, the second term remains small compared to the first (except at very low temperatures), and the ideal diode law is obeyed. In Si and GaAs p-n junctions, I_{Ro} is large compared to I_s near room temperature, and the depletion region recombination current dominates the diffusion current. However, since the diffusion current varies as $\exp(V_F/V_T)$, whereas the recombination current varies as $\exp(V_F/2V_T)$, at sufficiently higher values of the bias voltage, the diffusion current becomes higher than the recombination current. Figure 7.6 shows the measured $\log I$ versus V_F characteristics of Si and GaAs p-n junction diodes at room temperature [3]. It is seen that the recombination current dominates at low values of I, and the slope of the characteristic is $1/2V_T$. At higher current levels, the diffusion component dominates and the slope approaches a value $1/V_T$.

The forward current rises exponentially with the applied bias V_F only as long as the IR_s drop in Eq. (7.33) remains negligible compared to the junction voltage V_j. When this drop is significant, the diode current is given by Eq. (7.34). This explains the deviation of the plots in Fig. 7.6 from linearity for high currents in excess of about 5 mA. At high currents, the limit of low-level injection is exceeded, and the series resistance R_s becomes bias dependent.

It is often convenient to add the diffusion and the depletion region generation-recombination currents together in a single expression by writing the diode equation as

$$I = I_o[\exp(V_a/\eta V_T) - 1] \tag{7.37}$$

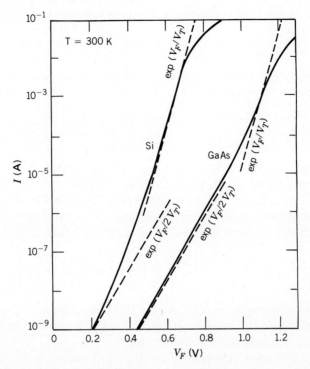

FIGURE 7.6 Measured forward I-V characteristics of silicon and gallium arsenide p-n junction diodes at 300 K. (From A. S. Grove, reference 3, p. 190. Copyright © 1967. Reprinted by permission of John Wiley & Sons, Inc., New York.)

where $I_o = I_s$ when the diode current is dominated by the diffusion process, and $I = I_{Ro}$ when the current is dominated by the depletion region generation-recombination process. The parameter η, known as the *ideality factor*, has a value of 1 for the diffusion current and is approximately 2 for the recombination current. When the two currents are comparable, η lies between 1 and 2. The current I_o can be determined experimentally by extrapolating the forward log I versus V_F plot to zero bias.

7.3 TEMPERATURE DEPENDENCE OF THE I-V CHARACTERISTIC

The temperature dependence of the I-V characteristic is important in many applications of diodes. Two temperature coefficients are commonly associated with a junction diode. The first is that of reverse current, and it is expressed as a fractional change $(1/I)(dI/dT)$ in the diode current. In the case of Ge diodes, the reverse current saturates to a value I_s, giving

$$I = -I_s = -qAn_i^2\left[\frac{D_p}{N_d L_p} + \frac{D_n}{N_a L_n}\right]$$

In this equation D_p, D_n, L_p, and L_n are all temperature dependent, but the temperature dependence of I is dominated by n_i^2, which rises exponentially with temperature. Substituting for n_i^2, and assuming that all other quantities remain invariant with temperature, I_s can be expressed as

$$I_s = A_3 T^3 \exp(-E_{go}/kT) \tag{7.38}$$

where E_{go} is the extrapolated band gap at 0° K and A_3 is a constant of proportionality. Differentiating Eq. (7.38) with respect to T, we obtain the following relation:

$$\frac{1}{I_s}\frac{dI_s}{dT} = \frac{1}{T}\left[3 + \frac{E_{go}}{kT}\right] \tag{7.39}$$

For Ge, $E_{go} = 0.78$ eV, and this equation predicts an 11 percent increase in I_s for each degree rise in temperature near 300 K.

In Si diodes, the reverse current is dominated by the carrier generation in the depletion region and surface leakage. The generation current varies as n_i, and a similar analysis, as was carried out for I_s, shows that

$$\frac{1}{I_{Ro}}\left(\frac{dI_{Ro}}{dT}\right) = \frac{1}{2T}\left[3 + \frac{E_{go}}{kT}\right] \tag{7.40}$$

Substituting $E_{go} = 1.16$ eV, we obtain a change of about 8 percent per degree Kelvin in the reverse current at 300 K. The surface leakage current is generally less sensitive to temperature than I_{Ro}.

The other temperature coefficient of interest is that of the forward voltage at a constant current. Assuming that the current is dominated by the diffusion process and that the forward bias is sufficiently large, we can write

$$I = I_s \exp(qV_F/kT)$$

and

$$\frac{1}{I}\frac{dI}{dT} = \frac{1}{I_s}\frac{dI_s}{dT} - \frac{qV_F}{kT^2} + \frac{q}{kT}\left(\frac{dV_F}{dT}\right)$$

Substituting for $(1/I_s)(dI_s/dT)$ from Eq. (7.39), we can write the above equation as

$$\frac{1}{I}\frac{dI}{dT} = \frac{3}{T} + \frac{q}{kT^2}(V_{go} - V_F) + \frac{q}{kT}\frac{dV_F}{dT} \tag{7.41}$$

A similar expression can be obtained for the depletion region recombination current, but it is not important because the diode is usually operated at currents on the order of 1 mA where the diffusion current dominates the recombination current. The temperature coefficient for the forward voltage for a constant current is obtained by putting $dI/dT = 0$ in Eq. (7.41), giving

$$\frac{dV_F}{dT} = -\left(\frac{3k}{q} + \frac{V_{go} - V_F}{T}\right) \tag{7.42}$$

where V_{go} is the voltage corresponding to the band gap E_{go}. Substituting for I_s from Eq. (7.38), we can express the forward current as

$$I = A_3 T^3 \exp\left[-\frac{q(V_{go} - V_F)}{kT}\right]$$

When this equation is combined with Eq. (7.42), the following relation is obtained [4]:

$$\frac{dV_F}{dT} = -P(T) + G \log_{10} I \tag{7.43}$$

where $P(T)$ is a slowly varying function of temperature and I is expressed in mA. For diodes with a maximum current rating spanning 40 mA to 200 mA, the constant P has a value of 2 mV/ °C near 300 K and $G = 0.198$. These values are valid for both Si and Ge diodes. The above equation gives $dV_F/dT = -1.86$ mV/ °C for a 5 mA current.

7.4 HIGH-LEVEL INJECTION EFFECTS [5]

A p-n junction diode is said to be operating under high-level injection when the injected minority carrier concentration on the low doped side of the junction becomes comparable to the majority carrier concentration. The low-level injection treatment is no longer valid at a high injection level, and the following changes occur:

1. The minority carrier concentration at the depletion region edge is no longer given by Eq. (7.7), and the dependence of the diode current on junction voltage is changed from that given by the ideal diode law.
2. The neglect of electric field in the neutral regions is no longer justified, and the minority carrier current in these regions has both a drift and a diffusion component.
3. Since the majority and minority carrier concentrations in the neutral regions are comparable, the mobility and diffusion constants are different from their values under low-level injection.

To simplify our analysis, we assume a p^+-n junction diode in which the high-level injection is attained on the n-side, while the p-side remains under low-level injection. Since there is an electric field in the neutral n-region, the electron and hole current densities in this region are given by

$$J_p = q\mu_p p_n \mathscr{E} - qD_p \left(\frac{dp_n}{dx}\right) \tag{7.44a}$$

$$J_n = q\mu_n n_n \mathscr{E} + qD_n \left(\frac{dn_n}{dx}\right) \tag{7.44b}$$

Since practically all the current in the depletion region is carried by holes, we assume $J_n = 0$ at $x = x_n$ and

$$\mathscr{E}(x) = -\frac{D_n}{\mu_n}\left(\frac{1}{n_n}\right)\frac{dn_n}{dx}\bigg|_{x=x_n} \tag{7.45}$$

From the quasi-neutrality approximation, we have $p_e \simeq n_e$ and $dp_n/dx \simeq dn_n/dx$. Substituting for $\mathscr{E}(x)$ from Eq. (7.45) into Eq. (7.44a) and eliminating μ_n and

μ_p with the help of the Einstein relation, we obtain

$$J = J_p(x_n) = -qD_p\left(1 + \frac{p_n}{n_n}\right)\frac{dp_n}{dx}\bigg|_{x=x_n} \quad (7.46)$$

where

$$p_n = p_{no} + p_e(x)$$

and

$$n_n = n_{no} + p_e(x) \quad (7.47)$$

If $p_e(x_n)$ remains negligibly small in comparison with n_{no}, the ratio p_n/n_n is practically zero, and Eq. (7.46) reduces to the low-level injection case. However, when $p_e(x_n)$ is large compared to n_{no}, the ratio p_n/n_n is nearly unity and

$$J = -2qD_p\left(\frac{dp_n}{dx}\right)_{x=x_n} \quad (7.48)$$

Thus, at a very high-level injection, the diode current can be expressed as a diffusion current with D_p replaced by $2D_p$.

Let us now examine how the law of junction is modified at high-level injection. We have seen that under low-level injection the pn-product at the depletion region edge is given by Eq. (7.24). This relation is valid for any arbitrary injection level as can be seen from the following discussion. For a nondegenerate semiconductor, the electron and hole concentrations at x_n can be expressed by the relations

$$n_n(x_n) = n_i \exp[(E_{Fn} - E_i)/kT]$$

and

$$p_n(x_n) = n_i \exp[(E_i - E_{Fp})/kT]$$

where E_{Fn} and E_{Fp} represent the quasi-Fermi levels for electrons and holes, respectively, and E_i is the intrinsic Fermi level. Multiplying these two relations yields:

$$p_n(x_n)n_n(x_n) = n_i^2 \exp[(E_{Fn} - E_{Fp})/kT] \quad (7.49)$$

Under thermal equilibrium conditions $E_{Fn} = E_{Fp}$, and the pn-product equals to n_i^2 everywhere. However, when the junction is forward biased by a voltage $V_j = (V_a - IR_s)$, the electron quasi-Fermi level in the neutral n-region will be displaced from the quasi-Fermi level for the holes in the neutral p-region by qV_j. Since the pn-product is constant throughout the depletion region, it follows that $(E_{Fn} - E_{Fp}) = qV_j$ at $x = x_n$ and

$$p_n(x_n)n_n(x_n) = n_i^2 \exp(V_j/V_T) \quad (7.50)$$

At very high-level injection $p_n(x_n) = n_n(x_n)$ which gives

$$p_n(x_n) = n_i \exp(V_j/2V_T) \quad (7.51)$$

As in the case of low-level injection, in steady state we have $dp_n/dx|_{x_n} = -p(x_n)/L_p$. Substituting this value in Eq. (7.48), we can write the diode current as

$$I = AJ = \frac{2qAD_p}{L_p} n_i \exp(V_j/2V_T) \qquad (7.52)$$

Here D_p and L_p will be different from their values under low-level injection because both mobility and lifetime are functions of the majority carrier concentration in the n-region. It is evident from Eq. (7.52) that at high-level injection, the slope of the log I versus junction voltage plot becomes $1/2V_T$ instead of $1/V_T$. This change in slope is difficult to observe experimentally. Because of the presence of the voltage drop across the series resistance, which changes in a nonlinear manner with the current, there is no simple way to determine the voltage V_j at high currents.

7.5 EXAMPLE OF A p-n JUNCTION DIODE

To illustrate the various points discussed in this chapter, we will consider a Si abrupt p-n junction diode with $N_d = 10^{15}$ cm^{-3}, $N_a = 10^{18}$ cm^{-3}, and a diode area $A = 10^{-3}$ cm^2. We assume both the n- and p-regions to be 2×10^{-2} cm long so that the diode can be treated as a long-base diode.

From the plots of mobility versus impurity concentration data, we obtain the values of μ_n and μ_p, and the respective diffusion constants are then calculated using the Einstein relation. Let us take the following values at 300 K (Fig. 4.5).

$$\left. \begin{array}{ll} \mu_p = 480 \text{ cm}^2 \text{ V}^{-1} \text{ sec}^{-1}, & \mu_n = 1280 \text{ cm}^2 \text{ V}^{-1} \text{ sec}^{-1} \\ D_p = 12.4 \text{ cm}^2 \text{ sec}^{-1}, & D_n = 33 \text{ cm}^2 \text{ sec}^{-1} \end{array} \right\} \text{ on n-side}$$

$$\left. \begin{array}{ll} \mu_p = 110 \text{ cm}^2 \text{ V}^{-1} \text{ sec}^{-1}, & \mu_n = 280 \text{ cm}^2 \text{ V}^{-1} \text{ sec}^{-1} \\ D_p = 2.8 \text{ cm}^2 \text{ sec}^{-1}, & D_n = 7.25 \text{ cm}^2 \text{ sec}^{-1} \end{array} \right\} \text{ on p-side}$$

We further assume that $\tau_n = \tau_p = \tau_o = 1 \times 10^{-6}$ sec, so that

$$L_p = \sqrt{D_p \tau_p} = \sqrt{12.4 \times 10^{-6}} = 3.5 \times 10^{-3} \text{ cm on n-side}$$

and

$$L_n = \sqrt{D_n \tau_n} = \sqrt{7.25 \times 10^{-6}} = 2.7 \times 10^{-3} \text{ cm on p-side}$$

(a) CALCULATION OF THE DIODE CURRENT

Let us consider the diode with a reverse bias of 4 V. From Eq. (6.7) we get $V_i \simeq 0.75$ V, and taking $\varepsilon_s = 1.05 \times 10^{-12}$ F/cm for Si, we obtain a depletion region width $W = 2.5 \times 10^{-4}$ cm for a total voltage of 4.75 V. The diode reverse current, from Eq. (7.35), can be written as

$$I = (I_s + I_{Ro})$$

From Eq. (7.17a) we get

$$I_s = 1.6 \times 10^{-19} \times 10^{-3} \times (1.5 \times 10^{10})^2 \left[\frac{12.4}{10^{15} \times 3.5 \times 10^{-3}} + \frac{7.25}{10^{18} \times 2.7 \times 10^{-3}} \right]$$

$$= 3.6 \times 10^{-14} [3.54 + 0.0027] = 1.27 \times 10^{-13} \text{ A}$$

The second term in the bracket represents the electron current, and it is negligible compared to the first term. From Eq. (7.32a), we have $I_{Ro} = 3 \times 10^{-10}$ A, which is three orders of magnitude higher than I_s. If the surface conditions are not ideal, the diode may have a surface leakage current of a few nanoampers which will be the dominant reverse current component. As the temperature is increased gradually above the room temperature, both I_s and I_{Ro} will increase and the two currents will become equal when

$$n_i = \frac{N_d L_p W}{2 D_p \tau_p} = \frac{N_d W}{2 L_p} = \frac{10^{15} \times 2.5 \times 10^{-4}}{2 \times 3.5 \times 10^{-3}} = 3.57 \times 10^{13} \text{ cm}^{-3}$$

This value corresponds to a temperature of about 165°C. Above this temperature, I_s will dominate the other components and the ideal diode equation will be valid.

Under forward bias in excess of 0.1 V, the total current is given by Eq. (7.36), which, after substituting the values of I_s and I_{Ro}, can be written as

$$I = 1.27 \times 10^{-13} \exp(V_F/V_T) + 1.2 \times 10^{-6} W \exp(V_F/2V_T)$$

where W is the function of the applied bias V_F. These two currents are plotted in Fig. 7.7 by dashed lines, and the total current is shown by the solid line. The two

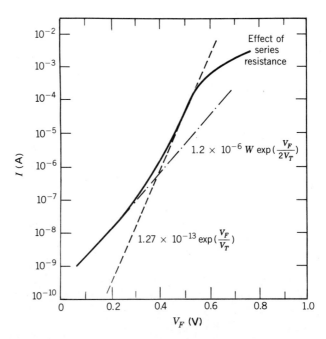

FIGURE 7.7 Calculated forward I-V characteristics of the Si diode of Sec. 7.5.

EXAMPLE OF A p-n JUNCTION DIODE

components became equal in magnitude at $V_F = 0.39$ V. Above this voltage, the diffusion current dominates the depletion region recombination current.

(b) MINORITY CARRIER DISTRIBUTIONS

Let us consider the diode at a forward bias $V_F = 0.5$ V. At this voltage, the depletion region recombination current I_R is negligible compared to the diffusion current, and the total current is low enough to ensure low-level injection. Using Eq. (7.8), we get

$$p_e(x_n) = \frac{n_i^2}{N_d} \exp(V_F/V_T) = \frac{2.25 \times 10^{20}}{10^{15}} \exp\left(\frac{0.5}{0.02586}\right) = 5.6 \times 10^{13} \text{ cm}^{-3}$$

and from Eq. (7.21)

$$p_e(y) = 5.6 \times 10^{13} \exp\left(-\frac{10^3}{3.5}y\right)$$

which is plotted in Fig. 7.8a. The excess electron concentration in the quasi-neutral p-region is obtained as

$$n_e(y') = 5.6 \times 10^{10} \exp\left(-\frac{10^3}{2.7}y'\right)$$

which is negligibly small compared to $p_e(y)$ and has been ignored in Fig. 7.8a.

(c) MINORITY AND MAJORITY CARRIER CURRENTS

From the excess hole distribution discussed above, we obtain

$$\frac{dp_e}{dy} = -\frac{5.6 \times 10^{13}}{3.5 \times 10^{-3}} \exp\left(-\frac{10^3}{3.5}y\right) = -1.6 \times 10^{16} \exp\left(-\frac{10^3}{3.5}y\right)$$

and

$$I_p(y) = -qAD_p\frac{dp_e}{dy} = 1.6 \times 10^{-19} \times 10^{-3} \times 12.4 \times 1.6 \times 10^{16} \exp\left(-\frac{10^3 y}{3.5}\right)$$

$$= 3.17 \times 10^{-5} \exp\left(-\frac{10^3}{3.5}y\right) \text{ A}$$

This current has its maximum value of 3.17×10^{-5} A at $y = 0$. The diffusion current of electrons on the p-side is similarly obtained from the slope of $n_e(y')$ giving

$$I_n(y') = 1.6 \times 10^{-19} \times 10^{-3} \times 7.25 \times 2.07 \times 10^{13} \exp\left(-\frac{10^3}{2.7}y'\right)$$

$$= 2.4 \times 10^{-8} \exp\left(-\frac{10^3}{2.7}y'\right) \text{ A}$$

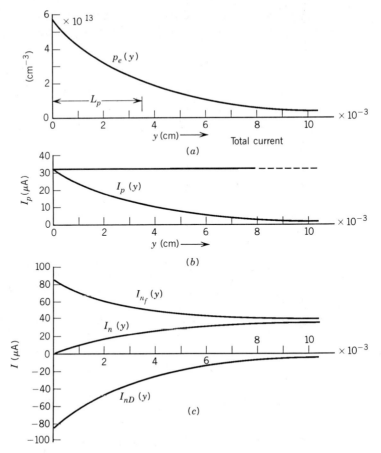

FIGURE 7.8 (a) Minority carrier distribution, (b) hole, and (c) electron currents in the neutral n-region of the p-n junction diode of Sec. 7.5.

which has a maximum value of 2.4×10^{-8} A. Thus, the total diode current is given by

$$I = I_p(0) + I_n(0) \simeq 3.17 \times 10^{-5} + 2.4 \times 10^{-8} \simeq 3.17 \times 10^{-5} \text{ A}$$

As explained in Sec. 7.1.4, the electron current in the quasi-neutral n-region is obtained by subtracting the hole current $I_p(y)$ from the total current I. Thus

$$I_n(y) = I - I_p(y) = 3.17 \times 10^{-5}\left[1 - \exp\left(-\frac{10^3}{3.5}y\right)\right]$$

This current has a diffusion and a drift component. The diffusion component $I_{nD}(y)$ is given by the relation $I_{nD}(y) = qAD_n(dn_n/dy)$. From the quasi-neutrality assumption, we have $dn_n/dy = dp_e/dy$, and since $D_n = 33$ cm^2 sec^{-1} in the n-region, we obtain

$$I_{nD}(y) = -1.6 \times 10^{-19} \times 10^{-3} \times 33 \times 1.6 \times 10^{16} \exp\left(-\frac{10^3}{3.5}y\right)$$

$$= -8.45 \times 10^{-5} \exp(-10^3 y/3.5) \text{ A}$$

EXAMPLE OF A p-n JUNCTION DIODE

The drift component $I_{nf}(y)$ of the electron current in the n-region is obtained by subtracting the diffusion current $I_{nD}(y)$ from the total current $I_n(y)$. Thus, we have

$$I_{nf}(y) = I_n(y) - I_{nD}(y)$$
$$= 3.17 \times 10^{-5}[1 - \exp(-10^3 y/3.5)] + 8.45 \times 10^{-5} \exp(-10^3 y/3.5)$$
$$= 3.17 \times 10^{-5}[1 + 1.67 \exp(-10^3 y/3.5)] \text{ A}$$

The current components $I_p(y)$, $I_{nD}(y)$, $I_n(y)$, and $I_{nf}(y)$ are plotted in Figs. 7.8b and c.

(d) VALIDITY OF APPROXIMATIONS

In the derivation of the ideal diode equation, we have assumed the electric field in the quasi-neutral regions to be sufficiently low to ensure that (1) the voltage drops across these regions are negligibly small; (2) the minority carrier drift currents are negligible; and (3) the space-charge neutrality is maintained. We will now examine the validity of these assumptions.

The electric field $\mathscr{E}(y)$ in the quasi-neutral n-region is obtained from the relation $I_{nf}(y) = qA\mu_n n_n(y)\mathscr{E}(y)$. Since the variation in $n_n(y)$ is negligibly small, we take $n_n(y) = N_d$, $\mu_n = 1280$ cm^2 V^{-1} sec^{-1} and obtain

$$\mathscr{E}(y) = \frac{I_{nf}(y)}{qAN_d\mu_n} = 0.15[1 + 1.67 \exp(-10^3 y/3.5)] \text{ V/cm}$$

The potential drop in the quasi-neutral n-region is obtained by integrating $\mathscr{E}(y)$ from $y = 0$ to $y = 2.0 \times 10^{-2}$ cm, and it is found to be less than 3 mV. Since the minority carrier current in the p-region is negligibly small, the voltage drop across this region can be obtained by multiplying its resistance with the diode current. This region has a resistivity of $(qN_a\mu_p)^{-1} = (1.6 \times 10^{-19} \times 10^{18} \times 110)^{-1} = 0.057$ Ω-cm. Hence, its resistance is

$$\frac{2 \times 10^{-2}}{10^{-3}} \times 0.057 = 1.14 \text{ }\Omega$$

and the voltage drop across this region is $1.14 \times 3.17 \times 10^{-5} = 0.036$ mV. Thus, the voltage appearing across the neutral base regions is negligibly small compared to 0.5 V, and practically all the applied voltage appears across the depletion region.

The presence of the electric field in the neutral n-region causes a drift current of holes $I_{pf}(y) = qA\mu_p p_e(y)\mathscr{E}(y)$. This current has its maximum value at $y = 0$. That is

$$I_{pf}(0) = 1.6 \times 10^{-19} \times 10^{-3} \times 480 \times 5.6 \times 10^{13} \times 0.15 \times 2.67 = 1.72 \times 10^{-6} \text{ A}$$

which is less than 6 percent of the hole diffusion current at this point. Hence, the assumption that the hole current in the n-region flows by diffusion is approximately correct.

Finally, the assumption of space-charge neutrality can be checked by using the Poisson equation ($\rho = \varepsilon_s (d\mathscr{E}/dy)|_{y=0}$). Thus

$$\rho(y=0) = \frac{1.05 \times 10^{-12} \times 1.67 \times 0.15}{3.5 \times 10^{-3}} = 7.51 \times 10^{-11} \text{ coulomb cm}^{-3}.$$

This space-charge will cause an imbalance of 4.69×10^8 electronic charges cm^{-3} in the electron concentration. Since the equilibrium electron concentration is 10^{15} cm^{-3}, only one out of 2.13×10^6 electrons needs to be removed to produce the space charge at $y = 0$. Therefore, the assumption of space-charge neutrality is well justified.

(e) DIODE BEHAVIOR AT HIGH CURRENTS

The resistance of the n-region under low-level injection is obtained by

$$\frac{W_n}{q\mu_n N_d A} = \frac{2 \times 10^{-2}}{1.6 \times 10^{-19} \times 1280 \times 10^{15} \times 10^{-3}} = 97.65 \text{ } \Omega.$$

Since the resistance of the p-region is 1.14 Ω, we take $R_s = 98.8$ Ω. At low currents, the voltage drop IR_s remains negligible compared to the junction voltage V_j. However, when I exceeds 0.2 mA, the IR_s drop becomes significant, and the I-V characteristic in Fig. 7.7 starts deviating from the straight line. The transition from low- to high-level injection occurs when $p_e(0) = N_d$, and this happens at $V_j = 0.575$ V. The diode current at this value of V_j is 0.56 mA, and the resulting IR_s drop is 0.055 V, making $V_a = V_j + IR_s = 0.63$ V. As the current is increased further, the electron concentration in the neutral n-region increases above n_{no}, and its resistance is decreased causing a reduction in R_s.

Finally, let us consider the situation in which $p_e(0) = 10^{16}$ cm^{-3}. The diode is now operating under high-level injection. From Eq. (7.51), we obtain $V_j = 0.69$ V. Substituting this value of V_j in Eq. (7.52) we get $I = 10$ mA. The total voltage drop V_a will now be significantly higher than V_j because of the large IR_s drop. In the example we have considered, R_s is rather high. In many diodes R_s is only a few ohms.

REFERENCES

1. W. SHOCKLEY, *Electrons and Holes in Semiconductors,* Van Nostrand, Princeton, N.J., 1950.

2. J. L. MOLL, S. KRAKAUER, and R. SHEN, "P-N junction charge-storage diodes," *Proc. IRE.* 50, 43 (1962).

3. A. S. GROVE, *Physics and Technology of Semiconductor Devices,* John Wiley, New York, 1967.

4. H. J. SIMPSON, "Temperature dependence of forward characteristics of p-n junction diodes," *SCP and Solid State Technology* 7, 22 (Sept. 1964) and 36 (Oct. 1964).

5. A. NUSSBAUM, "The theory of semiconducting junctions," Chapter 2 in *Semiconductors and Semimetals,* R. K. WILLARDSON and A. C. BEER (eds.), Academic Press, New York, Vol. 15, 39 (1981).

ADDITIONAL READING LIST

1. J. L. MOLL, *Physics of Semiconductors,* McGraw-Hill, New York, 1964, Chapter 7.
2. M. J. O. STRUTT, *Semiconductor Devices,* Academic Press, New York, 1966.
3. S. K. GHANDHI, *Semiconductor Power Devices,* John Wiley, New York, 1977, Chapters 2 and 3.
4. R. S. MULLER and T. I. KAMINS, *Device Electronics for Integrated Circuits,* 2d ed., John Wiley, New York, 1986, Chapter 5.
5. E. S. YANG, *Microelectronic Devices,* McGraw-Hill, New York, 1988.

PROBLEMS

7.1 Using the procedure described in Sec. 7.1.2, obtain (7.11).

7.2 A long-base Si abrupt p-n junction diode with a junction area of 10^{-2} cm^{-2} has $N_d = 10^{18}$ cm^{-3}, $N_a = 10^{17}$ cm^{-3}, $\tau_p = 10^{-8}$ sec, $\tau_n = 10^{-6}$ sec, $D_p = 5.2$ cm^2 sec^{-1}, and $D_n = 20$ cm^2 sec^{-1}. At 300 K, calculate the diode current with a forward bias of 0.5 V, and then with a reverse bias of 5 V. Include the generation-recombination current and assume $\tau_o = 10^{-7}$ sec.

7.3 In a long-base diode, show that the probability for hole recombination in the neutral n-region within the interval y and $(y + \Delta y)$ is $\exp(-y/L_p)\Delta y/L_p$. From this result show that the average distance that a hole diffuses before recombining is L_p.

7.4 The end contact on the n-siode of a p$^+$-n junction short-base diode is characterized by the surface recombination velocity $J_p(W_1) = qSp_e(W_1)$. Determine the expression for the saturation current I_s. Discuss the cases for $S = 0$ and $S = \infty$.

7.5 Show that the saturation current I_s in a long-base diode is due to a collection of minority carriers generated within one diffusion length away from the depletion region edge on each side.

7.6 Show that the depletion region width in a forward-biased, short-base p$^+$-n junction diode is given by

$$W = \left[\frac{2\varepsilon_s V_T}{qN_d} \ln \frac{qAN_a D_p}{W_n I}\right]^{1/2}$$

Calculate the value of W for a Ge diode for a forward current of 1 mA with the following parameters: $N_d = 10^{16}$ cm^{-3}, $N_a = 2.4 \times 10^{18}$ cm^{-3}, $D_p = 44$ cm^2 sec^{-1}, $\tau_p = 10^{-5}$ sec, and $A = 2 \times 10^{-4}$ cm^2. Assume $W_n = 0.1\, L_p$ and $T = 300$ K.

7.7 A long-base Ge p-n junction diode has an abrupt junction with uniformly doped regions. The p-side has a resistivity of 1 Ω-cm and the n-side has a resistivity of 0.2 Ω-cm. (a) Calculate the concentrations of minority carriers at the edges of the depletion region with a forward bias of 0.207 V, and sketch the majority and minority carrier current densities as functions of distance from the edges of the depletion region on each side of the junction. (b) Calculate the locations of the planes at which the majority and minority carrier currents are equal in magnitude. Assume $\tau_p = 10^{-7}$ sec, $\tau_n = 10^{-5}$ sec, and $T = 300\,\text{K}$ in your calculations.

7.8 Sketch the electron and hole currents as a function of position in a forward-biased Si p-n junction diode with $N_d = 5 \times 10^{16}$ cm^{-3} and $\tau_p = 10^{-6}$ sec on the n-side and $N_a = 10^{17}$ cm^{-3} and $\tau_n = 6 \times 10^{-7}$ sec on the p-side. The length of the neutral region on each side is half of the minority carrier diffusion length. Assume $T = 300\,\text{K}$ and a forward bias of 15 V_T. Neglect carrier recombination in the depletion region. Repeat the calculations for a reverse bias of 2 V.

7.9 A Ge p-n junction diode has $N_d = 2 \times 10^{16}$ cm^{-3} on the n-side and $N_a = 3 \times 10^{19}$ cm^{-3} on the p-side. Calculate the forward voltage at which the injected hole concentration at the edge of depletion region on the n-side becomes equal to the majority carrier concentration. Assuming $T = 300\,\text{K}$, $D_p = 42$ cm^2 sec^{-1}, and $\tau_p = 3 \times 10^{-7}$ sec, calculate the current density at this voltage and compare it with the thermal equilibrium diffusion current density.

7.10 The temperature dependence of the reverse saturation current of a p-n junction diode is given by the relation

$$I_s(T) = I_s(T_o) \exp[0.057(T - T_o)]$$

where $T_o = 300\,\text{K}$. Determine the temperature at which the current becomes twice its value at 300 K and estimate the band gap of the semiconductor.

7.11 A Si diode forward biased at a fixed current is being used as a thermometer. The diode voltage is 0.62 V at 300 K. (a) How much will this voltage change when the temperature is varied by ± 2 percent? (b) Estimate the diode voltage at 50° C.

7.12 Two ideal p-n junction diodes are connected in series across a 1 V battery such that both of them are forward biased. One diode has $I_s = 10^{-5}$ A and another $I_s = 10^{-8}$ A. Calculate the current through the circuit and voltage drop across each diode at 300 K.

7.13 A Ge p-n junction diode has a reverse saturation current of 10^{-6} A at 300 K. A parallel combination of two such diodes is connected to a 1 V battery in series with a 50-Ω resistance such that both the diodes are forward biased. Calculate the current through the circuit and voltage drop across each diode. Use graphical means if necessary.

7.14 A p-n junction diode with $I_s = 10^{-6}$ A at 300 K is forward biased in series with a 60-Ω resistor. Plot the I-V characteristic of the combination on I versus V_F linear plot when V_F is varied from 0 to 1.2 V.

8
ELECTRICAL BREAKDOWN IN p-n JUNCTIONS

INTRODUCTION

In the previous chapter we saw that the ideal diode theory predicts a small, voltage-independent saturation current I_s, when the junction is sufficiently reverse biased. In junction diodes where the reverse current is dominated by a carrier generation in the space-charge region or surface leakage, the current does not saturate but is found to increase with the reverse bias. However, this increase is too small to cause any perceptible change in the rectifying properties of the diode. Even in junctions in which the space-charge generation and surface leakage currents are negligible, it is observed that the reverse current increases gradually with the bias until a critical voltage is reached at which the current increases abruptly with the voltage. This phenomenon is known as *breakdown of the p-n junction*.

Figure 8.1 shows the I-V characteristic of a typical p-n junction diode including the reverse breakdown region. Two features of the breakdown are noteworthy. First, the breakdown is only electrical, and there is no mechanical damage to the diode. Second, the breakdown is reversible and nondestructive as long as the power dissipation at the junction is kept below the value allowed by thermal considerations.

There are two basically different mechanisms of junction breakdown: the Zener mechanism (also known as internal field emission), due to tunneling, and avalanche breakdown. The Zener mechanism involves the direct excitation of electrons from the valence band to the conduction band under the action of the high electric field in the junction depletion region, and it occurs only in p-n junctions whose two sides are heavily doped so that the width of the depletion layer is small. Avalanche breakdown is caused by a process known as secondary multiplication, and it occurs in junctions having thicker depletion regions, which makes tunneling less probable.

In this chapter, a brief account of the two breakdown mechanisms is presented. Methods to calculate the breakdown voltages for the abrupt and the lin-

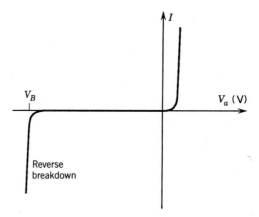

FIGURE 8.1 Current-voltage characteristic of a p-n junction diode showing reverse breakdown.

early graded junctions are given, and the effects of temperature and junction imperfections on the breakdown voltage are qualitatively described. Finally, applications of junction diodes in the breakdown region are mentioned.

8.1 PHENOMENOLOGICAL DESCRIPTION OF BREAKDOWN MECHANISMS [1]

For a qualitative understanding of the breakdown mechanisms in p-n junctions, it is helpful to consider the relationships between the impurity concentration, the depletion region width, and the electric field in the junction region. For simplicity, let us consider an abrupt p^+-n junction. Neglecting the depletion width in the p^+-region, the following relations are obtained

$$W^* = \sqrt{\frac{2\varepsilon_s}{qN_d}}$$

$$W = W^*\sqrt{(V_i + V_R)} \quad \text{and} \quad |\mathscr{E}_m| = \frac{2\sqrt{(V_i + V_R)}}{W^*} \tag{8.1}$$

where $V_R = -V_a$ is the applied reverse voltage. It is possible to understand the breakdown mechanisms with the help of these relations.

(a) INTERNAL FIELD EMISSION OR ZENER BREAKDOWN

Breakdown because of internal field emission is called *Zener breakdown* after C. Zener who initially proposed the idea in 1934. Equation (8.1) shows that when N_d is large (on the order of 10^{18} cm^{-3} in Si and Ge), W^* is small and $\mathscr{E}_m$ is high even for small voltages. Under these conditions, the depletion layer width W is small, and the electrons from the valence band on the p-side of the junction can tunnel into the empty states in the conduction band on the n-side as shown in Fig 8.2a. If W^* and V_R are such that $\mathscr{E}_m$ exceeds 10^6 V/cm, the tunnel current becomes sufficiently high to cause the junction breakdown.

(b) AVALANCHE BREAKDOWN

As the impurity concentration N_d is decreased, both W^* and W increase, and $\mathscr{E}_m$ decreases. A point is reached where $\mathscr{E}_m$ has a value smaller than that required for

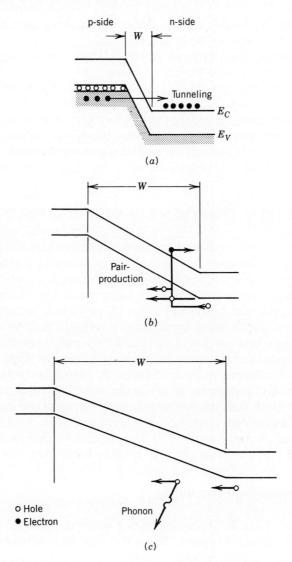

FIGURE 8.2 Breakdown mechanisms in p-n junction diodes. (a) Tunneling, (b) avalanche multiplication, and (c) carriers lose the gained energy to phonons and remain too cold to ionize. (From M. S. Tyagi, reference 1, p. 101. Copyright © 1968. Reprinted by permission of Pergamon Press, Oxford.)

tunneling. In a reverse-biased junction, the thermally generated minority carriers (mainly holes in a p^+-n junction) injected into the space-charge region are accelerated by the field and gain kinetic energy from the field. When the junction voltage is large enough to produce a sufficiently high field, some of the accelerated carriers gain energy higher than the gap energy E_g. A carrier that has energy appreciably higher than E_g can produce a secondary electron–hole pair by knocking out an electron from the covalent bond of a lattice atom in the depletion region, as shown in Fig. 8.2b. The secondary carriers produced by this process can produce further electron–hole pairs. Avalanche breakdown occurs when the number of carriers produced by this ionization process becomes very large.

For an accelerated carrier, the ionization collision is not the only event by which the gained energy can be lost. A carrier may also lose its energy to the crystal lattice by emitting phonons. This energy loss to phonons obviously does not contribute to the ionization process.

Besides the above two processes, there is also a possibility of breakdown occurring by punch-through as shown in Fig. 8.2c. When N_d is very low, both W^* and W are large and $\mathscr{E}_m$ becomes very low. For Si and Ge p-n junctions, if $\mathscr{E}_m$ remains lower than 10^5 V/cm, then the accelerated carriers lose their excess energy by emitting phonons and cannot gain enough energy to produce secondary electron–hole pairs. In such a case breakdown occurs when the space-charge region extends through the neutral region on the n-side up to the ohmic contact. This has already been discussed in Sec. 7.1 for short-base diodes.

8.2 THEORETICAL TREATMENT OF INTERNAL FIELD EMISSION [2]

Tunneling is a quantum mechanical phenomenon that has no classical analogue. Consider an electron with energy E incident on a potential barrier of height V_o as shown in Fig. 8.3. If one treats the electron as a particle and applies the laws of classical mechanics to describe its motion, it then becomes clear that the electron can pass from region I to region III only when its energy E is higher than V_o. If this is not the case, the electron is reflected back to region I after striking the barrier. However, when we consider the wave nature of the electron and employ the results of quantum mechanics, we find that there exists a finite probability for the electron to tunnel through the barrier into region III even when E is less than V_o. Calculations show that the tunneling probability is significant only when the width W_t of the potential barrier is small in comparison with the wavelength of the incident electron. For thermal electrons at room temperature, the wavelength is on the order of 10^{-4} cm, so that for significant tunneling to occur, the width W_t should be orders of magnitude less than this value.

8.2.1 The Tunneling Probability

When large fields are applied to a semiconductor, the energy band diagram may be represented as shown in Fig 8.4. The slope of the band edges in this figure is

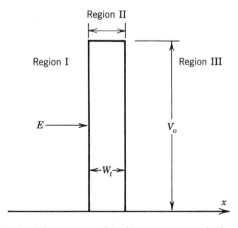

FIGURE 8.3 Electron with an energy E incident on a potential barrier of height V_o.

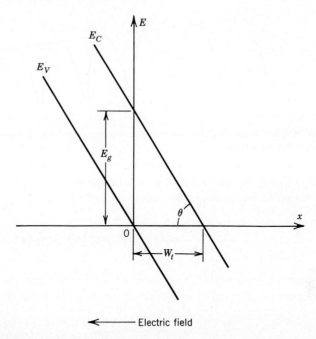

FIGURE 8.4 Electron energy band diagram for a semiconductor in the presence of a high electric field.

a measure of the applied electric field, and the width W_t of the potential barrier is given by

$$q\mathscr{E} = -\frac{E_g}{W_t}$$

or

$$W_t = \frac{V_g}{|\mathscr{E}|} \qquad (8.2)$$

where V_g represents the voltage corresponding to the gap energy E_g and $\mathscr{E}$ is the electric field. For Si with $E_g = 1.12$ eV, the tunnel current becomes significant only when $|\mathscr{E}| \simeq 10^6$ V/cm and W_t is on the order of 10^{-6} cm.

The fields that cause electron tunneling are much lower than those produced by the interatomic forces of lattice atoms. Thus, the applied field may be assumed to produce only a small perturbation on the crystal periodic potential as shown in Fig. 8.5. Zener concluded that for these small perturbations, the wave function ψ of an electron can be described by the equation [3]

$$-\frac{\hbar^2}{2m^*} \cdot \frac{d^2\psi}{dx^2} + [qV(x) - q\mathscr{E}x]\psi = E\psi \qquad (8.3)$$

where $V(x)$ is the electron potential due to lattice forces, and $q\mathscr{E}x$ is the energy due to the presence of the electric field. For a field strength of 10^6 V/cm, the potential change over a lattice spacing is only on the order of 10^{-2} V/cm and is negligible in comparison with $V(x)$. Hence, the term $q\mathscr{E}x$ in Eq. (8.3) may be treated as practi-

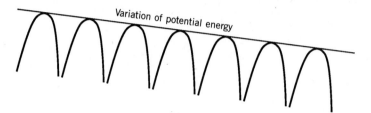

FIGURE 8.5 Tunneling field shown as producing a small perturbation on the periodic potential of the crystal lattice.

cally invariant over a distance of a lattice constant. Under these conditions, the solution of Eq. (8.3) can be approximated by

$$\psi(x) = u_k(x) \exp\left[j \int_0^x k(x)\,dx\right] \tag{8.4}$$

where the wave vector $k(x)$ is now a function of x because it depends in the same manner on $(E + q\mathscr{E}x)$ as k depends on E in the force-free case. Thus

$$[k(x)]^2 = \frac{2m^*}{\hbar^2}[E + q\mathscr{E}x - V(x)]$$

$$= \frac{2m^*}{\hbar^2}[E'(x) - V(x)] \tag{8.5}$$

As long as the energy $E'(x)$ falls in the allowed band, the wave vector $k(x)$ is real and has the character of a lattice modulated wave with $k(x)$ varying slowly in space. This will be the case up to the point x_V in Fig. 8.6, which represents the upper edge of the valence band. When $E'(x)$ increases above its value at x_V, it enters the forbidden band in which $k(x)$ becomes imaginary and the resulting wave function decays exponentially. As $E'(x)$ is gradually increased, a point x_C is

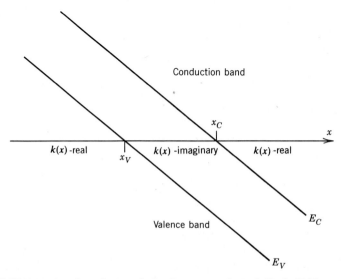

FIGURE 8.6 Wave vector of an electron during its passage through the forbidden energy band.

reached at which the electron enters the conduction band. Above this point $k(x)$ again becomes real and assumes the character of a lattice-modulated wave.

The probability of tunneling of electrons P_t through the forbidden gap is given by the ratio of the squares of the amplitudes of the wave function at the two points x_C and x_V. Thus,

$$P_t = \frac{|\psi(x_C)|^2}{|\psi(x_V)|^2} \tag{8.6}$$

Substituting for $\psi(x_C)$ and $\psi(x_V)$ from Eq. (8.4), we can write the above equation as

$$P_t = \left(\frac{|u_k(x_C)| \exp\left[j \int_0^{x_C} k(x)\,dx\right]}{|u_k(x_V)| \exp\left[j \int_0^{x_V} k(x)\,dx\right]} \right)^2 \tag{8.7}$$

Zener assumed that the lattice periodic factor did not influence the value of P_t significantly, and could be omitted, giving

$$P_t = \exp\left[2j \int_{x_V}^{x_C} k(x)\,dx\right] \tag{8.8}$$

In the forbidden band, $k(x) = jf(x)$ is imaginary, so that P_t becomes

$$P_t = \exp\left[-2 \int_{x_V}^{x_C} f(x)\,dx\right] = \exp\left[-2 \int_{x_V}^{x_C} |k(x)|\,dx\right] \tag{8.9}$$

where $|k(x)|$ is the absolute value of $k(x)$ in the forbidden band. The points x_V and x_C are the so-called classical turning points. At these points the energy $E'(x) - V(x) = 0$. Although Eq. (8.9) is based on several simplifying assumptions, more involved calculations using the *WKB* (Wentzel, Kramers, and Brillouin) *method* [2] lead to the same result as in Eq. (8.9).

The detailed form of the potential barrier in the forbidden gap of a semiconductor is seldom known accurately. However, it turns out that the tunneling probability is not very sensitive to the actual form of the potential barrier. The two most commonly used forms of potential barriers, namely, the triangular and the parabolic barriers are shown in Fig. 8.7. For both these barrier shapes, the tunneling probability may be written as

$$P_t = \exp\left[-\frac{\beta^* E_g^{3/2}}{|\mathscr{E}|}\right] \tag{8.10}$$

with

$$\beta^* = \frac{4\sqrt{2m^*}}{3q\hbar} \tag{8.11}$$

for the triangular barrier and

$$\beta^* = \frac{\pi\sqrt{m^*}}{2\sqrt{2}\,q\hbar} \tag{8.12}$$

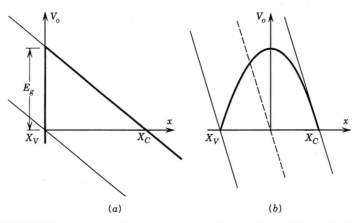

FIGURE 8.7 Two forms of potential barrier: (a) triangular and (b) parabolic.

for the parabolic barrier. In these expressions, q represents the electronic charge, and m^* is the average effective mass of a tunneling electron, which is given by

$$\frac{1}{m^*} = \frac{1}{2}\left(\frac{1}{m_e^*} + \frac{1}{m_h^*}\right) \tag{8.13}$$

where m_e^* and m_h^* represent the density of states, effective masses of the electron, and the hole, respectively. Tunnel currents are more readily observed in abrupt p-n junctions than in the linearly graded junctions. The potential distribution in the depletion region of an abrupt junction is parabolic, and the value of β^* given by Eq. (8.12) has been commonly used to analyze the tunnel current.

Let us estimate P_t for a Si abrupt p-n junction with $(V_i + V_R) = 2.25$ V and $W = 300$ Å. The maximum field in the depletion region is

$$\mathcal{E}_m = \frac{2 \times 2.25}{300 \times 10^{-8}} = 1.5 \times 10^6 \text{ V/cm}$$

From Table 4.2 we have $m_e^* = 1.18\, m_o$ and $m_h^* = 0.81\, m_o$ giving $m^* = 0.96\, m_o$. From Eq. (8.12)

$$\beta^* = \frac{3.14 \times (0.96 \times 9.1 \times 10^{-31})^{1/2}}{2\sqrt{2} \times 1.6 \times 10^{-19} \times 1.055 \times 10^{-34}} = 6.15 \times 10^{37} \text{ V cm}^{-1}(eV)^{-3/2}$$

and

$$\beta^* E_g^{3/2} = 6.15 \times 10^{37} (1.12 \times 1.6 \times 10^{-19})^{3/2} = 4.66 \times 10^9 \text{ V/m} = 4.66 \times 10^7 \text{ V/cm}$$

Taking $\mathcal{E} = \mathcal{E}_m$ in Eq. (8.10), we get

$$P_t = \exp\left(-\frac{4.66 \times 10^7}{1.5 \times 10^6}\right) = 3.22 \times 10^{-14}$$

Thus, it is clear that P_t is very small, and only an insignificantly small fraction of electrons from the valence band can tunnel into the conduction band.

8.3 ZENER BREAKDOWN IN p-n JUNCTIONS

The tunnel current in a p-n junction can be expressed by

$$I = A^* \mathcal{E}^{3/2} D(V) P_t \tag{8.14}$$

where A^* is a constant of proportionality. In general, the density of states function $D(V)$, is too complicated to be described by a simple expression. However, for tunneling in a reverse-biased junction, $D(V)$ is proportional to the reverse voltage V_R and Eq. (8.14) becomes

$$I = A' V_R \mathcal{E}^{3/2} \exp\left(-\frac{\beta^* E_g^{3/2}}{\mathcal{E}}\right) \tag{8.15}$$

This expression will be used to study the Zener current in an abrupt and a linearly graded p-n junction.

8.3.1 Abrupt p-n Junction

The electric field in the depletion region of an abrupt p-n junction varies with position, and the average field $\overline{\mathcal{E}}$ is given by

$$|\overline{\mathcal{E}}| = \left(\frac{V_i + V_R}{W}\right) = \frac{1}{W^*}\sqrt{(V_i + V_R)} \tag{8.16}$$

In a p-n junction where both sides are doped to degeneracy, the voltage V_i can be approximated by [1]

$$V_i = V_g + 0.6(\Phi_n + \Phi_p) \tag{8.17}$$

where $q\Phi_n = (E_F - E_C)$ and $q\Phi_p = (E_V - E_F)$. In a degenerate semiconductor, Φ_n is approximately given by

$$\Phi_n = \frac{kT}{q}\left[\ln\frac{N_d}{N_C} + 0.35\left(\frac{N_d}{N_C}\right)\right] \tag{8.17a}$$

and a similar relation can be written for Φ_p.

The correct field that one should substitute in Eq. (8.15) to obtain the measured current is smaller than the maximum field $\mathcal{E}_m$ but is larger than the average field $\overline{\mathcal{E}}$. When the applied reverse bias is small compared to V_i, it is a good approximation to substitute $\mathcal{E} = (2/3)\mathcal{E}_m$ in Eq. (8.15). However, when the applied reverse bias is larger than V_i, a better approximation is $\mathcal{E} = \mathcal{E}_m$.

In Eq. (8.15), most of the variation of I with respect to $\mathcal{E}$ is contained in the exponential factor, and after substituting $\mathcal{E} = \mathcal{E}_m$, it becomes obvious that a plot of $\log(I/V_R)$ versus $(V_i + V_R)^{-1/2}$ should give a straight line whose slope can be used to determine the value of β^*, and hence the effective mass m^*. Such plots for two heavily doped germanium p^+-n junctions of different dopings are shown in Fig. 8.8. It is seen that, for an applied voltage as large as 2.8 V, the measured current is in good agreement with Eq. (8.15). However, for larger values of reverse bias, the current becomes larger than that predicted by Eq. (8.15) and the mea-

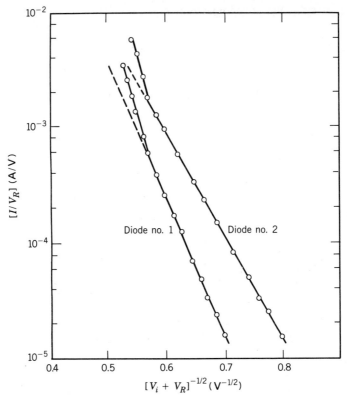

FIGURE 8.8 (I/V_R) plotted as a function of $(V_i + V_R)^{-1/2}$ for two Ge p-n junction diodes.

sured points deviate from the straight line. A similar deviation is observed in the case of heavily doped silicon p-n junctions when the applied bias becomes larger than about $4V_g$. This increase in current is caused by the avalanche multiplication of the field-emitted electrons in the high field of the junction depletion region. Thus, it is clear that true Zener breakdown is observed only for a breakdown voltage less than about $4V_g$.

8.3.2 Linearly Graded Junctions

The maximum and the average fields in a linearly graded junction at a reverse bias V_R are given by

$$|\mathscr{E}_m| = \frac{3}{2}\left(\frac{V_i + V_R}{W}\right)$$

and

$$|\overline{\mathscr{E}}| = \frac{(V_i + V_R)}{W} \tag{8.18}$$

where

$$V_i \simeq V_g$$

It is obvious that the difference between the average and the maximum field does not cause a large change in the exponent of Eq. (8.15) for a linearly graded junction as was the case for an abrupt p-n junction. For a linearly graded junction in Si, appreciable tunneling is observed when the grading constant is on the order of 10^{26} to 10^{27} cm^{-4}. However, tunneling can occur more readily in an abrupt p-n junction than in a linearly graded junction because the depletion layer width in the abrupt p-n junction can be reduced by heavily doping the two sides of the junction. When both the n- and p-sides of the junction are doped to degeneracy, the breakdown voltage is reduced to zero, and at still higher dopings, tunneling also occurs in a forward-biased junction. These effects are considered in Chapter 10.

The reverse I-V characteristics of a p-n junction in which breakdown occurs by tunneling are shown in Fig. 8.9 at two temperatures. It is seen that the current rises slowly with the voltage, and a sharp breakdown point, the knee point, is not observed. Such characteristics are known as *soft characteristics.* In this case, the breakdown voltage is customarily defined as the voltage corresponding to a reverse current of 5 mA.

8.3.3 Temperature Dependence of the Zener Breakdown Voltage

For p-n junctions in which breakdown occurs by the Zener mechanism, the breakdown voltage decreases with temperature (Fig. 8.9). A quantity γ_T, called the *temperature coefficient of breakdown voltage,* is defined by the relation

$$\gamma_T = \frac{1}{V_{Bo}}\left(\frac{dV_B}{dT}\right) \tag{8.19}$$

where V_{Bo} is the breakdown voltage at a reference temperature (e.g., 300 K). Obviously, γ_T is negative for junctions showing the Zener breakdown.

The decrease in Zener breakdown voltage with temperature can be related to a decrease in the band gap E_g. In Eq. (8.15), the voltage dependence of I is largely

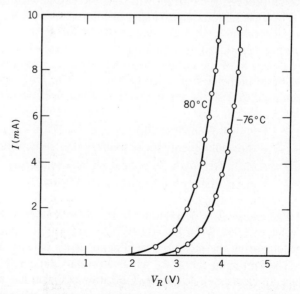

FIGURE 8.9 Measured reverse characteristics of a silicon p-n junction with Zener breakdown.

contained in the exponential factor. Thus, to maintain a constant current, the term $\beta^* E_g^{3/2}/\mathscr{E}$ must remain invariant with temperature. Since E_g decreases with temperature, to keep this term constant, $\mathscr{E}$ must be decreased by reducing the reverse bias across the junction (see Prob. 8.5).

8.4 SECONDARY MULTIPLICATION IN SEMICONDUCTORS [2]

In our discussion of nonlinear conductivity in Sec. 4.6, we saw that, when a semiconductor is subjected to an increasing electric field, a point is reached where the electron and hole drift velocities approach their saturation values. Any further increase in the electric field above this value simply increases the random thermal velocity of carriers and, hence, causes an increase in their average kinetic energy. Since the kinetic energy is a measure of temperature, this means that, in the presence of a high electric field, electrons become "hot" in the sense that their temperature becomes considerably higher than the lattice temperature. If these hot electrons are assumed to have a Maxwellian distribution, then their average thermal velocity $\bar{v}$ can be written as (see Sec. 1.7)

$$\bar{v} = \left[\frac{8kT_e}{\pi m_e}\right]^{1/2} \tag{8.20}$$

where m_e and T_e represent the electron conductivity effective mass and electron temperature, respectively. A similar expression can be written for holes. At a field of about 2×10^5 V/cm in Ge and Si, the average kinetic energy of a hot electron becomes as high as 0.2 eV and a temperature T_e of about 2500 K is reached. When fields of this order of magnitude are applied to crystals of these semiconductors, a significant number of hot carriers acquire energy higher than the gap energy. These hot carriers collide with lattice atoms and impart enough energy to valence band electrons to excite them to the conduction band. This results in secondary electron–hole pair generation. Because of the electron–hole pair generation process, the current will start increasing with the electric field.

The most important parameter in connection with the secondary electron–hole pair generation is the ionization coefficient α_i defined as the number of electron–hole pairs produced by a single electron (or hole) when it moves a unit distance in the direction of the electric field. Efforts have been made to determine α_i from the basic properties of a semiconductor. The four basic parameters that have been used to describe the interaction of hot electrons with the crystal lattice are:

E_r = optical phonon energy
E_i^* = threshold energy for secondary pair generation
lr = mean free path for optical phonon scattering
li = mean free path for ionization collision.

The energy E_r is determined from the optical absorption experiments and has a value of 0.063 eV for Si and 0.037 eV for Ge. The values of lr and li are deduced by comparing the measured field dependence of α_i with the theory.

The minimum energy E_i^* which a primary carrier should have in order to produce an electron–hole pair by ionization collision has to be at least equal to the gap energy E_g. For a semiconductor in which the electron and hole effective masses are equal and isotropic, the minimum ionization energy that satisfies the momentum conservation condition is $3E_g/2$ (see Prob. 8.7). For semiconductors like Si and Ge the above conditions are not satisfied, and E_i^* may be different from $3E_g/2$. The measured value of E_i^* in Si ranges from 1.8 eV to 3.6 eV for electrons and from 2.4 eV to 5.0 eV for holes [2, 4].

Different approaches have been used to study the distribution function of hot electrons and holes in order to calculate the field dependence of α_i. All these efforts, as well as the experimental results, have shown that over a wide range of electric fields, $\alpha_i(\mathscr{E})$ can be expressed as

$$\alpha_i(\mathscr{E}) = a_o \exp(-b/\mathscr{E})^m \tag{8.21}$$

where a_o, b, and m are constants whose values for Ge and Si are given in Table 8.1 along with E_r and lr. The situation for GaAs is more complicated, and ionization coefficients are found to depend on crystal orientation [5, 6].

8.5 AVALANCHE BREAKDOWN IN p-n JUNCTIONS

8.5.1 The Ionization Integral

Figure 8.10 illustrates the manner in which the avalanche breakdown occurs in the depletion region of a p-n junction. For simplicity, we consider an abrupt p^+-n junction such that the primary current essentially consists of holes. We further assume that each of the electrons and holes, on the average, produces N_s secondary pairs. Let a current I_o of holes be injected into the space-charge layer at $x = 0$. The first generation of carriers produces a current $I_o N_s$ after the secondary multiplication. This current, in turn, generates a current $I_o N_s^2$ and so on. After successive generations of secondary carriers, the final current I is given by

$$I = I_o(1 + N_s + N_s^2 + \ldots) = \frac{I_o}{1 - N_s} \tag{8.22}$$

Thus, the avalanche multiplication factor $M = I/I_o$ can be written as

$$M = \frac{1}{(1 - N_s)}$$

TABLE 8.1
Secondary Ionization Data for Ge and Si*

Semiconductor	E_r(eV)	Charge Carrier	lr(300 K) (Å)	Ionization coefficient (300 K)		
				a_o(cm^{-1})	b(V/cm)	m
Ge	0.037	Electron	65±10	1.55×10^7	1.56×10^6	1
		Hole	—	1.0×10^6	1.28×10^6	1
Si	0.063	Electron	62±5	3.8×10^6	1.75×10^6	1
		Hole	45±5	2.25×10^7	3.26×10^6	1

*From S. M. Sze, Physics of Semiconductor Devices, p. 62, John Wiley, New York, 1969. Reprinted by permission.

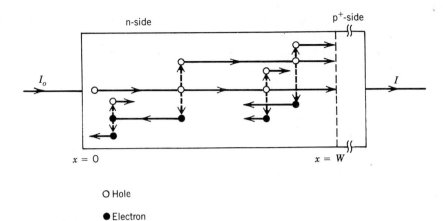

FIGURE 8.10 Schematic diagram showing avalanche multiplication in the depletion region of a p^+-n junction.

It is evident that M tends to infinity leading to avalanche breakdown when $N_s = 1$, which means that each of the electrons and holes produces one secondary carrier on the average.

The multiplication factor M can be expressed in terms of the ionization coefficients of electrons and holes, which, as can be seen from Table 8.1, are generally different. Let α_n and α_p represent the ionization coefficients of electrons and holes, respectively. In a p^+-n junction, secondary ionization is initiated by holes and the avalanche multiplication factor M is given by [2]

$$\left(1 - \frac{1}{M}\right) = \int_0^W \alpha_p \exp\left[-\int_0^x (\alpha_p - \alpha_n)\, dx'\right] dx \tag{8.23}$$

Avalanche breakdown occurs when M becomes infinitely large and the ionization integral on the right-hand side of Eq. (8.23) becomes unity. Thus, the breakdown condition in a p^+-n junction is expressed by

$$\int_0^W \alpha_p \exp\left[-\int_0^x (\alpha_p - \alpha_n)\, dx'\right] dx = 1 \tag{8.24a}$$

In n^+-p junctions, the avalanche is initiated by electrons, and the corresponding relation becomes

$$\int_0^W \alpha_n \exp\left[-\int_0^x (\alpha_n - \alpha_p)\, dx'\right] dx = 1 \tag{8.24b}$$

In a p-n junction, where the current I_o consists of both electrons and holes, the breakdown voltage is calculated using the simultaneous solutions of these two equations.

It is not possible to integrate the above equations directly to obtain a closed form expression for the breakdown voltage V_B. However, both Eq. (8.24a) and Eq. (8.24b) can be simplified considerably if an average coefficient α_i is used instead of the two separate coefficients. Thus, writing $\alpha_n = \alpha_p = \alpha_i$ in the above equations, we obtain

$$\int_0^W \alpha_i(\mathscr{E}) \, dx = 1 \tag{8.25}$$

8.5.2 Avalanche Breakdown Voltage of Abrupt and Linearly Graded Junctions

Based on knowledge of the electric field $\mathscr{E}(x)$ and potential distribution $\phi(x)$ in the junction depletion region, it is possible to determine the width $W = W_b$ at breakdown using either Eq. (8.24) or (8.25). The breakdown voltage V_B can then be calculated by substituting $x = W_b$ in the expression for $\phi(x)$. In general, both $\mathscr{E}(x)$ and $\phi(x)$ can be obtained by solving the Poisson equation in the depletion region. The evaluation of W_b from Eq. (8.24), however, requires the use of numerical methods.

In the case of a two-sided abrupt p-n junction, it is often convenient to replace it with an equivalent one-sided abrupt junction with maximum field $\mathscr{E}_m$ occurring at $x = 0$, as shown by the dashed lines in Fig. 8.11. The dopant concentration on the heavily doped side of the equivalent junction is assumed to be very high, whereas that on the low doped side is replaced by an effective value $N_I = N_d N_a / (N_d + N_a)$. The electric field $\mathscr{E}(x)$ in the depletion region can then be written as

$$\mathscr{E}(x) = \mathscr{E}_m (1 - x/W)$$

with

$$|\mathscr{E}_m| = \sqrt{2qN_I(V_i + V_R)/\varepsilon_s} \tag{8.26}$$

Avalanche breakdown voltage V_B as a function of N_I is plotted in Fig 8.12 for Ge, Si, $\langle 100 \rangle$ GaAs, and GaP abrupt p-n junctions. Figure 8.13 shows V_B for linearly graded junctions in these semiconductors plotted as a function of the grading constant K. These plots have been obtained from the numerical integration of Eq. (8.24) using the known values of a_o and b. The dashed lines in both figures indicate the upper limits of N_I and K for which the avalanche breakdown calculations are valid. To the right of these lines, the Zener mechanism dominates the breakdown process.

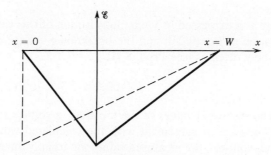

FIGURE 8.11 Representation of the electric field in a two-sided abrupt p-n junction (solid lines) by an equivalent one-sided abrupt junction (dashed lines).

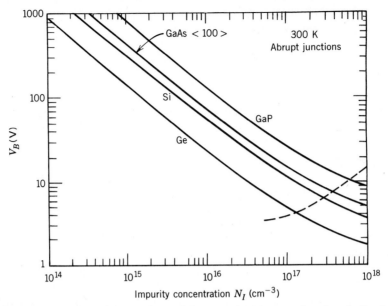

FIGURE 8.12 Avalanche breakdown voltage of abrupt p-n junctions as a function of effective dopant concentration N_I. The dashed line indicates the upper limit beyond which Zener breakdown dominates. (From S. M. Sze and G. Gibbons, Appl. Phys. Lett 8, p. 112, (1966) and S. M. Sze, reference 5, p. 101, Copyright © 1981. Reprinted by permissions of American Institute of Physics and John Wiley & Sons, Inc., New York.)

The following approximate universal expression has been obtained for the breakdown voltage of an abrupt p-n junction in all four semiconductors [5].

$$V_B(\text{volts}) \simeq 60(E_g/1.1)^{3/2}(N_I/10^{16})^{-3/4} \qquad (8.27)$$

where E_g is expressed in eV and N_I in cm^{-3}. Similarly for linearly graded junctions, V_B is approximately given by

$$V_B(\text{volts}) \simeq 60(E_g/1.1)^{6/5}(K/3 \times 10^{20})^{-2/5} \qquad (8.28)$$

where K is expressed in cm^{-4}.

It is sometimes useful to obtain an analytical expression for the breakdown voltage assuming $\alpha_n = \alpha_p = \alpha_i$. For example, in the case of Si, a power series approximation for $\alpha_i(\mathscr{E})$ is obtained [7].

$$\alpha_i(\mathscr{E}) \simeq 1.8 \times 10^{-35} \mathscr{E}^7 \text{ cm}^{-1} \qquad (8.29)$$

where $\mathscr{E}$ is expressed in V/cm. Substitution of this value of $\alpha_i(\mathscr{E})$ allows direct integration of Eq. (8.25), and the breakdown voltage of an abrupt Si p-n junction can be written as (see Prob. 8.11).

$$V_B(\text{volts}) \simeq 5.3 \times 10^{13} N_I^{-3/4} \qquad (8.30)$$

The measured values of the breakdown voltage in abrupt and linearly graded junctions are in agreement with the calculated values up to nearly 200 V. For higher voltages, the measured values are usually lower than the calculated values. The difference increases with increasing values of V_B, and it is caused by crystal imperfections and junction curvature effects that induce premature junction breakdown.

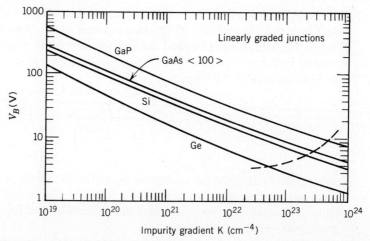

FIGURE 8.13 Avalanche breakdown voltage of linearly graded junctions as a function of the grading constant K. (From S. M. Sze and G. Gibbons, Appl. Phys. Lett 8, p. 112, (1966) and S. M. Sze, reference 5, p. 102, Copyright © 1981. Reprinted by permissions of American Institute of Physics and John Wiley & Sons, Inc., New York.)

8.5.3 Temperature Dependence of Avalanche Breakdown Voltage

Figure 8.14 shows the reverse characteristics of a p-n junction with avalanche breakdown at two temperatures. In constrast to the curves of Fig. 8.9, these characteristics are hard in the sense that they have a well-defined *knee of breakdown*, above which the current increases very rapidly with voltage. Moreover, the breakdown voltage is seen to increase with temperature, making the temperature coefficient of breakdown voltage positive. These are the characteristic features of

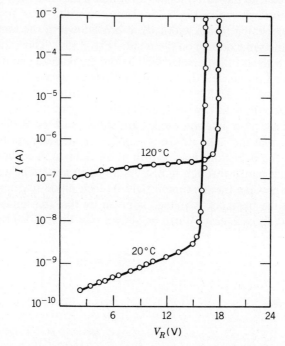

FIGURE 8.14 Reverse I-V characteristics of a p-n junction with avalanche breakdown. After reference 1.

avalanche breakdown. In Fig. 8.14 the pre-breakdown current at 20°C is dominated by carrier generation in the space-charge layer. However, at 120°C, the thermal current I_s dominates, and the slow increase in the pre-breakdown current is caused by the avalanche multiplication of electrons and holes injected into the space-charge region. An empirical relation

$$M = \frac{1}{\left[1 - \left(\frac{V_R}{V_B}\right)^m\right]} \quad (8.31)$$

is often used to describe the avalanche multiplication factor as a function of junction reverse voltage where m ranges from 4 to 6 in silicon and from 2 to 6 in germanium. This relation has no theoretical justification, and Eq (8.23) must be used for accurate values of M.

The increase in avalanche breakdown voltage with temperature results from a decrease in the ionization coefficients (α_n and α_p) and is to be understood as follows: As the temperature of the semiconductor increases, the optical phonon scattering of carriers in the space-charge layer becomes more intense, and the mean free path lr is reduced. Therefore, these carriers now lose more energy to the lattice and, to compensate for this energy loss, a higher voltage has to be applied across the junction to achieve the condition for M to become infinite.

8.5.4 The Punched-Through Diode [8]

For junction diodes that have to operate at high reverse voltages, extremely thick depletion layers are required. It is undesirable to have such thick layers, because when the diode is forward biased, most of the thick region remains undepleted and adds to the diode series resistance. This effect can be reduced in the p^+-n-n^+ structure shown in Fig. 8.15c. Here the width W of the n-region is smaller than the depletion region width W_b at breakdown in the long base p^+-n diode, which has the same doping on the n-side (Fig. 8.15a). Thus, the n-region in the diode of Fig. 8.15c is fully depleted well before breakdown occurs. The breakdown voltage of such a punched-through (PT) diode is always lower than that of the normal long-base diode. This will become clear from the following considerations.

Electric field distributions in the n-region of the normal and the punched-through p^+-n junction diodes are shown in Figs. 8.15b and d, respectively. It is clear that the slope $d\mathscr{E}/dx = -qN_d/\varepsilon_s$ is the same for both diodes. Let W_b represent the depletion region width at breakdown in a normal diode and $F = W/W_b$ be the punch-through factor. As an approximation, we assume that breakdown occurs when the maximum field in each diode reaches a critical value $\mathscr{E}_{mb}$, and that the breakdown voltage is given by the area under the $\mathscr{E}$ versus distance x curve in each case. It can be shown (see Prob. 8.15) that

$$\frac{V_{BPT}}{V_B} \simeq F(2 - F) \quad (8.32)$$

where

$$V_{BPT} = \frac{W}{2}(2 - F)\mathscr{E}_{mb}$$

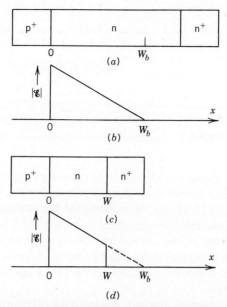

FIGURE 8.15 Schematics of (a) long-base and (c) punched-through diode near reverse breakdown; (b) and (d) show the electric field distributions for these diodes. After reference 8.

is the breakdown voltage of the punched-through diode, and V_B is that of the normal long-base diode. Since F is less than unity, V_{BPT} is always lower than V_B.

As an example, consider a Si punched-through p^+-n-n^+ diode with $N_d = 10^{14}$ cm^{-3} and $W = 85$ μm. Using Eq. (8.30) and writing N_d for N_I, we obtain

$$V_B = 5.3 \times 10^{13}(10^{14})^{-3/4} = 1675 \text{ V}$$

The depletion region width at breakdown W_b is given by

$$W_b = \left(\frac{2\varepsilon_s}{qN_d}VB\right)^{1/2} = \left(\frac{2.1 \times 10^{-12} \times 1675}{1.6 \times 10^{-19} \times 10^{14}}\right)^{1/2} = 1.48 \times 10^{-2} \text{ cm}$$

The punch-through factor is $F = 85/148 = 0.57$, and from Eq. (8.32), we obtain

$$V_{BPT} = 0.57(2 - 0.57)1675 = 1365 \text{ V}.$$

8.6 EFFECT OF JUNCTION CURVATURE AND CRYSTAL IMPERFECTIONS ON THE BREAKDOWN VOLTAGE

8.6.1 Junction Curvature

Thus far, we have considered only a plane structure obtained by making a junction at a uniform depth. When a junction is fabricated by the planar process (see Sec. 19.4), performing diffusion through a window in the SiO$_2$ layer, the impurities diffuse not only vertically downward but also sideways. This situation results in a junction with curved boundaries as shown in Fig. 8.16. Depending on the shape of the window in the SiO$_2$, the curved boundaries may be either cylindrical or spherical in shape. In both cases, the electric field strength in the depletion

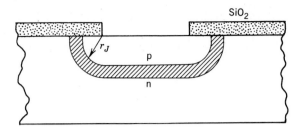

FIGURE 8.16 Junction curvature formation in the planar diffusion process.

region is higher than that in a plane p-n junction, and the avalanche breakdown voltage is determined by the junction curvature [8].

Figure 8.17 shows the normalized breakdown voltage $V_B(C)/V_B$ plotted as a function of r_J/W_b for cylindrical and spherical junctions. Here, $V_B(C)$ is the breakdown voltage of the curved junction with a radius of curvature r_J, and V_B is the breakdown voltage of the plane junction. As seen in Fig. 8.17, for a given value of r_J/W_b, the breakdown voltage is lower for a spherical junction than for a cylindrical junction.

In the case of linearly graded and diffused p-n junctions, the breakdown voltage remains almost independent of the junction curvature provided that the electric field in the depletion region is symmetrical on both sides of the metallurgical junction plane.

8.6.2 Crystal Imperfections

The breakdown voltage of a p-n junction is usually limited by crystal imperfections. It is observed that in most cases the current at the onset of avalanche breakdown does not flow uniformly across the junction, but rather funnels through narrow discrete channels known as microplasmas. These microplasmas

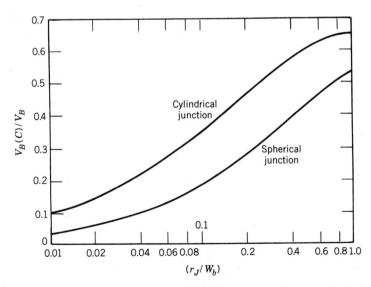

FIGURE 8.17 Normalized avalanche breakdown voltages of cylindrical and spherical silicon abrupt p-n junctions as a function of radius of curvature. (From S. K. Ghandhi, reference 8, p. 61. Copyright © 1977. Reprinted by permission of John Wiley & Sons, Inc., New York.)

occur at those places in the depletion region where the electric field is increased above its average value owing to the presence of imperfections, as shown in Fig. 8.18. Because the ionization coefficients (α_n and α_p) are strong functions of the electric field, they have higher values in the microplasma region than in the rest of the junction. This causes the multiplication factor M to become larger in this region. Thus, when breakdown occurs at a microplasma site, the neighboring regions may carry only a negligibly small pre-breakdown current.

A p-n junction with microplasma breakdown is characterized by an instability at the onset of breakdown. When the junction current remains below a critical value of about 100 μA, the breakdown remains unstable and the microplasma exhibits current fluctuations. In this unstable region, practically all the current flows in the form of current pulses of constant amplitude and random duration. The frequency of these pulses increases with the applied voltage until at a sufficiently large voltage the pulses disappear and a continuous current flow is obtained. The random current pulses are observed because the minority carriers contributing to the pre-breakdown current have a statistical distribution over the junction. Whenever a carrier passes through the microplasma region, an avalanche is triggered and a current pulse is observed. However, at times when no carrier passes through this region, an avalanche is not triggered, and the diode passes only the small pre-breakdown current. This on-off behavior of the microplasma corresponds to a bistable switch and results in a small region of apparent negative resistance in the static I-V characteristic of the junction near the knee of the breakdown.

Microplasmas are caused by the presence of certain types of crystal defects in the junction depletion region. The following defects are found to cause microplasma formation.

(a) DISLOCATIONS

Dislocations reduce the breakdown voltage because during p-n junction formation, enhanced diffusion occurs along their lengths and metallic impurities tend to segregate at these sites causing a local enhancement of the electric field.

(b) PRECIPITATES IN THE JUNCTION REGION

The presence of small metallic or dielectric precipitates in the junction depletion region causes an increase in the electric field above its average value. This in-

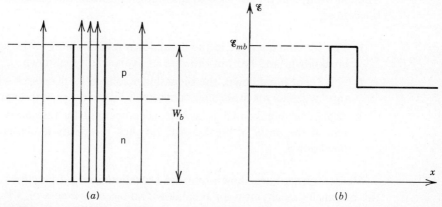

FIGURE 8.18 Breakdown condition in a microplasma; (a) current and (b) electric field distributions.

crease occurs because the dielectric constant of the precipitated particles is different from that of the semiconductor. Small particles of SiO_2 and SiC in silicon are found to be particularly harmful.

(c) DOPANT IMPURITY STRIATIONS

The nonhomogeneous distribution of dopant atoms either during crystal growth or during the junction formation may also produce localized regions with higher than average dopant concentrations in the junction. The electric field in such regions is enhanced above its average value, causing a premature avalanche breakdown in a reverse-biased junction.

The effect of imperfections on Zener breakdown is to cause more field emission current at the sites of imperfection. The effect, however, is not as significant as in case of avalanche breakdown.

8.6.3 Structurally Perfect Junctions

Even in a structurally perfect p-n junction, the statistical spatial fluctuations in the donor and acceptor atoms cause a nonuniformity in the breakdown voltage. Thus, a truly uniform breakdown of large-area junctions is not easy to achieve. However, in the case of structurally perfect junctions, the fluctuation in the impurity distribution is small, and the reduction in the breakdown voltage from its ideal value is much less than that in junctions with microplasmas. In junctions with no microplasmas, the preferred sites of breakdown are the regions where the junction meets the semiconductor surface. Because of the presence of surface charges and junction curvature, the electric field at the surface of the junction is higher than that in the bulk, and this field concentration causes a localized reduction in the breakdown voltage. A common method for reducing the tangential field concentration at the surface is to shape the semiconductor surface near the p-n junction so that the width of the space-charge region is increased. (See Sec. 18.1).

8.7 DISTINCTION BETWEEN THE ZENER AND AVALANCHE BREAKDOWN

The following criteria are used to distinguish between the Zener and avalanche breakdown.

1. The I-V characteristic in the breakdown region is soft in the case of Zener breakdown, and hard in the case of avalanche breakdown.
2. For Zener breakdown, the breakdown voltage V_B decreases with temperature, whereas for avalanche breakdown V_B increases with temperature.
3. Avalanche breakdown is always accompanied by instability in the current at the onset of breakdown. No such instability is observed in Zener breakdown.

Measurements have shown that the breakdown in p-n junctions is caused by the avalanche mechanism for breakdown voltages in excess of 8 V_g, and by the Zener mechanism for values of V_B up to about 4 V_g. Between these two limits, both of these mechanisms operate, and the breakdown is a mixed type.

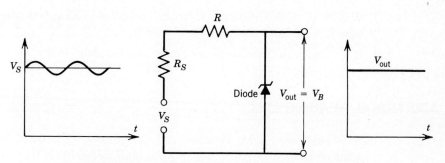

FIGURE 8.19 A voltage regulator circuit using a Zener diode.

8.8 APPLICATIONS OF BREAKDOWN DIODES

Breakdown diodes are used as voltage regulators in power supply circuits. A simple voltage regulator circuit is shown in Fig. 8.19. In this circuit, V_s and R_s represent the power supply voltage and its resistance, respectively, and R is the resistance placed in series to limit the diode current to a safe value. The voltage V_s is the rectified filtered voltage having a dc value and ripple component. The output voltage (V_{OUT}) is a nearly constant voltage equal to the breakdown voltage of the diode. Variations in V_{OUT} are small compared to those of V_s. More complicated voltage regulator circuits have been designed depending on the type of the signal V_s and the nature of the load.

Breakdown diodes can also be used as sources of a reference voltage. Since the diode breakdown voltage has a well-defined value, it can be used as a reference in circuits that require a known value of voltage. To obtain a temperature-independent reference source, two diodes, one with Zener breakdown and the other with avalanche breakdown, can be placed in series. Since the avalanche breakdown voltage increases with temperature while the Zener breakdown voltage decreases with temperature, proper matching of the two temperature coefficients will produce a constant reference voltage that is independent of temperature.

Voltage regulator diodes are popularly called *Zener diodes*. Although the majority of these diodes show avalanche breakdown, the Zener diode terminology is in widespread use.

REFERENCES

1. M. S. TYAGI, "Zener and avalanche breakdown in silicon alloyed p-n junctions," *Solid-State Electron.* 11, 99 (1968).
2. J. L. MOLL, *Physics of Semiconductors*, McGraw-Hill, New York, 1964.
3. E. SPENKE, *Electronic Semiconductors*, McGraw-Hill, New York, 1958, Chapter 7.
4. S. M. SZE, *Semiconductor Devices: Physics and Technology*, John Wiley, New York, 1985, Chapter 2.
5. S. M. SZE, *Physics of Semiconductor Devices*, John Wiley, New York, 1981, Chapters 1 and 2.
6. M. H. LEE and S. M. SZE, "Orientation dependence of breakdown voltage in GaAs," *Solid-State Electron.* 23, 1007 (1980).

7. W. Fulop, "Calculation of avalanche breakdown voltages of silicon p-n junctions," *Solid-State Electron.* 10, 39 (1967).

8. S. K. Ghandhi, *Semiconductor Power Devices,* John Wiley, New York, 1977.

ADDITIONAL READING LIST

1. S. Mahadevan, S. M. Hardas, and G. Suryan, "Electrical breakdown in semiconductors," *Physica Status Solidi* (a) 8, 335–374 (1971).

PROBLEMS

8.1 Determine the width of the potential barrier when a uniform field of 10^6 V/cm is applied to a sample of (a) Ge, (b) Si, and (c) GaAs.

8.2 The tunneling probability for a triangular barrier is given by Eqs. (8.10) and (8.11). For $m^* = m_o$ and $|\mathscr{E}| = 1.2 \times 10^6$ V/cm, calculate the value of E_g that will give a tunneling probability of 5×10^{-12}.

8.3 Consider a Si abrupt p-n junction diode with a dopant concentration of 2×10^{18} cm^{-3} on both the sides. Assume a reverse bias of 2 V applied to the junction. Determine the value of A' in Eq. (8.15) when a tunnel current of 1 mA is flowing through the diode. Take $\mathscr{E} = \mathscr{E}_m$, $V_i = 1.0$ V, and $T = 300$ K.

8.4 The Ge diode No. 2 in Fig. 8.8 has a donor concentration of 2.7×10^{17} cm^{-3} on the n-side and an acceptor concentration of 2×10^{19} cm^{-3} on the p-side. Substituting $\mathscr{E} = \mathscr{E}_m$, determine the value of $\beta^* E_g^{3/2}$ from the slope of the plot and estimate the value of m^* at 300 K. Assume $E_g = 0.8$ eV, the direct gap for Ge.

8.5 In the diode of Prob. 8.3, determine the temperature coefficient γ_T of the voltage $(V_i + V_R)$ at a reverse current of 1 mA, assuming that this change is caused by a change in the band gap of the semiconductor. Take $(dE_g/dT)/E_g = 2.5 \times 10^{-4}\,°\text{K}^{-1}$.

8.6 Using Eq. (8.20), determine the average random thermal velocity of an electron at $T_e = 2000$ K in (a) Ge, (b) Si, and (c) GaAs. Use Table 4.2 for values of conductivity effective mass m_e.

8.7 Consider a semiconductor with spherical constant energy surfaces and $m_e = m_h$. Prove that for such a semiconductor the threshold energy for secondary electron–hole pair production is $3E_g/2$.

8.8 A symmetrical abrupt Ge p-n junction has an impurity concentration of 10^{15} atoms cm^{-3} on both sides. Calculate the avalanche breakdown voltage if the maximum field at breakdown is 2.5×10^5 V cm^{-1}. Calculate the breakdown voltage if the impurity concentration on the n-side remains the same, but on the p-side it becomes 10^{19} cm^{-3}.

8.9 The electric field $\mathscr{E}(x)$ in an abrupt p$^+$-n junction is given by $\mathscr{E}(x) = \mathscr{E}_m(1 - x/W)$. Assuming $\alpha_n = \alpha_p = \alpha_i = a_o \exp(-b/\mathscr{E})$, show that V_B can be expressed as $V_B \simeq (b/2a_o) \exp(b/\mathscr{E}_{mb})$.

PROBLEMS

8.10 Consider a Si abrupt p^+-n junction diode at 300 K. Using the result of Prob. 8.9 and assuming $a_o = 1.2 \times 10^7 \text{ cm}^{-1}$ and $b = 1.5 \times 10^6 \text{ V cm}^{-1}$, calculate the dopant concentration in the n-region that will make $V_B = 36$ V.

8.11 Using the value of $\alpha_i(\mathscr{E})$ in Eq. (8.29), show that the avalanche breakdown voltage of an abrupt silicon p-n junction is given by Eq. (8.30).

8.12 A Si abrupt p^+-n junction has a breakdown voltage of 950 V. Calculate the dopant concentration on the n-side using Eq. (8.30) and determine the depletion layer width W_b at breakdown.

8.13 The ionization coefficient in GaAs can be approximated by $\alpha_n = \alpha_p = (\mathscr{E}/4)^6 \times 10^{-26} \text{ cm}^{-1}$ where $\mathscr{E}$ is expressed in V/cm. Calculate the breakdown voltage of a GaAs abrupt p^+-n junction with a doping of $6 \times 10^{15} \text{ cm}^{-3}$ on the lightly doped n-side.

8.14 Consider a linearly graded Si p-n junction with $K = 2 \times 10^{19} \text{ cm}^{-4}$. Using the value of $\alpha_i(\mathscr{E})$ in Eq. (8.29), calculate the breakdown voltage.

8.15 Show that the breakdown voltage V_{BPT} in a punched-through diode is given by Eq. (8.32).

8.16 A p^+-n-n^+ diode has an n-region width of 45 μm. Calculate the doping in the n-region that will be required to obtain a breakdown voltage of half the ideal voltage given by Eq. (8.30).

9
DYNAMIC BEHAVIOR OF p-n JUNCTION DIODES

INTRODUCTION

The static I-V characteristic of a p-n junction diode considered in Chapter 7 is obtained when the diode voltage changes so slowly that the excess carrier distributions in n- and p-regions always remain in a dc steady state. When the voltage changes rapidly with time, the electron-hole distributions cannot follow it. Consequently, the current and voltage can no longer be described by the static characteristic and the diode shows capacitive behavior.

Two effects give rise to the capacitive behavior in p-n junction diodes. First, we have already seen that when the voltage across a p-n junction is varied, there is a change in the width of the depletion region. This change is caused by the flow of majority carriers in the quasi-neutral regions and appears as a displacement current in the center of the depletion region. Second, whenever the junction voltage is varied, a change takes place in the minority carrier concentrations in the neutral regions. Thus, charging and discharging of these regions also occurs with the change in the diode voltage. The behavior of the current resulting from this charging and discharging process is governed by the diffusion constant and the lifetime of minority carriers. The first effect is of more importance in a reverse-biased diode, while the second is dominant in the forward-biased case.

In this chapter we first consider the high-frequency performance of junction diodes (see Sec. 9.1). The charge control equation of the diode is presented in Sec. 9.2. The turn-on transient and diode reverse recovery are discussed in Sec. 9.3. The chapter closes with a brief account of high-speed switching diodes in Sec. 9.4.

9.1 SMALL-SIGNAL ac IMPEDANCE OF A JUNCTION DIODE

9.1.1 Differential Resistance of a Forward-Biased Diode

The problem we are going to consider is the following: Suppose that a p-n junction diode is carrying a constant current at some forward-bias voltage V_o (Fig. 9.1). Let a small sinusoidal voltage $\tilde{v}$ be superimposed on the dc bias. We now want to know the impedance presented by the diode to the small-signal excitation. For reasons mentioned above, the impedance is composed of both ohmic and capacitive parts and thus, will be a function of the frequency of the ac signal. At very low frequencies, the capacitive effects may be neglected, and the diode resistance r_j can be obtained by simply differentiating the static I-V characteristic

$$I = I_s[\exp(V_a/V_T) - 1]$$

at the bias voltage, giving

$$r_j = \frac{dV_a}{dI} = \frac{V_T}{(I + I_s)} \tag{9.1}$$

In the case of a sufficiently large reverse bias, $I = -I_s$ and r_j will become infinite. However, when surface leakage and carrier generation in the depletion region dominate the reverse current, it will increase slowly with voltage and will not saturate. The diode resistance will then be finite but very large. For a forward bias in excess of about $3\,V_T$, the current I_s becomes negligible in comparison with I, and at 300 K we obtain

$$r_j = \frac{25.86}{I(\text{mA})}\,\Omega \tag{9.1a}$$

If the diode has an ideality factor $\eta > 1$, the resistance r_j will be η times higher than the above value.

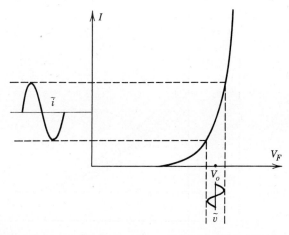

FIGURE 9.1 Voltage and current waveforms when a small sinusoidal signal is superimposed on a forward-biased diode.

At higher frequencies, Eq. (9.1) is not correct and capacitive effects must be taken into consideration. This can be done by writing the minority carrier continuity equation for the small-signal ac excitation and solving it with appropriate boundary conditions.

9.1.2 Diffusion Capacitance and Diode Impedance [1]

Let us consider a forward-biased, long-base p^+-n junction diode. Assume that the diode has a dc voltage V_o on which a small-signal ac voltage $\tilde{v} = \hat{v} \exp(j\omega t)$ is superimposed. The hole and electron distributions in the quasi-neutral n-region are shown in Fig. 9.2. When the ac voltage reaches its positive peak, the junction voltage becomes $(V_o + \hat{v})$, and the excess hole concentration increases above its steady state value $\bar{p}_e$ corresponding to the dc voltage V_o. Similarly, when the ac voltage goes to its negative peak, the excess hole concentration decreases below $\bar{p}_e$. Thus, as the ac voltage completes its one full cycle, the change in the excess hole charge shown by the hatched area in Fig. 9.2 takes place. Because of the quasi-neutrality in this region, the majority carrier (i.e., electron) charge changes by the same amount as the hole charge. This change in the minority and majority carrier charges is equivalent to a diffusion capacitance whose value will now be calculated.

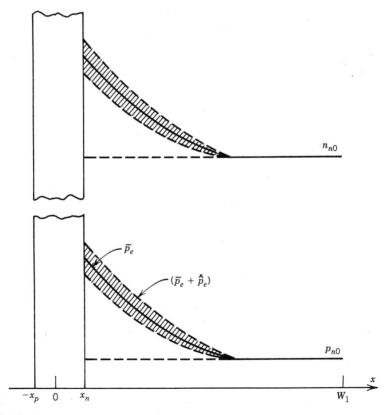

FIGURE 9.2 Variations in the minority and majority carrier stored charges in neutral n-region with forward bias.

Let us assume that the excess hole concentration p_e in the quasi-neutral n-region can be expressed as

$$p_e = \bar{p}_e + \tilde{p}_e \qquad (9.2)$$

where $\bar{p}_e$ is the dc part of p_e associated with the bias voltage V_o and $\tilde{p}_e$ is the ac component which varies sinusoidally

$$\tilde{p}_e = \hat{p}_e \exp(j\omega t)$$

with the additional assumption that the hole current in the n-region flows only by diffusion, the hole continuity equation (7.1) can be written as

$$D_p \frac{\partial^2}{\partial x^2}(\bar{p}_e + \tilde{p}_e) - \frac{\bar{p}_e + \tilde{p}_e}{\tau_p} = \frac{\partial}{\partial t}(\bar{p}_e + \tilde{p}_e) \qquad (9.3)$$

In the steady state,

$$\frac{\partial \bar{p}_e}{\partial t} = 0 \quad \text{and} \quad \frac{\partial \tilde{p}_e}{\partial t} = j\omega \tilde{p}_e.$$

Separating the dc and ac components in Eq. (9.3) leads to

$$\frac{d^2 \bar{p}_e}{dx^2} - \frac{\bar{p}_e}{L_p^2} = 0 \qquad (9.4)$$

and

$$\frac{d^2 \tilde{p}_e}{dx^2} - \left(\frac{1 + j\omega \tau_p}{L_p^2}\right)\tilde{p}_e = 0 \qquad (9.5a)$$

where $L_p^2 = D_p \tau_p$. Note that Eq. (9.4) is the same as Eq. (7.3), which was solved to obtain the static I-V characteristic in Chapter 7. Therefore, it needs no further consideration. Equation (9.5a) can be rewritten as

$$\frac{d^2 \tilde{p}_e}{dx^2} = \frac{\tilde{p}_e}{L_p^{*2}} \qquad (9.5b)$$

where

$$\tau_p^* = \frac{\tau_p}{1 + j\omega \tau_p} \qquad (9.6a)$$

and

$$L_p^* = \sqrt{D_p \tau_p^*} \qquad (9.6b)$$

Equation (9.5b) has the solution

$$\tilde{p}_e(x) = C_1 \exp(-x/L_p^*) + C_2 \exp(x/L_p^*) \qquad (9.7)$$

The constants C_1 and C_2 are determined by the boundary conditions at $x = x_n$ and at $x = W_1$. From the Boltzmann relation at $x = x_n$, we have

$$p_e(x_n) = p_{no}\left[\exp\left(\frac{V_o + \tilde{v}}{V_T}\right) - 1\right]$$

If the magnitude of ac voltage is small compared to the thermal voltage V_T, then $\exp(\tilde{v}/V_T) \simeq (1 + \tilde{v}/V_T)$. This leads to the following relations:

$$\bar{p}_e(x_n) = p_{no}[\exp(V_o/V_T) - 1]$$

$$\tilde{p}_e(x_n) = p_{no} \exp\left(\frac{V_o}{V_T}\right) \cdot \frac{\tilde{v}}{V_T} \qquad (9.8)$$

We assume a perfect ohmic contact at $x = W_1$ (see Fig. 7.1) so that

$$\tilde{p}_e(W_1) = 0 \qquad (9.9)$$

Substituting the boundary conditions (9.8) and (9.9) into (9.7), we obtain

$$\tilde{p}_e(x) = p_{no} \exp(V_o/V_T)\frac{\tilde{v}}{V_T}\left[\sinh\left(\frac{W_1 - x}{L_p^*}\right) \Big/ \sinh\left(\frac{W_1 - x_n}{L_p^*}\right)\right] \qquad (9.10)$$

The ac current $\tilde{i}_p = -qAD_p(\partial \tilde{p}_e/\partial x)$ is obtained from Eq. (9.10):

$$\tilde{i}_p(x) = \frac{qAD_p n_i^2}{N_d L_p^*}\frac{\tilde{v}}{V_T}\exp\left(\frac{V_o}{V_T}\right)\left[\cosh\left(\frac{W_1 - x}{L_p^*}\right) \Big/ \sinh\left(\frac{W_1 - x_n}{L_p^*}\right)\right]$$

where the relation $N_d p_{no} = n_i^2$ has been used. This current has its maximum value at $x = x_n$

$$\tilde{i}_p(x_n) = \left(\frac{qAD_p n_i^2}{N_d L_p^*}\right)\frac{\tilde{v}}{V_T}\coth\left[\frac{W_1 - x_n}{L_p^*}\right]\exp(V_o/V_T) \qquad (9.11)$$

A similar expression can be obtained for the current $\tilde{i}_n(-x_p)$ due to electron injection into the p-region. The total current $\tilde{i}_o$ is obtained by adding the two components together. Thus

$$\tilde{i}_o = \tilde{i}_p(x_n) + \tilde{i}_n(-x_p)$$

$$= qAn_i^2\left[\frac{D_p}{N_d L_p^*}\coth\left(\frac{W_n}{L_p^*}\right) + \frac{D_n}{N_a L_n^*}\coth\left(\frac{W_p}{L_n^*}\right)\right]\frac{\tilde{v}}{V_T}\exp(V_o/V_T) \qquad (9.12)$$

where L_n^* is the complex diffusion length similar to L_p^*, $W_n = (W_1 - x_n)$, $W_p = |W_2 - x_p|$, and all other symbols have their usual meanings.

Equation (9.12) gives only the minority carrier diffusion component of the current and does not include the displacement current associated with the junction capacitance. This current must be added to the diffusion current to obtain the total current. For a diode with junction area A, the displacement current $\tilde{i}_d$ is given by

$$\tilde{i}_d = -A\varepsilon_s \frac{d\mathscr{E}}{dt}$$

where $\mathscr{E} = [V_i - (V_o + \tilde{v})]/W$ represents the average field in the depletion region. The negative sign in the expression for $\tilde{i}_d$ appears because the current is taken positive when the forward bias on the junction is increased, and this causes a decrease in $\mathscr{E}$. If the ac voltage $\tilde{v}$ is small compared to V_o, the change in the depletion region width W can be neglected, and we obtain

$$\tilde{i}_d = j\omega C_j \tilde{v} \qquad (9.13)$$

where $C_j = A\varepsilon_s/W$ is the junction capacitance. Thus, writing $\tilde{i} = \tilde{i}_o + \tilde{i}_d$, we can express the diode admittance y as

$$y = \frac{\tilde{i}}{\tilde{v}} = \frac{qAn_i^2}{V_T}\left[\frac{D_p}{N_d L_p^*}\coth\left(\frac{W_n}{L_p^*}\right) + \frac{D_n}{N_a L_n^*}\coth\left(\frac{W_p}{L_n^*}\right)\right]\exp\left(\frac{V_o}{V_T}\right) + j\omega C_j \qquad (9.14)$$

We will now limit our discussion to the two cases of a long-base and a short-base diode.

(a) THE LONG-BASE DIODE

Let us consider a long-base p^+-n junction diode. The electron current component in Eq. (9.14) is then negligible, $\coth(W_n/L_p^*) = 1$, and we get

$$y = \frac{qAD_p n_i^2}{V_T N_d L_p}\exp(V_o/V_T)\sqrt{1 + j\omega\tau_p} + j\omega C_j$$

$$= \frac{I}{V_T}\sqrt{1 + j\omega\tau_p} + j\omega C_j \qquad (9.15)$$

For low frequencies, such that $\omega\tau_p \ll 1$, the above relation can be simplified to yield

$$y = \frac{I}{V_T} + j\omega\left[\frac{I}{2V_T}\tau_p + C_j\right] \qquad (9.16a)$$

$$= g + j\omega(C_D + C_j) \qquad (9.16b)$$

where

$$C_D = \frac{I\tau_p}{2V_T} = \frac{g\tau_p}{2} \qquad (9.17)$$

represents the diffusion capacitance of the diode.

The small-signal equivalent circuit of the diode resulting from Eq. (9.16b) is shown in Fig. 9.3. The diode conductance g is shunted by a parallel combination

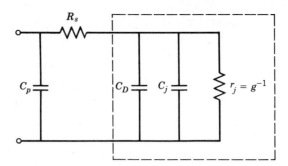

FIGURE 9.3 Small-signal low-frequency ac equivalent circuit of a packaged diode. The dashed boundary shows the equivalent circuit of the p-n junction.

of C_j and C_D. The capacitance C_D increases linearly with the diode current, and at a sufficiently large forward bias, C_j becomes negligible in comparison with C_D. On the other hand, for a reverse-biased diode, C_D can be ignored in comparison to C_j. In the circuit of Fig. 9.3 the series resistance R_s is the ohmic resistance of the neutral n- and p-regions, and C_p is the capacitance of the diode package. Although R_s and C_p are not predicted by Eq. (9.16), they have to be included for a complete description of a packaged diode.

For high frequencies, such that $\omega\tau_p \gg 1$, the diode admittance in Eq. (9.15) can again be expressed as

$$y = \frac{I}{V_T}\sqrt{j\omega\tau_p} + j\omega C_j$$

$$= g\sqrt{\frac{\omega\tau_p}{2}} + j\omega(g\sqrt{\tau_p/2\omega} + C_j) \tag{9.18}$$

At sufficiently large forward bias, C_j is small, and the admittance in Eq. (9.18) can still be regarded as a parallel resistance capacitance (R-C) combination in which both the resistance and the capacitance decrease as $1/\sqrt{\omega}$.

(b) THE SHORT-BASE DIODE

When x is small, $\coth x \simeq 1/x + x/3$. Thus, for a short-base p^+-n junction diode, we obtain

$$y = \frac{qAD_p n_i^2}{V_T N_d W_n} \exp(V_o/V_T)\left[1 + \frac{1}{3}\frac{W_n^2}{L_p^{*2}}\right] + j\omega C_j$$

Substituting L_p^* from Eq. (9.6b), we can write

$$y = \frac{I}{V_T}\left[1 + \frac{W_n^2}{3L_p^2}(1 + j\omega\tau_p)\right] + j\omega C_j \tag{9.19}$$

Separating the real and imaginary parts, and noting that $W_n^2/3L_p^2$ is small compared to unity, we can write the diode admittance as

$$y = g + j\omega\left[\frac{g}{3}\frac{W_n^2}{L_p^2}\tau_p + C_j\right] \tag{9.20}$$

where $g = I/V_T$ is the diode conductance, and the diffusion capacitance C_D is now given by

$$C_D = \frac{g}{3}\frac{W_n^2}{L_p^2}\tau_p \qquad (9.21)$$

The small-signal ac equivalent circuit for the short-base diode is the same as that shown in Fig. 9.3.

The value of the diffusion capacitance for a long-base diode is not the same as that obtained from the ratio of the incremental charge to the incremental voltage. Consider a long-base p^+-n junction diode with an applied voltage V_a. The excess hole concentration $p_e(x)$ is given by

$$p_e(x) = p_{no}[\exp(V_a/V_T) - 1]\exp\left(\frac{x_n - x}{L_p}\right)$$

and the stored charge Q_p becomes

$$Q_p = qA\int_{x_n}^{\infty} p_e(x)\,dx = I_s[\exp(V_a/V_T) - 1]\tau_p = I\tau_p$$

which gives

$$\frac{dQ_p}{dV_a} = \frac{I}{V_T}\tau_p \qquad (9.22)$$

The diffusion capacitance C_D obtained from Eq. (9.17) is just half of this value. The physical meaning of this result is that only half of the incremental minority carrier charge shown by the hatched area in Fig. 9.2 can be reclaimed through the junction during a cycle of the sine wave. The remaining half recombines in the neutral n-region. A similar analysis for a short-base diode shows that two-thirds of the incremental charge ΔQ_p can be reclaimed through the junction (see Prob. 9.4).

As an example, consider a long-base p^+-n junction diode carrying a forward current of 1 mA and $C_j = 50$ pF. We assume $\tau_p = 10^{-6}$ sec and $V_T = 25$ mV. Let us calculate the diode admittance at 100 KHz.

The diode conductance is given by

$$g = \frac{I}{V_T} = \frac{1}{25} = 0.04 \text{ A/V}$$

and from Eq. (9.17)

$$C_D = \frac{0.04 \times 10^{-6}}{2} = 2 \times 10^{-8} \text{ F}$$

which is large compared to C_j. Thus, neglecting C_j, the diode admittance at 100 KHz becomes

$$y = 0.04 + j2\pi \times 10^5 \times 2 \times 10^{-8} = (0.04 + j0.0126) \text{ A/V}.$$

9.2 THE CHARGE CONTROL EQUATION OF A JUNCTION DIODE

Thus far we have described the diode current as a function of the diode voltage. We will now consider a different perspective in which the current is expressed in terms of the excess minority carrier charges stored in the neutral p- and n-regions of the diode. This approach is more suited for studying the transient behavior of junction diodes.

We have already seen that the distribution and flow of excess holes in the n-region of a p-n junction diode are described by the hole continuity equation

$$D_p \frac{\partial^2 p_e}{\partial x^2} - \frac{p_e}{\tau_p} = \frac{\partial p_e}{\partial t}$$

and the hole diffusion current is given by

$$i_p(x,t) = -qAD_p \frac{\partial p_e}{\partial x} \tag{9.23}$$

Furthermore, we have also seen that under quasi-equilibrium and steady state conditions, the excess hole concentration at the edge x_n of the depletion region is related to the thermal equilibrium hole concentration p_{no} by the Boltzmann relation. The question that now arises is whether the Boltzmann relation is also valid under transient conditions. To answer this question, we note that a typical value of the junction depletion region width is on the order of 10^{-4} cm, and the electrons and holes may be assumed to move through this region with the mean thermal velocity, which is about 10^7 cm/sec at room temperature. Thus, the transit time of mobile electrons and holes through the depletion region is on the order of 10^{-11} sec. Hence, if the time during which the changes in the diode voltage occur is large compared to this value, the Boltzmann relation remains valid, and we can write

$$p_e(x_n, t) = p_{no}\left[\exp\left(\frac{v(t)}{V_T}\right) - 1\right] \tag{9.24}$$

where $v(t)$ is the instantaneous voltage across the junction. As $v(t)$ changes with time, the edge concentration $p_e(x_n, t)$ changes in accordance with Eq. (9.24).

Multiplying the continuity equation by $qAdx$ and integrating it over the neutral region from x_n to W_1, using Eq. (9.23) gives us

$$-\int_{x_n}^{W_1} \frac{di_p}{dx} dx = \frac{qA}{\tau_p} \int_{x_n}^{W_1} p_e(x)\, dx + \frac{d}{dt} qA \int_{x_n}^{W_1} p_e(x)\, dx$$

which can be written as

$$i_p(x_n, t) - i_p(W_1, t) = \frac{Q_p(t)}{\tau_p} + \frac{dQ_p}{dt} \tag{9.25}$$

where

$$Q_p(t) = qA \int_{x_n}^{W_1} p_e(x,t)\, dx$$

represents the excess stored charge in the n-region at any time t, $i_p(x_n, t)$ is the hole current at x_n, and $i_p(W_1, t)$ is the same at $x = W_1$. The first term on the right-hand side of Eq. (9.25) represents the rate of excess carrier charge loss due to recombination, while the second term represents the rate at which the charge is being stored. Therefore, this equation just states the principle of conservation of charge. The excess charge injected per unit of time in the neutral n-region by the flow of hole current $i_p(x_n, t)$ is either lost by recombination or gets stored in this region; the leftover charge flows out causing the current $i_p(W_1, t)$.

We now define a time constant τ_N by the relation

$$i_p(W_1, t) = \frac{Q_p(t)}{\tau_N} \tag{9.26}$$

The hole distribution in the neutral n-region of the diode is given by Eq. (7.11), which can be rewritten as

$$p_e(x, t) = p_e(x_n, t) \frac{\sinh[(W_1 - x)/L_p]}{\sinh[(W_1 - x_n)/L_p]}$$

Using this expression to obtain the values of $i_p(W_1, t)$ from Eq. (9.23) and $Q_p(t)$ from Eq. (9.25), we get

$$\tau_N = \frac{Q_p(t)}{i_p(W_1, t)} = \tau_p \left[\cosh\left(\frac{W_n}{L_p}\right) - 1 \right] \tag{9.27}$$

It can be easily verified (see Prob. 9.5) that τ_N is closely related to the hole transit time through the neutral n-region of width W_n. Substituting for $i_p(W_1, t)$ from Eq. (9.26) into Eq. (9.25), and defining an effective lifetime τ_F, we obtain

$$i_p(x_n, t) = \frac{Q_p(t)}{\tau_F} + \frac{dQ_p}{dt} \tag{9.28}$$

where

$$\tau_F^{-1} = \tau_p^{-1} + \tau_N^{-1} \tag{9.29}$$

A similar relation can be obtained for the electron current $i_n(-x_p, t) = i_n(x_n, t)$. Adding the two currents, we have

$$i_o(x_n, t) = i_p(x_n, t) + i_n(x_n, t) = \frac{Q_p(t)}{\tau_F} + \frac{Q_n(t)}{\tau_F'} + \frac{dQ}{dt} \tag{9.30}$$

where $Q_n(t)$ represents the stored charge of excess electrons in the neutral p-region, τ_F' is the effective lifetime similar to τ_F, and $Q(t) = Q_p(t) + Q_n(t)$. The total diode current consists of the current given by Eq. (9.30) plus the current caused by the charging and discharging of the depletion region capacitance. If we write the average depletion region capacitance as $\overline{C}_j$, then the current through this capacitance is given by $\overline{C}_j \, dv/dt$ and Eq. (9.30) is modified to

$$i(t) = \frac{Q_p(t)}{\tau_F} + \frac{Q_n(t)}{\tau_F'} + \frac{dQ}{dt} + \overline{C}_j \frac{dv}{dt} \tag{9.31}$$

This is the charge control equation of a p-n junction diode and expresses the current in terms of the stored charges $Q_p(t)$, $Q_n(t)$, and $\overline{C}_j v(t)$. Note that in a steady state the third and fourth terms on the right-hand side of Eq. (9.31) are zero, and the diode current is determined by the recombination rates of excess electron–hole pairs in the neutral n- and p-regions.

In the rest of this chapter, we will be concerned with the transient behavior of p-n junction diodes. To concentrate on essentials we will assume a p^+-n junction diode in which the electron current $i_n(x_n, t)$ can be neglected. Consequently, the charge control equation becomes

$$i(t) = \frac{Q_p(t)}{\tau_F} + \frac{dQ_p}{dt} + \overline{C}_j \frac{dv}{dt} \tag{9.32}$$

Note that for a long-base diode τ_N is infinite and $\tau_F = \tau_p$. Although Eq. (9.32) is not a general equation, its solution will illustrate all the major features we want to know about the transient behavior of p-n junction diodes.

9.3 SWITCHING TRANSIENTS IN JUNCTION DIODES

In practical circuits, the voltage used to switch a diode from one state to another is obtained from a source whose output impedance has some finite value. The rise (or fall) of the diode current and voltage with time will largely depend on the output impedance of the source. However, the exact behavior of the diode can always be deduced from the study of the two extreme cases involving a constant current source and a constant voltage source.

9.3.1 Turn-On Transient [2]

(a) CONSTANT CURRENT SOURCE

Let a constant current step I_F be applied to a p^+-n junction diode through the circuit shown in Fig. 9.4. For $t < 0$, the switch remains open and no current flows. The switch is closed abruptly at $t = 0$. The switch cannot be closed manually but rather through a fast switching electronic device. The resistance R_1 in the circuit is very large, so that $I_F \approx V_1/R_1$ at all time for $t > 0$. As soon as the switch is closed, the holes from the p-side of the junction start diffusing to the n-side. Since the current remains constant at I_F, the hole concentration must ad-

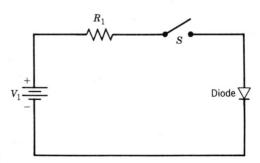

FIGURE 9.4 A basic circuit to study the turn-on behavior of a diode.

just in such a way that its slope at plane $x = x_n$ corresponds to the current I_F, which means that

$$\left.\frac{\partial p_e}{\partial x}\right|_{x=x_n} = \frac{-I_F}{qAD_p} \qquad (9.33)$$

The holes cannot diffuse into the n-region instantaneously after the switch is closed. Hence, there will be no recombination current at $t = 0$, and the diode current will be given by the second and the third terms in Eq. (9.32). However, if I_F is sufficiently large, the $\overline{C}_j dv/dt$ term will have little effect on the current and can be ignored. Note that the voltage across the junction cannot change abruptly without causing an infinite current. Thus, the diode voltage will start building up slowly from its zero value at $t = 0$, and so will be the excess hole concentration $p_e(x_n, t)$. With the passage of time, the injected holes diffuse into the neutral n-region, and the concentration $p_e(x_n, t)$ increases gradually as the voltage across the junction builds up. The injected holes will now start recombining with electrons, and a part of the current I_F is replaced by the first term in Eq. (9.32), thereby causing a decrease in dQ_p/dt. With a further elapse of time, the charge storage current will be gradually replaced by the recombination current and a steady state will be reached when the excess hole concentration $p_e(x_n, t)$ attains its equilibrium value with $v(t) = V_F$ in Eq. (9.24). The hole distribution in the n-region at various time intervals after the switch is closed is shown in Fig. 9.5a (assuming a long-base diode). The diode current and voltage as a function of time are sketched in Fig. 9.5b and Fig. 9.5c, respectively.

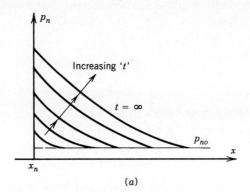

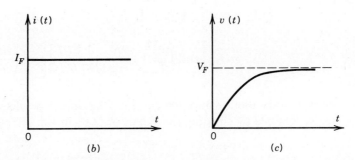

FIGURE 9.5 Turn-on transient in a long-base p^+-n junction diode with a *constant current source*. (a) Hole distribution in the neutral n-region, (b) the diode current, and (c) the diode voltage as a function of time.

The diode voltage as a function of time can be obtained by solving Eq. (9.32). However, to solve this equation we must know how $p_e(x, t)$ develops with distance and time. Therefore, we have to formulate and solve a time-dependent continuity equation that can be obtained simply by stating the conditions, which the excess carrier distribution must satisfy for every value of x and t. This approach leads to a partial differential equation that can be solved only by advanced mathematical techniques. We abandon this approach and will use a simpler method known as the *stored charge approximation*. In this approximation, it is assumed that the actual hole distribution at any time can be represented by a steady state distribution corresponding to that time. The time variation of $p_e(x, t)$ is then reflected only in the variation of the excess hole density at x_n, which means that

$$p_e(x, t) = p_{no}\left[\exp\left(\frac{v(t)}{V_T}\right) - 1\right] \frac{\sinh[(W_1 - x)/L_p]}{\sinh(W_n/L_p)} \quad (9.34)$$

This assumption, though not valid for rapidly varying transients, leads to results that are not radically different from those obtained from a more accurate analysis.

Let us now solve charge control equation (9.32) after neglecting the $\bar{C}_j dv/dt$ term. Since I_F is constant, we have a solution of the form

$$Q_p(t) = A + B \exp(-t/\tau_F) \quad (9.35)$$

The constants A and B are determined by the conditions of charge continuity at $t = 0$. Thus

$$Q_p(0) = 0 \quad \text{and} \quad \left.\frac{dQ_p}{dt}\right|_{t=0} = I_F \quad (9.36)$$

Substituting these values in Eq. (9.35) gives

$$Q_p(t) = I_F \tau_F [1 - \exp(-t/\tau_F)] \quad (9.37)$$

From Eq. (9.34), we can express the stored charge $Q_p(t)$ as

$$Q_p(t) = qA \int_{x_n}^{W_1} p_e(x, t)\, dx$$

$$= I_s \tau_p \left[\coth\left(\frac{W_n}{L_p}\right) - \frac{1}{\sinh(W_n/L_p)}\right]\left[\exp\left(\frac{v(t)}{V_T}\right) - 1\right] \quad (9.38)$$

where I_s is the diode saturation current. The time dependence of the diode voltage is obtained by equating the values of $Q_p(t)$ in Eq. (9.37) and (9.38). That is

$$v(t) = V_T \ln\left\{\frac{I_F \tau_F [1 - \exp(-t/\tau_F)]}{I_s \tau_p \left[\coth\left(\frac{W_n}{L_p}\right) - 1 \Big/ \sinh\left(\frac{W_n}{L_p}\right)\right]} + 1\right\} \quad (9.39)$$

In the case of a long-base diode,

$$\coth\left(\frac{W_n}{L_p}\right) \simeq 1, \quad 1 \Big/ \sinh\left(\frac{W_n}{L_p}\right) \to 0, \quad \tau_F = \tau_p$$

and Eq. (9.39) reduces to

$$v(t) = V_T \ln\left[\frac{I_F}{I_s}(1 - \exp(-t/\tau_p)) + 1\right] \tag{9.40}$$

For a long-base p^+-n junction diode with $I_s = 1$ μA and $I_F = 1$ mA, the voltage $v(t)$ as a function of t/τ_p given by Eq. (9.40) is plotted in Fig. 9.6 along with the variation obtained from a more accurate solution of the partial differential equation [2, 5]. According to the stored charge approximation, the transient is completed in a time of about $3\tau_p$, while the same point is reached in a time of $\tau = \tau_p$ in the accurate model.

In a more general case of switching from a constant forward current I_1 to a higher current I_2, similar considerations can be used, and for a long-base diode, $v(t)$ can be written as (see Prob. 9.8).

$$v(t) = V_T \ln\left[\frac{I_1}{I_s} + \frac{I_2 - I_1}{I_s}(1 - \exp(-t/\tau_p)) + 1\right] \tag{9.41}$$

In the above analysis of the turn-on transient, we have considered only the hole (i.e., the minority carrier) curent. This does not mean that the electron current does not flow. In fact, as we move away from the depletion region edge x_n into the n-region, the hole current is gradually replaced by a current of electrons flowing into this region from the ohmic contact (see Sec. 7.1 for details). However, because the electron current at $x = x_n$ is zero at all times, it is not necessary to include the electron motion in the analysis of the diode transient behavior.

(b) CONSTANT VOLTAGE SOURCE

A constant voltage source results if the resistance R_1 in Fig. 9.4 is reduced to zero. When a constant voltage step V_F is applied to the diode, the excess hole concentration at $x = x_n$ abruptly changes from zero to $p_{no}[\exp(V_F/V_T) - 1]$. Since

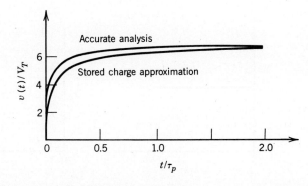

FIGURE 9.6 The normalized voltage $v(t)/V_T$ as a function of time for a long-base p-n junction diode. (From J. F. Gibbons, *Semiconductor Electronics*, p. 287. McGraw-Hill, New York, 1966.)

holes cannot diffuse into the neutral n-region just after the switch is closed, the slope of excess hole concentration $\partial p_e/\partial x\,|_{x=x_n}$ at $t = 0$ will be large; consequently, a very high current will flow through the diode. In fact, the current will be limited only by the diode series resistance. As time goes on, holes diffuse into the neutral n-region to the right of x_n causing a decrease in the slope of the hole concentration at x_n. Thus, the hole current continuously decreases with time until the steady state is reached with current I_F, required by the diode static characteristic. Fig. 9.7a shows the hole distribution in the neutral n-region for various times, and Fig. 9.7c depicts the diode current as a function of time.

To obtain the exact variation of current with time is rather difficult. However, an approximate expression for $i(t)$ can be obtained from Eq. (9.40) if $v(t)$ is replaced by V_F, and $i(t)$ is substituted for I_F. We then get

$$i(t) = I_s \frac{[\exp(V_F/V_T) - 1]}{[1 - \exp(-t/\tau_p)]} \tag{9.42}$$

This expression, though approximate, shows the two essential features of the transient, namely, the current tends to infinity at $t = 0$ and approaches the value I_F when t goes to infinity.

9.3.2 Turn-Off Transient and Diode Recovery [3, 4]

When a diode carrying a steady forward current is suddenly reverse biased, it will not respond to the reverse voltage until the excess electrons and holes stored in

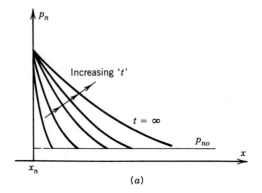

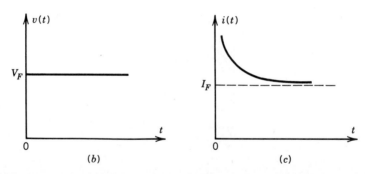

FIGURE 9.7 Turn-on transient in a long-base p^+-n junction diode with a *constant voltage source*. (a) Hole distribution in the neutral n-region, (b) the diode voltage, and (c) the diode current as a function of time.

the neutral n- and p-regions have been withdrawn. Consequently, on application of the reverse bias, the diode will pass a reverse current appreciably higher than the saturation current I_s for some time. The current then starts falling as the stored charge of excess carriers is withdrawn, eventually reaching its reverse saturation value. The way in which the stored charge is removed depends on the source impedance of the reverse drive.

Let us consider an abrupt p^+-n junction diode connected in the circuit shown in Fig. 9.8. Suppose that the switch is in position A and the diode is carrying a steady forward current I_F that can be measured by measuring the voltage drop across the small resistance R_3. We assume that the voltages V_1 and V_2 and the resistances R_1 and R_2 are sufficiently large so that the diode forward voltage V_F is negligibly small compared to V_1 and V_2. The current I_F is then given by

$$I_F = \frac{V_1 - V_F}{R_1 + R_3} \simeq \frac{V_1}{R_1 + R_3}$$

At $t = 0$, the switch is suddenly moved over to position B. This initiates the recovery phase of the diode, and the current is abruptly changed from its forward value I_F to a constant reverse value $I_R \simeq -V_2/(R_2 + R_3)$ as shown in Fig. 9.9b. The constant current $-I_R$ is maintained as long as there is sufficient stored charge of excess holes in the neutral n-region. During this period, the diode remains forward biased and thus acts as a short circuit with a small voltage drop.

Figure 9.10 illustrates the hole distribution in the neutral n-region of the diode. Just before $t = 0$, the diode is carrying a steady forward current I_F and the slope of excess hole distribution at x_n is given by

$$\left.\frac{\partial p_e}{\partial x}\right|_{x=x_n} = -I_F/qAD_p$$

At $t = 0$ the current becomes $-I_R$ and the slope at x_n changes to a positive value

$$\left.\frac{\partial p_e}{\partial x}\right|_{x=x_n} = I_R/qAD_p \tag{9.43}$$

As time passes, the slope given by Eq. (9.43) remains almost constant until the excess hole concentration $p_e(x_n) = [p_n(x_n) - p_{no}]$ reduces to zero. Once this point

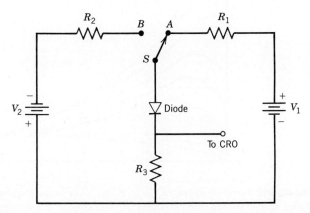

FIGURE 9.8 Basic circuit to study the turn-off transient in a p-n junction diode.

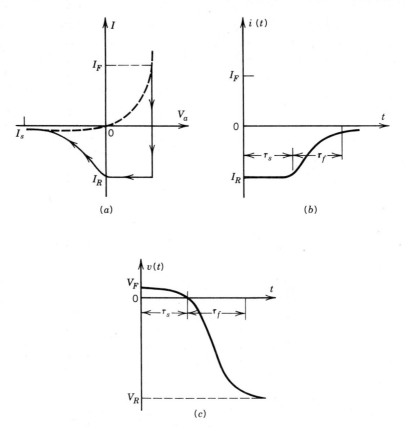

FIGURE 9.9 Turn-off transient in a p^+-n junction diode. (a) Switching trajectory, (b) the current, and (c) the voltage as a function of time.

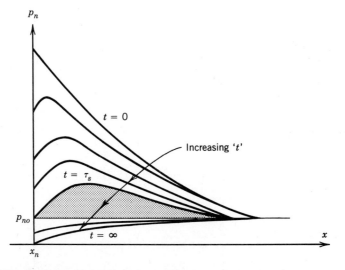

FIGURE 9.10 Decay of stored charge of excess holes in the long-base p^+-n junction diode. The shaded area shows the stored charge at $t = \tau_s$.

is reached, the stored hole charge will no longer be sufficient to support the constant current $-I_R$. Thus, the slope of the hole concentration and the reverse current start decreasing, and the junction voltage becomes negative. The excess hole concentration also becomes negative. Eventually, a point is reached when the diode has fully recovered to its nonconducting state and the slope of the hole concentration at x_n attains a constant value corresponding to the reverse saturation current $-I_s$. Under this condition, the hole concentration $p_n(x_n)$ approaches zero making $p_e(x_n) \simeq -p_{no}$.

The diode switching trajectory (i.e., the path followed by the current and the voltage) and the time dependence of current and voltage are shown in Fig. 9.9. Note that the diode remains forward biased until $t = \tau_s$. Beyond this point, the diode starts reverse biasing itself. The reverse bias tends to attain a constant value as the diode current approaches its reverse saturation value $-I_s$. The time during which the reverse current decreases from its value I_R to 0.1 I_R is called the *fall time* τ_f of the diode. The total time $\tau_r = \tau_s + \tau_f$ is called the *diode reverse recovery time*.

Expressions for the storage and fall times can be obtained from the solution of the charge control equation making suitable approximations. During the storage phase, the diode voltage changes only by a small amount (i.e., from V_F to 0). Hence, the current through the depletion region capacitance can be ignored, and Eq. (9.32) reduces to

$$\frac{dQ_p}{dt} + \frac{Q_p}{\tau_F} = i(t)$$

Since $i(t) = -I_R$ remains almost constant during the storage phase, the above equation has the solution

$$Q_p(t) = C_1 \exp(-t/\tau_F) - I_R \tau_F \tag{9.44}$$

The constant of integration C_1 is determined by the initial condition, $Q_p(0) = I_F \tau_F$ which gives $C_1 = \tau_F(I_F + I_R)$ and

$$Q_p(t) = (I_F + I_R)\tau_F \exp(-t/\tau_F) - I_R \tau_F \tag{9.44a}$$

It now becomes necessary to make some assumption about the charge remaining in the n-region at $t = \tau_s$. Referring to Fig. 9.11, let us assume that the excess holes are stored over a distance $(x_1 + x_2)$ which is on the order of L_p in a long-base diode but is equal to W_n in the case of a short-base diode. We assume a triangular shape for the hole concentration with the slopes

$$\tan(-\theta_1) \propto -\frac{\partial p_e}{\partial x}\bigg|_{t=0} = I_F/qAD_p \tag{9.45a}$$

and

$$\tan \theta_2 \propto \frac{\partial p_e}{\partial x}\bigg|_{t=\tau_s} = -I_R/qAD_p \tag{9.45b}$$

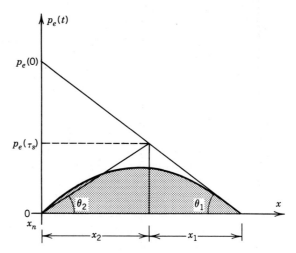

FIGURE 9.11 Triangular approximation for the excess hole concentration at $t = \tau_s$.

giving

$$\frac{\tan \theta_1}{\tan \theta_2} = I_F/I_R$$

If $p_e(0)$ and $p_e(\tau_s)$ represent the maxima in the excess hole concentration at $t = 0$ and $t = \tau_s$, respectively, then from the similar triangles of Fig. 9.11 we have

$$\frac{p_e(\tau_s)}{p_e(0)} = \frac{1}{1 + \tan \theta_1/\tan \theta_2} = \frac{1}{1 + I_F/I_R} \quad (9.46a)$$

The stored charge is determined by the area of the triangle whose base is $(x_1 + x_2)$, and the height at any time is given by the maximum in the electron concentration. This situation allows us to write

$$\frac{Q_p(\tau_s)}{Q_p(0)} = \frac{p_e(\tau_s)}{p_e(0)} = \frac{1}{1 + I_F/I_R} \quad (9.46b)$$

Substituting for $Q_p(\tau_s)$ from Eq. (9.46b) into Eq. (9.44a), writing $Q_p(0) = I_F \tau_F$, and solving for τ_s yield

$$\tau_s = \tau_F \left[\ln\left(1 + \frac{I_F}{I_R}\right) - \ln\left(1 + \frac{I_F}{I_F + I_R}\right) \right] \quad (9.47)$$

Note that in Eq. (9.47) both I_F and I_R are positive numbers. The storage time of a long-base diode is obtained by substituting $\tau_F = \tau_p$ in the above equation giving

$$\tau_s = \tau_p \left[\ln\left(1 + \frac{I_F}{I_R}\right) - \ln\left(1 + \frac{I_F}{I_F + I_R}\right) \right] \quad (9.47a)$$

It is evident from Eq. (9.47) that τ_s dpends on the effective lifetime τ_F and is a function of both I_F and I_R. Since $\tau_F < \tau_p$, for a given ratio I_F/I_R a short-base diode has a lower value of τ_s than a long-base diode.

A simpler approximation is to take $Q_p(\tau_s) = 0$ instead of that given by Eq. (9.46b). Solving Eq. (9.44a) with $Q_p(\tau_s) = 0$ and assuming a long-base diode lead to the relation

$$\tau_s = \tau_p \ln\left(1 + \frac{I_F}{I_R}\right) \qquad (9.48)$$

A more accurate determination of τ_s involves the solution of a time-dependent continuity equation with appropriate boundary conditions. For a long-base diode the result is [5]

$$\tau_s = \tau_p \left[\operatorname{erfc}\left(\frac{I_F}{I_F + I_R}\right)\right]^2 \qquad (9.49)$$

where erfc is the complementary error function. In general, Eqs. (9.47a) and (9.48) are poor approximations of the accurate expression (9.49). However, for $I_F = I_R$ we obtain $\tau_s = 0.7\tau_p$ from Eq. (9.48) but Eq. (9.47a) gives $\tau_s = 0.29\tau_p$ which is close to the value $\tau_s = 0.25\tau_p$ obtained from Eq. (9.49).

The diode gets reverse biased after $t = \tau_s$, and the current through the depletion region capacitance can no longer be ignored. Let us define a time constant τ_R for the reverse current such that

$$i(t) = -\frac{Q_p(t)}{\tau_R} \quad \text{for } t > \tau_s \qquad (9.50)$$

It is possible to determine τ_R from the measurements of τ_s. However, an approximate expression for τ_R is obtained by writing the second term in the bracket of Eq. (9.47) as $\ln(1 + \tau_R/\tau_F)$ giving

$$\tau_R = \tau_F \left(\frac{I_F}{I_F + I_R}\right) \qquad (9.51)$$

From the circuit in Fig. 9.8 we have $V_2 = v(t) + Ri(t)$ where $R = (R_2 + R_3)$. Thus, we obtain

$$\frac{dv}{dt} = -\frac{R\,di}{dt}$$

Substituting this value of dv/dt and for $Q_p(t)$ from Eq. (9.50) into Eq. (9.32), we obtain the following relation:

$$\frac{di}{dt}(\tau_R + R\bar{C}_j) + i(t)\left(1 + \frac{\tau_R}{\tau_F}\right) = 0 \qquad (9.52)$$

This equation has the solution

$$i(t) = A_1 \exp(-t/\tau)$$

At $t = \tau_s$ we have $i(\tau_s) = -I_R$ and

$$i(t) = -I_R \exp\left[-\left(\frac{t - \tau_s}{\tau}\right)\right] \qquad (9.53)$$

where

$$\tau = \frac{\tau_R + R\overline{C}_j}{1 + \tau_R/\tau_F} \qquad (9.54)$$

The fall time τ_f, during which the current falls from I_R to $0.1I_R$, is obtained from Eq. (9.53) as

$$\tau_f \approx 2.3\left[\frac{\tau_R + R\overline{C}_j}{1 + \tau_R/\tau_F}\right] \qquad (9.55)$$

In many instances τ_R is small compared to $R\overline{C}_j$, and we obtain $\tau_f \approx 2.3R\overline{C}_j$. Note that $\overline{C}_j$ is the average capacitance of the diode between the two extreme voltage limits 0 to $-V_2$ and can be approximated to the average of the capacitances at zero and at the full reverse bias.

As an illustration, consider a forward-biased diode with $I_F = 1.5$ mA. Let a reverse bias of 10 V be applied to the diode through a 5 KΩ resistance at $t = 0$. Let the junction capacitance be 18 pF at zero bias, and 6 pF at full reverse bias, so that $\overline{C}_j \simeq 12$ pF. Let us assume $\tau_F = 2 \times 0^{-7}$ sec and estimate the values of τ_s and τ_f.

Neglecting the voltage drop across the diode, we get

$$|I_R| = \frac{10}{5} = 2 \text{ mA}$$

and from Eq. (9.47)

$$\tau_s = 2 \times 10^{-7}\left[\ln\left(1 + \frac{1.5}{2}\right) - \ln\left(1 + \frac{1.5}{3.5}\right)\right] = 4 \times 10^{-8} \text{ sec}.$$

From Eq. (9.51), we get

$$\tau_R = \frac{1.5}{3.5}\tau_F = \frac{3}{7}\tau_F$$

and substitution in Eq. (9.55) gives

$$\tau_f = 2.3\left[\frac{\frac{3}{7} \times 2 \times 10^{-7} + 5 \times 10^3 \times 12 \times 10^{-12}}{1 + \frac{3}{7}}\right] = 2.35 \times 10^{-7} \text{ sec}$$

Thus, $\tau_r = \tau_s + \tau_f = 2.75 \times 10^{-7}$ sec.

9.4 HIGH-SPEED SWITCHING DIODES

A finite recovery time limits the use of diodes for rectification and switching. The diode can be used as a rectifier only for frequencies whose time period is large compared to the recovery time. When the time period of the ac voltage to be rectified is of the same order of magnitude as the reverse recovery time, the rectifying ability of the diode is lost. This happens because during the positive half cycle of the ac voltage, the excess carrier charges will be stored in the neutral n- and p-regions, while during the negative half-cycle these charges will be withdrawn. Consequently, the diode will act as a capacitor rather than a rectifier. Thus, depending on the value of its reverse recovery time, each diode has an upper frequency limit for which it can be used as a rectifier.

In switching applications, the diode is turned on and off by fast pulses, and the finite reverse recovery time limits the rate of pulses that can be applied, thus limiting the diode switching speed. Diodes intended for high-frequency rectification and fast switching speeds must have a very short recovery time. This can be achieved by reducing the storage time by making τ_F as small as possible. The effective lifetime τ_F can be reduced either by decreasing the minority carrier lifetime or by using a short-base diode for which τ_F can be made almost equal to the transit time τ_N. A short-base diode may have an undesirably large value of reverse current and a low reverse breakdown voltage because of punch-through. A more practical approach then is to use a p^+-n-n^+ structure (Fig. 9.12) in which the n-type midregion thickness W_1 is kept small compared to the hole diffusion length. The electric field at the n-n^+ junction prevents hole injection in the n^+-region during forward bias and aids their rapid removal during the reverse bias. For this structure the hole transit time is $\tau_N = W_1^2/2D_p$, and it can be made small compared to τ_p. Note that a n^+-p-p^+ structure with the same W_1 is still better because of the larger value of the electron diffusion constant.

The minority carrier lifetime is reduced either by doping the semiconductor with a deep level impurity such as gold in Si, or by electron bombardment as discussed in Sec. 5.4. The price one has to pay for reducing the minority carrier lifetime is an increased reverse saturation current. The typical storage time of a switching diode ranges from 1 to 10 nsec.

The reverse recovery characteristics of a p-n junction can be improved by incorporating a large dopant concentration gradient near the junction as shown in Fig. 9.13a. The electric field resulting from the sharp dopant gradient keeps the injected minority carriers confined very close to the junction region when the diode is forward biased. Upon application of the reverse bias, the built-in field accelerates the removal of the injected minority carriers whose concentration at the junction reaches zero at the time the stored charge has been removed. Moreover, because of the sharp dopant concentration gradient, the difference between the depletion region width at zero bias and at full reverse bias is small, and

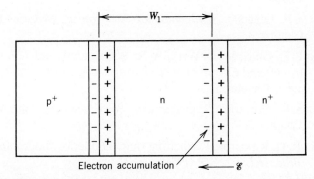

FIGURE 9.12 Schematic diagram of a p^+-n-n^+ fast switching diode.

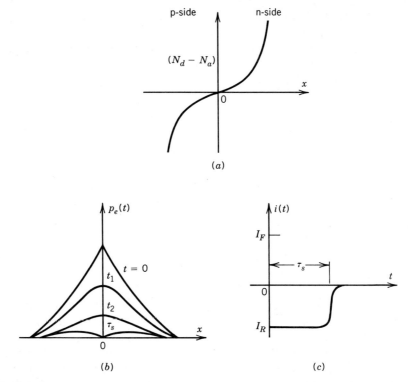

FIGURE 9.13 The step recovery diode: (a) impurity distribution near the junction, (b) excess minority carrier distributions, and (c) the current waveform after switching from forward to reverse bias.

charging time of the depletion region capacitance is reduced. Such diodes are known as *step recovery diodes* [6]. Figures 9.13b and c show the excess carrier distribution and the current waveform after the reverse current pulse is applied.

Another solution to reduce the reverse recovery time is to eliminate minority carriers and to use diodes in which current is carried by majority carriers. These devices are considered in the next chapter.

REFERENCES

1. J. L. Moll, *Physics of Semiconductors,* McGraw-Hill, New York, 1964, Chapter 7.
2. J. F. Gibbons, *Semiconductor Electronics,* McGraw-Hill, New York, 1966, Chapter 8.
3. P. E. Gray, D. DeWitt, A. R. Boothroyd, and J. F. Gibbons, *Physical Electronics and Circuit Models of Transistors,* SEEC Vol. 2, John Wiley, New York, 1964, Chapter 5.
4. S. K. Ghandhi, *Semiconductor Power Devices,* John Wiley, New York, 1977, Chapter 3.
5. R. H. Kingston, "Switching time in junction diodes and junction transistors," *Proc. IRE.* 42, 829 (1954).
6. J. L. Moll, S. Krakauer, and R. Shen, "p-n junction charge-storage diodes," *Proc. IRE.* 50, 43 (1962).

ADDITIONAL READING LIST

1. A. BAR-LEV, *Semiconductors and Electronic Devices*, 2d ed., Prentice-Hall International, London, 1984, Chapter 11.
2. A. FORTINI and S. MUNOGLU, "Influence of series resistance and internal capacitance on the forward voltage decay in a p-n junction," *Solid-State Electron* 30, 357–360 (1987).
3. V. K. TEWARY and S. C. JAIN, "Open-circuit voltage decay in solar cells," *Advances in Electronics and Electron Physics,* Vol. 67, 329–414 (1986).

PROBLEMS

9.1 Consider a Si abrupt p^+-n junction long-base diode with a forward current of 2 mA at 300 K. The hole lifetime in the n-region is 10^{-7} sec. Neglecting the depletion region capacitance, calculate the diode impedance at frequencies of 140 KHz, 1.2 MHz, and 350 MHz.

9.2 (a) Ignoring the depletion region capacitance, calculate the impedance of a Si abrupt p^+-n junction diode at 500 MHz. The diode is carrying a forward current of 1 mA and has the following parameters: $W_n = 5$ μm, $\tau_p = 10^{-6}$ sec, $D_p = 12$ cm^2/sec; the junction area is 3×10^{-4} cm^2 and $T = 300$ K. (b) Taking $N_d = 2 \times 10^{15}$ cm^{-3} and the built-in voltage $V_i = 0.7$ V, calculate the voltage at which the depletion region capacitance becomes equal to the diffusion capacitance.

9.3 The forward current of a long-bse n^+-p junction diode is given by the ideal diode equation. Show that the low-frequency diffusion capacitance C_D of the diode varies linearly with the forward current I_F. If the slope of the C_D versus I_F plot is 1.6×10^{-5} F/A, calculate the electron lifetime, the stored charge, and the value of C_D at $I_F = 2$ mA.

9.4 Consider a p^+-n junction short-base diode of neutral n-region width W_n. The diode is forward biased at a dc voltage V_o, and a low-frequency small-signal ac voltage is superimposed on the dc bias. Calculate the change in the stored charge ΔQ_p during a full cycle of the ac signal and show that only two thirds of this charge can be reclaimed through the junction.

9.5 Show that τ_N in Eq. (9.27) is related to the hole transit time through the neutral n-region of width W_n and prove that for a short-base p^+-n junction diode the transit time is given by $W_n^2/2D_p$.

9.6 A p^+-n junction short-base diode has the following parameters: $W_n = 25$ μm; the junction area is 10^{-3} cm^2, and $D_p = 40$ cm^2/sec. Determine the value of the hole lifetime in the neutral n-region, so that 95 percent of the injected hole current diffuses through the base and reaches the ohmic contact. Calculate the hole transit time through the base and determine the value of τ_F for the diode.

9.7 A Ge abrupt p^+-n junction long-base diode has a saturation current of 1 μA at 300 K. A 20 mA forward current step is applied to this diode. The time during which the diode voltage rises from 10 percent to 90 percent of its steady state value is 10^{-7} sec. Using stored charge analysis, estimate the diffusion capacitance of the diode at 20 mA.

9.8 A p^+-n junction long-base diode with a saturation current I_s is carrying a steady forward curent I_1 for $t < 0$. At $t = 0$ the diode current is abruptly increased to I_2. Using stored charge approximation, show that the diode voltage as a function of time is given by Eq. (9.41). Assuming $I_s = 1\ \mu A$, $I_1 = 1\ mA$, $I_2 = 10\ mA$, $\tau_p = 10^{-6}$ sec, and $T = 300\ K$, calculate the stored charge and the diode voltage for $t = 0.5\ \mu sec$.

9.9 An abrupt p^+-n junction long-base diode is carrying a steady forward current I_F for $t < 0$. For $t > 0$, the diode current decays as $I_F \exp(-t/\tau_p)$. Determine the stored charge $Q_p(t)$, and using the stored charge approximation, calculate the diode voltage as a function of time.

9.10 A p^+-n junction long-base diode is carrying a steady forward current of 10 mA for $t < 0$. At $t = 0$ the diode is abruptly switched to a reverse current of 10 mA. After $t = 1\ \mu sec$, the diode is again switched to a forward current of 10 mA.
(a) Determine the voltage as a function of time for $t > 1\ \mu sec$.
(b) Calculate the hole charge Q_p stored in the diode at $t = 0$ and also at $t = 1\ \mu sec$. Assume $\tau_p = 2\ \mu sec$ and $I_s = 10^{-8}$ A.

9.11 A p-n^+ junction long-base diode has a saturation current of $0.2\ \mu A$ at $300\ K$. The diode is carrying a steady forward current of 5 mA for $t < 0$. At $t = 0$ the biasing battery is suddenly disconnected, and it is observed that the diode voltage reduces to two thirds of its steady-state value after $10\ \mu sec$. Calculate the electron lifetime in the neutral p-region of the diode.

9.12 A p^+-n junction diode is carrying a steady forward current for $t < 0$. At $t = 0$ the diode is abruptly reverse biased by a 20 V battery in series with a 1 KΩ resistor. Using the stored charge approximation, estimate as accurately as you can the storage and fall times of the diode assuming $\tau_F = 0.2\ \mu sec$ and $\overline{C}_j = 15\ pF$.

10
MAJORITY CARRIER DIODES

INTRODUCTION

In the previous chapter we saw that the response time of p-n junction diodes is limited by the storage and removal of the minority carriers in the neutral n- and p-regions of the diode. Techniques for improving the switching speed of the diodes were also discussed. This chapter is concerned with semiconductor diodes in which the current is carried by majority carriers. These include tunnel, backward, and Schottky barrier diodes.

A brief account of the tunnel diode is presented in Sec. 10.1, whereas the backward diode is the subject matter of Sec. 10.2. The Schottky barrier diode is discussed in Sec. 10.3, where our main concern is to understand the origin of barrier height and the mechanisms of current flow in metal–semiconductor contacts. Furthermore, since low-resistance ohmic contacts are used in almost all devices, a brief account of them is given in Sec. 10.4. The last section, 10.5, describes heterojunctions, which are formed between two dissimilar semiconductors. Not every heterojunction acts as a majority carrier diode. However, the way in which the energy band diagram of a Schottky barrier is constructed is also applicable to heterojunctions. This is why they are taken up in this chapter.

10.1 THE TUNNEL DIODE

The tunnel diode (also known as the Esaki diode) consists of a p-n junction whose two sides are doped so heavily that the Fermi level lies within the conduction band on the n-side and the valence band on the p-side. Because of the heavy dopings on the two sides, the zero-bias depletion region width of the junction is less than 100 Å, and the electric field, therefore, reaches a value higher than 10^6 V/cm. At such high fields, a large tunnel current is observed even at a small reverse bias, causing the Zener breakdown of the junction. Thus, the tunnel diode can be considered a Zener diode with a zero reverse breakdown voltage. However, the forward characteristic of the diode is more complex.

10.1.1 Principle of Operation

The thermal equilibrium energy band diagram of a tunnel diode is shown in Fig. 10.1. The two sides of the junction are heavily doped, and the Fermi level is located inside the allowed bands on both sides. Consequently, the built-in voltage of the junction is larger than the voltage V_g corresponding to the band gap of the semiconductor. At a finite temperature, the distribution of filled states is shown by the shaded area in Fig. 10.1. In the conduction band on the n-side, there is a small number of occupied states but a large number of empty ones. The situation is just the reverse in the valence band on the p-side where a small number of empty states face the occupied states of the conduction band on the n-side. The electrons from the occupied states in the conduction band on the n-side can tunnel into the empty valence band states on the p-side through the forbidden gap, causing a flux of electrons from the n- to the p-side. However, the conventional current I_{CV} flows in the opposite direction as shown in Fig. 10.1. Similarly, electrons from the occupied states of the valence band on the p-side tunnel into the empty conduction band states on the n-side constituting a current I_{VC}. Under thermal equilibrium, I_{VC} exactly balances I_{CV}. Figure 10.2a shows the energy band diagram when a small reverse bias is applied to the diode. In this situation a large number of valence band electrons on the p-side face a large number of empty conduction band states on the n-side. On the other hand, the electrons in the filled states of the conduction band on the n-side face only filled states in the valence band on the p-side. Consequently, electrons tunnel from the p- to the n-side, while the tunneling in the opposite direction becomes impossible. Thus, the current I_{VC} becomes large and increases rapidly with the bias, while I_{CV} tends to zero. This causes a net current from the n- to the p-side.

The situation for a small forward bias is shown in Fig. 10.2b. The conduction band electrons on the n-side now face empty states in the valence band on the p-side. Thus, these electrons can tunnel from the n- to the p-side, causing the tunnel current I_{CV} to increase. Since the electrons in the valence band on the p-side are placed opposite the forbidden energy gap on the n-side, the tunneling of electrons from the p- to the n-side is not possible. Therefore, a current flows across the junction from the p- to the n-side. As the forward bias is increased further,

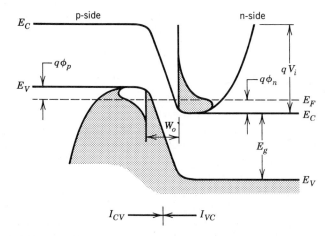

FIGURE 10.1 Energy band diagram of a tunnel diode in thermal equilibrium. The shaded areas show the states occupied by electrons.

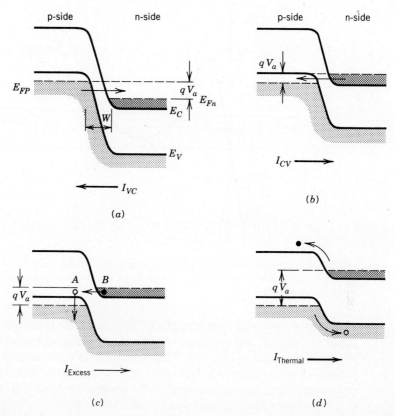

FIGURE 10.2 Simplified energy band diagrams of a tunnel diode for various biasing conditions: (a) reverse bias, (b) small forward bias, (c) cross band condition, and (d) large forward bias, which causes minority carrier injection over the barrier. After R. N. Hall, IRE Trans. Electron Devices, ED-7, 1 (1960).

a situation is reached where a maximum number of empty states in the valence band on the p-side is placed opposite to a maximum number of occupied conduction band states on the n-side. The tunnel current attains its maximum value for this bias. At still higher voltages, the bands on the two sides of the junction begin to pass each other, causing a decrease in the tunnel current. Finally, at a forward bias $V_a = (\phi_n + \phi_p)$, the bottom edge of the conduction band on the n-side comes exactly opposite to the top edge of the valence band on the p-side (Fig. 10.2c). Direct tunneling of electrons is now made impossible, and, ideally, the tunnel current should reduce to zero. However, the measured current is found to be considerably in excess of its theoretical value. This excess current is believed to arise from the tunneling of electrons by way of energy levels within the forbidden energy gap. In a semiconductor doped to degeneracy, electron states are found in the forbidden gap because of the band tails and also because of the spread in the dopant impurity ionization energy (see Sec. 3.8). The postulated tunneling mechanism, responsible for the excess current, is also shown in Fig. 10.2c. An electron at some point B in the conduction band on the n-side can tunnel to a local level at A from which it could drop into the valence band on the p-side. The excess current is found to increase exponentially with the voltage as shown in Fig. 10.3. With a still further increase in the forward bias, the minority carriers are injected over the barrier causing the usual thermal current of holes and electrons (Fig. 10.2d).

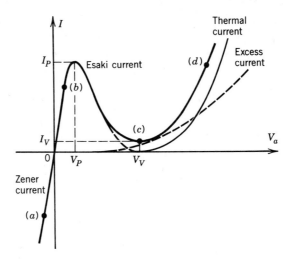

FIGURE 10.3 Current-voltage characteristic of a typical tunnel diode showing the various current components. The points (a), (b), (c), and (d) on the characteristic correspond to the biasing conditions in Fig. 10.2.

The current-voltage characteristic of a tunnel diode is shown in Fig. 10.3. The points (a), (b), (c), and (d) shown on the characteristic correspond to the respective band diagrams in Fig. 10.2. The characteristic has a number of current components. The monotonically increasing reverse current is caused by the tunneling of electrons from the valence band to the conduction band (Fig. 10.2a). This process has already been discussed in Sec. 8.3 and is known as *Zener tunneling*. For low forward voltages, the current is due to the tunneling of electrons from the conduction band on the n-side to the valence band on the p-side. This current is called the *Esaki current*. The forward current initially increases with the bias, reaches its peak value I_P at voltage V_P, and then decreases, passing through a minimum at the valley voltage V_V. For voltages higher than V_V the current consists of two components, both of which rise exponentially with the bias voltage. However, the excess current has a weaker dependence on the bias voltage than the thermal current. Between V_P and V_V the characteristic shows a region of differential negative resistance.

10.1.2 I-V Characteristic

The above discussion shows that the current that flows through the tunnel diode is the difference between the Esaki current I_{CV} and the Zener current I_{VC}. Let us first consider I_{CV}. This current obviously would be determined by (1) the number of electrons in the conduction band on the n-side, (2) the number of unoccupied states in the valence band on the p-side, and (3) the tunneling probability P_t of an electron through the barrier. Consequently,

$$I_{CV} = A_C \int_{E_C}^{E_V} N(E) F_C(E) N'(E) [1 - F_V(E)] P_t dE \qquad (10.1)$$

where $N(E)$ and $N'(E)$ are the densities of states in the conduction and valence bands, respectively. $F_C(E)$ and $F_V(E)$ represent the respective probabilities of occupancy of the state by an electron and are given by the FD distribution function. A_C is a constant that involves the diode area. Obviously, the product $N(E)F_C(E)$ gives the density of occupied states in the conduction band on the n-side, and $N'(E)[1 - F_V(E)]$ gives the density of empty states in the valence band on the p-side. Similarly, the Zener current I_{VC} can be written as

$$I_{VC} = -A_C \int_{E_C}^{E_V} N'(E) F_V(E) N(E) [1 - F_C(E)] P_t \, dE \qquad (10.2)$$

where the significance of the terms $N'(E)F_V(E)$ and $N(E)[1 - F_C(E)]$ can be drawn from the above description. In Eqs. (10.1) and (10.2) the limits of integration are always from E_C on the n-side to E_V on the p-side as this is the only available energy range in which tunneling occurs. The total current I is obtained by adding the two components giving

$$I = A_C \int_{E_C}^{E_V} [(F_C(E) - F_V(E)] N(E) N'(E) P_t \, dE \qquad (10.3)$$

The integral in Eq. (10.3) has been evaluated to obtain a closed form expression for I under the following assumptions: (1) P_t is given by Eq. (8.10) and is taken as a constant over the small voltage range from 0 to V_V; (2) $N(E)$ varies as $(E - E_C)^{1/2}$ in the conduction band and $N'(E)$ varies as $(E_V - E)^{1/2}$ in the valence band; and (3) both $(E_V - E_F)$ and $(E_F - E_C)$ are less than $2kT$. With these assumptions we obtain [1]

$$I = \frac{A'_C P_t q V_a}{kT} [(\phi_n + \phi_p) - V_a]^2 \qquad (10.4)$$

where A'_C is a constant different from A_C, V_a is the forward voltage, and $q\phi_n = (E_F - E_C)$ and $q\phi_p = (E_V - E_F)$ represent the penetrations of the Fermi level into the conduction and valence bands, respectively. For a degenerately doped semiconductor, ϕ_n can be approximated by [2]

$$\phi_n \approx \frac{kT}{q} \left[\ln\left(\frac{N_d}{N_C}\right) + 0.35\left(\frac{N_d}{N_C}\right) \right] \qquad (10.5)$$

A similar expression can be obtained for ϕ_p after replacing N_d and N_C by N_a and N_V, respectively. Equation (10.4) is valid only up to the valley voltage $V_V = (\phi_n + \phi_p)$. Differentiating Eq. (10.4) with respect to V_a and equating $dI/dV_a = 0$, we can show (see Prob. 10.2) that

$$V_P = \frac{(\phi_n + \phi_p)}{3} \quad \text{and} \quad V_V = (\phi_p + \phi_n) \qquad (10.6)$$

The diode conductance remains zero at both these voltages.

A more convenient expression for the diode characteristic valid for all values of forward bias has been obtained in terms of I_P and V_P as [3]

$$I = I_P \left(\frac{V_a}{V_P}\right) \exp\left(1 - \frac{V_a}{V_P}\right) + I_s \left[\exp\left(\frac{qV_a}{kT}\right) - 1\right] + I_{\text{excess}} \qquad (10.7)$$

Here the first term gives the tunnel current, and the second term is the thermal current of minority carriers.

The tunnel diode characteristic is relatively insensitive to temperature. The peak current I_P may increase or decrease with temperature, depending on doping on the two sides of the junction, but both V_P and I_V decrease with an increase in temperature. The valley voltage V_V also decreases at a rate of about 1 mV/°C. Most of these effects are explained by the decrease in E_g with temperature.

10.1.3 The Tunnel Diode as a Circuit Element

The I-V characteristic of a tunnel diode is reproduced in Fig. 10.4a. Note that between I_P and I_V the diode has three values of voltage for a given current, but for a given voltage the current has only one value. Thus, the dynamic resistance exhibited by the diode is a voltage-controlled negative resistance. The negative differential conductance dI/dV_a becomes zero at V_P as well as at V_V and has its maximum negative value ($-1/R_{min}$) at the inflection point P, where

$$R_{min} \simeq 2\, V_P/I_P \tag{10.8}$$

The simple description of a tunnel diode as a pure negative resistance is correct only at relatively low frequencies. At microwave frequencies the junction capacitance and the inductance of lead wires also have to be considered. The small-signal ac equivalent circuit of a packaged diode biased in the negative resistance region is shown in Fig. 10.5. Here C_j is the junction capacitance, $-R$ is the junction resistance, R_s is the series resistance, and L_s is the lead inductance. From the equivalent circuit, the input impedance of the diode Z_{in} is obtained as

$$Z_{in} = \left[R_s + \frac{-R}{1 + (\omega R C_j)^2} \right] + j \left[\omega L_s - \frac{\omega R^2 C_j}{1 + (\omega R C_j)^2} \right] \tag{10.9}$$

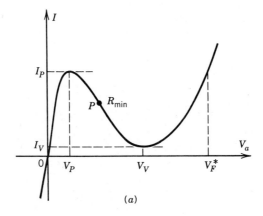

(a)

(b)

FIGURE 10.4 The tunnel diode (a) I-V characteristic and (b) circuit symbol. The negative conductance has its maximum value at the inflection point P.

THE TUNNEL DIODE

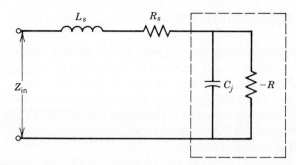

FIGURE 10.5 Small-signal equivalent circuit of a packaged tunnel diode.

The real part of Z_{in} is zero at a frequency f_{r0} given by

$$f_{r0} = \frac{1}{2\pi RC_j}\sqrt{\left(\frac{R}{R_s} - 1\right)} \qquad (10.10\text{a})$$

The imaginary part becomes zero at a frequency f_{x0}, where

$$f_{x0} = \frac{1}{2\pi}\sqrt{\frac{1}{L_s C_j} - \frac{1}{(RC_j)^2}} \qquad (10.10\text{b})$$

In typical circuit applications $f_{x0} > f_{r0} > f_0$, where f_0 is the operating frequency.

10.1.4 Materials and Device Performance

A large majority of commercially available tunnel diodes are fabricated from Ge or GaAs. Si and GaSb have also been used, but it is difficult to manufacture Si tunnel diodes with a high I_P/I_V ratio. The alloying process is used to fabricate the device, and the resulting p-n junction is an abrupt junction. Linearly graded junctions do not result in good tunnel diodes. Typical values of important parameters for Si, Ge, and GaAs tunnel diodes are given in Table 10.1. Here V_F^* is the diode voltage at which the forward current I_F equals the peak current I_P. The voltage V_V given in this table is generally higher than the value obtained from Eq. (10.6). The negative resistance of a tunnel diode can be used to achieve switching, amplification, oscillation, and other circuit functions. The diode can be used in high speed switching circuits and in microwave amplifiers and oscillators.

Besides its high speed, the other advantages of a tunnel diode are its low cost, simplicity, and immunity to ambient. The disadvantages are its low-output voltage swing and hence its low-power handling capability.

TABLE 10.1
Typical Parameters of Tunnel Diodes

Semiconductor	I_P/I_V	$V_P(\text{V})$	$V_V(\text{V})$	$V_F^*(\text{V})$
Ge	8	0.055	0.35	0.5
Si	3.5	0.065	0.42	0.7
GaAs	15	0.15	0.50	1.1

10.2 THE BACKWARD DIODE

In a tunnel diode the two sides of the junction are doped so that the Fermi level lies in the allowed bands. Now consider that the dopings on the two sides are such that the Fermi level on each side lies very close to the respective band edge in the forbidden gap. When this is the case, a large tunnel current is obtained even at small values of reverse bias, but no direct tunneling is possible for a forward bias. Since this diode conducts in the backward (i.e., reverse) direction, it is called a *backward diode*.

Figure 10.6 shows the thermal equilibrium energy band diagram and the current-voltage characteristic of a backward diode. Note that the tunneling of electrons is not possible for a forward bias and that no significant current flows until the minority carriers are injected over the barrier at a sufficiently large bias. However, at a small reverse bias, electrons from the valence band on the p-side tunnel into the empty states in the conduction band on the n-side and a large current is observed. The I-V characteristic of the diode can be described by Eq. (8.15) giving

$$I = A'V_a \mathscr{E}^{3/2} \exp(-\beta E_g^{3/2}/\mathscr{E}) \qquad (10.11)$$

Since $V_i \gg V_a$ in this case, the above expression can be rewritten as (see Prob. 10.5)

$$I = A_1 V_a \exp(\beta_1 V_a) \qquad (10.11a)$$

where A_1 and β_1 are slowly varying functions of V_a. Thus, the current in a backward diode rises exponentially with the applied reverse bias.

10.3 THE SCHOTTKY BARRIER DIODE

A Schottky barrier is a metal–semiconductor contact that possesses rectifying properties. Rectification at metal–semiconductor contact was first observed in 1874 by F. Braun. The early metal–semiconductor diodes were point contact

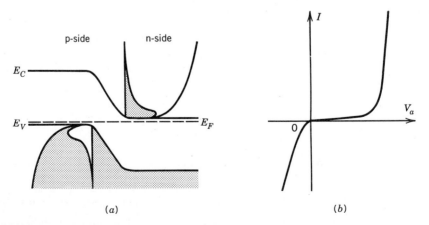

FIGURE 10.6 The backward diode. (*a*) Energy band diagram at thermal equilibrium and (*b*) the static I-V characteristic.

THE SCHOTTKY BARRIER DIODE

diodes that employed a sharpened metallic wire in contact with an exposed surface of a semiconductor. These devices proved highly unreliable and were eventually replaced by rectifiers obtained by the deposition of a thin film of a metal on the surface of a semiconductor. Our present understanding of the behavior of metal–semiconductor (m-s) contacts derives from studies on these devices. In order to understand whether a m-s contact is rectifying or nonrectifying, we have to investigate the mechanism of barrier formation at the metal–semiconductor interface.

10.3.1 Formation of a Barrier

(a) THE SCHOTTKY-MOTT THEORY

Figure 10.7 illustrates the process of barrier formation according to the Schottky-Mott theory. We assume the semiconductor to be the n-type and uniformly doped. Figure 10.7a shows the two substances isolated from each other. Here the vacuum level (which represents the energy of an electron at rest outside the solid) is taken as the reference level. The work function $q\phi_m$ of the metal is assumed to be larger than the semiconductor work function $q\phi_{sc}$. As discussed in Sec. 2.6, the work function of a substance is the energy required to bring an electron from the Fermi level to the vacuum level. The electron affinity $q\chi$ of the semiconductor is the energy difference between the conduction band edge E_C and the vacuum level. When the metal is brought into contact with the semiconductor electrons from the semiconductor conduction band (which have higher energy than the metal electrons) flow into the metal till the Fermi levels in the two substances are aligned. This results in the upward bending of energy bands in the semiconductor as shown in Fig. 10.7b. The electrons that flow into the metal leave behind a positive charge of ionized donors, which extends up to a thickness W_o in the semi-

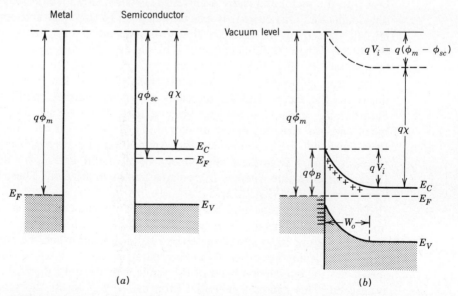

FIGURE 10.7 Energy band diagrams of metal-n-type semiconductor contact with $\phi_m > \phi_{sc}$. (a) Two materials isolated from each other and (b) thermal equilibrium situation after the contact is made.

conductor. However, the electron charge on the metal side is contained within a distance of about an atomic layer and is essentially a surface charge. The dipole layer of charges establishes an electric field from the semiconductor to the metal.

In the energy band diagram of Fig. 10.7b, we have assumed that ϕ_m, χ and the band gap E_g of the semiconductor remain unchanged after the contact is made between the two materials. Thus, we can draw this diagram by drawing the equilibrium Fermi level E_F and then locating the relative positions of E_C, E_V, and the vacuum level with respect to E_F in the neutral semiconductor. Finally, in the transition region the vacuum level from the semiconductor side is made to approach the vacuum level on the metal side gradually to preserve its continuity. The band edges E_C and E_V follow the same variation as the vacuum level, because $q\chi$ remains unchanged. This procedure is quite general and can be used to draw the energy band diagram of any two materials in contact. It is evident from Fig. 10.7b that the band bending in the semiconductor is equal to the difference between the two work functions. The difference $V_i = (\phi_m - \phi_{sc})$ is called the built-in voltage, and the barrier height as viewed from the metal toward the semiconductor is given by

$$q\phi_B = q(\phi_m - \chi) \tag{10.12a}$$

$$= qV_i + (E_C - E_F) \tag{10.12b}$$

since $q\phi_{sc} = q\chi + (E_C - E_F)$. Equation (10.12a) was stated by W. Schottky and independently by N. F. Mott; henceforth it will be referred to as the Schottky approximation. In most cases, the potential ϕ_B is orders of magnitude higher than kT/q, and the space-charge region in the semiconductor becomes a depletion region similar to that of a p-n junction.

It can now be easily seen that the contact in Fig. 10.7b is a rectifying contact. In thermal equilibrium at a temperature T, a small fraction of conduction band electrons will have sufficient energy to surmount the barrier. These electrons flow into the metal and cause a current I_{ms}, which flows from the metal to the semiconductor. This current is exactly balanced by an equal and opposite current I_{sm} caused by the electron flow from the metal into the semiconductor Fig. 10.8a). When the semiconductor is biased negative with respect to the metal by a voltage V_F (Fig. 10.8b), the barrier for electrons in the semiconductor decreases from qV_i to $q(V_i - V_F)$. More electrons can now flow from the semiconductor to the metal, and I_{ms} increases above its thermal equilibrium value. However, I_{sm} remains unchanged because there is no voltage drop across the metal and $q\phi_B$ remains almost unaltered. Thus, there is a net flow of current from the metal to the semiconductor. Application of a reverse bias V_R reduces the electron flow from the semiconductor to the metal, and I_{ms} is reduced below its equilibrium value whereas I_{sm} remains almost unaffected. Thus, a small reverse current flows (Fig. 10.8c).

The above discussion applies to the n-type semiconductor with $\phi_m > \phi_{sc}$. The energy band diagrams for the n-type semiconductor with $\phi_m < \phi_{sc}$ are shown in Fig. 10.9. Figure 10.9a shows the energy band diagram when the two materials are isolated from each other. After the contact is made, electrons flow from the metal into the conduction band of the semiconductor until thermal equilibrium is reached. This causes a potential drop $(\phi_{sc} - \phi_m)$ across the semiconductor (Fig. 10.9b). The negative charge of electrons that accumulate in the semiconductor is confined to a thickness on the order of the Debye length and is es-

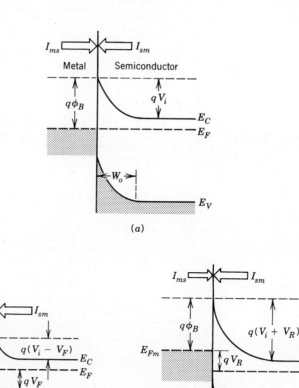

FIGURE 10.8 Energy band diagrams of rectifying metal-n-type semiconductor contact at (a) thermal equilibrium, (b) forward bias, and (c) reverse bias.

sentially a surface charge. The same is also true for the positive charge in the metal. Since there is no depletion region in the semiconductor, no barrier exists for the electron flow either from the semiconductor into the metal or in the opposite direction. When a bias is applied to this system, practically all the applied voltage appears across the neutral semiconductor as shown in Figs. 10.9c and d. This type of nonrectifying contact is often referred to as the *ohmic contact*.

From the above discussion it is seen that a metallic contact on the *n*-type semiconductor is rectifying when $\phi_m > \phi_{sc}$ and is nonrectifying when $\phi_m < \phi_{sc}$. The opposite is true for a metal–*p*-type semiconductor contact. For example, Fig. 10.10a shows the energy band diagrams of the isolated metal and *p*-type semiconductor having $\phi_{sc} > \phi_m$. When the two substances are brought into intimate contact, electrons flow from the metal into the semiconductor until E_F is the same throughout. Each electron flowing into the semiconductor removes a hole from the valence band, leaving behind an unneutralized charge of ionized acceptor in the semiconductor. Thus, a depletion region is formed in the semiconductor as shown in Fig. 10.10b. Since the current in a *p*-type semiconductor is carried mainly by holes, it is not difficult to see that the contact of Fig. 10.10b is a rectify-

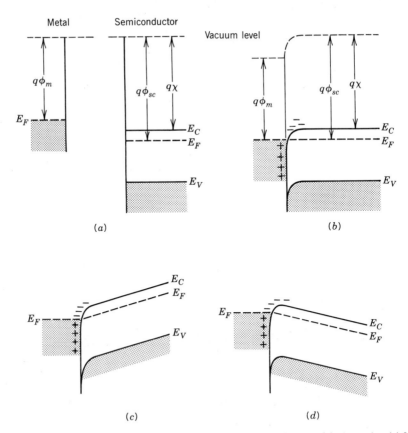

FIGURE 10.9 Energy band diagrams of a metal-n-type semiconductor with $\phi_m < \phi_{sc}$. (a) Materials isolated from each other, (b) contact at thermal equilibrium, (c) negative bias on the semiconductor, and (d) positive bias on the semiconductor.

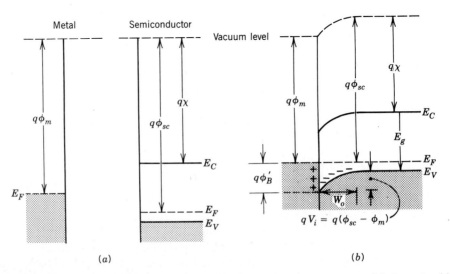

FIGURE 10.10 Energy band diagrams of metal-p-type semiconductor contact with $\phi_m < \phi_{sc}$. (a) Materials isolated from each other and (b) situation after the contact is made and thermal equilibrium is reached.

ing contact and the barrier height $q\phi'_B$ to the hole flow is

$$q\phi'_B = (q\chi + E_g - q\phi_m) \qquad (10.13)$$

From Eqs. (10.12a) and (10.13) we obtain $q(\phi_B + \phi'_B) = E_g$, which shows that for a given m-s system, the sum of electron and hole barrier heights equals the band gap of the semiconductor. Schottky barrier contacts on *p*-type semiconductors are less frequently used. Consequently, our subsequent discussion will be confined to metal–*n*-type semiconductor systems only.

(b) MODIFICATIONS TO THE SCHOTTKY-MOTT THEORY

Practical m-s contacts do not obey the guidelines provided by the Schottky-Mott theory. It is observed that regardless of the values of ϕ_m and ϕ_{sc}, a large majority of m-s combinations form rectifying contacts with potential barriers. Moreover, Eq. (10.12a) predicts that ϕ_B varies linearly with ϕ_m. Although ϕ_B depends on ϕ_m in some semiconductors, it is almost independent of ϕ_m in covalently bonded semiconductors like Ge and Si.

One of the first explanations for the insensitivity of barrier height to ϕ_m was given by John Bardeen who pointed out the importance of localized surface states. In a covalently bonded crystal, the surface atoms have no neighbors on the vacuum side with whom they can make covalent bonds. Thus, each surface atom has one broken covalent bond known as the *dangling bond*. Dangling bonds give rise to surface states that are continuously distributed in energy within the forbidden gap. These states pin the Fermi level at the surface and thus influence the barrier height.

Figure 10.11 illustrates the process of barrier formation in the presence of surface states. Figure 10.11a shows the energy band diagram of an *n*-type semiconductor under the flat-band condition. Here no net charge exists either in the surface states or in the bulk semiconductor. The surface states are characterized by a neutral level $q\phi_0$ such that all states below $q\phi_0$ are occupied while those above it are empty. This situation is one of nonequilibrium. Equilibrium is reached when electrons from the semiconductor adjacent to the surface occupy states above $q\phi_0$ and the Fermi level becomes constant throughout (Fig. 10.11b). Thus, the surface becomes negatively charged, and a depletion region is created within the semiconductor near the surface. If a metal is now brought into contact with the semiconductor, exchange of electrons takes place largely between the metal and the semiconductor surface states and the depletion region charge remains practically unaffected (Fig. 10.11c). In the limit that the density of surface states becomes infinitely large, we obtain

$$q\phi_B = (E_g - q\phi_0) \simeq \frac{2}{3} E_g \qquad (10.14)$$

since in covalently bonded semiconductors $q\phi_0$ is estimated to occur $E_g/3$ above the valence band. Equation (10.14) will be called the *Bardeen approximation*.

C. A. Mead [4] has suggested that, from the viewpoint of barrier formation, semiconductors can be classified into two categories. In the first category are covalently bonded semiconductors like Si and GaAs. These materials have a large density of surface states in the band gap, and the barrier height is thus given by Eq. (10.14). In the second category are the semiconductors with ionic bonding

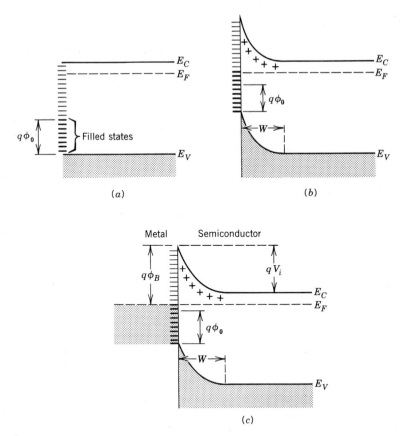

FIGURE 10.11 Energy band diagrams illustrating the barrier-formation process on n-type semiconductor with a large density of surface states. (a) Flat-band condition, (b) surface in thermal equilibrium with the bulk, and (c) semiconductor in contact with a metal.

like ZnS. In these materials there are no surface states in the band gap, and ϕ_B is determined primarily by the difference $(\phi_m - \chi)$. Since the work function is not known accurately for many metals, Mead replaced it by the electronegativity $q\chi_m$ of the metal, which has a well-defined value and is linearly related to $q\phi_m$. An analysis of a number of m-s systems showed that ϕ_B can be expressed by the empirical relation

$$\phi_B = S\chi_m + \phi_0(s) \tag{10.15}$$

where $\phi_0(s)$ represents the contribution of the surface states and $S = d\phi_B/d\chi_m$ gives the dependence of barrier height on $q\chi_m$. For a semiconductor with two different types of atoms, the electronegativity difference between the anion and cation of the semiconductor is a measure of ionicity. A plot of S as a function of this electronegativity difference is shown in Fig. 10.12. Note that semiconductors with ionic bonding have a large value of S, but S is small for covalently bonded materials.

In recent years there has been growing awareness that, even if Eq. (10.15) is true in some cases, it does not provide a complete description of m-s contacts. A precise knowledge of the microscopic structure of the m-s interface and the inter-

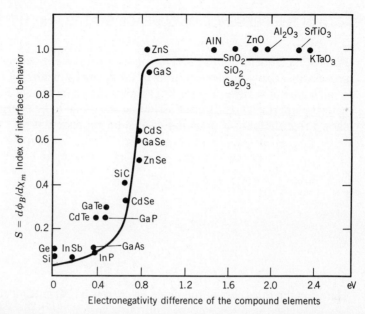

FIGURE 10.12 Plot showing variation of the interface index S as a function of the electronegativity difference between cations and anions of compound semiconductors. (From S. Kurtin, T. C. McGill, and C. A. Mead, Fundamental transitions in electronic nature of solids, *Phys. Rev. Letters*, vol. 22, p. 1434. Copyright © 1969. Reprinted by permission of the American Physical Society, New York.)

facial reaction of the metal–semiconductor atoms is necessary for understanding the behavior of metal contacts on semiconductors.

10.3.2 Classification of Metal–Semiconductor Interfaces

Four different types of interfaces between metals and semiconductors can be delineated: (1) A metal is only physisorbed on the surface of a semiconductor without making any chemical bond. (2) A nonreactive metal forms a weak chemical bond with a highly polarizable semiconductor (dielectric constant > 7) but does not form any compound with it. (3) A highly polarizable semiconductor reacts with a metal to form one or more chemical compounds. (4) A thin insulating film of native oxide prevents intimate contact between the metal and the semiconductor.

The type 1 interface represents an ideal Schottky barrier, while the type 2 interface approximates to the Bardeen barrier, with ϕ_B given by Eq. (10.14). Type 3 and 4 interfaces are more common and will be considered below in some detail.

(a) CONTACT ON REACTIVE INTERFACES

The most extensively studied reactive interfaces are the metal silicides, which are formed as a result of chemical reaction between the transition metal and silicon. The silicides exhibit metallic conductivity, and the barrier is formed at the silicide–Si interface, which is located in the interior of the semiconductor. Thus, the barrier height does not depend on the surface property but is determined by a parameter related to metallurgical reaction. Studies on a number of transition metal silicides-Si systems have revealed [5] that ϕ_B decreases almost linearly with the eutectic temperature.

(b) CONTACTS IN THE PRESENCE OF SURFACE STATES AND THE INTERFACIAL LAYER

Prior to metal deposition, the semiconductor surface is chemically cleaned. This process invariably leaves a thin (5 to 20 Å thick) insulating oxide layer on the semiconductor surface. Figure 10.13 shows the energy band diagram of a contact with the interfacial oxide layer assuming the oxide to be charge free. It has been shown that the barrier potential ϕ_B can be expressed as [6]

$$\phi_B = C_1(\phi_m - \chi) + (1 - C_1)(E_g/q - \phi_0)$$
$$= C_1\phi_m + C_2 \tag{10.16}$$

where

$$C_1 = \frac{\varepsilon_i}{\varepsilon_i + q^2\delta D_s} \tag{10.16a}$$

Here δ is the thickness of the oxide layer, ε_i is its permittivity, and D_s is the density of surface states per eV per unit area within the band gap. If $D_s = 0$, then $C_1 = 1$, and Eq. (10.16) reduces to (10.12). On the other hand, if D_s is very large, C_1 goes to zero and ϕ_B approaches the value given by Eq. (10.14).

Measurement of barrier height on the chemically cleaned surfaces of a large number of m-s combinations indicates that ϕ_B depends on the method of surface preparation. Hence, the values of C_1 and C_2 are not unique for a metal–semiconductor system. Nevertheless, some investigators have fitted straight lines to the measured ϕ_B versus ϕ_m data, determined C_1 and C_2, and obtained the values of ϕ_0 and D_s (see Prob. 10.10).

As an example, consider a metal–Si Schottky barrier having an SiO$_2$ layer with $\delta = 15$ Å and $\varepsilon_i = 3.9\varepsilon_o$. Let $qD_s = 3.5 \times 10^{13}$ cm^{-2}, $(E_g - q\phi_0) = 0.82$ eV and

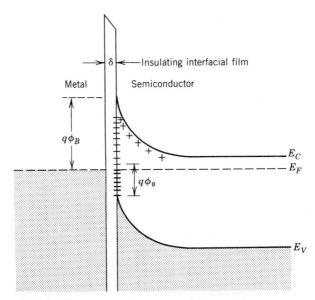

FIGURE 10.13 Energy band diagram of a metal-semiconductor contact with surface states and interfacial oxide layer.

$\chi = 4.05$ V. Using Eq. (10.16a), we get

$$C_1 = \frac{3.9 \times 8.85 \times 10^{-14}}{3.9 \times 8.85 \times 10^{-14} + 1.6 \times 10^{-19} \times 1.5 \times 10^{-7} \times 3.5 \times 10^{13}} = 0.29$$

and from Eq. (10.16) we obtain

$$\phi_B = 0.29\phi_m + (1 - 0.29) \times 0.82 - 0.29 \times 4.05 = 0.29\phi_m - 0.587$$

10.3.3 Contacts on Vacuum-Cleaved Surfaces

Metallic contacts on vacuum-cleaved surfaces have been fabricated mainly to understand the mechanism of barrier formation [7, 8]. These contacts are prepared by cleaving the semiconductor in a vacuum just before the metal deposition. Hence, there is no insulating oxide layer between the metal and semiconductor in these intimate contacts. Let us investigate what factors determine the barrier height in these clean contacts.

In covalently bonded semiconductors, Mead has assumed [4] that surface states intrinsic to the semiconductor pin the Fermi level at the surface. These states are supposed to remain unchanged even after the semiconductor comes into contact with the metal and thus determine the barrier height. This viewpoint is not acceptable for clean contacts because the interface states can be altered by the presence of the metal for two reasons. First, as pointed out by V. Heine [9], electrons from the metal can tunnel into the forbidden gap of the semiconductor. The wave functions of these electrons decay exponentially into the semiconductor, leaving tails that penetrate the semiconductor up to a depth of about 10 Å. These tail states can become more important than the intrinsic surface states and may play a dominant role in determining the barrier height. Second, when metal is evaporated onto the semiconductor surface, the intermixing of atoms that occurs across the boundary creates chemical defects near the m-s interface. These defects give rise to donor and acceptor states that may pin the Fermi level at the surface. Considerable evidence exists which shows that, in a large majority of covalently bonded semiconductors, the barrier height at the m-s interface is determined by the interface states produced by the intermixing of metal semiconductor atoms. This is true not only for clean m-s contacts, but also for contacts with an interfacial oxide layer [7].

10.3.4 The Roles of the Interfacial Oxide and Ionicity

The presence of native oxide on the surface of the semiconductor produces interface states whose nature and density depend only on the oxide–semiconductor combination. The oxide layer suppresses the tunneling of metal electrons into the forbidden gap of the semiconductor, and more significantly, it tends to reduce the importance of interdiffusion. If the density of interface states at the oxide-semiconductor interface is sufficiently large, the barrier height becomes insensitive to the metal work function.

As mentioned earlier, in most cases the barrier height is determined by the chemical reactivity of the constituent atoms at the m-s interface. The chemical reactivity may be expressed in terms of heat of reaction of the metal–semiconductor system. Covalently bonded semiconductors have a low-heat of reaction and tend to react readily with the metal to produce chemical defects in the interfacial region. However, semiconductors with ionic bonding have a high-heat of reaction and do not readily react with the metal. Consequently, the barrier height in semiconductors with predominantly ionic bonding is strongly dependent on the metal work function. In the light of these observations, the plot in Fig. 10.12 can be readily interpreted in terms of the heat of reaction of the metal–semiconductor system rather than the existence of intrinsic surface states. The heat of reaction can be directly correlated with the electronegativity difference and appears to be largely responsible for the occurrence of interface states.

10.3.5 The Voltage Dependence of Barrier Height

ϕ_B is dependent on voltage largely because of the image force and the presence of the interfacial oxide. Here we will investigate only the barrier lowering owing to the image force.

The image force barrier lowering can be understood by referring to Fig. 10.14. An electron at a distance x from the metal experiences an electric field perpendicular to the metal surface. This field may be calculated by assuming a hypothetical image charge q located at a distance $(-x)$ inside the metal. The force of attraction between the electron and its image charge is $q^2/4\pi\varepsilon_d(2x)^2$, and the electron has a potential energy $E(x) = -q^2/16\pi\varepsilon_d x$ relative to that of an electron at infinity as shown by the dashed curve in Fig. 10.14. This energy must be added to the barrier energy $-q\mathscr{E}x$ to obtain the total potential energy $PE(x)$ of the electron. Thus

$$-PE(x) = \frac{q^2}{16\pi\varepsilon_d x} + q\mathscr{E}x \qquad (10.17)$$

The maximum in $PE(x)$ occurs at a distance x_m from the metal surface. From Eq. (10.17) it can be shown (see Prob. 10.12) that

$$x_m = \sqrt{q/16\pi\varepsilon_d \mathscr{E}} \qquad (10.18)$$

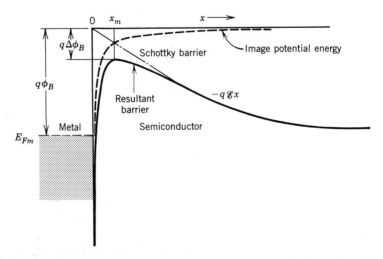

FIGURE 10.14 Energy band diagram showing the barrier lowering due to image force.

and the barrier lowering $\Delta\phi_B$ is given by

$$\Delta\phi_B = 2\mathscr{E}x_m = (q\mathscr{E}/4\pi\varepsilon_d)^{1/2} \tag{10.19a}$$

The maximum field strength $\mathscr{E}_m$ for a bias voltage V_a is given by Eq. (10.23), and substituting this value for $\mathscr{E}$ in Eq. (10.19a), we get

$$\Delta\phi_B = \left[\frac{q^3 N_d}{8\pi^2 \varepsilon_d^2 \varepsilon_s}(V_i - V_a)\right]^{1/4} \tag{10.19b}$$

The image force permittivity ε_d may be different from the static permittivity ε_s because when the electron transit time through the barrier region is small compared to the dielectric relaxation time, the semiconductor does not get fully polarized. However, in most cases the transit time is sufficiently large and we can write $\varepsilon_d = \varepsilon_s$.

10.3.6 Capacitance Voltage Characteristics

The electric field and potential distributions in the space-charge region of a Schottky barrier are obtained from the solution of a one-dimensional Poisson equation. Because there is practically no voltage drop in the metal, the results for a Schottky barrier on the n-type semiconductor are identical to those for a one-sided abrupt p^+-n junction. Thus, using the depletion approximation, we can write

$$\mathscr{E}(x) = \mathscr{E}_m\left(1 - \frac{x}{W}\right) \tag{10.20}$$

$$\phi(x) = \frac{qN_d}{\varepsilon_s}[Wx - x^2/2] \tag{10.21}$$

$$W = \left[\frac{2\varepsilon_s}{qN_d}(V_i - V_a)\right]^{1/2} \tag{10.22}$$

Here V_a is positive for forward bias and negative for reverse bias. The maximum in $\mathscr{E}$ occurs at the m-s boundary $x = 0$ and is given by

$$\mathscr{E}_m = \frac{-2(V_i - V_a)}{W} = -\sqrt{2qN_d(V_i - V_a)/\varepsilon_s} \tag{10.23}$$

The depletion region charge Q_d per unit area and the depletion region capacitance C are obtained as

$$Q_d = qN_d W = \sqrt{2q\varepsilon_s N_d(V_i - V_a)} \tag{10.24}$$

and

$$C = A\left|\frac{dQ_d}{dV_a}\right| = A\sqrt{\frac{q\varepsilon_s N_d}{2(V_i - V_a)}} = \frac{\varepsilon_s A}{W} \tag{10.25}$$

where A is the cross-sectional area of the Schottky barrier contact. From Eq. (10.25) we obtain

$$\frac{1}{C^2} = \frac{2(V_i - V_a)}{A^2 q \varepsilon_s N_d} \qquad (10.26a)$$

A more accurate analysis that takes into account the contribution of majority carriers (i.e., electrons in this case) to the depletion region charge gives a slightly different result [10]:

$$\frac{1}{C^2} = \frac{2(V_i - V_a - kT/q)}{A^2 q \varepsilon_s N_d} \qquad (10.26b)$$

Thus, a plot of $1/C^2$ versus V_a gives a straight line from which it is possible to obtain both V_i and N_d.

Let us consider a metal–silicon Schottky barrier with a (A^2/C^2) versus a V_a slope of -3.3×10^{15} (cm/F)^2 V^{-1}. Using Eq. (10.26b), we obtain

$$N_d = \frac{2}{q \varepsilon_s} \left(\frac{1}{3.3 \times 10^{15}} \right)$$

$$= \frac{2}{1.6 \times 10^{-19} \times 1.05 \times 10^{-12} \times 3.3 \times 10^{15}} = 3.6 \times 10^{15} \text{ cm}^{-3}$$

When N_d varies with distance into the semiconductor, a plot of $1/C^2$ versus V_a is not a straight line. However, the slope of the characteristic at any point W, at the edge of the depletion region, is still given by $-2/A^2 q \varepsilon_s N_d(W)$ and W is obtained from the relation $C = \varepsilon_s A/W$. Thus, by measuring the slope of $1/C^2$ versus V_a plot at a given depth W, it is possible to determine the dopant concentration $N_d(W)$. This provides a convenient method of measuring the dopant impurity distribution in a semiconductor.

10.3.7 Current Flow Mechanisms Through the Barrier [10,11]

The various ways in which charge carriers can be transported from semiconductor to metal in a forward-biased Schottky barrier contact are shown in Fig. 10.15.

Figure 10.15a illustrates the process of *thermionic emission* (TE). Here an electron that has energy higher than $q(V_i - V_F)$ is emitted into the metal over the barrier. If the semiconductor is doped to degeneracy, the depletion region becomes very thin and electrons can tunnel into the metal through the barrier. The two ways in which tunneling can occur are shown in Fig. 10.15b. At low temperature, electrons near the Fermi level in the semiconductor can tunnel into the metal. This process is known as *field emission* (FE). As the temperature increases, electrons are excited to higher energies where they "see" a thinner and lower barrier. These electrons can tunnel into the metal before reaching the top of the barrier. This is known as *thermionic field emission* (TFE). In a forward-biased Schottky barrier diode, electrons are injected into the depletions region from the neutral semiconductor, and holes are injected from the metal. These electron–hole pairs recombine in the depletion region to cause a forward current (Fig. 10.15c). When the barrier height in a metal–n-type semiconductor contact

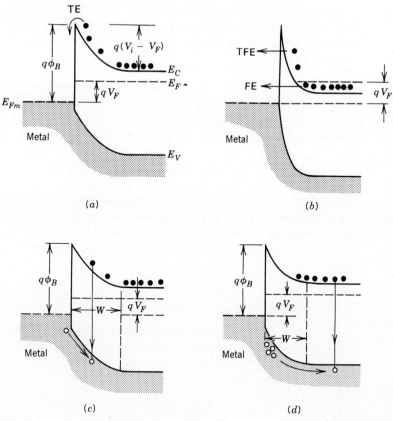

FIGURE 10.15 Energy band diagrams showing different current transport processes in metal-*n*-type semiconductor Schottky barrier diode. (*a*) Thermionic emission, (*b*) tunneling, (*c*) electron–hole pair recombination in the depletion region, and (*d*) minority carrier injection.

becomes larger than $E_g/2$, the semiconductor adjacent to the metal becomes *p*-type. Under a forward bias holes from the p-region diffuse into the neutral n-region. These holes recombine with electrons in the neutral n-region, causing a minority carrier current similar to that in a forward-biased p-n junction (Fig. 10.15*d*).

Thermionic emission is usually the dominant mechanism in Si and GaAs Schottky barrier diodes and leads to the ideal diode characteristic. Tunneling and carrier generation and recombination in the depletion region cause a departure from the ideal behavior. A brief account of these processes follows.

(a) THERMIONIC EMISSION

Before an electron is emitted into the metal over the barrier, it has to pass through the depletion region in the semiconductor. The electron motion within the depletion region is governed by the drift and diffusion processes, while its emission into the metal is determined by the density of available states in the metal. The two processes are essentially in series, and the one that causes the larger impediment to the electron flow determines the current. In high-mobility semiconductors like Ge, Si, and GaAs, the current is limited by thermionic emission over the barrier. Consequently, we will ignore the effect of drift and diffusion in the barrier region in the following discussion.

It is obvious from Fig. 10.15a that only those electrons that have energy higher than $q(V_i - V_a)$ can reach the top of the barrier. Assuming a Maxwellian distribution of velocities, we observe that the electron concentration n_{to} at the top of the barrier is given by

$$n_{to} = n_o \exp[-(V_i - V_a)/V_T] \qquad (10.27a)$$

where n_o is the electron concentration in the neutral semiconductor. For a nondegenerate semiconductor n_o is given by Eq. (3.20), and we obtain

$$n_{to} = N_C \exp[-(\phi_B - V_a)/V_T] \qquad (10.27b)$$

If the electrons have an isotropic distribution of velocities, then from the kinetic theory of gases the electron flux incident on the barrier is $\bar{v}n_{to}/4$. Assuming that all the electrons incident on the barrier cross into the metal, we see that the current I_{ms} flowing from the metal to semiconductor becomes

$$I_{ms} = \frac{qA\bar{v}}{4} N_C \exp[-(\phi_B - V_a)/V_T] \qquad (10.28)$$

Here $\bar{v}$ is the average thermal velocity of electrons in the semiconductor.

Under thermal equilibrium, I_{ms} is still given by Eq. (10.28) with $V_a = 0$. Since no net current flows in this situation, we must have

$$I_{sm} = -\frac{qA\bar{v}}{4} N_C \exp(-\phi_B/V_T) \qquad (10.29)$$

When a voltage V_a is applied, the barrier height $q\phi_B$ remains practically unchanged, and thus I_{sm} becomes constant. Writing $I_{sm} = I_s$ and combining Eq. (10.28) and (10.29) lead to

$$I = I_s\left[\exp\left(\frac{V_a}{V_T}\right) - 1\right] \qquad (10.30)$$

For a Maxwellian distribution $\bar{v} = (8kT/\pi m_e)^{1/2}$. Substituting for N_C from Eq. (3.21), we can write the saturation current I_s as

$$I_s = AR^*T^2 \exp(-\phi_B/V_T) \qquad (10.31)$$

where

$$R^* = 4\pi m_e q k^2/h^3$$

is the Richardson constant for the thermionic emission of electrons from the metal into the semiconductor having electron effective mass m_e.

A more rigorous analysis that combines the processes of drift and diffusion and thermionic emission into a single theory has shown that the diode current can still be described by Eq. (10.30), but I_s is now given by [6]

$$I_s = \frac{qAN_C v_R}{1 + v_R/v_d} \exp(-\phi_B/V_T) \qquad (10.32)$$

where v_d is the effective diffusion velocity of electrons through the depletion region and $v_R = \bar{v}/4$. It is evident that for $v_d \gg v_R$, Eq. (10.32) reduces to Eq. (10.31) and thermionic emission theory applies. If we assume that the electron velocity v_d through the barrier is constant equal to $\mu_n \mathscr{E}_m$, the condition for the validity of thermionic emission becomes $\mu_n \mathscr{E}_m \gg \bar{v}/4$. Writing $\mu_n = q\tau_c/m_e$ and $\bar{l} = \bar{v}\tau_c$, we obtain

$$\bar{l} \gg 2kT/\pi q\mathscr{E}_m \tag{10.33}$$

where $\bar{l}$ and τ_c represent the electron mean free path and relaxation time, respectively. The right-hand side of Eq. (10.33) represents the distance within which the barrier decreases by an amount $2kT/\pi$ from its maximum value. Thus, if $\bar{l}$ is large compared to this distance, the thermionic emission theory is valid. The inequality in Eq. (10.33) is well satisfied for Schottky barriers on Ge, Si, and other high-mobility semiconductors.

(b) TUNNELING

Tunneling can occur either as field emission (FE) or as thermionic field emission (TFE). FE is possible only in semiconductors that are doped to degeneracy, but TFE can occur even in nondegenerate materials. Tunneling would cause a significant increase in the diode current because it represents a process in parallel with thermionic emission. In Si and GaAs Schottky barriers, TFE becomes significant only when the dopant concentration $N_d > 10^{17}$ cm^{-3}.

(c) CARRIER RECOMBINATION IN THE DEPLETION REGION

The expression for the depletion region generation recombination current derived for a p-n junction diode is also valid for a Schottky barrier diode. Thus, using Eq. (7.32)

$$I_{rg} = I_{Ro}\left[\exp\left(\frac{V_a}{2V_T}\right) - 1\right] \tag{10.34}$$

where

$$I_{Ro} = qAn_iW/2\tau_o \tag{10.34a}$$

Here τ_o is the carrier lifetime in the depletion region. From Eqs. (10.31) and (10.34a), we have

$$\frac{I_{Ro}}{I_s} = \frac{qn_i}{R^*T^2}\left(\frac{W}{2\tau_o}\right)\exp(\phi_B/V_T) \tag{10.35}$$

This relation shows that I_{Ro} is likely to become important only in the case of large barrier height, low temperature, and a lightly doped semiconductor (i.e., large value of W).

(d) MINORITY CARRIER INJECTION

The hole current I_p in a long-base Schottky barrier diode is the same as in a long-base p-n junction diode and can be written as (see Eq. (7.17)):

$$I_p = \frac{qAD_pn_i^2}{N_dL_p}[\exp(V_a/V_T) - 1] \tag{10.36}$$

From Eqs. (10.30), (10.31), and (10.36) we have

$$\frac{I_p}{I} = \frac{qD_p n_i^2}{R^*T^2 L_p N_d} \exp(\phi_B/V_T) \tag{10.37}$$

In a Schottky barrier made on Si, ϕ_B is typically 0.8 V and the above ratio remains lower than 10^{-3} for $N_d = 10^{15}$ cm^{-3} and below.

10.3.8. Current-Voltage Characteristics

(a) FORWARD CHARACTERISTIC

Schottky barrier diodes have been made on a number of semiconductors, and most of the devices follow the I-V relation of the form

$$I = I_s \left[\exp\left(\frac{V_a}{\eta V_T}\right) - 1 \right] \tag{10.38}$$

where η is called the *ideality factor*. For an ideal Schottky barrier, diode $\eta = 1$. Factors that make η to exceed unity are: (1) the bias dependence of barrier height, (2) the tunneling of electrons through the barrier, and (3) the electron–hole pair recombination in the depletion region. The depletion region recombination in Si and GaAs diodes becomes significant only for $N_d < 10^{15}$ cm^{-3}, and tunneling occurs for concentrations in excess of 10^{17} cm^{-3}. A large majority of Schottky barrier diodes are made in the dopant concentration range of 10^{15} cm^{-3} to 10^{17} cm^{-3}. The departure of η from unity in these devices occurs mainly from the bias dependence of the barrier height. It has been shown that [10]

$$\frac{1}{\eta} = \left(1 - \frac{\partial \phi_B}{\partial V_F}\right) \tag{10.39}$$

Since ϕ_B increases with the forward bias, $(\partial \phi_B/\partial V_F)$ is positive and η is larger than unity.

(b) REVERSE CHARACTERISTIC

According to thermionic emission theory, the reverse current of a Schottky barrier diode should saturate to a value I_s given by Eq. (10.31). However, in actual devices the reverse current is always found to increase with the bias. In good diodes where proper design eliminates the surface leakage, the most common cause of an increase in current is the decrease in the barrier height with an increase in the reverse bias. Substituting $(\phi_B - \Delta\phi_B)$ for ϕ_B in Eq. (10.31) shows that the reverse current increases as $\exp(\Delta\phi_B/V_T)$. A primary cause of barrier lowering is the image force on the electron emitted from the metal into the semiconductor. The image force barrier lowering can be obtained from Eq. (10.19b) after replacing V_a by the reverse voltage $-V_R$.

Other mechanisms that cause the reverse current to increase with bias are the tunneling of electrons from the metal into the semiconductor and electron–hole pair generation in the depletion region.

THE SCHOTTKY BARRIER DIODE

10.3.9 Measurement of Barrier Height

(a) C-V MEASUREMENT

In this method, the diode is reverse biased at a dc voltage, and its capacitance is measured by superimposing a small ac signal on the bias voltage. These measurements are made for a number of values of reverse bias, and $1/C^2$ is plotted as a function of the bias voltage. As seen in Eq. (10.26b), the plot represents a straight line with an intercept $V_o = (V_i - kT/q)$ on the voltage axis, and we obtain

$$\phi_B = V_o + kT/q + (E_C - E_F)/q - \Delta\phi_B \tag{10.40}$$

The C-V method measures the barrier height under flat-band conditions. To obtain the zero-bias barrier height, we have to subtract the barrier lowering that is present at zero bias.

As an example, consider a metal–silicon Schottky barrier with $N_d = 3 \times 10^{15}$ cm^{-3}. At 300 K we have $(E_C - E_F)/q = 0.236$ V and $kT/q = 0.0258$ V. If the measured intercept is $V_o = 0.67$ V, the value of $\Delta\phi_B$ at zero bias obtained from Eq. (10.19b) is 0.0173 V and from Eq. (10.40) we obtain $\phi_B = 0.915$ V.

(b) CURRENT-VOLTAGE MEASUREMENT

In these measurements the diode current I is measured as a function of the forward bias and $\ln I$ is plotted against the applied voltage. The saturation current I_s is obtained by extrapolating the straight line to $V_a = 0$. If the diode area A and the Richardson constant R^* are known, ϕ_B can be obtained from Eq. (10.31).

(c) PHOTOELECTRIC MEASUREMENT

If a monochromatic light with $q\phi_B < h\nu < E_g$ is made incident on the Schottky barrier from the metal side, the incident photons will excite some electrons from the metal over the barrier. For $(h\nu - q\phi_B) > 3kT$ the resulting photocurrent I_{ph} is given by the relation $I_{ph} = B_1(h\nu - q\phi_B)^2$ where B_1 is a constant of proportionality. Thus, a plot of $\sqrt{I_{ph}}$ against $h\nu$ gives a straight line whose intercept on the $h\nu$ axis directly gives the barrier height.

Results of various measurements have shown that the barrier height depends on the method of semiconductor surface preparation. As a result, there is considerable scatter in the values of ϕ_B measured by different workers for the same m-s combination. The average values of barrier height for some metal-semiconductor combinations are given in Table 10.2.

10.3.10 Comparison with the p-n Junction Diode

A Schottky barrier diode has a number of advantages over a p-n junction diode. First, the thermionic emission process leads to a current density J_s of the order of 10^{-7} A/cm^2 in Si and GaAs. This is more than four orders of magnitude higher than the saturation current density of p-n junction diodes made in these materials. Thus, for the same forward current a Schottky barrier diode will have about 0.3 V less voltage drop than a p-n junction diode. Second, the ideality factor η of a Schottky barrier diode lies in the range 1.02 to 1.15, which is smaller than that

TABLE 10.2
Measured Schottky Barrier Heights of Various Metals on Some Elemental and Compound Semiconductors (in eV) at 300 K

Semiconductor	E_g (eV)	Type	Ag	Al	Au	Cu	In	Mg	Ni	Pb	Pt	W
Ge	0.67	n	0.54	0.48	0.59	0.52	0.64	—	0.49	0.38	—	0.48
Ge		p	0.50	—	0.30	—	0.55	—	—	—	—	—
Si	1.12	n	0.78	0.72	0.80	0.58	—	0.4	0.61	—	0.90	0.67
Si		p	0.54	0.58	0.34	0.46	—	—	0.51	0.55	—	0.45
GaAs	1.42	n	0.88	0.80	0.90	0.82	0.83	0.77	—	—	0.84	0.80
GaAs		p	0.63	—	0.42	—	—	—	—	—	—	—
GaP	2.24	n	1.20	1.07	1.30	1.20	—	1.04	1.27	—	1.45	—
CdS	2.43	n	0.56	—	0.78	0.50	—	—	0.45	0.59	1.10	—
CdSe	1.70	n	0.43	—	0.49	0.33	—	—	—	0.37	—	—
ZnS	3.60	n	1.65	0.80	2.00	1.75	1.50	0.82	—	—	1.84	—
ZnSe	2.7	n	1.21	0.76	1.47	1.10	0.91	—	—	1.16	1.40	—

of a p-n junction diode. Both features make the Schottky barrier diode a preferred device in low-voltage, high-current rectifiers. However, because of its large reverse current, a Schottky barrier diode is not suitable for use in high-voltage, low-current applications.

Another advantage of a Schottky barrier diode is its fast transient response. Since the hole current in a Schottky barrier diode forms only a negligibly small fraction of the total current, there is practically no minority carrier storage. Consequently, the diode has no storage time, and its recovery time is determined by the product CR_s where R_s is the diode series resistance and C is the depletion region capacitance. The reverse recovery time of a Schottky barrier diode may be as low as 10^{-11} sec, which is orders of magnitude smaller than the recovery time of a p-n junction diode. At a constant forward current the temperature dependence of the forward voltage of a Schottky barrier diode is lower than that of a p-n junction diode. This also favors the use of the Schottky barrier diode in some applications.

The disadvantage of a Schottky barrier diode is its large reverse current. Moreover, Schottky barrier diodes have a poor reproducibility, though they are easier to fabricate.

10.3.11 Device Structures and Technology

Two semiconductors that are widely used to make Schottky barrier diodes are Si and GaAs. Gold, platinum, and tungsten are the most commonly used metals, but aluminum is also used for Schottky barriers in integrated circuits.

Four different structures of Si-Schottky barrier diodes are shown in Fig. 10.16. Each of these devices employs an n-type epitaxial layer on n^+-substrate to reduce the diode series resistance. The simple structure shown in Fig. 10.16a is fabricated by depositing a metal dot of suitable size onto the semiconductor surface. This device shows a large reverse leakage current and has a low breakdown voltage because of the concentration of the electric field near the periphery. To

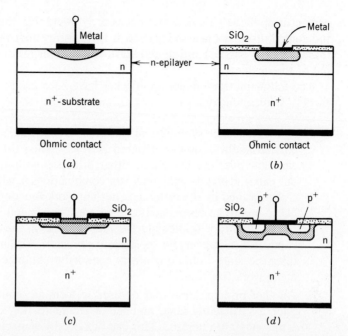

FIGURE 10.16 Four different types of Schottky barrier diode structures: (a) Simple metal contact onto an exposed surface of the semiconductor. Planar structure (b) without metal overlap, (c) with metal overlap, and (d) with a diffused p^+-guard ring. The shaded areas show the depletion region.

overcome these problems, most of the present-day Schottky barrier diodes are constructed using planar technology (see Sec. 19.4). A thin SiO_2 layer is grown on the semiconductor surface, and windows are opened in the oxide using standard photolithography. The three common structures fabricated in this way are shown in Figs. 10.16b to d. The structure of Fig. 10.16b does not give ideal characteristics because of the relatively sharp edge and a positive charge that exists in the oxide. These conditions create a high electric field in the depletion region near the diode periphery, leading to similar effects as in the structure shown in Fig. 10.16a. The edge effects can be eliminated by allowing the metal layer to overlap the oxide as shown in Fig. 10.16c. The depletion region under the oxide is now rounded off, avoiding the occurrence of the sharp edge. Another way to avoid the edge effects is to fabricate a guard ring structure as shown in Fig. 10.16d. This structure, however, has a p-n junction and suffers from long reverse recovery times.

After metal deposition, most of the metal–semiconductor contacts are subjected to heat treatment to promote the adhesion of the metal to the semiconductor. The temperature at which the contact is subjected to heat treatment has to be kept below the eutectic temperature of the metal–semiconductor system. Even at these low temperatures, the migration of semiconductor atoms through the metal may occur. As a consequence, both the barrier height and I-V characteristic may undergo a change.

10.4 OHMIC CONTACTS [12]

A metal–semiconductor contact is defined as ohmic if its resistance is negligibly small compared with the resistance of the semiconductor specimen to which the contact is applied. This does not necessarily imply that the I-V characteristic of the contact itself is linear. However, it is necessary that the contact should not inject minority carriers and should serve only as a means of taking current in and out of the device.

The following three major approaches have been used to achieve ohmic contacts to semiconductors:

1. A low-resistance symmetrical contact to a semiconductor is obtained if the barrier height is small compared with kT. When this is the case, carriers can flow over the barrier in either direction without any impediment. If one can find metal–semiconductor combinations in which the barrier height is determined by the difference in their work functions, it should be possible to create an ohmic contact by choosing a metal with $\phi_m < \phi_{sc}$ in the case of an n-type semiconductor and $\phi_m > \phi_{sc}$ in the case of a p-type semiconductor. However, for most semiconductors, because of the presence of interface states, ohmic contacts cannot be obtained by proper choice of metal work function.

2. Another principle that may be used to form an ohmic contact is to introduce a sufficiently large number of recombination centers at the metal–semiconductor interface. This can be done by damaging the semiconductor surface either by sandblasting or by ion bombardment. When a metal is deposited on a damaged surface, electron–hole pair recombination in the depletion region dominates the current transport and the contact presents a very low resistance to the current flow.

3. Low-resistance contact to a semiconductor is obtained when the depletion region of the contact barrier is reduced in thickness by heavy doping of the semiconductor. Since a very high field is present in a thin barrier even at zero bias, the reduction in thickness also causes a decrease in the barrier height because of a lowering of the barrier due to image force. Thus, for a thin depletion region the barrier becomes transparent to electrons in both directions and FE and TFE dominate the current transport. Although the I-V characteristic of the contact itself is nonlinear, its resistance is very low in comparison to that of the semiconductor sample.

The most widely used method for making ohmic contacts to semiconductors is the third method. Ohmic contact to a lightly doped semiconductor is formed by creating a thin, heavily doped semiconductor region of the same conductivity type between the metal and the bulk semiconductor. For example, ohmic contact to an n-type semiconductor is made by producing an n^+-region near the surface so as to form an n^+-n junction either before or after the metal deposition. Similarly, formation of a p^+-p junction is required for making ohmic contact to a p-type semiconductor. The n^+-n and p^+-p high-low junctions have very high leakage currents showing linear I-V relation for both directions of current flow. Figure 10.17 shows the energy band diagram of a metal-n^+-n semiconductor contact. The contact barrier is very thin and is essentially transparent to electrons in both directions.

The high-low junction may be formed by the incorporation of an appropriate dopant by diffusion, ion implantation, or epitaxy before metal deposition. The heavy doping in the interfacial region can also be achieved by depositing an alloy containing an element that acts as a donor or acceptor and by a subsequent heat treatment of the alloy above its melting point.

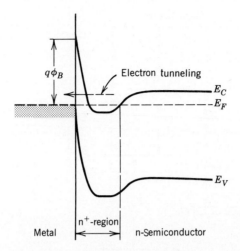

FIGURE 10.17 Energy band diagram of a metal -n^+-n (high-low) ohmic contact. (From S. M. Sze, reference 6, p. 306. Copyright © 1981. Reprinted by permission of John Wiley & Sons, Inc., New York.)

A contact is generally characterized by its specific contact resistance, which is the product of contact resistance and its area. A good ohmic contact should have a specific contact resistance of less than 10^{-4} Ωcm^2.

10.5 HETEROJUNCTIONS [13,14]

A junction formed between two dissimilar semiconductors is called a *heterojunction*. It can either be a p-n junction or an n-n (or a p-p) junction. The p-n junction is called an *anisotype junction* and carries current due to minority carrier injection, while the n-n junction is called an *isotype heterojunction* and is a majority carrier device.

10.5.1 Energy Band Diagram

Our first concern in approaching the heterojunction problem is to construct the energy band diagram. We will first consider an ideal heterojunction that has no interfacial layer of any kind between the two semiconductors as well as no interface states. The ideal junction model was first proposed by R. L. Anderson and is considered below in Fig. 10.18. One of the most widely investigated semiconductor heterojunctions that closely conforms to the Anderson model is the Ge-GaAs heterojunction. Here the lattice constant mismatch between the two semiconductors is less than 1 percent, so that the interface states density is low.

Figure 10.18a shows the energy band diagram of two isolated semiconductors. Semiconductor 1 is p-type and has a smaller band gap E_{g1} but a larger electron affinity than the n-type semiconductor 2. When the two materials are brought into intimate contact, electron–hole flow occurs across the junction until thermal equilibrium is reached. This process creates a depletion region of ionized donors on the n-side and of ionized acceptors on the p-side in the same manner as in a p-n homojunction. To draw the equilibrium energy band diagram of the system we draw a common E_F and locate the positions of respective band edges and the vacuum level relative to E_F in the neutral n- and p-regions. In the junction deple-

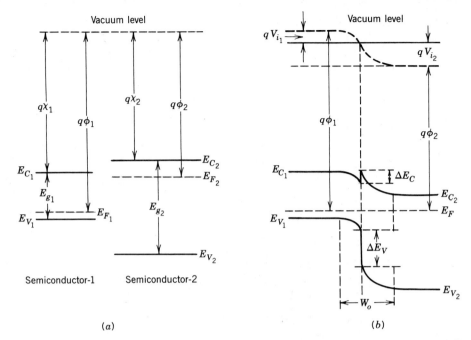

FIGURE 10.18 Energy band diagrams showing (a) two neutral isolated semiconductors and (b) heterojunction after the contact is made and thermal equilibrium is reached.

tion region, the vacuum level from semiconductor 2 is made to gradually approach that of semiconductor 1. Finally, the band edges E_{C1}, E_{C2}, E_{V1}, and E_{V2} are located parallel to the vacuum level. The energy band diagram of the heterojunction obtained using this procedure is shown in Fig. 10.18b. Note that the exact nature of the band bending will be determined by the dopant distributions on the two sides of the junction.

The built-in voltage V_i of the junction in Fig. 10.18b is given by

$$V_i = (V_{i1} + V_{i2}) = \phi_1 - \phi_2 \tag{10.41}$$

where V_{i1} and V_{i2} are the voltages that appear across the semiconductors 1 and 2, respectively. Since E_{C1} and E_{C2} remain parallel to the vacuum level, there is an abrupt discontinuity ΔE_C in the conduction band edge at the planar boundary. From Fig. 10.18b it is seen that

$$\Delta E_C = q(\chi_1 - \chi_2) \tag{10.42}$$

Similarly, the abrupt discontinuity ΔE_V in the valence band edge is the difference between the hole affinities of the two semiconductors. Thus

$$\Delta E_V = (E_{g2} - E_{g1}) - q(\chi_1 - \chi_2) \tag{10.43}$$

From the above two relations we obtain

$$\Delta E_C + \Delta E_V = E_{g2} - E_{g1} \tag{10.44}$$

We will now consider practical heterojunctions where cross-diffusion, chemical interaction, and surface damage may be present giving rise to interfacial layers, interface states, and surface dipoles. These nonideal situations modify the idealized energy band diagram. In Fig. 10.19a, an interfacial layer present in semiconductor 1 contains interface dipoles causing a voltage drop $(-\Delta V)$ across this layer. Since the potential drop from the neutral region in semiconductor 1 to that in semiconductor 2 is still $(\phi_1 - \phi_2)$, we must have

$$V_{i1} + V_{i2} - \Delta V = (\phi_1 - \phi_2) \qquad (10.45)$$

which shows that V_{i1} and V_{i2}, and the band bending will now be different from that predicted by Eq. (10.41).

The interdiffusion of atoms across the semiconductor boundary or surface damage may give rise to interface states that may contain a net negative or positive charge. This modifies the band shape across the interface as shown in Fig. 10.19b. Note that the sum of V_{i1} and V_{i2} is again $(\phi_1 - \phi_2)$, but their individual values are now different from those obtained from Fig. 10.18b.

Finally, Fig. 10.19c shows the energy band diagram when there is an interfacial region over which gradual transition occurs from semiconductor 1 to semiconduc-

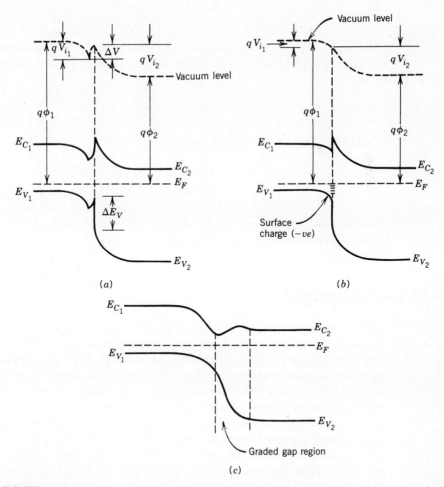

FIGURE 10.19 Energy band diagrams of heterojunction with nonideal situations. (a) Interfacial dipole layer, (b) surface states with a net negative charge, and (c) transition region with graded band gap and electron affinity.

tor 2. Consequently, the band discontinuities ΔE_C and ΔE_V do not occur at a plane but are developed over the graded band gap transition region.

For the ideal heterojunction (Fig. 10.18b) the depletion region widths can be obtained by solving the Poisson equation on two sides of the junction. One of the boundary conditions is the continuity of electric flux across the interface, which means that

$$\varepsilon_1 \mathscr{E}_{m1} = \varepsilon_2 \mathscr{E}_{m2} \tag{10.46}$$

where ε_1 and ε_2 are the permittivities and $\mathscr{E}_{m1}$ and $\mathscr{E}_{m2}$ are the maximum electric field strengths in the two regions. For an abrupt p-n junction we have

$$\mathscr{E}_{m1} = -\frac{2(V_{i1} - V_{a1})}{x_1} \tag{10.47}$$

and

$$\mathscr{E}_{m2} = -\frac{2(V_{i2} - V_{a2})}{x_2} \tag{10.48}$$

where $(V_{a1} + V_{a2}) = V_a$ and x_1 and x_2 are the depletion region widths in semiconductors 1 and 2, respectively. In addition, the condition of space-charge neutrality leads to

$$N_{a1} x_1 = N_{d2} x_2 \tag{10.49}$$

Combining Eqs. (10.46) through (10.49), we obtain

$$\frac{V_{i1} - V_{a1}}{V_{i2} - V_{a2}} = \frac{\varepsilon_2 N_{d2}}{\varepsilon_1 N_{a1}} \tag{10.50}$$

This relation indicates that, if the dopings on the two sides of the junction are significantly different, most of the applied voltage appears across the lightly doped side.

10.5.2 Current Transport

Under ideal conditions, the main current component in a forward-biased p-n heterojunction is caused by the minority carrier injection over the barrier. It is evident from Fig. 10.18b that the electrons in the wide band gap (n-type) semiconductor see a much lower barrier than the holes in the narrow band gap (p-type) semiconductor. Thus, the wide band gap semiconductor injects more minority carriers across the junction than the narrow band gap semiconductor. If the band gaps of the two semiconductors are substantially different, the injection from the narrow band gap semiconductor can be neglected. The current then flows because of minority carrier injection from the wide band gap semiconductor into the narrow band gap semiconductor, even when the dopant concentration in the wide band gap semiconductor is lower than that in the narrow band gap semiconductor. This offers a potential advantage of using heterojunctions in some devices such as bipolar transistors and injection lasers.

For the p-n junction shown in Fig. 10.18b, the current voltage characteristic can be approximated as [13]

$$I = \frac{qAD_n n_{po}}{L_n}\left[\exp\left(\frac{K_2 V_a}{V_T}\right) - 1\right] \quad (10.51)$$

where n_{po} is the thermal equilibrium electron concentration in the p-type semiconductor and $K_2 V_a = V_{a2}$. It turns out that Eq. (10.51) is not in agreement with the experimental results because it ignores several other current components that might be present in actual p-n heterojunctions. For example, depletion region generation-recombination current may become important in some situations. However, in most cases, the current is dominated by the tunneling of electrons through the spikes formed due to discontinuities in the band edges and through the interface states in the junction transition region. The resulting I-V characteristic is discussed by A. G. Milnes and D. L. Feucht [13].

In an isotype (n-n or p-p) heterojunction, the dominant current transport mechanism is the thermionic emission of carriers over the barrier, and the I-V characteristic takes the form

$$I = I_s\left(1 - \frac{V_a}{V_i}\right)\left[\exp\left(\frac{V_a}{V_T}\right) - 1\right] \quad (10.52)$$

where

$$I_s = \frac{qAR^* T V_i}{k}\exp(-V_i/V_T) \quad (10.52\text{a})$$

R^* is the effective Richardson constant, and all other symbols have their usual meanings.

Heterojunctions have shown great promise in a number of devices such as light-emitting diodes, semiconductor lasers, solar cells, photodetectors, and bipolar transistors. Some of these applications will be discussed in subsequent chapters.

REFERENCES

1. J. KARLOVSKY, "Simple method for calculating the tunneling current of an Esaki diode," *Phys. Rev.* 127, 419 (1962).

2. H. KROEMER, "The Einstein relation for degenerate carrier concentrations," *IEEE Trans. Electron Devices* ED-25, 850 (1978).

3. T. A. DEMASSA and D. P. KNOTT, "The prediction of tunnel diode voltage-current characteristics," *Solid-State Electron.* 13, 131 (1970).

4. C. A. MEAD, "Metal-semiconductor surface barriers," *Solid-State Electron.* 9, 1023 (1966).

5. G. OTTAVIANI, K. N. TU, and J. W. MAYER, "Interfacial reaction and Schottky barrier in metal-silicon systems," *Phys. Rev. Lett.* 44, 284 (1980).

6. S. M. SZE, *Physics of Semiconductor Devices,* John Wiley, New York, 1981, Chapter 5.

7. W. E. SPICER, I. LINDAU, P. SKEATH, C.Y. SU and P. CHYE, "Unified mechanism for Schottky barrier formation and III-V oxide interface states," *Phys. Rev. Lett.* 44, 420 (1980).
8. N. NEWMAN, T. KENDELEWICZ, D. THOMSON, S. H. PAN, S. J. EGLASH, and W. E. SPICER, "Schottky barriers on atomically clean cleaved GaAs," *Solid-State Electron.* 28, 307 (1985).
9. V. HEINE, "Theory of surface states," *Phys. Rev.* A138, 1689 (1965).
10. E. H. RHODERICK, *Metal-Semiconductor Contacts*, Clarendon Press, Oxford, 1978.
11. K. SHENAI and R.W. DUTTON, "Current transport mechanisms in atomically abrupt metal-semiconductor interfaces," *IEEE Trans. Electron Devices* ED-35, 468 (1988).
12. E. H. RHODERICK, "Metal-semiconductor contacts," *IEE Proc.* 129 (part I), 1 (1982).
13. A. G. MILNES and D. L. FEUCHT, *Heterojunctions and Metal-Semiconductor Junctions*, Academic Press, New York, 1972.
14. A. G. MILNES, "Semiconductor heterojunction topics: introduction and overview," *Solid-State Electron.* 29, 99 (1986).

ADDITIONAL READING LIST

1. C. N. DUNN, "Tunnel Diodes," in *Microwave Semiconductor Devices and Their Circuit Applications*, H. A. Watson (ed.), McGraw-Hill, New York, 1969, Chapter 13.
2. B. L. SHARMA (ed.), *Metal-Semiconductor Schottky Barrier Junctions and Their Applications*, Plenum Press, New York, 1984, Chapters 1, 2, and 3.
3. V. L. RIDEOUT, "A review of the theory, technology and applications of metal-semiconductor rectifiers," *Thin Solid Films* 48, 261–291 (1978).
4. S. S. COHEN and G. SH. GILDENBLAT, *Metal-Semiconductor Contacts and Devices*, Vol. 13 in *VLSI Electronics*, N. G. Einspruch (ed.), Academic Press, Orlando, Florida, 1986, Chapters 2, 3, 4, and 5.
5. A. G. MILNES, "Heterojunctions: some knowns and unknowns," *Solid-State Electron.* 30, 1099–1105 (1987).

PROBLEMS

10.1 The n-side of a Ge tunnel diode is doped with 10^{20} donor atoms cm^{-3} and the p-side with 4×10^{19} acceptor atoms cm^{-3}. Using the values of N_C and N_V from Table 4.2, calculate the penetration of the Fermi level in the allowed bands on each side and determine the built-in voltage of the diode assuming a band gap of 0.56 eV for degenerate Ge. Also determine the voltage at which the tunnel current will have its minimum value.

10.2 Differentiate Eq. (10.4) with respect to V_a and making $dI/dV_a = 0$ obtain the relations in Eq. (10.6).

PROBLEMS

10.3 Consider a tunnel diode with I-V characteristic given by Eq. (10.7). Take $I_P = 10$ mA, $I_s = 10^{-11}$ A, $V_P = 0.1$ V, and $kT/q = 25$ mV. (a) Plot the diode characteristic and determine the valley voltage. (b) Determine the maximum value of resistance which should be placed in series with the diode to obtain stable characteristics in the negative resistance region. (c) Calculate the diode voltage at which the forward current is $I_F = I_P$. Ignore the excess current.

10.4 The typical equivalent circuit parameters of a Ge tunnel diode at $I_P = 10$ mA are: $R = -30\ \Omega$, $Cj = 20$ pF, $R_s = 1\ \Omega$, and $L_s = 5$ nH. Calculate the values of f_{r0}, f_{x0}, and the diode input impedance at 800 MHz.

10.5 In Eq. (10.11) assume an abrupt p-n junction and take the average electric field $|\mathscr{E}| = \sqrt{(V_i + V_a)}/W^*$ where W^* is the width constant of the junction. Substituting this value of $\mathscr{E}$, show that for small reverse biases such that $V_i \gg V_a$ the current is given by Eq. (10.11a).

10.6 A Schottky barrier diode is fabricated on n-type Si with $N_d = 1.5 \times 10^{15}$ cm^{-3} by evaporating a metal with $q\phi_m = 4.9$ eV. Neglecting the effect of interface states, calculate the built-in voltage, the barrier height, and the depletion region width at zero bias. Assume $T = 300$ K.

10.7 Draw the thermal equilibrium energy band diagram of a metal–p-type semiconductor system with $\phi_m > \phi_{sc}$ and show that according to the Schottky-Mott theory the contact should be ohmic.

10.8 A silicon sample has a uniform donor concentration of 4.5×10^{15} cm^{-3} and a uniform density of surface states 4×10^{12} cm^{-2} eV^{-1}. The neutral level at the surface is 0.3 eV above the valence band. Determine the surface potential of the free surface and the width of the surface depletion region when the surface is in thermal equilibrium with the bulk. (*Hint:* equate the depletion region charge with the surface state charge.)

10.9 A Schottky barrier diode is fabricated on the Si substrate of Prob. 10.8 by evaporating gold ($\phi_m = 5$ V). There is a 10 Å thick SiO$_2$ layer with a dielectric constant of 3.9 between the gold and Si. Calculate the barrier height, the built-in potential, and the depletion region width at a reverse bias of 3 V.

10.10 Schottky barrier diodes are fabricated by evaporating a number of metals on the chemically cleaned surface of n-type ZnSe. The barrier height data can be approximated by the straight line $\phi_B = 0.56\phi_m - 1.27$. Assume a 20 Å thick insulating oxide layer of dielectric constant 3 between ZnSe and the metal. Calculate the density of interface states and the position of the neutral level at the ZnSe surface. Take $q\chi = 4$ eV and $E_g = 2.7$ eV.

10.11 The capacitance of a Au-n-type GaAs Schottky barrier diode is given by the relation $1/C^2 = 1.74 \times 10^5 - 2.12 \times 10^5 V_a$ where C is expressed in μF and V_a is in volt. Taking the diode area to be 10^{-1} cm^2, calculate the built-in voltage, the barrier height, the dopant concentration, and the work function of Au.

10.12 Using Eq. (10.17), show that the maximum in potential energy occurs at x_m given by Eq. (10.18) and that the image force barrier lowering is given by Eq. (10.19a).

10.13 In a metal–Si Schottky barrier contact, the barrier height is 0.8 eV and $R^* = 110\ A\ \text{cm}^{-2}\ {}^\circ K^{-2}$. Assume $N_d = 1.5 \times 10^{16}\ \text{cm}^{-3}$ and $T = 300\ \text{K}$. (a) Calculate the values of V_i and ϕ_m. (b) Calculate the ratio of the injected hole current to the electron current assuming $D_p = 12\ \text{cm}^2\ \text{sec}^{-1}$ and $L_p = 1 \times 10^{-3}\ \text{cm}$.

10.14 The current-voltage characteristic of a Schottky barrier is given by Eq. (10.38). The measured forward current at 300 K is 3×10^{-8} A at 0.2 V and 1×10^{-6} A at 0.3 V. The diode area is $0.2\ \text{cm}^2$ and $\phi_B = 1$ V. Calculate the saturation current I_s, the ideality factor η, and the value of R^* assuming that the current is caused by thermionic emission.

10.15 Draw the thermal equilibrium energy band diagram of a heterojunction formed between n-type Ge with $N_d = 1.5 \times 10^{16}\ \text{cm}^{-3}$ and p-type GaAs with $N_a = 8.5 \times 10^{15}\ \text{cm}^{-3}$. Calculate the built-in voltage of the junction.

10.16 Repeat Prob. 10.15 for an n-type Ge and n-type GaAs isotype heterojunction with $N_d = 3.6 \times 10^{16}\ \text{cm}^{-3}$ on each side of the junction.

PART 4
SEMICONDUCTOR DEVICES

11
MICROWAVE DIODES

INTRODUCTION

The microwave frequency range extends from 1 GHz to about 10^3 GHz and corresponds to wavelengths ranging from 30 cm to 0.3 mm. A number of semiconductor devices have found applications in this region. These include two-terminal devices like the Schottky barrier, tunnel and backward diodes, and three-terminal devices like bipolar transistors, MESFETs, and MOS transistors. However, the use of these devices is not exclusively confined to microwave frequencies.

In this chapter we will consider a number of special devices that are used mainly at microwave frequencies. Although the operating principles of these devices differ, they are all two-terminal devices. Section 11.1 provides a brief account of the varactor diode, and the p-i-n diode is considered in Sec. 11.2. The next two sections describe avalanche injection devices, namely, the IMPATT and TRAPATT diodes. The BARITT diode forms the subject matter of Sec. 11.5, and the chapter ends with a brief reference to the transferred electron device.

11.1 THE VARACTOR DIODE

The word "varactor," coined from the phrase *variable reactor,* means a device whose capacitance can be varied in a controlled manner by the application of a bias voltage. Varactors have found use in voltage-controlled tuning, harmonic generation, parametric amplification, mixing, detection, and frequency modulation.

The basic element of a varactor diode is the capacitance of its junction space-charge region. Thus, a varactor can be either a p-n junction, a Schottky barrier, or even a metal insulator semiconductor (MIS) diode. However, our discussion here will be limited to a reverse-biased p-n junction diode.

11.1.1 Equivalent Circuit and Device Parameters

The steady state small-signal equivalent circuit of a p-n junction diode was obtained in Fig. 9.3. The equivalent circuit of a packaged reverse biased varactor diode derived from that circuit is shown in Fig. 11.1. Here, r_j is the small-signal junction resistance resulting from the generation recombination and leakage current, C_j is the junction capacitance, R_s is the diode series resistance, L_s is the inductance of the leads, and C_p is the capacitance of the diode package. Note that L_s and C_p are external to the semiconductor wafer and hence are shown by dashed lines. Since there is practically no diffusion capacitance at reverse bias, C_D does not appear in the circuit. For a reverse-biased junction, r_j is very high. It can, therefore, be left out, and by neglecting the effect of L_s and C_p, the quality factor Q of the diode can be written as

$$Q = \frac{1}{\omega C_j R_s} \tag{11.1}$$

The intrinsic cutoff frequency f_c is defined as the frequency at which $Q = 1$ and is given by

$$f_c = \frac{1}{2\pi R_s C_j} \tag{11.2}$$

This relation shows that to obtain a high value of f_c, we must decrease R_s and C_j. To reduce R_s, we should use a semiconductor with high carrier mobility. Hence, n-type material is preferred to the p-type. Further reductions in R_s require a low-resistivity semiconductor and a large cross-sectional area for the diode. These features tend to increase C_j, and thus, a compromise is necessary. Note that an increase in the reverse bias causes a decrease in C_j, and hence, an increase in f_c. Practical varactor diodes are fabricated on n-type epitaxial layers grown on low-resistivity n^+-semiconductor substrates. The thickness of the epitaxial layer is adjusted with the requirement that the desired value of f_c is obtained by optimizing the values of R_s and C_j.

As an example, consider a varactor diode with a zero-bias capacitance $C_j = 1$ pF and $f_c = 50$ GHz. Let us calculate the series resistance R_s and the quality factor at 200 MHz.

From Eq. (11.2), we get $R_s = 1/2\pi f_c C_j = (2\pi \times 5 \times 10^{10} \times 1 \times 10^{-12})^{-1} = 3.18\,\Omega$. Also from Eq. (11.1), $Q = (2\pi \times 2 \times 10^8 \times 3.18 \times 10^{-12})^{-1} = 250$. Thus, Q is high at 200 MHz.

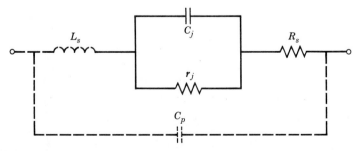

FIGURE 11.1 Equivalent circuit of a packaged reverse-biased p-n junction varactor diode.

THE VARACTOR DIODE

Besides the cutoff frequency, the other design parameters of a varactor are the voltage sensitivity of the junction capacitance C_j and the breakdown voltage. The voltage sensitivity of C_j is expressed by the C-V index $s(V)$ given by the relation

$$s(V) = -\frac{d(\log C_j)}{d(\log V)} \tag{11.3}$$

where $V = (V_i + V_R)$ is the total voltage across the junction. In situations where the impurity profile on the two sides of the junction is such that $s(V)$ remains constant (independent of voltage), Eq. (11.3) can be integrated to yield

$$C_j(V) = \frac{K_1}{(V_i + V_R)^s} \tag{11.4}$$

where K_1 is a constant of integration. Note that Eq. (11.4) reduces to the C-V relation of an abrupt p-n junction for $s = 1/2$, and to that of a linearly graded junction for $s = 1/3$.

In many applications of varactor diodes, a more rapid variation of C_j with applied bias is required. This is achieved by making p-n junctions with s-values higher than $1/2$. Such diodes are known as *hyperabrupt varactor diodes*. A peculiar feature of a hyperabrupt diode is that the dopant impurity concentration decreases with distance from the junction. Practical hyperabrupt varactors mostly employ a p^+-n junction like the one shown in Fig. 11.2. Here the p-side is heavily doped, and the n-side forms a retrograded region in which the donor concentration $N_d(x)$ decreases with the distance x from the junction. Note that the retrograded n-region is backed by a heavily doped n^+-region to reduce the series resistance. The thickness W_1 of the n-region is kept small enough to obtain the desired values of f_c and the breakdown voltage. With increasing reverse bias, the depletion layer moves into a more lightly doped region. As a result, the junction capacitance decreases more rapidly with the reverse bias than that of an abrupt

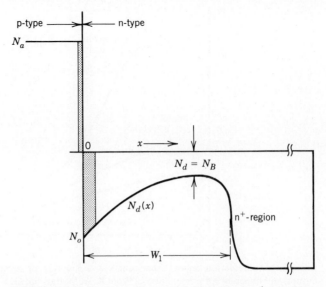

FIGURE 11.2 Impurity profile across a hyperabrupt p^+-n junction.

p-n junction diode. Figure 11.3 compares the C-V characteristics for the linearly graded, abrupt, and hyperabrupt junctions [1]. The hyperabrupt junction is observed to have the highest sensitivity to the bias voltage and gives rise to the largest capacitance variation.

The retrograded region can be formed by diffusion, epitaxial growth, or ion implantation. Depending on the fabrication process, this region may assume an impurity distribution that may be either erfc, Gaussian, power law, or a combination of these three. A voltage invariant value of s is obtained only in the case of a power law distribution in which $N_d(x)$ can be expressed as

$$N_d(x) = N_o x^m \quad \text{for } x > 0 \tag{11.5}$$

where N_o is the concentration at the junction and m is the grading exponent. Solving the one-dimensional Poisson equation with the depletion approximation for the distribution given by Eq. (11.5) yields [1]

$$C_j = A \left[\frac{qN_o(\varepsilon_s)^{m+1}}{(m+2)(V_i + V_R)} \right]^{-1/(m+2)} \tag{11.6}$$

where A is the junction area. From Eq. (11.6) we have

$$s(V) = -\frac{d(\log C_j)}{d(\log V)} = \frac{1}{m+2} \tag{11.7}$$

The hyperabrupt junction is obtained only when m is negative; for $m = 0$ the junction is abrupt, and for $m = 1$ it is linearly graded. Any desired value of s may be obtained by a proper choice of the exponent m. For example, in voltage-controlled R-C oscillators (where the frequency of oscillation is controlled by varying bias across the junction), a value of $s = 1$ is desirable, and this can be achieved by making $m = -1$ in Eq. (11.5).

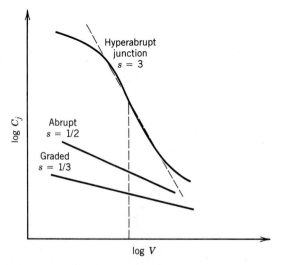

FIGURE 11.3 Depletion region capacitance as a function of reverse bias for linearly graded, abrupt, and hyperabrupt p-n junctions. (From R. A. Moline and G. F. Foxhall, reference 1, p. 268. Copyright © 1972 IEEE. Reprinted by permission.)

Retrograded profiles with power law distributions are difficult to realize in practice, and these are not the only profiles of interest. A simple method for making hyperabrupt diodes is the double diffusion process. In this process, an n-type epitaxial layer with concentration N_B is first grown on an n$^+$-substrate. The retrograded region is then formed by the diffusion of a donor impurity in the n-region with surface concentration a few orders of magnitude higher than N_B. Finally, a p$^+$-n junction is made by performing a shallow p-diffusion into the retrograded region. The retrograded impurity profile is either a Gaussian or an erfc, which in most cases can be approximated by a decaying exponential of the form [2]

$$N_d(x) \simeq (N_o - N_B) \exp(-x/\lambda) + N_B \tag{11.8}$$

where λ is a characteristic length related to the time and temperature of the diffusion.

The C-V index of the exponentially retrograded junction is a function of the ratio N_B/N_o. For a given value of this ratio, s does not remain constant but varies with the reverse bias, reaching its maximum value at a voltage $V = qN_o\lambda^2/\varepsilon_s$. It has been shown [3] that s remains almost constant over small changes in voltage around the maximum point. Figure 11.4 shows a plot of s_{max} as a function of N_B/N_o for Si hyperabrupt p$^+$-n junctions. It is seen that s_{max} values as large as 10 can be obtained by a proper choice of N_B/N_o.

The last parameter, namely, the breakdown voltage V_B of a varactor, is important because it determines the range of the available variable capacitance. Breakdown voltages of abrupt and linearly graded junctions have already been considered in Chapter 8. As for the hyperabrupt junctions, detailed calculations of V_B have been made for exponentially retrograded p$^+$-n junctions, assuming that the n-region is thicker than the depletion region width at breakdown [2]. This condition is not always satisfied, because to minimize the diode series resistance R_s, the n-region is often made thin enough to allow the depletion layer to extend completely through this region before the junction breaks down. The breakdown voltage of such a punched-through diode is reduced below its value given in [2].

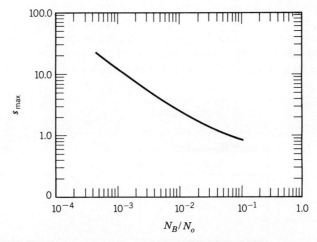

FIGURE 11.4 The maximum value (S_{max}) of C-V index for exponentially retrograded junction as a function of the ratio N_B/N_o. (From A. K. Gupta and M. S. Tyagi, reference 3, p. 509. Copyright © 1978. Reprinted by permission of Pergamon Press, Oxford.)

11.1.2 Device Fabrication and Performance

Varactor diodes have been fabricated using diffusion, epitaxial growth, and ion implantation techniques. Figure 11.5a shows the cross-sectional view of a p-n junction varactor diode fabricated using planar technology. The structure has a low-resistivity n^+-substrate on which the n-layer is epitaxially grown. The thickness and resistivity of this layer are determined from the requirement of cutoff frequency and the range of capacitance variation. Retrograded doping, if needed, is performed in the n-region. The wafer is then oxidized, a window is etched in the oxide, and a p^+-diffusion is made. The device fabrication is completed by making ohmic contacts to the top p^+-layer and the n^+-substrate. The fabrication technology of the Schottky barrier varactor shown in Fig. 11.5b is simpler. Here the p^+-region is replaced by a rectifying metal–semiconductor contact, and the contact area is defined by etching the structure in the form of a mesa.

Microwave varactors generally employ a linearly graded junction. Hyperabrupt varactors usually have a lower cutoff frequency than the linearly graded devices. This is because the retrograded region of the hyperabrupt diode offers a high series resistance when not fully depleted.

As for the choice of the semiconductor material, only Si and GaAs have been used for varactor diodes. GaAs has an advantage over Si because of its higher electron mobility. High-performance GaAs varactors have been fabricated with a zero-bias capacitance as low as 0.12 pF and a cutoff frequency in excess of 700 GHz.

11.2 THE P-I-N DIODE

11.2.1 General Considerations

A p-i-n diode consists of an intrinsic region of semiconductor material sandwiched between end regions of heavily doped p^+- and n^+-type material. The device was orginally proposed [4] as a low-frequency, high-power rectifier because it could support a high reverse voltage. The resistance of a reverse-biased p-i-n diode is very high. However, under forward bias, a large number of electrons and holes are injected into the central intrinsic zone and modulate the conductivity of this region. As a result, the diode resistance becomes very low and can be controlled by the forward current. Because of this property, the diode can be used as a switch or a variable attenuator at microwave frequencies.

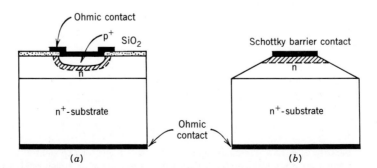

FIGURE 11.5 Schematic cross-sections of (a) p-n junction and (b) Schottky barrier varacter diodes.

THE P-I-N DIODE

Figure 11.6a shows the schematic cross-section of a p-i-n diode. These diodes are invariably made in Si, and it is difficult to produce a central region that is truly intrinsic. In practical devices, the central region is either weakly doped p-type (called π) or weakly doped n-type (called ν). Typical dopant concentrations in the central region may range from 10^{12} to 10^{13} cm^{-3}, so that this region is fully depleted only at a small reverse bias, and in extreme cases, even by virtue of the built-in voltage. Sketches (b) through (d) in Fig. 11.6 show the dopant concentration, space charge, and electric field distribution in a p-i-n structure as well as in a p$^+$-π-n$^+$ structure, assuming that the central region is fully depleted.

11.2.2 Current-Voltage Characteristics

(a) REVERSE CHARACTERISTIC

Application of a reverse bias causes the depletion region to extend mainly in the lightly doped central region, and at a reverse bias of a few volts, this region becomes fully depleted. The device capacitance then reaches a substantially constant value given by

$$C_j = \frac{\varepsilon_s A}{W} \tag{11.9}$$

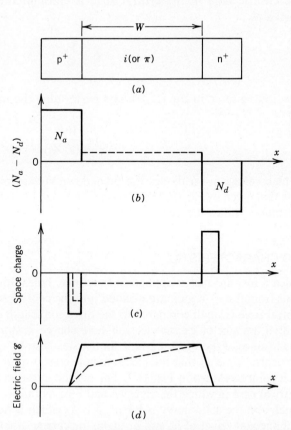

FIGURE 11.6 The p-i-n diodes: (a) schematic diagram, (b) impurity distribution, (c) space-charge, and (d) electric field distributions. The solid lines depict the various distributions for a p-i-n diode and the dashed lines are for a fully depleted p$^+$-π-n$^+$ structure. (From H. S. Veloric and M. B. Prince, reference 4, p. 983. Copyright © 1957 AT&T. Reprinted by special permission.)

where A is the diode cross-sectional area, and W is the width of the central π (or ν) zone. As the reverse bias is increased further, the depletion layer extends slowly in the n^+- and p^+-end regions and C_j may decrease slightly with increasing reverse voltage.

The dominant current component in the reverse-biased diode is the current resulting from the electron-hole pair generation in the depletion region, and from Eq. (7.28) we have

$$I = -\frac{qAn_i W}{2\tau_o} \tag{11.10}$$

where τ_o is the carrier generation lifetime in the depletion region.

The breakdown voltage V_B of the diode is obtained by evaluating the ionization integral across the depletion region of width W. Assuming an effective ionization coefficient α_i, we can obtain the breakdown condition from Eqs. (8.25) and (8.21), giving

$$a_o \int_0^W \exp(-b/\mathscr{E})\,dx = 1 \tag{11.11}$$

The electric field in the central zone is constant throughout and can be expressed as

$$\mathscr{E} = \frac{V_B}{W} \tag{11.12}$$

Substituting for $\mathscr{E}$ in Eq. (11.11), and performing the integration yield

$$V_B = \frac{Wb}{\ln(a_o W)} \tag{11.13}$$

In high-voltage p-i-n diodes the breakdown voltage may be substantially lower than that given by Eq. (11.13) because of junction curvature and electric field concentration at the surface.

(b) FORWARD CHARACTERISTIC

When a forward bias is applied to the diode, holes from the p^+-region and electrons from the n^+-region are injected into the central zone. If the width W of the central zone is small compared to the diffusion length of injected carriers, a substantial amount of excess electron–hole charge is stored in this region even at small values of the diode current. In the steady state, quasi-neutrality prevails in the central zone so that the electron–hole concentrations remain nearly equal at each point as shown in Fig. 11.7. The dashed curve in this figure is for the symmetrical case in which the electron and hole mobilities have been assumed to be equal, and the solid curve is for $\mu_n = 3\,\mu_p$ which is nearly the case for Si. Note that unequal values of μ_n and μ_p make the carrier distribution asymmetrical. Figure 11.7 also shows the electron–hole concentrations injected into the end regions from the central zone. These concentrations generally remain negligibly small unless the current density through the diode is very high. In most p-i-n diodes, the onset of high-level injection occurs at a small forward bias of about 0.2 V.

THE P-I-N DIODE

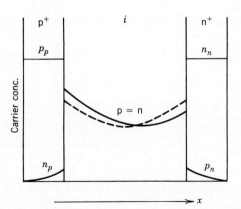

FIGURE 11.7 Typical electron-hole distributions in the various regions of a p-i-n diode under forward bias. The dashed curve is for $\mu_n = \mu_p$ whereas the solid curve is for $\mu_n = 3\mu_p$.

This situation necessitates the use of ambipolar mobility and diffusion constants in the analysis of the diode behavior.

Derivation of the current-voltage relation for a p-i-n diode using a self-consistent theory is difficult. However, a simplified expression for the diode current is easily obtained by making the following assumptions:

1. Minority carrier injection in the p^+- and n^+-regions is negligible.
2. High-level injection prevails in the i-region, and the ambipolar diffusion constant D_a and the high-level injection lifetime τ_a are uniform throughout this region.
3. The excess carrier concentration in the i-region is constant throughout, so that the diffusion currents of electron and holes may be neglected.

Let $\bar{p}_e$ denote the excess hole concentration in the midregion, which we assume to be large compared to the intrinsic carrier concentration. In the steady state, $d\bar{p}_e/dt = 0$, and the current density can be written as:

$$J = q \int_0^W \frac{\bar{p}_e}{\tau_a} dx = \frac{q\bar{p}_e W}{\tau_a} \tag{11.14}$$

where $\tau_a = (\tau_{po} + \tau_{no})$ is the high-level injection lifetime in the midregion. Since the diffusion current is neglected, J can be expressed as

$$J = q(\mu_n + \mu_p)\bar{p}_e \overline{\mathscr{E}} \tag{11.15a}$$

where $\overline{\mathscr{E}}$ represents the average field in the i-region and $\bar{p}_e = \bar{n}_e$. By making use of the Einstein relation $qD_p = \mu_p kT$ and writing $b_o = \mu_n/\mu_p$, the above relation becomes

$$J = \frac{q}{kT}(1 + b_o)qD_p\bar{p}_e\overline{\mathscr{E}} \tag{11.15b}$$

The ambipolar diffusion constant D_a is related to D_p by the relation

$$D_a = \frac{2D_p D_n}{D_p + D_n} = \frac{2b_o D_p}{(b_o + 1)} \tag{11.16}$$

Substituting for D_p from Eq. (11.16) into Eq. (11.15b) leads to

$$J = \frac{q}{kT} \frac{(b_o + 1)^2}{2b_o} q D_a \bar{p}_e \overline{\mathscr{E}} \qquad (11.17)$$

Eliminating J between Eq. (11.14) and Eq. (11.17), we obtain

$$\overline{\mathscr{E}} = \frac{kT}{q} \frac{2b_o W}{(1 + b_o)^2 D_a \tau_a} \qquad (11.18)$$

Finally, the voltage V_M across the mid i-region is given by

$$V_M = \overline{\mathscr{E}} W$$

Substituting for $\overline{\mathscr{E}}$ from Eq. (11.18) and writing the ambipolar diffusion length $L_a = \sqrt{D_a \tau_a}$, we obtain the following expression for V_M:

$$V_M = \frac{kT}{q} \frac{2b_o}{(1 + b_o)^2} \left(\frac{W}{L_a}\right)^2 \qquad (11.19a)$$

In the case of silicon, $b_o \simeq 3$ and

$$V_M = \frac{3kT}{8q} \left(\frac{W}{L_a}\right)^2 \qquad (11.19b)$$

This expression shows that the voltage drop is independent of current. This rather surprising result is a direct consequence of the assumption that all the forward current is caused by an increase in the electron and hole concentrations. The increase in the excess carrier charge decreases the resistance of the i-region so that the product of the current and the resistance remains constant.

A more complete theory of a p-i-n diode has been developed by solving the ambipolar transport equation in the i-region with appropriate boundary conditions, and taking into account the effect of voltage drops across the two junctions. This approach leads to the following equation for the diode current [5]:

$$I = \frac{4qAD_a n_i}{W} F_L \exp\left(\frac{qV_a}{2kT}\right) \qquad (11.20)$$

where V_a is the applied voltage. The parameter F_L is independent of I but is a function of the ratio W/L_a and is plotted in Fig. 11.8. It is seen that F_L has its maximum value for $W/2L_a \simeq 1$, and the diode voltage drop for a given current is minimized when this is the case. The condition $W = 2L_a$ marks the transition between a short and a long p-i-n diode. For a short structure with $W < 2L_a$, there is an excess carrier buildup in the i-region which needs higher junction voltages associated with the p^+-i and i-n^+ junctions to support the forward current. For the long structure with $W > 2L_a$, the conductivity modulation of the i-region is insufficient; this increases the voltage drop across this region, which causes a large forward drop across the diode.

When the injected carrier concentration in the i-region exceeds 10^{17} cm^{-3}, additional effects like minority carrier recombination in the p^+ and n^+-end regions,

THE IMPATT DIODE

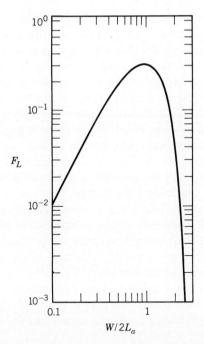

FIGURE 11.8 The parameter F_L plotted as a function of the ratio $W/2L_a$ for a p-i-n diode. (From S. K. Ghandhi, reference 5, p. 117. Copyright © 1977. Reprinted by permission of John Wiley & Sons, Inc., New York.)

Auger recombination, and electron-electron and hole-electron scattering become important and should be considered. Although analytical models have been developed to predict the I-V characteristics of p-i-n diodes by taking into account some of these effects, numerical methods have to be used to obtain close agreement between theoretically predicted and experimentally observed characteristics.

As an example, consider a silicon p-i-n diode at 300 K with $W = 100\ \mu\mathrm{m}$, $(W/2L_a) = 2$, $D_a = 12\ \mathrm{cm^2\ sec^{-1}}$, and $V_a = 1$ V. Let us estimate the current density.

From the plot of Fig. 11.8 we get $F_L = 3 \times 10^{-2}$, and from Eq. (11.20)

$$I/A = \frac{4 \times 1.6 \times 10^{-19} \times 12 \times 1.5 \times 10^{10} \times 3 \times 10^{-2}}{100 \times 10^{-4}} \exp\left(\frac{1 \times 10^3}{2 \times 25.86}\right)$$

$$= 86.2\ \mathrm{A/cm^2}$$

11.3 THE IMPATT DIODE

An IMPATT (impact ionization avalanche transit time) diode employs the avalanche and carrier drift processes in a semiconductor to produce a dynamic negative resistance at microwave frequencies. The diode is one of the most powerful solid-state sources of microwave power.

11.3.1 Principle of Operation

In principle, any p-n junction diode at avalanche breakdown can be used as an IMPATT diode. However, to understand how the device operates, we consider the p^+-n-i-n^+ structure originally proposed by W.T. Read [6]. The schematic of the device is shown in Fig. 11.9a, and the doping profile and electric field distribution at reverse breakdown are shown in Fig. 11.9b and c, respectively. At breakdown, the n-region is punched-through and forms the avalanche region of the diode. The high-resistivity i-region is the drift zone through which the avalanche-generated electrons move toward the anode.

Now consider a dc bias V_B, just short of that required to cause breakdown, applied to the diode in Fig. 11.9. Let an ac voltage $\tilde{v} = V_1 \sin \omega t$ of sufficiently large magnitude be superimposed on the dc bias, such that during the positive cycle of the ac voltage, the diode is driven deep into the avalanche breakdown. At $t = 0$, the ac voltage is zero, and only a small pre-breakdown current flows through the diode. As t increases, the voltage goes above the breakdown voltage, and secondary electron–hole pairs are produced by impact ionization. As long as the field in the avalanche region is maintained above the breakdown field $\mathscr{E}_b$, the electron–hole concentrations grow exponentially with t. Similarly, these concentrations decay exponentially with time when the field is reduced below $\mathscr{E}_b$ at the negative swing of the ac voltage. The holes generated in the avalanche region disappear in the p^+-region and are collected by the cathode. The electrons are injected into the i-zone where they drift toward the n^+-region. At $t = T/4$ (i.e., $\omega t = \pi/2$) the field in the avalanche region reaches its maximum value ($\mathscr{E}_b + |\tilde{\mathscr{E}}|$), and

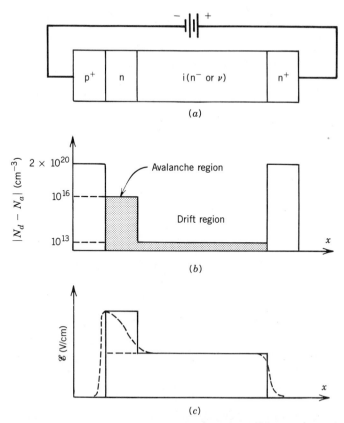

FIGURE 11.9 The Read diode: (a) schematic diagram, (b) doping profile, and (c) electric field distribution at breakdown. The dashed lines in (c) show the actual and the solid lines the idealized field distributions. The figures given in (b) are typical for a silicon diode. (From W.T. Read, reference 6, p. 402. Copyright © 1958 AT&T. Reprinted by special permission.)

the population of the electron–hole pairs starts building up. At this time, the ionization coefficients α_n and α_p have their maximum values. Although α_n and α_p follow the electric field instantaneously, the generated electron concentration does not, because it also depends on the number of electron–hole pairs already present in the avalanche region. Hence, the electron concentration at $t = T/4$ will have a small value (Fig. 11.10a). Even after the field has passed its maximum value, the electron–hole concentrations continue to grow because the secondary carrier generation rate still remains above its average value. For this reason, the electron concentration in the avalanche region attains its maximum value at $t = T/2$ when the field has dropped to its average value (Fig. 11.10b). Thus, it is clear that the avalanche region introduces a 90° phase shift between the ac signal and the electron concentration in this region.

With a further increase in t, the ac voltage becomes negative, and the field in the avalanche region drops below its critical value $\mathscr{E}_b$. The bunch of electrons in the avalanche region is then injected into the drift zone. The movement of electrons through this zone induces a current in the external circuit which has a phase opposite to that of the ac voltage. The ac field, therefore, absorbs energy from the drifting electrons as they are deaccelerated by the decreasing field. Figure 11.10c shows the electron bunch at $t = 3T/4$. It is clear that an ideal phase shift between the diode current and the ac (or rf) signal is achieved if the thickness of the drift zone is such that the bunch of electrons is collected at the n^+-anode at

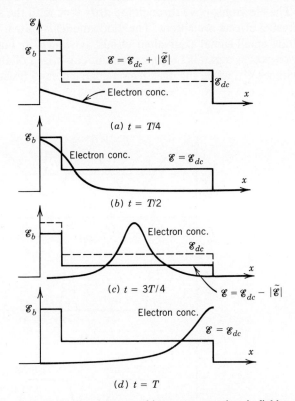

FIGURE 11.10 Successive time development of instantaneous electric field and electron charge concentration as a function of distance in a Read diode biased in avalanche breakdown. (From B. C. De Loach Jr., Avalanche transit-time microwave diodes, Chapter 15, in *Microwave Semiconductor Devices and their Circuit Applications*, H. A. Watson (ed.), p. 478. Copyright © 1969. Reprinted by permission of McGraw-Hill, Inc., New York.)

$t = T$ when the ac voltage goes to zero, as shown in Fig. 11.10d. This condition is achieved by making $\tau_d = T/2$ where τ_d is the electron transit time through the drift zone. This situation produces an additional phase shift of 90° between the ac voltage and the diode current. The waveforms of the ac voltage, the injected electron charge, and the current induced in the external circuit are shown in Fig. 11.11. Note that the induced current starts at a time when the electron bunch is injected into the drift zone, and continues until the bunch is collected at the anode.

The electric field in the drift zone must remain sufficiently high at all times to ensure that the electrons move through it with the saturated drift velocity v_s. If the field drops below this value, the terminal current decreases and causes a power loss. Moreover, the bunch of electrons spreads out owing to diffusive motion, causing a less steep fall in the diode current, and thus decreases the overall phase shift.

11.3.2 Small-Signal Impedance of the Read Diode

A small-signal theory has been developed for the Read diode under the following assumptions [7]: (1) the diode can be divided into separate avalanche and drift zones, and the current flow is one-dimensional. (2) Both the electrons and the holes have the same ionization coefficient α_i and the same saturation velocity v_s. (3) The electrons move through the drift zone with the saturation velocity v_s, and diffusion effects are absent. The diode model is shown in Fig. 11.12a and the resulting small-signal equivalent circuit is shown in Fig. 11.12b. The avalanche region acts as a parallel resonant circuit comprised of an inductance L_A and a capacitance C_A, while the drift region contributes to resistance R. The values of L_A

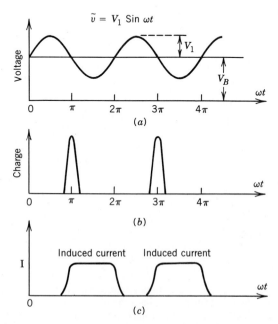

FIGURE 11.11 Waveforms of voltage, injected electron charge, and the induced current in a Read diode.

THE IMPATT DIODE

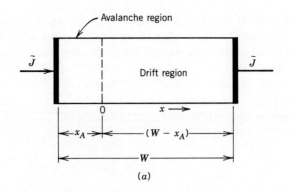

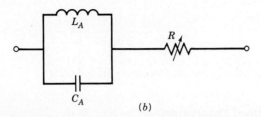

FIGURE 11.12 (*a*) Model of a Read diode with avalanche and drift zones and (*b*) the small-signal equivalent circuit of the diode.

and C_A are given by the relations [7]

$$L_A = \frac{x_A}{2\alpha_i' I_{dc} v_s} \tag{11.21}$$

and

$$C_A = \frac{\varepsilon_s A}{x_A} \tag{11.22}$$

Here A is the diode cross-sectional area, x_A is the thickness of the avalanche zone, I_{dc} is the bias point dc current, and $\alpha_i' = d\alpha_i/d\mathcal{E}_A$ where $\mathcal{E}_A$ represents the electric field in the avalanche region. The avalanche resonant frequency ω_A of the diode is given by

$$\omega_A = \frac{1}{\sqrt{L_A C_A}} = \left[\frac{2\alpha_i' v_s I_{dc}}{A \varepsilon_s}\right]^{1/2} \tag{11.23}$$

It is clear that ω_A can be varied over a wide range by changing the bias current I_{dc}. It turns out that the real part of the diode impedance is positive if the operating frequency $\omega < \omega_A$. It changes sign at $\omega = \omega_A$ and becomes negative for $\omega > \omega_A$. Similarly, the imaginary part remains inductive for $\omega < \omega_A$ and becomes capacitive for $\omega > \omega_A$. Although the range of negative resistance is very wide, the diode is usually designed to have the total transit time (W/v_s) equal to one-half the oscillation period. This means that $W/v_s = \pi/\omega$, or

$$f = \frac{v_s}{2W} \tag{11.24}$$

where W is the total width of the diode and f is the operating frequency. The small-signal analysis gives a qualitative description of the diode behavior but does not predict the correct values of conversion efficiency and output power.

As an example, consider a Si IMPATT diode with $x_A = 2\ \mu$m, $I_{dc} = 200$ mA, $v_s = 10^7$ cm sec^{-1}, $A = 10^{-3}$ cm^2, and $\alpha_i' = 0.5\ V^{-1}$. Let us calculate L_A, C_A, and the resonant frequency.

From Eqs. (11.21) and (11.22) we obtain

$$L_A = \frac{2 \times 10^{-4}}{2 \times 0.5 \times 200 \times 10^{-3} \times 10^7} = 1 \times 10^{-10}\ H$$

and

$$C_A = \frac{1.05 \times 10^{-12} \times 10^{-3}}{2 \times 10^{-4}} = 5.25 \times 10^{-12}\ F$$

which gives a resonant frequency of 6.9 GHz.

11.3.3 Large-Signal Operation

In order to obtain a high value of output power and power conversion efficiency, an IMPATT diode has to be operated under large-signal conditions. The large-signal performance of a Read diode can be obtained from the numerical solution of device equations with appropriate boundary conditions. The phase relations obtained for the diode voltage and current are close to those shown in Fig. 11.11.

(a) POWER CONVERSION EFFICIENCY

To calculate the diode conversion efficiency from dc to microwave rf power, we assume an ac voltage $\tilde{v} = V_1 \sin \omega t$ superimposed on the dc bias V_B. We further assume that the charge injected from the avalanche region into the drift zone can be represented by a sharp pulse that corresponds to the situation shown in Fig. 11.11. A Fourier analysis of the induced current waveform $I(t)$ in Fig. 11.11 gives

$$I(t) = \frac{I_o}{2} + \frac{2I_o}{\pi} \sin \omega t + \text{higher order harmonics} \qquad (11.25)$$

where I_o denotes the amplitude of the current pulse. Thus, we have $I_{dc} = I_o/2$ and the ac current $\tilde{I} = (2I_o/\pi) \sin \omega t$. The ac output power P_{ac} is

$$P_{ac} = \tilde{v}\tilde{I} = \frac{2I_o V_1}{\pi} \frac{1}{2\pi} \int_0^{2\pi} \sin^2 \theta\, d\theta = \frac{I_o V_1}{\pi}$$

The conversion efficiency η of the diode is given by the relation

$$\eta = \frac{\tilde{v}\tilde{I}}{V_B I_{dc}} \qquad (11.26)$$

and substituting the values of various quantities gives

$$\eta = \frac{2}{\pi} \frac{V_1}{V_B} \tag{11.27}$$

The maximum allowable value of the rf voltage swing V_1 depends on the swing in the electric field $\Delta\mathscr{E}$ in the avalanche and drift regions. If $\mathscr{E}_A$ and $\mathscr{E}_d$ represent the electric fields in the avalanche and the drift zones, respectively, then at the breakdown voltage V_B we have

$$V_B = x_A \mathscr{E}_A + (W - x_A)\mathscr{E}_d \tag{11.28}$$
$$V_1 = \Delta\mathscr{E} W$$

and we obtain

$$\frac{V_1}{V_B} = \frac{1}{\dfrac{\mathscr{E}_d}{\Delta\mathscr{E}} + \dfrac{x_A}{W}\left(\dfrac{\mathscr{E}_A - \mathscr{E}_d}{\Delta\mathscr{E}}\right)} \tag{11.29}$$

The most optimistic situation is the one in which the voltage across the drift zone swings to zero, which means that $\mathscr{E}_d$ goes to zero at the maximum negative swing of the rf voltage, making $\Delta\mathscr{E} = \mathscr{E}_d$. In addition, if we assume $\mathscr{E}_A = 2\mathscr{E}_d$, then Eq. (11.29) leads to

$$\frac{V_1}{V_B} = \frac{1}{1 + \dfrac{x_A}{W}} \tag{11.30}$$

Thus, V_1 may be made to approach V_B in the limit when the ratio x_A/W approaches zero. In actual devices, this ratio ranges from about 0.15 to 0.25 leading to a (V_1/V_B) ratio of about 0.8 and a conversion efficiency in excess of 50 percent. Such high values of η have not been obtained because the assumptions made in the above analysis are not realistic. First, the field in the drift zone must not be allowed to fall below that required for velocity saturation. If the field does fall below this value, the electrons will travel with a field-dependent velocity, and their transit time will depend on the field. The second condition on the field is imposed by the space charge of electrons injected into the drift zone. In fact, for satisfactory operation of the diode, $\Delta\mathscr{E}$ should not exceed $\mathscr{E}_d/2$. The allowable swing in the diode voltage is then limited to less than 50 percent of the breakdown voltage. Other factors that reduce the conversion efficiency include the series resistances of inactive p^+- and n^+-regions, saturation of the ionization coefficient, and tunnelling at high electric fields [7]. The practical values of efficiency may be as high as 30 percent in Si and 35 percent in GaAs devices.

11.3.4 Diode Structures and Device Fabrication

(a) DIODE STRUCTURES

Besides the Read diode considered above, there are other members of the IMPATT family. The dopant concentrations and field distributions at breakdown for four different types of structures are shown in Fig. 11.13. The drift zone in each case is assumed to be punched-through.

The structure in Fig. 11.13a is the simple one-sided abrupt p^+-n junction. Here the avalanche region, where 95 percent of the avalanche multiplication occurs, forms only a small part of the total depletion region width W. The structure (b) is an improved form of the Read diode. Here the width of the n-region is so chosen that this region closely corresponds to the avalanche zone. The low-high-low structure of (c) is another solution to the problem of increasing efficiency. The clump charge in the central n^+-region keeps the field $\mathscr{E}$ almost constant in the n^--region, and the field in the ν-region is kept below that required for avalanching. Thus, the field in the avalanche region is no longer limited by the space charge. The structure of Fig. 11.13d is the p-i-n diode, which is also known as the Misawa diode. In this device, avalanche occurs throughout the full central region, and the avalanche and drift zones cannot be separated. The ac equivalent circuit of this diode is similar to that shown in Fig. 11.12b, but L_A is replaced with a frequency-independent negative resistance in series with the inductance [8].

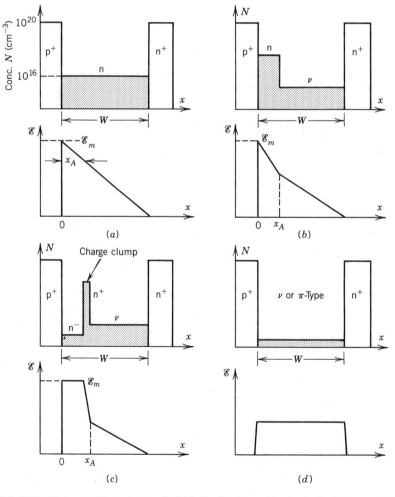

FIGURE 11.13 Doping profile and electric field distributions of basic single-drift zone IMPATT diodes. (a) One-sided abrupt p^+-n junction diode, (b) and (c) improved high-low type Read diodes, and (d) p^+-i-n^+ diode.

Higher values of efficiency and power output have been obtained in structures with two drift zones. Figure 11.14 shows doping profiles and electric fields in two structures that have a double-drift zone. In these devices, the avalanche zone is centrally located, and electrons and holes drift in opposite directions from this zone. A double-drift structure improves the conversion efficiency and increases the ac power output close to 100 percent with respect to the single-drift zone case, provided that all the other conditions remain the same.

(b) DEVICE FABRICATION

IMPATT diodes have been fabricated using epitaxial growth, diffusion, and ion implantation, or a combination of all three processes. The design is optimized to obtain maximum power output and efficiency. Figure 11.15 depicts the cross-sections of two typical devices. The configuration of Fig. 11.15a is formed by epitaxial growth of an n-layer on an n^+-substrate, and subsequent diffusion of a thin p^+-layer into the n-region. Figure 11.15b shows a modified Read-type Schottky barrier diode. Here a ν-layer is grown on an n^+-substrate. Next, an n-layer is diffused in the ν-layer, and a rectifying metal contact is made on the n-layer to form a Schottky barrier. In this device, the maximum field occurs at the metal–semiconductor contact. Heat generated there can be easily conducted away by the metal contact, which has a superior thermal conductivity.

11.3.5 State of the Art

IMPATT diodes have been fabricated in Ge, Si, and GaAs. Silicon has been most commonly used, but GaAs has also been used for frequencies below 30 GHz. Because of almost equal values of α_n and α_p for GaAs, the avalanche zone is nar-

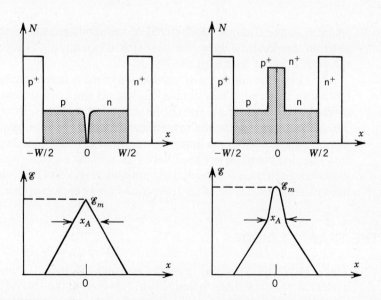

FIGURE 11.14 Doping profiles and electric field distributions in double-drift zone IMPATT diodes: (a) p-n junction diode, and (b) high-low-high type double drift diode. (From S. Teszner and J. L. Teszner, reference 11, pp. 323, 324 and 325. Copyright © 1975. Reprinted by permission of Academic Press, Orlando, Florida.)

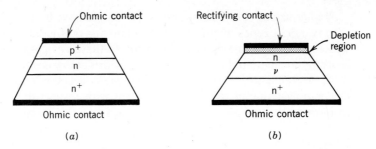

FIGURE 11.15 Two typical IMPATT diode structures: (a) single-drift zone p^+-n diode and (b) improved Read-type Schottky barrier diode.

rower and efficiencies as large as 36 percent can be obtained. However, above 30 GHz, the GaAs diodes become inferior to the Si devices.

The upper frequency limit of an IMPATT diode is determined by the transit time condition given by Eq. (11.24). An IMPATT diode with distinct avalanche and drift zones cannot operate above 200 GHz, but the uniform avalanche p-i-n diode can be operated at frequencies as high as 600 GHz. The main disadvantage of an IMPATT diode is its high noise. The reason for the high noise is the statistical nature of the impact ionization process, which introduces a time jitter in the leading edge of the current pulse.

Two factors limit the rf power output from an IMPATT diode; one is thermal, and the other is electronic. At lower frequencies the output is limited by the thermal power that can be dissipated in the semiconductor wafer. In the steady state the output power at a given value of the diode reactance X_c is given by the relation [7]

$$P_{\text{out}} = \frac{K_s \Delta T}{2\pi X_c \varepsilon_s f} \tag{11.31}$$

where K_s is the thermal conductivity of the semiconductor and ΔT is the rise in junction temperature above the heat sink temperature. This relation shows that the power output decreases as $1/f$.

The electronic limitation arises because of the inherent limitations of the semiconductor material. These limitations are (1) the critical field $\mathscr{E}_b$ at which avalanche breakdown occurs, and (2) the saturation velocity v_s through the drift zone. These factors determine the maximum voltage that can be applied across the depletion region and the maximum current that can be carried by the semiconductor. The maximum power then varies as $1/f^2$ for a given impedance level. The electronic limitation is expected to dominate above 30 GHz, whereas the thermal limitation dominates at lower frequencies for continuous operation.

11.4 THE TRAPATT DIODE

The TRAPATT (trapped plasma avalanche-triggered transit) mode of an avalanche injection diode is basically different from the IMPATT mode. To understand the principle of operation of the device, let us consider a highly punched-through p^+-p-n^+ diode that is often a preferred structure for the TRAPATT mode. If the reverse bias across the diode is increased above the breakdown voltage V_B, the diode current will initially increase with voltage. However, when

THE TRAPATT DIODE

the current becomes sufficiently high, the p-region is filled with an electron–hole plasma produced by the secondary ionization process, and the diode voltage is reduced to a low value. The diode thus exhibits a dynamic negative resistance between a high-voltage, low-current state and a low-voltage, high-current state. If proper circuit loading is provided, the diode is switched back and forth periodically, and it generates microwave power.

To understand the process of diode voltage collapse at high current, it is useful to invoke the concept of an avalanche shock front. Let us consider that a sharply rising reverse current pulse is applied to a nonconducting p^+-p-n^+ diode. As long as the diode voltage remains below V_B, the current remains largely capacitive and the field everywhere in the p-region rises at the rate

$$\frac{d\mathscr{E}}{dt} = \frac{J(t)}{\varepsilon_s} \tag{11.32}$$

where $J(t)$ is the time-dependent current density.

The current will remain capacitive, and Eq. (11.32) will be valid as long as $\mathscr{E}$ is everywhere less than the breakdown field $\mathscr{E}_b$. This is the case until $t = t_3$ in Fig. 11.16. After the maximum field has increased above $\mathscr{E}_b$, a large number of electron–hole pairs will be produced for the current to become conductive with a very small field. If the rise in current is very fast, such that the carrier drift during the current rise is negligible, then the point at which $\mathscr{E} = \mathscr{E}_b$ will move rapidly toward the p^+-region, as shown in Fig. 11.16 for $t > t_3$. Thus, for sufficiently large values of $J(t)$, the avalanche shock front moves rapidly to fill the p-region with an electron–hole plasma. To obtain the velocity of propagation v_z of the leading edge of the avalanche zone, we note that no electron–hole pairs are produced to the right of the plane at $\mathscr{E} = \mathscr{E}_b$, and we have

$$\frac{d\mathscr{E}}{dx} = \frac{qN_a}{\varepsilon_s} \tag{11.33}$$

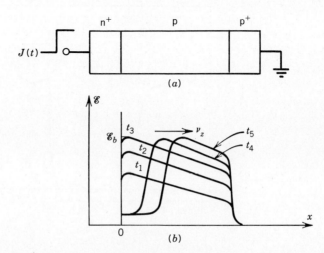

FIGURE 11.16 TRAPATT mode of operation of a highly punched-through reverse-biased n^+-p-p^+ diode: (a) schematic cross-section and (b) field rise before (t_1 to t_3) and during electron–hole plasma formation (t_4 and t_5).

where N_a represents the acceptor concentration in the p-region. From Eqs. (11.32) and (11.33) we get

$$v_z = \frac{dx}{dt} = \frac{J}{qN_a} \tag{11.34}$$

The avalanche zone will quickly sweep across the diode if $v_z \gg v_s$, and this leads to the condition

$$J \gg qN_a v_s \tag{11.35}$$

For Si at room temperature, $v_s = 10^7$ cm/sec, and if we take $v_z = 5 \times 10^7$ cm/sec and $N_a = 10^{15}$ cm^{-3}, we obtain $J = 8 \times 10^3$ A cm^{-2}. After collapse of the diode voltage, the field is reduced below that required for velocity saturation. Thus, the plasma will be trapped and its extraction will take place with a velocity $\mu\mathcal{E} < v_s$.

Figure 11.17 shows the waveforms of the diode voltage and current as a function of time. During period 1, the diode voltage rises significantly above V_B, followed by a sudden drop in the voltage and rise in the current as the electron–hole plasma fills the depletion region. Period 2 begins with the recovery of the diode voltage. Initially, the field is low and carriers move out of the depletion region with a velocity lower than v_s. However, as the plasma is extracted, the diode voltage recovers with a subsequent increase in $\mathcal{E}$. The carrier velocity then increases

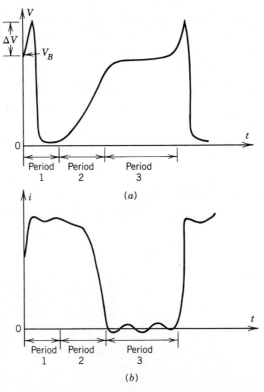

FIGURE 11.17 Voltage and current waveforms during the TRAPATT mode of a highly punched-through n$^+$-p-p$^+$ diode. (From S. Teszner and J.L. Teszner, reference 11, p. 331. Copyright © 1975. Reprinted by permission of Academic Press, Orlando, Florida.)

simultaneously and causes an increase in $d\mathscr{E}/dt$ and the diode voltage. During this period, the current stays close to its maximum value, but suddenly drops to a very low value at the end when no carriers are left in the p-zone. In period 3, the current remains low and the voltage stays at a high-level slightly below V_B until, at the end of the period, an overvoltage pulse restores the situation to that at the start of period 1. It is possible to assure the periodic generation of overvoltage and current by an external generator of appropriate design.

The reason for interest in TRAPATT oscillators is their high-output power and high efficiencies. The efficiency is high because the voltage is very low during the plasma extraction period. At the same time, this mode requires very high power densities (on the order of 10^6 W cm^{-2}) for its operation. For this reason, TRAPATT oscillators are considered mostly for pulse operation. The maximum efficiency achievable in the TRAPATT mode is about 80 percent. However, it is not easy to establish an optimum waveform to achieve the maximum efficiency. Silicon is the only material that has been used, and in a pulsed operation in the 1-2 GHz range, 60 percent efficiency with 1 KW of output power is possible.

Because of the large extraction time of the electron–hole plasma, TRAPATT diodes operate at comparatively low frequencies below 10 GHz. The noise behavior of a TRAPATT diode is even worse than that of the IMPATT diode. This is because extremely large fluctuations occur in the startup of the current by avalanche multiplication.

11.5 THE BARITT DIODE

A BARITT (barrier-injection transit-time) diode utilizes the injection and transit time delay property of minority carriers to produce a negative conductance at microwave frequencies. In this device, the minority carriers are injected over the barrier of a forward-biased junction, and no avalanche is involved. Consequently, the available phase delay is much less, and the power output and conversion efficiencies are considerably smaller than those of the IMPATT diode.

11.5.1 Principle of Operation

To understand the operation of the device, let us first consider the symmetrical metal–semiconductor–metal (MSM) structure shown in Fig. 11.18. The semiconductor is uniformly doped n-type with rectifying contacts on the two ends. Figures 11.18b to d show the space charge, electric field distribution, and the energy band diagram of the device at zero bias. Let a dc bias V_B be applied to the structure, making the left-hand contact positive with respect to the right-hand contact. The junction J_1 is then forward biased and J_2 is reverse biased. As the bias voltage is increased gradually, practically all the voltage appears across J_2, and a point is reached when the depletion region width W_2 associated with junction J_2 reaches the depletion region associated with J_1. The voltage corresponding to this situation is called the *punch-through voltage* V_{PT} and is given by [7]

$$V_{PT} \simeq \frac{qN_d W^2}{2\varepsilon_s} - W\left(\frac{2qN_d V_i}{\varepsilon_s}\right)^{1/2} \tag{11.36}$$

where $W = (W_1 + W_2)$, W_1 is the depletion region width of the junction J_1, and N_d is the donor concentration in the semiconductor. If the bias is increased still fur-

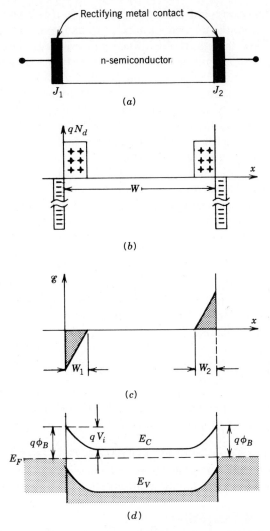

FIGURE 11.18 A metal-semiconductor-metal (MSM) structure at thermal equilibrium: (a) schematic diagram, (b) space-charge distribution, (c) electric field distribution, and (d) energy band diagram.

ther, W_2 keeps on widening, and when it becomes equal to W, the bands at J_1 become flat. The flat-band voltage V_{FB} is obtained from the relation

$$V_{FB} = \frac{qN_d W^2}{2\varepsilon_s} \tag{11.37}$$

Let us now consider the MSM structure biased with a dc voltage higher than V_{PT} but less than V_{FB}. The energy band diagram for this situation is shown in Fig. 11.19. Let an ac voltage $\tilde{v} = V_1 \sin \omega t$ be superimposed on the dc bias. When the voltage $\tilde{v}$ is positive, additional holes will be injected from the metal into the semiconductor. Since the hole injection rises exponentially with the applied voltage, the injected current will have its maximum value at the positive peak of the rf voltage. Movement of holes through the semiconductor drift zone induces a current in the external circuit which lasts until the holes are collected at J_2. Fig-

THE BARITT DIODE

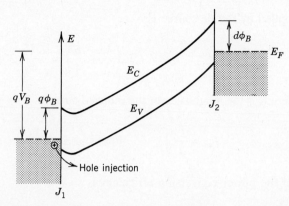

FIGURE 11.19 Energy band diagram of a reverse-biased MSM structure with dc bias higher than V_{PT} but lower than V_{FB}.

ure 11.20 shows the waveforms of the ac voltage, the injected hole current, and the current induced in the external circuit. Note that holes are injected almost as a delta function when the phase angle of $\tilde{v}$ is $\pi/2$. From $\pi/2$ to π, the ac voltage and the induced current are in phase and holes absorb energy from the field. For $\pi \leq \omega t < 2\pi$, the voltage and current are out of phase, and the ac field absorbs energy from the holes. It is clear that to obtain the maximum ac power, the thickness W should be such that the injected charge bunch gets collected at J_2 slightly before the end of the ac cycle.

11.5.2 Power Output and Efficiency

From the waveform of Fig. 11.20, it is seen that the external current flows for a time $(T/2 + T\delta/4)$ where $T = 1/\omega$ is the time period of the ac voltage, and δ is a parameter that determines the length of the induced current pulse. The power

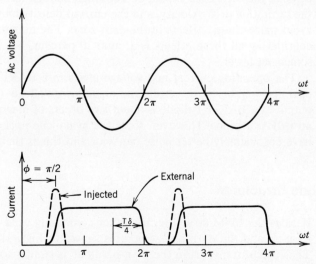

FIGURE 11.20 Waveforms of ac voltage, injected charge, and induced current in a BARITT diode.

supplied by the dc source is

$$P_{dc} = I_o V_B \left(\frac{1}{2} + \frac{\delta}{4} \right) = \frac{I_o V_B}{4}(2 + \delta) \qquad (11.38)$$

where I_o is the magnitude of the induced current. The ac power output is given by

$$P_{ac} = \frac{I_o}{T} \int_0^{(T/2 + T\delta/4)} V_1 \cos \omega t \, dt = \frac{I_o V_1}{2\pi} \sin \frac{\pi}{2}(2 + \delta) \qquad (11.39)$$

and the power conversion efficiency is

$$\eta = \frac{1}{2\pi} \frac{V_1}{V_B} \frac{4}{2 + \delta} \sin \frac{\pi}{2}(2 + \delta) \qquad (11.40)$$

The maximum value of η occurs for $\delta = 0.86$, giving

$$\eta_{max} = 0.22 \frac{V_1}{V_B} \qquad (11.41)$$

Thus, the efficiency can swing to 22 percent as V_1 approaches V_B. The optimum width of the drift zone W for maximum efficiency is obtained from the relation $\omega \tau_d = (2 + \delta)\pi/2$, and writing $\tau_d = W/v_s$ and $\delta = 0.86$, we obtain

$$W_{opt} = 0.72 \frac{v_s}{f} \qquad (11.42)$$

There are a number of limitations on the amplitude V_1. First, the total voltage across the diode should not exceed the flat-band voltage V_{FB}. If this happens, the current will saturate to a constant value before $\tilde{v}$ reaches its peak. The other limitations are similar to those considered in the case of IMPATT diodes. The field through the drift zone should be kept high enough for the carriers to move with the saturation drift velocity, and the current density must be kept low enough to avoid space-charge effects in the drift zone. The maximum efficiency estimated considering all these effects is around 10 percent, but the measured values are somewhat lower.

The operating current and voltage levels in a BARITT are low because it is operated well below the avalanche breakdown voltage. Consequently, the power output of a BARITT diode is about two orders of magnitude lower than that of an IMPATT diode. However, since the avalanche noise is absent, these devices have considerably better noise behavior and this is their main advantage.

11.5.3 Diode Structures

Besides the MSM structure, the other commonly used BARITT diodes are the p^+-n-p^+ and the p^+-n-i-p^+ structures shown in Figs. 11.21a and b, respectively. The current transport in the p^+-n-p^+ diode is similar to that in the MSM diode, but there is one important difference. In the case of the MSM diode, holes are injected from the metal into the semiconductor over the potential barrier which, in

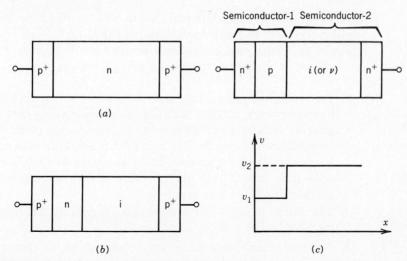

FIGURE 11.21 Common device structures for BARITT diodes: (a) p^+-n-p^+ diode, (b) p^+-n-i-p^+ diode, and (c) DOVETT diode with its velocity distribution.

a Pt-Si Schottky diode, is about 0.2 eV. No such barrier exists in a p^+-n-p^+ diode, so that at a given voltage, the current in this diode is about three orders of magnitude higher than that in the MSM diode. The advantage of the p^+-n-i-p^+ structure is that the electric field in the i-zone remains constant, which leads to a higher value of η compared to that of the p^+-n-p^+ diode.

The conversion efficiency can be further improved in the double velocity transit time (DOVETT) diode shown in Fig. 11.21c. The velocity distribution in the device at punch-through is also shown. Here, the central p- and ν-regions are of two different semiconductors such that the saturated drift velocity v_2 in the ν-region is higher than that in the p-region. Thus, the central p-ν junction is a heterojunction between semiconductors 1 and 2, while the remaining two junctions are homojunctions. The device is biased such that both sides of the p-ν junction are punched-through and a large current is injected over the forward-biased n^+-p junction. Because the velocity v_1 in the p-zone is lower than v_2, an injection delay is achieved, and losses are reduced resulting in high efficiency and power output. One combination that satisfies the condition $v_2 > v_1$ is when semiconductor 1 is AlGaAs and semiconductor 2 is GaAs. Efficiencies in excess of 20 percent have been predicted for this combination.

11.6 TRANSFERRED-ELECTRON DEVICES

Transferred-electron devices are bulk effect devices that utilize hot electrons to produce a voltage-controlled differential negative resistance (DNR). The phenomenon that gives rise to DNR is the decrease in electron mobility with the electric field. This decrease is caused by the transfer of conduction electrons from a high-mobility, low-energy state to a low-mobility, high-energy state under the action of high electric fields in excess of a few KV/cm. This transfer is made possible by the peculiar band structure of semiconductors like GaAs and InP. The first experimental observation of microwave oscillations resulting from the transferred electron effect was observed by J. B. Gunn [9] in GaAs. Although the effect has also been observed in other semiconductors like InAs, ZnSe, and CdTe, our discussion will be limited to GaAs and InP.

Figure 11.22 shows a cross-section through the E-k diagram of the GaAs conduction band. The diagram has the following features:

1. The lowest minimum occurs at **k** = 0 (valley 1). Here the E-k diagram has a sharp curvature, and the electrons in this lower valley have a low effective mass m_1 and hence, a high mobility μ_1. Besides the lower valley, there are energy minima along the [111] axis known as *satellite valleys,* each of which is separated from the lower valley by an energy $\Delta E = 0.31$ eV. In a satellite valley (i.e., valley 2), the electron mass m_2 is higher, and the mobility μ_2 is considerably lower than μ_1.
2. The density of states in a satellite valley is considerably higher than that in the lower valley.
3. The energy difference ΔE is large compared to the thermal energy kT of the electrons at room temperature. Thus, the transfer of electrons from the lower to the satellite valley by thermal agitation is not very likely. Furthermore, since ΔE is small compared to E_g, electron transfer can occur at fields much lower than those required for avalanche breakdown.

Consider now that an electric field $\mathscr{E}$ is applied to a sample of GaAs. At low fields, all the conduction electrons are located in the lower valley 1. However, as $\mathscr{E}$ is increased, electrons gain energy from the field and make a transition to the low-mobility satellite valleys. As a result, the conductivity of the material decreases at high fields. The electron transfer sets in abruptly after the field has reached the threshold value $\mathscr{E}_T$. This causes a decrease in the average drift velocity of the electron and results in a region of negative differential mobility (NDM).

The conductivity in a two-valley semiconductor can be written as

$$\sigma = q[n_1(\mathscr{E})\mu_1 + n_2(\mathscr{E})\mu_2] = \frac{qn_o v_n(\mathscr{E})}{\mathscr{E}} \tag{11.43}$$

where n_1 and n_2 are the electron concentrations in the lower and the upper valley, respectively, and $n_o = (n_1 + n_2)$. From Eq. (11.43), the average drift velocity $v_n(\mathscr{E})$

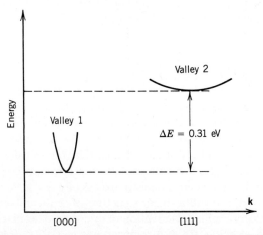

FIGURE 11.22 *E*-k diagram of GaAs conduction band along the [111] direction.

TRANSFERRED-ELECTRON DEVICES

is obtained as

$$v_n(\mathcal{E}) = \frac{\{n_1(\mathcal{E})\mu_1 + n_2(\mathcal{E})\mu_2\}\mathcal{E}}{n_o} \quad (11.44)$$

It can be seen from this relation that the differential mobility $dv_n/d\mathcal{E}$ can become negative at a sufficiently high field since n_1 decreases and n_2 increases with $\mathcal{E}$ (see Prob. 11.14). At very high fields when all the electrons have made a transition to the upper valleys, $v_n(\mathcal{E})$ again starts increasing, with $\mathcal{E}$ making $dv_n/d\mathcal{E}$ positive. Figure 11.23 shows [7] the experimentally determined velocity versus $\mathcal{E}$ curves for GaAs and InP. The threshold field $\mathcal{E}_T$ is about 3.2 KV/cm for GaAs and about 10.5 KV/cm for InP.

11.6.1 Static I-V Characteristics of a Differential Negative Mobility (DNM) Crystal

Let us consider a sample of length L of a semiconductor whose velocity field characteristic is shown in Fig. 11.23. Let voltage V be applied to the sample. If V is increased gradually, we can expect to obtain the I-V characteristic similar to that of a tunnel diode, provided the field is uniform throughout. However, in real crystals, local fluctuations in the mobile carrier concentration occur due to doping inhomogeniety. This causes the field to become nonuniform, and as a result, the negative resistance characteristic is not observed. The one-dimensional equations governing the I-V characteristic of the sample are the Poisson equation

$$\frac{\partial \mathcal{E}}{\partial x} = \frac{q}{\varepsilon_s}[n(x) - n_o] \quad (11.45)$$

and the current density equation

$$J(x) = qn(x)v_n(x) \quad (11.46)$$

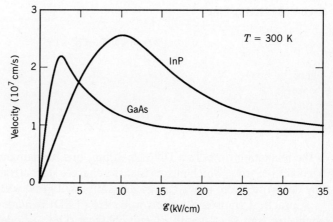

FIGURE 11.23 Velocity field characteristics of hot electrons in GaAs and InP. (From S. M. Sze, *Semiconductor Devices: Physics and Technology*, p. 239. Copyright © 1985. Reprinted by permission of John Wiley & Sons, Inc., New York.)

where n_o is the thermal equilibrium electron concentration and $n(x)$ is the concentration at any point x. From the above two relations, we obtain

$$\frac{\partial \mathcal{E}}{\partial x} = \frac{q}{\varepsilon_s}\left[\frac{J(x)}{qv_n(x)} - n_o\right] \quad (11.47)$$

A computer solution of this nonlinear differential equation reveals that in the steady state, $\mathcal{E}(x)$ increases monotonically with x and no negative resistance characteristic is observed. This is because an internal space charge exists in the sample, and $n(x)$ varies with $\mathcal{E}$ such that $n(x)v_n(x)$ always increases with $\mathcal{E}$.

(a) HIGH FIELD DOMAIN FORMATION

Under the nonsteady state condition, high field domain formation occurs and the steady state solutions are not necessarily stable with respect to small fluctuations. Referring to Fig. 11.24, let us consider that a small increase in $n(x)$ occurs at some region A of the bar as shown in Fig. 11.24b. Since the current density J is constant throughout, an increase in n causes v_n to decrease. Thus, the electrons in this region are slowed down and it causes an increase in the negative charge. At the front end of this perturbation, more electrons are leaving than are entering, and this creates a depletion region having positive charges. Thus, a dipole layer of charges is formed in the crystal in which the electric field is higher than the field outside it. The resulting electric field and potential distributions in the bar are shown in Figs. 11.24c and d, respectively. If the device is initially biased in the negative resistance region, the space charge will grow until the field inside the domain rises to $\mathcal{E}_2$ and that outside the domain decreases to $\mathcal{E}_1$. When this happens, the currents in the two regions become equal and further growth of the domain is stopped [10].

11.6.2 Electron Dynamics in a DNM Medium

A medium in which the electron drift velocity increases monotonically with $\mathcal{E}$ is a positive differential mobility medium. In this medium, any space charge $Q(0)$ injected at $t = 0$ decays exponentially with time because of mutual repulsion between electrons. Thus,

$$Q(t) = Q(0)\exp(-t/\tau_c) \quad (11.48)$$

where

$$\tau_c = \frac{\varepsilon_s}{\sigma} = \left(\frac{qn_o}{\varepsilon_s}\frac{dv_n}{d\mathcal{E}}\right)^{-1} \quad (11.49)$$

is the relaxation time. In a DNM medium, this fundamental law is violated as τ_c becomes negative, which means that the charges start attracting each other. Thus, any departure from space-charge neutrality will grow with time.

A practical transferred electron device (TED) is made from an n-type semiconductor (say, GaAs) with n^+-end contacts. The cathode contact of the device injects excess electrons into the interior of the crystal, and the high field domain

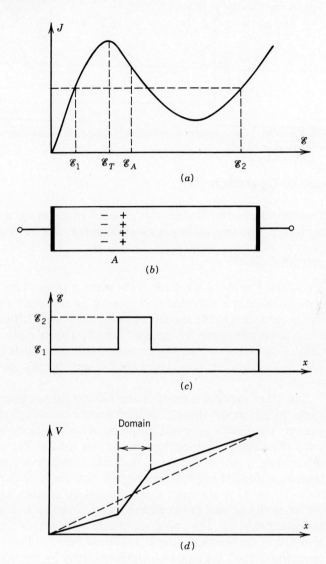

FIGURE 11.24 High field domain formation in a bulk differential negative resistance device. (a) J-$\mathscr{E}$ characteristic, (b) device cross-section, (c) electric field, and (d) potential distributions. After reference 10.

discussed above is nucleated at the cathode. The domain traverses to the anode, and during its transit, the space charge increases exponentially with time and the domain continues to grow until it attains maturity. When a fully matured domain is collected at the anode, the field rises above $\mathscr{E}_T$ and a new domain is nucleated at the cathode. The cycle repeats itself, and the current through the device oscillates as shown in Fig. 11.25. For small lengths, the frequency of oscillation lies in the microwave region.

A matured domain will be formed when its transit time L/v_n is larger than τ_c, and for GaAs, this condition requires that $n_o L \geqslant 1.6 \times 10^{12}$ cm^{-2}. Samples with $n_o L < 10^{12}$ cm^{-2} cannot support fully developed dipole domains, but they can support a growing space-charge wave and can be operated as linear microwave amplifiers.

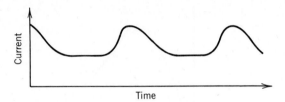

FIGURE 11.25 Typical current waveforms resulting from nucleation, propagation, and collection of domains in a TED.

11.6.3 Modes of Operation [11]

Transferred electron devices can be used either as microwave amplifiers or as oscillators. The important modes of operation will be described in this section.

(a) OSCILLATORS

The simplest mode of operation is the domain transit time mode in which a fully matured domain is collected at the anode as explained above. When a resistive load is connected to the anode, the power conversion efficiency is very low. However, the efficiency can be increased considerably by replacing the resistance with a high Q resonant circuit. The resonant loading modifies the oscillation process, and the frequency is now determined mainly by the resonant behavior of the circuit.

The other important mode is the limited space-charge accumulation (LSA) mode. In this mode, the $n_o L$ product should be such that a matured domain is formed. The device is biased in the negative resistance region by a dc voltage, and a rf voltage is superimposed on the bias voltage. The rf voltage swing is such that, during a part of the cycle, the field in the device remains below $\mathcal{E}_T$. The frequency of the rf signal is kept so high that the time during which the field remains above $\mathcal{E}_T$ is too small to form a mature domain. Under these conditions, the space charge that grows during a part of the cycle is quenched subsequently as $\mathcal{E}$ goes below $\mathcal{E}_T$. Thus, no space-charge instability is created, and the I-V characteristic of the sample is similar to the v-$\mathcal{E}$ curve in Fig. 11.23. The frequency of operation in the LSA mode is determined only by the external circuit and is independent of the transit time. This mode is the most efficient mode for dc-to-rf power conversion at a single frequency.

(b) AMPLIFIERS

As stated above, a TED biased above the threshold field shows a positive resistance in the dc steady state. However, when a rf field of frequency higher than the transit time frequency is applied to such a device, the space-charge fluctuations induced by the ac field do not have time to stabilize. Thus, the space charge grows during its transit from the cathode to the anode. In this case, the diode behaves as an amplifier with gain increasing exponentially with the length of the sample. However, this amplifier has to be stabilized. Some methods of stabilization are given below:

1. A growing domain does not reach maturity if the transit time of the rf perturbation across the sample is small compared to the electron relaxation time. In the case of GaAs, this condition requires $n_o L < 10^{12}$ cm^{-2}. Diodes

satisfying this condition are called *subcritically doped diodes* and act as amplifiers.

2. If the rf signal in the LSA mode is large enough to bring the field in the device below $\mathcal{E}_T$ for a certain fraction of the period and, if the frequency is sufficiently high, no domain growth occurs. Under this condition, the I-V characteristic is similar to the v-$\mathcal{E}$ curve, and the diode can be used as a negative resistance amplifier.

3. In a diode in which the $n_o L$ product is such that a mature domain formation is possible, the n$^+$-cathode injects electrons into the n-region and, if sufficient bias is applied to the device, the high field region fills the whole of the sample and a stable negative resistance operation with gain is obtained [11].

11.6.4 Device Fabrication and Performance

The transferred electron effect has been observed in GaAs, InP, CdTe, ZnSe, GaAsP, and some quarternary semiconductors like InGaAsP [12]. However, practical devices have been fabricated mainly in GaAs and to a lesser extent in InP.

Successful device fabrication requires extremely pure and uniform materials with a minimum of deep impurities. The desired value of n_o in GaAs ranges from 10^{13} cm^{-3} to 10^{16} cm^{-3}. Present-day devices are fabricated on n-material grown on a n$^+$-substrate approximately 0.2 mm thick. Finally, a thin n$^+$-layer is epitaxially grown on the n-layer to make a n$^+$-n-n$^+$ structure. After making ohmic contacts on the two n$^+$-sides, the wafer is divided into individual units. Each unit is then mounted on a gold-plated heat sink as shown in Fig. 11.26. Note that the n-region forms the active part of the device.

The theoretically predicted power conversion efficiency of a GaAs TED is 26 percent in the transit time mode and 30 percent in the LSA mode. However, the experimentally obtained values are less than 20 percent. The main reason for this poor performance is the temperature rise due to power dissipation in the device. The high temperature during operation causes changes in μ and n_o, and thus decreases the peak-to-valley current ratio. An efficient heat sink, therefore, is essential for good efficiency.

The main areas of application for TEDs are in microwave amplification and power generation. In these applications, they have to compete with IMPATT diodes and with bipolar and field-effect transistors. Recent progress in bipolar transistors and MESFETs has shown the superiority of transistors over TEDs at frequencies below 20 GHz. At higher frequencies IMPATT diodes are preferred because of their large power-handling capability. The only advantage of a TED is its low noise.

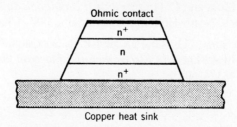

FIGURE 11.26 Cross-sectional view of a typical TED mounted on a heat sink.

REFERENCES

1. M. H. NORWOOD and E. SHATZ, "Voltage variable capacitor tuning: A review," *Proc. IEEE* 56, 788 (1968). Also R. A. MOLINE and G. F. FOXHALL, *IEEE Trans. Electron Devices* ED-19, 267 (1972).
2. A. K. GUPTA and M. S. TYAGI, "Avalanche breakdown voltage of hyperabrupt silicon p-n junctions," *Solid-State Electron.* 19, 342 (1976).
3. A. K. GUPTA and M. S. TYAGI, "C-V index of hyperabrupt p-n junctions," *Solid-State Electron.* 21, 507 (1978).
4. H. S. VELORIC and M. B. PRINCE, "High-voltage conductivity-modulated silicon rectifier," *Bell System Tech. J.* 36, 975 (1957).
5. S. K. GHANDHI, *Semiconductor Power Devices,* John Wiley, New York, 1977.
6. W. T. READ, "A proposed high-frequency negative-resistance diode," *Bell System Tech. J.* 37, 401 (1958).
7. S. M. SZE, *Physics of Semiconductor Devices,* John Wiley, New York, 1981.
8. T. MISAWA, "Negative resistance in p-n junctions under avalanche breakdown conditions," *IEEE Trans. Electron Devices* ED-13, 137 (1966).
9. J. B. GUNN, "Microwave oscillations of current in III-V semiconductors," *Solid State Commun.* 1, 88 (1963). See also *IEEE Trans. Electron Devices* ED-23, 705 (1976).
10. H. KROEMER, "Negative conductance in semiconductors," *IEEE Spectrum* 5, 47 (Jan. 1968).
11. S. TESZNER and J. L. TESZNER, "Microwave power semiconductor devices. I," in *Advances in Electronics and Electron Physics,* Academic Press, Vol. 39, 291 (1975).
12. W. KOWALSKY and A. SCHLACHETZKI, "Analysis of the transferred-electron effect in the InGaAsP system," *Solid-State Electron.* 30, 161 (1987).

ADDITIONAL READING LIST

1. H. A. WATSON (ed.), *Microwave Semiconductor Devices and Their Circuit Applications,* McGraw-Hill, New York, 1969.
2. M. J. HOWES and D. V. MORGAN (eds.), *Microwave Devices,* John Wiley, London, 1977, Chapters 2 and 3.
3. T. S. MOSS and C. HILSUM (eds.), *Handbook on Semiconductors,* Vol. 4, Device Physics, North-Holland, Amsterdam, 1981, Chapters 4A and 4B.
4. M. P. SHAW, H L. GRUBIN, and P. R. SOLOMON, "Gunn-Hilsum effect electronics," in *Advances in Electronics and Electron Physics,* Academic Press, Vol. 51, 309–433 (1980).
5. P. A. BLAKEY, "A reassessment of TRAPATT theory-I. Charge production and device design," *Solid-State Electron.* 27, 751–761 (1984).

PROBLEMS

11.1 A varactor diode has a junction capacitance of 1.2 pF and a series resistance of 0.8 Ω at zero bias. Determine the zero-bias cutoff frequency and the frequency at which the quality factor becomes 100.

11.2 In a certain application of a GaAs varactor diode it is required to have $s = 2$ independent of voltage. Consider a p$^+$-n diode with $N_o = 10^{17}$ cm^{-3}. Assuming the diode area to be 10^{-2} cm^2, determine the exponent of the power law distribution and the diode capacitance at $(V_i + V_R) = 3$ V.

11.3 Consider a Si p$^+$-n junction varactor diode fabricated on an epitaxial layer with $N_d = 6 \times 10^{15}$ cm^{-3} grown on n$^+$-substrate of 0.002 Ω-cm resistivity. The epitaxial layer thickness is 5 μm, and the junction depth from the surface is 1 μm. The dopant concentration on the p$^+$-side is 10^{19} cm^{-3}, and the junction area is 10^{-3} cm^2. Calculate the diode cutoff frequency at (a) zero bias, (b) a reverse bias of 3 V, and (c) a reverse bias of 25 V. Ignore the resistance of the n$^+$-substrate.

11.4 In Prob. 3 assume that the n-side of the junction is linearly graded but the total number of impurities in the n-epitaxial layer remains the same. Calculate the cutoff frequency for the three bias voltages. Assume $V_i = 0.8$ V.

11.5 The central π-region of a Si p-π-n diode is 200 μm wide and has a resistivity of 1 K Ω-cm. The dopant concentration in each of the p$^+$- and n$^+$-regions is 10^{18} cm^{-3}. Calculate the zero-bias capacitance cm^{-2} and the voltage at which the capacitance becomes 25 percent of its value at zero bias. Also determine the R-C time constant of the junction under these two conditions assuming $\tau_o = 5\mu$ sec.

11.6 A Si p-i-n diode has the avalanche breakdown voltage of 250 V. Using $a_o = 1.07 \times 10^6$ cm^{-1} and $b = 1.65 \times 10^6$ V/cm, calculate the width of the i-region of the diode.

11.7 A Si p-i-n diode has a junction area of 10^{-2} cm^2 and an i-region width of 100 μm. The diode is carrying a 60-mA forward current at 300 K. Taking the values of μ_n and μ_p for the intrinsic silicon, determine the voltage drop across the i-region. Also calculate the diode voltage using Eq. (11.20) assuming $\tau_a = 3\mu$ sec.

11.8 The variation in electric field in the depletion region due to avalanche-generated space-charge gives rise to an incremental resistance R_{sc} for a p$^+$-n junction diode. Show that for a junction having a depletion region width W the resistance R_{sc} is given by $R_{sc} = (W - x_A)^2/2A\varepsilon_s v_s$.

11.9 A silicon Read diode, biased in the avalanche breakdown region, has an area of 2.6×10^{-3} cm^2 and is carrying a dc bias current of 0.5 A. The width of depletion region is 15 μm, and significant multiplication occurs only in 10 percent of this width. The diode is biased at a voltage of 168 V, and the field in the avalanche region is five times that in the drift zone. Taking the values of a_o and b from Prob. 11.6, calculate: (a) the capacitance and the inductance of the avalanche region, and (b) the resonant frequency f_A and the avalanche region impedance at $f = 2f_A$.

11.10 A silicon IMPATT diode is operating at a frequency of 10 GHz. The diode reactance at this frequency is 50 Ω and $\Delta T = 25°\text{C}$. Taking $K_s = 1.5 \ W\,\text{cm}^{-1}\,\text{K}^{-1}$, calculate the maximum obtainable value of the output power.

11.11 A MSM diode on n-Si has an active region thickness of 7 μm and a punch-through voltage of 72 V. Determine the dopant concentration in the semiconductor and calculate the flat-band voltage assuming $V_i = 0.6$ V.

11.12 The MSM structure of Prob. 11.11 is used as a BARITT diode. Assume that the negative conductance is given by $G = (h/\omega\tau_d)\sin\omega\tau_d$ where τ_d is the minority carrier transit time through the drift zone and h is a constant. The peak value of G is found to be $-0.05\ \Omega^{-1}$. Estimate the frequency at which this maximum will occur and the diode conductance at 22 GHz.

11.13 A Si BARITT diode is to be operated at a frequency of 7.5 GHz with a dc bias of 60 V and a bias current of 5 mA. Estimate the width of the active region for obtaining the maximum efficiency and calculate the power supplied by the dc source. Determine the amplitude of the ac voltage if the measured efficiency is 5 percent.

11.14 Differentiating Eq. (11.44) with respect to the electric field, determine the condition for which the differential mobility becomes negative.

11.15 The active length of a GaAs Gunn diode is 10 μm. It is desired to operate the device at 6 GHz and at 12 GHz. Discuss, giving reasons what will be the modes of operation at each frequency.

11.16 If the carrier concentration in the sample of Prob. 11.15 is 10^{15} cm^{-3}, determine the domain growth rate constant. Take a negative differential mobility of -3000 cm^2 V^{-1} sec^{-1} and a low field mobility of 7500 cm^2 V^{-1} sec^{-1}.

12
OPTOELECTRONIC DEVICES

INTRODUCTION

Optoelectronic devices convert optical energy into electrical energy or vice versa. The mechanism of converting optical radiation into electrical energy is known as the *photovoltaic effect*. Two important devices based on the photovoltaic effect are the solar cell and the photodiode. The converse of the photovoltaic effect is the phenomenon of emission of optical radiation by converting electrical energy into light. This phenomenon is called *electroluminescence*. The light-emitting diode and laser are examples of electroluminescent devices.

This chapter presents a brief account of photovoltaic and electroluminescent devices. Section 12.1 discusses solar cells which are designed to convert sunlight into electrical power. In Sec. 12.2 we consider photoconductors and photodiodes which are used to detect optical signals. Light-emitting diodes have assumed considerable importance in recent years and are described in Sec. 12.3. The chapter ends with a brief reference to semiconductor lasers in Sec. 12.4.

12.1 THE SOLAR CELL

In a solar cell, electron-hole pairs are produced by absorption of light. These pairs are then separated by the electric field present in the cell. Broadly speaking, there are four principal types of solar cells: (1) the p-n homojunction, (2) the heterojunction, (3) the Schottky barrier junction, and (4) the semiconductor–electrolyte junction. The p-n junction cell has been the focus of substantial effort during the last several years, and a Si p-n junction cell serves as a reference device for all other types of solar cells. Besides the p-n homojunction solar cell, we will also briefly discuss the heterojunction and the Schottky barrier cells. The semiconductor–electrolyte junction solar cell is beyond the scope of this book and will not be considered here.

12.1.1 Photovoltaic Effect in a p-n Junction [1]

The sunlight incident on a p-n junction solar cell causes the absorption of photons in the semiconductor, and electron–hole pairs are generated on the n- and p-sides of the junction. A fraction of the light-generated minority carriers diffuse to the edges of the junction depletion region and are swept across the junction by the electric field. Thus, the holes are collected on the p-side and electrons on the n-side. This collection of charges gives rise to a photocurrent through a load connected across the cell.

Not all the light that falls on the surface of the semiconductor is absorbed; a fraction of it is reflected, and some light is transmitted through the material. Only the light absorbed in the semiconductor can produce electron–hole pairs. The ability of a material to absorb light is measured by its absorption coefficient α (see Sec. 5.1). Figure 12.1 shows α plotted as a function of photon energy for a number of semiconductors. Note that for a given photon energy the larger the value of E_g, the lower the absorption coefficient.

We now consider the p-n junction in Fig. 12.2a to be illuminated with light of photon energy higher than E_g. Let F_{ph} denote the incident photon flux (number of photons/area-sec) at the surface and R the reflection coefficient. If we assume that each of the absorbed photons generates one electron–hole pair (unit quantum efficiency), the pair generation rate G_L at any distance x' from the surface of the semiconductor is given by

$$G_L = \alpha F_{ph}(1 - R) \exp(-\alpha x') \tag{12.1}$$

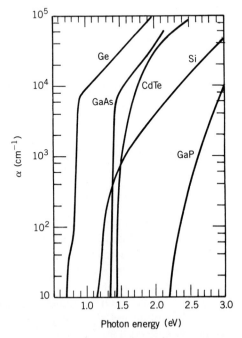

FIGURE 12.1 Absorption coefficient as a function of photon energy for some semiconductors used in solar cells. (From J. J. Loferski, Recent research on Photovoltaic solar energy converters, *Proc. of IEEE*, vol. 51, p. 669. Copyright © 1963 IEEE. Reprinted by permission.

THE SOLAR CELL

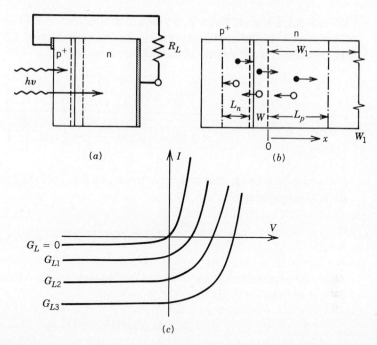

FIGURE 12.2 Photovoltaic effect in a p-n junction solar cell: (*a*) absorption of light by the cell, (*b*) diffusion of electrons and holes to produce photocurrent, and (*c*) I-V characteristics of the illuminated cell.

Under low-level injection, the steady state continuity equation for holes on the n-side of the junction can be written as (see Sec. 5.6)

$$\frac{1}{q}\frac{dJ_p}{dx} + \frac{p_e}{\tau_p} - G_L = 0 \tag{12.2}$$

We consider the simplest model of the solar cell in which the n- and p-sides of the junction are uniformly doped. The hole current density J_p in the neutral n-region can be expressed as $J_P = -qD_p(dp_e/dx)$, and Eq. (12.2) becomes

$$D_p\frac{d^2p_e}{dx^2} - \frac{p_e}{\tau_p} + G_L = 0 \tag{12.3}$$

From Eq. (12.1), it is seen that G_L is a decreasing function of x' and, hence, also of x. However, as an approximation, we assume G_L to be constant and independent of x. This approximation is valid if the thickness of the semiconductor is small compared to $1/\alpha$. If we assume a bias V_j across the junction, the excess hole concentration at the depletion region edge ($x = 0$) in Fig. 12.2b is given by

$$p_e(0) = p_{no}[\exp(V_j/V_T) - 1] \tag{12.4a}$$

We further assume that the contact at $x = W_1$ is perfectly ohmic so that

$$p_e(W_1) = 0 \tag{12.4b}$$

We can solve Eq. (12.3) for $p_e(x)$ with the boundary conditions Eqs. (12.4a) and (12.4b), and the following expression for the hole current $I_p(0)$ at the depletion region edge ($x = 0$) can be obtained (see Prob. 12.2):

$$I_p(0) = \frac{qAD_p}{L_p} p_{no} \coth\left(\frac{W_1}{L_p}\right)\left[\exp(V_j/V_T) - 1\right]$$
$$- qAG_L L_p\left[\coth\left(\frac{W_1}{L_p}\right) - \left(\sinh\frac{W_1}{L_p}\right)^{-1}\right] \quad (12.5)$$

For a long-base diode, the coth term tends to unity, and the last term in Eq. (12.5) goes to zero giving

$$I_p(0) = \frac{qAD_p p_{no}}{L_p}[\exp(V_j/V_T) - 1] - qAG_L L_p \quad (12.6)$$

A similar expression can be obtained for the electron current in the p-region, and adding the two currents together, we obtain

$$I = I_s[\exp(V_j/V_T) - 1] - I_L \quad (12.7)$$

where I_s is the saturation current and I_L is the light-generated current and is given by

$$I_L = qAG_L(L_p + L_n) \quad (12.8)$$

where A is the junction area and L_p and L_n are the minority carrier diffusion lengths in the n- and p-regions, respectively.

The I-V characteristics expressed by Eq. (12.7) are shown in Fig. 12.2c for different values of G_L. It is clear that the total current is just the diode dark current lowered by the current I_L which is directed from the n- to the p-side of the junction.

The diode can be operated either in the third or the fourth quadrant of the I-V characteristics. When operated in the third quadrant, the junction is reverse biased by a few volts, and the diode carries a reverse current

$$I = -(I_s + I_L) \quad (12.9)$$

The diode operated in this mode acts as a photodiode and can be used to measure the light intensity. Operation in the fourth quadrant is achieved by connecting a load resistance across the diode terminals as shown in Fig. 12.2a. The diode operated in this manner is a photovoltaic cell. The appearance of a forward voltage across an illuminated cell is called the *photovoltaic effect*.

12.1.2 The p-n Junction Solar Cell

(a) BASIC DEVICE

It is observed from Eq. (12.8) that I_L can be increased by increasing the electron–hole pair generation rate G_L and making the minority carrier diffusion lengths L_p and L_n large. Large values of L_n and L_p require that the minority carrier diffusion constants and lifetimes on both sides be as large as possible. The top layer in a solar cell is produced by solid-state diffusion with a high dopant concentration. Lattice damage in this layer introduces a high density of recombination centers, which makes the lifetime in this region very short. This short lifetime creates a so-called dead zone near the surface which makes little contribution to light-generated current I_L. Thus, in a solar cell, the top layer is kept rather thin, and most of the contribution to I_L arises from the substrate, which is called the base of the cell.

Figure 12.3 shows the schematic of a basic p-n junction solar cell. It is an n^+-p junction device made on a p-type base. The switchover from the p^+-n to the n^+-p structure has resulted in an improvement in the overall performance of the cell because of the larger diffusion constant of electrons in the p-type base region. A large value of D_n increases the diffusion length L_n, and, hence, the current I_L. The n^+-region is produced by diffusion of phosphorus. In conventional diffusion, the surface concentration of phosphorus is about 5×10^{20} cm^{-3} and the junction depth may be around 0.4 μm. In an effort to avoid the dead zone created by the high concentration of P atoms, the surface concentration is kept below 10^{20} atoms cm^{-3} with a junction depth of less than 0.2 μm.

(b) ELECTRICAL CHARACTERISTICS

In a solar cell, the current flows in the reverse direction (i.e., from n- to p-side) and Eq. (12.7) can be written as

$$I = I_L - I_s[\exp(V_j/V_T) - 1] \tag{12.10}$$

In the derivation of this relation, we assumed the ideal situation in which the dark current results from the injection of thermally generated minority carriers into the neutral n- and p-regions. Besides the thermal current, Si and GaAs p-n junction solar cells also show depletion region recombination current. When the doping on the two sides of the junction is high ($\simeq 10^{18}$ cm^{-3}), an additional current component due to tunneling may also be present. We will neglect these components, however, and keep our analysis confined to the ideal case.

A plot of Eq. (12.10) is shown in Fig. 12.4 for an experimental device [2] under air mass 1 (AM1) illumination. The AM1 spectrum represents the sunlight falling on the device placed at sea level under a clear sky with the sun at the zenith. The power reaching the device under this condition is about 92.5 mW cm^{-2}. The solar spectrum just outside the earth's atmosphere has an incident power of

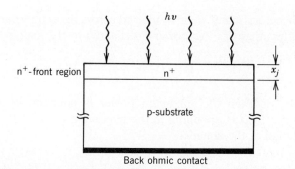

FIGURE 12.3 Schematic of a diffused n^+-p junction Si solar cell.

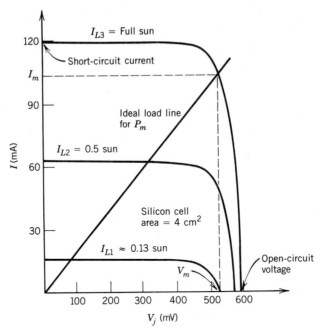

FIGURE 12.4 I-V characteristics of a typical silicon solar cell under AM1 illumination condition. (From E. S. Yang, reference 2, p. 151. Copyright © 1978. Reprinted by permission of McGraw-Hill, Inc., New York.)

about 135 mW cm^{-2} and is known as the *AM0 spectrum*. For the sun at an angle of 60° from the vertical, the path length through the atmosphere is doubled, and this is called the *AM2 spectrum*.

The three most important parameters of a solar cell are the short-circuit current I_{sc}, the open-circuit voltage V_{oc}, and the fill factor (*FF*). For $V_j = 0$ in Eq. (12.10), we obtain

$$I_{sc} = I_L \tag{12.11}$$

and for $I = 0$

$$V_{oc} = V_T \ln\left(1 + \frac{I_L}{I_s}\right) \tag{12.12}$$

When a resistance R_L is connected across the cell terminals, V_j is less than V_{oc} and I is less than I_L. The power P delivered to the load is

$$P = IV_j = I_L V_j - I_s V_j[\exp(V_j/V_T) - 1]$$

The condition for maximum power is obtained by making $\partial P/\partial V_j = 0$. The voltage V_m corresponding to the maximum value of P is then obtained from the solution of the equation

$$\left(1 + \frac{V_m}{V_T}\right)\exp(V_m/V_T) = \left(1 + \frac{I_L}{I_s}\right) \tag{12.13}$$

and the current I_m at this voltage is

$$I_m = (I_s + I_L)\left[\frac{V_m}{V_m + V_T}\right] \tag{12.14}$$

The maximum obtainable power P_m is then given by

$$P_m = I_m V_m = (I_s + I_L)\left[\frac{V_m^2}{V_m + V_T}\right] \tag{12.15}$$

and the optimum load resistance R_{opt} which achieves the maximum power is obtained as

$$R_{opt} = \frac{V_m}{I_m} = \frac{(V_m + V_T)}{(I_s + I_L)} \tag{12.16}$$

The fill factor (FF) is defined by the relation

$$FF = \frac{V_m I_m}{V_{oc} I_L} \tag{12.17}$$

and is a measure of the squareness of the I-V curve in Fig. 12.4. For a well-designed cell, the FF lies between 0.7 to 0.8. The ideal power conversion efficiency is given by

$$\eta = \frac{V_m I_m}{P_{in}} = FF\frac{V_{oc} I_L}{P_{in}} \tag{12.18}$$

Thus, it is clear that high values of I_L, V_{oc}, and FF lead to high efficiencies in the p-n junction solar cell. I_L can be increased by increasing G_L and the minority carrier diffusion lengths L_p and L_n.

The open-circuit voltage V_{oc} is increased by decreasing the dark current, and this indicates the desirability of decreasing the depletion region recombination current. A conventional way to reduce the dark current I_s is to dope the diffused front layer heavily and keep it thin so that the minority carrier injection in this region remains negligibly small compared to that in the base region. When this is the case, recombination in the base determines the dark current and I_s can be written as

$$I_s \simeq \frac{qAD_n n_i^2}{L_n N_a} \tag{12.19}$$

It is evident from this relation that V_{oc} will increase with increasing N_a. In Si solar cells, V_{oc} reaches its maximum value at a dopant concentration of about 10^{17} cm^{-3}. Beyond this point, tunnel currents become important and V_{oc} starts decreasing.

Even when every effort has been made to reduce the dark current, the measured values of V_{oc} in Si devices are found to be only 75 to 80 percent of the expected value. It has been pointed out that heavy doping effects in the top diffused layer are responsible for the decrease in V_{oc}.

The first effect of heavy doping is to reduce the band gap in the top n^+-layer. This reduction causes an increase in the minority carrier concentration in this region. The second effect of heavy doping is to lower the minority carrier lifetime in the n^+-region. This lowering is caused by the generation of additional recombination centers at heavy doping, as well as by the band-to-band Auger recombination. Both effects cause an increase in the component of the dark current resulting from the top n^+-region.

A high value of FF is obtained when the diode ideality factor is close to unity; this again emphasizes the desirability of reducing the depletion region recombination and surface leakage currents.

As an example, let us consider a silicon n^+-p junction solar cell with a 1 cm^2 surface area and $N_a = 10^{15}$ cm^{-3} in the base region. We neglect the contribution of the n^+-region and assume $D_n = 35$ cm^2 sec^{-1}, $\tau_n = 2.57$ μsec, and $G_L = 2.7 \times 10^{19}$ cm^{-3} sec^{-1}. Let us determine I_L and V_{oc}. We have $L_n = \sqrt{D_n \tau_n} = 9.48 \times 10^{-3}$ cm. Neglecting the contribution of n^+-layer, we get

$$I_s = \frac{qAD_n n_i^2}{L_n N_a} = \frac{1.6 \times 10^{-19} \times 35 \times (1.5 \times 10^{10})^2}{9.48 \times 10^{-3} \times 10^{15}} = 1.33 \times 10^{-10} \text{ A}$$

and from Eq. (12.8), $I_L = qAG_L L_n = 40.95$ mA. Now from Eq. (12.12), $V_{oc} = 25.86 \times 10^{-3} \ln(1 + 40.95 \times 10^{-3}/1.33 \times 10^{-10}) = 0.505$ V.

(c) ELECTRICAL EQUIVALENT CIRCUIT

Figure 12.5 shows the electrical equivalent circuit of an illuminated solar cell. The photocurrent is represented by a current source I_L which opposes the forward current of the diode. The shunt resistance R_{sh} may be caused by surface leakage along the edges of the cell or by crystal defects along the junction depletion region. The series resistance R_s results from the resistances of neutral n- and p-regions, as well as the resistance of the ohmic contacts. From Fig. 12.5 it is seen that

$$I\left(1 + \frac{R_s}{R_{sh}}\right) = I_L - I_{\text{dark}} - \frac{V_o}{R_{sh}} \tag{12.20}$$

which shows that both the diode current I and the output voltage V_o are affected by the presence of R_s and R_{sh}.

A low value of R_s requires a heavily doped top layer and deep junction. Both features reduce the minority carrier lifetime and, hence, tend to increase the dark

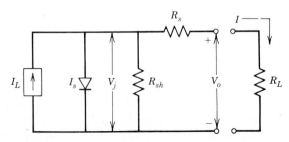

FIGURE 12.5 Electrical equivalent circuit of an illuminated p-n junction solar cell including series and shunt resistances.

current. Thus, for an optimum design, a compromise has to be reached between these conflicting requirements. A practical method to reduce the part of R_s associated with the n$^+$-region is to make contact using a grid pattern (Fig. 12.6) with 5 to 10 percent of the surface covered with the contact metal [3].

(d) MATERIALS AND DESIGN CONSIDERATIONS

Figure 12.7 shows the calculated ideal efficiency as a function of E_g at 300 K for 1 sun ($C = 1$) and for 1000 sun ($C = 1000$) concentrations [4]. Semiconductors with band gaps between 1 and 2 eV have a theoretical efficiency in excess of 20 percent, and all of them can be considered solar cell materials. However, because of technological and other considerations, the choice has to be made between Si and GaAs. Although GaAs offers the possibility of higher efficiency and power output, Si is widely used in p-n junction cells because of its more advanced technology and lower cost.

12.1.3 Device Fabrication and Further Developments in p-n Junction Solar Cell

(a) SILICON CELLS

Figure 12.8 shows the schematic diagram of a solar cell made on a p-silicon base by diffusion of phosphorus. After diffusing the n$^+$-region, ohmic contact is made to the back surface of the base region. The contact to the top layer is made in the form of a grid pattern. An antireflection (AR) coating is then used to reduce the amount of light lost by reflection from the Si surface. The ideal AR coating material is the one whose refractive index is the geometric mean of the refractive indices of air and the semiconductor. Practical coating materials include SiO, SiO_2, TiO_2, and Ta_2O_5. The typical thickness of an AR coating is 1000 Å. Finally, a cover glass is bonded to the cell using a transparent adhesive to prevent high-energy particles from reaching and degrading the device.

Over the years, efforts have been made to increase the efficiency and power output of the Si cells. The increase in power output has been achieved largely by an increase in the short-circuit current I_L, which can be increased by increasing the amount of light entering the cell and by increasing the collection of photogenerated carriers. Considerable improvement in carrier collection is obtained by doping the back side of the base heavily p-type so as to make a p-p$^+$ junction. This p-p$^+$ junction provides a back surface field in the base. Figure 12.9 shows the energy band diagram of the n$^+$-p-p$^+$ solar cell. It is evident that the potential

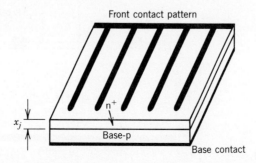

FIGURE 12.6 Cross-section of an n$^+$-p junction solar cell showing a front contact grid pattern.

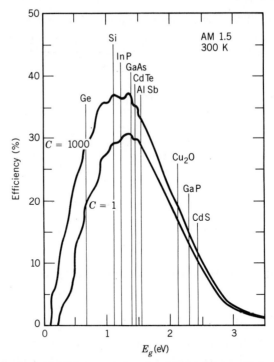

FIGURE 12.7 Ideal solar cell efficiency as a function of the band gap. (From S. M. Sze, Semiconductor device development in the 1970's and 1980's—a perspective, *Proc. of IEEE*, vol. 69, p. 1130. Copyright © 1981 IEEE. Reprinted by permission. Also reference 4, p. 798.)

barrier at the p-p$^+$ junction tends to keep the photogenerated electrons confined to the p-region, which enhances their collection by the p-n junction.

The performance of solar cells has also been improved by reducing the reflection of the incident light from the surface. This has been achieved by surface texturing using chemical etchants. Some slow etchants are found to selectively etch the Si surface and produce pyramidical tetrahedra of high density that are uniformly distributed over the surface (Fig. 12.10a). Surface texturing has two advantages. First, the multiple reflections reduce the amount of light reflected back from the surface. Second, the light gets refracted as it enters Si and travels obliquely through the cell causing its absorption closer to the junction (Fig. 12.10b).

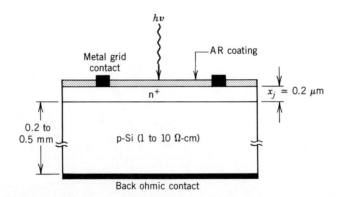

FIGURE 12.8 Cross-sectional view of a typical Si p-n junction solar cell.

THE SOLAR CELL

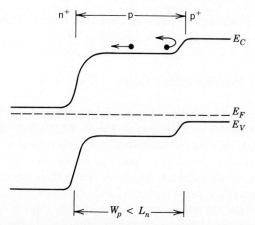

FIGURE 12.9 Energy band diagram of an n^+-p-p^+ back-surface field solar cell.

By employing surface texturing and a back surface field, it has become possible to increase the open-circuit voltage above 650 mV, whereas a typical value of I_L is 40 mA cm^{-2}. Cells with 15 percent AM1 efficiency are commercially available, and laboratory units have been reported with 20 percent efficiencies.

The most important use of Si solar cells has been in space satellites where no other satisfactory energy sources are available. The final barrier to the widespread use of solar cells for terrestrial applications is their cost. At present, the cost is too high to compete with conventional energy sources. However, cost may be reduced substantially by reducing the cost of crystal growth and wafer preparation. Low-cost crystal growth technologies will go a long way toward reducing the cost.

(b) GaAs SOLAR CELLS

The conversion efficiency of a Si p-n junction solar cell decreases rapidly with increasing temperature. This decrease is caused mainly by a rapid increase in the reverse saturation current I_s with rising temperature. The conversion efficiency of a GaAs p-n junction solar cell decreases less rapidly with temperature because of the higher band gap of GaAs. However, a serious problem with GaAs is that because of its direct band gap the value of α is high. As a consequence, the electron–hole pairs are created very near the surface and are lost by surface recombination

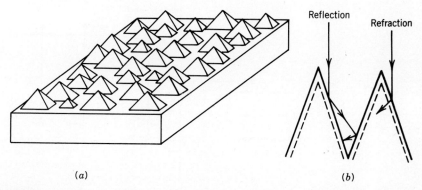

FIGURE 12.10 (a) Textured Si surface with pyramidical structue, and (b) reflection and refraction of light from the surface.

before reaching the junction depletion region. Thus, a very thin n-region is required on the *p*-type base.

12.1.4 Schottky Barrier Solar Cells

In a Schottky barrier solar cell, a thin (less than 100 Å thick) metal film is deposited on the semiconductor. Figure 12.11 shows the energy band diagram of a cell made on a *p*-type semiconductor. When light is incident on the front surface, photons with energy $h\nu > q\phi_B$ can excite holes from the metal over the barrier into the semiconductor. However, for a thin metal film, a large fraction of the light enters the semiconductor, and photons with $h\nu > E_g$ produce electron–hole pairs in the neutral semiconductor and the depletion region. The photogenerated electrons are collected by the junction, while the holes move toward the back contact causing a photocurrent across the diode. The photocurrent is opposed by the dark current of the diode, which arises from the thermionic emission of holes from the semiconductor into the metal. Since the dark current in a Schottky barrier diode is a few orders of magnitude higher than that in a p-n junction diode with the same area, V_{oc} is reduced. As a consequence, the conversion efficiency is much lower. The main advantage of Schottky barrier solar cells is that they do not require high temperature processing like diffusion, and thus, the processing cost is reduced.

The dark current may be reduced and V_{oc} increased by inserting a thin insulating layer (usually SiO_2) between the metal and the semiconductor. The resulting structure is known as the MIS (metal insulator-semiconductor) solar cell. Figure 12.12 shows the cross-section and energy band diagram of a MIS cell formed on a p-semiconductor. It is evident from Fig. 12.12*b* that the thin SiO_2 layer is transparent to electrons (i.e., minority carriers), but forms a barrier to the hole flow. Thus, the photogenerated electrons flow to the metal, but the dark current of thermionically emitted holes is considerably reduced. As a consequence, the V_{oc} of a MIS solar cell will be larger than that of a Schottky barrier solar cell. An optimum thickness of SiO_2 for the Si MIS cell is about 20 Å, and efficiencies as large as 17 percent can be obtained. A thicker SiO_2 layer reduces the short-circuit current and lowers the efficiency.

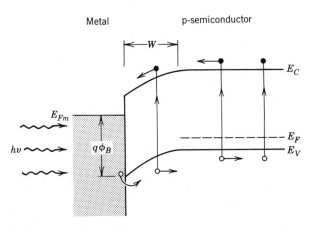

FIGURE 12.11 Energy band diagram of an illuminated Schottky barrier solar cell made on a *p*-type semiconductor.

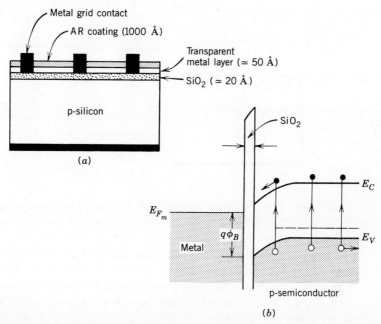

FIGURE 12.12 The MIS Schottky barrier solar cell. (a) Schematic cross-section and (b) energy band diagram under illuminated condition.

12.1.5 Heterojunction and Thin Film Solar Cells

Heterojunction solar cells have some advantages over the conventional homojunction cells. If the top layer semiconductor has a larger band gap E_{g1} than the band gap E_{g2} of the base region, then photons with energy $h\nu > E_{g1}$ are largely absorbed in the top layer. The top layer, however, acts as a window for the low-energy photons that will be absorbed by the second semiconductor. This will enhance the short wavelength response. The main difficulty in obtaining a heterojunction solar cell is to find semiconductors that have a good lattice match. Two such semiconductors are AlAs and GaAs. Figure 12.13 shows the cross-section and energy band diagram of the AlAs/GaAs heterojunction solar cell. AlAs has a wider band gap (2.2 eV) than GaAs. The absorption coefficient of AlAs for sunlight is relatively small, and a large fraction of the incident light is absorbed in the GaAs base. The light-generated carriers are collected in a similar manner as in an n^+-p homojunction solar cell. Power conversion efficiencies are as high as 18 percent.

Thin film solar cells are fabricated using thin films of semiconducting materials deposited on electrically active or passive substrates like glass, ceramic, graphite, or a metal. A variety of processes such as vacuum deposition, sputtering, growth from the vapor phase, and plating have been used to deposit films of CdS, CdTe, ZnSe, InP, GaAs, Si, and other semiconductors. These deposited films are either polycrystalline or amorphous, and thus have a highly disordered structure with properties much inferior to single crystals. Because of the availability of inexpensive materials and lower processing costs, thin film solar cells have the advantage of low cost. However, these cells have low efficiencies and suffer from long-term instability that results from the chemical reaction of the deposited semiconductor with the ambient.

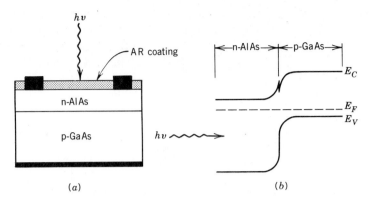

FIGURE 12.13 The AlAs-GaAs heterojunction solar cell: (a) schematic cross-section and (b) the energy band diagram under thermal equilibrium.

One of the important thin film solar cells is the Cu_2S-CdS heterojunction solar cell [5]. This cell is fabricated by vacuum deposition of about 20 μm thick CdS film on either a metal or a metal-coated insulating substrate. Dipping the unit into a hot solution of cuprous chloride for a few seconds forms a thin Cu_2S layer. The cell is completed by applying an AR coating and ohmic contacts.

Figure 12.14 shows the cross-sectional view and the energy band diagram of the Cu_2S-CdS heterojunction cell. As shown in Fig. 12.14b the heterojunction interface has a large density of interface states caused by lattice mismatch between Cu_2S and CdS. These states act as recombination centers and reduce the cell efficiency. With sunlight incident on the cell, most of the photocurrent is generated in the thin Cu_2S layer ($E_g = 1.2$ eV). Typical power conversion efficiencies range from 6 to 10 percent.

Growth of thin films of polycrystalline Si on a variety of inexpensive substrates is another way of obtaining low-cost thin film solar cells [5]. Either p-n junction or Schottky barrier solar cells can be made on these films. However, the grain size of the deposited Si film is very small, and the cell efficiency is only 1 to 2 percent.

The most promising Si thin film technology is based on an amorphous silicon-hydrogen alloy [6]. The hydrogenated amorphous Si films can be produced by the glow-discharge decomposition of silane (SiH_4). These films contain approximately 5 to 10 percent hydrogen (which results from the decomposition of SiH_4) and can be doped n- or p-type. It is postulated that the hydrogen saturates the

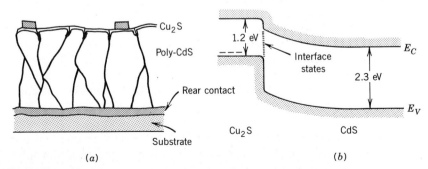

FIGURE 12.14 The thin-film Cu_2S-CdS solar cell: (a) schematic view and (b) the energy band diagram under thermal equilibrium. (From M. A. Green, *Solar Cells,* Prentice-Hall, Inc., Englewood Cliffs, New Jersey, 1982, p. 197. Reprinted by permission of the author.)

dangling bonds on the internal microvoids and other defects. Hydrogenated amorphous Si films have an absorption coefficient on the order of 10^5 cm^{-1} across the visible light spectrum, which means that 1 μm thick films are sufficient to absorb most of the solar energy in the visible region. Schottky-barrier and p-n junction solar cells fabricated on these films have shown efficiencies on the order of 6 percent.

12.2 PHOTODETECTORS

The detection of optical radiation with a semiconductor photodetector involves three processes: (1) Generation of carriers by the incident radiation, (2) transport and/or multiplication of carriers to produce an electric current, and (3) interaction of the photocurrent with an external circuit to yield an output signal. Photodetectors have found use in infrared sensors and as detectors for optical-fiber communication systems. Next, we will describe the operating principles and important characteristics of some common types of photodetectors.

12.2.1 Photoconductor

The most fundamental photoconductor may be fabricated by making ohmic contacts to the two ends of a slab of a semiconductor, as shown in Fig. 12.15a.

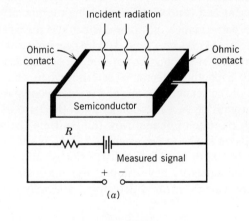

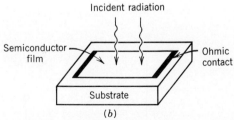

FIGURE 12.15 Schematic diagrams of photoconductors: (a) slab of a semiconductor with ohmic contacts on two opposite faces and (b) a thin film photoconductor deposited on an insulating substrate. (From C.T. Elliot, Infrared detectors, Chapter 6B in *Handbook on Semiconductors*, vol. 4, T.S. Moss and C. Hilsum (eds.), North-Holland, Amsterdam, p. 732. Copyright © 1981. Reprinted by permission of Elsevier Science Publishers, Physical Sciences & Engineering Div., and the author.)

Alternatively, one can deposit a thin film of a semiconductor on an insulating substrate with ohmic contacts provided at the two ends of the film (Fig. 12.15b). The incident light produces free carriers in the semiconductor by two different processes as shown in Fig. 12.16. Radiation with $h\nu > E_g$ produces electron–hole pairs by band-to-band transitions (a). This is the intrinsic photoexcitation process. Transitions (b) and (c) are the extrinsic excitation processes. In transition (b), the incident photon excites an electron from a donor level into the conduction band, whereas in (c) an electron is excited from the valence band to an acceptor level creating a hole in the semiconductor. Both the intrinsic and extrinsic excitation processes cause an increase in the conductivity, that can be measured by measuring the change in the voltage across the photoconductor when a constant current is passed through the circuit (Fig. 12.15a).

For the intrinsic excitation process, $hc/\lambda > E_g$, and the long wavelength cutoff λ_c is given by

$$\lambda_c = \frac{hc}{E_g} \tag{12.21}$$

where c is the velocity of light. In the case of extrinsic excitation, λ_c is determined by the depth of the donor or acceptor level in the band gap.

Extrinsic photoconductors have low optical absorption and must be cooled to low temperatures to avoid the thermal ionization of centers which occur at high temperatures. Intrinsic photoconductors have a much higher optical absorption coefficient and can be operated at high temperatures.

The performance of a photodetector is measured in terms of its (1) quantum efficiency, (2) response time, and (3) sensitivity of detection. Quantum efficiency is expressed by the number of free carriers generated per photon. The response time of a photoconductor is determined by the carrier transit time along the length of the semiconductor, and it is long compared to that of photodiodes. The current gain of the photoconductor is the ratio of the charge produced per unit time to the rate of charge flow (i.e., current) between the electrodes. This gain depends on the ratio τ/τ_t where $\tau_t = L/v_d$ is the transit time across the semiconductor of length L, v_d is the drift velocity, and τ is the carrier lifetime.

Photoconductors have limited use for high-frequency operations. However, they are used as infrared detectors, especially beyond a few μm wavelengths.

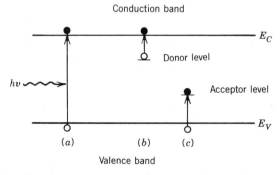

FIGURE 12.16 Optical excitation processes in a photoconductor: (a) band-to-band (intrinsic) excitation and extrinsic excitation involving (b) a donor level and (c) an acceptor level.

12.2.2 Photodiode [4]

A p-n junction photodiode is shown in Fig. 12.17. The incident optical signal produces electron–hole pairs that are collected by the junction producing a photocurrent in the external circuit. During its operation the photodiode is usually reverse biased with a relatively large voltage that is kept substantially below the avalanche breakdown voltage. The large reverse bias lowers the carrier transit time through the depletion region and also reduces the depletion region capacitance. Both features improve the diode capability for high-frequency operation.

The quantum efficiency of the diode is determined mainly by the absorption coefficient α of the semiconductor. Since α is a strong function of the wavelength, appreciable photocurrent is observed only in a limited wavelength range. The long wavelength cutoff λ_c is given by Eq. (12.21). For short wavelengths, α is large, and most of the absorption occurs in the upper dead-layer region that limits the photocurrent to a low value. In the near infrared region, Si photodiodes (with AR coating) can reach 100 percent quantum efficiency near the 0.8 μm to 0.9 μm region.

Three factors that limit the response speed of a photodiode are: (1) the diffusion of carriers, (2) the carrier transit time through the junction depletion region, and (3) the depletion region capacitance. To reduce the diffusion time of carriers from the neutral regions to the depletion region, the incident radiation must be absorbed close to the junction. This situation can be achieved by using a shallow junction and choosing the dopant concentration in the base region, such that the depletion region width is on the order of $1/\alpha$. However, a too wide depletion region can increase the carrier transit time through it, and a compromise has to be reached between the two conflicting requirements. The detection sensitivity of the photodiode is increased by increasing the quantum efficiency and reducing the dark current.

Besides the p-n homojunction diode, the other members of the photodiode family are the p-i-n, Schottky barrier, and heterojunction diodes.

The schematic cross-section and energy band diagram of a reverse biased p-i-n photodiode are shown in Fig. 12.18. The front surface of the p-zone has an AR coating to reduce reflection losses and increase the quantum efficiency. The field in the i-zone is constant, and for high quantum efficiency, the i-region thickness $W \gg 1/\alpha$. If W is made too thick, the transit time delay through this zone becomes

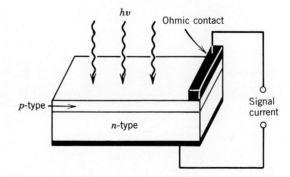

FIGURE 12.17 Schematic diagram of a p-n junction photodiode. (From C.T. Elliot, in *Handbook on Semiconductors,* vol. 4, T. S. Moss and C. Hilsum (eds.), North-Holland, Amsterdam, p. 732. Copyright © 1981. Reprinted by permission of Elsevier Science Publishers, Physical Sciences and Engineering Div., and the author.)

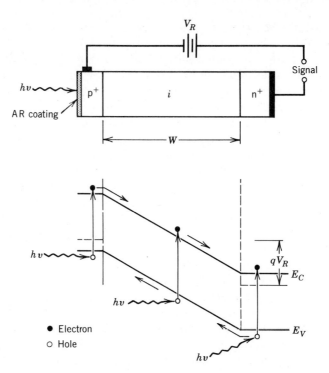

FIGURE 12.18 Cross-section and energy band diagram of p-i-n photodiode under reverse bias.

considerable, which adversely affects the response time. A compromise between the high quantum efficiency and low response time is achieved by keeping W between $1/\alpha$ and $2/\alpha$. Since W can be tailored to optimize the quantum efficiency and the response time, the p-i-n diode is one of the most common photodetectors.

Schottky barrier photodiodes are useful in the visible and ultraviolet portion of the spectrum and can have higher efficiencies compared to the p-n junction diode [7]. Figure 12.19 shows the energy band diagram of an illuminated reverse-biased Schottky barrier diode on an n-type semiconductor. The radiation is made incident from the metal contact, and to keep absorption losses low, its thickness

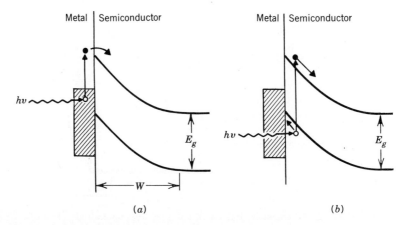

FIGURE 12.19 Energy band diagrams of a Schottky barrier diode showing (*a*) photoemission of electrons from the metal into the semiconductor and (*b*) band-to-band excitation of electron–hole pairs.

is restricted to approximately 100 Å. Moreover, an AR coating is used to reduce the reflection losses. Two modes of operation of the photodiode are shown in Fig. 12.19. When $h\upsilon > q\phi_B$ but less than E_g, the incident radiation can excite electrons from the metal over the barrier into the semiconductor (Fig. 12.19a). However, for photons with energy $h\upsilon > E_g$, electron–hole pairs are produced in the semiconductor. The generated holes are collected by the junction which constitutes the photocurrent. This mode of operation is shown in Fig. 12.19b. The resulting characteristics of the photodiode are similar to those of a conventional p-n junction or a p-i-n photodiode.

In a homojunction photodiode, light is absorbed throughout, and the quantum efficiency depends critically on the distance of the junction from the surface. This limitation is removed in heterojunction photodiodes. With this technology, the band gap of the semiconductor (through which the radiation enters) can be made sufficiently large so that the incident light is not absorbed. Consequently, the quantum efficiency is determined by the lower band gap semiconductor, and the loss of carriers due to surface recombination is avoided.

A heterojunction photodiode tends to have a large value of dark current. To obtain a low dark current, the lattice constants of the two semiconductors must be closely matched. This limits the choice of the semiconductors that can be used to make an efficient heterojunction photodiode. An $In_xGa_{1-x}As$ epitaxial layer grown on an InP substrate forms a heterojunction with a nearly perfect lattice match. Figure 12.20 shows the structure of a back-side illuminated InP-InGaAs heterojunction photodiode. [7]. InP has a band gap of 1.3 eV and remains transparent to the incident radiation which is absorbed largely in the high-resistivity, n-type $In_xGa_{1-x}As$ ($x = 0.53$) which has a band gap of 0.73 eV.

12.2.3 Avalanche Photodiode [7]

Avalanche photodiodes are operated at reverse biases close to avalanche breakdown voltage to enable avalanche multiplication of photogenerated carriers.

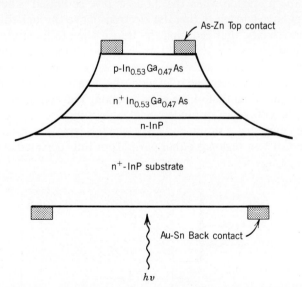

FIGURE 12.20 Schematic cross-section of a back illuminated InP-InGaAs heterojunction photodiode. (From R. G. Smith, reference 7, p. 298. Copyright © 1982. Reprinted by permission of Springer-Verlag, Heidelberg and the author.)

Thus, the current is higher than the primary current generated by the incident radiation, and this increases the sensitivity of the diode to the incident radiation.

Figure 12.21 shows the schematic representation of a typical avalanche photodiode (APD) biased in the avalanche multiplication region. The depletion region width W has an avalanche zone of width x_A and a drift zone of width $(W - x_A)$. The incident radiation creates electron–hole pairs in the various regions of the device. The photogenerated carriers then diffuse to the avalanche zone, and the photocurrent is multiplied by the secondary multiplication of the electron–hole pairs. Because of the statistical nature of the avalanche multiplication process, there is a fluctuation in gain, and this fluctuation produces an avalanche multiplication noise in excess of the normal shot noise of the diode.

The avalanche noise is a function of the avalanche multiplication factor M, and it also has a strong dependence on the electron to hole ionization coefficient ratio α_n/α_p. In order to achieve a low value of noise, α_n and α_p should be as different as possible, and the avalanche process should be initiated by carriers with a higher ionization coefficient. The structure of Fig. 12.21 is designed to ensure these features. Suppose that the diode is made in Si. In this material, $\alpha_n \simeq 3\alpha_p$. Thus, the n^+ and p-regions are kept thin so that little absorption occurs in these regions, and more than 95 percent of the incident radiation is absorbed in the thick π-zone. The photogenerated electrons then move to the avalanche region to initiate avalanche while holes are collected in the p^+-region.

An important requirement of an APD is that the multiplication factor M be uniform over the area of the junction through which the photocurrent flows. This criterion requires dislocation free material with minimal resistivity fluctuations to avoid the occurrence of microplasmas. Furthermore, the active area of the device is kept no larger than necessary to accommodate the beam. A small area has a lower probability of occurrence of defects that may cause microplasmas.

Two representative device configurations for APDs are shown in Fig. 12.22. The structure shown in Fig. 12.22a is similar to that in Fig. 12.21 and is called a *reach-through diode*. Here, the depletion region reaches all the way from the n^+ to the p^+-region when the diode is reverse biased. The field in the π-region is kept large enough so that the electrons move through this region with the saturation velocity, thus reducing the transit time to its minimum possible value. The n-guard rings are diffused to reduce the premature breakdown of the n^+-p junction by the field concentration at the edges. Since the n-region has a lower donor concentration than the n^+-region, the breakdown voltage of the n-p junction will be higher than that of the n^+-p junction. The configuration shown in Fig. 12.22b is a Schottky barrier diode made on an *n*-type semiconductor. The active area of the device has a thin metal layer which is largely transparent to the incident radiation. The n-region is either partially or fully depleted at the reverse bias, and the

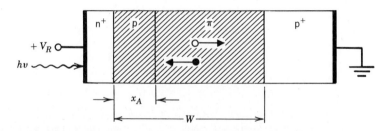

FIGURE 12.21 Schematic cross-section of a typical avalanche photodiode.

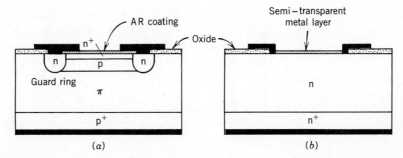

FIGURE 12.22 Schematic representation of two common types of avalanche photodiodes: (a) the reach-through diode and (b) Schottky barrier diode.

photogenerated electrons produce secondary electron–hole pairs in the high field depletion region. Note that in this diode no holes are collected by the junction, and the avalanche multiplication is initiated by electrons. Thus, in a material like Si this results in a very low avalanche noise. Schottky barrier APDs made on n-type Si are particularly useful in the visible and ultraviolet (UV) range.

Avalanche photodiodes have been fabricated in many semiconductors including Ge, Si, various III-V compounds, and their alloys. In the 0.8 to 1 μm wavelength range, the material used is mainly Si, while Ge and some III-V compounds have been used in the wavelength range from 1 to 1.6 μm. Most of the current research efforts are focused on developing III-V compound semiconductor APDs employing heterojunctions. These devices have separate absorbing and avalanche multiplications regions [7]. Typical examples are AlGaAs grown on a GaAs substrate and GaInAsP or GaInAs grown on an InP substrate.

12.3 LIGHT EMITTING DIODES

Semiconductor electroluminescent diodes can be classified into two categories: *light emitting diodes* (LEDs), diodes that emit incoherent spontaneous light, and *laser diodes,* diodes that emit coherent radiation.

The main areas of application for LEDs are in visible displays and indicators, and in infrared optical communication with optical fibers. In this section, we will be concerned with LEDs that emit in the visible region.

12.3.1 Radiative and Nonradiative Transitions

Figure 12.23 schematically illustrates the important radiative (a) and nonradiative (b) transitions in semiconductors [8]. In Fig. 12.23a, path 1 depicts the intrinsic band-to-band transition involving the recombination of an electron–hole pair. This process has a high probability of occurrence only in a direct gap semiconductor, and the emitted photon energy is then $h\nu = E_g$ (see Sec. 5.3). Path 2 shows an extrinsic recombination between a hole bound to a neutral acceptor and a free electron in the conduction band. Path 3 depicts the exciton recombination involving an acceptor-type trap. This type of recombination is important in some III-V compounds doped with an isovalent impurity. A typical example is nitrogen in GaP. When a P atom is replaced by an N atom in GaP, it creates an acceptor-type trap level in the band gap very close to the conduction band edge E_C. This level captures an electron from the conduction band and becomes nega-

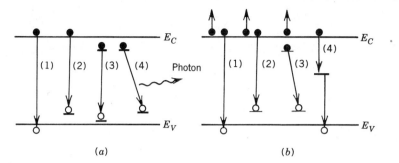

FIGURE 12.23 Schematic illustrations of some basic (a) radiative and (b) nonradiative transitions in semiconductors.

tively charged. Subsequently, through Coulombic attraction, a hole from the valence band gets bound to the trap level, and the electron–hole pair remaining in the excited state is known as a *bound exciton*. The pair then recombines by emitting a photon.

Path 4 shows the extrinsic process of donor-acceptor pair recombination. Here an electron from the conduction band is captured by the ionized donor, and a hole is captured by the ionized acceptor. The subsequent transition of the electron to the acceptor state emits a photon equal to the difference between the two energy levels.

In the nonradiative transitions shown in Fig. 12.23b, path 1 shows the band-to-band Auger recombination involving two electrons and one hole (eeh process). Trap-assisted Auger recombination can also occur. Path 2 depicts an Auger transition involving two free electrons and a trapped hole, and path 3 shows the same involving an occupied donor-acceptor pair and a free electron. Finally, path 4 depicts the nonradiative recombination process via a deep-level recombination center.

12.3.2 Basic Processes in LEDs

The light emission from a semiconductor involves basically three processes [9]. The first is an excitation process in which electron–hole pairs are produced. The second process is the recombination process in which the excited carriers give up their energy either through a radiative or a nonradiative process. The third process is the extraction or passage of emitted photons from the active region of the semiconductor to the observer. An efficiency can be assigned to each of these processes, and the overall efficiency of the LED is the product of these three efficiencies.

(a) EXCITATION MECHANISMS

The most common method of excitation for invoking electroluminescence in semiconductors is injection of minority carriers through a forward-biased junction, and virtually all commercial LEDs rely on this method. The other methods of excitation include [4] avalanche excitation and tunneling. In avalanche excitation, the excess carriers are produced by an impact ionization process that causes avalanche breakdown in the semiconductor. Here we will be concerned only with carrier injection via a forward-biased p-n junction.

In a forward-biased p-n junction, the hole-to-electron current ratio I_p/I_n can be expressed as (see Sec. 7.1)

$$\frac{I_p}{I_n} = \frac{D_p L_n N_a}{D_n L_p N_d} \tag{12.22}$$

In the majority of LEDs, dominant radiative recombination occurs only on one side of the junction, and this condition is achieved by a proper choice of N_a and N_d. For example, if the excess carriers are to recombine predominantly on the n-side of the junction, then holes from the p-side have to be preferentially injected into the n-side, making I_p large compared to I_n. This is achieved by making a p^+-n junction with $N_a \gg N_d$.

Electron–hole pairs also recombine in the junction depletion region causing a recombination current I_{rec} given by Eq. (7.31). When this current is added to the diode current, the overall injection efficiency η_i in a p^+-n junction LED is given by

$$\eta_i = \frac{I_p}{I_p + I_n + I_{\text{rec}}} \tag{12.23}$$

Here the denominator represents the total current flowing through the diode. Thus, it is clear that η_i can be increased by reducing I_n and I_{rec}. Two guidelines can be used to reduce I_{rec} in GaP and GaAs p-n junctions. Since most of the recombination occurs where the depletion region meets the surface, the diode perimeter should be reduced to a minimum that is still compatible with the device area necessary for obtaining the desired value of the current. Second, I_{rec} may be reduced by keeping the density of surface states and recombination centers in the depletion region as low as possible. Typical values of η_i range from 30 to 60 percent.

(b) RECOMBINATION OF EXCESS CARRIERS

After the excess electron–hole pairs have been injected on the appropriate side of the junction, the next process is the recombination of these pairs. The efficiency η_r of radiative recombination is written as

$$\eta_r = \frac{U_r}{U_r + U_{nr}} \tag{12.24}$$

Here $U_r = p_e/\tau_r$ represents the radiative recombination rate, and $U_{nr} = p_e/\tau_{nr}$ is the nonradiative recombination rate. The variables τ_r and τ_{nr} are the radiative and the nonradiative lifetimes, respectively. Substituting the values of U_r and U_{nr} in Eq. (12.24), we obtain

$$\eta_r = \frac{\tau_{nr}}{\tau_{nr} + \tau_r} \tag{12.25}$$

It is apparent that η_r can be increased either by increaseing τ_{nr} or by reducing τ_r. In a direct gap semiconductor like GaAs, τ_r is on the order of 10^{-9} sec at a dopant concentration of about 10^{17} cm^{-3}, and the main problem in increasing η_r is to increase τ_{nr} by reducing the nonradiative recombination paths. In the case of an indirect gap semiconductor like GaP, radiative recombination occurs via the acceptor levels of the isovalent dopant nitrogen, and η_r can be increased by increasing the N atom concentration and, at the same time, by closing the nonradiative recombination paths.

(c) EXTRACTION OF LIGHT FROM THE SEMICONDUCTOR

In a typical LED, photons are generated within a few μm of the p-n junction, and the emitted light must pass through the semiconductor to reach the surface from where it is emitted to reach the observer's eye. There are three loss mechanisms that reduce the amount of light reaching the observer. First, the amount of light transmitted to the surface depends on the absorption coefficient α of the semiconductor and decreases with the distance x from the junction as $\exp(-\alpha x)$. Second, when the light passes from the semiconductor material of refractive index $\bar{n}_2$ to the outside medium (i.e., air) of refractive index $\bar{n}_1$, a fraction of it is reflected into the semiconductor. This loss is known as the Fresnel loss. A third loss mechanism results from the total internal reflection of the light rays which reach the semiconductor surface at an angle greater than the critical angle θ_c given by the relation

$$\sin \theta_c = \frac{\bar{n}_1}{\bar{n}_2} \qquad (12.26)$$

The angle θ_c is 16° for GaAs and 17° for GaP. In order to obtain a high extraction efficiency η_o, these losses must be minimized.

For a semiconductor, α depends on the band structure and the wavelength of the light. In GaAs, it is on the order of 10^4 cm^{-1} at the emission wavelength. Thus, for every μm of material that the photon traverses, approximately two-thirds of the light is lost by reabsorption. This is a major limitation in achieving a good extraction efficiency in a direct gap semiconductor. For indirect gap materials such as GaP and GaAs$_y$P$_{1-y}$ (with $y = 0.6$), α as a function of wavelength is shown [9] in Fig. 12.24. For the case of GaP doped with isovalent nitrogen, α rises almost abruptly around 0.56 μm corresponding to the band gap energy. Near the band edge, α is lower than 10^2 cm^{-1}, which is considerably smaller than the α of a direct gap semiconductor resulting in higher values of η_o.

As an example, consider a p-n junction LED with an injection efficiency of 60 percent and a light extraction efficiency $\eta_o = 50$ percent. Let us calculate τ_r if $\tau_{nr} = 10^{-8}$ sec, and the overall efficiency is 1.5 percent.

Since $\eta = 0.015 = \eta_i \eta_o \eta_r$ or $\eta_r = 0.05$, from Eq. (12.25)

$$\frac{1}{1 + (\tau_r/\tau_{nr})} = 0.05 \quad \text{or} \quad \tau_r = 19\tau_{nr} = 1.9 \times 10^{-7} \text{ sec}$$

12.3.3 LED Materials

The requirements of a good LED material are the ability to fabricate a p-n junction, to achieve good injection and luminescence efficiencies, and at the same time, extract a reasonable fraction of the generated light. The wavelength λ of the emitted radiation is related to the photon energy $h\nu$ by the relation

$$\lambda = \frac{hc}{h\nu} \simeq \frac{1.24}{h\nu(\text{eV})} \mu\text{m} \qquad (12.27)$$

LIGHT EMITTING DIODES

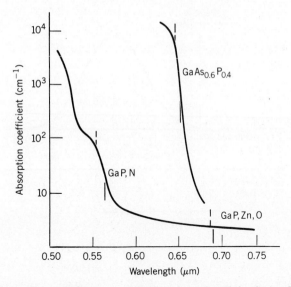

FIGURE 12.24 Absorption coefficient α as a function of wavelength for $GaAs_{0.6}P_{0.4}$ and GaP containing 5×10^{18} cm^{-3} of N atoms. (From A. R. Peaker, reference 9, p. 205. Copyright © 1980. Reprinted by permission from IEE Publishing Department (U.K.) and the author.)

For band-to-band radiative recombination, $h\nu = E_g$ and the materials for a visible LED (0.44 to 0.7 μm wavelength) must have a band gap between 1.8 and 2.8 eV. Alternatively, the band gap can be larger than 2.8 eV if we can introduce an appropriate level of localized states in the band gap through which the recombination process will be radiative. The choice is mainly confined to GaP and GaAsP, although LEDs from other materials have also been produced.

(a) GaP

GaP has an indirect band gap of 2.26 eV at 300 K. It can be made p- and n-type by using appropriate dopants, and p-n junctions in GaP have shown good injection and extraction efficiencies. A green-colored LED with $\lambda = 0.565$ μm has been achieved by using a recombination path via a trap level created by the isovalent nitrogen impurity with which both the p- and n-sides of the junction are doped. A more complex isovalent state obtained by doping with Zn and oxygen, as the nearest neighbor pairs, gives rise to red emission with $\lambda = 0.69$ μm. However, the main problem with these LEDs is that, at this wavelength, the eye is very insensitive. Figure 12.25 shows the relative sensitivity of the human eye to different colors. The corresponding photon energy and the band gaps of some important materials are also shown. It can be observed that the sensitivity of the human eye reaches its maximum at $\lambda = 0.555$ μm and falls nearly to zero at the two ends of the visible region. The radiative recombination efficiency of green-colored GaP LEDs is rather low, and the only effective way to increase it is to reduce recombination through the nonradiative paths.

(b) $GaAs_{1-y}P_y$

The $GaAs_{1-y}P_y$ compound behaves as a direct gap semiconductor for $y \leq 0.45$ and has an indirect gap for larger values of y. Any composition can be grown using vapor-phase epitaxial growth. However, commercially available LEDs employ compositions that give red, yellow, and orange emissions. The compositions emit-

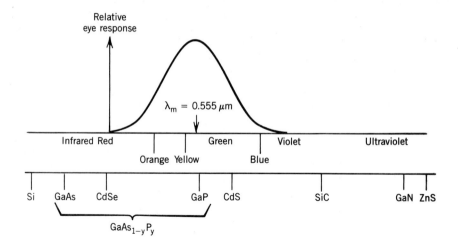

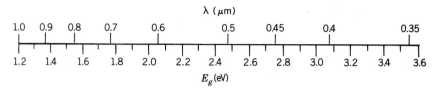

FIGURE 12.25 Relative sensitivity of the human eye for different colors. The corresponding photon energy and band gaps of some semiconductors are also shown. (From S. M. Sze, reference, 16, p. 259. Copyright © 1985. Reprinted by permission of John Wiley & Sons, Inc., New York.)

ting at yellow and orange have indirect gap and behave much the same way as GaP. Nitrogen as an isovalent dopant is used to obtain the radiative recombination. For red emission with a peak at 0.65 μm, $y = 0.4$ and the band gap is direct. However, the situation is rather complex as both the direct and indirect radiative transitions occur. The radiative efficiency η_r of this material is lower than that of GaP because of the larger density of nonradiative recombination centers.

(c) OTHER MATERIALS

The II-VI compounds ZnSe, CdS, and ZnTe have direct band gap and have potential for emission in the visible region. These materials have not been used in producing LEDs because they are unipolar. Hence, p-n junctions cannot be made in these semiconductors. Two other materials on which work has been done are SiC and GaN, both of which are candidates for blue emission. The technology of SiC requires very high temperatures and tedious processing. The major problem with GaN is that it is n-type and cannot be doped to p-type. Light emission has been obtained [9] using metal insulator-GaN (MIS) structures, but it offers no reliable technology for the commercial production of GaN LEDs.

12.3.4 LED Structures

The light-emitting diodes of GaP and GaAs$_{1-y}$P$_y$ are fabricated by the successive growth of appropriately doped n- and p-layers of these compounds onto a GaAs or a GaP substrate. These layers are grown either by liquid phase epitaxy (LPE) or by vapor phase epitaxy (VPE) (see Sec. 19.2). The device structure may vary depending on the application and the material used. The basic LED structures have the flat-diode configuration shown in Fig. 12.26. $GaAs_{1-y}P_y$ LEDs with direct band gap are fabricated on a GaAs substrate (Fig. 12.26a). To avoid lattice mismatch at the GaAs-GaAsP interface, a graded region of about 10 μm thickness is grown on the GaAs. The composition is gradually changed from $y = 0$ to $y = 0.4$. The n-type $GaAs_{1-y}P_y$ layer with $y = 0.4$ is then grown, and a thin p-region is produced either by diffusion of Zn or by epitaxial growth. Light generated at the p-n junction is emitted in all directions. The light emitted downward is absorbed in the graded region and in the GaAs, both of which are opaque to the emitted radiation. In addition, the light emitted upward is absorbed partly in the p-$GaAs_{1-y}P_y$ layer. Thus, only a small fraction of light can emerge from the surface.

In the case of nitrogen doped indirect gap material LEDs (i.e., GaP or GaAsP), the substrate is GaP (Fig. 12.26b). Since GaP has a higher band gap than GaAsP, it is transparent to the emitted radiation. As a consequence, the light emitted downward can be reflected from the back-side metal contact as well as from the sides, and this results in increased light output.

The performance of a LED can be further improved by designing the diode geometry so that more of the emitted light arrives at an angle lower than the critical angle θ_c, and coating the surface with a medium of refractive index $\bar{n} = \sqrt{\bar{n}_1 \bar{n}_2}$. Figure 12.27 shows [10] the cross-sections of some of the geometries that have been considered. However, these shapes are difficult to fabricate, and the resulting device becomes very expensive.

12.4 SEMICONDUCTOR LASERS

The word "laser" is an acronym for *l*ight *a*mplification by *s*timulated *e*mission of *r*adiation. The basic mechanism of light emission from a laser is the same as

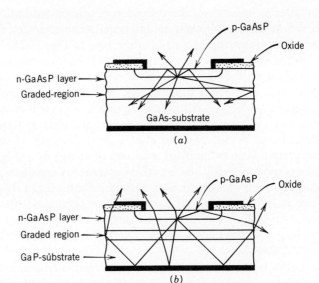

FIGURE 12.26 Schematic cross-sections of $GaAs_{1-y}P_y$ LEDs fabricated (a) on a GaAs substrate and (b) on a GaP substrate. (After R. Kniss, *Electronics May 2*, p. 34, 1974, McGraw-Hill, New York.)

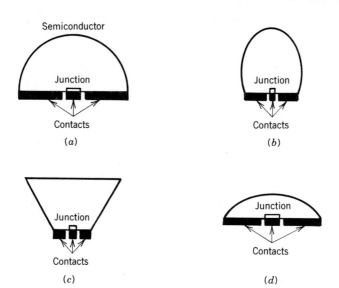

FIGURE 12.27 Some LED geometries to increase the light extraction: (a) hemisphere, (b) truncated ellipsoid, (c) paraboloid, and (d) truncated cone. (From W. N. Carr, reference 10, pp. 9, 11, 12, 13, 15. Copyright © 1966. Reprinted by permission of Pergamon Press, Oxford.)

that from a LED. However, in contrast to the incoherent light of LEDs, laser light has spatial and temporal coherence. Spatial coherence means that all the points on the wave front have the same phase, and this results in a highly directional beam of light. Temporal coherence means that, at any point in space, the wave amplitude varies sinusoidally with time; hence, the emitted radiation is monochromatic.

12.4.1 Laser Action

In order to understand the operation of a laser, we have to understand the processes of stimulated emission and population inversion, in addition to the usual processes occurring in a LED. Referring to Fig. 12.28, let us consider an atom with an electron that can occupy one of the two energy levels E_1 and E_2. The atom is in ground state when the electron is at E_1 and in its excited state when the electron is at E_2. Now suppose that the atom is in its ground state. A photon of energy $h\nu_{12} = (E_2 - E_1)$ impinging on the atom will be absorbed and will cause the atom to make a transition to the excited state as shown in Fig. 12.28a. The excited state is unstable, and after a short time, the atom makes a transition to the ground state by emitting a photon of energy $h\nu_{12}$ (Fig. 12.28b). This process occurs spontaneously and is called *spontaneous emission*. An atom in the excited state need not wait for spontaneous emission to occur. If a photon of energy $h\nu_{12}$ falls on the atom while it is still in its excited state, the atom is stimulated to make a transition to the ground state by emitting a photon of energy $h\nu_{12}$ which is in phase with the impinging radiation. This process is called *stimulated emission* (Fig. 12.28c).

Let us now consider a system consisting of a large number of atoms in thermal equilibrium at a temperature T having concentrations n_1 and n_2 at E_1 and E_2, respectively. For $(E_2 - E_1) > 3kT$ the Boltzmann distribution gives

$$\frac{n_2}{n_1} = \exp[-(E_2 - E_1)/kT] \qquad (12.28)$$

SEMICONDUCTOR LASERS

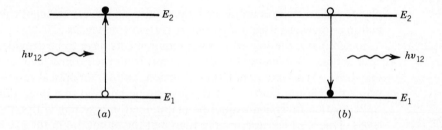

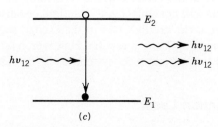

FIGURE 12.28 The three basic processes that occur between two energy levels of an atom when electromagnetic radiation interacts with it: (a) absorption, (b) spontaneous emission, and (c) stimulated emission.

This equation shows that, in thermal equilibrium, there are more atoms in the ground state than in the excited state. If radiation of energy $h\nu_{12}$ is continuously falling on the system and the atoms are in a radiation field of energy density $\rho(\nu_{12})$, then stimulated emission can also occur along with the absorption and spontaneous emission processes. The stimulated emission is proportional to the energy density $\rho(\nu_{12})$ and the concentration n_2, and can be written as $B_{21}n_2\rho(\nu_{12})$ where B_{21} is a constant of proportionality. The absorption rate is proportional to n_1 and can be expressed as $B_{12}n_1\rho(\nu_{12})$. The spontaneous emission rate is not influenced by $\rho(\nu_{12})$ and can be written as $A_{21}n_2$ where A_{21} is also a constant of proportionality. In the steady state, the rate of absorption must be balanced by the two emission rates, so that

$$B_{12}n_1\rho(\nu_{12}) = A_{21}n_2 + B_{21}n_2\rho(\nu_{12}) \tag{12.29}$$

From this relation, we observe that

$$\frac{\text{Stimulated emission rate}}{\text{Spontaneous emission rate}} = \frac{B_{21}}{A_{21}}\rho(\nu_{12}) \tag{12.30}$$

Thus, the stimulated emission rate can be made large compared to the spontaneous emission rate by having a large photon field density $\rho(\nu_{12})$. In addition, from Eq. (12.29) we have

$$\frac{\text{Stimulated emission rate}}{\text{Absorption rate}} = \frac{B_{21}}{B_{12}} \cdot \frac{n_2}{n_1} \tag{12.31}$$

which shows that the stimulated emission will dominate the absorption of a photon when there are more atoms in the excited state compared to the ground state. This condition is called *population inversion*. In practical laser systems, a large photon field density is produced by providing an optical resonant cavity in which multiple internal reflections occur at certain frequencies.

In summary, the following basic requirements have to be provided to achieve laser action: (1) A method to excite atoms from the ground level to a higher energy level, (2) a large population inversion, and (3) an optical resonant cavity in which the photon field energy density can be built up to a large value.

Population inversion cannot be maintained in a system with only two energy levels. This is so because if more than half the atoms are in the excited state initially, the impinging radiation will cause stimulated emission until half the atoms remain in the excited state and the remaining half are in the ground state. Further radiation will not change the situation, because for each atom that makes the transition from the ground state to the excited state, another will be making the transition in the opposite direction. However, a large population inversion can be easily achieved in three-level system, as is the case with the Ruby laser [11].

12.4.2 The p-n Junction Laser

In semiconductor lasers, transitions are associated with the electron states in the valence and conduction bands. This situation is in contrast to other types of lasers where transition occurs between discrete energy levels. Semiconductor lasers are compact in size, require low power to operate, and have high efficiencies. However, they do not have the narrow spectral line widths or the degree of coherence of competing lasers. The first semiconductor lasers were the p-n junction diode lasers made in GaAs, our subsequent discussion in this section will be limited to these lasers only.

(a) BASIC STRUCTURE

Figure 12.29 shows the basic structure of a GaAs p-n junction laser diode. A heavily doped p-region is produced by diffusion of Zn into n-type GaAs. The doping in the n- and p-regions is such that both are degenerate. One pair of faces perpendicular to the junction plane is cleaved and polished so that they act as reflecting mirrors. The remaining two faces are purposefully roughened to eliminate lasing in these directions. Thus, this structure acts as a resonant cavity and is known as a *Fabry-Perot cavity*.

(b) POPULATION INVERSION

In a laser diode, the doping levels on the two sides of the junction are kept lower than those in a tunnel diode (see Sec. 10.1). Thus, the I-V characteristic of a laser diode does not exhibit a negative resistance. Figure 12.30a shows the energy band diagram of a p-n junction at thermal equilibrium, and Fig. 12.30b shows the same structure with a forward bias that is larger than the band gap. Figure 12.30b shows that the region near the junction is not depleted of mobile carriers but rather contains a high concentration of electrons in the conduction band and vacant states in the valence band. Thus, a condition of population inversion occurs in a thin region about the junction. When photons of appropriate energy impinge on this region, they will induce electron transitions from the conduction to the valence band and cause stimulated emission.

The meaning of population inversion in Fig. 12.30b can be understood by referring to quasi-Fermi levels E_{Fn} and E_{Fp}, which are shown by the dashed lines in

SEMICONDUCTOR LASERS

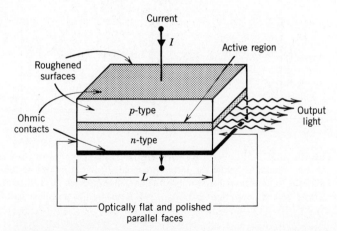

FIGURE 12.29 Basic structure of a GaAs p-n junction laser diode.

the figure. Note that in the neutral n- and p-regions, far away from the junction, E_{Fn} and E_{Fp} coincide. However, it is evident from Fig. 12.30b that, in the population inversion region near the junction, E_{Fn} and E_{Fp} are separated from each other by an energy equal to $E_g + q(\phi_n + \phi_p)$. Population inversion has occurred in this region in the sense that, in the energy range between E_{Fn} and E_{Fp}, more states are occupied in the conduction band than in the valence band. It is clear that the incident photon that can induce stimulated emission should have an energy $E_g < h\nu < [E_g + q(\phi_n + \phi_p)]$.

In a forward-biased p-n junction, the injected electron–hole pairs can also recombine spontaneously and cause spontaneous emission of photons in energy ranging from E_g to $E_g + q(\phi_n + \phi_p)$. Spontaneous radiation dominates before the occurrence of strong population inversion at a large forward bias. Figure 12.31 shows the typical plots of the intensity of the emitted radiation as a function of $h\nu$. At low bias currents, a spontaneous emission spectrum with energies between E_g and $E_g + q(\phi_n + \phi_p)$ is obtained (Fig. 12.31a). As the diode current is gradually increased, a point is reached where significant population inversion exists near the junction region. Consequently, stimulated emission occurs at frequencies cor-

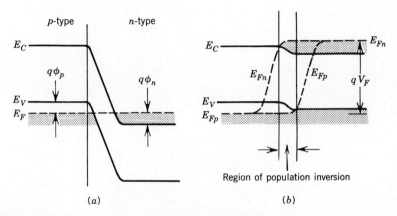

FIGURE 12.30 Energy band diagram of a p-n junction laser diode (a) under thermal equilibrium and (b) at a large forward bias in excess of the band gap voltage.

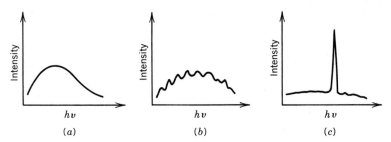

FIGURE 12.31 Light intensity as a function of photon energy for a p-n junction laser diode: (a) incoherent emission below the laser threshold, (b) laser modes at the onset of laser threshold, and (c) dominant laser mode above the threshold. (From B. G. Streetman, *Solid State Electronic Devices,* 2nd ed., copyright © 1980, p. 393. Reprinted by permission of Prentice Hall, Inc., Englewood Cliffs, New Jersey.)

responding to the normal modes of the Fabry-Perot cavity as shown in plot (b). The wavelengths of the normal modes are given by the relation

$$\lambda = \frac{2L}{n} \tag{12.32}$$

where L is the length of the cavity and $n = 1, 2, 3 \ldots$. Finally, at sufficiently high currents, strong population inversion occurs, and the most preferred mode dominates the intensity as shown in plot (c). The laser output is now composed of almost monochromatic radiation superimposed on a relatively weak background radiation level. The intense monochromatic radiation is the main output from the device. Thus, a laser diode is characterized by a threshold current I_{th}. Below I_{th}, only spontaneous emission occurs and the radiation is incoherent. Above I_{th}, the diode emits stimulated radiation and the spectral emission narrows to a few Å. Figure 12.32 compares the power outputs of a laser diode and a LED as a function of the diode forward current. A sharp rise in the power output above the knee of the threshold current is observed in the case of the laser diode. However, the power from a LED increases continuously with the bias current, and no threshold is observed.

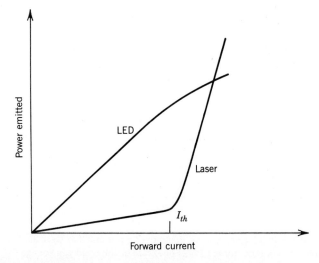

FIGURE 12.32 Power emitted from a laser diode and from a LED as a function of the bias current.

12.4.3 Heterojunction Lasers [12, 15]

In a GaAs p-n junction laser, excess carrier recombination is more efficient on the p-side than on the n-side, and laser action occurs mainly on the p-side of the junction. The electrons injected into the p-region diffuse away from the junction, and the thickness of the active region in which population inversion occurs is only a small fraction of the total thickness in which the excess carriers are present. As a consequence, GaAs p-n homojunction lasers have a large value of the threshold current density J_{th} which is on the order of 50 KA cm^{-2} at room temperature. Such a large value of J_{th} imposes serious problems for continuous wave (CW) operation of the laser.

The threshold current density can be reduced considerably if the injected electrons in the p-GaAs are kept confined to a narrow zone near the p-n junction. This situation can be achieved by keeping the p-GaAs layer thin and enclosing it with a *p*-type semiconductor of wider band gap, thus making a p-p heterojunction as shown schematically in Fig. 12.33. The wider band gap semiconductor should have a direct band gap and a good lattice match with GaAs. The ternary compound $Al_xGa_{1-x}As$ satisfies both requirements. AlAs has an indirect gap of 2.2 eV at 300 K. The band gap of $Al_xGa_{1-x}As$ varies with the fraction x of Al atoms. Up to $x = 0.35$, the gap is direct and can be expressed as

$$E_g(x) = 1.424 + 1.247x \tag{12.33}$$

The lattice constant match between GaAs and AlAs is excellent, and the GaAs/AlGaAs junction has a low interface state density.

Further improvement in the confinement of injected electrons can be obtained if the n-GaAs in Fig. 12.33 is replaced by n-$Al_xGa_{1-x}As$. This structure is known as a *double heterojunction* (DH) *laser*. Let us look into the salient features of this device.

Figure 12.34*a* shows the thermal equilibrium energy band diagram of n-$Al_xGa_{1-x}As$/p-GaAs heterojunction for $x = 0.3$ based on the Anderson model (see Sec. 10.5). The hole affinities of the two semiconductors are assumed to be nearly the same,* while the electron affinity of $Al_xGa_{1-x}As$ is taken to be 0.4 eV higher than that of GaAs. Both semiconductors are doped to degeneracy. Figure 12.34*b* shows the energy band diagram for a large forward-bias V_F. It is evident from this figure that current flows predominantly through the injection of electrons from n-AlGaAs to p-GaAs. The electrons are unable to return to AlGaAs because they face a potential barrier ΔE_C which is higher than the bar-

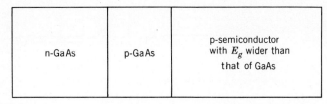

FIGURE 12.33 Schematic diagram employing p-p heterojunction for confinement of injected electrons.

---

*A. G. Milnes, Semiconductor Devices and Integrated Electronics, p. 849, Van Nostrand Reinhold, 1980.

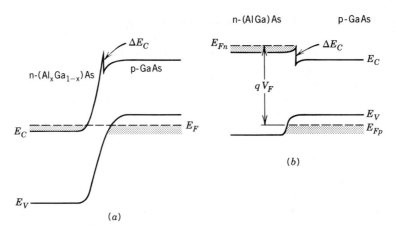

FIGURE 12.34 Energy band diagrams of n-Al_xGa_{1-x}As/p-GaAs heterojunction; (a) at thermal equilibrium and (b) at a large forward bias V_F.

rier for electrons from the n-side of the junction. Therefore, the injected electrons remain confined to the p-side.

The energy band diagram of p-GaAs and p-Al_xGa_{1-x}As heterojunction at thermal equilibrium is shown in Fig. 12.35a. It is observed that electrons in the conduction band of p-GaAs see a potential barrier ΔE_C. The energy band diagram of an n-AlGaAs/p-GaAs/p-AlGaAs heterostructure laser can be envisaged by putting Fig. 12.35a to the right of Fig. 12.34b, as is shown in Fig. 12.35b. It is evident that the energy ΔE_C acts as a confinement barrier for the electrons in the conduction band of p-GaAs. The width of the confinement region can be made small enough to keep the whole p-GaAs region under strong population inversion. Thus, this region becomes the active region where excess electron–hole pairs recombine to emit the laser light.

Besides the excess carrier confinement considered above, a laser structure also needs optical confinement. This means that the photons emitted during the electron–hole pair recombination should remain confined to the active region, and not be allowed to escape to the adjacent n- and p-AlGaAs regions. Optical confinement is provided by refractive index changes at the GaAs/AlGaAs interfaces. The wider band gap semiconductor has a lower value of refractive index, and this keeps the light confined to the active region. The excess carrier and optical confinements reduce the value of J_{th}. Carrier confinement mainly controls the population inversion, and hence, the gain of the Fabry-Perot cavity. Optical confinement reduces the photon losses from the cavity, and thus decreases its effective absorption coefficient.

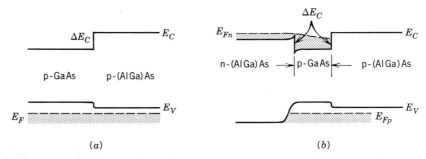

FIGURE 12.35 Energy band diagrams showing (a) a p-GaAs/p-AlGaAs heterojunction at thermal equilibrium and (b) a forward biased n-AlGaAs/p-GaAs/p-AlGaAs DH laser diode.

SEMICONDUCTOR LASERS

Figure 12.36 shows a comparison of the DH and the p-n junction GaAs lasers [13]. The band shapes under operating bias, the refractive index steps, and the optical power distributions are shown. The small changes in the refractive index in the case of the GaAs p-n junction laser results from the doping differences. Both the carrier and optical confinements are poor in the case of the GaAs p-n junction laser. However, for the DH laser, both the carriers and the light are confined to a narrow region, and this enhances the stimulated emission, causing a lower value of the threshold current density J_{th}.

Besides the DH laser described above, there are several other configurations of heterostructure lasers [4]. In all these devices, the optical feedback necessary for lasing is obtained by reflecting light from the polished faces of the Fabry-Perot cavity. Figure 12.37 shows the cross-section of a distributed feedback (DFB) laser in which the feedback is obtained from a periodic perturbation of the refractive index. This perturbation is obtained by etching periodic corrugations at the p-GaAs and p-AlGaAs interface. The output of this type of laser has a narrow linewidth at a single spectral line [14] (Fig. 12.37b).

12.4.4 Laser Diode Materials

Laser diodes have been fabricated with emission wavelengths extending from the visible to the infrared region using various semiconductors, and virtually all the

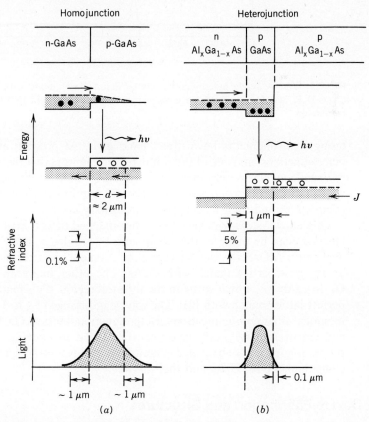

FIGURE 12.36 Comparison of p-n homojunction and double heterostructure laser diodes. (After I. Hayashi et al, reference 13, p. 1930. Copyright © 1971. Reprinted by permission of American Institute of Physics, New York.)

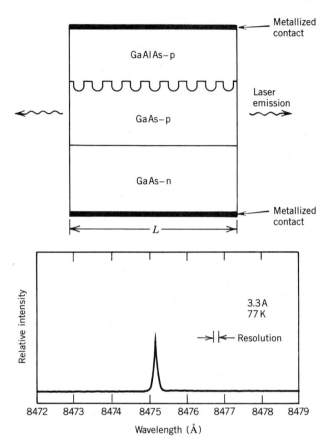

FIGURE 12.37 A single heterojunction GaAs/AlGaAs laser diode with distributed feedback. (From D. R. Scriffers et al, reference 14, pp. 610 and 611. Copyright © 1975 IEEE. Reprinted by permission.)

lasing semiconductors have direct band gaps. At present, all laser diodes use heterojunctions, and one of the prime requirements is that the lattice between the two semiconductors must match closely. Only a limited number of semiconductor combinations have been able to satisfy this requirement; most of these are III-V compounds, and their ternary and quarternary alloys.

One alloy system with very low strain-induced defects is $Al_xGa_{1-x}As$. Heterojunctions can be made in this compound by varying x, and laser diodes can be made with AlGaAs as the active region. As mentioned earlier, this material has a very good lattice match with GaAs. The other important semiconductor is $Ga_xIn_{1-x}As_yP_{1-y}$ which emits in the infrared region. This material has an almost perfect lattice match with InP. The wavelength range of 1 to 1.3 μm is of interest in optical-fiber communications. Heterojunction lasers of $Ga_xIn_{1-x}As/Ga_xIn_{1-x}P$ and $Ga_xIn_{1-x}As_yP_{1-y}/InP$ have been fabricated to emit laser radiation in this range. Single and double heterojunction lasers have also been produced in IV-VI compounds PbTe, PbSe, and their alloys.

12.4.5 Device Fabrication and Structures

Heterojunction laser diodes are fabricated using vapor phase epitaxy (VPE) or liquid phase epitaxy (LPE). LPE is more widely used. Diodes are designed either for continuous or pulse power operation, and the two commonly fabricated struc-

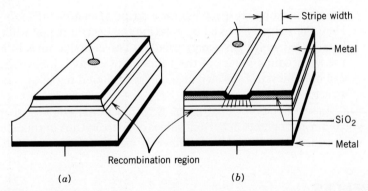

FIGURE 12.38 Schematic diagrams of (a) broad area and (b) stripe contact laser diodes. (From H. Kressel et al, reference 15, p. 15. Copyright © 1982. Reprinted by permission of Springer-Verlag, Heidelberg.)

tures are shown [15] in Fig. 12.38. The stripe contact structure (Fig. 12.38b) is superior to the broad area diode and offers the possibility of an extended CW operating range.

Fabrication of the broad area diode begins with a polished single-crystal semiconductor wafer that serves as the substrate. The desired layers are grown on the substrate using LPE, and the substrate is then thinned by lapping. Ohmic contacts are made on the two sides, after which the wafer is cut into individual diodes. Fabrication of stripe contact devices involves additional steps. For example, in the structure of Fig. 12.38b the various layers are grown on the substrate in the same way as in case of the broad area diode. The SiO_2 layer is then deposited on the top surface, and the stripe is defined by window etching in the oxide using standard photolithographic processes. Finally, the contact metal is deposited all over the top surface. Several other types of stripe contact structures have been fabricated [4]. The stripe widths typically range from 2 to 30 μm. A stripe geometry laser is not a spectrally pure light source, and the radiation can be emitted in the form of emission lines spread over a range of about 50 Å. Figure 12.39

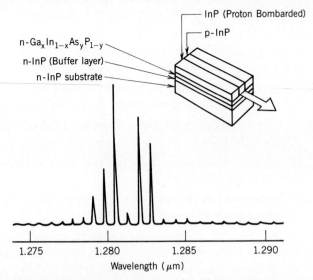

FIGURE 12.39 High-resolution emission spectra of a InP/GaInAsP DH laser. After A. G. Foyt, IEEE Device Research conference, Colo., 1979. (From S. M. Sze, reference 16, p. 227. Copyright © 1985. Reprinted by permission of John Wiley, Inc., New York.)

shows the emission spectrum for a stripe geometry InP/GaInAsP DH laser [16]. Here the stripe is formed by masking the central region while the top surface is bombarded with high-energy protons. The emission lines in Fig. 12.39 belong to the longitudinal modes of the Fabry-Perot cavity [4]. Single-frequency operation can be achieved if the stripe geometry is altered to obtain a distributed-feedback structure.

Laser diodes for continuous wave operation employs stripe widths of 10 to 20 μm and operate at current values of 50 to 300 mA at room temperature. Diodes designed for pulse operation have J_{th} values in excess of 10 KA cm^{-2}.

REFERENCES

1. R. H. BUBE and A. L. FAHRENBRUCH, "Photovoltaic effect," in *Advances in Electronics and Electron Physics,* Vol. 56, C. Marton (ed.), 163–217 (1981).

2. E. S. YANG, *Fundamentals of Semiconductor Devices,* McGraw-Hill, New York, 1978.

3. D. K. SCHRODER and D. L. MEIER, "Solar cell contact resistance—A review," *IEEE Trans. Electron Devices* ED-31, 637 (1984).

4. S. M. SZE, *Physics of Semiconductor Devices,* John Wiley, New York, 1981.

5. H. J. HOVEL, "Photovoltaic materials and devices for terrestrial solar energy applications," *Solar Energy Materials* 2, 277 (1980).

6. P. G. LECOMBER, "Preparation and applications of amorphous silicon," Chapter 16 in *Crystalline Semiconducting Materials and Devices,* P. N. Butcher, N. H. March, and M. P. Tosi (eds.), Plenum Press, New York, 1986.

7. D. P. SCHINKE, R. G. SMITH, and A. R. HARTMAN, "Photodetectors," Chapter 3, in *Semiconductor Devices for Optical Communication,* H. Kressel (ed.), *Topics in Applied Physics,* Vol. 39, Springer-Verlag, Heidelberg, 1982, Also R. G. SMITH, Chapter 11.

8. M. H. PILKUHN, "Light emitting diodes," Chapter 5A, in *Handbook on Semiconductors,* Vol. 4, T. S. Moss and C. Hilsum (eds.), North-Holland, Amsterdam, 1981.

9. A. R. PEAKER, "Light emitting diodes," *IEE Proc. Part A* (U.K.) 127, 202 (1980).

10. W. N. CARR, "Photometric figures of merit for semiconductor luminescent sources operating in spontaneous mode," *Infrared Phys.* 6, 1 (1966).

11. B. G. STREETMAN, *Solid State Electronic Devices,* Prentice-Hall, Englewood Cliffs, N.J., 1979.

12. R. BAETS, "Heterostructures in III-V optoelectronic devices," *Solid-State Electron.* 30, 1175 (1987).

13. I. HAYASHI, M. B. PANISH, and F. K. REINHART, "GaAs-Al$_x$Ga$_{1-x}$As double heterostructure injection lasers," *J. Appl. Phys.* 42, 1929 (1971).

14. D. R. SCIFRES, R. D. BURNHAM, and W. STREIFER, "Longitudinal and radiation modes in GaAs single heterojunction distributed-feedback injection lasers," *IEEE Trans. Electron Devices* ED-22, 609 (1975).

15. H. KRESSEL, M. ETTENBERG, J. P. WITTKE, and I. LADANY, "Laser diodes and LEDs for fiber optical communication," Chapter 2 in reference 7.

16. S. M. SZE, *Semiconductor Devices: Physics and Technology,* John Wiley, New York, 1985.

ADDITIONAL READING LIST

1. M. A. GREEN, *Solar Cells: Operating Principles, Technology and System Applications,* Prentice-Hall, Englewood Cliffs, N.J., 1982.

2. E.W. WILLIAMS and R. HALL, *Luminescence and the Light Emitting Diode,* Pergamon Press, Oxford, 1978.

3. H. KRESSEL and J. K. BUTLER, *Semiconductor Lasers and Heterojunction LEDs,* Academic Press, New York, 1977.

4. G. H. B. THOMPSON, *Physics of Semiconductor Laser Devices,* John Wiley, Chichester, 1980.

5. R. C. GOODFELLOW, "Semiconductor materials and structures for optical-communication devices," Chapter 14 in *Crystalline Semiconducting Materials and Devices,* P. N. Butcher, N. H. March, and M. P. Tosi (eds.), Plenum Press, New York, 1986.

6. K. A. JONES, *Introduction to Optical Electronics,* Harper and Row, New York, 1987.

7. Special Issue, "Optoelectronics," *Solid-State Electron.* 30, January, 1987.

PROBLEMS

12.1 Consider an abrupt p-n junction solar cell with uniformly doped n- and p-regions. Draw the energy band diagram of the illuminated cell under (a) the short-circuit condition and (b) the open-circuit condition.

12.2 Solve Eq. (12.3) for the excess carrier concentration $p_e(x)$ and using the boundary conditions (12.4a) and (12.4b) obtain Eq. (12.5). Write $I_p(0) = -qAD_p(dp_e/dx)_{x=0}$.

12.3 Consider a Si n$^+$-p junction solar cell with 1 cm^2 area having $N_a = 2 \times 10^{18}$ cm^{-3}, $\tau_n = 4 \times 10^{-7}$ sec, $D_n = 5$ cm^2 sec^{-1}, and $G_L = 2.7 \times 10^{19}$ cm^{-3} sec^{-1}. Calculate the open-circuit voltage V_{oc}. If the measured value of V_{oc} is 0.63 V, calculate the band gap narrowing in the p-base. Neglect the contribution of the n$^+$-layer. Assume $T = 300$ K.

12.4 A 1 cm^2 area Si p-n junction solar cell has its dark current given by $1 \times 10^{-8} [\exp(V_j/1.5V_T) - 1]$ A. The cell has $I_L = 38$ mA under the AM0 condition at 300 K. Determine the maximum output power and the optimum value of the load resistance. Also determine the fill factor and the cell efficiency assuming a 20 percent power loss due to reflection at the surface.

12.5 One p-n junction solar cell has $V_{oc} = 0.5$ V and $I_L = 1$ A. A second cell has $V_{oc} = 0.6$ V and $I_L = 0.8$ A. Assuming that both cells obey the ideal diode equation, calculate the values of V_{oc} and I_{sc} when the two are connected in (a) series and (b) parallel.

12.6 Consider two n$^+$-p junction solar cells made on a 150-μm thick p-Si substrate that has $L_n = 600$ μm. One cell exhibits an ohmic contact on the back, while the other has a perfectly reflecting back contact. The first cell has $V_{oc} = 0.55$ V and $I_L = 1.8$ A. If the second cell has $I_L = 2$ A, calculate the value of V_{oc} for this cell. Take $T = 300$ K.

12.7 A 2 cm² Schottky barrier solar cell is fabricated by depositing an Au film on a 150-μm thick p-Si substrate having $L_n = 100$ μm. The cell is illuminated through the Au film at AM1 solar spectrum. Neglecting the reflection and absorption losses in the Au film and assuming $FF = 0.75$, calculate I_{sc}, V_{oc}, and the cell efficiency for the maximum power output. The barrier height of the Au-Si barrier is 0.8 eV at 300 K and $R^* = 32$ cm^{-2}K^{-2}. Assume $F_{ph} = 3.7 \times 10^{17}$ cm^{-2} sec^{-1}, $\alpha = 100$ cm^{-1}, and unit quantum efficiency.

12.8 Show that the power conversion efficiency of a Schottky barrier solar cell increases with increasing barrier height and reaches its maximum value when the barrier height equals the band gap of the semiconductor.

12.9 A photoconducting bar of n-CdSe is 0.5 cm long, 0.2 cm wide, and 0.02 cm thick. The material has $E_g = 1.7$ eV, $\mu_n = 400$ cm² V^{-1} sec^{-1}, and negligible hole mobility. One of the largest area faces of the bar is uniformly illuminated with 0.45 μm light with an incident power of 1.5 mW cm^{-2}. Assuming unit quantum efficiency and $\tau_n = 10^{-3}$ sec, calculate the photocurrent when 5 V is applied across the bar.

12.10 A Si Schottky barrier photodiode with a junction area of 0.5 cm² has a 100 Å thick Au film to form the barrier. The Si sample is 100-μm thick n-type and has a resistivity of 200 Ω-cm. The back side of the diode employs a transparent ohmic contact. The front surface of the diode is uniformly illuminated with a 0.95-μm light, and the incident power is 80 mW. Neglecting losses in the Au film and assuming unit quantum efficiency and $\alpha = 150$ cm^{-1}, calculate the diode photocurrent at a reverse bias of 200 V.

12.11 In a Si p⁺-n-n⁺ avalanche photodiode, the n-region is 25 μm thick and has a donor concentration of 10^{14} cm^{-3}. What reverse bias will be required to obtain a multiplication factor of 10? Assume that $\alpha_i(\varepsilon)$ is given by Eq. (8.29).

12.12 A GaAs p-n junction LED has a dopant concentration of 10^{18} cm^{-3} on each side. The n-region of the diode is very long, whereas the p-region is only one diffusion length long. The radiative recombination lifetime is given by $\tau_r = [7.2 \times 10^{-10}(N_d + N_a)]^{-1}$ sec. The Auger coefficient for GaAs is 10^{-31} cm⁶ sec^{-1}. A deep-level recombination center with a lifetime of 6.67×10^{-9} sec is also present. Calculate the injection and radiative recombination efficiencies at a forward bias of 1 V. Assume $\mu_n = 4500$ cm² V^{-1} sec^{-1} and $\mu_p = 0.03 \mu_n$.

12.13 In Prob. 12.12 determine the light extraction efficiency and estimate the absorption coefficient of GaAs if the overall efficiency of the LED is 0.2 percent.

12.14 A GaAsP diode that emits at 0.65 μm has an overall efficiency of 0.5 percent. The diode is replaced by a GaP diode with an overall efficiency of 0.12 percent. Which of the two diodes will appear brighter to the human eye? Why?

12.15 Consider a two-level system in thermal equilibrium at a sufficiently high temperature. Using Boltzmann statistics and Planck's distribution law, show that $B_{21} = B_{12}$.

12.16 The nonradiative lifetime in a DH laser is given by $\tau_{nr} = d/2S$ where d is the active region thickness and S is the surface recombination velocity. A GaAs/InGaP DH laser has $d = 0.3$ μm and a radiative lifetime of 10^{-9} sec. Calculate the value of S and the lattice mismatch factor $\Delta a/a$ that will lead to a radiative recombination efficiency of 50 percent. Assume that $S = 2 \times 10^7 \, \Delta a/a$ cm/sec.

12.17 The forward voltage drop across a laser diode is given by $V_F = (E_g/q) + IR_s$. An AlGaAs/GaAs DH laser diode is operated at $I = 0.4$ A with a power conversion efficiency of 5 percent and an output power of 68 mW. Determine the series resistance R_s.

13
BIPOLAR JUNCTION TRANSISTORS I: FUNDAMENTALS

INTRODUCTION

The bipolar transistor represents one of the most significant developments in the field of semiconductor devices. As discrete devices or as part of integrated circuits, bipolar transistors are used in all modern communication systems, in vehicles and satellites, and in high-speed computers. Although field-effect transistors, considered in Chapters 15 and 16, are competing with the bipolar transistor in many applications, the bipolar transistor still occupies an important place in modern electronic and communication circuits. Therefore, we will present a detailed discussion of this device.

This chapter is devoted to the basic physics and current-voltage characteristics of bipolar transistors. We begin with a description of transisitor action in Sec. 13.1. The next section describes the structure and doping profiles of some common transistor designs. In Sec. 13.3, the intrinsic model of a diffusion transistor is analyzed. Real transistors differ from intrinsic transistor in several respects which are considered in Sec. 13.4. The current-voltage characteristics in the active region of operation of the device are considered in Sec. 13.5, and the chapter ends with the charge control description of transistor behavior in Sec. 13.6.

13.1 PRINCIPLE OF OPERATION

13.1.1 The Transistor Action

The bipolar transistor is a two-junction, three-layer device in which the current flow properties of one p-n junction can be modulated by interaction with another nearby p-n junction. The structure may be either p-n-p or n-p-n, but in our dis-

PRINCIPLE OF OPERATION

cussion we will concentrate on the p-n-p transistor. We made this choice because the direction of charge flow in a p-n-p device is the same as the direction of the current flow, which makes the charge transport mechanisms somewhat easier to understand.

In order to understand the meaning of interaction between two p-n junctions, let us first consider the p-n-p structure of Fig. 13.1a in which two p-n junctions (J_1 and J_2) are produced in a semiconductor sample. The distance between J_1 and J_2 is several times larger than the hole diffusion length L_p in the mid n-region. Under this condition, practically all the holes injected into the n-region at the forward-biased junction J_1 get lost by recombination with electrons, and practically none is able to reach the reverse-biased junction J_2. Thus, the two junctions are independent of each other, and the structure is equivalent to the two back-to-back diodes shown in Fig. 13.1b.

Now let us assume that the width of the n-region is decreased gradually by moving J_1 closer to J_2. When the width becomes small compared to L_p, the structure along with the hole distribution in the n-region is shown in Fig. 13.2. This figure shows that now most of the holes injected at J_1 are able to reach J_2, and only a few are lost by recombination with electrons. Hence, practically all the hole current emitted from J_1 is collected by J_2, and transistor action is obtained.

The structure shown schematically in Fig. 13.2 is a p-n-p transistor. The p-layer to the left of J_1 emits holes into the n-region when J_1 is forward biased and is called the *emitter,* the narrow n-region is called the *base,* and the p-region to the right of J_2 is called the *collector.*

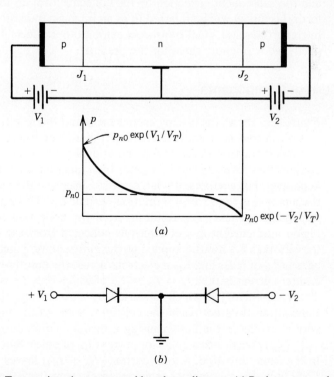

FIGURE 13.1 Two p-n junctions separated by a large distance. (*a*) Basic structure along with the minority carrier distribution in the mid n-region with a forward bias on J_1, and (*b*) the electrical equivalent circuit.

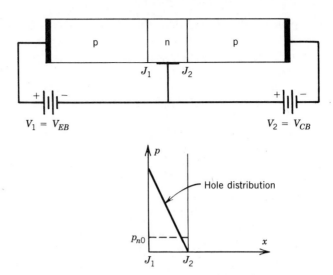

FIGURE 13.2 Basic structure of a p-n-p transistor and hole distribution in the mid n-region showing interaction between two closely spaced p-n junctions.

In the operation of the transistor as an active device (i.e., an amplifier), the emitter-base junction is forward biased while the collector-base junction is reverse biased. When operated with this combination of biases, the transistor is said to be operating in its normal active mode. The three basic processes, namely, the injection of minority carriers from the emitter, their transport through the base, and the subsequent collection by the collector form an idealized model. Henceforth this model will be referred to as an intrinsic transistor model, or simply, the intrinsic transistor. Real transistors exhibit other effects, but for the sake of simplicity we will ignore them for the present.

13.1.2 Current Components

Figure 13.3 shows the biasing arrangement and the resulting current components in a p-n-p transistor under normal operation. The electron–hole pair generation and recombination in the two junction depletion regions is ignored. Let us first consider the reverse-biased collector-base junction when the emitter terminal is kept open. For a sufficiently large reverse bias V_{CB}, the current that flows across the junction is the reverse saturation current I_{CO}. This current is the sum of the current components I_{pCO} resulting from the collection of holes from the base region, and current I_{nCO} of electrons collected from the collector region. Let a forward-bias V_{EB} now be applied to the emitter-base junction. This bias causes a current I_{pE} of holes and I_{nE} of electrons across the emitter-base junction. The total emitter current is then $I_E = I_{pE} + I_{nE}$ of which only I_{pE} is emitted into the base. The component I_{nE}, which is caused by electron injection from the base into the emitter, does not reach the collector. Some of the injected holes recombine with electrons in the base, causing a recombination current I_{pr}. The difference $(I_{pE} - I_{pr})$ constitutes a collector current I_{pC} of holes. Since both I_{pC} and I_{CO} flow in the same direction, a total current $(I_{pC} + I_{CO})$ leaves the collector terminal. Each hole that recombines removes an electron from the base. Thus, the negative terminal of the battery V_{EB} is supplying electrons to maintain the current components I_{pr} and I_{nE}, and a current $I_{nb} = I_{nE} + I_{pr}$ flows from the emitter to

PRINCIPLE OF OPERATION

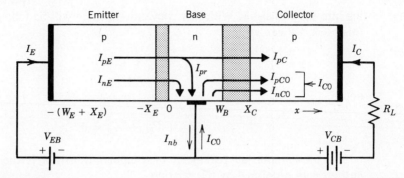

FIGURE 13.3 Biasing arrangement and current components in a p-n-p transistor under normal active operation.

the base. Note that I_{nb} and I_{CO} oppose each other at the base contact. The magnitude of I_{nb} is generally larger than that of I_{CO}. Consequently, a net current $(I_{nb} - I_{CO})$ will be leaving the base of the p-n-p transistor depicted in Fig. 13.3. One important aspect of transistor design is to keep the base current as low as possible so that practically all the emitter current will flow through the collector terminal. By making the reverse voltage V_{CB} and the load resistance R_L high, the voltage drop across R_L can be made substantially larger than the emitter-base voltage. This explains how voltage amplification is achieved using a bipolar transistor.

13.1.3 Circuit Symbol and Basic Parameters

From the foregoing discussion we have seen that the transistor is a three-terminal device with three currents and three node pair voltages. Figure 13.4 shows the circuit symbols along with the terminal variables for p-n-p and n-p-n transistors. The arrow at the emitter indicates the direction of the emitter current. For the p-n-p transistor, the current is flowing into the emitter and the arrow is pointing inward. However, the direction of the arrow is reversed in the n-p-n transistor because the current is flowing out of the emitter. For the purpose of making calculations, it is more convenient to assume that all the terminal currents are flow-

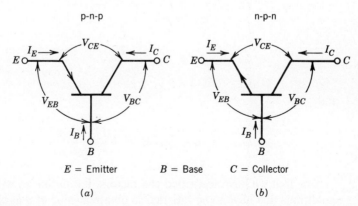

FIGURE 13.4 Circuit symbols and terminal variables for (a) a p-n-p transistor and (b) an n-p-n transistor.

ing into the transistor. With this convention, which will be followed throughout this book, we can write

$$I_E + I_B + I_C = 0 \tag{13.1}$$

and

$$V_{EB} + V_{BC} + V_{CE} = 0 \tag{13.2}$$

The most important parameter of a bipolar transistor is the common base-current gain α defined by the relation

$$\alpha = \frac{I_{pC}}{I_E} = -\frac{I_C - I_{CO}}{I_E} \tag{13.3}$$

The negative sign in Eq. (13.3) arises because of our assumption that all the terminal currents are flowing into the transistor. For a p-n-p transistor, I_E is positive, but the currents I_C and I_{CO} are negative because the actual current flows out of the terminal. Similarly, in an n-p-n transistor, I_E is negative, but both I_C and I_{CO} are positive. Thus, α is positive in either case. For a small-signal ac operation, we can define an ac current gain $\tilde{\alpha}$ by the relation

$$\tilde{\alpha} = -\frac{\partial(I_C - I_{CO})}{\partial I_E} \approx -\frac{\partial I_C}{\partial I_E} \tag{13.4}$$

The two alphas given by Eqs. (13.3) and (13.4) are, in general, not equal. They will be equal only if I_C varies linearly with I_E, which is usually not the case. Note that for a transistor, α and $\tilde{\alpha}$ should be as close to unity as possible.

The alpha of a transistor is composed of three factors: the emitter efficiency γ, the base transport factor α_T, and the collector multiplication factor M. For static operation of the p-n-p transistor, these parameters are defined as

$$\gamma = \frac{I_{pE}}{I_E} = \frac{I_{pE}}{I_{pE} + I_{nE}} = \frac{1}{1 + \frac{I_{nE}}{I_{pE}}} \tag{13.5a}$$

$$\alpha_T = \frac{I_{pC}}{I_{pE}} \tag{13.5b}$$

$$M = -\frac{(I_C - I_{CO})}{I_{pC}} \tag{13.5c}$$

and

$$\alpha = \gamma \alpha_T M = -\frac{I_C - I_{CO}}{I_E} \tag{13.6}$$

Now that we have identified the factors that define α, we can determine the conditions that have to be satisfied to obtain a value of α as close to unity as possible. To be specific, let us consider the transistor structure shown in Fig. 13.5, which is fabricated by alloying p-type emitter and collector regions to a thin slice

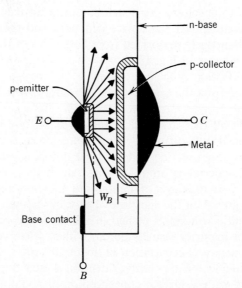

FIGURE 13.5 Schematic cross-section of an alloyed p-n-p transistor. The lines with arrows show the direction of hole motion when the device is biased for normal active operation.

of n-type semiconductor. This type of device is no longer in use today, but it illustrates all the essential features of transistor design. In order to obtain a large value of α, both γ and α_T should be made to approach unity. It is evident from Eq. (13.5a) that γ approaches unity when the ratio I_{nE}/I_{pE} is made vanishingly small, and this is achieved by doping the p-emitter heavily compared to the base. To improve α_T, the recombination current I_{pr} should be reduced to its lowest possible value. I_{pr} is reduced, first, by reducing the base width W_B, and, second, by keeping the ohmic contact to the base away from the edge of the emitter-base junction. When the base width W_B is made small, only a small fraction of injected holes will recombine in the base region. As shown in Fig. 13.5, some of the injected holes also diffuse toward the base contact, and if this contact is near the emitter-base junction, most of these holes will be drained through the contact and cause an increase in I_{pr}. Therefore, it is desirable to keep the base contact more than a diffusion length away from the edge of the emitter-base junction. Another condition for keeping I_{pC} large is also evident from Fig. 13.5. As shown by the arrows, holes from the emitter are emitted in all possible directions of the half plane to the right of the emitter. To collect as many of them as possible, the collector-base junction area should be made large compared to the emitter-base junction area.

13.2 FABRICATION METHODS AND DOPING PROFILES

Bipolar transistors have been fabricated in several different ways. An early process, used mainly for Ge transistors, was that of alloying the emitter and collector regions to a thin slice of an appropriately doped semiconductor. This process did not allow good control of the emitter and collector geometries and resulted in base widths on the order of several microns. Consequently, in the beginning Si transistors were made by growing the emitter, base, and collector regions during the process of crystal growth from the melt (see Sec. 19.2.1). The next significant step was to develop a solid-state impurity diffusion process that allowed much tighter control over the device geometry and the base width. Diffusion could be simultaneously performed from both sides of the slice resulting in the impurity distribution of Fig. 13.6a. Alternatively, transistors could be produced by a double diffusion process in which p- and n-type diffusions were performed in succession from the same face of the wafer giving the impurity profile shown in Fig. 13.6b.

Nearly all the present-day transistors are n-p-n Si transistors fabricated by using the double-diffused planar process (see Sec. 19.4). Figure 13.7 shows a schematic cross-section and the typical doping profile of such a transistor. The starting material is an n^+-Si substrate with a donor concentration of about 10^{19} cm^{-3}. An n-Si layer of appropriate thickness and dopant concentration is epitaxially grown on the substrate. The wafer is then oxidized to grow a thin SiO_2 layer, and a window is etched in the oxide for performing the base diffusion. Acceptor impurity (usually boron) is then selectively diffused into the n-region with a surface concentration of about 10^{18} cm^{-3}. The wafer is then reoxidized, and a window for the emitter diffusion is etched into the oxide. Next, an n-type diffusion with a surface concentration of about 10^{21} donor atoms cm^{-3} is performed to define the emitter-base junction. Finally, ohmic contacts are made to the n^1-substrate as well as to the base and emitter regions as shown in Fig. 13.7a. The doping profile of the device is shown in Fig. 13.7b. Because of the sharp impurity gradient in the base region, there exists a built-in electric field in which the minority carriers injected from a forward-biased emitter are accelerated toward the collector. Consequently, a drift motion is superimposed over the normal diffusive motion of the minority carriers during their transit through the base region. This type of transistor, therefore, is called a *drift transistor*. On the other hand, the transistor structure of Fig. 13.5 has abrupt p-n junctions with uniformly doped emitter, base, and collector regions. No drift field exists in the base region of this transistor, and the minority carriers injected from the emitter move across the base only by the process of diffusion. Such a device is called a *diffusion transistor*.

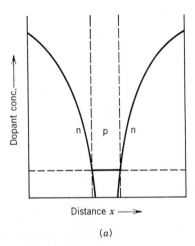
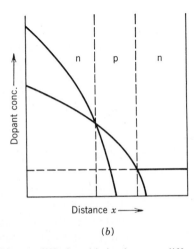

FIGURE 13.6 Doping profiles in transistors made by solid-state diffusion: (a) simultaneous diffusion from both sides of a wafer and (b) successive diffusion of p- and n-type dopants from the same face.

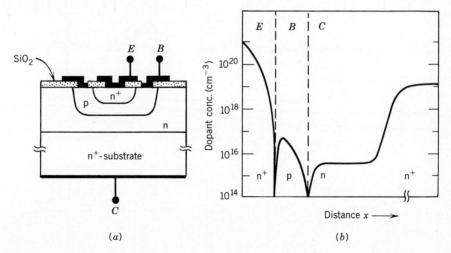

FIGURE 13.7 The double diffused planar transistor: (a) schematic cross-section and (b) doping profile.

A diffusion transistor is the simplest form of a transistor, and its theory is easier to understand than the theory of a drift transistor. In this chapter, we will deal mainly with the diffusion transistor because it permits a clear focus on many phenomena that are important in transistor operation without mathematical complication.

13.3 ANALYSIS OF THE IDEAL DIFFUSION TRANSISTOR

13.3 Calculation of Terminal Currents

Figure 13.8a shows the schematic cross-section of a p-n-p diffusion transistor biased for normal active operation. The emitter-base and collector-base junction areas are assumed to be equal, and it is further assumed that the device has abrupt junctions with uniformly doped emitter, base, and collector regions as shown in Fig. 13.8b. The energy band diagram of the biased transistor is shown in Fig. 13.8c where the directions of the electrons and holes flowing across the two junctions are also indicated. For a forward bias V_{EB} on the emitter-base junction, holes are injected from the emitter into the base, and electrons are injected from the base into the emitter. The holes injected into the base diffuse toward the collector. After reaching the depletion region edge of the reverse-biased collector-base junction, they fall through a large potential gradient and are collected by the collector.

In order to calculate the hole and electron currents across the emitter-base and the collector-base junctions, the continuity equations for minority carriers in the emitter, base, and collector regions have to be solved. Making a one-dimensional approximation, and following the approach of Sec. 7.1.2, we can write the excess hole concentration p_e in the neutral n-base region in the steady state as

$$\frac{d^2 p_e}{dx^2} - \frac{p_e}{L_B^2} = 0 \qquad (13.7)$$

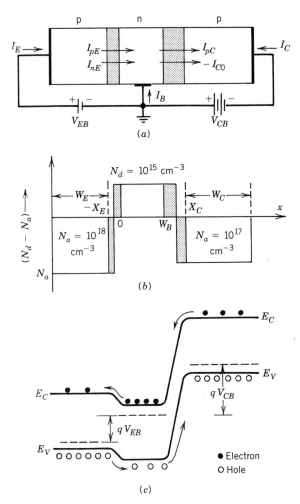

FIGURE 13.8 A p-n-p diffusion transistor under normal active operation. (a) Biasing arrangement, (b) doping profiles and depletion regions (hatched area), and (c) energy band diagram showing electron and hole motion across the junctions.

This equation was solved in Sec. 7.1.2 to obtain the static I-V characteristics of a p-n junction diode. We will now solve it for the p-n-p transistor with reference to the coordinate system shown in Fig. 13.8b. Note that the origin ($x = 0$) is now taken to be at the edge of the emitter-base junction depletion region. As previously discussed, Eq. (13.7) has the solution

$$p_e(x) = A_1 \exp(x/L_B) + A_2 \exp(-x/L_B) \tag{13.8}$$

Here L_B represents the hole diffusion length in the base, and the constants A_1 and A_2 are determined from the boundary conditions

$$p_e(0) = p_{no}\left[\exp\left(\frac{V_{EB}}{V_T}\right) - 1\right] = \frac{n_i^2}{N_d}\left[\exp\left(\frac{V_{EB}}{V_T}\right) - 1\right] \equiv p_E \quad \text{at } x = 0 \tag{13.9}$$

$$p_e(W_B) = \frac{n_i^2}{N_d}\left[\exp\left(\frac{V_{CB}}{V_T}\right) - 1\right] \equiv p_C \quad \text{at } x = W_B \tag{13.10}$$

where V_{EB} and V_{CB} represent the emitter-base and collector-base junction voltages, respectively, $V_T = kT/q$, and W_B is the base width. Substituting Eqs. (13.9) and (13.10) into Eq. (13.8) yields

$$p_e(x) = p_E \left[\frac{\sinh([W_B - x]/L_B)}{\sinh(W_B/L_B)} \right] + p_C \left[\frac{\sinh(x/L_B)}{\sinh(W_B/L_B)} \right] \quad (13.11)$$

It is evident from Eq. (13.11) that the excess hole concentration $p_e(x)$ at any point in the base region is a linear superposition of p_E and p_C which depend exponentially on V_{EB} and V_{CB}, respectively. Note that Eqs. (13.9) and (13.10) have been obtained from the Boltzmann relation and are valid only under low-level injection conditions. For a field-free base region, the hole current I_{pE} is given by

$$I_{pE} = -qAD_B \frac{dp_e}{dx}\bigg|_{x=0} = \frac{qAD_B}{L_B}\left[p_E \coth\left(\frac{W_B}{L_B}\right) - \frac{p_C}{\sinh(W_B/L_B)}\right]$$
$$(13.12)$$

where D_B is the hole diffusion coefficient in the base, and A is the junction area. A similar procedure can be used to obtain the electron current I_{nE} across the emitter-base junction giving

$$I_{nE} = qAD_E \frac{dn_e}{dx}\bigg|_{x=-X_E} = \frac{qAD_E n_E}{L_E} \coth\left|\frac{W_E}{L_E}\right| \quad (13.13)$$

where

$$n_E = \frac{n_i^2}{N_a(E)}[\exp(V_{EB}/V_T) - 1] \quad (13.14)$$

denotes the excess electron concentration at the emitter edge $(-X_E)$ of the emitter-base junction depletion region, D_E and L_E represent the electron diffusion constant and diffusion length in the emitter, respectively, W_E is the width of the neutral emitter region, and $N_a(E)$ is the acceptor concentration in the emitter region. The emitter current $I_E = (I_{pE} + I_{nE})$ is obtained from Eqs. (13.12) and (13.13). After substituting for $p_E, p_C,$ and n_E from Eqs. (13.9), (13.10) and (13.14), we obtain the following expression for I_E:

$$I_E = qAn_i^2 \left[\frac{D_B}{L_B N_d} \coth\left(\frac{W_B}{L_B}\right) + \frac{D_E}{L_E N_a(E)} \coth\left|\frac{W_E}{L_E}\right| \right] \left[\exp\left(\frac{V_{EB}}{V_T}\right) - 1 \right]$$
$$- \frac{qAn_i^2 D_B}{L_B N_d \sinh(W_B/L_B)}\left[\exp\left(\frac{V_{CB}}{V_T}\right) - 1\right] \quad (13.15)$$

Similarly, the collector current I_C is obtained from

$$I_C = -(I_{pC} + I_{nC}) = qA\left[D_B \frac{\partial p_e}{\partial x}\bigg|_{x=W_B} - D_C \frac{\partial n_e}{\partial x}\bigg|_{x=X_C}\right] \quad (13.16)$$

The electron current I_{nC} is obtained from Eq. (13.13) after substituting W_C for W_E, and L_C for L_E, and replacing n_E in Eq. (13.14) by n_C where

$$n_C = \frac{n_i^2}{N_a(C)}\left[\exp\left(\frac{V_{CB}}{V_T}\right) - 1\right]$$

For a wide collector region such that $W_C \gg L_C$, the $\coth(W_C/L_C)$ term in the expression of I_{nC} becomes unity, and from Eq. (13.16) we obtain

$$I_C = -\frac{qAD_B n_i^2}{L_B N_d \sinh(W_B/L_B)}\left[\exp\left(\frac{V_{EB}}{V_T}\right) - 1\right]$$
$$+ qAn_i^2 \left[\frac{D_B}{L_B N_d}\coth\left(\frac{W_B}{L_B}\right) + \frac{D_C}{L_C N_a(C)}\right]\left[\exp\left(\frac{V_{CB}}{V_T}\right) - 1\right] \quad (13.17)$$

Here D_C and L_C represent the electron diffusion coefficient and diffusion length in the collector region, and $N_a(C)$ is the acceptor concentration in this region. Equations (13.15) and (13.17) form the basis of the Ebers-Moll equations and will be discussed further.

13.3.2 Calculation of dc Parameters

The emitter efficiency of the p-n-p transistor can be directly obtained from Eqs. (13.12), (13.13), and (13.14). For a large reverse bias applied to the collector-base junction, it is seen from Eq. (13.10) that $p_C \simeq -n_i^2/N_d$. Hence, in Eq. (13.12), the p_C term can be neglected to obtain

$$\frac{I_{nE}}{I_{pE}} = \frac{N_d}{N_a(E)}\frac{D_E L_B}{D_B L_E}\left[\frac{\coth|W_E/L_E|}{\coth(W_B/L_B)}\right]$$

With the assumption that $W_B \ll L_B$ and $|W_E| \ll L_E$, and substituting for I_{nE}/I_{pE} in Eq. (13.5a) we get

$$\gamma = \frac{1}{1 + \dfrac{N_d}{N_a(E)}\dfrac{D_E}{D_B}\left|\dfrac{W_B}{W_E}\right|} = \frac{1}{1 + K_1\dfrac{N_d}{N_a(E)}} \quad (13.18)$$

It is evident from Eq. (13.18) that γ can be made to approach unity if the ratio $N_a(E)/N_d$ is made large (about 10^3).

The base transport factor is given by

$$\alpha_T = \frac{I_{pC}}{I_{pE}} = \left.\frac{dp_e}{dx}\right|_{x=W_B} \bigg/ \left.\frac{dp_e}{dx}\right|_{x=0}$$

After putting $p_C = 0$ in Eq. (13.11) and evaluating the differentials, we get

$$\alpha_T = \frac{1}{\cosh(W_B/L_B)} \quad (13.19a)$$

which, for $W_B \ll L_B$, can be written as

$$\alpha_T \simeq \frac{1}{1 + W_B^2/2L_B^2} \tag{13.19b}$$

The parameter M defined by Eq. (13.5c) is a function of the collector-base reverse bias V_{CB} because of the avalanche multiplication of carriers in the depletion region of this junction. The dependence of M on V_{CB} can be expressed by the empirical relation

$$M = \frac{1}{\left[1 - \left(\dfrac{V_{CB}}{V_{Br}}\right)^m\right]} \tag{13.20}$$

where V_{Br} represents the avalanche breakdown voltage of the collector-base junction and m varies between 4 and 6. It is seen from Eq. (13.20) that M is almost unity at low values of V_{CB} but becomes very large as V_{CB} approaches V_{Br}.

The common base-current gain α of the transistor is obtained by multiplying Eqs. (13.18), (13.19b), and (13.20), giving

$$\alpha = \frac{M}{\left[\left(1 + K_1 \dfrac{N_d}{N_a(E)}\right)\left(1 + \dfrac{W_B^2}{2L_B^2}\right)\right]} \tag{13.21}$$

In the normal operation of the transistor, the reverse voltage V_{CB} is kept so small that M is essentially unity and α of the transistor is less than unity. The extent to which α deviates from unity represents the alpha-defect of the transistor, which is a measure of the current I_{nb} supplied from the base contact.

In most circuit applications of bipolar transistors, the emitter is used as a common terminal with the base as the input and the collector as the output terminal. A relation between the base and the collector currents is obtained by eliminating I_E between Eqs. (13.1) and (13.3), giving

$$I_C = \frac{\alpha}{1 - \alpha} I_B + \frac{I_{CO}}{1 - \alpha} \equiv \beta I_B + (\beta + 1)I_{CO} \tag{13.22}$$

where we have defined

$$\beta = \frac{\alpha}{1 - \alpha} \tag{13.23}$$

as the common emitter current gain. This gain is usually written as h_{FE}. Thus, writing $\beta = h_{FE}$ in Eq. (13.23) and substituting the value of α from Eq. (13.21) and taking $M = 1$, we obtain

$$\frac{1}{h_{FE}} \simeq \frac{W_B^2}{2L_B^2} + K_1 \frac{N_d}{N_a(E)} \tag{13.24}$$

In many transistors, the second term on the right-hand side of Eq. (13.24) is small, and we can write $h_{FE} \simeq 2(L_B/W_B)^2$. Note that α is close to unity and h_{FE} is a large number. For example, if $L_B/W_B = 10$, then $h_{FE} = 200$.

13.3.3 The Ebers-Moll Equations [1]

We have derived Eqs. (13.15) and (13.17) for the emitter and the collector currents of an intrinsic p-n-p transistor. Similar equations can also be obtained for an n-p-n transistor. Thus, it is possible to express the emitter and the collector currents of a bipolar transistor by the following general relations:

$$I_E = -I_{ES}\left[\exp\left(\frac{V_{EB}}{V_T}\right) - 1\right] + \alpha_I I_{CS}\left[\exp\left(\frac{V_{CB}}{V_T}\right) - 1\right] \quad (13.25)$$

$$I_C = \alpha_N I_{ES}\left[\exp\left(\frac{V_{EB}}{V_T}\right) - 1\right] - I_{CS}\left[\exp\left(\frac{V_{CB}}{V_T}\right) - 1\right] \quad (13.26)$$

Here I_{ES} and I_{CS} are proportionality factors and are taken to be positive for an n-p-n transistor and negative for a p-n-p transistor. The parameters α_N and α_I represent the common base current gains of the transistor under normal active and inverse active operations, respectively. In the inverse active operation of a transistor, the roles of the emitter and the collector are interchanged which means that the collector-base junction is forward biased and the emitter-base junction is reverse biased. Equations (13.25) and (13.26), known as *the Ebers-Moll (E-M) equations*, are valid for all combinations of V_{EB} and V_{CB}. It is obvious that for a p-n-p transistor, V_{EB} is positive when the emitter-base junction is forward biased and negative when it is reverse biased. Similar considerations also apply for V_{CB}. In the case of an n-p-n transistor, the situation is reversed and E-M equations are obtained by writing V_{BE} for V_{EB} and V_{BC} for V_{CB}. However, in this book we will use Eqs. (13.25) and (13.26), with the convention that V_{EB} and V_{CB} are positive for a forward bias and negative for a reverse bias.

For the p-n-p transistor considered above, the parameters I_{ES}, I_{CS}, α_N, and α_I can be evaluated in terms of the junction area, the base and emitter region widths, the diffusion constants and dopant concentration, and so on, by comparing Eq. (13.25) with Eq. (13.15) and Eq. (13.26) with Eq. (13.17). Similar calculations can also be made for the case of an n-p-n transistor. For a symmetrical transistor in which the emitter and the collector areas are equal, and the emitter and collector region dopings are also identical, $I_{ES} = I_{CS}$ and $\alpha_N = \alpha_I$. For the case of the intrinsic p-n-p transistor considered in Sec. 13.3.1, it is seen that I_{ES} and I_{CS} are not equal unless the acceptor concentration $N_a(E)$ is the same as $N_a(C)$. However, a comparison of Eqs. (13.15) and (13.17) with Eqs. (13.25) and (13.26) reveals that

$$\alpha_N I_{ES} = \alpha_I I_{CS} = -\frac{qAn_i^2 D_B}{L_B N_d \sinh(W_B/L_B)} \quad (13.27)$$

Even when the emitter and collector junction areas are unequal, and the transistor is not intrinsic, the reciprocity relation $\alpha_N I_{ES} = \alpha_I I_{CS}$ is always valid. In fact, it has been found that the reciprocity relation is valid for any transistor in which the parameters I_{ES}, I_{CS}, α_N, and α_I are obtainable from measurements [2].

The four parameters of the E-M equations can be determined by making a set of measurements on the transistor. For example, if we put a short between the collector and base to make $V_{CB} = 0$, then from Eq. (13.25) we obtain

$$I_E = -I_{ES}[\exp(V_{EB}/V_T) - 1] \quad (13.28)$$

ANALYSIS OF THE IDEAL DIFFUSION TRANSISTOR

This equation shows that the external behavior of the emitter-base diode with the collector shorted to the base is similar to a p-n junction diode having a saturation current I_{ES}. However, the internal behavior of the diode is different from a normal p-n junction diode, as can be seen in Fig. 13.9. Here, the collector-base junction collects most of the hole current injected from the emitter because of the presence of the built-in voltage. The current I_{ES} is determined by fitting Eq. (13.28) to the measured forward characteristics of the diode in Fig. 13.9.

If the collector current I_C is also measured for $V_{CB} = 0$, then from Eqs. (13.26) and (13.28), we get

$$\alpha_N = -\frac{I_C}{I_E}\bigg|_{V_{CB}=0} \tag{13.29}$$

The parameters I_{CS} and α_I can be determined in a similar manner as I_{ES} and α_N by making $V_{EB} = 0$ and putting a forward bias on the collector-base junction. Once these four parameters are known, it becomes possible to calculate I_E and I_C for any combination of V_{EB} and V_{CB}.

The currents I_{ES} and I_{CS} can also be expressed in terms of the saturation currents I_{EO} and I_{CO} of the emitter-base and the collector-base junctions, respectively. To do so let us eliminate V_{EB} between the two E-M equations (13.25) and (13.26), and write

$$I_C = -\alpha_N I_E - I_{CS}(1 - \alpha_N \alpha_I)[\exp(V_{CB}/V_T) - 1] \tag{13.30}$$

If we keep the emitter lead open to make $I_E = 0$, then the collector-base diode current can be expressed as

$$I_C = -I_{CO}[\exp(V_{CB}/V_T) - 1] \tag{13.31}$$

Comparing Eq. (13.31) with (13.30) after making $I_E = 0$, we obtain

$$I_{CS} = \frac{I_{CO}}{1 - \alpha_N \alpha_I} \tag{13.32}$$

and similarly

$$I_{ES} = \frac{I_{EO}}{1 - \alpha_N \alpha_I} \tag{13.33}$$

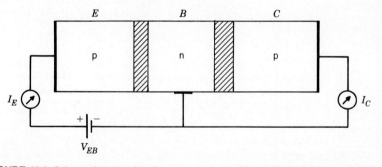

FIGURE 13.9 Schematic setup for the measurement of I_{ES} and α_N of a p-n-p transistor.

where I_{EO} is the saturation current of the emitter-base junction with $I_C = 0$. Like I_{ES} and I_{CS}, the currents I_{EO} and I_{CO} are positive for an n-p-n transistor but negative for a p-n-p transistor.

13.3.4 Regions of Operation

The Ebers-Moll equations for the intrinsic transistor are valid for all combinations of junction voltages. Corresponding to the four possible combinations of V_{EB} and V_{CB}, a bipolar transistor has four regions of operation. Figure 13.10 shows these regions of operation for a p-n-p device. The corresponding idealized minority carrier distributions in the base region are also shown. The same figure can also be used for an n-p-n transistor if we write V_{BE} for V_{EB} and V_{BC} for V_{CB}. The normal active region of operation discussed thus far falls in the fourth quadrant where V_{EB} is positive and V_{CB} is negative. This is the most important region of operation because here the transistor is used as an active device.

In the saturation region, both V_{EB} and V_{CB} are positive. Thus, both junctions are forward biased and are injecting holes into the base region. Because of enhanced carrier concentration in the base, there will be a large recombination current that will flow through the base terminal. Under this condition, the voltage drop across the transistor is very low, and the device acts as a closed (or ON) switch.

In the inverse-active region, the collector-base junction is forward biased, and the emitter-base junction is reverse biased. Therefore, the operation of the device is similar to that in the normal active region except that the roles of the emitter and collector have been interchanged. As mentioned earlier, the collector area in a transistor is kept larger than the emitter area. Thus, not all the holes injected

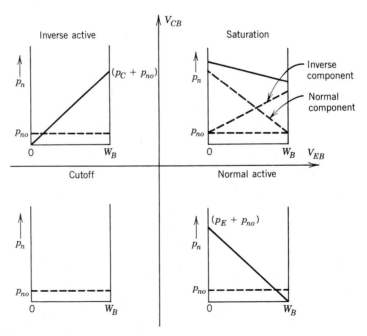

FIGURE 13.10 Regions of operation and minority carrier distributions in the base region of a p-n-p transistor.

into the base at the collector-base junction are collected by the emitter, many of them are lost by recombination in the remote parts of the base region. As a result, the current gain α_I for the inverse-active region is lower than the current gain α_N for the normal active region.

In the cutoff region, both junctions are reverse biased and the base region is virtually free of minority carriers. The transistor, therefore, corresponds to an open switch with a negligibly small emitter, base, and collector current. Note that for sufficiently large reverse voltages (V_{EB} and V_{CB}), the values of I_E and I_C obtained from Eqs. (13.25) and (13.26) are $I_E = I_{ES} - \alpha_I I_{CS}$ and $I_C = I_{CS} - \alpha_N I_{CS}$, which are very low.

13.4 REAL TRANSISTORS

In our discussion thus far, we have considered an intrinsic transistor in which only three basic processes, namely, the injection, transport, and collection of minority carriers, are emphasized. A real transistor differs from an intrinsic transistor in several respects. These differences arise first, because the geometry of a real transistor is more complex than the simple structure shown in Fig. 13.8a and second because several second-order effects have to be included in the description of a real transistor. This section investigates mechanisms that must be considered to obtain a realistic picture of transistor operation.

13.4.1 Carrier Recombination in the Emitter-Base Junction Depletion Region

In the intrinsic transistor, we assumed that the junctions follow the ideal diode law and we neglect the carrier generation and recombination in the junction depletion regions. This assumption is not always justified in real transistors. In the normal active region of operation, the current due to carrier recombination in the emitter-base junction depletion region becomes important and must be added to the total emitter current I_E. As seen in Sec. 7.2.1, the depletion region recombination current I_R can be expressed as

$$I_R = I_{Ro}[\exp(V_{EB}/2V_T) - 1] \qquad (13.34)$$

where

$$I_{Ro} = \frac{qAn_i X_E}{2\tau_o}$$

Here X_E represents the width of the emitter-base junction depletion region, and τ_o is the carrier lifetime in this region. The current I_R flows only in the base-emitter leads and does not affect the collector current. Consequently, the expression for emitter efficiency in Eq. (13.5a) is no longer valid, and we get

$$\gamma = \frac{I_{pE}}{I_{pE} + I_{nE} + I_R} \qquad (13.35)$$

Neglecting I_{nE} and writing $I_{pE} \simeq I_{EO} \exp(V_{EB}/V_T)$, we can write the above equation as

$$\gamma = \frac{1}{[1 + (I_{Ro}/I_{EO}) \exp(-V_{EB}/2V_T)]} \quad (13.36)$$

Since the ratio I_{Ro}/I_{EO} can be 10^3 or larger for Si and GaAs p-n junctions at room temperature and below, it is evident from Eq. (13.36) that γ is small for low values of forward bias but approaches unity for $V_{EB} \gg V_T$. The variation in γ with V_{EB} in real transistors makes the current gains α and h_{FE} strongly dependent on V_{EB} and, hence, also on the collector current I_C. Figure 13.11 shows a typical plot of h_{FE} as a function of I_C. The low values of I_C in this figure correspond to the low values of V_{EB} where the emitter efficiency γ is low; thus, h_{FE}, which is almost equal to $\gamma/(1 - \gamma)$, is also low. As I_C is increased by increasing V_{EB}, the emitter efficiency improves, and h_{FE} increases gradually, reaching its maximum value for $V_{EB} \gg V_T$. At still higher collector currents, h_{FE} starts decreasing. This decrease is caused by high-injection effects considered in Sec. 14.5. An important high injection effect is a decrease in γ, which occurs when the injected hole concentration in the base approaches the base doping N_d. This situation causes an increase in the electron concentration in the base and, hence, in the electron current component I_{nE}.

Besides reducing the emitter efficiency, carrier recombination in the emitter junction depletion region also causes an increase in the small-signal ac resistance r_e of the emitter-base junction. This can be seen by writing I_E as

$$I_E \simeq I_{EO} \exp(V_{EB}/V_T) + I_{Ro} \exp(V_{EB}/2V_T) \quad (13.37)$$

Differentiating this relation with respect to V_{EB}, we can show that (see Prob. 13.7)

$$r_e = \frac{\partial V_{EB}}{\partial I_E} = \frac{V_T}{I_E(1 - I_R/2I_E)} \quad (13.38)$$

It is evident from this relation that $r_e = V_T/I_E$ when $I_R = 0$, but increases above this value when I_R becomes comparable to I_E.

13.4.2 Effect of Collector-Bias Variation

In the idealized model of the last section, the base-width was assumed to be constant and independent of the junction voltages. However, the base-width of real

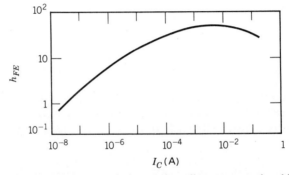

FIGURE 13.11 Dependence of current gain h_{FE} on the collector current in a bipolar transistor.

transistors varies with the collector-base junction voltage V_{CB}. As the reverse bias on the collector-base junction is increased, the depletion region of the junction widens, causing a decrease in the effective base-width W_B. This decrease in W_B can be expressed by the relation

$$W_B = W_{B0} - W_B^* f(V_{CB}) \tag{13.39}$$

where W_{B0} represents the total width of the base region and W_B^* is the width constant of the base-collector junction on the base side. The mathematical form of the function $f(V_{CB})$ can be established if the impurity distributions on both sides of the junction are specified. When the base region doping is uniform, and the collector is heavily doped compared to the base, Eq. (13.39) can be written as

$$W_B = W_{B0} - \left[\frac{2\varepsilon_s}{qN_d}(V_i - V_{CB})\right]^{1/2} \tag{13.40}$$

Base-width modulation resulting from the variations in the collector-base bias was first analyzed by J. M. Early and is known as the *Early effect* [3].

Because of base-width modulation, both I_E and I_C increase with increasing reverse bias on the collector-base junction, as can be seen in Fig. 13.12. This figure shows the hole distribution in the base plotted for two values of the reverse bias. When the reverse voltage is increased from V_{CB} to $(V_{CB} + \Delta V_{CB})$, the base-width decreases from W_B to $(W_B - \Delta W_B)$, and the slope of the hole concentration in the base is also changed. Since the currents I_E and I_C are proportional to this slope, both currents increase with increasing reverse bias. The decrease in the base-width also causes an increase in the base transport factor α_T, which will cause an increase in the current gain α.

Base narrowing also places an upper limit on the reverse voltage that can be applied to the collector-base junction. If the reverse bias on the collector is increased, and avalanche breakdown does not occur, the maximum collector voltage will be the one at which the collector-base junction depletion region will reach the emitter-base junction and W_B becomes zero. This voltage is called the *punch-*

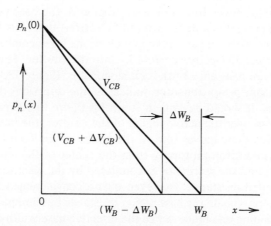

FIGURE 13.12 Influence of base-width modulation on the minority carrier distribution in the base region of a p-n-p transistor.

through voltage V_{PT}. For a uniformly doped p$^+$-n-p$^+$ transistor, the voltage V_{PT} is obtained from Eq. (13.40) by making $W_B = 0$, giving

$$V_{PT} = \frac{qN_d W_{B0}^2}{2\varepsilon_s} \tag{13.41}$$

13.4.3 Avalanche Multiplication in the Collector-Base Junction Depletion Region

Another effect that causes an increase in the collector current with reverse bias V_{CB} is avalanche multiplication of minority carriers in the depletion region of the collector-base junction. When the reverse voltage V_{CB} is sufficiently large, the minority carriers diffusing into the collector-base junction depletion region from either side create secondary electron–hole pairs. Consequently, the ratio of the current flowing out of the depletion region to that which enters it is given by the avalanche multiplication factor M. Thus, the current I_C in Eq. (13.30) becomes

$$I_C = -\gamma\alpha_T M I_E + M I_{CO} \tag{13.42}$$

where M is the avalanche multiplication factor given by Eq. (13.20). The current gain $\alpha = \gamma\alpha_T M$ of the transistor can now be expressed as

$$\alpha = \frac{\gamma\alpha_T}{[1 - (V_{CB}/V_{Br})^m]}$$

Since m is greater than 3, the avalanche multiplication factor M remains almost unity unless V_{CB} gets close to the avalanche breakdown voltage V_{Br}. As V_{CB} approaches V_{Br}, the current gain α, and hence, the collector current I_C becomes very large.

13.4.4 Base Resistance

A real transistor is a three-dimensional structure in which a lateral current $I_B(\simeq I_{nb})$ flows from the active region to the base contact. In a p-n-p transistor, this current is a drift current of electrons that flow from the base ohmic contact to the active region to supply electrons for recombination with holes, as well as to inject an electron current I_{nE} into the emitter. This base current flows through a finite resistance of the semiconductor. Figure 13.13 shows the cross-section of a planar p-n-p transistor, indicating the directions of the base and collector currents. It should be evident from this figure that since the emitter-base current is spread over the active region of the emitter, the base current decreases continuously as one moves toward the center of the emitter. Consequently, the voltage does not drop uniformly from the center of the emitter to the base ohmic contact, and the resistance encountered by the electrons in the base region is a distributed resistance. However, over a considerable range of current, it is possible to approximate the base region resistance by a lumped resistance r_B called the base spreading resistance. Since the effective base-width decreases with reverse bias at the collector-base junction, r_B will also increase with the reverse bias. We can calculate r_B if the transistor geometry and the base region resistivity are known [4, 5].

STATIC I-V CHARACTERISTICS IN THE NORMAL ACTIVE REGION

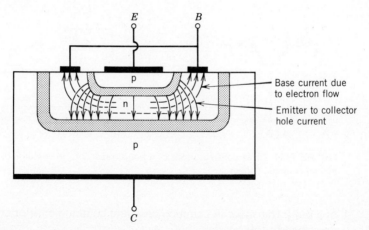

FIGURE 13.13 Cross-sectional view of p-n-p planar transistor under normal active operation showing the base current and edge crowding.

Two main effects are associated with the voltage drop across the base spreading resistance. The first is that the voltages appearing across the emitter-base and the collector-base junctions are not the same as the externally applied voltages. This will cause a modification in the E-M equations by subtracting a voltage drop $I_B r_B$ from both V_{EB} and V_{CB}. The second effect is that the potential drop in the active region under the emitter reduces the forward bias across the emitter-base junction in such a way that the forward bias is larger at the emitter periphery near the contact than at the center of the emitter. Because the emitter current depends exponentially on the forward bias, the hole current density injected from the emitter will have its highest value in the outer portion near the emitter edge and a minimum value at the center of the emitter. This phenomenon, which causes current concentration near the edge of the emitter, is known as *edge crowding* and is depicted in Fig. 13.13 by the crowding of current lines near the emitter periphery.

13.4.5 Consequences of Structural Asymmetries

In the intrinsic transistor in Sec. 13.3, we had assumed that the collector and the emitter areas were equal. As mentioned earlier, in real transistors the collector area is generally larger than the emitter area. In addition, the collector region is lightly doped compared to the emitter region. As a result of these asymmetries, I_{CS} in a real transistor is larger than I_{ES}, and α_N is larger than α_I, but the reciprocity relation $\alpha_N I_{ES} = \alpha_I I_{CS}$ still holds.

13.5 STATIC I-V CHARACTERISTICS IN THE NORMAL ACTIVE REGION

The I-V characteristics of intrinsic transistor are described by the E-M equations, and those of real transistors can be derived from these equations after making appropriate corrections. A bipolar transistor can be operated in any one of the three configurations shown in Fig. 13.14. In each of these configurations, the transistor is characterized by relating the input current to the input voltage and the output current to the output voltage. The input and output characteristics

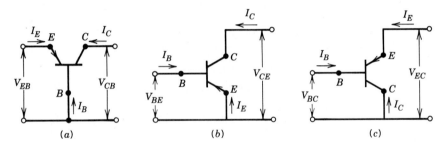

FIGURE 13.14 Three configurations of a p-n-p transistor: (*a*) common base, (*b*) common emitter, and (*c*) common collector.

of a p-n-p transistor in common base and common emitter configurations will be described in this section.

13.5.1 Common Base Configuration

(a) INPUT CHARACTERISTICS

In describing the input current I_E as a function of V_{EB}, either the collector current I_C or the voltage V_{CB} may be held constant during the measurements. In practice, it becomes difficult to hold I_C constant, and the input characteristics are obtained with V_{CB} as a parameter. From Eq. (13.25), it is clear that for a sufficiently large reverse bias on the collector we obtain

$$I_E = -I_{ES}[\exp(V_{EB}/V_T) - 1] - \alpha_I I_{CS} \tag{13.43}$$

In the case of a p-n-p transistor, both I_{ES} and I_{CS} are negative so that I_E is positive. Equation (13.43) gives the characteristics of a p-n junction diode except that the current I_E is not zero when $V_{EB} = 0$. Using the relation $\alpha_N I_{ES} = \alpha_I I_{CS}$, we obtain the emitter voltage for $I_E = 0$ as

$$V_{EB}(I_E = 0) = V_T \ln(1 - \alpha_N) \tag{13.44}$$

This is a small negative voltage. Equation (13.43) shows that I_E is independent of V_{CB}. However, in real transistors it is not so, and because of the base-width modulation, I_E is found to increase with increasing reverse bias on the collector-base junction.

(b) OUTPUT CHARACTERISTICS

I_C as a function of reverse voltage V_{CB} is generally measured with the emitter current I_E as parameter, and not the voltage V_{EB}. This is because I_C varies almost linearly with I_E while its dependence on V_{EB} is exponential. Eliminating V_{EB} in the two E-M equations, (13.25) and (13.26), we obtain

$$I_C = -\alpha_N I_E - I_{CO}[\exp(V_{CB}/V_T) - 1] \tag{13.45}$$

where I_{CO} is given by Eq. (13.32). Figure 13.15 shows the I_C versus V_{CB} characteristics of a p-n-p transistor for various values of I_E. For $I_E = 0$ the above equation represents a p-n junction diode whose current saturates to I_{CO} for a sufficiently

STATIC I-V CHARACTERISTICS IN THE NORMAL ACTIVE REGION

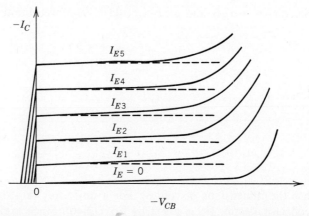

FIGURE 13.15 Common base output characteristics of a p-n-p transistor in the normal active region. The dashed lines show the characteristics for an ideal transistor, and the solid lines are for a real transistor.

large reverse bias V_{CB}. In Si transistors I_{CO} is typically of the order of a few nanoamperes. For a finite value of I_E, a current $-\alpha_N I_E$ is added to the diode current, and because of the presence of this term, a small forward bias is required on the collector-base junction to make $I_C = 0$. Thus, substituting $I_C = 0$ in Eq. (13.45) gives

$$V_{CB}(I_C = 0) = V_T \ln\left[1 - \frac{\alpha_N I_E}{I_{CO}}\right] \tag{13.46}$$

For a p-n-p transistor, this voltage is positive because I_{CO} is negative while I_E is positive.

In the active region of operation, V_{CB} is a large negative value, and Eq. (13.45) reduces to

$$I_C = -\alpha_N I_E + I_{CO} \tag{13.47}$$

The characteristics predicted by this relation are shown by the dashed lines in Fig. 13.15. Since I_C is constant and independent of V_{CB}, the slope $dI_C/dV_{CB} = 0$. The characteristics of a real transistor differ from this ideal behavior as shown by the solid curves in Fig. 13.15. Here I_C is seen to increase with increasing reverse bias, and it results in a finite small-signal output conductance dI_C/dV_{CB}. This behavior is a consequence of the Early effect already discussed in Sec. 13.4.2. Finally, at large values of reverse bias, I_C increases further because of the avalanche multiplication of current carriers in the collector-base junction depletion region. This phenomenon eventually leads to avalanche breakdown of the junction at high reverse voltages.

The dynamic output conductance g_{cb} resulting from the Early effect can be readily estimated for the p-n-p transistor shown in Fig. 13.8. For a constant value of I_E, the conductance is given by

$$g_{cb} = \frac{\partial I_C}{\partial V_{CB}} = \frac{\partial I_C}{\partial W_B} \frac{\partial W_B}{\partial V_{CB}}$$

Since $I_C \simeq -\alpha_N I_E$ and $\alpha_N = 1/\cosh(W_B/L_B)$, we obtain

$$g_{cb} = -I_E \frac{\partial \alpha_N}{\partial W_B} \frac{\partial W_B}{\partial V_{CB}} = -\frac{|I_C|}{V_{AF}} \quad (13.48)$$

where

$$V_{AF} = \frac{L_B}{\tanh\left(\dfrac{W_B}{L_B}\right)\left(\dfrac{\partial W_B}{\partial V_{CB}}\right)} \quad (13.49)$$

has the dimensions of voltage. For a uniformly doped base and heavily doped collector region, W_B is given by Eq. (13.40). Since $\tanh(W_B/L_B) \simeq W_B/L_B$ for a narrow base transistor, we obtain

$$V_{AF} = -\frac{L_B^2}{W_B}\left[\frac{2qN_d(V_i - V_{CB})}{\varepsilon_s}\right]^{1/2} \quad (13.50a)$$

and

$$g_{cb} = \frac{|I_C|W_B}{L_B^2}\left[\frac{\varepsilon_s}{2qN_d(V_i - V_{CB})}\right]^{1/2} \quad (13.50b)$$

where V_{CB} is the reverse voltage (negative for a p-n-p transistor). The negative sign in Eq. (13.50a) indicates that W_B decreases as the reverse bias on the collector is increased. It is evident from Eq. (13.50b) that the Early effect conductance varies as $1/V_{CB}^{1/2}$. As an example, consider a silicon p^+-n-p^+ transistor having $N_d = 10^{16}$ cm^{-3} in the base region, with $W_B = 1$ μm and $L_B = 10$ μm. If the transistor is operated at $I_C = -1$ mA and $(V_i - V_{CB}) = 10$ V, then from Eq. (13.50b) we have

$$g_{cb} = \frac{1 \times 10^{-3} \times 1 \times 10^{-4}}{(10 \times 10^{-4})^2}\left[\frac{8.85 \times 10^{-14} \times 11.9}{2 \times 1.6 \times 10^{-19} \times 10^{16} \times 10}\right]^{1/2}$$

which gives $1/g_{cb} = 1.74 \times 10^6$ Ω.

13.5.2 Common Emitter Configuration

A bipolar transistor is more frequently used in the common emitter configuration than in the other two configurations. The input and output characteristics of this configuration are of considerable interest.

(a) INPUT CHARACTERISTICS

The input characteristics of the common emitter configuration represents the plot of the base current I_B as a function of the base-emitter voltage V_{BE}. Since $I_B = -(I_E + I_C)$, substituting the values of I_E and I_C from the two E-M equations and assuming a large reverse bias on the collector-base junction, we obtain the following relation for I_B:

$$I_B = I_{ES}(1 - \alpha_N)[\exp(V_{EB}/V_T) - 1] - I_{CS}(1 - \alpha_I) \quad (13.51)$$

STATIC I-V CHARACTERISTICS IN THE NORMAL ACTIVE REGION

This equation is similar to that of a p-n junction diode except for the term $I_{CS}(1 - \alpha_I)$. For a p-n-p transistor, both I_{ES} and I_{CS} are negative, so that I_B is negative in the normal active region and reaches zero at a small forward voltage

$$V_{EB}(I_B = 0) = V_T \ln\left[1 + \frac{I_{CS}(1 - \alpha_I)}{I_{ES}(1 - \alpha_N)}\right] \quad (13.52)$$

According to Eq. (13.51), I_B appears to be independent of V_{CB}. However, for real transistors this is not true, and I_B decreases with increasing reverse bias on the collector-base junction. This decrease is caused by a decrease in the base width, which reduces the recombination current I_{pr} in the base region.

(b) OUTPUT CHARACTERISTICS

The I_C versus V_{CE} output characterisitics of the common emitter configuration are somewhat more complex than the common base output characteristics. To understand the behavior of a transistor in this configuration, let us consider the p-n-p transistor in Fig. 13.16 whose emitter-base junction is forward biased by a sufficiently large voltage V_{EB}. Let a voltage V_{CE} be applied which biases the collector negative with respect to the emitter, and let this voltage be changed gradually from zero to a large negative value. Since $V_{CE} + V_{BC} + V_{EB} = 0$, we have

$$V_{CE} = (V_{CB} - V_{EB}) \quad (13.53)$$

Initially, when $V_{CE} = 0$, the voltages V_{CB} and V_{EB} are equal, and the collector-base junction is heavily forward biased. In this situation, the collector is injecting holes into the base, and the injected hole current is larger than the fraction $\alpha_N I_E$ of the emitter current I_E which would be collected by the collector. Consequently, the collector current I_C is positive. As V_{CE} is made increasingly negative, V_{CB} becomes less than V_{EB}, and the forward bias is reduced which causes a reduction in the collector current. Eventually, a point is reached where the forward bias on the collector is reduced to the extent that the forward current is just bal-

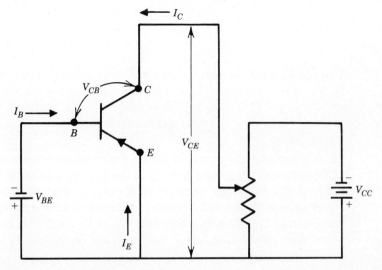

FIGURE 13.16 A p-n-p transistor with forward-biased emitter and gradually increasing negative voltage V_{CE}.

anced by the reverse current component $\alpha_N I_E$, and $I_C = 0$. Making V_{CE} more negative reduces the forward voltage V_{CB} still further, and the reverse current $\alpha_N I_E$ exceeds the forward current so that I_C becomes negative. As long as V_{CB} remains positive, the transistor remains in saturation and I_C increases with bias, but remains less than $-\alpha_N I_E$. At $V_{CE} = -V_{EB}$, the voltage V_{CB} reduces to zero and $I_C = -\alpha_N I_E$. For more negative values of V_{CE}, the transistor operates in the normal active region. Finally, when V_{CE} is a large negative value, the collector-base junction is reverse biased with $V_{CB} \simeq V_{CE}$ and I_C becomes independent of V_{CE}. Thus, the I_C versus V_{CE} characteristic can be divided into a saturation region and a normal active region.

To calculate I_C as a function of V_{CE} in the saturation region, we solve Eq. (13.45) for V_{CB} to obtain

$$V_{CB} = V_T \ln\left[1 - \frac{\alpha_N I_E + I_C}{I_{CO}}\right] \qquad (13.54)$$

Similarly

$$V_{EB} = V_T \ln\left[1 - \frac{\alpha_I I_C + I_E}{I_{EO}}\right] \qquad (13.55)$$

Substituting these values in Eq. (13.53) yields

$$V_{CE} = V_T \ln\left\{\frac{[I_{CO} - \alpha_N I_E - I_C]\alpha_I}{[I_{EO} - \alpha_I I_C - I_E]\alpha_N}\right\} \qquad (13.56)$$

where we have used the relation $I_{EO}/I_{CO} = \alpha_I/\alpha_N$. Substituting $I_E = -(I_C + I_B)$ in Eq. (13.56) and neglecting I_{EO} and I_{CO}, the following expression for V_{CE} is obtained (see Prob. 13.12):

$$V_{CE} = V_T \ln \alpha_I + V_T \ln \frac{\left[1 - \frac{I_C}{I_B}\left(\frac{1 - \alpha_N}{\alpha_N}\right)\right]}{\left[1 + \frac{I_C}{I_B}(1 - \alpha_I)\right]} \qquad (13.57)$$

This equation can be used to calculate V_{CE} as a function of I_C for a given value of I_B. The offset voltage for $I_C = 0$ is obtained from Eq. (13.57):

$$V_{CE}(I_C = 0) = V_T \ln \alpha_I$$

which is a small negative voltage for a p-n-p transistor because α_I is less than unity.

Equation (13.57) is valid only up to the knee of saturation in Fig. 13.17 where $V_{CE} = -V_{EB}$. For larger negative values of V_{CE}, the collector-base junction becomes reverse biased, and I_C tends to saturate to a constant value. When the reverse voltage V_{CE} becomes a few volts $V_{CB} \simeq V_{CE}$, and I_C is given by Eq. (13.47), which after substituting $I_E = -(I_C + I_B)$ can be written as

$$I_C = h_{FE}I_B + (1 + h_{FE})I_{CO} \qquad (13.58)$$

STATIC I-V CHARACTERISTICS IN THE NORMAL ACTIVE REGION

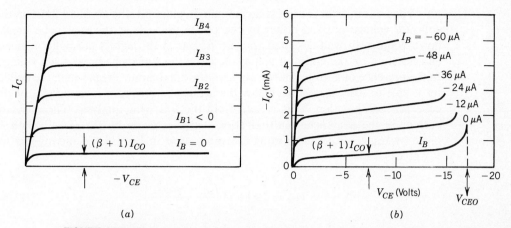

FIGURE 13.17 The common emitter output characteristics of a p-n-p transistor in the normal active mode. (a) Idealized characteristics obtained from Eqs. (13.57) and (13.58) and (b) characteristics of a real transistor.

where

$$h_{FE} = \frac{\alpha_N}{1 - \alpha_N}.$$

Note that Eq. (13.58) is the same as Eq. (13.22) if we write $\alpha = \alpha_N$.

One interesting result from Eq. (13.58) is that, when the base is open-circuited ($I_B = 0$), the current I_C becomes $(h_{FE} + 1)I_{CO}$ instead of the common base saturation current I_{CO}. This happens because the electron current I_{nCO} injected from the collector into the base cannot leave through the base lead. The electrons associated with this current produce a negative charge in the base region. To neutralize this charge, holes from the emitter diffuse into the base and the emitter-base junction becomes slightly forward biased. The electrons then leave through the emitter, and the collector current rises to $(1 + h_{FE})I_{CO}$ because of the hole injection from the emitter into the base.

The transistor characteristics obtained from Eqs. (13.57) and (13.58) are shown in Fig. 13.17a. The measured characteristics of a real transistor are shown in Fig. 13.17b. The finite slope in the characteristics of a real transistor is a consequence of the Early effect. It is evident that the slopes of the $I_C - V_{CE}$ characteristics in Fig. 13.17b are considerably larger than those of the I_C versus V_{CB} characteristics in Fig. 13.15. It can easily be shown that the conductance $g_{ce} = dI_C/dV_{CE}$ is h_{FE} times larger than g_{cb} given by Eq. (13.50b) (see Prob. 13.15).

13.5.3 Limitation on Junction Voltages

In the normal operation of a transistor, the emitter-base junction is forward biased and the voltage across this junction remains less than a volt. In situations where a reverse bias is to be applied to the emitter, the maximum voltage is limited by the breakdown voltage of the junction. In a typical graded-base transistor, this voltage is on the order of 6 to 8 V.

The maximum reverse voltage that can be applied to the collector of a transistor depends on the circuit configuration and operating conditions of the transistor. When a transistor is used in a common base circuit, the absolute limit on the collector voltage is fixed either by the punch-through or the avalanche breakdown voltage. In well-designed transistors, punch-through is normally avoided, and the limit on the collector voltage is set by the avalanche breakdown voltage. In the presence of a finite emitter current, the maximum voltage which the collector is able to sustain remains below the avalanche breakdown voltage V_{Br} of the collector-base junction. This is seen from the following analysis. The collector current in the avalanche multiplication region is given by Eq. (13.42) with M given by Eq. (13.20). Substituting the value of M in Eq. (13.42) and solving for V_{CB} gives

$$V_{CB} = V_{Br}\left[1 - \frac{I_{CO} - \gamma\alpha_T I_E}{I_C}\right]^{1/m} \tag{13.59}$$

The ratio I_{CO}/I_C is essentially zero at breakdown, and the maximum value of V_{CB} is V_{Br} which is obtained for $I_E = 0$. This maximum value is denoted by V_{CBO} and is shown in Fig. 13.18. Note that in the normal active region of operation, the currents I_C and I_E have opposite signs, so that for a finite value of I_E, the voltage V_{CB} in Eq. (13.59) is lower than V_{CBO}. This behavior is evident from the $I_C - V_{CB}$ characteristics shown in Fig. 13.15.

In the common emitter configuration, current amplification combines with avalanche multiplication to impose additional limitations on the collector current, and the voltage that the collector can sustain is lower than V_{CBO}. If we include the effect of avalanche multiplication and write $\alpha_N = \gamma\alpha_T M$, the collector current in Eq. (13.58) can be written as

$$I_C = \frac{\gamma\alpha_T M}{1 - \gamma\alpha_T M}I_B + \frac{I_{CO}}{1 - \gamma\alpha_T M}$$

Substituting the value of M from Eq. (13.20) and solving for $V_{CB} \simeq V_{CE}$ leads to

$$V_{CE} = V_{CBO}\left[1 - \frac{h_{FE}}{1 + h_{FE}}\left(1 + \frac{I_B}{I_C}\right)\right]^{1/m} \tag{13.60}$$

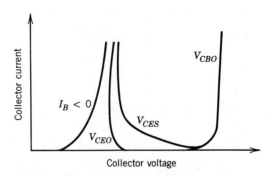

FIGURE 13.18 Collector current versus collector voltage characteristics in the avalanche breakdown region of a transistor for different circuit conditions. (From P. Leturcq, Power junction devices, Chapter 7A in *Handbook on Semiconductors*, vol. 4, T.S. Moss and C. Hilsum (eds.), North-Holland, Amsterdam, p. 884. Copyright © 1981. Reprinted by permission of Elsevier Science Publishers, Physical Science & Engineering Div., and the author.)

where I_{CO} has been neglected, and we have used the relation

$$h_{FE} = \frac{\gamma \alpha_T}{1 - \gamma \alpha_T}$$

as already defined in Eq. (13.24). It is evident from Eq. (13.60) that when the base is open-circuited, $I_B = 0$ and the emitter-collector reverse voltage V_{CE} is given by

$$V_{CEO} = V_{CBO}/(1 + h_{FE})^{1/m} \qquad (13.61)$$

This voltage is considerably lower than the voltage V_{CBO}. For example, for a transistor with $h_{FE} = 80$ and $m = 4$, the voltage V_{CEO} is only 33 percent of V_{CBO}. The voltage V_{CEO} is also shown in Fig. 13.18. The lowering of this voltage at high values of I_C is caused by an increase in h_{FE}.

It is also evident from Eq. (13.60) that V_{CE} is lower than V_{CEO} when the emitter-base junction is forward biased, and I_B has the same sign as the collector current I_C. A case of special interest occurs when the collector-base junction is reverse biased, but the emitter is shorted to the base. The characteristic is then specified by V_{CES} and is also shown in Fig. 13.18. In this case, the emitter-base junction offers a high resistance at low currents, and most of the current flows through the base lead making $V_{CB} \simeq V_{CBO}$. However, as the current through the device is increased, the voltage drop across the internal base resistance r_B forward biases the emitter-base junction so that I_B becomes negligibly small, and the reverse voltage approaches V_{CEO}.

13.6 CHARGE CONTROL EQUATIONS [6]

To this point, we have described transistor currents in terms of junction voltages. We will now use a different approach in which the terminal currents will be expressed in terms of charges stored within the various regions of the device.

Let us consider a p-n-p transistor with the emitter-base junction forward biased and the collector terminal shorted to the base. The excess hole distribution in the base region for this situation is shown by the dashed curve in Fig. 13.19. The excess hole concentration is $p_E = p_{no}[\exp(V_{EB}/V_T) - 1]$ at $x = 0$ and is zero

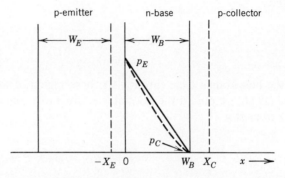

FIGURE 13.19 Excess hole distribution in the base region of a p-n-p transistor under the normal active region of operation. The dashed line shows the actual and the solid line the idealized distribution.

at $x = W_B$. Since very few holes are lost by recombination in the base region, the hole distribution can be approximated by the solid straight line shown in Fig. 13.19. The charge Q_N stored in the quasi-neutral base region in normal operation of the transistor, as calculated from the area of the triangle, is given by

$$Q_N = \frac{qAp_E W_B}{2} \tag{13.62}$$

where A is the junction area. The component I_{pr} of the base current which results from the hole recombination in the base region can now be written as

$$I_{pr} = \frac{Q_N}{\tau_B} \tag{13.63}$$

where τ_B represents the hole lifetime in the base region. The other components of the base current are the electron current I_{nE} injected into the emitter and the current I_{CO} flowing from the base to collector. If we neglect the component I_{CO}, the base current I_B is given by

$$-I_{BN} = I_{pr} + I_{nE} = \frac{Q_N}{\tau_B}\left(1 + \frac{I_{nE}}{I_{pr}}\right) = \frac{Q_N}{\tau_{BN}} \tag{13.64}$$

where

$$\tau_{BN} = \frac{\tau_B}{1 + I_{nE}/I_{pr}} \tag{13.65}$$

Note that in Eq. (13.64) I_{BN} is written as a negative current because both I_{pr} and I_{nE} flow from the emitter to base and, thus, are leaving the base terminal. From Eq. (13.65), we see that when I_{nE} is small compared to I_{pr}, $\tau_{BN} \simeq \tau_B$.

Let us now determine the collector current I_C. From the hole distribution in Fig. 13.19, we obtain

$$-I_{CN} = \frac{qAD_B p_E}{W_B} = \frac{qAW_B p_E}{2}\left(\frac{2D_B}{W_B^2}\right) = \frac{Q_N}{\tau_N} \tag{13.66}$$

where

$$\tau_N = \frac{W_B^2}{2D_B} \tag{13.67}$$

is the hole transit time through the base region of width W_B. From Eqs. (13.64) and (13.66) we obtain the normal (i.e., forward) common emitter current gain of the transistor

$$\frac{I_{CN}}{I_{BN}} = h_{FE} = \beta_N = \frac{\tau_{BN}}{\tau_N} \tag{13.68}$$

Since τ_{BN} is given by Eq. (13.65), we see that if I_{nE} is negligible compared to I_{pr}, then β_N is simply the ratio of the hole lifetime τ_B to the hole transit time τ_N

through the base region. Thus, it is evident that the current gain β_N can be increased by increasing the ratio τ_B/τ_N.

Equations (13.64) and (13.66) represent the static situation in which the junction voltages and the terminal currents are invariant with time. Let us now extend the charge control equations to time-varying situations. We will denote the instantaneous values of terminal voltages and currents by lower case variables with upper case subscripts. For example, the instantaneous emitter-base voltage will be denoted by v_{EB}, the instantaneous collector current by i_C and so on. In the p-n-p transistor under discussion, when the voltage v_{EB} varies with time, the stored charge of holes Q_N in the base region will also vary with time. Clearly, if Q_N increases with time, there will be a component of base current equal to dQ_N/dt, and the instantaneous base-current i_{BN} can be written as

$$-i_{BN} = \frac{Q_N}{\tau_{BN}} + \frac{dQ_N}{dt} \qquad (13.69a)$$

In writing the expression for the collector current i_C, we assume that the time rate of change of terminal voltages and currents is so small that the so-called quasi-static approximation holds. This means that the hole distribution in the quasi-neutral base region at any instant of time corresponds to a steady-state situation coinciding with that time. This implies that the triangular shape of the excess hole distribution similar to that of Fig. 13.19 is maintained at all times. When this is the case, the collector current at any time will depend only on the slope of the excess hole distribution, and hence, only on the charge Q_N. Thus, following Eq. (13.66) the instantaneous collector current i_{CN} is given by

$$-i_{CN} = \frac{Q_N}{\tau_N} \qquad (13.69b)$$

Note that in Eqs. (13.69a) and (13.69b) the subscript N denotes the normal operation in which the emitter-base junction is forward biased and $v_{CB} = 0$. The emitter current i_{EN} is obtained from the relation $i_{EN} = -(i_{BN} + i_{CN})$, giving

$$i_{EN} = \frac{dQ_N}{dt} + \frac{Q_N}{\tau_{BN}} + \frac{Q_N}{\tau_N} \qquad (13.69c)$$

We now use the principle of superposition that was discussed in Sec. 13.3.1. Since the hole concentration in the base region is a linear combination of the two boundary values p_E and p_C (see Eq. 13.11), the total hole charge stored in the quasi-neutral base region can be resolved into the normal charge Q_N given by Eq. (13.62) and an inverse charge Q_I given by the relation

$$Q_I = \frac{qAW_B p_C}{2} \qquad (13.70)$$

where p_C is given by Eq. (13.10). The various inverse current components associated with Q_I can be written by interchanging the roles of the collector and the emitter. Thus

$$-i_{BI} = \frac{Q_I}{\tau_{BI}} + \frac{dQ_I}{dt} \qquad (13.71a)$$

$$-i_{EI} = Q_I/\tau_I \tag{13.71b}$$

and

$$i_{CI} = \frac{dQ_I}{dt} + \frac{Q_I}{\tau_{BI}} + \frac{Q_I}{\tau_I} \tag{13.71c}$$

where τ_I is the hole transit time through the base region in the inverse operation, and τ_{BI} is a time constant similar to τ_{BN}. The total instantaneous terminal currents are obtained by the addition of the normal and inverse components. However, any dynamic model should also account for the components of the terminal currents resulting from the change in charge in the depletion regions of the two junctions. These components of the terminal currents can be taken into account by adding two terms, namely, dQ_{VE}/dt and dQ_{VC}/dt, in the expression for the base current. We thus write

$$i_B = i_{BN} + i_{BI} = -\frac{Q_N}{\tau_{BN}} - \frac{dQ_N}{dt} - \frac{Q_I}{\tau_{BI}} - \frac{dQ_I}{dt} - \frac{dQ_{VE}}{dt} - \frac{dQ_{VC}}{dt} \tag{13.72a}$$

$$i_C = i_{CN} + i_{CI} = -\frac{Q_N}{\tau_N} + \frac{dQ_I}{dt} + Q_I\left(\frac{1}{\tau_I} + \frac{1}{\tau_{BI}}\right) + \frac{dQ_{VC}}{dt} \tag{13.72b}$$

$$i_E = i_{EN} + i_{EI} = \frac{dQ_N}{dt} + Q_N\left(\frac{1}{\tau_N} + \frac{1}{\tau_{BN}}\right) - \frac{Q_I}{\tau_I} + \frac{dQ_{VE}}{dt} \tag{13.72c}$$

In these equations dQ_{VE}/dt represents the current that charges the emitter-base junction depletion region, and dQ_{VC}/dt has the same significance for the collector-base junction depletion region. Equations (13.72) a, b, and c are the charge control equations for a p-n-p transistor. The same equations can be used for an n-p-n transistor by changing the signs of the base, emitter, and collector currents. We will find important applications of these equations in studying the dynamic behavior of transistors in the next chapter and in obtaining circuit models of bipolar transistor in Chapter 17.

Note that the charge control equations are based on the quasi-static approximation that assumes the existence of direct proportionality between the quantity of charge and current. This situation implies that the time constants entering these equations are independent of the stored charge or the junction voltages. Such a simple assumption is justified only when the time for which the dynamic response of the device being studied is large compared to the base transit times (τ_N and τ_I). When the time scale of interest is comparable to τ_N (or τ_I), the quasi-static approximation is no longer valid, and the charge control equations will fail to predict the correct behavior of the transistor.

REFERENCES

1. J. J. EBERS and J. L. MOLL, "Large-signal behavior of junction transistors," *Proc. IRE* 42, 1761 (1954).

2. H. J. DE MAN, M.Y. GHANNAM, and R. P. MERTENS, "Mathematical proof of the validity of reciprocity in one-dimensional bipolar transistors with arbitrary base parameters," *IEEE Trans. Electron Devices* ED-31, 1720 (1984).

3. J. M. EARLY, "Effects of space-charge layer widening in junction transistors," *Proc. IRE* 40, 1401 (1952).

4. R. M. WARNER and J. N. FORDEMWALT, *Integrated Circuits,* McGraw-Hill, New York, 1965, pp. 108–109.

5. F. HEBERT and D. J. ROULSTON, "Base resistance of bipolar transistors from layout details including two dimensional effects at low currents and low frequencies," *Solid-State Electron.* 31, 283 (1988).

6. J. TE WINKEL, "Past and present of the charge-control concept in the characterization of the bipolar transistor," in *Advances in Electronics and Electron Physics,* L. Marton (ed.), Academic Press, New York, Vol. 39, 253 (1975).

ADDITIONAL READING LIST

1. J. L. MOLL, *Physics of Semiconductors,* McGraw-Hill, New York, 1964, Chapter 8.
2. P. E. GRAY and C. L. SEARLE, *Electronic Principles: Physics, Models, and Circuits,* John Wiley, New York, 1969, Chapters 7 and 8.
3. R. L. PRITCHARD, *Electrical Characteristics of Transistors,* McGraw-Hill, New York, 1967.
4. R. S. MULLER and T. I. KAMINS, *Device Electronics for Integrated Circuits,* 2d ed., John Wiley, New York, 1986, Chapters 6 and 7.
5. W. SHOCKLEY, "The path to the conception of the junction transistor," *IEEE Trans. Electron Devices* ED-31, Centennial special issue, 1523–1546 (1984).

PROBLEMS

13.1 Draw the energy band diagrams for an n-p-n transistor when it is biased in (a) the saturation region and (b) the cutoff region. In both cases draw the majority and the minority carrier distributions in the emitter, base, and collector regions.

13.2 Consider a p-n-p transistor biased in the normal active region of operation at room temperature. In this situation, both I_B and I_C are negative. Now if I_C is held constant and the temperature is raised gradually, I_B will decrease and ultimately become positive. Explain this behavior in terms of physical phenomena that occur in the device.

13.3 A symmetrical Ge p-n-p transistor with emitter-base and collector-base junctions, each 1 mm in diameter, has an impurity concentration of 5×10^{15} cm^{-3} in the base and 10^{18} cm^{-3} in the emitter and the collector. The base-width is 10 μm, $\tau_B = 4 \times 10^{-6}$ sec, $\tau_E = 10^{-8}$ sec, and the emitter region is much longer than the diffusion length L_E. Calculate the current gains α and h_{FE} of the transistor. Take $D_B = 47$ cm^2 sec^{-1} and $D_E = 52$ cm^2 sec^{-1}.

13.4 Consider a p-n-p transistor with uniformly doped emitter, base, and collector regions. If $h_{FE}(\gamma)$ denotes the common emitter current gain when α_T is unity, and $h_{FE}(\alpha_T)$ denotes the same when γ is unity, show that when both α_T and γ are nonunity, the current gain h_{FE} is given by $(h_{FE})^{-1} = [h_{FE}(\alpha_T)]^{-1} + [h_{FE}(\gamma)]^{-1}$.

13.5 A Si n-p-n transistor has the following parameters at 300 K: $N_a = 5 \times 10^{16}$ cm^{-3}, $N_d(E) = 1 \times 10^{18}$ cm^{-3}, $W_B = 2$ μm, $W_E = 0.2$ μm, $\mu_B = 1000$ cm^2 V^{-1} sec^{-1}, $\mu_p(E) = 150$ cm^2 V^{-1} sec^{-1}, $\tau_B = 10^{-6}$ sec, and $\tau_E = 10^{-8}$ sec. The emitter-base junction area is 0.01 cm^2, $I_E = 1$ mA, and the collector-base junction is reverse biased by 2 V. Neglect carrier generation and recombination in the two junction depletion regions. (a) Calculate the emitter-base junction voltage and the excess electron concentration in the base at the edge of the emitter-base junction depletion region. (b) Calculate γ, α_T, and h_{FE} for the transistor.

13.6 A p-n-p transistor with uniformly doped base, emitter, and collector regions has $I_E = 1.2$ mA. Sketch the minority carrier distribution in the base when (a) the collector is shorted to the base, (b) the collector is shorted to the emitter, and (c) the collector terminal is kept open.

13.7 Differentiating Eq. (13.37) with respect to V_{EB}, obtain Eq. (13.38).

13.8 A Si n$^+$-p-n$^+$ power transistor has a base-width of 150 μm and a base doping of 10^{14} cm^{-3}. A reverse bias of 500 V is applied to the collector-base junction. Calculate the minority carrier lifetime in the base region that will give a common emitter current gain of 40.

13.9 The punch-through voltage of a Ge alloyed p-n-p transistor is 25 V, the base doping is 10^{15} cm^{-3}, and the emitter and the collector dopant concentrations are 10^{19} cm^{-3} each. Calculate the zero-bias base width and α of the transistor at a 10 V reverse bias across the collector-base junction. Assume $\tau_B = 10^{-6}$ sec and $T = 300$ K.

13.10 The emitter and collector regions of a Si alloyed p-n-p transistor are heavily doped, and the impurity concentration in the base is 10^{15} cm^{-3}. Calculate the base-width that will make the avalanche breakdown voltage equal to the punch-through voltage. Assume that avalanche breakdown occurs when the maximum field strength in the depletion region becomes 5×10^5 V/cm. What minority carrier diffusion length and lifetime in the base are required to obtain a value of $\alpha = 0.95$? Assume $\gamma = 0.99$ and a reverse bias of 15 V at the collector-base junction.

13.11 An n-p-n transistor with uniformly doped emitter, base, and collector regions has a collector-base junction punch-through voltage of 60 V, and $h_{FE} = 40$ at a reverse bias of 5 V. Determine the reverse-bias voltage at which h_{FE} becomes 80.

13.12 Using Eq. (13.56) and making appropriate approximations, show that for a transistor operating in the saturation region, V_{CE} is given by Eq. (13.57). Calculate the value of V_{CE} for $\alpha_N = 0.985$, $\alpha_I = 0.72$, and $I_C/I_B = 10$. Assume $T = 300$ K.

13.13 The emitter current of a p-n-p transistor with $\alpha_N = \alpha_I$ is 0.5 mA when the emitter-base junction is forward biased and the collector is left open. When the collector is shorted to the base, the current rises to 25 mA. Calculate h_{FE} and the base-width of the transistor assuming a minority carrier diffusion length of 20 μm in the base and the emitter efficiency to be unity.

PROBLEMS

13.14 A Ge alloyed p-n-p transistor has $I_{ES} = -2\ \mu A$, $I_{CS} = -3\ \mu A$, and $\alpha = 0.95$. The transistor is connected to a 5 V battery in series with a 1 KΩ resistor such that the positive of the battery is connected to the emitter, the negative to the collector, and the base is open-circuited. Calculate the current through the circuit and the voltage drops across each of the two junctions.

13.15 Show that the conductance $g_{ce} = \partial I_C/\partial V_{CE}$ in the normal active region is h_{FE} times larger than the conductance g_{cb} given by Eq. (13.50b).

13.16 The reverse breakdown voltage V_{CBO} of a p-n-p transistor is 70 V and $h_{FE} = 50$. The transistor is operated in the common base configuration with an emitter current of 1 mA. Taking $m = 4$, calculate the collector-base junction reverse voltage at which the collector current becomes -5 mA.

13.17 Consider the transistor in Prob. 13.16 to be operated in a common emitter configuration with $I_C = 80 I_B$. Calculate the value of V_{CE}. Also determine the voltage V_{CEO}.

13.18 A p-n-p transistor has the following parameters: $\tau_N = 1.2 \times 10^{-8}$ sec, $\tau_I = 3.6 \times 10^{-8}$ sec, $\beta_N = 100$, and $\beta_I = 10$. Evaluate the steady-state base charge components Q_N and Q_I if $I_B = -0.5$ mA and $I_C = -2$ mA. Determine the signs of the bias voltages V_{CB} and V_{EB}.

14
BIPOLAR JUNCTION TRANSISTORS II: DEVICES

INTRODUCTION

In the previous chapter, our treatment of the bipolar transistor was based on a diffusion transistor with abrupt junctions and uniformly doped emitter, base, and collector regions. As mentioned in Sec. 13.2, modern-day transistors are produced by impurity diffusion process and have emitter and base regions with graded impurity distributions. These devices have a built-in electric field in the base that results in the drift of minority carriers in addition to the usual diffusion transport.

In this chapter, we will first point out the high-frequency limitations of the uniformly doped base transistor. Next, we will discuss the drift transistor in some detail to explain how impurity grading in the base improves the high-frequency performance of the device. This is followed by a brief account of some common types of transistors, their requirements, structures, and design considerations. In our discussion, we are concerned mainly with high-frequency, high-power, and high-speed switching transistors.

14.1 THE DIFFUSION TRANSISTOR AT HIGH FREQUENCIES

One important function of a transistor is voltage amplification. The device parameter that ultimately limits the voltage gain of an amplifier is the small-signal current gain $\tilde{\alpha}$ of the transistor. A first-order approximation for the frequency dependence of $\tilde{\alpha}$ can be obtained from the charge control equations (Sec. 13.6). Let us assume that the total charge Q_N stored in the normal operation can be expressed as

$$Q_N = \bar{Q}_N + q_N \tag{14.1}$$

where $\bar{Q}_N$ is the charge associated with the dc bias and q_N represents the charge associated with the small-signal ac voltage that is superimposed on the dc bias. We further assume that

$$q_N = q_{NO} \exp(j\omega t) \tag{14.2}$$

Now consider a p-n-p transistor in the normal operating region with a sufficiently large bias on the collector. For this situation $Q_I \approx 0$, and neglecting the depletion region charges Q_{VE} and Q_{VC}, from Eq. (13.72) we can write

$$i_E = I_E + i_e = \frac{\bar{Q}_N + q_N}{\tau_N^*} + \frac{d}{dt}(\bar{Q}_N + q_N) \tag{14.3a}$$

and

$$i_C = I_C + i_c = -\frac{\bar{Q}_N + q_N}{\tau_N} \tag{14.3b}$$

Here I_E and I_C represent the emitter and collector currents at the bias point, i_e and i_c are the respective small-signal ac currents, and

$$\tau_N^* = \left(\frac{1}{\tau_N} + \frac{1}{\tau_{BN}}\right)^{-1} \tag{14.4}$$

In the steady state $d\bar{Q}_N/dt = 0$, and separating out the ac components, we obtain

$$i_e = \frac{q_N}{\tau_N^*}(1 + j\omega\tau_N^*) \tag{14.5}$$

$$i_c = -\frac{q_N}{\tau_N}$$

and

$$\tilde{\alpha}(\omega) = -\frac{i_c}{i_e} = \frac{\alpha_o}{1 + j\omega/\omega_\alpha} \equiv \frac{\alpha_o}{1 + jf/f_\alpha} \tag{14.6}$$

where

$$\alpha_o = \frac{\tau_N^*}{\tau_N} = \frac{\tau_{BN}}{\tau_N + \tau_{BN}} \tag{14.7a}$$

and

$$\omega_\alpha = \frac{1}{\tau_N} + \frac{1}{\tau_{BN}} \simeq \frac{1}{\tau_N} \tag{14.7b}$$

since in a transistor, τ_{BN} is large compared to τ_N. The frequency $f_\alpha = \omega_\alpha/2\pi$ is known as the *α-cutoff frequency* of the transistor. It is evident from Eq. (14.5) that the emitter current i_e consists of two components: a conductive component q_N/τ_N^*

and a capacitive component associated with the change in the stored charge q_N. The two components become equal in magnitude at the α-cutoff frequency f_α.

From Eq. (14.7b) it is seen that ω_α can be increased by reducing the normal transit time τ_N. This is understandable because the phase difference between the emitter and collector currents arises from the fact that the minority carriers injected from the emitter require a time τ_N to reach the collector. When the time period of the ac signal becomes comparable to τ_N, the emitter and collector currents will be out of phase, making $\tilde{\alpha}$ a complex quantity. Since from Eq. (13.67) $\tau_N = W_B^2/2D_B$, we can reduce τ_N by decreasing W_B and using semiconductors with a high value of carrier mobility. Since the electron mobility is larger than the hole mobility in Ge, Si, and GaAs, for the same value of W_B, an n-p-n transistor will have a larger value of ω_α compared to an analogous p-n-p transistor. This is one reason why n-p-n transistors are preferred over p-n-p transistors.

In the above discussion we have assumed that the minority carriers move through the base only by diffusion. In a graded-base transistor, the built-in drift field accelerates the injected minority carriers toward the collector so that their transit time through the base is reduced, and the frequency f_α is increased.

14.2 THE DRIFT TRANSISTOR

14.2.1 Current-Voltage Characteristics

A typical doping profile for an n-p-n graded-base transistor is shown is Fig. 13.7. For a p-n-p graded base transistor, the profile will look like the one shown in Fig. 14.1a. The dopant concentration in the base resulting from two successive diffusion steps has a maximum at some point x_1. The holes injected from the emitter are subjected to a retarding field to the left of x_1 and to an accelerating field in the remainder of the base region. In most cases, the retarding field region is less than 20 percent of the total base width W_B, and the effect of the retarding field is ignored in the calculations. The idealized impurity profile for which calculations are made is shown in Fig. 14.1b. Here the base is assumed to be exponentially graded, and this results in a constant electric field in the base. This assumption considerably simplifies the calculations.

We will first present calculations for a general case in which the exact form of $N_d(x)$ in the base is not known. An important observation about the currents in the device in Fig. 14.1 is that, in the normal active region of operation, the electron current remains negligibly small. This is because the emitter is heavily doped compared to the base so that the electron current density J_{nE} is negligible compared to the hole current density J_{pE}. In addition, there is negligible electron flow between the base and the collector. Consequently, we can write

$$J_{nE} = q\mu_n n_n \mathcal{E}(x) + qD_n \frac{dn_n}{dx} = 0$$

or

$$\mathcal{E}(x) = -\frac{D_n}{\mu_n}\frac{1}{n_n}\frac{dn_n}{dx} = -\frac{kT}{q}\frac{1}{n_n}\frac{dn_n}{dx} \qquad (14.8)$$

Here D_n and μ_n represent the diffusion constant and mobility of the electrons in the base region, and n_n is the electron concentration in this region. Note that

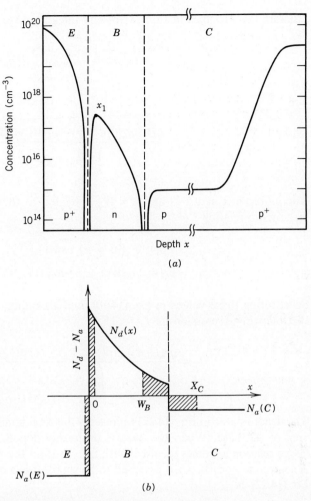

FIGURE 14.1 Impurity profile of a graded-base p-n-p transistor. (*a*) Actual profile resulting after a two-step diffusion process and (*b*) idealized profile with exponentially decaying donor concentration in the base.

Eq. (14.8) is valid under thermal equilibrium, as well as under low-level injection, as long as $J_{nE} \ll J_{pE}$.

The hole current density $J_p(x)$ in the base can be written as

$$J_p(x) = q\mu_B p_n \mathscr{E}(x) - qD_B \frac{dp_n}{dx} \tag{14.9a}$$

where $p_n = p_{no} + p_e$ is the total hole concentration in the base. Substituting for $\mathscr{E}(x)$ from Eq. (14.8) into Eq. (14.9a) and writing $D_B = (kT/q)\mu_B$, we obtain

$$J_p(x) = -\frac{qD_B}{n_n(x)} \frac{d}{dx}(p_n n_n) \tag{14.9b}$$

In a high-performance transistor, minority carrier recombination in the base may be ignored, and $J_p(x)$ may be taken as a constant (independent of x). We can now separate the variables in Eq. (14.9b) and integrate over the base-width W_B

$$J_p \int_0^{W_B} n_n(x)\, dx = -qD_B \int_0^{W_B} d(p_n n_n)$$

and

$$J_p = \frac{qD_B[p_n(0)n_n(0) - p_n(W_B)n_n(W_B)]}{\int_0^{W_B} n_n(x)\, dx} \qquad (14.10)$$

The pn-product at $x = 0$ and $x = W_B$ is related to the junction voltages by the Boltzmann relation

$$p_n(0)n_n(0) = n_i^2 \exp(V_{EB}/V_T) \qquad (14.11)$$
$$p_n(W_B)n_n(W_B) = n_i^2 \exp(V_{CB}/V_T)$$

Substituting these values in Eq. (14.10), and assuming $n_n(x) = N_d(x)$, we obtain the following expression for $I_p = AJ_p$:

$$I_p = \frac{qAD_B n_i^2 [\exp(V_{EB}/V_T) - \exp(V_{CB}/V_T)]}{\int_0^{W_B} N_d(x)\, dx} \qquad (14.12)$$

The denominator in Eq. (14.12) represents the total number of impurities per unit area of the base. When the base is uniformly doped, N_d is constant, and the above relation becomes identical with Eq. (13.15) if the term corresponding to I_{nE} is neglected, and W_B/L_B is assumed to be small. It is a common practice to write Eq. (14.12) as

$$I_p = \frac{qA n_i^2 [\exp(V_{EB}/V_T) - \exp(V_{CB}/V_T)]}{G_B} \qquad (14.13)$$

where

$$G_B = \int_0^{W_B} (N_d(x)/D_B)\, dx \qquad (14.14)$$

is called the *base Gummel number* [1]. Note that we have treated D_B as constant, although it varies with the dopant concentration in the base. The emitter current I_{pE} is obtained by substituting $V_{CB} = 0$ in Eq. (14.13), giving

$$I_{pE} = \frac{qA n_i^2}{G_B} [\exp(V_{EB}/V_T) - 1] \qquad (14.15)$$

14.2.2 Calculation of Device Parameters

In the derivation of Eq. (14.13), the base transport factor has been assumed to be unity. The emitter current I_{nE} may be calculated in a similar manner as Eq. (14.15) and can be written as

$$I_{nE} = \frac{qAn_i^2}{G_E}[\exp(V_{EB}/V_T) - 1] \tag{14.16}$$

The emitter Gummel number G_E will depend on the doping profile in the emitter, and the emitter region width W_E. For a thin emitter with negligible recombination and perfect ohmic end-contact, G_E will have the same form as G_B. Thus

$$G_E = \int_0^{W_E} \frac{N_a(x)}{D_E} dx$$

It is evident that G_E will be small in thin emitters. However, it can be increased by reducing the surface recombination velocity S at the ohmic contact. One way to accomplish this objective is to use polysilicon emitters [2]. In this technology, a heavily arsenic-doped polysilicon layer is deposited on the top of the base where the emitter is to be constructed. This polysilicon layer is then used as a diffusion source to create an extremely shallow emitter of an n-p-n transistor. The emitter-polysilicon interface is characterized by a low value of S, and this feature causes a reduction in the base current of minority carriers injected into the emitter (see Prob. 7.4).

To obtain an analytical expression for G_E, we assume the emitter region concentration N_a to be uniform. For the perfect ohmic contact condition, G_E is given by

$$G_E = \frac{N_a L_E}{D_E} \coth\left|\frac{W_E}{L_E}\right| \tag{14.17a}$$

where D_E is the minority carrier diffusion coefficient in the emitter and L_E is the diffusion length. For a wide emitter with $|W_E| \gg L_E$, we have

$$G_E = \frac{N_a L_E}{D_E} \tag{14.17b}$$

The total emitter current is obtained from Eqs. (14.15) and (14.16). Thus,

$$I_E = qAn_i^2\left(\frac{1}{G_B} + \frac{1}{G_E}\right)[\exp(V_{EB}/V_T) - 1]$$
$$= -I_{ES}[\exp(V_{EB}/V_T) - 1] \tag{14.18}$$

which is identical to the Ebers-Moll equation (13.25) when $V_{CB} = 0$.

The emitter efficiency γ is obtained from Eqs. (14.15) and (14.16) giving

$$\gamma = \frac{1}{1 + \dfrac{I_{nE}}{I_{pE}}} = \frac{1}{1 + \dfrac{G_B}{G_E}} \tag{14.19}$$

and the common emitter current gain becomes

$$h_{FE} \approx \frac{\gamma}{1-\gamma} = \frac{G_E}{G_B} \tag{14.20}$$

A large value of h_{FE} can be obained by making G_E large compared to G_B.

14.2.3 Excess Carrier Distribution in the Base Region

The excess hole distribution in the base can be obtained from Eq. (14.9b). Regarding J_p as a constant and writing $n_n = N_d$ and $p_n = p_e$, we obtain

$$N_d \frac{dp_e}{dx} + p_e \frac{dN_d}{dx} + \frac{I_p}{qAD_B} N_d = 0$$

which can be solved for $p_e(x)$ to yield

$$p_e(x) = \frac{I_p}{qAD_B} \frac{\int_x^{W_B} N_d(x)\, dx}{N_d(x)} \tag{14.21}$$

This equation can be used to obtain $p_e(x)$ at any location x for a given value of I_p when $N_d(x)$ is known. As an example, we consider the specific case of the exponential impurity profile shown in Fig. 14.1b. Let us write

$$N_d(x) = N_d(0) \exp(-\eta x/W_B) \tag{14.22}$$

Note that the grading factor η is given by

$$\eta = \ln \frac{N_d(0)}{N_d(W_B)} \tag{14.23}$$

Substituting $n_n = N_d$ in Eq. (14.8) and using Eq. (14.22), the field $\mathscr{E}(x)$ in the base region can be expressed as

$$\mathscr{E}(x) = \frac{kT}{q} \frac{\eta}{W_B} \tag{14.24}$$

which is constant independent of x. Substituting for $N_d(x)$ from Eq. (14.22) into Eq. (14.21) gives

$$p_e(x) = \left(\frac{I_p W_B}{qAD_B}\right) \left\{\frac{1 - \exp[-\eta(1 - x/W_B)]}{\eta}\right\} \tag{14.25}$$

For known values of I_p and η, the concentration $p_e(x)$ can be obtained at any desired depth in the base region.

The base transit time τ_N is given by

$$\tau_N = \int_0^{W_B} \frac{dx}{v_p(x)} = \frac{qA}{I_p} \int_0^{W_B} p_e(x)\, dx$$

and substituting for $p_e(x)$ from Eq. (14.25), we obtain the following expression for τ_N (see Prob. 14.3):

$$\tau_N = \frac{W_B^2}{D_B} \left\{ \frac{\eta + \exp(-\eta) - 1}{\eta^2} \right\} \qquad (14.26)$$

Note that for $\eta = 0$ this expression reduces to $\tau_N = W_B^2/2D_B$.

14.2.4 Heavy Doping Effects in the Emitter

It appears from Eq. (14.19) that the emitter efficiency can be made to approach unity and h_{FE} in Eq. (14.20) can be made very large by increasing the emitter Gummel number G_E. This, however, is not true because, as the doping in the emitter becomes very high, two additional effects have to be considered: the band gap narrowing effect, and the decrease in the minority carrier lifetime [3].

Band gap narrowing was considered in Sec. 3.8. There we noted that the electrical band gap narrowing is the relevant parameter for device design. It causes an increase in the effective intrinsic concentration n_{ie} which is related to n_i by Eq. (3.59). A plot of n_{ie} as a function of dopant concentration is given in Fig. 3.18.

The decrease in the minority carrier lifetime is caused by generation of additional lattice defects at heavy doping, as well as by Auger recombination. The effective minority carrier lifetime is obtained by the reciprocal addition of the SRH lifetime given by Eq. (5.34) or Eq. (5.35) and the Auger lifetime τ_A given by Eq. (5.41). As the dopant concentration is increased, the concentration of recombination centers N_t increases, and this causes a decrease in the SRH lifetime. At the same time the Auger lifetime is also decreased. Both effects cause a sharp decrease in the minority carrier lifetime at heavy doping.

Band gap narrowing increases the minority carrier concentration in the emitter and hence lowers the emitter Gummel number. The reduction in the carrier lifetime causes a decrease in the minority carrier diffusion length from L_E to some lower value L_E'. As a result, the Gummel number in the emitter is reduced to a new value G_E' given by the relation

$$G_E' = G_E \frac{L_E'}{L_E} \left(\frac{n_i}{n_{ie}}\right)^2 = G_E \frac{L_E'}{L_E} \exp\left(-\frac{\Delta E_g^{\text{elect}}}{kT}\right) \qquad (14.27)$$

Since $L_E' < L_E$ and $n_{ie} > n_i$, an important conclusion can be drawn from Eq. (14.27). If the dopant concentration in the emitter region is increased beyond a certain level, the emitter Gummel number does not increase proportionately but rather levels off. It even starts decreasing if the reduction in the ratio (n_i/n_{ie}) and the decrease in L_E' have a more pronounced effect than the increase in the dopant concentration. Figure 14.2 shows the calculated values of the emitter Gummel number as a function of dopant impurity concentration [4] in a uniformly doped emitter of an n-p-n transistor for two different values of SRH lifetime. The results of some measurements on commercial transistors are also shown (see the hatched area). It is seen that G_E' reaches its maximum value for dopant concentrations between 10^{18} and 10^{19} cm^{-3}, and starts decreasing thereafter. Thus, the band gap reduction and Auger recombination put a fundamental limit on h_{FE} which is reduced when the emitter is very heavily doped.

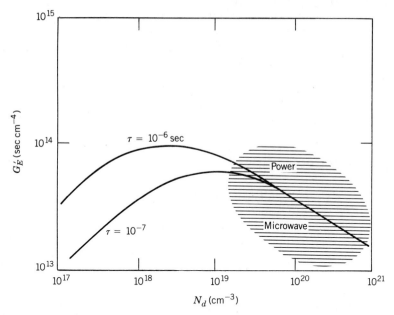

FIGURE 14.2 Emitter Gummel number as a function of dopant concentration in the uniformly doped emitter of an n-p-n transistor. (From P. A. H. Hart, reference 4, p. 114. Copyright © 1981. Reprinted by permission of Elsevier Science Publishers, Physical Science and Engineering Div., and the author.)

Variation in h_{FE} with temperature is also of some interest. When heavy doping effects are absent in the emitter, the change in h_{FE} results from an increase in the diffusion length in the base and the emitter regions with temperature. However, in the presence of band gap narrowing, a larger increase in h_{FE} occurs with temperature because of the exponential factor in Eq. (14.27).

14.3 HIGH-FREQUENCY PERFORMANCE

14.3.1 The Alpha and Beta Cutoff Frequencies

For a transistor with a uniformly doped base region, the frequency dependence of $\tilde{\alpha}$ is given by Eq. (14.6) with $\omega_\alpha = 2D_B/W_B^2$. It is evident from Eq. (14.6) that at $\omega = \omega_\alpha$, the phase angle is 45° and $|\tilde{\alpha}| = \alpha_o/\sqrt{2}$. These predictions are not in agreement with the measured results.

A better approximation for $\tilde{\alpha}(\omega)$ can be obtained from the solution of the time-dependent continuity equation for minority carriers in the base. At all frequencies of interest, the frequency dependence of $\tilde{\alpha}$ is determined by that of the transport factor α_T. Since $\alpha_T = \text{sech}(W_B/L_B)$, the frequency dependence of $\tilde{\alpha}$ is obtained by replacing the base diffusion length L_B by its complex value $L_B/\sqrt{1 + j\omega\tau_B}$, where τ_B is the minority carrier lifetime in the base region. Thus,

$$\tilde{\alpha}(\omega) = \text{sech}\left[\frac{W_B}{L_B}\sqrt{1 + j\omega\tau_B}\right] \quad (14.28)$$

When this expression is expanded in a power series, and only the first three terms are retained, we obtain Eq. (14.6).

It can be seen from Eq. (14.28) that the relation $|\tilde{\alpha}| = \alpha_o/\sqrt{2}$ is obtained at a frequency higher than $2D_B/W_B^2$, and the phase angle at this new frequency is 58° instead of 45°. These features are incorporated by defining an excess phase factor m and writing

$$\tilde{\alpha}(\omega) = \frac{\alpha_o}{1 + j\omega/\omega_\alpha} \exp\left(-j\frac{m\omega}{\omega_\alpha}\right) \qquad (14.29)$$

This equation is also valid for a graded-base transistor with a constant field in the base, and the values of m and ω_α are given by [5]

$$m = (0.22 + 0.1\eta)$$

and

$$\omega_\alpha = \frac{2.43 D_B}{W_B^2}[1 + (\eta/2)^{4/3}] \qquad (14.30)$$

The case for a uniformly doped base transistor is obtained for $\eta = 0$. Note that ω_α can be increased considerably by increasing the grading factor η. The practical values of η are in the range 2 to 5 and seldom exceed 8 because of technological limitations [4].

In most cases, a transistor is used for frequencies that are low compared to ω_α, and for this situation, Eq. (14.6) is fairly accurate. Defining the common emitter current gain $\beta(\omega)$ by the relation

$$\beta(\omega) = \frac{\tilde{\alpha}(\omega)}{1 - \tilde{\alpha}(\omega)}$$

and substituting for $\tilde{\alpha}(\omega)$ from Eq. (14.6), we obtain

$$\beta(\omega) = \frac{\beta_o}{1 + jf/f_\beta} \qquad (14.31)$$

where

$$\beta_o = \frac{\alpha_o}{1 - \alpha_o} \qquad (14.32a)$$

and

$$f_\beta = f_\alpha(1 - \alpha_o) \qquad (14.32b)$$

is called the β-cutoff frequency. From these two relations, it follows that

$$\alpha_o f_\alpha = \beta_o f_\beta \qquad (14.33)$$

This is an important result and shows that, for a transistor with a given geometry and impurity profile, the product of the current gain and the cutoff frequency is an intrinsic property of the transistor and does not depend on the configuration in which the device is used.

14.3.2 The Current Gain Bandwidth Frequency f_T

In order to compare and evaluate the ability of different transistors to function at high frequencies, a frequency f_T is defined at which the magnitude of β becomes unity. This frequency is an important figure of merit for high-frequency and microwave transistors. From Eq. (14.31) the frequency f_T at which $|\beta| = 1$ is given by

$$f_T = f_\beta \sqrt{\beta_o^2 - 1} \approx \beta_o f_\beta = \alpha_o f_\alpha \tag{14.34}$$

and substituting for f_β from Eq. (14.34) into Eq. (14.31) yields

$$\beta(\omega) = \frac{\beta_o}{1 + j\beta_o f/f_T} \tag{14.35}$$

Our discussion to this point has been limited to an ideal transistor in which junction capacitances and series resistances have been ignored, and the base width W_B has been assumed to be large. In such a device, the base transit time τ_N is large and determines the frequency dependence of the current gains. However, in a transistor with a short base width, other time constants become comparable to τ_N, and f_T is given by the relation

$$f_T = \frac{1}{2\pi \tau_{ec}} \tag{14.36}$$

where τ_{ec} is the emitter to collector delay time and is given by

$$\tau_{ec} = \tau_e + \tau_N + \tau_{cd} + \tau_{cb} \tag{14.37}$$

Here τ_e is the emitter-base junction capacitance charging time, τ_{cd} is the collector depletion region delay-time constant, and τ_{cb} is the collector-base junction capacitance charging time. The significance of the various terms in Eq. (14.37) can be understood by considering the p-n-p transistor in Fig. 14.3 to which a base current step $-I_B$ is applied at $t = 0$. This step injects electrons into the base re-

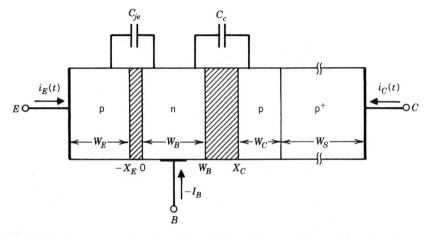

FIGURE 14.3 Diagram showing the origins of the emitter-to-collector delay time in a p-n-p transistor.

gion, which causes a hole current $i_E(t)$ to flow from the emitter. As the holes move from the emitter to the collector, the four delay times in Eq. (14.37) are encountered in sequence and their sum represents the average delay time between the application of the current step at the base and the collector current.

The time constant τ_e is the product of the emitter-base junction resistance $r_e = kT/qI_E$ and the junction capacitance C_{je}. Thus

$$\tau_e = C_{je}\frac{kT}{qI_E} \quad (14.38)$$

Since C_{je} and I_E depend on the emitter base junction voltage, τ_e is bias dependent. The base transit time τ_N is given by Eq. (14.26) and can be reduced by increasing the grading factor η.

For a sufficiently large reverse bias on the collector, the field in the collector-base junction depletion region is high, and holes can be assumed to move through this region with the saturation velocity v_s. It has been shown that the time constant τ_{cd} associated with the transit of holes through the depletion region of width $(X_C - W_B)$ is given by [6] (see Prob. 14.9)

$$\tau_{cd} = \frac{(X_C - W_B)}{2v_s} \quad (14.39)$$

The collector junction capacitance C_c charges through the collector series resistance r_c, making

$$\tau_{cb} = r_c C_c \quad (14.40)$$

τ_{cb} can be minimized by reducing r_c. The resistance r_c is reduced by employing a heavily-doped substrate on which the appropriately doped collector layer is grown epitaxially (Fig. 14.1).

The cutoff frequency f_T is a function of the operating current I_E (or I_C). At low currents, the emitter-base junction resistance is high and τ_e dominates in determining the value of f_T. As the operating current is increased, r_e decreases causing f_T to increase. The increase in f_T comes to an end when the other time constants in Eq. (14.37) become comparable to τ_e. Finally, at sufficiently high currents, high-level injection prevails in the base, and the width W_B is increased because of the Kirk effect (see Sec. 14.5). At the same time, the electric field in the collector-base junction depletion region is reduced, and carriers move through this region with a velocity lower than v_s. As a consequence, both τ_N and τ_{cd} begin to increase and f_T starts decreasing with the current. Figure 14.4 shows [4] the typical behavior of f_T as a function of the collector current I_C.

Apart from f_T, there exists a maximum frequency of oscillation f_{max} at which the power gain of the transistor becomes unity. This frequency is given by the relation [7]

$$f_{max} \simeq \left[\frac{f_T}{8\pi r_B C_c}\right]^{1/2} \quad (14.41)$$

Above f_{max} the transistor is no longer able to act as an active device.

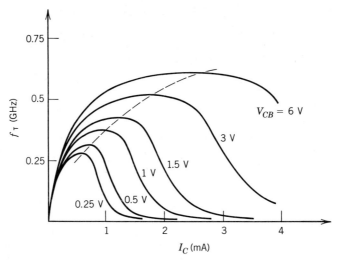

FIGURE 14.4 The cutoff frequency f_T plotted as a function of the collector current for different values of V_{CB}. (From P. A. H. Hart, reference 4, p. 133. Copyright © 1981. Reprinted by permission of Elsevier Science Publishers, Physical Science and Engineering Div., and the author.)

14.3.3 An Illustration

To illustrate the various points discussed in this section, let us consider a Si transistor at 300 K with a bias current $I_E = 1$ mA and the following set of parameters:

$$(X_C - W_B) = W_B = 1 \times 10^{-4} \text{ cm}, \quad D_B = 12 \text{ cm}^2/\text{sec}, \quad \eta = 5$$

$$C_{je} = 10 \text{ pF}, \quad C_c = 3 \text{ pF}, \quad \alpha_o = 0.98, \quad r_c = 2 \text{ } \Omega, \quad \text{and} \quad r_B = 25 \text{ } \Omega$$

From Eq. (14.30) we obtain

$$f_\alpha = \frac{2.43 \times 12}{2\pi \times 10^{-8}}[1 + (5/2)^{4/3}] = 2.04 \text{ GHz}$$

and $f_\beta = (1 - \alpha_o)f_\alpha = 40.8$ MHz. From Eq. (14.34), the ideal cutoff frequency is $f_T = 2$ GHz. However, the actual value of f_T is considerably lower because of the presence of C_{je} and C_c.

We will now evaluate the various time constants in Eq. (14.37). Since $I_E = 1$ mA, from Eq. (14.38) we get

$$\tau_e = 10 \times 10^{-12} \times \frac{25.86 \times 10^{-3}}{10^{-3}} = 2.58 \times 10^{-10} \text{ sec}$$

and from Eq. (14.26), $\tau_N = 1.33 \times 10^{-10}$ sec.

Taking $v_s = 10^7$ cm/sec in Eq. (14.39) gives $\tau_{cd} = 5 \times 10^{-12}$ sec. Finally, from Eq. (14.40) $\tau_{cb} = 2 \times 3 \times 10^{-12} = 6 \times 10^{-12}$ sec. Substituting these values in Eq. (14.37) gives $\tau_{ec} = 4.02 \times 10^{-10}$ sec and $f_T = 1/2\pi\tau_{ec} = 396$ MHz. Also from Eq. (14.41), we obtain

$$f_{\max} = \left(\frac{396 \times 10^6}{8\pi \times 25 \times 3 \times 10^{-12}}\right)^{1/2} = 458.2 \text{ MHz}$$

14.4 HIGH-FREQUENCY AND MICROWAVE TRANSISTORS

All present-day high-frequency transistors are Si planar n-p-n transistors fabricated in an n-epitaxial layer grown on an n^+-Si substrate. Ge and GaAs bipolar transistor technology has not reached the stage of perfection to compete with Si devices.

The difference between a low- and a high-frequency transistor lies in the dimensions of the active areas and in the magnitudes of junction and package capacitances and the series resistances. In a high-frequency transistor, the active area and the parasitic elements must be decreased to obtain a low value of τ_{ec} and a high value of f_{max}. In order to minimize τ_{ec} the individual time constants in Eq. (14.37) should be decreased as far as possible. The emitter time constant τ_e is reduced by operating the transistor with a sufficiently high emitter current I_E which reduces r_e. However, I_E is kept below the value that will cause high-level injection into the base. The base transit time τ_N is decreased by having a sharp impurity grading in the base and by making W_B small. Reduction in τ_{cb} is achieved by using a thin n-epitaxial layer on an n^+-substrate to reduce the collector series resistance r_c. A low value of τ_{cd} requires that the depletion region width $(X_C - W_B)$ be made small. This requirement is met if both sides of the collector base junction are heavily doped. Heavy doping on both sides reduces the junction breakdown voltage, and compromises have to be made between the high-frequency and high-voltage operational capability.

Figure 14.5 shows the basic geometry for a Si n-p-n planar transistor. Operation of the device at high frequency requries that f_T and f_{max} be large. If all other

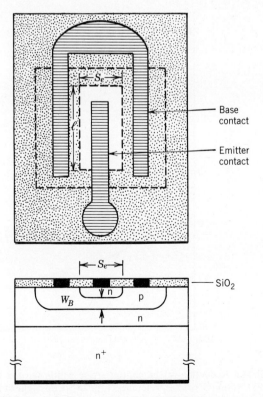

FIGURE 14.5 Basic device geometry for a high-frequency n-p-n transistor.

time constants except τ_N are optimized, then f_T is increased by reducing W_B, and f_{max} (for a given value of f_T) is increased by reducing r_B and C_c. For the structure shown in Fig. 14.5 to a first approximation, we can write [5]

$$r_B = \frac{\bar{\rho}_B S_e}{12 W_B l} \quad (14.42)$$

and

$$C_c \simeq C_o S_e l$$

Here $\bar{\rho}_B$ is the average resistivity of the base layer, l is the emitter stripe length, S_e is the stripe width, and C_o is the collector-base junction capacitance per unit area. Substituting for r_B and C_c from Eq. (14.42) into Eq. (14.41), we can see that f_{max} varies inversely with S_e. In addition, f_T varies inversely with W_B. Thus, S_e and W_B are the two critical parameters in Fig. 14.5. The reduction in these dimensions over the years (since 1952) is shown in Fig. 14.6. In the early 1960s, the stripe width S_e was well above 1 μm, W_B was above 0.1 μm, and the limitations on high-frequency performance were set by these dimensions. However, with the advent of ion implantation doping and electron beam lithography, the S_e and W_B limitations have become less important, and the high-frequency performance has moved closer to the fundamental limits imposed by the properties of the semiconductor material. As discussed in Sec. 11.3, these limits are set by the critical field $\mathcal{E}_b$ at the onset of avalanche breakdown, and by the saturation velocity v_s of

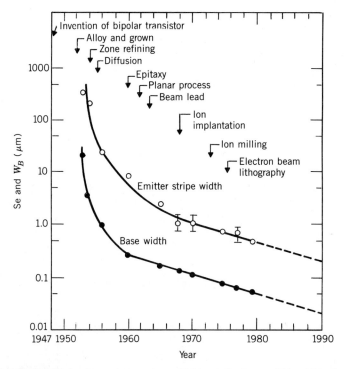

FIGURE 14.6 Reduction in the emitter stripe width and the base width of bipolar transistors since 1952. (From S. M. Sze, reference 7, p. 157. Copyright © 1981. Reprinted by permission of John Wiley & Sons, Inc., New York.)

the carriers. These limitations predict an upper limit on the power output varying as $1/f_T^2$ for a given value of the reactance $X_c = 1/2\pi f_T C_c$.

DEVICE GEOMETRY

The stripe geometry transistor shown in Fig. 14.5 is the basic structure. In order to obtain a large output power, the emitter area must be increased. With an increase in area, the emitter periphery should also be increased to minimize edge crowding. The two most commonly used geometries for microwave transistors are shown [8] in Fig. 14.7. Both structures have a large emitter periphery to area ratio. The geometry in Fig. 14.7a shows the emitter and base interdigitated with emitter fingers enclosed by base fingers and vice versa. In the overlay transistor (Fig. 14.7b), a number of emitter diffusions are made which are connected together by emitter metallization.

14.5 POWER TRANSISTORS

Power transistors are designed for power amplification, which means that the operating voltage and current must be large. In order to support a large reverse bias on the collector-base junction, this junction must have a high breakdown voltage. This condition requires a thick, high-resistivity zone on at least one side of the junction. Because of the occurrence of the high-resistivity region and high current levels in the device, high-level injection conditions are easily reached in a power transistor. A brief account of high-level injection effects will now be presented.

14.5.1 High-Level Injection Effects [9]

In a Si graded-base transistor, high-level injection typically occurs at a forward emitter-base junction voltage of about 0.7 V. Some of the high-level injection effects are associated with the emitter-base junction and will be referred to as

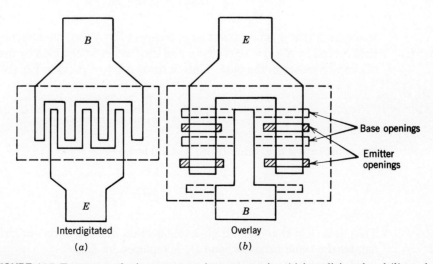

FIGURE 14.7 Two types of microwave transistor geometries: (a) interdigitated and (b) overlay. (From H. F. Cooke, reference 8, p. 1164. Copyright © 1971 IEEE. Reprinted by permission.)

emitter effects. The others associated with the collector-base junction will be referred to as collector effects.

(a) EMITTER EFFECTS

Important high-level injection effects on the emitter side are: a decrease in emitter efficiency at high currents, modifications introduced in the law of junction and minority carrier transport in the base, and the edge crowding effect. We will now examine these effects in more detail.

Let us first consider the decrease in emitter efficiency. At a sufficiently large value of I_E, the concentration of injected minority carriers in the base becomes comparable to the dopant concentration. Quasi-neutrality in the base requires that the majority carrier concentration also increase by the same amount as the minority carrier concentration. Thus, if $p_e(x)$ represents the excess hole concentration in the base of a p-n-p transistor, then $n_n(x) = N_d(x) + p_e(x)$ and the integral in Eq. (14.14) becomes

$$\int_0^{W_B} \frac{n_n(x)\,dx}{D_B} = \int_0^{W_B} \left[\frac{N_d(x) + p_e(x)}{D_B}\right] dx \tag{14.43}$$

When $p_e(x)$ is comparable to $N_d(x)$, the integral in Eq. (14.43) is increased, which in turn causes an increase in the base Gummel number G_B. Since the emitter is heavily doped, high-level injection does not occur in the emitter and the emitter Gummel number G_E remains unaffected. Hence, an increase in G_B reduces emitter efficiency and the current gain h_{FE}.

A second effect of high-level injection is the modification of carrier transport in the base region and a change in the law of junction which expresses the dependence of the collector current on the emitter-base junction voltage. Let us consider a p-n-p transistor. Assuming that I_{nE} is negligibly small compared to I_{pE} and substituting $n_n(x) = N_d(x) + p_e(x)$ in Eq. (14.8) give

$$\mathcal{E}(x) = -\frac{kT}{q}\frac{1}{[N_d(x) + p_e(x)]}\frac{d}{dx}[N_d(x) + p_e(x)] \tag{14.44a}$$

In the limit that $p_e(x)$ becomes large compared to $N_d(x)$, the effect of the built-in field caused by $N_d(x)$ is washed out and the field is determined by the distribution of excess carriers in the base. Thus, writing $p_e(x) = p_n(x)$ in Eq. (14.44a), we get

$$\mathcal{E}(x) = -\frac{kT}{q}\frac{1}{p_n(x)}\frac{dp_n}{dx} \tag{14.44b}$$

and substituting this value in Eq. (14.9a) gives

$$J_p \simeq -2qD_B\frac{dp_n}{dx} \tag{14.45}$$

Thus, it is clear that under high-level injection, the diffusion and drift transistors exhibit the same behavior, and D_B is replaced by $2D_B$.

The modification in the law of junction can be obtained by determining the boundary value of excess carrier concentration $p_e(0)$. As discussed in Sec. 7.4, the *pn*-product at the edge of the emitter-base junction depletion region can be written as

$$p_n(0)n_n(0) = n_i^2 \exp(V_{EB}/V_T) \tag{14.46}$$

Since at high-level injection $n_n(0) = p_n(0) \simeq p_e(0)$, we get

$$p_e(0) = n_i \exp(V_{EB}/2V_T) \tag{14.47}$$

Thus, the collector current at high-level injection rises as $\exp(V_{EB}/2V_T)$ instead of $\exp(V_{EB}/V_T)$.

Another high-level injection effect associated with the emitter-base junction is the edge-crowding effect already discussed in Sec. 13.4. Edge crowding becomes important when the base current causes a transverse voltage drop on the order of (kT/q) across the distributed base resistance beneath the emitter. Because of edge crowding, the current density near the edge of the emitter becomes substantially higher than that in the central region. As a consequence, high-level injection tends to occur near the edge of the emitter earlier than in the regions further away from the emitter periphery. Thus, edge crowding causes localized high-level injection at a current density considerably lower than that at which high-level injection will occur for a uniform current flow across the emitter.

Edge crowding causes more power dissipation at the reverse-biased collector-base junction beneath the emitter periphery. As a result, this region becomes hotter than the surrounding regions, and the power-handling capability of the transistor is reduced. Therefore, power transistors are designed to have a large ratio of emitter periphery to emitter area to minimize edge crowding.

(b) COLLECTOR EFFECTS

High-level injection at the collector-base junction causes an increase in the effective base width. To understand this effect, let us consider the collector-base junction depletion region of a p-n-p transistor. Holes enter the depletion region from the base side. They drift through this region with a finite velocity and disappear in the p-collector. At low values of the collector current I_C, the hole concentration in the depletion region remains small compared to the dopant concentration, and their influence on the depletion region charge may be neglected. However, as I_C is increased, the excess hole concentration becomes comparable with the concentration of dopant atoms and causes a change in the depletion region charge. We will consider high-level injection effects in a p^+-n-p^+ transistor.

Figure 14.8 shows the collector-base junction depletion region of a p^+-n-p^+ transistor with uniform doping N_d in the base. Since the injected holes have a positive charge, their charge is added to ionized donor charge and subtracted from the ionized acceptor charge. Thus, the charge concentration in the depletion region on the base side becomes $(N_d + p_e)$, and on the collector side it is $(N_a - p_e)$. For a sufficiently large bias on the collector-base junction, the holes may be assumed to move through the depletion region with the saturation velocity v_s, and the collector current density J_C can be expressed as

$$J_C = qp_e v_s \tag{14.48}$$

If the concentration p_e remains small compared to the dopant concentration N_a in the collector, then the depletion layer thickness on the collector side remains

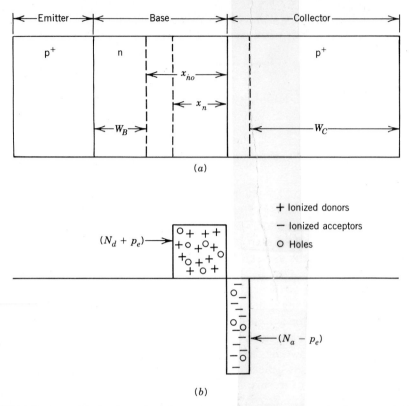

FIGURE 14.8 Base widening in a p^+-n-p^+ transistor showing (a) a schematic diagram and (b) the depletion layer charges.

practically unchanged, but on the base side the depletion region thickness x_n is given by [9]

$$x_n = \frac{x_{no}}{\left(1 + \dfrac{J_C}{J_1}\right)^{1/2}} \tag{14.49}$$

where x_{no} is the depletion region width for $J_C = 0$ and $J_1 = qv_s N_d$ is the critical current density at which the base widening becomes significant. The increase in the base width is given by $\Delta W_B = (x_{no} - x_n)$. Base widening in a n^+-p-n^+ transistor can be obtained in a similar manner. The situation is more complex in the case of a n^+-p-ν-n^+ transistor; details are given by S. K. Ghandhi [9].

The change in the base width as a result of the modification in the space charge at the collector-base junction was first examined by C.T. Kirk, and it is known as the Kirk effect [10].

14.5.2 The Fall in Current Gain at High Currents

Three factors are responsible for the drop in current gain h_{FE} at high currents. First, the high-level injection in the base increases the base Gummel number, and thus causes a decrease in the emitter efficiency. Second, when high-level injection occurs in the base, the collector current I_C increases as $\exp(V_{EB}/2V_T)$, whereas the base current continues to rise as $\exp(V_{EB}/V_T)$ because low-level injection prevails in the emitter. Finally, an increase in the base width at high currents decreases the base transport factor of the transistor, which further reduces I_C but increases the base current I_B. A simple analysis that neglects edge crowding has shown that h_{FE} can be written as [11]

$$h_{FE} \simeq \frac{K_2 A_E}{W_{BT}^2 I_C J_{BEO}} \tag{14.50}$$

Here K_2 is a constant of proportionality, and J_{BEO} is the saturation current density of minority carriers injected from the base into the emitter. Thus, from a design perspective, the variables that affect h_{FE} are the emitter area A_E, the collector current I_C, and the effective base width W_{BT}. Typically, J_{BEO} has a value lower then 10^{-12} A cm^{-2} at room temperature, and it is of little consequence.

14.5.3 Device Structures

Presently, all power transistors are made in Si. Typical doping profiles for the two most commonly used devices are shown in Fig. 14.9.

The homogeneous base n$^+$-p-n$^+$ transistor (Fig. 14.9a) is fabricated using a single deep diffusion step that is simultaneously made into both sides of a lightly doped p-type Si wafer. The device has a wide base region, and its I-V characteristics are similar to those of the transistor described in Chapter 13. The main advantage of this type of transistor is its low saturation voltage which results from the heavily doped collector. However, high-voltage homogeneous base devices require a wide base region to accommodate the depletion region of the collector-base junction. Since a wide base region is detrimental to high-frequency performance, the use of homogeneous base devices is limited to low frequencies and low-speed switching applications.

The impurity profile of the graded-base transistor in Fig. 14.9b has a thick n-type collector with a donor concentration of 10^{14} cm^{-3} or less. This lightly doped region (ν-region) serves as the starting material, and the device is fabricated using three diffusions that are performed sequentially. When the collector-base junction is reverse biased, the depletion region extends mainly in the ν-region. The main advantage of this device is that the voltage-handling capability can be made independent of the base width. In this way, a high-voltage operation can be combined with a considerably improved dynamic performance.

Figure 14.10 shows the common emitter output characteristics of a typical high-voltage n$^+$-p-ν-n$^+$ transistor [9]. The characteristics can be divided into three regions that are referred to as the hard-saturation region (I), the quasi-saturation region (II), and the active region (III). In the hard-saturation region (I), the p-ν junction is heavily forward biased and the ν-layer is filled with holes injected from the p-base. Since the charge neutrality is maintained in the ν-zone, the electron concentration equals the hole concentration, and the resistance of this zone decreases, resulting in a very low-voltage drop.

In the quasi-saturation region II, the p-ν junction is less heavily forward biased, so that the electron–hole concentrations in the ν-layer are reduced and the resistance is increased. Finally, as the collector-emitter voltage V_{CE} is increased, the voltage across the p-ν junction becomes zero at some point. At still higher voltages, the junction is reverse biased, and the transistor enters the active re-

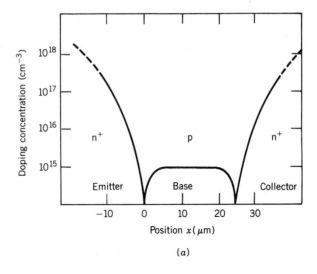

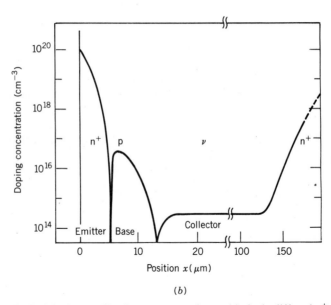

FIGURE 14.9 Typical doping profiles for power transistors: (a) single diffused n^+-p-n^+ transistor and (b) triple diffused n^+-p-ν-n^+ transistor. (From P. Leturcq, reference 11, p. 880. Copyright © 1981. Reprinted by permission of Elsevier Science Publishers, Physical Science and Engineering Div., and the author.)

gion (III). From this point onward the device characteristics are similar to those of a normal transistor.

With respect to device geometry, a power transistor is designed to achieve the best current handling capability per unit area. Since edge crowding occurs for normal operating conditions, the emitter periphery to area ratio must be large. Thus, the interdigitated and overlay structures (Fig. 14.7) have also been used for power transistors. To improve the power-handling capability of the device, the heat generated at the collector-base junction must be conducted away efficiently. This condition requires that the collector be mounted on a heat sink to keep the junction temperature at its lowest value.

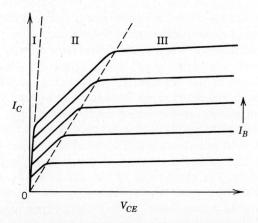

FIGURE 14.10 Common emitter output characteristics of a high-power n^+-p-ν-n^+ transistor showing hard saturation (I), quasi-saturation (II), and active (III) region of operation.

14.5.4 Second Breakdown and Safe Operation Area

The use of power transistors is often limited because of a second breakdown that occurs when the device is biased at high voltage near the breakdown region and the current is relatively high. The second breakdown is marked by an abrupt decrease in the voltage and a corresponding increase in the collector current which often leads to the destruction of the device owing to excessive heating. Figure 14.11 shows the I_C versus V_{CE} characteristic of a power transistor under the second breakdown condition. It is seen that as V_{CE} reaches its breakover value given by Eq. (13.61), and I_C is correspondingly increased, a region of unstable operation occurs, and the device switches to a low-voltage, high-current mode of operation. Subsequent operation in this region usually results in permanent damage to the transistor.

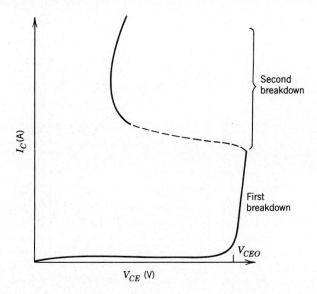

FIGURE 14.11 Typical current-voltage characteristics of a transistor showing second breakdown.

The second breakdown occurs from hot-spot formation in the semiconductor because of nonuniform current distribution across the area and a positive feedback mechanism. When a sufficiently large amount of power is dissipated in the transistor, a hot spot usually develops near the center of the device. The power dissipation in the hot-spot region becomes excessive, and the current density reaches its maximum value. Because of its high temperature, the intrinsic carrier concentration n_i in this region begins to increase. This increase in n_i increases the current through this area and thus causes a further rise in temperature. The device is triggered into a second breakdown when the hot-spot temperature becomes high enough to make n_i in this region equal to the collector doping concentration. The exact cause of hot-spot formation is not known. However, it is believed that crystal defects, doping fluctuations, and other nonuniformities that give rise to microplasmas in the collector-base junction depletion region are responsible for the hot-spot formation.

To safeguard a transistor from permanent damage, a safe operating area (SOA) in the I_C-V_{CE} plane is specified as important information for the circuit designer. Figure 14.12 shows the SOA of a Si power transistor. For safe operation, the operating point must fall in the area bounded by the solid lines.

The line AB represents the upper limit of the device current. The maximum value of I_C is limited by the junction heating effects and the melting of the bond leads. However, a transistor is normally operated below a value of I_C for which a sharp drop in h_{FE} occurs (see Fig. 13.11). The line BC indicates the thermal limit imposed by the maximum allowable junction temperature T_j. The power P dissipated in the device is related to T_j by the relation

$$P = \frac{T_j - T_0}{R_{th}} \tag{14.51}$$

where T_0 is the ambient temperature and R_{th} is the thermal resistance (usually expressed in °C/watt). For Si devices, the maximum temperature of operation is

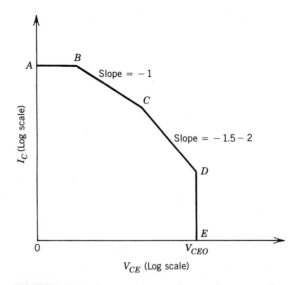

FIGURE 14.12 Safe operating area for transistor operation.

about 175 °C. If R_{th} is assumed to be independent of temperature, then for a given value of T_0

$$P_{max} = V_{CE}I_C = \frac{T_j(\text{max}) - T_0}{R_{th}} = \text{constant} \quad (14.52)$$

Thus, a plot of log I_C versus log V_{CE} is a straight line with a slope of -1. The line CD corresponds to the limit imposed by the second breakdown and has a slope ranging from -1.5 to -2. Finally, the vertical line DE is associated with the maximum value of $V_{CE} = V_{CEO}$ given by Eq. (13.61). Note that SOA is reduced with an increase in the ambient temperature.

14.6 THE SWITCHING TRANSISTOR

A switching transistor is designed to function as an on-off switch that can change its state from a high-voltage, low-current state to a low-voltage, high-current state and vice versa in a short time. Switching is a large-signal transient process, and thus, the operation of a switching transistor is different from that of a transistor used for small-signal amplification. However, the device requirements for obtaining high-frequency response and high-switching speed are almost identical.

In order to be used as a current switch with significant gain in a digital circuit, the transistor is operated in the common emitter configuration. The voltage drop across the transistor in its high-current on-state should be as small as possible, and this is achieved by driving the device into saturation. The low-current off-state, on the other hand, corresponds to the cutoff region. Figure 14.13 shows the circuit connections for a p-n-p transistor used to switch the current in a load R_L along the load-line drawn on the $I_C - V_{CE}$ characteristics. The load-line is given by the relation

$$I_C = \frac{V_{CC} - V_{CE}}{R_L} \quad (14.53)$$

In the cutoff region, $I_C \simeq 0$ and $V_{CE} \simeq V_{CC}$. This is the off-state of the switch. In the on-state, I_C is large, which makes $I_C R_L \simeq V_{CC}$ and $V_{CE} \simeq 0$. The transition between the on- and off-states is obtained by controlling the base current I_B. For a positive bias on the base, the emitter-base junction is reverse biased and the transistor is in cutoff. For a sufficiently large negative bias on the base, the emitter-base junction becomes forward biased, and the transistor operates in the saturation region and the product $I_B h_{FE}$ exceeds I_C.

The characteristic parameters describing the behavior of a switching transistor are: current carrying capability, maximum off-state voltage, on- and off-state impedances, and switching time. The current-carrying capability is determined by thermal considerations and is limited by the maximum allowable power dissipation at the collector-base junction. The maximum voltage in the off-state is limited either by punch-through or by the breakover condition Eq. (13.61). The on- and off-resistances can be obtained from the Ebers-Moll equations. In the off-state, both junctions are reverse biased, and from Eq. (13.26) we obtain

$$I_C(\text{off}) = -\alpha_N I_{ES} + I_{CS} = I_{CS}(1 - \alpha_I)$$

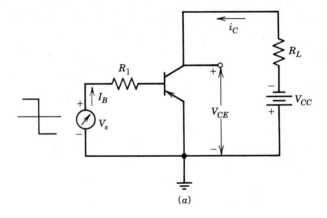

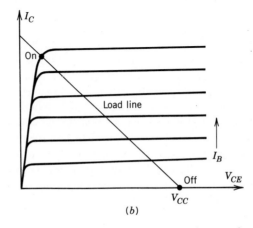

FIGURE 14.13 Circuit connections and load-line for a p-n-p transistor used to switch current in a load R_L.

The off-state resistance is

$$\frac{V_{CE}}{I_C(\text{off})} \simeq \frac{V_{CC}}{I_{CS}(1 - \alpha_I)} \quad (14.54)$$

Since I_{CS} is on the order of 10^{-10} A in Si transistors, and V_{CC} is several volts, the off-state resistance is very high.

The on-state voltage is obtained from Eq. (13.57), which can be rewritten as

$$V_{CE}(\text{on}) = V_T \ln \left[\frac{1 - \dfrac{I_C}{I_B}\left(\dfrac{1 - \alpha_N}{\alpha_N}\right)}{\dfrac{1}{\alpha_I} + \dfrac{I_C}{I_B}\left(\dfrac{1 - \alpha_I}{\alpha_I}\right)} \right] \quad (14.55)$$

Here V_{CE} is on the order of 0.1 V. For example, if we take $I_C/I_B = 20$, $\alpha_N = 0.98$, and $\alpha_I = 0.25$, we get $V_{CE} \simeq 0.12$ V, at 300 K, and for $I_C = 10$ mA, the resistance is 12 Ω.

THE SWITCHING TRANSISTOR

We will now consider the switching operation. Figure 14.14 shows a typical switching cycle for the circuit in Fig. 14.13a. We assume the resistance R_1 and voltage V_s to be large enough, so that in the on-state, the base current can be taken as constant. Let us assume that initially the transistor is in the cutoff region. At $t = 0$, a negative current step $(-I_{B1})$ is applied which is large enough to bring the transistor into the saturation region. Since the collector current does not respond to the base current instantaneously, a finite time is required for I_C to reach its final value as shown in Fig. 14.14b. Note that the collector current is caused by injection of holes from the emitter into the base in response to the negative charge provided by the base current step. It is obvious that the emitter cannot inject holes into the base unless the emitter-base junction becomes slightly forward biased. Since V_{CE} remains constant, the change in the emitter-base junction voltage will cause the same change in the collector-base junction voltage. These changes reduce the depletion region widths of the two junctions, and the charge necessary for this purpose is supplied by the base current during the delay time τ_d. At the end of τ_d, the emitter-base junction becomes forward biased, and the transistor enters the active region. Consequently, the emitter injects holes into the base. Next, the collector current begins to rise, and excess holes are stored in the base reigon. As time passes, the forward bias on the emitter-base junction increases until the current reaches its maximum value (V_{CC}/R_L). The rise

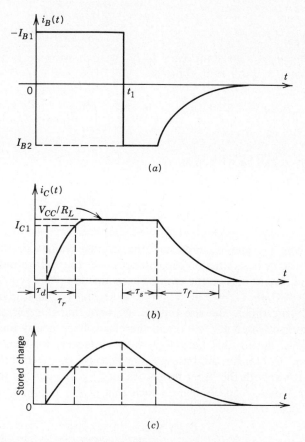

FIGURE 14.14 Switching response of the transistor shown in Fig. 14.13. (a) Base drive current, (b) the collector current, and (c) the stored charge in the base region.

time τ_r is the time during which I_C reaches the value $I_{C1} = 0.9V_{CC}/R_L$. The sum $(\tau_d + \tau_r)$ is called the on-time of the transistor.

After the collector current has reached its maximum value, the transistor enters the saturation region, and the excess hole charge in the base continues to increase (Fig. 14.14c). Eventually, steady state is reached when the transistor goes into hard saturation. Let us now assume that the transistor is in hard saturation and steady state has been achieved. At $t = t_1$, the input voltage V_s abruptly reverses its polarity. The base current does not reduce to zero, but a reverse current flows through the base until the stored charge of excess carriers has been removed. The storage and fall times (τ_s and τ_f) are then defined from the criteria discussed in Sec. 9.3. The sum $(\tau_s + \tau_f)$ is called the turn-off time of the transistor.

The turn-on and turn-off times of a transistor can be obtained from the solution of the charge control equations. However, the problem of switching between the states involves operating the transistor in a nonlinear manner. Consequently, a closed form solution of the charge control equations is not possible, and numerical analysis of the device has to be carried out [12]. Approximate expressions for τ_d and τ_r can be obtained from the simple considerations explained below.

As noted above, during the delay time τ_d, the emitter-base and the collector-base junctions are reverse biased. Therefore, $Q_N = Q_I \approx 0$, and from Eq. (13.72), we get

$$i_B = -\frac{d}{dt}(Q_{VE} + Q_{VC}) \tag{14.56}$$

Since i_B is a constant equal to I_{B1}, we can estimate τ_d by integrating Eq. (14.56) from $t = 0$ to $t = \tau_d$. That is

$$I_{B1}\int_0^{\tau_d} dt = -\int_0^{\tau_d} d(Q_{VE} + Q_{VC}) \tag{14.57a}$$

which can be written as

$$\tau_d = -\frac{V_{\text{off}}}{I_{B1}}(\overline{C}_{je} + \overline{C}_c) \tag{14.57b}$$

where $\overline{C}_{je}$ and $\overline{C}_c$ represent the average values of the emitter-base and the collector-base junction capacitances over the voltage range of interest, and V_{off} is the reverse bias on the emitter-base junction. As a typical example, we may have $\overline{C}_{je} = 3$ pF, $\overline{C}_c = 2$ pF, $V_{\text{off}} = 2$ V, and $I_{B1} = 1$ mA, giving $\tau_d = 10^{-8}$ sec.

To estimate the rise time τ_r, we note that the collector-base junction is fully reverse biased at $t = \tau_d$, and the bias voltage nearly reaches zero at $t = \tau_d + \tau_f$. Thus, during this period $Q_I = 0$. Moreover, we neglect the dQ_{VE}/dt term in Eq. (13.72a) for the base current because this term remains negligible in comparison with the other terms. As for Q_{VC}, we assume that it can be represented by a linear capacitance $\overline{C}_c$ such that $Q_{VC} = \overline{C}_c v_{CB}$. Making these assumptions, we can write

$$i_B = -\frac{Q_N}{\tau_{BN}} - \frac{dQ_N}{dt} - \overline{C}_c\frac{dv_{CB}}{dt} = I_{B1} \tag{14.58a}$$

$$i_C = -\frac{Q_N}{\tau_N} + \bar{C}_c \frac{dv_{CB}}{dt} \qquad (14.58b)$$

Also, from the circuit in Fig. 14.13a, we have

$$i_C R_L + v_{CE} = V_{CC}$$

Since $v_{CE} \simeq v_{CB}$, this relation gives

$$\frac{dv_{CB}}{dt} = -R_L \frac{di_C}{dt} \qquad (14.59)$$

We can further simplify Eq. (14.58b) by neglecting the $C_c(dv_{CB}/dt)$ term. Neglect of this term in the expression for i_C is justified because the collector current is usually much higher than the base current, and this term has much less influence on i_C than it has on the base current. Thus, we write

$$i_C = -\frac{Q_N}{\tau_N} \qquad (14.60)$$

Combining Eqs. (14.58a), (14.59), and (14.60) yields

$$\frac{di_C}{dt}\left(1 + \frac{\bar{C}_c R_L}{\tau_N}\right) + \frac{i_C}{\tau_{BN}} = \frac{I_{B1}}{\tau_N} \qquad (14.61)$$

This equation can be solved for i_C to give

$$i_C(t) = I_{B1} h_{FE} \left\{ 1 - \exp\left[-t \bigg/ \left(1 + \frac{\bar{C}_c R_L}{\tau_N}\right) \tau_{BN} \right] \right\} \qquad (14.62)$$

where $h_{FE} = \tau_{BN}/\tau_N$. The rise time τ_r is obtained from Eq. (14.62) as the time for which $i_C(\tau_r) = I_{C1}$, giving

$$\tau_r = \tau_{BN}\left(1 + \frac{\bar{C}_c R_L}{\tau_N}\right) \ln\left[\frac{1}{1 - (I_{C1}/h_{FE} I_{B1})}\right] \qquad (14.63)$$

where both I_{C1} and I_{B1} have the same sign (i.e., negative for a p-n-p transistor). τ_r is often related to the current-gain bandwidth frequency $\omega_T = 1/\tau_N$ under normal operation. Writing $\omega_T = \omega_N$ in Eq. (14.63), we get

$$\tau_r = h_{FE}\left(\frac{1}{\omega_N} + \bar{C}_c R_L\right) \ln\left[\frac{1}{1 - (I_{C1}/h_{FE} I_{B1})}\right] \qquad (14.64)$$

For a transistor with $h_{FE} = 40$, $\bar{C}_c = 10$ pF, $R_L = 10^3 \,\Omega$, $V_{CC} = -10$ V, $\omega_N = 10^8$ rad/sec, and $I_{B1} = -1$ mA, we have $I_{C1} = -9$ mA, and Eq. (14.64) gives $\tau_r = 2.04 \times 10^{-7}$ sec which is considerably larger than τ_d.

The calculation of the storage time is more involved, and the following approximate expression has been obtained [13]

$$\tau_s = \frac{\omega_N + \omega_I}{\omega_N \omega_I (1 - \alpha_N \alpha_I)} \ln\left[\frac{1 + |I_{B1}/I_{B2}|}{1 + |I_{C1}/h_{FE} I_{B2}|}\right] \qquad (14.65)$$

where ω_I is the current-gain bandwidth product in the inverse operation, and I_{B2} is the constant base drive that results when the voltage across the emitter base junction is reversed to turn the transistor off (see Fig. 14.14a). Note that I_{B2} is positive in a p-n-p transistor.

It is evident from the above equations that the turn-on and turn-off times of a transistor are inversely proportional to the current-gain bandwidth frequency ω_N. Thus, the basic requirements for high-speed switching transistors are the same as those for high-frequency transistors.

REFERENCES

1. H. K. GUMMEL, "Measurement of the number of impurities in the base layer of a transistor," *Proc. IRE* 49, 834 (1961).
2. G. L. PATTON, J. C. BRAVMAN, and J. D. PLUMMER, "Physics, technology, and modeling of polysilicon emitter contacts for VLSI bipolar transistors," *IEEE Trans. Electron Devices* ED-33, 1754 (1986).
3. J. A. DEL ALAMO and R. M. SWANSON, "Modelling of minority-carrier transport in heavily doped silicon emitters," *Solid-State Electron.* 30, 1127 (1987).
4. P. A. H. HART, "Bipolar transistors and integrated circuits," Chapter 2, in *Handbook on Semiconductors,* T. S. Moss and C. Hilsum (eds.), Vol. 4, North-Holland, Amsterdam, 1981.
5. S. K. GHANDHI, *The Theory and Practice of Microelectronics,* John Wiley, New York, 1968.
6. J. L. Moll, *Physics of Semiconductors,* McGraw-Hill, New York, 1964.
7. S. M. SZE, *Physics of Semiconductor Devices,* John Wiley, New York, 1981, Chapter 3.
8. H. F. COOKE, "Microwave transistors: theory and design," *Proc. IEEE* 59, 1163 (1971).
9. S. K. GHANDHI, *Semiconductor Power Devices,* John Wiley, New York, 1977.
10. C. T. KIRK, "A theory of transistor cutoff frequency (f_T) falloff at high current density," *IEEE Trans. Electron Devices* ED-9, 164 (1962).
11. P. LETURCQ, "Power junction devices," Chapter 7A, in reference 4.
12. P. E. GRAY and C. L. SEARLE, *Electronic Principles: Physics, Models, and Circuits,* John Wiley, New York, 1969, Chapter 22.
13. A. G. MILNES, *Semiconductor Devices and Integrated Electronics,* Van Nostrand, New York, 1980.

ADDITIONAL READING LIST

1. R. S. MULLER and T. I. KAMINS, *Device Electronics for Integrated Circuits,* 2d ed., John Wiley, New York, 1986.
2. T. S. MOSS and C. HILSUM (eds.), *Handbook on Semiconductors,* Vol. 4, North-Holland, Amsterdam, 1981, Chapters 2 and 7A.

PROBLEMS

14.1 An n-p-n transistor has a base-width of 3 μm and $D_B = 10$ cm^2 sec^{-1} at 300 K. The minority carrier lifetime in the base is 0.1 μsec, and γ is nearly unity. Determine f_α and the magnitude and phase of α at a frequency of 15 MHz.

14.2 The donor concentration in the base region of a Si p-n-p transistor can be approximated by $N_d(x) = 2 \times 10^{18} \exp(-x/\lambda)$ where $\lambda = 1.2$ μm. The transistor has $W_B = 4.5$ μm, $W_E = 2.5$ μm, and a uniform doping concentration of $N_a = 5 \times 10^{19}$ cm^{-3} in the emitter. Assume $D_B = 8$ cm^2 sec^{-1}, $\tau_B = 2 \times 10^{-7}$ sec, $D_E = 2.5$ cm^2 sec^{-1}, $\tau_E = 1.5 \times 10^{-8}$ sec, and $T = 300$ K. Calculate the base and the emitter Gummel numbers and h_{FE}. Take the band gap narrowing in the emitter into account.

14.3 For a graded-base transistor with a constant electric field in the base, show that τ_N is given by Eq. (14.26).

14.4 In a p-n-p transistor with a constant built-in field in the base, show that the stored charge Q_s of excess holes in the base, in the normal active mode of operation, can be expressed as $Q_s \simeq I_p W_B^2 / \eta D_p$.

14.5 The common emitter current gain of a transistor with constant doping in the base is given by Eq. (14.28). Expanding this relation and assuming $\gamma = 1$, obtain Eq. (14.6).

14.6 A drift transistor with a constant field in the base has $W_B = 1.5$ μm, $D_B = 7$ cm^2 sec^{-1}, and $f_\alpha = 450$ MHz. Calculate the excess phase factor and the magnitude and the phase of the common base current gain $\tilde{\alpha}$ at 300 MHz.

14.7 An epitaxial planar n-p-n transistor with a constant field in the base has $W_B = 1$ μm. The grading in the base decreases the base transit time τ_N by a factor of 4 from its value in the situation where the base is homogeneously doped. If $N_a(0) = 10^{19}$ cm^{-3}, calculate the value of $N_a(W_B)$. Explain how τ_N will be affected at high collector currents.

14.8 An n-p-n transistor at $I_C = 5$ mA and $T = 300$ K has $\beta_o = 100$ and $|\beta| = 10$ at 10 MHz. Assuming that γ is unity and Eq. (14.6) is valid, calculate f_α, f_β, f_T, and the base width of the transistor. Assume the base to be homogeneously doped and $D_B = 12$ cm^2 sec^{-1}.

14.9 Consider that the collector-base junction of a p-n-p transistor is sufficiently reverse biased so that the holes move through the junction depletion region of width $(X_C - W_B)$ with saturation velocity v_s. Show that when an ac signal is superimposed on the dc bias the associated time constant is given by Eq. (14.39).

14.10 An n^+-p-n^+ power transistor has a base width of 125 μm and a base dopant concentration of 1.3×10^{14} cm^{-3}. The transistor has a current gain $h_{FE} = 50$ at a collector reverse bias of 750 V at low currents. Calculate the current gain for current densities of 50 A cm^{-2} and 150 A cm^{-2}.

14.11 Consider a p-n-p transistor with $W_B = 2$ μm, $h_{FE} = 40$, and the impurity profile in the base given in Prob. 14.2. The transistor has a single stripe emitter geometry shown in Fig. 14.5. With $S_e = 2$ μm and $l = 50$ μm estimate the collector current at the onset of edge crowding. Assume an average electron mobility of 250 cm^2/V sec in the base.

14.12 An n-p-n transistor having $\alpha_N = 0.98$, $I_{CO} = 0.2$ μA, and $I_{EO} = 0.16$ μA is used in the common emitter configuration with a collector series resistance of 4 K Ω and $V_{CC} = 12$ V. Determine the minimum base current for the transistor to enter the saturation region.

14.13 In the circuit of Fig. 14.13a assume that $R_L = 0$. The base current is abruptly raised to -0.5 mA at $t = 0$. The steady-state collector current is -10 mA, and the initial slope of the collector current is -25 mA/μsec. Determine the values of τ_N and τ_{BN}. Use the charge control equations and neglect the charges associated with the junction depletion regions.

14.14 An n-p-n power transistor is operated with $V_{CE} = 100$ V and $I_C = 0.5$ A at 300 K. The maximum allowable junction temperature is 175 °C. Calculate the thermal resistance.

14.15 A p-n-p transistor has $\tau_N = 1.2 \times 10^{-8}$ sec, $\tau_I = 3.6 \times 10^{-8}$ sec, $\alpha_N = 0.99$, and $\alpha_I = 0.9$. The transistor is connected in the circuit of Fig. 14.13a. The base current is varied from $I_{B1} = -0.5$ mA to $I_{B2} = 0.2$ mA and $R_1 = 10$ K Ω. The load resistance is 2 K Ω and $V_{CC} = -12$ V. Assume $\overline{C}_{je} = 4$ pF, $\overline{C}_c = 2$ pF, and $T = 300$ K. Calculate (a) the on resistance of the transistor and (b) the time constants τ_d and τ_s.

15
JUNCTION AND METAL–SEMICONDUCTOR FIELD-EFFECT TRANSISTORS

INTRODUCTION

The term *field effect* as used here describes the change in conductance of a sample that occurs as a result of a capacitively applied field normal to the sample surface. Field-effect transistors (FETs) are devices in which transistor action (signal amplification, etc.) is achieved by the field effect. These devices are also classified as unipolar transistors because the current flow involves only one type of carrier, namely, the majority carriers. This feature is in contrast to the bipolar transistors where the current is carried by both the majority and minority carriers.

In order to understand how the conductance modulation is achieved, let us consider a region of n-semiconductor with a cross-sectional area A and length L. The conductance G of this region can be written as

$$G = \frac{q\mu_n n A}{L} \tag{15.1}$$

where n represents the electron concentration and μ_n is the electron mobility. The conductance G in Eq. (15.1) may be changed either by altering the dimensions A and L of the region or by changing the carrier concentration n. Based on these two methods of conductance modulation, we have two classes of field-effect transistors: the junction field-effect transistor (JFET) and the surface field-effect transistor. In the JFET, one or two reverse-biased p-n junctions are used to change the cross-sectional area of the conducting region (called the *channel*). A metal–semiconductor field-effect transistor (MESFET) is similar to a JFET, except that a rectifying metal–semiconductor contact is used in place of the p-n junction. In the surface FET, the carrier concentration n in the channel is modulated by applying an electric field at the semiconductor's surface using a metal electrode separated by an insulator. Modulation of the channel conductance by either of the two methods causes a variation in the channel current which is responsible for the transistor action. Since the electric field that modulates the channel conductance is varied by changing the voltage on a third terminal called the *gate*, the FET is a voltage-controlled device.

Field-effect transistors have many advantages over bipolar transistors. They have a considerably higher input impedance than bipolar transistors, and this prevents feedback from the output to the input. Since the current in a FET is due to transport of majority carriers, the current voltage characteristics of FETs are relatively insensitive to temperature change and radiation damage. Field-effect transistors are also less noisy than bipolar transistors because of the absence of generation-recombination noise (see Appendix G). In addition, FETs are simpler to fabricate and require less space in integrated circuits than bipolar transistors.

The main disadvantage of FETs has been their lower gain-bandwidth product compared to that of bipolar transistors. However, this is true mainly for silicon FETs fabricated using conventional technology. During the last several years, high-mobility semiconductors like GaAs and InP have been employed, and new fabrication techniques have been used to reduce the device dimensions to the submicron range. These developments have led to MESFETs with high-frequency performance comparable to, and in some cases, even better than that of bipolar transistors.

This chapter is devoted to the JFET and MESFET, not only because the theory of these devices is simple compared to that of the surface FET, but also because, historically, the JFET was the first functional device. The initial three sections of this chapter are concerned with JFET structures fabricated using conventional silicon technology, while the remaining four are devoted to short-channel MESFETs.

15.1 PRINCIPLE OF OPERATION

Figure 15.1 shows a schematic diagram of an n-channel JFET along with the biasing arrangement. The device consists of a thin bar of an *n*-type semiconductor sandwiched between two p^+-type regions of the same semiconductor. The thin n-region is isolated from the p^+-regions by the junction depletion regions and is called the *channel*. The channel is provided with ohmic contacts at the two ends. During the operation of the device, one of these contacts is connected to the negative terminal of the supply voltage and is known as the *source electrode* because it serves as the source of electrons for the current in the channel. The other contact is connected to the positive terminal of the supply, and it acts as a "sink" for the electrons flowing out of the channel. It is called the *drain electrode*, or simply the drain. Each of the two p^+-regions is called a *gate*. These p^+-regions are shown connected together in Fig. 15.1. When the gate is reverse biased with respect to the source, the widths of the two depletion regions increase. Consequently, the cross-sectional area of the channel is reduced, and its resistance is increased. Thus, the JFET is basically a voltage-controlled resistor whose resistance can be varied by changing the bias between the gate and the source.

In order to understand the principle of operation of the device, let us consider the idealized structure shown in Fig. 15.2. We assume that the channel is uniformly doped with a donor concentration N_d and that the p^+-gate regions are

PRINCIPLE OF OPERATION

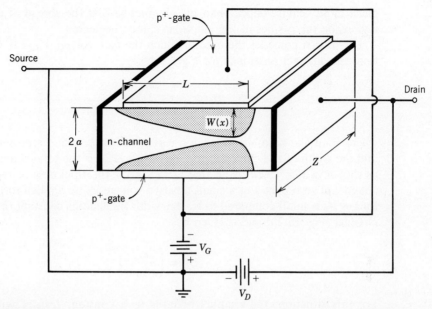

FIGURE 15.1 Schematic diagram showing the basic structure and biasing arrangement for an n-channel JFET.

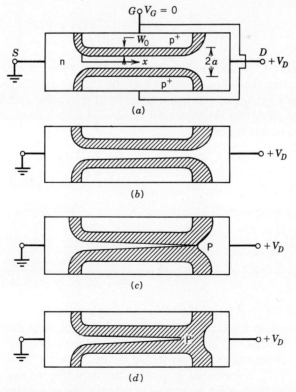

FIGURE 15.2 The channel thickness profile for various values of V_D with $V_G = 0$. (a) V_D is small and the channel thickness is uniform; (b) V_D is large but remains below pinch-off; (c) situation at pinch-off; and (d) beyond pinch-off.

much more heavily doped than the channel so that the spread of the depletion regions on the p-sides of the junctions can be neglected.

Let us first consider the case in which the gate voltage $V_G = 0$. In this situation, the channel resistance R_0 is given by

$$R_0 = \frac{L}{2q\mu_n N_d Z(a - W_0)} \tag{15.2}$$

where W_0 is the zero-bias depletion region width, L and Z represent the length and the width of the channel, respectively, and $2a$ is the total channel thickness as shown in Fig. 15.2a. Note that $2(a - W_0)$ is the thickness of the conducting channel at zero bias. Let a small positive voltage V_D be applied to the drain. As long as V_D is small compared to V_i, the width W_0 remains constant throughout the channel (Fig. 15.2a) and is given by

$$W_0 = \left(\frac{2\varepsilon_s}{qN_d} V_i\right)^{1/2} \tag{15.3}$$

For this situation, the channel behaves as a constant resistor with resistance given by Eq. (15.2). However, when V_D becomes comparable to V_i, then because of the voltage rise from source to drain, the channel becomes increasingly reverse biased toward the drain. If $V(x)$ represents the voltage at a distance x from the source, then the depletion region width $W(x)$ becomes

$$W(x) = \left[\frac{2\varepsilon_s}{qN_d}(V_i + V(x))\right]^{1/2} \tag{15.4}$$

Figure 15.2b shows the variation of $W(x)$ with distance x along the channel. Neglecting the voltage drops across the regions outside the narrow channel, we see that $V(x)$ has its maximum value V_D at the drain, and $W(x)$ is a maximum there. The elemental resistance dR associated with a finite length (dx) of the channel can be written as

$$dR = \frac{dx}{2q\mu_n N_d Z[a - W(x)]} \tag{15.5}$$

The total resistance R of the channel is obtained by integrating Eq. (15.5) from $x = 0$ to $x = L$, giving

$$R = \frac{1}{2q\mu_n N_d Z} \int_0^L \frac{dx}{[a - W(x)]} \tag{15.6}$$

Since $W(x)$ is larger than W_0 at all points except at $x = 0$, the resistance R is larger than R_0 in Eq. (15.2). Thus, it is clear that the channel resistance R increases with an increase in V_D. This increase in R causes the drain current to fall below the initial resistance line, as shown in the curve for $V_G = 0$ in Fig. 15.3.

As V_D is increased gradually, the depletion region width also increases, and eventually a point is reached at which V_D becomes large enough to cause the two depletion regions to meet near the drain (Fig. 15.2c). When this happens, the channel near the drain gets "pinched-off," and the source and drain are isolated

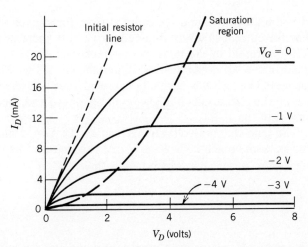

FIGURE 15.3 Calculated current-voltage characteristics of an n-channel JFET. (From A. S. Grove, reference 2, p. 249. Copyright © 1967. Reprinted by permission of John Wiley & Sons, Inc., New York.)

from each other through the depletion region of the reverse-biased p$^+$-n junction. The pinched-off region has a very high electric field directed along the negative x direction (away from the tip P of the channel in Fig. 15.2c). This field originates on the ionized donors in the depletion region and terminates on the electrons in the channel. Such an intense field causes the electrons to flow from the tip P of the channel into the pinched-off drain region. This situation is similar to that existing at the reverse-biased collector-base junction of a bipolar transistor. The reverse current of this junction is very small, but a large current can flow through the depletion region if minority carriers are injected into it from the emitter-base junction. Thus, the current flowing through the pinched-off region to the right of P is limited by the number of electrons that are injected into the depletion region from the tip of the channel. Therefore, the magnitude of this current is determined only by the voltage drop across the undepleted part of the channel.

The drain voltage at the onset of pinch-off is called the *saturation voltage* V_{Dsat}. Its value can be determined from Eq. (15.4) by substituting V_{Dsat} for $V(x)$, and a for $W(x)$ giving

$$V_{Dsat} = \frac{qN_d}{2\varepsilon_s}a^2 - V_i \qquad (15.7)$$

When V_D is raised above V_{Dsat}, practically all the additional voltage appears across the depletion region. Consequently, the depletion region widens, and the point P moves slightly toward the source as indicated in Fig. 15.2d. Even a relatively large increase in V_D above V_{Dsat} causes only a small increase in the depletion region width so that the channel length L is not significantly affected. Moreover, the voltage at the pinch-off point P does not change; it remains at V_{Dsat}. This means that the electrical conditions in the channel have changed only slightly. Thus, the drain current saturates to a value I_{Dsat} when V_D is increased above V_{Dsat}. This behavior is shown in Fig. 15.3.

When a reverse-bias V_G is applied to the gate, the depletion region of each p$^+$-n junction becomes wider, and the cross-sectional area of the channel is reduced. For small values of V_D, the channel will behave as a resistor, but its resistance will be higher now, and the drain current for a given value of V_D will be lower than the current that was obtained for $V_G = 0$. Increase in V_D will cause the current to saturate in the same way as explained above. The current-voltage characteristics of an n-channel JFET are shown in Fig. 15.3 for a number of gate voltages. The JFET is normally operated with a reverse-biased gate. A forward bias on the gate results in a low-input resistance and is undesirable in most applications of the JFET.

The gate-channel voltage that causes the two depletion regions to meet is known as the pinch-off voltage V_P. At this voltage, the channel disappears making $a = \sqrt{(2\varepsilon_s/qN_d)V_P}$, which means that

$$V_P = \frac{qN_d}{2\varepsilon_s}a^2 \tag{15.8}$$

The pinch-off voltage is an important parameter for a JFET and ranges in value from a few volts to a few tens of volts.

The saturation voltage V_{Dsat} of a JFET at a gate bias voltage V_G can be obtained by substituting $(V_i - V_G)$ for V_i in Eq. (15.7) making

$$V_{Dsat} = \frac{qN_d a^2}{2\epsilon_s} - (V_i - V_G)$$

which in view of Eq. (15.8) can be written as

$$V_{Dsat} = V_P - (V_i - V_G) \tag{15.9}$$

which is identical to Eq. (15.7) except that V_i is now replaced by $(V_i - V_G)$.

From the above discussion, we have seen that the current versus voltage characteristics of a JFET have two distinct regions. When V_D is small compared to $(V_i - V_G)$, the I-V characteristics are linear, and for $V_D > V_{Dsat}$, the current saturates at I_{Dsat}. The former region is called the *linear region* and the latter the *saturation region*.

15.2 STATIC I-V CHARACTERISTICS OF THE IDEALIZED MODEL

15.2.1 I-V Characteristics

In deriving the $(I_D - V_D)$ characteristics of a JFET, we need to make the following assumptions:

1. The transistor can be represented by the idealized model of Fig. 15.2 with uniformly doped channel and p^+-gate regions that are very heavily doped when compared to the dopant concentration in the channel.
2. The so-called gradual channel approximation is valid [1]. This statement needs explanation. When no bias is applied to the drain, the gate-to-channel junction depletion region width is constant along the channel and can be calculated using a relation similar to Eq. (15.3) which is obtained from the solution of the one-dimensional Poisson equation. However, when the source and the drain are at different potentials, the reverse voltage across the

STATIC I-V CHARACTERISTICS OF THE IDEALIZED MODEL

gate-to-channel junction varies with the distance x along the channel. An exact analysis of this situation requires a solution of the two-dimensional Poisson equation. In the gradual channel approximation, it is assumed that the depletion region width varies slowly enough along the channel so that it can be calculated from the solution of the one-dimensional Poisson equation. This assumption is justified if the electric field component $\mathscr{E}_y$ normal to the channel is large compared to the component $\mathscr{E}_x$ at all points along the edge of the depletion region. The gradual channel approximation is valid in all devices where the channel length L is large compared to the thickness a.

3. The electron mobility is constant along the channel. This assumption, though incorrect, is made because it considerably simplifies the analysis.

Let us consider the idealized JFET structure (Fig. 15.2) with appropriate biases on the gate and the drain. Since the channel is uniformly doped, diffusion effects are absent, and the drain current I_D can be written as a drift current of electrons.

$$I_D = 2qN_d v_d(x) Z [a - W(x)] \tag{15.10}$$

Here Z is the channel width and $v_d(x)$ is the electron drift velocity, which for the constant mobility case is given by

$$v_d(x) = -\mu_n \mathscr{E}_x = \mu_n \frac{dV}{dx} \tag{15.11}$$

Making the gradual channel approximation and assuming the gate to channel junctions as one-sided abrupt junctions, the depletion region thickness at any point x can be obtained from the relation

$$W(x) = \sqrt{\frac{2\varepsilon_s}{qN_d}[V_i - V_G + V(x)]} \tag{15.12}$$

Substituting for $v_d(x)$ and $W(x)$ from Eqs. (15.11) and (15.12) into Eq. (15.10), and integrating the resulting expression from the source ($x = 0$, $V = 0$) to the drain ($x = L$, $V = V_D$) lead to the following expression for the drain current (see Prob. 15.1):

$$I_D = G_o \left\{ V_D - \frac{2}{3} V_P \left[\left(\frac{V_i - V_G + V_D}{V_P} \right)^{3/2} - \left(\frac{V_i - V_G}{V_P} \right)^{3/2} \right] \right\} \tag{15.13}$$

where

$$G_o = \frac{2q\mu_n N_d Z a}{L} \tag{15.14}$$

is the conductance of the channel including the depletion regions. Figure 15.3 shows the I-V characteristics of a Si n-channel JFET calculated from Eqs. (15.13) and (15.14) using known values of μ_n, N_d, and other parameters [2]. Eq. (15.13) describes the characteristics correctly for $0 \le V_D < V_{D\text{sat}}$. Beyond this point the channel is pinched-off, and the drain current is assumed to be independent of V_D.

When V_D exceeds V_{Dsat}, Eq. (15.13) is no longer valid because it is based on the gradual channel approximation. Once the channel is pinched-off, the potential distribution in the depletion region near the tip of the channel can no longer be obtained using the one-dimensional Poisson equation, and the gradual channel approximation loses its validity.

We now proceed to examine the characteristics in the two regions of particular interest, namely, the linear and the saturation regions. For small drain voltages such that $V_D \ll (V_i - V_G)$, it is possible to simplify Eq. (15.13) by expanding the bracketed term containing V_D in a binomial series and retaining only the first two terms. This leads to

$$I_D = G_o \left[1 - \left(\frac{V_i - V_G}{V_P} \right)^{1/2} \right] V_D \qquad (15.15)$$

which represents the I-V characteristics in the linear region near the origin.

From Eq. (15.15), it is seen that for a given value of V_D, as V_G is made more and more negative, the drain current I_D continues to decrease until a value of V_G is reached at which I_D becomes zero and the transistor is turned off. This *turn-off voltage*, also known as the threshold voltage, is obtained by putting $V_G = V_{th}$ and $I_D = 0$ in Eq. (15.15) giving

$$V_{th} = V_i - V_P \qquad (15.16)$$

For an n-channel JFET, the threshold voltage V_{th} is negative and represents the voltage that must be applied to the gate to deplete the channel completely, that is, to make $W = a$. For the JFET characteristics shown in Fig. 15.3, $V_{th} = -4$ V.

The drain current in the saturation region is obtained by substituting $V_D = V_{Dsat}$ from Eq. (15.9) into Eq. (15.13). This gives

$$I_{Dsat} = G_o \left\{ \frac{V_P}{3} + \left[\frac{2}{3} \left(\frac{V_i - V_G}{V_P} \right)^{1/2} - 1 \right] (V_i - V_G) \right\} \qquad (15.17)$$

This current is independent of V_D, as it should be from the considerations of Sec. 15.1.

15.2.2 Device Parameters

Two important small-signal parameters of a JFET are the *drain conductance* (also called channel conductance) g_D defined by

$$g_D = \left. \frac{\partial I_D}{\partial V_D} \right|_{V_G = \text{constant}} \qquad (15.18)$$

and the *transconductance* g_m which is given by the relation

$$g_m = \left. \frac{\partial I_D}{\partial V_G} \right|_{V_D = \text{constant}} \qquad (15.19)$$

The drain conductance below the saturation region can be directly obtained from Eq. (15.13). In the linear region where Eq. (15.15) is valid, we have

$$g_D = G_o \left[1 - \left(\frac{V_i - V_G}{V_P} \right)^{1/2} \right] \quad (15.20)$$

Note that for an n-channel JFET, V_G is negative when the gate is reverse biased. Equation (15.20) shows that g_D decreases with increasing reverse bias on the gate, and it becomes zero when V_G approaches the threshold voltage V_{th}. In the saturation region, I_D is independent of V_D and $g_D = 0$.

The transconductance g_m is obtained by differentiating Eq. (15.13) with respect to V_G and treating V_D as constant. The result is

$$g_m = G_o \left[\left(\frac{V_i - V_G + V_D}{V_P} \right)^{1/2} - \left(\frac{V_i - V_G}{V_P} \right)^{1/2} \right] \quad (15.21)$$

In the linear region near the origin $V_D \ll (V_i - V_G)$, and Eq. (15.21) reduces to

$$g_m = \frac{G_o V_D}{2\sqrt{V_P(V_i - V_G)}} \quad (15.22)$$

The transconductance in the saturation region is obtained by inserting $V_D = V_{Dsat} = V_P - (V_i - V_G)$ in Eq. (15.21) yielding

$$g_m = G_o \left[1 - \left(\frac{V_i - V_G}{V_P} \right)^{1/2} \right] \quad (15.23)$$

A comparison of this relation with Eq. (15.20) reveals that g_m in the saturation region is the same as g_D in the linear region.

A similar analysis can be made for a p-channel JFET. However, in the case of the p-channel device, V_D will be negative and V_G will be positive for a reverse bias on the gate. Since the electron mobility is higher than the hole mobility for all the semiconductors of interest, an n-channel JFET has a larger value of g_m and a better high-frequency response than a p-channel device. For this reason, JFETs are fabricated primarily as n-channel devices.

15.3 JFET STRUCTURES

In the early 1950s, JFETs were fabricated from Ge and Si either by a *rate-grown* process (see Sec. 19.2.1) or by alloying, and these structures were closer to that shown in Fig. 15.1. Present-day devices are produced from Si using planar technology (Sec. 19.4) and differ considerably from the idealized structure discussed above. Figure 15.4 shows the cross-sectional view of two typical structures. In order to fabricate the double-gate structure of Fig. 15.4a, we start with a p^+-silicon substrate that is approximately 0.2 mm thick. A thin layer of n-Si is then epitaxially grown on the substrate and the wafer is oxidized. Next, windows are etched in the oxide, and an n-diffusion is performed to obtain the regions for making the source and drain contacts. The p^+-type gate region is then formed by diffusion from the top surface after etching an additional window through the oxide.

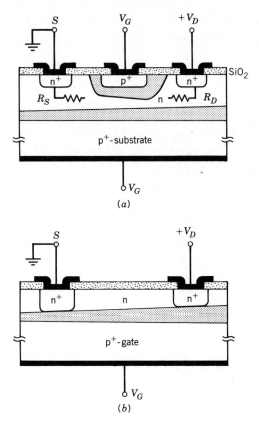

FIGURE 15.4 Cross-sectional views of n-channel JFETs: (*a*) double-gate structure and (*b*) single-gate structure.

Finally, ohmic contacts are made to the various regions. The fabrication process of the single-gate device (Fig. 15.4*b*) is similar to that of the double-gate structure except that the p^+-diffusion performed from the top surface is omitted.

A real JFET differs from the ideal device in several respects, as discussed in this section.

15.3.1 Graded Channel

In our discussion of the idealized structure, we have assumed a uniformly doped channel and a one-sided abrupt gate-to-channel junction. In a real JFET, because of out diffusion from the p^+-substrate during the epitaxial growth of the *n*-type layer, the impurity distribution in the channel becomes graded. However, it has been shown [3] that the I-V characteristics of a JFET are not very sensitive to the impurity distribution in the channel. Thus, the simple analysis of the previous section provides a fairly good approximation of the behavior of real JFETs.

15.3.2 Drain Current in the Saturation Region

Figure 15.5 shows the I-V characteristics of a typical n-channel JFET. These characteristics differ from the calculated characteristics of Fig. 15.3 in two impor-

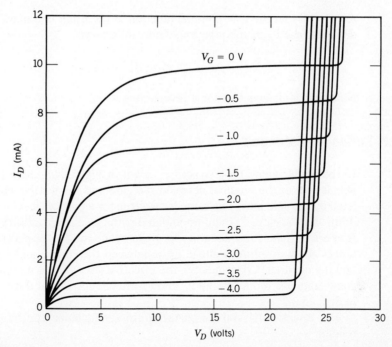

FIGURE 15.5 Measured common source I-V characteristics of a JFET. (From J. Millman and C. Halkias, *Integrated Electronics,* p. 313. Copyright © 1972. Reprinted by permission of McGraw-Hill, Inc., New York.)

tant respects. First, for values of V_D above V_{Dsat}, the drain current does not saturate but increases slowly with the drain voltage, giving rise to a small but finite drain conductance in the saturation region. Second, the maximum drain voltage that can be applied to the transistor is limited by the breakdown voltage of the gate-channel junction.

Two factors are responsible for the observed increase in I_D for $V_D > V_{Dsat}$. One of these is the reduction in the effective channel length with an increase in the drain voltage. As seen in Fig. 15.2d, as V_D is gradually increased above V_{Dsat}, the depletion region of the pinched-off channel widens and the tip moves toward the source. The voltage at P remains fixed at V_{Dsat}, but the effective length of the channel is reduced. A decrease in L causes an increase in G_o and a corresponding increase in the drain current.

The other factor, responsible for the observed increase in I_D, is the avalanche multiplication of the electron current in the pinched-off region. Because the electric field in this region is high and increases rapidly with V_D, it becomes intense enough to produce secondary electron–hole pairs at some value of V_D above V_{Dsat}. Consequently, the primary drain current gets multiplied by the avalanche multiplication factor M in the depletion region. This effect is similar to the multiplication of the collector current in the collector-base junction depletion region of a bipolar transistor.

As V_D is increased gradually, a point is reached at which the avalanche breakdown of the gate-to-channel p^+-n junction occurs. This causes an abrupt increase in the drain current as seen in Fig. 15.5. The magnitude of the drain voltage at which the avalanche breakdown occurs is obtained from the condition that the reverse bias on the gate-channel junction at the drain end must equal the breakdown voltage V_B of the junction. Thus, we obtain

$$V_D - V_G = V_B \tag{15.24}$$

Note that V_G is negative for a reverse bias.

15.3.3 Series Resistances

In the double-gate JFET structure shown in Fig. 15.4a, there exists a considerable length of the channel between the gate and the source which is not modulated by the gate bias. The same is also true for a part of the channel between the drain and the gate. These unmodulated paths give rise to source and drain series resistances that are denoted in Fig. 15.4a by R_S and R_D, respectively. These resistances cause an ohmic voltage drop between the source and the drain contacts and the channel. Consequently, the effective voltage that modulates the channel conductance is reduced, and g_D and g_m are changed from their values calculated in Sec. 15.2.2.

The effect of R_S and R_D on the drain conductance can be expressed by the relation

$$\frac{1}{g_D'} = \frac{1}{g_D} + R_S + R_D \tag{15.25}$$

where g_D' is the actually observed drain conductance and g_D is its value in the absence of the series resistances. In the linear region of the characteristic, all the terms on the right-hand side of Eq. (15.25) are comparable, and the observed conductance may be significantly reduced below g_D. However, in the saturation region, $1/g_D$ is large compared to $(R_S + R_D)$, and these resistances have little effect.

In the saturation region, the source resistance R_S reduces the value of g_m. Because of the presence of R_S, the potential at the beginning of the channel will be $I_D R_S$ instead of zero, as assumed in the treatment of Sec. 15.2.1. Thus, if V_G is the applied gate voltage, then the effective gate voltage at the entrance point of the channel will be $V_G - I_D R_S$, and it can be shown (see Prob. 17.8) that

$$\frac{1}{g_m'} = \frac{1}{g_m} + R_S \tag{15.26}$$

where g_m' is the measured value of the transconductance and g_m is its value when $R_S = 0$.

The voltage drop $I_D R_D$ across this resistance will increase the drain voltage required to bring about saturation of the drain current. However, R_D will have no effect on the g_m in the saturation region.

15.3.4 Effect of Temperature

Three parameters that may cause a change in the $I_D - V_D$ characteristics of a JFET with temperature are V_P, V_i, and the conductance G_o. If all the dopant atoms are assumed to be ionized, V_P is almost independent of temperature, while

V_i decreases slowly with increasing temperature. The most important factor that determines the temperature sensitivity of the characteristics is the carrier mobility μ_n which appears in the expression for G_o. Since μ_n is controlled by thermal scattering, it decreases with rising temperature and causes a small decrease in I_D for a given value of the gate voltage. This behavior is in contrast to that of the bipolar transistor where the collector current increases with temperature.

15.3.5 Voltage, Current, and Power Limitations

The highest voltage that can be applied to the drain of a JFET is limited by the avalanche breakdown voltage of the gate-channel diode. As evident from Eq. (15.24), this voltage is a function of V_G.

The current rating of a JFET is determined by the heating effects on the ohmic contacts. If the drain current is allowed to exceed its maximum permissible limit, excess heating at the contacts causes them to melt.

When high power is dissipated in a semiconductor device, heat is generated to raise the temperature of the semiconductor material. In a JFET, the highest voltage drop occurs in the pinched-off region near the drain, and it is this region that has the highest temperature. If excessive heat is generated in this part of the device, the temperature may rise to the point where the semiconductor becomes intrinsic. Consequently, the majority carrier concentration will increase and cause an increase in the drain current, which causes a further increase in temperature. If steady state is not reached, the process may result in thermal runaway of the transistor. This problem may be avoided using a design with low thermal resistance such that the heat is quickly removed from the high temperature region.

15.4 BASIC TYPES OF MESFETS

Because of their low noise and relative insensitivity to cosmic radiation, JFETs are ideally suited for amplification in the GHz frequency range in satellite communication systems. However, as mentioned earlier, the main disadvantage of JFETs has been their low gain bandwidth product. Since 1970, considerable work has been done to produce field-effect transistors using Si and high-mobility semiconductors like GaAs and InP. These devices employ a Schottky barrier metal gate instead of the p-n junction and consequently are known as metal–semiconductor field-effect transistors (MESFETs).

Figure 15.6 shows the cross-section of a GaAs MESFET. The device is fabricated by growing an *n*-type epitaxial layer on a high-resistivity GaAs substrate

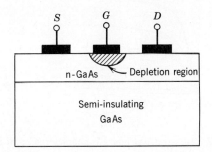

FIGURE 15.6 Schematic cross-section of a GaAs MESFET.

that may be either *n*-type or *p*-type. The source and drain ohmic contacts are made by depositing a suitable metal by vacuum evaporation and by alloying in a subsequent heat treatment. The metal gate offers a number of advantages. First, the formation of a Schottky barrier contact can be achieved at much lower temperatures than those required for p-n junction formation. Second, the gate length can be reduced to submicron dimensions by controlling the length of the metal electrodes. It is not possible to obtain such small values of gate and channel length using the conventional ion implantation and diffusion technologies because of limitations on the diffusion process.

Depending on the thickness of the epitaxial layer, a MESFET may be designed to operate either as a depletion mode (normally on) device or an enhancement mode (normally off) device. In a depletion mode device, the thickness of the epitaxial layer is more than the zero-bias depletion region width of the Schottky barrier gate, and the transistor has a conducting channel at $V_G = 0$. Thus, the gate is biased negative to deplete the channel, as shown in Fig. 15.7a. In the enhancement-type device (Fig. 15.7b), the epitaxial layer is kept thin, and the built-in voltage of the metal gate Schottky barrier junction is sufficient to deplete the channel completely at $V_G = 0$. Conduction in the channel occurs only for small positive values of V_G (Fig. 15.7b). Enhancement-type devices are useful in high-speed, low-power applications, but the majority of MESFETs are the depletion type.

15.5 MODELS FOR I-V CHARACTERISTICS OF SHORT-CHANNEL MESFETS

Thus far, our discussion of JFET was confined to long-channel devices where the gate length is tens of microns. In such devices, the drift field in the channel may

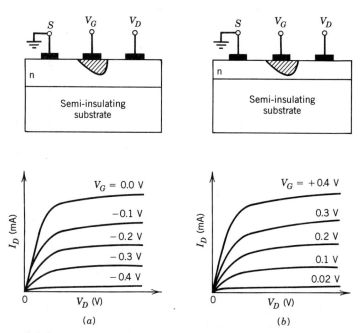

FIGURE 15.7 Cross-sections and I-V characteristics of two types of MESFET: (*a*) Depletion mode device and (*b*) enhancement mode device. (From S. M. Sze, reference 3, p. 324. Copyright © 1981. Reprinted by permission of John Wiley & Sons, Inc., New York.)

remain small enough for the constant mobility approximation to be valid. However, for short-channel JFETs and MESFETs having a small channel length to thickness (L/a) ratio the measured characteristics show considerable departure from the theory presented above. In these devices, the mobility does not remain constant but varies along the channel, and the gradual channel approximation is no longer valid. Therefore, the analysis requires a more refined treatment.

Depending on the gate length and the L/a ratio, several approaches have been used to describe the behavior of the short-channel devices. We will present only a brief, and largely qualitative, account of these developments. For details, the reader is referred to S. M. Sze [3] and the references mentioned there.

15.5.1 Field-Dependent Mobility Model

In the field-dependent mobility model, the drift velocity $v_d(x)$ is assumed to follow the relation

$$v_d(x) = \frac{\mu_n \mathscr{E}_x}{1 + \frac{\mu_n \mathscr{E}_x}{v_s}} \tag{15.27}$$

where v_s is the electron saturation velocity. Substituting for $v_d(x)$ from Eq. (15.27) into Eq. (15.10) and replacing $\mathscr{E}_x$ by dV/dx, the drain current can be written as

$$I_D = qN_d Z\mu_n a \left[1 - \sqrt{\frac{V_i - V_G + V(x)}{V_P}} \right] \left(\frac{dV}{dx} \frac{1}{1 + \frac{\mu_n}{v_s}\frac{dV}{dx}} \right) \tag{15.28}$$

Here we have assumed a single-gate structure, and hence, the factor 2 is missing from Eq. (15.28). The variables in the above equation can be easily separated. When the resulting expression is integrated from the source to the drain, the following expression of I_D is obtained (see Prob. 15.8):

$$I_D = \frac{G_o \left\{ V_D - \frac{2}{3} V_P \left[\left(\frac{V_i - V_G + V_D}{V_P} \right)^{3/2} - \left(\frac{V_i - V_G}{V_P} \right)^{3/2} \right] \right\}}{1 + \mu_n V_D/v_s L} \tag{15.29}$$

where

$$G_o = \frac{qN_d \mu_n Z a}{L}$$

Comparing this expression with Eq. (15.13), we see that the field-dependent mobility reduces the drain current by a factor $(1 + \mu_n V_D/v_s L)$. Note that in Eq. (15.29), (V_D/L) is the average field in the channel and $(\mu_n V_D/L)$ represents the average drift velocity of electrons. If this velocity remains small compared to v_s, the second term in the denominator of Eq. (15.29) may be neglected, and this equation becomes identical with Eq. (15.13).

The reduction in the drain current because of velocity saturation in the channel produces the following effects:

1. The drain saturation voltage $V_{D\text{sat}}$ becomes smaller than the pinch-off voltage.
2. The transconductance g_m is reduced below its value for the constant mobility case, and if $\mu_n V_P \gg v_s L$, g_m becomes almost independent of the gate bias.
3. The drain conductance in the saturation region at a given value of drain current becomes lower than the value it would have for a constant electron mobility.

15.5.2 Two-Region Model

Measured drain currents in short-channel Si MESFETs have been found to be in reasonably good agreement with the prediction of the field-dependent mobility model. However, the situation is more complex in GaAs, and a two-region model has been proposed for the current versus voltage characteristics of short-channel MESFETs in this material [4]. The difference between the field-dependent mobility model and the two-region model is illustrated by the plots in Fig. 15.8. Here the solid curve represents the velocity field relation as given by Eq. (15.27), and the dashed lines show the situation that corresponds to the two-region model. The channel of the device is separated into two longitudinal sections corresponding to two piece-wise linear segments of the assumed velocity-field characteristic. In region I, which extends from $x = 0$ to some distance $x = L_1$, Ohm's law is assumed to be valid, and the mobility μ_n is constant. In region II (from $x = L_1$ to $x = L$), the drift velocity is saturated at v_s and is independent of the electric field. The boundary between the two regions at $x = L_1$ is obtained by the condition that, at this point, the longitudinal electric field $\mathscr{E}_x$ reaches its critical value $\mathscr{E}_c$ at which the drift velocity saturates. The analysis includes the presaturation region and the region beyond pinch-off. However, the onset of pinch-off is defined by the condition that $\mathscr{E}_x$ reaches its critical value $\mathscr{E}_c$, and full channel depletion is not to be expected for a finite value of $\mathscr{E}_c$. The negative differential mobility in the

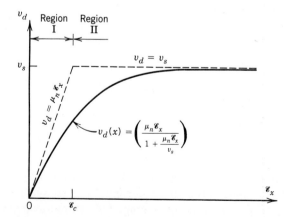

FIGURE 15.8 Two different approximations for the drift velocity electric field relation. The solid curve shows the field-dependent mobility case, and the dashed lines represent the two-region model approach.

velocity field characteristics of GaAs (see Sec. 11.6) has been ignored, since it was not thought to influence the characteristics of a short-channel GaAs MESFET to any significant extent.

15.5.3 Saturated Velocity Model

The saturated velocity model is another approximation to describe the behavior of short-channel MESFETs. In this model, it is assumed that the electrons move through the channel with the saturation velocity v_s. The drain current is then directly modulated by the difference between the depletion region width and the channel thickness a. Consequently, for a uniformly doped channel

$$I_D = qN_d v_s Z[a - W] \tag{15.30}$$

This model is expected to be valid in the limit of very narrow gates in which full current saturation may be assumed under the gate. In fact, Eq. (15.30) is found to agree quite well with experimental results for short-channel GaAs devices with $L \simeq 2$ μm. Note that the source-to-drain voltage does not appear in Eq. (15.30), since W is given by

$$W = \sqrt{\frac{2\varepsilon_s}{qN_d}(V_i - V_G)} \tag{15.31}$$

15.5.4 Two-Dimensional Models

When the gate length becomes shorter than about 2 μm, none of the above three models is valid, and two-dimensional effects dominate the device's operation.

Derivation of the I-V characteristics of a short-channel Si MESFET using two-dimensional analysis has revealed [5] that the current saturation occurs, not because of the channel pinch-off, but because of the electron velocity saturation in the channel. In fact, in short-channel devices, the channel is never pinched-off completely, and the departure from space-charge neutrality occurs in the conducting path of the channel before pinch-off. For a given value of the gate bias, the characteristics are linear when the drain voltage is small. However, as this voltage is increased, the field in the channel becomes high and the drift velocity does not increase linearly with the field. As a result, the current falls below the initial resistance line and the characteristics become nonlinear. Saturation occurs when the electric field in the channel reaches the critical value at which the drift velocity saturates. The situation in a short-channel GaAs MESFET is more complex because of the complicated electron velocity curve in this material (Fig. 11.23). Both the experimental data and two-dimensional calculations show that the formation of a stationary Gunn-domain at the drain side of the gate is responsible for current saturation in a short-channel GaAs MESFET. Operation of the device in the linear region of the characteristic is similar to that of a Si MESFET, but it is different in the saturation region. Figure 15.9 shows the situation for a GaAs-MESFET operated in the current saturation regime. Because of the widening of the depletion region resulting from the reverse bias on the gate, the channel is narrowed near the drain. At some point x_1, the electric field has reached its value $\mathscr{E}_p$ corresponding to the peak velocity $v_p \approx 2 \times 10^7$ cm/sec. To the right of x_1, the electric field is higher, and the drift velocity drops to the lower

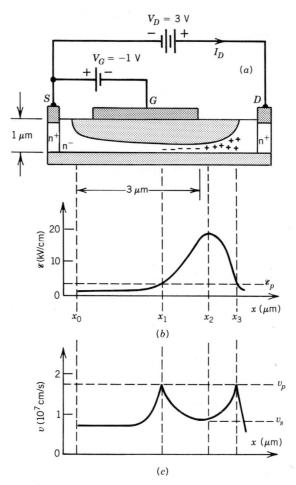

FIGURE 15.9 The GaAs MESFET operated in the saturated current region: (a) channel cross-section, (b) electric field, and (c) velocity distributions. (From C. A. Lichti, reference 5, p. 287. Copyright © 1976 IEEE. Reprinted by permission.)

saturation value of about 8×10^6 cm/sec because of electron transfer from the main valley to a satellite valley. Since the current

$$I_D = qn(x)v_d(x)Z[a - W(x)] \qquad (15.32)$$

remains constant throughout the channel, a reduction in $v_d(x)$ and the channel's cross-sectional area $Z[a - W(x)]$, must be compensated by an increase in the electron concentration $n(x)$. Consequently, electron accumulation occurs between x_1 and x_2 as shown in Fig. 15.9a. Between x_2 and x_3, the electron velocity increases and the channel becomes thicker. In order to keep I_D constant, electrons must be depleted from this region and give rise to a positive charge of ionized donors. The charges in the depletion and the accumulation regions form a dipole layer that supports most of the drain voltage after the current saturates.

Figure 15.10 shows the static I versus V characteristics of a GaAs MESFET. It is observed that there is a current drop back for $V_G = 0$ and $V_G = -1$ V which results from the decrease in the electron velocity in the channel, as shown in Fig. 15.9c. The current drop back occurs when the voltage across the dipole layer

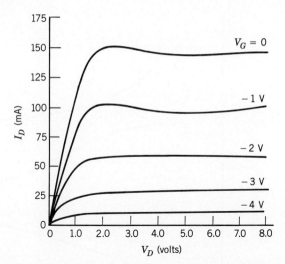

FIGURE 15.10 Static I-V characteristics of a short-channel GaAs MESFET. (From R. S. Pengelly, *Microwave Field-Effect Transistors Theory, Design and Applications,* p. 65. Copyright © 1982. Reprinted by permission of Research Studies Press Limited, Taunton, Somerset, England.)

increases at a faster rate than the source-to-drain voltage. It has been shown [6] that the current drop back can be suppressed either by a sufficient reduction of the epitaxial layer's thickness or by the addition of a high-resistivity substrate. Both effects increase the transverse field component $\mathscr{E}_y$, and if this field is kept large enough, no velocity decrease occurs along the channel to cause the negative resistance effect.

Some interesting results of the two-dimensional analysis of a short-channel GaAs MESFET are shown [7] in Fig. 15.11. As the gate length is decreased from 1.5 μm to 0.36 μm, the pinch-off voltage V_P, the transconductance g_m, and the saturation drain conductance g_D increase, but the gate-to-source capacitance (C_{GS}) does not decrease proportionately with the decrease in the gate length. While g_m increases largely because of the overall increase in the drain current, V_P increases mainly because of two-dimensional effects. All these effects are absent in any model that treats the gate control region as a one-dimensional diode.

15.6 HIGH-FREQUENCY PERFORMANCE

Two factors limit the high-frequency response of a FET: the transit time through the channel and the R-C time constant resulting from the input capacitance and the transconductance. For the constant mobility case, the transit time τ_t is given by

$$\tau_t = \frac{L}{\mu_n \mathscr{E}_x} \simeq \frac{L^2}{\mu_n V_D} \qquad (15.33)$$

and for the saturated drift velocity case

$$\tau_t = \frac{L}{v_s} \qquad (15.34)$$

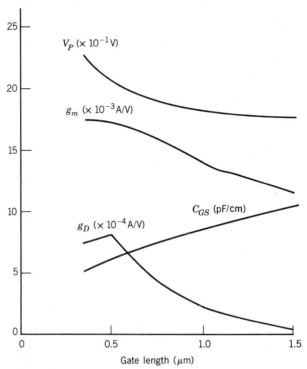

FIGURE 15.11 Plots showing the variations of V_P, g_m, g_D, and the gate-to-source capacitance C_{GS} as a function of gate length obtained from a two-dimensional analysis of a GaAs MESFET. (From W. R. Curtice, reference 7, p. 1695. Copyright © 1983 IEEE. Reprinted by permission.)

The transit time obtained either from Eq. (15.33) or (15.34) is usually small compared to the R-C time constant mentioned above.

The R-C time constant can easily be obtained from the small-signal equivalent circuit of the device. A brief account of the circuit models for JFET and MESFET is presented in Sec. 17.3. A simplified small-signal equivalent circuit of a MESFET is shown in Fig. 15.12 along with its circuit symbol. The arrow in Fig. 15.12a points from the p-to-n region. Thus, the symbol shown in Fig. 15.12a is for an n-channel FET; the direction of the gate arrow is reversed for a p-channel device.

In the simplified equivalent circuit of Fig. 15.12b, C_{GS} and C_{GD} represent the gate-to-source and the gate-to-drain capacitances, respectively, and C_{DS} is the drain-to-source capacitance. The current generator in the output is $g_m V_G$. A current cutoff frequency f_T is defined as the frequency at which current through C_{GS} is equal to the current generator $g_m V_G$, giving

$$f_T = \frac{g_m}{2\pi C_{GS}} \tag{15.35}$$

The R-C time constant $\tau_o = 1/2\pi f_T$ and the maximum frequency of oscillation $f_{\max}$ is directly related to f_T.

In order to obtain f_T in terms of device structure, we consider a single-gate FET in which the channel near the drain is just pinched-off. The average depletion region thickness is then $a/2$, and the capacitance C_{GS} is given by the relation

$$C_{GS} = \frac{2\varepsilon_s ZL}{a} \tag{15.36}$$

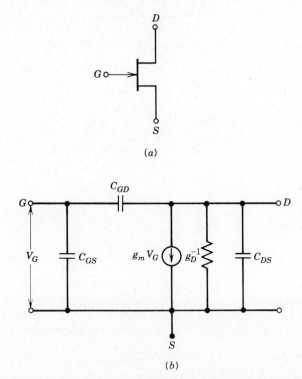

FIGURE 15.12 (a) Circuit symbol and (b) simplified small-signal equivalent circuit of a MESFET.

where ZL is the cross-sectional area of the capacitor. The maximum value that g_m can have is G_o, and substituting for G_o from Eq. (15.29) and C_{GS} from Eq. (15.36) into Eq. (15.35), we get the maximum possible value of f_T

$$f_T(\text{max}) = \frac{1}{2\pi}\left(\frac{qN_d a^2}{2\varepsilon_s}\right)\frac{\mu_n}{L^2} = \frac{1}{2\pi}\frac{V_P \mu_n}{L^2} \tag{15.37}$$

In practical devices, f_T will always be lower than that given by Eq. (15.37). It may be interesting to note that Eq. (15.37) becomes the same as Eq. (15.33) when V_P is replaced by V_D.

It is evident from Eq. (15.37) that the high-frequency limit of the JFET and MESFET is a natural limit caused by semiconductor properties and the physical dimensions of the device. For a transistor with a given value of V_P, the high-frequency performance can be improved by using semiconductors with higher carrier mobility and by decreasing the channel length L. Since the mobility of electrons in Si and GaAs is higher than the hole mobility in these semiconductors, an n-channel device has a higher operating frequency than a p-channel device with the same geometry. Moreover, since the electron mobility in GaAs is higher than in Si, we would expect a higher value of f_T in GaAs.

Apart from Si and GaAs, the other semiconductor that has been widely investigated for fabricating MESFETs is InP. This material has approximately a 50 percent higher value of the maximum drift velocity than GaAs. Therefore, the cutoff frequency f_T for InP MESFETs is expected to be higher than for GaAs devices. Figure 15.13 shows the calculated values of f_T plotted [8] as a function of the gate length for Si, GaAs, and InP at 300 K. The superiority of InP over the other two materials is evident. Besides the above three semiconductors, ternary compounds

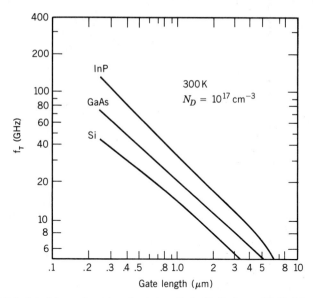

FIGURE 15.13 Calculated f_T as a function of gate length, in Si, GaAs, and InP. After T. J. Maloney and J. Frey, IEEE Trans. Electron Devices ED-23, 519 (1976). (From S. M. Sze, reference 3, p. 344. Copyright © 1981. Reprinted by permission of John Wiley & Sons, Inc., New York.)

InAsP and GaInAs and quarternary compound GaInAsSb have also been investigated for MESFET fabrication.

In a JFET structure, the smallest value of L is limited by the technological difficulties associated with the diffusion process. Ion implantation provides better control, but the gate length in a JFET can hardly be decreased below 1 μm. In MESFETs, electron beam lithography and plasma etching offer the possibility of submicron gate lengths. However, for the gate electrode to have adequate control of the current transport across the channel, the L/a ratio must be kept greater than unity. Thus, a reduction in channel length also means a reduction in the channel thickness a which implies a higher doping level to maintain a proper channel conductance. In Si and GaAs MESFETs, the highest doping that can be used without voltage breakdown in the channel is approximately 4×10^{17} cm^{-3}, and this limits the value of L to about 0.1 μm.

To illustrate the order of magnitude of the various characteristic times discussed above, let us consider an n-channel Si JFET with $V_P = 4$ V, $\mu_n = 1250$ cm^2/V-sec, and $L = 5$ μm. We assume $V_G = 0$ V. When the device is biased in the linear region at $V_D = 0.4$ V, μ_n may be assumed to be constant, and from Eq. (15.33), we obtain $\tau_t = 5 \times 10^{-10}$ sec. If V_D is now increased to cause velocity saturation in the channel, then with $v_s = 10^7$ cm/sec, Eq. (15.34) gives $\tau_t = 5 \times 10^{-11}$ sec. From Eq. (15.37), $f_T(\text{max}) = 10^{10}/\pi$ Hz, giving an R-C time constant $\tau_o = 5 \times 10^{-11}$ sec. Thus, τ_o is about the same as τ_t and limits the frequency response of the device. The maximum value of f_T is about 3.2 GHz. This is the most optimistic value we can expect. In a real device, g_m will be considerably lower than G_o, and hence, the value of f_T in a 5-μm gate length device will be lower than 3.2 GHz.

15.7 MESFET STRUCTURES

In order to obtain a large value of g_m and high power from the MESFET, the gate width Z should be as large as possible. The maximum gate width can be esti-

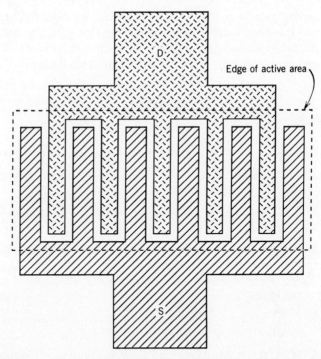

FIGURE 15.14 Top view of a power MESFET with interdigitated source and drain fingers. (From S. M. Sze, reference 3, p. 349. Copyright © 1981. Reprinted by permission of John Wiley & Sons, Inc., New York.)

mated from the criterion that the transmission line formed by the source and gate electrode should be a very small fraction of the wavelength. For a 1-μm gate length MESFET, the maximum gate width may be about 50 to 250 μm at 10 GHz. The gate width is usually increased by employing the interdigitated source and drain finger structure shown in Fig. 15.14. The parallel connection of gates can also be used to obtain a large value of gate width.

Single-gate MESFET structures have become most common and are widely used in microwave amplification. However, a dual-gate device has the advantage of a higher gain and a lower feedback capacitance, thus improving the gain stability. Figure 15.15 shows the schematic structure of a dual-gate MESFET. The

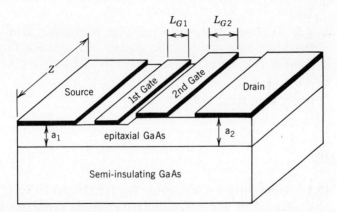

FIGURE 15.15 Structure of a dual-gate MESFET. (From S. Asai et al, reference 9, p. 898. Copyright © 1975 IEEE. Reprinted by permission.)

second gate near the drain can prevent the high field region between this gate and the drain reaching back under the first gate. By making the channel deeper under the second gate, a larger pinch-off voltage is obtained. This allows the second gate to operate with its maximum drain current.

REFERENCES

1. W. SHOCKLEY, "A unipolar field-effect transistor," *Proc. IRE* 40, 1365 (1952).
2. A. S. GROVE, *Physics and Technology of Semiconductor Devices,* John Wiley, New York, 1967, Chapter 8.
3. S. M. SZE, *Physics of Semiconductor Devices,* John Wiley, New York, 1981, Chapter 6.
4. R. A. PUCEL, H. A. HAUS, and H. STATZ, "Signal and noise properties of gallium arsenide microwave field-effect transistors," in *Advances in Electronics and Electron Physics,* L. Marten (ed.), Academic Press, New York, Vol. 38, 195 (1975).
5. C. A. LIECHTI, "Microwave field-effect transistors—1976," *IEEE Trans. Microwave Theory Tech.* MTT-24, 279 (1976).
6. M. A. R. AL-MUDARES and K. W. H. FOULDS, "Physical explanation of GaAs MESFET I/V characteristics," *IEE Proc.* 130 part I, 175 (1983).
7. W. R. CURTICE, "The performance of submicrometer gate length GaAs MESFET's," *IEEE Trans. Electron Devices* ED-30, 1693 (1983).
8. T. J. MALONEY and J. FREY, "Frequency limits of GaAs and InP field-effect transistors at 300 K and 77 K with typical active-layer doping," *IEEE Trans. Electron Devices* ED-23, 519 (1976).
9. S. ASAI, F. MURAI, and H. KODERA, "GaAs dual-gate Schottky-barrier FET's for microwave frequencies," *IEEE Trans. Electron Devices* ED-22, 897 (1975).

ADDITIONAL READING LIST

1. R. S. PENGELLY, *Microwave Field-Effect Transistors; Theory, Design and Applications,* John Wiley, Chichester, 1982.
2. J. A. TURNER, "Metal-semiconductor field effect transistors," Chapter 7 in *Metal-Semiconductor Schottky Barrier Junctions and Their Applications,* B. L. Sharma (ed.), Plenum Press, New York, 1984.
3. A. G. MILNES, *Semiconductor Devices and Integrated Electronics,* Van Nostrand, New York, 1980, Chapter 6.

PROBLEMS

15.1 Substituting for $W(x)$ from Eq. (15.12) and for $v_d(x)$ from Eq. (15.11) into Eq. (15.10) and integrating the resulting expression from $x = 0$ to $x = L$, obtain Eq. (15.13).

PROBLEMS

15.2 A double-gate n-channel Si JFET has $N_d = 5 \times 10^{15}$ cm^{-3}, $N_a = 10^{19}$ cm^{-3}, $a = 1$ μm, $L = 30$ μm, $Z = 0.1$ cm, and $\mu_n = 1050$ cm^2/V-sec. Determine (a) the pinch-off voltage, (b) the drain current for $V_D = V_P$ with the gate connected to the source, and (c) the drain current for $V_D = 0.5$ V and $V_G = -1$ V. Assume $T = 300$ K.

15.3 Consider a Si double-gate n-channel JFET with the following parameters: $N_a = 3 \times 10^{18}$ cm^{-3}, $N_d = 10^{15}$ cm^{-3}, $a = 2$ μm, $L = 20$ μm, and $Z/L = 5$. Assume $\mu_n = 1000$ cm^2/V-sec and $T = 300$ K. (a) Calculate the built-in voltage, the pinch-off voltage, and the value of the channel conductance with $V_G = 0$. (b) Calculate the drain conductance with $V_D = 0$ and $V_D = 0.5$ V for $V_G = -1$ V. (c) Calculate the transconductance in the linear region with $V_G = -1$ V and $V_D = 0.5$ V.

15.4 A uniformly doped single-gate p-channel Si JFET has the following parameters at 300 K: $N_a = 6 \times 10^{15}$ cm^{-3}, $N_d = 4.5 \times 10^{19}$ cm^{-3}, $L = 4$ μm, $Z = 50$ μm, and $a = 1$ μm. The hole mobility in the channel is 350 cm^2/V-sec. For this transistor, calculate (a) the pinch-off voltage and (b) the gain bandwidth product in the saturation region with a gate reverse bias of 2 V.

15.5 An n-channel JFET is being used as a controlled load by operating the transistor in saturation. The pinch-off voltage of the transistor is 3.5 V, and the built-in voltage of the gate-channel junction is 0.8 V. The gate is grounded. Assume $G_o = 1.44 \times 10^{-2}$ A/V and determine (a) the drain voltage V_{Dsat}, (b) the value of the load resistor with $V_D = V_{Dsat}$, and (c) the g_m of the transistor in the saturation region. Compare the value of g_m with that of a bipolar transistor with $I_C = I_{Dsat}$.

15.6 The avalanche breakdown voltage of the drain-substrate junction of an n-channel JFET is 30 V. Calculate the drain voltage that will produce avalanche breakdown when a reverse bias of 4 V is applied to the gate.

15.7 The dopant concentration in the channel of a single-gate n-channel JFET varies as $N_d(y) = Ky$ where $K = 10^{19}$ cm^{-4}. Calculate the pinch-off voltage, the drain current, and the values of g_D and g_m with $V_G = -2$ V and $V_D = 0.5$ V. Also calculate the value of g_m in the saturation region with $V_G = -2$ V. Assume $V_i = 0.6$ V in all your calculations and consider a single-gate device with the dimensions given in Prob. 15.3.

15.8 Starting from Eq. (15.28) and following the procedure described in the text, derive Eq. (15.29).

15.9 Consider an n-channel Si MESFET with $a = 1$ μm, and values of N_d, μ_n, V_P, and Z the same as in Prob. 15.3. Assume $V_i = 0.7$ V and $v_s = 10^7$ cm sec^{-1}. Using field-dependent mobility approximation, calculate the drain current for $\mu_n V_P/v_s L = 2$ with $V_G = -1$ V and $V_D = 2$ V.

15.10 Show that for a normally off n-channel MESFET, the saturated drain current can be expressed as

$$I_D = \frac{Z\mu_n \varepsilon_s}{2aL}(V_G - V_{th})^2$$

15.11 The current amplification factor μ of a JFET is defined by $\mu = \partial V_D / \partial V_G |_{I_D = \text{constant}}$. Show that $\mu = g_m g_D^{-1}$. Consider two identical JFETs and determine the values of μ and g_m for the composite unit when the two devices are connected (a) in series and (b) in parallel.

15.12 An InP MESFET has $N_d = 10^{17}$ cm^{-3}, $L = 1.5$ μm, $L/a = 5$, and $Z = 75$ μm. Assume $v_s = 6 \times 10^6$ cm sec^{-1} and $V_i = 0.7$ V. Using the saturated velocity model, calculate the values of I_D and g_m for $V_G = -1$ V and $V_D = 0.2$ V. Also estimate the electron transit time through the channel and the cutoff frequency f_T. Assume $\varepsilon_s = 12.4$ ε_o.

16
MOS TRANSISTORS AND CHARGE-COUPLED DEVICES

INTRODUCTION

An important feature of devices considered in this chapter is that their working is controlled by a capacitively coupled electric field acting perpendicular to the semiconductor surface. A large part of this chapter is devoted to metal–insulator–semiconductor transistors in which the perpendicular field modulates the conductance of the surface layer. These surface field-effect transistors are also known as *insulated gate field-effect transistors* (IGFETs) because the gate electrode in these devices is insulated from the source and drain by a thin insulator layer. Because of this insulation, an IGFET has a high-input resistance (of the order of 10^{13} ohms) that eliminates any possibility of feedback from output to input.

The idea of a solid-state amplifier based on surface field effect was conceived by J. E. Lilienfeld in 1930. However, a working device based on this principle could not be realized because a large density of electronic states existed at the surface of the semiconductor. It was only in 1960 that significant progress could be made in controlling the surface properties of silicon by thermal oxidation. Once this was achieved, the realization of a surface FET was simply the next step.

A surface field-effect transistor consists of one or more layers of metal and insulator deposited on a semiconductor and, therefore, can be classified as an MIS transistor where M stands for Metal, I for insulator, and S for semiconductor. A particular form of MIS transistor is the MOS transistor in which the insulator is the oxide, SiO_2. A charge-coupled device also consists of arrays of MOS structures. In this chapter we discuss the physical electronics of MOS devices in some detail. We start our study with a brief description of the electrical properties of the semiconductor surfaces. This is followed by a discussion of C-V characteristics of MOS capacitor and interface properties of the Si-SiO_2 system. Subsequently, the theory of the MOS transistor together with structural designs and short-channel effects is presented. The chapter ends with a brief account of charge-coupled devices.

16.1 SEMICONDUCTOR SURFACES

16.1.1 Electrical Properties of the Surface Space-Charge Region

The most important tool used in the study of semiconductor surfaces is the MOS capacitor shown in Fig. 16.1. It consists of a silicon substrate on which a thin SiO_2 layer is grown or deposited. A thin metal (usually Al) or a heavily doped polycrystalline Si-electrode is deposited on the top of the SiO_2 layer. This electrode is known as the *gate electrode* or simply the *gate*. The back side of the Si-substrate is provided with an ohmic contact. When a dc voltage is applied between the gate and the Si-substrate, the potential at the semiconductor surface under the gate electrode is altered, but no current can flow because of the presence of the SiO_2 layer. Consequently, the surface space-charge region in the semiconductor remains in thermal equilibrium, and the Fermi level will be constant throughout the semiconductor. However, the change in the potential at the semiconductor surface results in a relative shift between the Fermi level and the band edges. Since the electron and hole concentrations in the semiconductor depend on this shift, the mobile charge concentration in the surface space-charge layer changes with the bias on the gate. This is the process that we have called the *surface field effect*.

Figure 16.2 shows the energy band diagram of an ideal MOS structure made on a *p*-type semiconductor under thermal equilibrium. An ideal MOS system is one in which the following assumptions are valid: (1) The gate metal and the semiconductor have the same work function. (2) The oxide is a perfect insulator devoid of any charges. (3) The energy band diagram for the semiconductor is valid up to the surface, and no charges exist either on the oxide–semiconductor interface or in the bulk except those induced by the voltage at the gate.

In Fig. 16.2 the oxide is represented by an insulator with a conduction band edge lying below the vacuum level. The energy difference $q\phi_B$ between the lower edge of the oxide conduction band and the metal Fermi level is known as the *metal-to-oxide barrier energy*. In the Si-SiO_2 system this energy is smaller than the work function $q\phi_m$ of most of the metals.

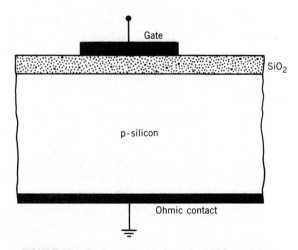

FIGURE 16.1 Cross-sectional view of a MOS capacitor.

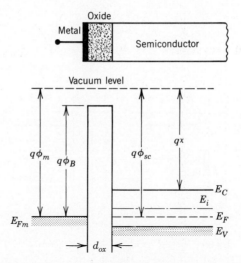

FIGURE 16.2 Energy band diagram of an ideal MOS capacitor on a p-type semiconductor under thermal equilibrium.

Plots in Fig. 16.3 show the energy band diagrams for three biasing conditions along with the idealized charge distribution for each situation [1]. These band diagrams can be obtained from considerations similar to those of Sec. 10.3.1. In Fig. 16.3a the metal gate is biased negative with respect to the semiconductor by a voltage V_G. The Fermi level on the metal side, therefore, rises above that on the semiconductor side by qV_G. Since ϕ_m and the electron affinity of the semiconductor remain unaffected by the bias voltage, the vacuum level on the metal side also rises by qV_G. In going from the semiconductor to the metal, the vacuum level must bend up gradually to accommodate the applied voltage V_G. Part of this bending occurs in the semiconductor; the rest occurs across the oxide. As the oxide is assumed to be charge free, the lower edge of the oxide conduction band will bend linearly as shown. The negative charge on the gate produces an equal and opposite charge in the semiconductor by attracting holes near the oxide-semiconductor interface. This causes an enhanced concentration of holes near the surface with a consequent upward bending of the energy levels E_C, E_V, and E_i. This condition is called *surface accumulation*.

When a small positive bias is applied to the gate, holes are pushed away from the vicinity of the oxide. This creates a depletion region in the semiconductor near the interface, consisting mainly of negatively charged acceptor ions. Since the hole concentration in the depletion region is considerably lower than that in the neutral semiconductor, the separation between the Fermi level and the valence band edge E_V is increased near the interface causing a downward bending of the energy levels E_C, E_V, and E_i. The energy band diagram and the corresponding charge distribution for this situation are shown in Fig. 16.3b. By using the depletion approximation, the charge Q_s contained per unit area of the semiconductor surface is given by

$$Q_s = -qN_a W \qquad (16.1)$$

Here W represents the width of the surface depletion region, and N_a is the acceptor concentration in the semiconductor which we have assumed to be uniform throughout.

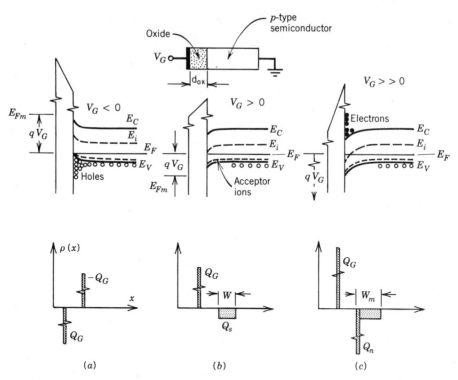

FIGURE 16.3 Energy band diagrams and charge distributions in an ideal MOS capacitor for different biasing conditions: (*a*) accumulation, (*b*) depletion, and (*c*) inversion. (From A. S. Grove, reference 1, p. 266. Copyright © 1967. Reprinted by permission of John Wiley & Sons, Inc., New York.)

If the positive bias on the gate is now increased gradually, the bands continue to bend downward and ultimately a point is reached at which E_i at the surface touches E_F. The semiconductor surface then becomes intrinsic. With a further increase in the gate voltage E_i crosses E_F and the conduction band edge E_C comes closer to E_F than the valence band edge E_V (Fig. 16.3c). In this situation the minority carriers (electrons) are attracted to the interface, and the semiconductor surface region contains more electrons than holes. The surface thus gets inverted from the *p*-type to the *n*-type. The thin zone in which the electron concentration exceeds the hole concentration is called the *inversion layer*. Once inversion occurs at the surface, a further increase in the gate bias will induce practically all the additional negative charge in the inversion layer. Consequently, the electron concentration in the surface space-charge region increases rapidly with the gate voltage, and the depletion region width reaches its maximum value W_m. When this happens, Q_s can be written as

$$Q_s = -Q_n - qN_a W_m \tag{16.2}$$

where Q_n represents the inversion layer charge per unit area. This charge distribution is shown in Fig. 16.3c. The inversion layer width is usually on the order of 50 to 100 Å and is small compared to W.

The n- and p-regions of the semiconductor in Fig. 16.3c are separated by a depletion layer. This situation is similar to that of a p-n junction. However, the junction in Fig. 16.3c is obtained by applying a field normal to the surface of the semiconductor, and thus is a field-induced junction.

16.1.2 Analysis of the Surface Space-Charge Region

We will now obtain relationships between the band bending and the electron-hole concentrations at the surface. To obtain these relations in their simplest form, we assume the semiconductor to be nondegenerate and uniformly doped. Under these assumptions, the electron and hole concentrations at any point in the semiconductor are given by (see Eqs. 3.27 and 3.28)

$$n = n_i \exp\left(\frac{E_F - E_i}{kT}\right)$$

and

$$p = n_i \exp\left(\frac{E_i - E_F}{kT}\right)$$

If p_o represents the hole concentration in the neutral semiconductor bulk, away from the surface, we can write

$$p_o = n_i \exp\left(\frac{E_{ib} - E_F}{kT}\right) \tag{16.3}$$

where E_{ib} is the intrinsic level in the neutral semiconductor. We define the Fermi potential ϕ_F by the relation $q\phi_F = (E_{ib} - E_F)$ and write

$$p_o = n_i \exp\left(\frac{q\phi_F}{kT}\right) \tag{16.4a}$$

Note that ϕ_F is the electrostatic potential at the Fermi level with respect to the intrinsic level E_{ib} and is positive for a p-type semiconductor. When all the acceptor atoms are ionized, from Eq. (16.4a) we obtain

$$\phi_F = \frac{kT}{q} \ln \frac{N_a}{n_i} \tag{16.4b}$$

Similarly, the hole concentration p_s at the surface of the semiconductor can be expressed as

$$p_s = n_i \exp\left[\frac{E_{is} - E_F}{kT}\right] \tag{16.5}$$

where E_{is} represents the intrinsic level at the semiconductor surface. From Eqs. (16.3) and (16.5) we obtain

$$p_s = p_o \exp\left[\frac{E_{is} - E_{ib}}{kT}\right] \tag{16.6}$$

Let us define the potential ϕ, deep in the semiconductor as zero and write $(E_{ib} - E_{is}) = q\phi_s$ where ϕ_s is the surface potential. We can then write

$$p_s = p_o \exp\left[\frac{-q\phi_s}{kT}\right]$$

and

$$n_s = n_o \exp\left[\frac{q\phi_s}{kT}\right] \quad (16.7)$$

The second relation in Eq. (16.7) follows from the fact that under thermal equilibrium $p_s n_s = p_o n_o = n_i^2$. Note that ϕ_s is the electrostatic potential at the surface of the semiconductor and designates the total bending of the energy level E_i (and also of E_C and E_V) from the bulk to the surface. The electron and hole concentrations at any point y in the surface space-charge region can be obtained simply by replacing $q\phi_s$ by $q\phi(y) = (E_{ib} - E_i(y))$. Thus

$$p(y) = p_o \exp\left[\frac{-q\phi(y)}{kT}\right]$$

$$n(y) = n_o \exp\left[\frac{q\phi(y)}{kT}\right] \quad (16.8)$$

The relations given by Eq. (16.7) are important as they relate the electron and hole concentrations at the surface to the bending of the energy bands. For example, when hole accumulation occurs near the surface $p_s > p_o$ and from Eq. (16.7), ϕ_s must be negative, which means that the intrinsic level E_{is} at the surface must be higher than E_{ib}. Therefore, the energy bands bend upward (Fig. 16.3a). In the case of depletion of holes from the surface, $p_s < p_o$ and ϕ_s must be positive, and energy bands bend as shown in Fig. 16.3b.

Figure 16.4 reproduces the energy band diagram of a MOS capacitor on the p-type semiconductor biased in the inversion regime. Note that all potentials are measured from the position of E_i in the bulk and are positive downwards. It is clear from this figure that the semiconductor surface becomes intrinsic when $\phi_s = \phi_F$. Thus, for a p-type semiconductor the onset of inversion is given by the condition $\phi_s > \phi_F$. Strong inversion is said to occur when the minority carrier concentration at the surface becomes equal to the majority carrier concentration in the bulk. For a p-type semiconductor this means that $n_s = p_o$, and the surface potential for this condition is given by

$$\phi_s(\text{strong inversion}) = 2\phi_F \quad (16.9)$$

This relation shows that strong inversion occurs when the energy level E_i at the surface comes below E_F as much as it is above E_F in the bulk.

16.2 C-V CHARACTERISTICS OF THE MOS CAPACITOR

16.2.1 The Ideal MOS System

The MOS structure in Fig. 16.1 would be acting as a simple parallel plate capacitor with SiO_2 as dielectric if the semiconductor below were a perfect conductor. How-

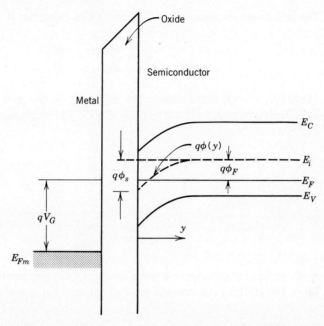

FIGURE 16.4 Energy band diagram of an inverted p-type semiconductor showing magnitudes and signs of ϕ_F and ϕ_s.

ever, the MOS capacitor is more complicated because the surface space-charge layer in the semiconductor under the gate metal is modified by the gate voltage. Let us analyze the effect of this layer on the capacitance of the MOS system.

When a voltage V_G is applied to the gate, a part V_{ox} of it will appear across the oxide so that

$$V_G = V_{ox} + \phi_s \tag{16.10}$$

If no charges are located at the oxide-semiconductor interface, the electric flux density will be continuous at the interface and from Gauss's law:

$$\varepsilon_{ox} \frac{V_{ox}}{d_{ox}} = \varepsilon_s \mathscr{E}_s = -Q_s \tag{16.11}$$

Here ε_{ox} and ε_s represent the oxide and semiconductor permittivities, respectively. $\mathscr{E}_s$ is the electric field at the surface of the semiconductor, and d_{ox} is the oxide thickness. Substituting for V_{ox} from Eq. (16.11) into Eq. (16.10) gives

$$V_G = -\frac{Q_s}{C_{ox}} + \phi_s \tag{16.12}$$

where $C_{ox}\ (=\varepsilon_{ox}/d_{ox})$ represents the oxide capacitance per unit area. Differentiating Eq. (16.12) with respect to V_G, we obtain

$$1 = -\frac{1}{C_{ox}} \frac{dQ_s}{dV_G} + \frac{d\phi_s}{dQ_s} \frac{dQ_s}{dV_G} \tag{16.13}$$

The small-signal capacitance C of the MOS capacitor is defined by the relation

$$C = \frac{dQ_G}{dV_G} = -\frac{dQ_s}{dV_G} \tag{16.14}$$

where $Q_G = -Q_s$ is the charge per unit area on the gate electrode. Substituting for dQ_s/dV_G from Eq. (16.14) into Eq. (16.13) and rearranging the terms, we obtain

$$\frac{1}{C} = \frac{1}{C_{ox}} + \frac{1}{C_s}$$

where

$$C_s = -\frac{dQ_s}{d\phi_s} \tag{16.15a}$$

is the semiconductor space-charge layer capacitance per unit area of the surface. From Eq. (16.15a) the normalized capacitance C/C_{ox} can be written as

$$\frac{C}{C_{ox}} = \frac{1}{1 + \dfrac{C_{ox}}{C_s}} \tag{16.15b}$$

For a given oxide thickness d_{ox}, C_{ox} is constant and independent of the gate voltage. Hence, if the voltage dependence of C_s is known, the ratio C/C_{ox} can be plotted as a function of V_G. A typical C-V plot of an ideal MOS capacitor made on the p-type Si is shown in Fig. 16.5. We will now explain how this curve is obtained.

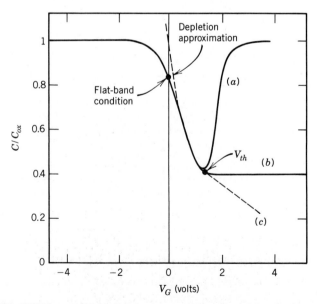

FIGURE 16.5 Ideal MOS capacitance-voltage curves: (*a*) low-frequency, (*b*) high-frequency, and (*c*) deep depletion. (From D.V. McCaughan and J.C. White, reference 12, p. 265. British Crown Copyright 1981. Reproduced by permission of the Controller of Her Britannic Majesty's Stationery Office.)

First consider that the MOS system is in thermal equilibrium at zero bias. Under this condition, the bands are flat and there is no charge in the semiconductor. Let a small ac signal now be applied to measure the capacitance of the system. The perturbation of the semiconductor surface by the small test signal produces an effective dynamic depletion region in the semiconductor whose thickness equals the extrinsic Debye length L_D given by

$$L_D = \sqrt{\frac{kT\varepsilon_s}{q^2(p_o + n_o)}} \quad (16.16)$$

For a strongly p-type semiconductor $p_o \simeq N_a, n_o \simeq 0$, and substituting $C_s = \varepsilon_s/L_D$ in Eq. (16.15b) the flat-band capacitance $C(\phi_s = 0) = C_{FB}$ can be written as

$$\frac{C_{FB}}{C_{ox}} = \frac{1}{1 + \dfrac{C_{ox} L_D}{\varepsilon_s}} \quad (16.17)$$

When the gate voltage is made negative, the hole concentration in the surface layer increases above p_o. This causes a decrease in the effective Debye length and a consequent increase in the capacitance C_s. Thus, the ratio C/C_{ox} starts increasing with the negative bias on the gate as shown by the solid curve in Fig. 16.5. Finally, at a sufficiently large negative gate voltage, a large number of excess holes are pulled very close to the oxide causing strong accumulation at the surface. As a result, C_s becomes very large and the ratio C/C_{ox} approaches unity.

For a positive value of V_G, holes are pushed away from the oxide–semiconductor interface creating a depletion region in the semiconductor. The depletion region capacitance in series with the oxide capacitance reduces the total capacitance well below C_{ox}. As the gate bias continues to rise, the depletion region widens and C_s goes on decreasing. This causes a gradual decrease in C/C_{ox} as V_G is made more and more positive. Finally, when V_G becomes sufficiently large to create strong inversion at the surface, the depletion region width reaches its maximum value. Then C_s has its minimum, and C/C_{ox} attains a constant value independent of the gate bias as shown by the solid curve (b) in Fig. 16.5.

The electric field and potential distributions in the surface space-charge region can be obtained from the solution of the one-dimensional Poisson equation, and C_s can be calculated [2] as a function of V_G. However, in the depletion regime of the MOS capacitor, conditions are similar to those in a reversed-biased p-n junction. This allows us to write (see Sec. 6.2.3)

$$\frac{d^2\phi}{dy^2} = \frac{qN_a}{\varepsilon_s} \quad (16.18)$$

Solving this equation with the boundary conditions $d\phi/dy = 0$ and $\phi = 0$ at $y = W$, we obtain

$$\phi(y) = \phi_s\left(1 - \frac{y}{W}\right)^2 \quad (16.19)$$

where ϕ_s is given by

$$\phi_s = \frac{qN_a W^2}{2\varepsilon_s} \quad (16.20)$$

The maximum width of depletion region W_m can be obtained from Eq. (16.20) after substituting the strong inversion condition $\phi_s = 2\phi_F$ giving

$$W_m = \sqrt{\frac{4\varepsilon_s}{qN_a}\phi_F} \qquad (16.21)$$

In the depletion approximation, the total charge $Q_s = -qN_aW$ and Eq. (16.12) becomes

$$V_G = \frac{qN_aW}{C_{ox}} + \phi_s \qquad (16.22)$$

The capacitance C_s can now be expressed as

$$C_s = \frac{\varepsilon_s}{W} \qquad (16.23)$$

Substituting for ϕ_s from Eq. (16.20) into Eq. (16.22) and solving for W gives

$$W = \frac{\varepsilon_s}{C_{ox}}\left[\sqrt{1 + \frac{2V_G}{qN_a\varepsilon_s}C_{ox}^2} - 1\right] \qquad (16.24)$$

Eliminating W between Eq. (16.24) and Eq. (16.23) and substituting for C_{ox}/C_s into Eq. (16.15b) leads to (see Prob. 16.4)

$$\frac{C}{C_{ox}} = \frac{1}{\sqrt{1 + \dfrac{2V_G}{qN_a\varepsilon_s}C_{ox}^2}} \qquad (16.25)$$

Calculations made using Eq. (16.25) are shown by the dashed curve in Fig. 16.5. Note that the depletion approximation is not valid for negative values of V_G including $V_G = 0$. If the inversion layer is not allowed to form, the MOS capacitor goes into deep depletion following the depletion approximation (curve c). We will now explain the low-frequency curve (a) and the part of the dotted curve denoted by (c).

As mentioned above, the MOS capacitance C is measured by superimposing a small ac signal (about 5 mV) on the dc bias. In the accumulation and depletion regimes of operation, the change in charge in the semiconductor in response to the applied signal requires the flow of majority carriers (i.e., holes in the p-type semiconductor) that either move out of the space-charge region or enter it. The time constant associated with this charge transport is the relaxation time τ_c which is on the order of 10^{-12} sec. Thus, for signal frequencies for which $\omega\tau_c \ll 1$, the C-V curve in the accumulation and the depletion regimes will be independent of the frequency. However, in the inversion regime the charge flow in response to the ac signal may also occur by movement of minority carriers between the inversion layer and the neutral semiconductor, and the MOS system shows a very strong frequency dependence.

To understand this frequency dependence, let us consider that the positive bias on the gate is increased by ΔV_G. This places an additional charge $+\Delta Q_G$ on the gate; consequently, a negative charge $-\Delta Q_G$ is induced in the semiconductor.

C-V CHARACTERISTICS OF THE MOS CAPACITOR

At high frequencies, the increase in the negative charge is caused by holes moving out of the depletion region edge as shown in Fig. 16.6a. The measured capacitance under this situation will be the oxide capacitance in series with the depletion region capacitance. Since the depletion region capacitance has reached its minimum value, the ratio C/C_{ox} also reaches its minimum as shown by the solid curve (b) in Fig. 16.5. Now if electron–hole pairs can be generated in the depletion region before ΔV_G goes to zero, the generated holes will replace the holes that had moved out of the depletion region edge while the electrons will move into the inversion layer (Fig. 16.6b). In this situation, the measured capacitance will be that of the oxide layer alone, and the C/C_{ox} will rise toward unity as shown by curve (a) in Fig. 16.5. Thus, it is clear that after the onset of strong inversion, curve (a) corresponds to low-frequency measurements, while curve (b) shows the high-frequency behavior of the MOS capacitor. For the Si-SiO$_2$ system near room temperature, the low-frequency behavior is exhibited below about 100 Hz. However, the transition frequency can be increased by an increase in the electron–hole pair generation rate (for example, by raising the temperature of the sample).

Part (c) of the dotted curve corresponds to the situation in which both the gate bias and the small-signal ac voltage vary at a faster rate than the rate of electron–hole pair generation in the depletion region. In this case, an inversion layer cannot form and the depletion layer becomes thicker than W_m. Consequently, the ratio C/C_{ox} keeps on decreasing with the gate bias. This situation occurs when a fast-rising voltage pulse is applied to the gate and is directly related to the operation of a charge-coupled device considered in Sec. 16.9.

Referring to the high-frequency curve (b) in Fig. 16.5, we recall that C/C_{ox} reaches its minimum value when strong inversion occurs at the semiconductor surface. The gate voltage at the onset of strong inversion is referred to as the *threshold voltage* V_{th} of the MOS system. This voltage can be obtained from Eq. (16.22) by substituting $\phi_s = 2\phi_F$ and $W = W_m$ where W_m is given by Eq. (16.21). Thus

$$V_{th} = -\frac{Q_{BO}}{C_{ox}} + 2\phi_F \tag{16.26}$$

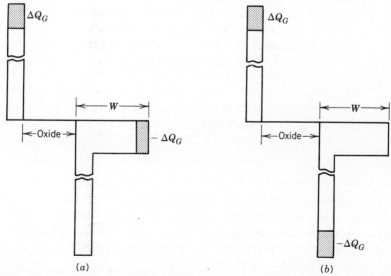

FIGURE 16.6 Variation of charge distribution in a MOS capacitor biased in the inversion regime: (a) high frequencies and (b) low frequencies.

where

$$Q_{BO} = -qN_aW_m = -\sqrt{4\varepsilon_s qN_a\phi_F} \qquad (16.27)$$

is the depletion region charge at the onset of strong inversion. The threshold voltage V_{th} is shown in Fig. 16.5.

16.2.2 Real MOS Capacitor

In the above discussion of the ideal MOS capacitor, we have assumed that there is no band bending in the semiconductor for $V_G = 0$. This condition is not realized in practice. Differences between the metal and the semiconductor work functions and charges in the oxide lead to nonideal behavior in practical MOS structures.

The gate metal and the semiconductor in general have different work functions. Thus, a difference $\phi_{ms} = (\phi_m - \phi_{sc})$ exists between the gate and the semiconductor bulk at $V_G = 0$. Note that the semiconductor work function is given by $q\phi_{sc} = (q\chi + E_g/2 \pm q\phi_F)$ so that

$$\phi_{ms} = \phi_m - \left(\chi + \frac{E_g}{2q} \pm \phi_F\right) \qquad (16.28)$$

Here the plus sign is valid for the p-type and the minus sign for the n-type semiconductor. Figure 16.7 gives values of ϕ_{ms} as a function of doping in n- and p-silicon for Al, Au, and degenerately doped n^+- and p^+-silicon gate electrodes.

The oxide-silicon system usually contains two types of charges: the mobile and fixed charges in the oxide and an interface charge at the Si-SiO$_2$ interface. Although the origin of some of these charges is still being debated, the total charge in the oxide can be determined by using high-frequency C-V characteristics of the MOS capacitor and by several other techniques [2].

As a result of oxide and interface charges and the metal–semiconductor work function difference, a shift in the ideal C-V curve of the MOS capacitor takes place. The solid curve in Fig. 16.8 shows the measured high-frequency C-V plot of an MOS capacitor made on p-silicon having $N_a = 5.5 \times 10^{16}$ cm^{-3} with Al-gate and SiO$_2$ thickness of 1100 Å. The dotted curve is the ideal curve which has been calculated from the known values of N_a and the SiO$_2$ thickness. The measured curve is displaced from the ideal curve toward the negative value of V_G, indicating the presence of a positive charge in the oxide. The gate voltage required to bring about the flat-band condition in the semiconductor is called the *flat-band voltage* V_{FB} and is given by [1]

$$V_{FB} = \phi_{ms} - \frac{Q_{it}}{C_{ox}} - \frac{1}{C_{ox}d_{ox}}\int_0^{d_{ox}} y\rho(y)\,dy \qquad (16.29)$$

where Q_{it} represents the charge per unit area at the oxide–semiconductor interface and $\rho(y)$ is the volume charge density in the oxide. In Eq. (16.29) the second term represents the shift in the C-V curve owing to interface charges, and the third term gives the shift owing to oxide charges. These two terms can be combined, and V_{FB} can be rewritten as

$$V_{FB} = \phi_{ms} - \frac{Q_{ox}}{C_{ox}} \qquad (16.29a)$$

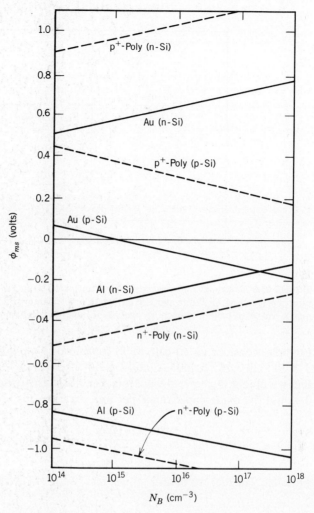

FIGURE 16.7 The metal-semiconductor work function difference ϕ_{ms} plotted as a function of dopant concentration in silicon for Al, Au, and polysilicon gates. (From S. M. Sze, reference 2, p. 397. Copyright © 1981. Reprinted by permission of John Wiley & Sons, Inc., New York.)

where Q_{ox} denotes the total charge. From the plot of Fig. 16.8, we obtain $V_{FB} = -4.5$ V. Figure 16.7 shows that for the Al-gate on p-silicon ϕ_{ms} is negative and its magnitude is less than 1 V. Consequently, the total oxide charge Q_{ox} must be positive because a negative value of V_{FB} requires Q_{ox} to be positive. The flat-band voltage causes a change in the threshold voltage. The new threshold voltage is the algebraic sum of V_{FB} and the voltage given by Eq. (16.26). Thus

$$V_{th} = V_{FB} - \frac{Q_{BO}}{C_{ox}} + 2\phi_F \qquad (16.30)$$

In the above discussion a p-type semiconductor has been assumed, but a similar analysis can be made for an n-type semiconductor. However, note that for the Si-SiO$_2$ system V_{FB} is negative in the case of p-silicon, but the remaining two terms in Eq. (16.30) are positive, whereas for n-silicon all the three terms are negative. Consequently, the magnitude of the threshold voltage of MOS transistors made on n-silicon is larger than that for transistors made on p-silicon. As an

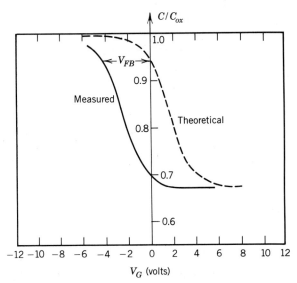

FIGURE 16.8 Calculated and measured high-frequency C-V plots of a MOS capacitor on p-silicon substrate. (From M. H. White and J. R. Cricchi, Complementary MOS transistors, *Solid-State Electronics*, vol. 9, p. 1000. Copyright © 1966. Reprinted by permission of Pergamon Press, Oxford.)

example, consider an Al-gate MOS capacitor on a p-silicon substrate with $N_a = 3 \times 10^{15}$ cm^{-3}, $d_{ox} = 1000$ Å, and 4×10^{10} charges cm^{-2} in the oxide. For this system we obtain $\phi_{ms} = -0.9$ V from Fig. 16.7. The dielectric constant of SiO$_2$ is 3.9, and the oxide capacitance per unit area is

$$C_{ox} = \frac{\varepsilon_{ox}}{d_{ox}} = \frac{3.9 \times 8.86 \times 10^{-14}}{10^{-5}} = 3.46 \times 10^{-8} \text{ F cm}^{-2}$$

From Eq. (16.29a) we obtain

$$V_{FB} = -0.9 - \frac{4 \times 10^{10} \times 1.6 \times 10^{-19}}{3.46 \times 10^{-8}} = -1.08 \text{ V}$$

From Eq. (16.4b) at 300 K, $\phi_F = 25.86 \times 10^{-3} \ln(3 \times 10^{15}/1.5 \times 10^{10}) = 0.316$ V and using this value of ϕ_F in Eq. (16.27) we get

$$Q_{BO} = -2.52 \times 10^{-8} \text{ C cm}^{-2}$$

Finally, from Eq. (16.30) we have

$$V_{th} = -1.08 + \frac{2.52}{3.46} + 2 \times 0.316 = 0.28 \text{ V}$$

For n-silicon with the same dopant concentration $\phi_{ms} = -0.25$ V and thus $V_{FB} = -0.43$ V. Note that now $\phi_F = -0.316$ V and $Q_{BO} = \sqrt{4\varepsilon_s q N_d \phi_F} = 2.52 \times 10^{-8}$ C cm^{-2} giving

$$V_{th} = -0.43 - \frac{2.52}{3.46} - 2 \times 0.316 = -1.79 \text{ V}.$$

16.3 THE Si-SiO$_2$ SYSTEM

Among all the insulator-semiconductor systems employed to fabricate IGFETs, the Si-SiO$_2$ system is by far the most important one. The most widely used method for growing SiO$_2$ layers on an Si surface is thermal oxidation (see Sec. 19.3.1). The following discussion will be confined to the thermally grown SiO$_2$.

The exact nature of the Si-SiO$_2$ interface, after thermal oxidation of Si, is not fully understood as yet. Perhaps the transition from Si to SiO$_2$ occurs over an incompletely oxidized monolayer of composition SiO$_x$ with x lying between 1 and 2. This monolayer is followed by a strained SiO$_2$ region of thickness 10-40 Å, and the remaining layer is strain-free amorphous SiO$_2$.

The four general types of charges associated with the Si-SiO$_2$ system are shown in Fig. 16.9. These charges may be classified as follows [3]: (1) Fixed oxide charge Q_f is a positive charge located in the oxide up to a distance of about 30 Å from the Si-SiO$_2$ interface. The origin of this charge is related to the oxidation process. It has been suggested that excess ionized silicon or loss of an electron from excess oxygen centers near the Si-SiO$_2$ interface are the probable causes of this charge. The charge Q_f is not in electrical communication with the underlying silicon. (2) The mobile oxide charge Q_m is primarily due to the presence of alkali ions like Na$^+$, K$^+$, and Li$^+$. Heavy metal ions may also contribute to this charge. These impurities are present in the cleaning solutions and other reagents as well as in the quartz tube used in the oxidation process. On application of gate bias, this charge can move through the oxide at elevated temperatures causing a change in the flat-band voltage with time. The charge Q_m can be reduced to a very low value by taking appropriate precautions at the time of oxide growth. (3) The oxide-trapped charge Q_{ot} may be either negative or positive due to electrons or holes trapped in the bulk of the oxide. The traps may result from ionizing radiation, from avalanche injection produced by hot electrons, or from other similar processes. (4) The interface trapped charge Q_{it} is mainly due to the breaking of covalent bonds at the Si-SiO$_2$ interface. The probable origin of this charge

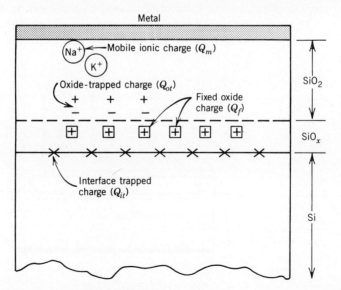

FIGURE 16.9 Charges associated with thermally oxidized silicon. (From B. E. Deal, reference 3, p. 607. Copyright © 1980 IEEE. Reprinted by permission.)

lies in oxidation-induced structural defects; metallic impurities diffusing to the interface; and other defects caused by radiation or similar bond breaking processes. This charge is in electrical communication with the underlying silicon and has often been referred to as fast surface states or interface states charge.

16.4 BASIC STRUCTURES AND THE OPERATING PRINCIPLE OF MOSFET

16.4.1 Types of MOSFETs

Two distinctly different types of MOS transistors are available commercially: n-channel and p-channel. The n-channel devices employ p-silicon substrate, whereas the p-channel devices are made on n-silicon substrate. On the basis of operating mode, each of these devices can be further classified into two categories, namely, the enhancment type and the depletion type. The enhancement-type transistor is normally off, and no current flows between the source and the drain for $V_G = 0$. The channel is then induced by applying a voltage of appropriate polarity to the gate. In a depletion-type device, a conducting channel already exists, and the device is on with no bias applied to the gate. The channel is depleted of mobile carriers by a gate voltage of polarity opposite to that of the drain voltage.

Figure 16.10 shows the basic structure of an n-channel enhancement type MOSFET. It consists of a p-Si substrate into which two heavily doped n^+-regions have been diffused. One of these serves as source and the other as drain. A thin SiO_2 layer grown by thermal oxidation of Si prior to diffusion separates the source and the drain. A metal contact is deposited on the oxide layer that completely covers the gap between the two n^+-diffusions. This contact serves as the gate. Metal contacts are also made to the two n^+-regions as shown in Fig 16.10. The surface of the underlying p-silicon can be inverted to produce a conducting channel between the source and the drain by biasing the gate positive with respect to the source. The source and the substrate are usually connected together, but in some applications the substrate is biased with respect to source.

The cross-section and the output characteristics of the n-channel enhancement-type MOSFET are shown in Fig. 16.11a. Note that initially the device is

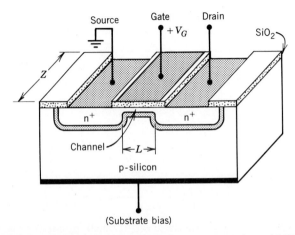

FIGURE 16.10 Schematic diagram of an enhancement-type n-channel MOS transistor.

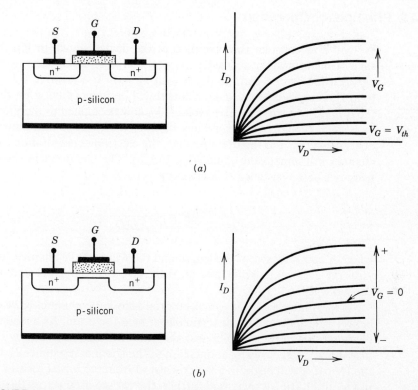

FIGURE 16.11 Schematic cross-sections and output characteristics of n-channel MOS transistors. (a) Enhancement-type device and (b) depletion-type device.

off and a positive gate voltage in excess of the threshold voltage V_{th} has to be applied to induce the channel and start conduction between the source and the drain. The electron concentration in the channel is enhanced by increasing the gate bias, and the device can be operated only in the enhancement mode. The n-channel depletion-type structure (Fig. 16.11b) is fabricated in a similar manner as the enhancement-type device. However, in the depletion-type device, the dopant concentration in the p-Si substrate is kept so low that the presence of ϕ_{ms} and the positive charge in the oxide creates an inversion layer at the Si surface below the gate electrode. Thus, the channel is already present with no bias on the gate, and the transistor is normally on. This device can now be operated either in the enhancement mode by applying a positive gate voltage or in the depletion mode with a negative bias on the gate.

The p-channel enhancement- and depletion-type devices can be produced in a similar manner on n-Si substrate. However, because of the existence of a positive charge in the oxide and the resulting negative value of the flat-band voltage V_{FB}, the p-channel MOSFETs usually tend to be of the enhancement type. The n-channel devices, on the other hand, can be made of either type. The electron mobility in Si is more than twice the hole mobility. So an n-channel MOSFET has a faster speed and better high-frequency capability than a p-channel device with the same geometry and dimensions. Consequently, an n-channel structure is preferred over a p-channel structure, and our subsequent discussion will be largely confined to an n-channel MOSFET. However, similar considerations can be applied to a p-channel device by changing the polarities of the gate and the drain voltages.

16.4.2 Principle of Operation

We will now consider the operation of the device shown in Fig. 16.10. For the sake of simplicity let us consider the situation when no conducting channel exists between the source and the drain with $V_G = 0$. If a positive voltage V_{G1} higher than V_{th} is applied to the gate, channel will be induced between the source and the drain. Let a small positive voltage V_D now be applied to the drain. As long as V_D is small enough not to cause any significant difference in the surface potential near the source and the drain regions, the electron concentration throughout the channel will remain the same (Fig. 16.12a). The channel will then behave as a resistor whose resistance R is given by

$$R = -\frac{L}{\mu_n Z Q_n} \qquad (16.31)$$

where L and Z represent the length and the width of the channel (see Fig. 16.10), μ_n is the electron mobility, and Q_n is the electron charge per unit area of the inversion layer.

As V_D is increased, the potential drop across the channel reduces the voltage between the gate and the inversion layer near the drain. As a result, the electron concentration in the channel near the drain will decrease. This will decrease the magnitude of Q_n, causing an increase in the channel resistance R. Consequently, the I-V curve will begin to bend downward from the initial resistor line as shown in Fig. 16.13. If V_D is increased further, the voltage drop across the oxide near the drain will continue to decrease until it falls below the value which is required to

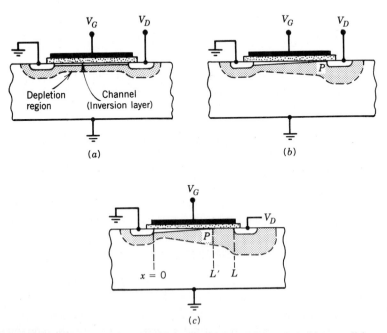

FIGURE 16.12 MOSFET cross-sections showing the effect of various bias conditions on the channel thickness. (a) The drain voltage V_D is small, and the depletion region is uniform along the channel; (b) $V_D = V_{Dsat}$, and the channel is pinched off near the drain; and (c) V_D exceeds V_{Dsat}, and the effective channel length is reduced.

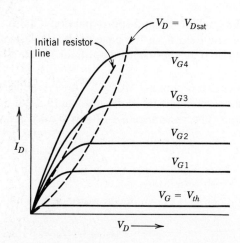

FIGURE 16.13 Idealized $I_D - V_D$ characteristics of an n-channel enhancement-type MOSFET.

maintain an inversion layer. When this happens, the channel near the drain gets pinched-off and the drain gets isolated from the conducting channel by a depletion region (Fig. 16.12b). The drain current will then saturate to a constant value. The current flow mechanism after the pinch-off is the same as that discussed for the JFET in Sec. 15.1.

An increase in V_D above the saturation voltage V_{Dsat} will cause the pinch-off point P to move toward the source, and the effective channel length will decrease slightly as shown in Fig. 16.12c. However, the potential at P will remain at V_{Dsat}, and the additional voltage in excess of V_{Dsat} will drop across the depletion region. Thus, the drain current will not be altered much when V_D is raised above V_{Dsat}.

If the gate bias is now increased to V_{G2}, it will cause an increase in the inversion layer charge Q_n. Hence, the channel resistance R will be reduced, causing a larger drain current for a given value of V_D. The pinch-off will now occur at a higher value of V_D. These features are evident from the $I_D - V_D$ characteristics of Fig. 16.13.

In the above discussion we have assumed an enhancement-type device. In a depletion-type MOSFET, the surface of the p-Si substrate is converted either to a high-resistivity p-type or a high-resistivity n-type using ion implantation. The positive charge in the SiO_2 and the presence of ϕ_{ms} then create conducting channel at the Si surface, causing an appreciable drain current even in the absence of any gate bias. The operating principle of this device is basically similar to that of the enhancement-type MOSFET discussed above.

16.5 CURRENT-VOLTAGE CHARACTERISTICS

16.5.1 Derivation of the Idealized Current-Voltage Relation

In deriving the $I_D - V_D$ characteristics of a MOSFET, we make the following assumptions:

1. The current flow in the channel is one-dimensional and no current flows from the source to the gate.
2. The electron mobility in the channel is constant, independent of the electric field.

3. The gradual channel approximation is valid. This means that the transverse electric field $\mathcal{E}_y$ at the edge of the depletion region in the channel is much higher than the longitudinal field $\mathcal{E}_x$ (see Sec. 15.2).

Consider an n-channel MOSFET of channel length L and width Z (Fig. 16.14). Let us take an elemental section of length dx of the channel at a distance x from the source. The resistance dR of this section is given by

$$dR = -\frac{dx}{Q_n(x)\mu_n Z} \tag{16.32}$$

The total charge $Q_s(x)$ induced in the semiconductor can be expressed as

$$Q_s(x) = Q_n(x) + Q_B(x) \tag{16.33}$$

where $Q_B(x)$ is the depletion region charge per unit area. From Eq. (16.12) we obtain Q_s in terms of V_G and the surface potential for an ideal MOS capacitor. In the case of a nonideal MOS capacitor with flat-band voltage V_{FB}, Eq. (16.12) at any point x becomes

$$V_G - V_{FB} = -\frac{Q_s(x)}{C_{ox}} + \phi_s(x) \tag{16.34}$$

Eliminating $Q_s(x)$ between Eq. (16.33) and (16.34), we obtain

$$Q_n(x) = C_{ox}[\phi_s(x) - (V_G - V_{FB})] - Q_B(x) \tag{16.35}$$

We will now determine the values of $\phi_s(x)$ and $Q_B(x)$ under the condition of strong inversion throughout the channel, assuming that the substrate is connected to the source. We have seen that in a MOS capacitor no drain current flows so that the system is in thermal equilibrium and $\phi_s = 2\phi_F$ at the onset of

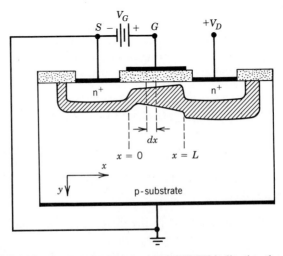

FIGURE 16.14 Cross-sectional view of an n-channel MOSFET indicating the differential length dx along the channel.

CURRENT-VOLTAGE CHARACTERISTICS

strong inversion. Under its actual operating condition, the space-charge region in a MOS transistor is not in thermal equilibrium. When the source and the substrate are grounded and the drain is biased positive, the surface potential will increase with increasing x, as will the amount of reverse bias across the channel-substrate junction. If $V(x)$ represents the channel potential at x for a drain voltage V_D, then to a good approximation the surface potential for strong inversion is given by

$$\phi_s(x) = 2\phi_F + V(x) \qquad (16.36)$$

Making use of depletion approximation, we can write the charge density $Q_B(x)$ at the onset of strong inversion as

$$Q_B(x) = -\sqrt{2qN_a\varepsilon_s[2\phi_F + V(x)]} \qquad (16.37)$$

Since $V(x)$ increases with x, the charge $Q_B(x)$ increases with position from the source end to the drain end. Substituting for $\phi_s(x)$ from Eq. (16.36) and for $Q_B(x)$ from Eq. (16.37) into Eq. (16.35) yields

$$Q_n(x) = C_{ox}[2\phi_F + V(x) - (V_G - V_{FB})] + \sqrt{2q\varepsilon_s N_a[2\phi_F + V(x)]} \qquad (16.38)$$

For a drain current I_D, the voltage drop across the element of resistance dR is

$$dV = I_D dR$$

and replacing dR by Eq. (16.32) leads to the following relation:

$$I_D dx = -\mu_n Z Q_n(x) dV \qquad (16.39)$$

Substituting for $Q_n(x)$ from Eq. (16.38) in Eq. (16.39) and integrating the resulting expression from $x = 0$ to $x = L$, we obtain the following expression for I_D (see Prob. 16.10)

$$I_D = \frac{Z\mu_n C_{ox}}{L} \left\{ \left[V_G - V_{FB} - 2\phi_F - \frac{V_D}{2} \right] V_D - \frac{2}{3}\Theta[(2\phi_F + V_D)^{3/2} - (2\phi_F)^{3/2}] \right\} \qquad (16.40)$$

where

$$\Theta = \sqrt{2\varepsilon_s q N_a}/C_{ox} \qquad (16.41)$$

is called the *body factor*. This relation is valid only below saturation. After the onset of saturation, I_D becomes independent of V_D and the above equation is no longer applicable. We will now consider two limiting cases of Eq. (16.40):

(a) LINEAR REGION

For very low values of V_D such that $V_D \ll (V_G - V_{FB} - 2\phi_F)$ and $V_D \ll 2\phi_F$, the term $V_D/2$ on the right-hand side in Eq. (16.40) can be neglected, and the equation simplifies to

$$I_D = \frac{Z\mu_n C_{ox}}{L}[V_G - V_{FB} - 2\phi_F - \Theta(2\phi_F)^{1/2}]V_D \qquad (16.42)$$

This relation describes the linear region of the $I_D - V_D$ characteristics near the origin in Fig. 16.13. Noting that

$$\Theta(2\phi_F)^{1/2} = \frac{\sqrt{4\varepsilon_s q N_a \phi_F}}{C_{ox}} = -\frac{Q_{BO}}{C_{ox}} \qquad (16.43)$$

Eq. (16.42) can be rewritten as

$$I_D = \frac{Z\mu_n C_{ox}}{L}(V_G - V_{th})V_D \qquad (16.44)$$

where V_{th} is given by Eq. (16.30). For an n-channel enhancement-type device, V_{FB} is small negative, but the remaining two terms in Eq. (16.30) are positive. Thus, V_{th} is a positive quantity and represents the turn-on voltage of the transistor. However, in a depletion-type device, the negative value of V_{FB} is large compared to the other two terms and V_{th} is negative. It then represents the voltage that must be applied to the gate to turn off the transistor by depleting the channel of mobile electrons.

(b) SATURATION REGION

The gradual channel approximation loses its validity when the channel gets pinched-off near the drain. The voltage V_{Dsat} at which I_D saturates can be obtained from the condition that at this voltage the charge density in the inversion layer near the drain becomes zero. Thus, substituting $V(x) = V_{Dsat}$ and making $Q_n(x) = 0$ in Eq. (16.38) and solving the resulting equation for V_{Dsat}, it can be shown that (see Prob. 16.11):

$$V_{Dsat} = V_G - V_{FB} - 2\phi_F + \frac{\Theta^2}{2} - \Theta\left(V_G - V_{FB} + \frac{\Theta^2}{4}\right)^{1/2} \qquad (16.45)$$

The drain current in saturation can be obtained after substituting $V_D = V_{Dsat}$ from Eq. (16.45) into Eq. (16.40). However, considerable simplification of Eq. (16.45) results when the dopant concentration N_a is low and the oxide layer is thin, so that the voltage drop across the oxide is negligibly small compared to $2\phi_F$. Under these conditions, the body factor Θ will be much less than unity and the following expression is obtained

$$V_{Dsat} = V_G - V_{FB} - 2\phi_F - \Theta(V_G - V_{FB})^{1/2}$$

Assuming that the voltage drop across the oxide is negligibly small compared to the surface potential at strong inversion and writing $(V_G - V_{FB}) \simeq 2\phi_F$, we get

$$V_{Dsat} = V_G - [V_{FB} + 2\phi_F + \Theta(2\phi_F)^{1/2}] = V_G - V_{th} \qquad (16.46)$$

CURRENT-VOLTAGE CHARACTERISTICS

where we have made use of Eqs. (16.43) and (16.30). The saturated drain current is now obtained after substituting for V_{Dsat} from Eq. (16.46) into Eq. (16.40) giving

$$I_{Dsat} = \frac{Z\mu_n C_{ox}}{L}\left\{[V_G - V_{FB} - 2\phi_F](V_G - V_{th}) - \frac{(V_G - V_{th})^2}{2}\right.$$
$$\left. - \frac{2}{3}\Theta[(2\phi_F + V_G - V_{th})^{3/2} - (2\phi_F)^{3/2}]\right\} \quad (16.47)$$

This current is independent of V_D as shown by the curves in Fig. 16.13. In the limit when $\Theta \ll 1$ and writing $V_{th} \approx V_{FB} + 2\phi_F$, the above equation reduces to

$$I_{Dsat} = \frac{Z\mu_n C_{ox}}{2L}(V_G - V_{th})^2 \quad (16.48)$$

This square law dependence of I_D on the gate voltage is valid for all values of $V_D > (V_G - V_{th})$.

16.5.2 Effect of Substrate Bias

When a bias voltage V_{BS} is applied to the substrate (body) with respect to the source, the surface potential for strong inversion at x changes from $[2\phi_F + V(x)]$ to $[2\phi_F + V(x) - V_{BS}]$. This causes the depletion region charge to change from $Q'_{BO}(x)$ to a new value $Q'_{BO}(x) = [2qN_a\varepsilon_s(2\phi_F + V(x) - V_{BS})]^{1/2}$. Consequently, $Q'_n(x)$ in Eq. (16.38) is changed to a new value $Q'_n(x)$ given by

$$Q'_n(x) = C_{ox}[2\phi_F + V(x) - V_{BS} - (V_G - V_{FB})] + \sqrt{2\varepsilon_s q N_a[2\phi_F + V(x) - V_{BS}]} \quad (16.49)$$

and the drain current I_D becomes

$$I_D = \frac{Z\mu_n C_{ox}}{L}\left\{\left(V_G - V_{FB} - 2\phi_F - \frac{V_D}{2}\right)V_D - \frac{2}{3}\Theta \right.$$
$$\left. \times [(2\phi_F + V_D - V_{BS})^{3/2} - (2\phi_F - V_{BS})^{3/2}]\right\} \quad (16.50)$$

In the linear region this equation simplifies to

$$I_D = \frac{Z\mu_n C_{ox}}{L}(V_G - V'_{th})V_D \quad (16.51)$$

where

$$V'_{th} = V_{FB} + 2\phi_F - \frac{Q_{BO}}{C_{ox}}\left(1 - \frac{V_{BS}}{2\phi_F}\right)^{1/2} \quad (16.52)$$

Thus, the effect of the substrate bias is to shift the threshold voltage from its value V_{th} to a new value V'_{th} given by Eq. (16.52). Similarly, in the saturation region

$$V_{Dsat} \simeq (V_G - V'_{th}) \quad (16.53)$$

and I_D to a good approximation can be rewritten as [4]

$$I_{Dsat} \simeq \frac{K'Z}{L}(V_G - V'_{th})^2 \qquad (16.54)$$

where K' is a constant of proportionality. It is seen from Eq. (16.52) that the substrate bias can be used to shift the threshold voltage of a MOSFET.

16.5.3 Device Parameters

As in the case of a JFET, the two important parameters of a MOSFET are the drain conductance g_D and the transconductance g_m. For given values of V_D and V_G, these parameters can be obtained from Eq. (16.40). In the linear region with source connected to the substrate, g_D is obtained from Eq. (16.44) and is given by

$$g_D = \left.\frac{\partial I_D}{\partial V_D}\right|_{V_G=\text{const}} = \frac{Z\mu_n C_{ox}}{L}(V_G - V_{th}) \qquad (16.55)$$

Similarly, g_m in the linear region is

$$g_m = \left.\frac{\partial I_D}{\partial V_G}\right|_{V_D=\text{const}} = \frac{Z\mu_n C_{ox}}{L}V_D \qquad (16.56)$$

which increases with increasing V_D. The transconductance in the saturation region can be obtained from Eq. (16.48) or simply by substituting $V_D = V_{Dsat} = (V_G - V_{th})$ in Eq. (16.56). In either case the result is

$$g_{m(sat)} = \frac{Z\mu_n C_{ox}}{L}(V_G - V_{th}) \qquad (16.57)$$

which is the same as Eq. (16.55).

For the idealized characteristics considered above, g_D in the saturation region is zero. But in real devices it will be finite because of the modulation of the channel length by the drain voltage, as is discussed next.

16.5.4 Real MOS Transistors

The I-V characteristics of real MOSFETs differ from the characteristics of the idealized model in several respects. These differences will now be considered.

(a) SOURCE AND DRAIN SERIES RESISTANCES

As explained for JFET in Sec. 15.3, unmodulated paths near the source and the drain give rise to source and drain series resistances R_S and R_D, respectively. The effect of these resistances is to reduce the value of g_D in the linear region and that of g_m in the saturation region.

(b) DRAIN CURRENT IN THE SATURATION REGION

The measured $I_D - V_D$ characteristics of an enhancement-type n-channel MOSFET are shown in Fig. 16.15. Note that for $V_D > V_{Dsat}$, the drain current I_D does

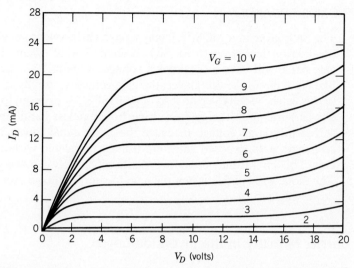

FIGURE 16.15 Experimental $I_D - V_D$ characteristics of a silicon n-channel enhancement-type MOSFET. (From D. L. Critchlow, R. H. Dennard, and S. E. Schuster, *IBM J. of Research and Development*, 17(5), p. 431. Copyright 1973 by International Business Machine Corporation; reprinted with permission.)

not saturate but increases slowly with V_D. This increase in I_D is caused partly by a decrease in the channel length with increasing V_D (Fig. 16.12) and partly by the avalanche multiplication of electrons in the high field depletion region near the drain. At still higher drain voltages (not shown in Fig. 16.15) avalanche breakdown will occur in the depletion region, causing a sharp increase in drain current after a critical value of V_D is reached. All these phenomena are similar to those described for JFET in Sec. 15.3.

(c) EFFECT OF INTERFACE STATES

When the mobile carrier concentration in the channel is changed by varying the gate voltage in a MOSFET, some of the induced charge is trapped by the states located at the Si-SiO$_2$ interface. As a result, the measured value of g_m will be lower than the theoretical value. However, this lowering in g_m will be observed only for low-frequency measurements. At high frequencies the interface states will be unable to follow the rapid changes in the gate voltage, and g_m will approach its theoretical value. Thus, in the presence of interface states the g_m of the transistor becomes frequency dependent. However, in the present-day MOS devices the interface state density can be reduced below 10^{10} cm^{-2}. Consequently, interface states will have little effect on the g_m of the device.

(d) SUBTHRESHOLD CURRENTS

In the above discussion, it was assumed that the conduction in MOSFET begins abruptly when V_G exceeds the threshold voltage V_{th}. The theory presented thus far fails to predict the subthreshold currents arising from the charge in the weak inversion regime. It has been shown [4] that in the subthreshold region the inversion layer charge density Q_n and the drain current increase exponentially with $(V_G - V_{th})$. However, at the onset of strong inversion Q_n becomes linearly dependent on V_G. Subthreshold currents can be made negligibly small by biasing the transistor about half a volt below V_{th}.

(e) GATE OXIDE BREAKDOWN

The SiO$_2$ layer in a MOSFET may have a thickness from 300 Å to 1500 Å. The breakdown field strength for SiO$_2$ is about 7×10^6 V/cm. Thus, a 1000 Å thick gate oxide will have a breakdown voltage of about 70 V. Because of weak spots like pin holes in the SiO$_2$ layer, the gate voltage is usually restricted to about half of this value. When the gate lead of a MOSFET is exposed to the ambient, a static charge buildup on the gate resulting from careless handling may easily produce sufficiently large voltage to cause the oxide breakdown. Therefore, MOSFET structures with exposed gates are usually fabricated with a gate protection circuit. This circuit may contain a Zener diode between the gate and the source. The diode will conduct when the gate-to-source voltage exceeds the diode breakdown voltage.

16.5.5 Carrier Mobility in the Inversion Layer

The electron mobility μ_n in the channel of the MOSFET has been assumed to be constant, independent of electric field. However, this is not the case, and two facts about the carrier mobility in the channel are noteworthy. First, the field-effect mobility characterizing the carrier transport in the inversion layer is considerably lower than the mobility in the bulk semiconductor. Second, this mobility is influenced by both the transverse and the longitudinal fields in the channel [5].

The reduction in mobility in the inversion layer in Si-MOSFETs is caused by scattering at the Si-SiO$_2$ interface [6]. The two extreme ways in which the free carriers can be scattered from the Si-SiO$_2$ interface are diffuse and specular scatterings. In the diffuse scattering, the carriers lose all memory of their velocity prior to collision. Therefore, this scattering leads to a reduction in the mobility of carriers present within a mean free path from the interface. In the specular scattering, the carrier momentum perpendicular to the interface undergoes a change, but the momentum component parallel to the interface remains unchanged. Thus, there is no mobility reduction for specular scattering. The electron mobility in the inversion layer of an n-channel MOSFET can be expressed as

$$\frac{1}{\mu_n} = \frac{1}{\mu_B} + \frac{1}{\mu_s} \tag{16.58}$$

where μ_s represents the surface scattering limited mobility and μ_B is the bulk mobility. A similar relation is valid for hole mobility in p-channel devices. A large concentration of majority carriers in the inversion layer screens the ionized impurity atoms so that the ionized impurity scattering becomes insignificant and μ_B is determined largely by thermal scattering.

The surface mobility μ_s is strongly influenced by the thickness of the inversion layer which is a function of the transverse electric field $\mathcal{E}_y$ in the channel and is given by [7]

$$\mu_s = \frac{2qd_i}{\bar{p}m_e v_{\text{rms}}} \tag{16.59}$$

where m_e and v_{rms} are the electron effective mass and the thermal velocity, respectively, $\bar{p}$ is the probability of diffuse scattering, and d_i is the thickness of the inversion layer. For Si at room temperature $\bar{p}$ can be written as [7]

$$\bar{p} = 0.09 + \frac{1.5 \times 10^{-8} N_f}{[Q_n/qd_i]^{1/4}} \tag{16.60}$$

where N_f represents the number of fixed charges per cm^2 at the Si-SiO$_2$ interface. The inversion layer thickness d_i can be expressed as

$$d_i(\text{cm}) = \frac{0.039}{\mathscr{E}_{\text{yeff}}} + \frac{1.24 \times 10^{-5}}{(\mathscr{E}_{\text{yeff}})^{1/3}} \tag{16.61}$$

where $\mathscr{E}_{\text{yeff}}$ is expressed in V/cm and is given by

$$\mathscr{E}_{\text{yeff}} = \frac{1}{\varepsilon_s} \left| Q_B + \frac{Q_n}{2} \right| \tag{16.62}$$

The first term in Eq. (16.61) is the classical thickness of the inversion layer, while the second term arises from the quantum mechanical broadening of the layer. For a Si n-channel MOSFET at 300 K, $\mu_B \simeq 1150$ cm^2/V-sec and μ_n becomes [7]

$$\frac{1}{\mu_n} = \frac{1}{\mu_B} + \frac{3.2 \times 10^{-9} \bar{p}}{d_i} \tag{16.63}$$

Besides its variation with the transverse field, the channel mobility is also a function of the longitudinal field $\mathscr{E}_x$. This dependence can be approximated by the relation

$$\mu_{\text{neff}} = \frac{\mu_n}{1 + \mathscr{E}_x/\mathscr{E}_c} \tag{16.64}$$

where μ_{neff} is the effective value and $\mathscr{E}_c$ is the critical field for velocity saturation.

16.5.6 Threshold Voltage Control

The control of threshold voltage is a major concern in the production of MOS transistors and integrated circuits. In many applications of MOSFETs, enhancement-type devices are desired, and in order to obtain such devices, V_{th} must be negative for p-channel and positive for n-channel devices. From Eq. (16.30) V_{th} is given by

$$V_{th} = V_{FB} + 2\phi_F - \frac{Q_{BO}}{C_{ox}}$$

As mentioned in Sec. 16.2.2, the flat-band voltage V_{FB} is negative for both p- and n-channel devices, but the remaining two terms are negative for p-channel and positive for n-channel transistors. Thus, the requirement of a negative value of V_{th} for a p-channel device is easily met. However, for an n-channel transistor V_{th} will be positive only if the sum of the two terms in the above expression is larger than V_{FB}. One requirement for this to happen is that ϕ_F be made large, which in turn requires that the p-type substrate must be heavily doped. Heavy doping of the substrate is undesirable because it leads to a high-substrate capacitance and to a low value of the drain-substrate breakdown voltage.

One way to increase V_{th} is to use a thick gate oxide. However, this reduces g_m of the transistor. Another way is to apply a substrate bias that will shift the threshold voltage to the desired value. This possibility is unattractive because it requires the provision of an additional power supply. Use of ion implantation permits the adjustment of the threshold voltage by changing the doping selectively in the channel.

Ion implantation can be used to decrease or increase the net dopant concentration in the channel. In the case of an n-channel MOSFET, the process is used to increase the channel doping. Figure 16.16 shows the distribution of implanted acceptors as a function of distance into the semiconductor for a typical implant. The implant causes a sharp peak in the distribution (dashed curve) which is broadened by thermal annealing. For the sake of calculations, the annealed profile is approximated by a box distribution with a constant acceptor concentration N_{ai} up to a distance y_i. Beyond y_i the acceptor concentration is the substrate concentration N_a. A substantial increase in the threshold voltage above its value without implantation can be obtained if y_i is kept smaller than the maximum depletion region width W_m at the onset of strong inversion.

16.5.7 Effect of Temperature on MOSFET Performance [5, 8]

Two parameters that affect the MOSFET performance with temperature are the carrier mobility in the channel and the threshold voltage V_{th}. As mentioned above, the carrier mobility is limited by scattering at the Si-SiO$_2$ interface and varies with temperature as T^{-a} where a lies between 0.5 and 1.

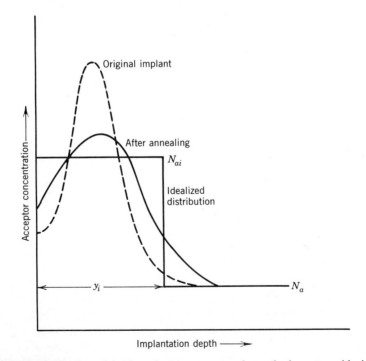

FIGURE 16.16 Distribution of implanted acceptor atoms beneath the gate oxide in an ion-implanted MOSFET.

The change in V_{th} is caused primarily by the change in the Fermi potential ϕ_F with temperature. The derivative dV_{th}/dT is always negative for both n-channel and p-channel devices. The combined effect of variations in the carrier mobility and V_{th} is to cause a decrease in the drain current with increasing temperature for both n- and p-channel devices. Accordingly, g_D and g_m also decrease with temperature. Note that this behavior is in contrast to that of the bipolar transistor where both the collector current and the current gain h_{FE} increase rapidly with temperature.

16.6 TRANSISTOR RATINGS AND FREQUENCY LIMITATIONS

16.6.1 Voltage Current and Power Limitations

The maximum voltage that can be applied to the drain of a MOSFET is limited by the avalanche breakdown of the drain-substrate diode. The drain voltage at which the avalanche breakdown occurs in a MOSFET is generally considerably lower than the value expected on the basis of the substrate resistivity. This lowering in the breakdown voltage is caused by the junction curvature of the drain diffusion, as well as by the proximity of the gate electrode to the depletion region formed near the drain.

The maximum current of a MOS transistor is limited by the same factors that were discussed in the case of JFET, namely, the melting of the source and the drain contacts by excessive Joule heating. The power rating of MOS transistors is a serious problem especially in devices with short-channel lengths. Although a second breakdown does not occur in a MOSFET, increased power dissipation may lead to the destruction of the device by excessive heating even at low-power levels. This happens because of the poor thermal conductivity of SiO_2.

16.6.2 High-Frequency Limitations

As in the case of JFET, there are two intrinsic limits on the high-frequency response of a MOSFET: (1) the carrier transit time through the channel, and (2) the charging time of capacitances associated with the device structure.

For a constant drift velocity $v_d = \mu_n \mathcal{E}$ of electrons in the channel, the transit time is $\tau_t = L/\mu_n \mathcal{E}$. However, when the MOSFET is operated in the saturation region, an approximate solution for the field $\mathcal{E}(x)$ can be obtained in terms of the gate voltage and the channel length. It has been shown [4] that

$$\tau_t = \frac{4}{3} \frac{L^2}{\mu_n(V_G - V_{th})} \tag{16.65}$$

For a MOSFET with $L = 10$ μm, $(V_G - V_{th}) = 5$ V, and $\mu_n = 600$ cm^2/V-sec. We obtain $\tau_t = 4.4 \times 10^{-10}$ sec, which is lower than the time required to charge the various capacitances inherent in a structure of these dimensions.

The capacitance charging time of a MOSFET can be readily obtained from its small-signal equivalent circuit (see Chapter 17). Figure 16.17 shows the circuit symbol and the small-signal equivalent circuit in the common source configuration. The circuit symbol is for an n-channel device, and the arrow in Fig. 16.17a represents the p-n junction between the substrate and the channel. The direction

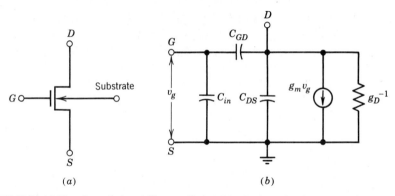

FIGURE 16.17 (*a*) Circuit symbol and (*b*) a small-signal equivalent circuit of a MOSFET in a common source configuration. The symbols *G*, *S*, and *D* refer to gate, source, and drain, respectively.

of the arrow is from *p* to *n* and will be reversed in a p-channel device. In the circuit of Fig. 16.17b the capacitance C_{GD} is the gate to drain capacitance and C_{DS} is the capacitance of the reverse-biased source-drain diode. The input capacitance C_{in} consists of the source-substrate diode capacitance and the capacitance from the gate electrode to the active area.

As in the case of JFET, a current cutoff frequency f_T is defined by the relation

$$f_T = \frac{g_m}{2\pi C_{in}} \qquad (16.66)$$

In general, silicon bipolar transistors are capable of operation at higher frequencies compared to MOSFETs. Two factors are responsible for the poor high-frequency performance of MOSFETs. First, for a given area and operating current, the g_m of a silicon MOSFET is less than that of a bipolar transistor (see Prob. 16.15). Second, because the gate is separated from the channel by a thin SiO_2 layer, appreciable input capacitance is observed in MOSFET structures. Any effort to improve on the frequency response must be directed to increase g_m and decrease C_{in}. In amplifying circuits, a MOSFET is operated in the saturation region where g_m is given by Eq. (16.57) and assuming that C_{in} essentially equals the gate to channel capacitance, we can write

$$C_{in} \simeq \frac{\varepsilon_{ox} ZL}{d_{ox}} = C_{ox} ZL$$

Substituting for C_{in} and g_m from Eq. (16.57) into Eq. (16.66) we obtain

$$f_T = \frac{\mu_n}{2\pi L^2}(V_G - V_{th}) \qquad (16.67)$$

Thus, for a given bias point, f_T can be increased by reducing *L* and employing semiconductor materials with higher carrier mobility. These same requirements were arrived at in the case of MESFETs in Chapter 15.

16.7 SHORT-CHANNEL EFFECTS [9,10]

Since the first fabrication of MOSFET in 1960, the minimum channel length has been shrinking continuously. The motivation behind this decrease has been an increasing interest in high-speed devices and in very large scale integrated circuits. The theory of MOSFET presented so far is valid only for long-channel devices. As the channel length decreases, departures from the long-channel behavior occur basically for two reasons. First, for a given doping in the channel a gradual reduction in the channel length leads to a situation in which the depletion region widths of the source and the drain diffusions become comparable to the channel length. When this happens, the potential distribution in the channel becomes two-dimensional, and the gradual channel approximation loses its validity. This results in large values of subthreshold current, a decrease in the threshold voltage, and nonsaturation of drain current due to punch-through.

The second cause of departure from the long-channel behavior is the presence of high electric fields in the channel. When a short-channel MOSFET is operated at a drain voltage $V_D > V_{D\text{sat}}$, the electric field near the drain may become large enough to cause avalanche multiplication of carriers. This leads to high substrate currents and parasitic bipolar transistor action. High fields may also cause hot carrier injection into the gate oxide, resulting in oxide charging and a shift in the threshold voltage. A brief account of these effects follows.

16.7.1 Subthreshold Conduction

It was pointed out in Sec. 16.5 that the subthreshold current in a MOSFET can be made negligible by keeping the gate voltage a few tenths of a volt below the threshold voltage. While this is true in the case of long-channel devices, short-channel MOSFETs show a substantially high subthreshold current.

Figure 16.18 shows drain current plotted as a function of V_G for three n-channel enhancement-type devices. It is seen that whereas the subthreshold current remains small for $L = 7$ μm, it increases for $L = 3$ μm, and for $L = 1.5$ μm the device fails to turn off. Also note that the drain voltage V_D has practically no effect on the long-channel device but has considerable effect on devices with $L = 3$ μm and $L = 1.5$ μm.

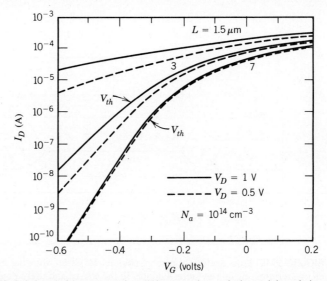

FIGURE 16.18 Subthreshold currents for different values of channel length in n-channel enhancement-type MOSFETs. (From S. M. Sze, reference 2, p. 470. Copyright © 1981. Reprinted by permission of John Wiley & Sons, Inc., New York.)

The boundary between a long- and a short-channel device is given by the following empirical relation:

$$L_{\min} = A_1[x_j d_{ox}(W_S + W_D)^2]^{1/3} \qquad (16.68)$$

Here $L_{\min}$ is the minimum-channel length for which the long-channel behavior is observed, x_j is the junction depth of source and drain diffusion in μm, d_{ox} is the gate oxide thickness in Å, and A_1 is 0.41 (Å)$^{-1/3}$. W_D is the drain depletion region width (in μm) which for a one-sided abrupt junction can be written as

$$W_D = \sqrt{\frac{2\varepsilon_s}{qN_a}(V_i + V_D)} \qquad (16.69)$$

where V_i is the built-in voltage, and it has been assumed that the source is connected to the substrate. The source depletion region width W_S is obtained from Eq. (16.69) with $V_D = 0$. Short-channel behavior is observed when L is less than $L_{\min}$ given by Eq. (16.68).

Figure 16.19 shows the cross-section of a short-channel MOSFET made on p-silicon. Note that some of the field lines from the charges in the channel terminate on ionized donor atoms in the source and the drain depletion regions rather than on the gate electrode. As a consequence, the full effect of the depletion region charge Q_{BO} on the threshold voltage is reduced, causing a decrease in the Q_{BO}/C_{ox} ratio. Since

$$V_G = V_{FB} - \frac{Q_{BO}}{C_{ox}} + \phi_s \qquad (16.70)$$

a reduction in Q_{BO}/C_{ox} for a given value of V_G increases ϕ_s. This increase in ϕ_s causes an increase in the inversion layer charge, leading to an increase in subthreshold current.

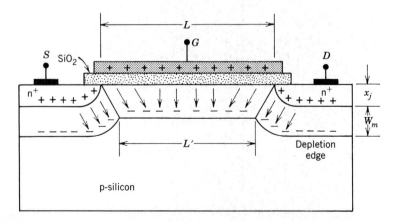

FIGURE 16.19 Cross-section of a short-channel MOSFET showing field lines from the channel charge terminating on the ionized donors in the source and drain depletion regions. (After L. D. Yau, *Solid-State Electronics*, vol. 17, p. 1060. Copyright © 1974. Reprinted by permission of Pergamon Press, Oxford.)

16.7.2 Threshold Voltage

The reduction in the threshold voltage in a short-channel MOSFET can be understood as follows [11]. It should be evident from Fig. 16.19 that the electric field lines in the oxide, as well as in the depletion region near the source and drain, are perturbed owing to the proximity of source and drain electrodes. In a short-channel device the electric field gets modified even in the middle of the channel. As a result, the electric field intensity in the channel will be less than what it will be in a long-channel device. Since the voltage is the line integral of the electric field intensity, a reduction in the electric field intensity implies a reduction in the gate voltage required to bring strong inversion in the channel.

A first-order estimate of the threshold voltage can be made from Fig. 16.19. The channel charge whose field lines terminate on the gate charge is contained inside the trapezoid of area $W_m(L + L')/2$. Consequently, the charge Q_B^* contained inside the trapezoid is

$$Q_B^* = -qN_a W_m Z\left(\frac{L + L'}{2}\right) = Q_{BO} Z\left(\frac{L + L'}{2}\right) \tag{16.71}$$

For a long-channel device $L' \simeq L$ and $Q_B^* = Q_{BO} ZL$. It has been shown that the decrease in the threshold voltage can be written as [2]

$$\Delta V_{th} = \frac{Q_{BO}}{2C_{ox}}\left[1 - \frac{L'}{L}\right] \tag{16.72}$$

16.7.3 Operation Under Punch-Through

In very short-channel devices, the sum of the source and the drain depletion widths may exceed the channel length even when $V_D = 0$. The channel is then punched through and upon application of the drain voltage, electrons are injected from the source into the depletion region, where they are swept by the electric field and are collected at the drain. The drain current is then dominated by the space-charge limited current which varies as V_D^2. This current flows in parallel to the inversion layer current, which increases linearly with the gate voltage but remains small compared to the space-charge limited current. Consequently, the drain current of such devices fails to saturate even at large values of the drain voltage V_D.

16.7.4 Operation in Avalanche Multiplication Regime

Consider an n-channel MOSFET operated in the saturation region. At a sufficiently large drain voltage, the electric field in the depletion region near the drain becomes large enough to produce electron–hole pairs by avalanche multiplication. The secondary electrons produced in this way flow to the drain, while the holes move into the substrate causing a substrate current I_{BS} (Fig. 16.20). This current causes an ohmic drop $I_{BS}R_B$ across the substrate. If I_{BS} is sufficiently large, this ohmic drop forward biases the source-substrate junction, causing electron injection from the source into the substrate. These electrons diffuse through the substrate, and in a short-channel device most of them get collected at the reverse-

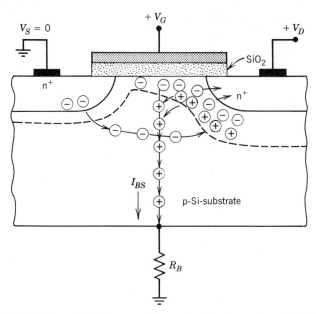

FIGURE 16.20 Parasitic bipolar transistor action in a short-channel MOSFET.

biased drain-substrate junction leading to a parasitic n-p-n (source-substrate-drain) bipolar transistor action. Because of this parasitic transistor action, the maximum drain voltage is considerably reduced. For example, when the resistance R_B is high, the substrate (i.e., base) of the parasitic transistor may be assumed to be open-circuited and following Eq. (13.60) we obtain $V_D(\text{max}) = V_B(1 - \alpha_{\text{npn}})^{1/m}$. Here V_B is the avalanche breakdown voltage of drain-substrate junction, α_{npn} is the alpha of the parasitic transistor, and $m \simeq 4$.

Another effect of high electric field in the channel is the oxide charging. As the field in the inversion layer becomes high, some of the hot electrons in the channel gain enough energy to surmount the Si-SiO$_2$ barrier and are injected into the SiO$_2$. This injection tends to reduce the positive charge in the oxide resulting in an increase in V_{th} and a decrease in the g_m of the device. Devices based on hot electron injection effects are considered in Sec. 19.10.3.

16.7.5 Downscaling of Device Dimensions

The short-channel effects mentioned above are undesirable and hence should be minimized. One way to maintain the long-channel behavior of a short-channel MOSFET is simply to scale down all dimensions and voltages of a long-channel device so that the internal fields remain the same. For example, consider a long-channel device with a channel length $L > L_{\min}$ given by Eq. (16.68). If this length is now reduced by a factor S, the oxide thickness, the channel width Z, junction depth, and depletion region width are also to be reduced by the same factor S. In order to reduce the depletion region width by S, the substrate doping is increased by S and all the voltages are reduced by S. The threshold voltage of the scaled-down device is also reduced nearly by S, and the power dissipation per device is decreased by S^2.

16.8 MOSFET STRUCTURES [12]

16.8.1 Self-Aligned Gate Devices

A cross-sectional view of a typical p-channel MOSFET fabricated using conventional technology is shown in Fig. 16.21a. In this device, the gate considerably overlaps with the source and the drain diffusions. If the gate length remains shorter than the distance between the two p^+-diffusions, a part of the channel will remain unmodulated and the device will not operate. To avoid this possibility, the conventional Al gate technology requires that sufficient overlap of the gate with source and drain diffusions be maintained to allow for mask registration tolerance. Because of this overlap, parasitic capacitances develop between the gate and source and between the gate and drain. These capacitances degrade the dynamic performance of the device and hence are undesirable.

One way to avoid the overlap capacitances and to obtain self-alignment between the gate and the source and drain regions is shown in Fig. 16.21b. Here the length of the gate metal is kept smaller than the spacing between the source and drain diffusions. Ion implantation is then performed using boron ions. The gate metal acts as a mask, but the boron ions penetrate the thin SiO_2 layer extending the source and drain regions to the edges of the gate. The main disadvantage of this method is that it produces radiation damage in Si and SiO_2 which may not be completely annealed out in a subsequent heat treatment because heating of Al to temperatures in excess of 570°C is undesirable. An alternative and commonly used process for obtaining gate alignment is the silicon gate technology [13].

16.8.2 DMOS

MOS transistors produced by the standard diffusion and oxidation processes cannot sustain high drain voltages. This is because in order to sustain a high drain voltage the channel must be lightly doped and should be long enough to avoid punch-through. With typical values of channel length (3 to 6 μm) and substrate doping ($N_a < 10^{15}$ cm^{-3}), V_D can hardly exceed 50 V.

The DMOS (double diffused MOS) structure shown in Fig. 16.22 is a high-voltage transistor. It has a lightly doped $\bar{p}$-Si substrate in which a p-diffusion is first selectively performed under the source region. This is followed by n^+-source

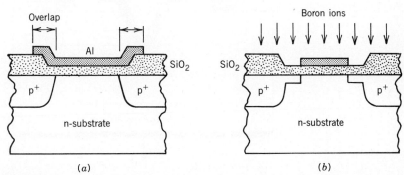

FIGURE 16.21 Reduction of parasitic capacitances due to gate overlap. (a) Conventional processing with Al-gate and (b) use of ion implantation to achieve self-aligned gate structure.

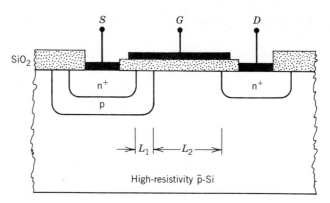

FIGURE 16.22 Cross-sectional view of a DMOS transistor. (From D.V. McCaughan and J.C. White, reference 12, p. 299. British Crown Copyright 1981. Reproduced by permission of the Controller of Her Britannic Majesty's Stationery Office.)

and drain diffusions. The channel region is thus linked to the n^+-source by the diffused p-island. The gate metal completely covers the p- and $\bar{p}$-regions and parts of the n^+-source and drain diffusions. Because of the presence of oxide charges and the ϕ_{ms} term, the high-resistivity $\bar{p}$-zone of length L_2 gets strongly inverted at very low gate voltages. Thus, the gate voltage modulates the conductivity only of the narrow region L_1 which is shorter than 1.5 μm. The device has a high drain breakdown voltage because of the presence of the $\bar{p}$-zone which is punched through at a low value of V_D. The electrons move through this high field zone with saturation velocity so that their transit time is minimized.

16.8.3 VMOS

Figure 16.23 shows the cross-section of a VMOS (vertical MOS) transistor. Fabrication of this device starts with an n^+-Si substrate on which an n-layer is grown epitaxially. The wafer is then oxidized, and p-diffusion is performed through

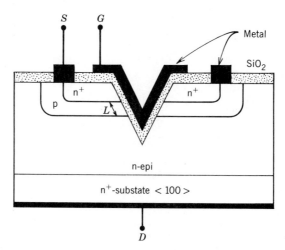

FIGURE 16.23 Schematic cross-section of a VMOS transistor. (From F. E. Holmes and C. A.T. Salama, *Solid-State Electronics*, vol. 17, p. 793. Copyright © 1974. Reprinted by Permission of Pergamon Press, Oxford.)

a window in the oxide into the n-epilayer. This is followed by the n^+-source diffusion through a smaller window in the oxide. The orientation of the n-Si layer is $\langle 100 \rangle$. In the middle of the source region a V groove is etched down to the n-epitaxial layer selectively using an anisotropic etchant like hydrazine. After groove etching the surface is oxidized, and Al contacts are evaporated to form source, drain, and gate. In the operation of the device the gate is driven positive, and the p-body forms the channel. When a drain voltage is applied, the n-epitaxial layer is punched through and gives a high-voltage capability to the device.

16.9 CHARGE-COUPLED DEVICES

In a charge-coupled device (CCD), the information is represented by a charge packet that is stored in potential wells created in the semiconductor by applying appropriate voltages at the gates of an array of MOS capacitors. The charge is transported in a controlled manner across a channel formed in the semiconductor. This way of representing information is different from that of conventional devices like transistors where current and voltage levels are used instead of charge.

16.9.1 Transient Characteristics of a MOS Capacitor

In order to understand the operation of a charge-coupled device, it is necessary to know the behavior of a MOS capacitor under transient conditions which drives it into deep depletion (curve c in Fig. 16.5). Consider a MOS capacitor made on a p-Si substrate. Let a positive voltage pulse of magnitude higher than the threshold voltage be applied to the gate. Figure 16.24a shows the energy band diagram of the system just after the application of the pulse. The device gets biased into deep depletion, and no inversion layer is formed because no minority carriers are immediately available. Under this condition, a large fraction of the applied bias appears across the semiconductor, and the surface potential ϕ_s is large. However, this is not a steady state situation. As time elapses, electron–hole pairs are generated thermally in the depletion region, and electrons diffuse to the Si-SiO$_2$ inter-

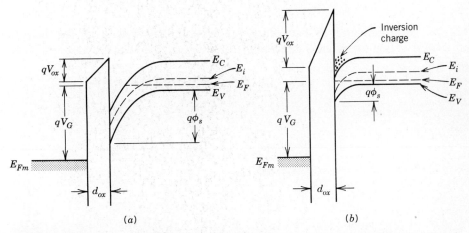

FIGURE 16.24 Energy band diagrams of a MOS capacitor (a) under deep depletion ($t = 0$) and (b) under thermal equilibrium ($t = \infty$).

face, forming an inversion layer at the Si-surface under the oxide. The generated holes neutralize a fraction of ionized acceptors and thus reduce the width of the depletion region. This process continues until thermal equilibrium is reached (Fig. 16.24b). Note that most of the applied voltage now appears across the oxide.

The time that elapses between the initial creation of a depletion region and the formation of a nearly full inversion layer is called the *thermal relaxation time*. This time is a function of the flat-band voltage, the gate voltage, and the density of generation recombination centers at the Si-SiO$_2$ interface and in the neighboring bulk. Its value may range from about a second to several minutes depending on the fabrication process of the CCD. Just after the application of the gate pulse, if a signal charge Q_{sig} of minority carriers is introduced by some means, the carriers collect at the surface of the semiconductor and cause a decrease in ϕ_s. It should be clear that charge Q_{sig} can be held in the potential well under the gate only for times that are small compared to the thermal relaxation time.

The surface potential as a function of gate bias V_G can be obtained following Eq. (16.34) and Eq. (16.20). Thus

$$V_G - V_{FB} = -\frac{Q_s}{C_{ox}} + \phi_s \tag{16.73}$$

and

$$\phi_s = \frac{qN_a W^2}{2\varepsilon_s} \tag{16.74}$$

When no signal charge is present, $Q_s = -qN_a W$. However, in the presence of a signal charge $-Q_{sig}$, we can write

$$Q_s = -qN_a W - Q_{sig} \tag{16.75}$$

Substituting for Q_s from Eq. (16.75) into Eq. (16.73) we obtain

$$V_G - V_{FB} = \frac{qN_a W}{C_{ox}} + \frac{Q_{sig}}{C_{ox}} + \phi_s \tag{16.76}$$

Eliminating W between Eqs. (16.74) and (16.76) and solving the resulting equation for ϕ_s yield (see Prob. 16.18)

$$\phi_s = V'_G + V_o - (2V'_G V_o + V_o^2)^{1/2} \tag{16.77}$$

where

$$V'_G = V_G - V_{FB} - \frac{Q_{sig}}{C_{ox}} \tag{16.77a}$$

$$V_o = \frac{qN_a \varepsilon_s}{C_{ox}^2} \tag{16.77b}$$

The above three relations show that, for a given value of V_G, the surface potential ϕ_s decreases almost linearly with the signal charge Q_{sig}.

16.9.2 Principle of Operation

Figure 16.25 shows a cross-section of a basic charge-coupled device. The structure consists of a p-Si substrate whose top surface is covered with a thermally grown SiO_2 layer. An array of closely spaced metallic electrodes is deposited on the surface of the oxide layer. Each of these electrodes forms a MOS capacitor that can be biased in deep depletion.

(a) STORAGE AND TRANSFER OF CHARGE

Let us consider the situation shown in Fig. 16.26a where a cutaway of a CCD with three metal gates is shown. The voltage V_2 on the central electrode has been made substantially more positive than the voltage V_1 existing on the two neighboring electrodes. Consequently, the surface potential will be higher, and the electron energy will be lower under the central electrode as compared to the other two electrodes. If electrons are now introduced into the depletion region, they will collect under the central electrode and the large potential on this electrode will keep them confined to the potential well formed there. It is obvious that the signal charge can be stored in the well only for times that are orders of magnitude smaller than the thermal relaxation time.

Figure 16.26b illustrates the charge transfer process. At time, say, $t = t_o$, the potential on the right electrode is raised to $V_3 > V_2$. The resulting potential variation is shown by the dotted line. The change in potential distribution causes the electrons to be transferred from the potential well under the second electrode to that under the third. Subsequently, the potentials on the electrodes can be readjusted so that the quiescent storage site is located under the third electrode. One way in which the charge transfer is realized along a linear array is the three-phase CCD whose two elements are shown in Fig. 16.27. The electrodes are connected in groups of three and are operated with a three-phase periodic waveform called the *clock voltage*. The curves on the left in Fig. 16.27 show the clock voltage waveforms for the three-phase operation. The composite waveform has a period T. Each of the parts ϕ_a, ϕ_b, and ϕ_c represents one phase and consists of three subintervals of duration $T/3$. During these subintervals the voltage is either at V_1 or at V_2 or is changing from V_1 to V_2. At $t = 0$, the voltage ϕ_a has a value V_2 and the charge is under the electrode a_1. There is no charge under b electrodes because the voltage ϕ_b is at V_1 and has been so for the full interval $T/3$. In addition, since ϕ_c has just completed its fall to V_1 there is no charge under the c electrode either. During the time interval from 0 to $T/3$, the voltage ϕ_b remains at V_2 while ϕ_a decreases from V_2 to V_1 and the charge is being transferred from the

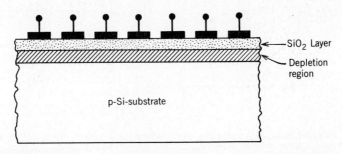

FIGURE 16.25 Cross-sectional view of a basic charge-coupled device.

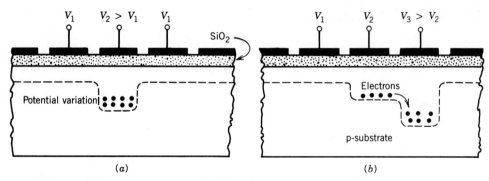

FIGURE 16.26 Cutaway view of a CCD illustrating the processes of (a) charge storage and (b) charge transfer.

region under the electrode a_1 to that under the electrode b_1. The transfer process is completed at $t = T/3$. At $t = 2T/3$ the charge gets transferred to the potential well under the c_1 electrode. Finally, at $t = T$ when the full time period of the clock wave is completed, the charge has moved to the potential well under the electrode a_2, two electrodes away from a_1. A fresh cycle of the clock voltage will move the charge under the next cell a_3. Thus, the charge can propagate along the linear array. If the time of propagation along the array is small compared to the thermal relaxation time, little charge is lost in the transfer, and the information represented by the signal charge is carried away from the input to the output of the device.

To this point in our discussion we have considered a p-Si substrate. The operation of a device made on an n-Si substrate is similar except that the voltatges V_1

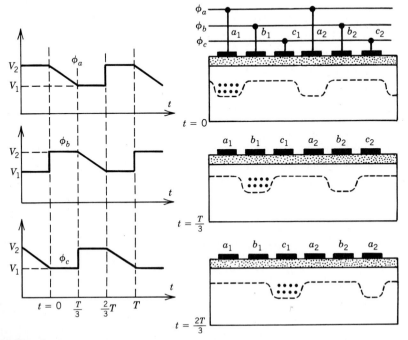

FIGURE 16.27 Schematic diagrams illustrating the operation of a three-phase CCD. (After W. S. Boyle and G. E. Smith, Charge coupled semiconductor devices, *Bell Syst. Tech. J.*, vol. 49, p. 589. Copyright © 1970 AT&T. Reprinted by special permission.)

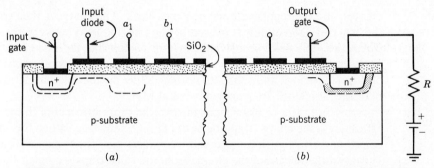

FIGURE 16.28 Schematics showing the input and output arrangements for a CCD.

and V_2 become negative. A number of geometries other than the three-phase structure utilizing the same basic concepts have been used. Details of these may be found in the literature mentioned at the end of this chapter.

(b) INPUT AND OUTPUT ARRANGEMENTS

The charge-coupling phenomenon in itself is not sufficient to construct a useful device unless we have an input to introduce the necessary charge carriers and an output where the stored charge could be detected. The minority carrier charge at the start of an array of electrodes can be generated using a variety of processes. For example, carriers can be generated by avalanche injection. In this process a large voltage pulse is applied to cause avalanche breakdown in the semiconductor. Another method is the optical injection where electron–hole pairs are generated by shining a light on the semiconductor.

The most commonly used method is the minority carrier injection from a p-n junction as shown in Fig. 16.28a. Here an n^+-type region is diffused into the p-substrate under an opening in the SiO_2 layer. Another gating electrode is added to serve as the input gate. The signal that is to be entered in the CCD is connected to the input diode which acts as a source of electrons. The input gate is kept at low voltage so that no electron enters the channel. When the clocking voltage is such that the a_1 electrode next to the input gate is in high-voltage condition, the gate is turned on by raising its voltage. Electrons will now fill the potential well formed under a_1 until the energy level of electrons in the well becomes equal to that in the source. When the desired amount of charge is stored under a_1, the voltage at the input gate is removed to isolate the source. The charge entered under the electrode a_1 can now be transferred laterally by the successive phases of the clock waveform.

At the output the charge may be detected by a reverse-biased p-n junction as shown in Fig. 16.28b. When the p-n junction is reverse biased, charge carriers are collected across it in the same way as minority carriers are collected by the collector-base junction of a bipolar transistor. An output voltage is then developed across the series resistor R.

(c) CHARGE TRANSFER EFFICIENCY

The process of charge transfer is not 100 percent efficient, and some charge is always lost during its transfer from the one electrode to the next. There are three basic charge transfer mechanisms. In the initial stages when the charge packet is very dense, a large fraction of the charge transfer occurs due to repulsive force between the electrons. As time passes, the electron density in the packet decreases, and the repulsive force becomes weak. The charge is then transferred by thermal diffusion and drift caused by the fringing field between the neighboring electrodes. The last two mechanisms are responsible for some loss of charge at every transfer. However, calculations have shown that by appropriate choice of electrode spacings and substrate doping, 99.99 percent of charge transfer can be obtained at a clock frequency of several MHz.

Apart from the transfer mechanisms, the other effect that limits the transfer efficiency is the charge trapping at the Si-SiO$_2$ interface. When the charge packet comes in contact with unoccupied interface traps, the traps are filled instantly. However, the release of the charge from the filled traps is a much slower process. Consequently, some charge is removed from the packet at each transfer, which is later released as noise. Thus, a prime requirement for fabrication of CCD is that the trap density at the Si-SiO$_2$ interface should be as low as possible. The effect of interface traps can be reduced substantially by propagating a constant background charge known as "fat zero" through the channel to fill the traps.

16.9.3 Surface and Buried Channel CCDs

A CCD must be designed in such a way that the surface potential has a smooth transition across the boundaries between the neighboring electrodes. This requires that the capacitors must be kept close together so that their depletion regions overlap. In order to realize this condition in the single metal gate structure of Fig. 16.25, the spacing between the gate electrodes has to be kept about 2 μm for a SiO$_2$ thickness of 1000 Å. The conventional photolithographic process is not suitable for obtaining such small gaps. A simple alternative to the single metal gate, three-phase CCD (Fig. 16.27) is the doped polysilicon structure shown in Fig. 16.29. In this structure, the SiO$_2$ layer is first covered uniformly by a nearly intrinsic polysilicon layer. The polysilicon is then doped heavily in the desired regions after masking the area where doping is not desired. Besides this simple structure, several other electrode structures and clocking schemes have been implemented.

In the device described thus far, the charge is stored and moved in potential wells at the Si surface under the SiO$_2$. Such a device is therefore known as a *surface channel CCD*. As mentioned above, traps at the Si-SiO$_2$ interface have a strong influence on the charge transfer efficiency of CCD. Obviously, the effect

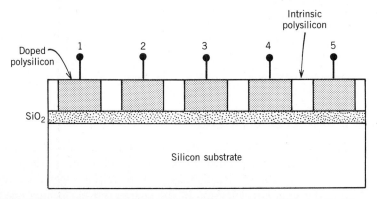

FIGURE 16.29 A doped polysilicon three-phase CCD structure. (From C. K. Kim and E. H. Snow, *Appl. Phys. Letters*, vol. 20(12), p. 515. Copyright © 1972. Reprinted by permission of American Institute of Physics, New York.)

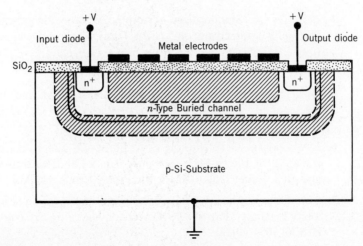

FIGURE 16.30 Cross-sectional view of a buried channel CCD with positive bias at the input and output diodes. (From R. H. Walden, R. H. Krambeck, R. J. Strain, J. McKenna, N. L. Schryer, and G. E. Smith, The buried channel charge coupled device, *Bell Syst. Tech. J.,* vol. 51, p. 1636. Copyright © 1972 AT&T. Reprinted by special permission.)

of surface traps will be eliminated if the channel is moved at some distance away from the Si-SiO$_2$ interface. This is achieved in the buried channel CCD (BCCD) shown in Fig. 16.30. This device employs a lightly doped p-Si substrate on the top of which an *n*-type layer is produced by ion implantation. The wafer is then oxidized, and n$^+$-diffusion is performed for the input and the output diodes through the windows etched in the oxide. Finally, ohmic contacts are made to the n$^+$-regions, and the SiO$_2$ layer is covered with a series of metal electrodes.

In order to understand the operation of the device, consider a large positive voltage applied to the n-layer through the input and the output diodes while the gate is held at the ground potential. The substrate to channel p-n junction is then reverse biased, and the depletion region of this junction will widen in both the n- and p-regions. Since the gate is also biased negative with respect to the n-region, a field-induced depletion layer forms in the region below the metal electrodes as shown in Fig. 16.30. As the reverse bias is increased, the two depletion regions will meet at some distance away from the Si-SiO$_2$ interface, and the potential will have its maximum value there. Therefore, the electrons introduced into the structure will be attracted to this potential maximum and will stay there in a potential well. This charge can then be transferred from one electrode to the next by application of suitable clock pulses in the same way as was done in the case of the surface CCD. Note that there is one important difference between the surface and the buried channel devices. In the surface device, the electrons forming the charge packet are minority carriers. These electrons form a very thin inversion layer near the Si-SiO$_2$ interface. On the other hand, in the buried channel CCD the stored charge is of majority carriers, which replaces some of the electrons that were removed during the formation of the depletion region.

REFERENCES

1. A. S. GROVE, *Physics and Technology of Semiconductor Devices,* John Wiley, New York, 1967.

2. S. M. SZE, *Physics of Semiconductor Devices,* John Wiley, New York, 1981, Chapters 7 and 8.

3. B. E. Deal, "Standardized terminology for oxide charges associated with thermally oxidized silicon," *IEEE Trans. Electron Devices* ED-27, 606 (1980).
4. R. S. Muller and T. I. Kamins, *Device Electronics for Integrated Circuits,* 2d ed., John Wiley, New York, 1986.
5. W. M. Soppa and H. G. Wagemann, "Investigation and modeling of the surface mobility of MOSFET's from −25 to +150°C," *IEEE Trans. Electron Devices* ED-35, 970 (1988).
6. J. A. Cooper, D. F. Nelson, S. A. Schwarz, and K. K. Thornber, "Carrier transport at the Si-SiO$_2$ interface," in VLSI Electronics, N. G. Einspruch and R. S. Bauer (eds.), Academic Press, New York, Vol. 10, 323–361 (1985).
7. S. A. Schwarz and S. E. Russek, "Semi-empirical equations for electron velocity in silicon: Part II-MOS inversion layer," *IEEE Trans. Electron Devices* ED-30, 1634 (1983).
8. F. M. Klaassen and W. Hes, "On the temperature coefficient of the MOSFET threshold voltage," *Solid-State Electron.* 29, 787 (1986).
9. D. K. Ferry, "Physics and modeling of submicron insulated-gate field-effect transistors. I," Chapter 6 in *VLSI Electronics,* N. G. Einspruch (ed.), Academic Press, New York, Vol. 1, 231-263 (1981).
10. D. G. Ong, *Modern MOS Technology: Processes, Devices, and Design,* McGraw-Hill, International Edition, 1986, Chapter 6.
11. C. R. Viswanathan, B. C. Burkey, G. Lubberts, and T. J. Tredwell, "Threshold voltage in short-channel MOS devices," *IEEE Trans. Electron Devices* ED-32, 932 (1985).
12. D. V. McCaughan and J. C. White, "MOS Transistors and Memories," Chapter 3A in *Handbook on Semiconductors,* T. S. Moss and C. Hilsum (eds.), Vol. 4, North-Holland, Amsterdam, 1981.
13. F. Faggin and T. Klein, "Silicon gate technology," *Solid-State Electron.* 13, 1125 (1970).

ADDITIONAL READING LIST

1. E. H. Nicollian and J. R. Brews, *MOS Physics and Technology,* John Wiley, New York, 1982.
2. E. S. Yang, *Microelectronic Devices,* McGraw-Hill, New York, 1988, Chapters 9–13.
3. D. G. Ong, *Modern MOS Technology: Processes, Devices, and Design,* McGraw-Hill, International Edition, 1986.
4. B. J. Baliga, *Modern Power Devices,* John Wiley, New York, 1987.
5. D. F. Barbe (ed.), *Charge-Coupled Devices,* Springer-Verlag, Berlin, 1980.
6. G. S. Hobson, *Charge-Transfer Devices,* Edward Arnold, London, 1978.
7. J. D. E. Beynon and D. R. Lamb (eds.), *Charge-Coupled Devices and Their Applications,* McGraw-Hill, London, 1980.

PROBLEMS

Unless stated otherwise, assume $T = 300\,\text{K}$ in the following problems.

16.1 An n-type silicon sample has a uniform donor concentration $N_d = 5 \times 10^{15}\,\text{cm}^{-3}$. Calculate the surface potential required (a) to make the surface intrinsic and (b) to bring strong inversion at the surface.

16.2 A thin oxide is grown on the surface of a uniformly doped p-silicon sample of 10 Ω-cm resistivity. A positive charge in the oxide raises the electron concentration at the surface to $3 \times 10^{15}\,\text{cm}^{-3}$. Calculate the magnitude and the sign of the surface potential.

16.3 A 0.5-μm thick oxide is grown on a uniformly doped n-silicon sample with $N_d = 1.5 \times 10^{15}\,\text{cm}^{-3}$. Assuming the oxide to be charge free, calculate the surface potential and the gate voltage for (a) making the surface intrinsic and (b) creating strong inversion at the surface. Also determine the maximum width W_m of the surface depletion region.

16.4 Combining Eqs. (16.23) and (16.24) and substituting for C_{ox} from Eq. (16.15b) obtain Eq. (16.25).

16.5 An Au gate MOS capacitor is fabricated on an n-silicon substrate with $N_d = 10^{15}\,\text{cm}^{-3}$. The thickness of the gate oxide is 1200 Å, and the charge density at the Si-SiO$_2$ interface is 3×10^{11} charges cm^{-2}. Calculate (a) the flat-band voltage, (b) the turn-on voltage, and (c) draw the energy band diagram of the system under thermal equilibrium and at the onset of strong inversion.

16.6 Consider an n-channel MOS capacitor made on p-silicon substrate with $N_a = 10^{15}\,\text{cm}^{-3}$, and the oxide thickness and surface charge densities the same as in Prob. 16.5. Calculate V_{FB}, V_{th} and compare these results with those of Prob. 16.5.

16.7 Figure 16.8 shows the (C-V) plot of a MOS capacitor made on p-silicon with $N_a = 5.5 \times 10^{16}\,\text{cm}^{-3}$, using Al as gate metal. Calculate the oxide thickness and the charge density at the Si-SiO$_2$ interface assuming there are no mobile charges in the oxide.

16.8 Consider a MOS capacitor of area 1 cm^2 made on n-silicon with $N_d = 1.5 \times 10^{14}\,\text{cm}^{-3}$ and Al gate. The SiO$_2$ layer is 0.2 μm thick. The Si is 20 μm thick and is epitaxially grown on n$^+$-silicon substrate having $N_d = 10^{19}\,\text{cm}^{-3}$. Neglecting any interface charge between Si and SiO$_2$, determine the flat-band capacitance, the zero-bias capacitance, $C_{\min}$ and $C_{\max}$ for the structure, and sketch the C-V plot.

16.9 Assume that the oxide in Prob. 16.8 contains 10^{12} mobile ions. Determine the flat-band and the threshold voltages when (a) all the ions are uniformly distributed in the oxide, (b) all the ions are at the Si-SiO$_2$ interface, and (c) all the ions are at the Al-SiO$_2$ interface.

16.10 Substituting for $Q_n(x)$ from Eq. (16.38) into Eq. (16.39) and integrating the resulting equation from $x = 0$ to $x = L$, derive Eq. (16.40).

16.11 Writing the charge $Q_n(x) = 0$ in Eq. (16.38) show that the saturation drain voltage is given by Eq. (16.45).

16.12 (a) An Al gate enhancement-type n-channel Si MOSFET has a substrate concentration $N_a = 6 \times 10^{15}$ cm^{-3} and an oxide thickness of 0.1 μm. The SiO$_2$ has 10^{10} charges cm^{-2}. Determine the threshold voltage of the transistor. When the transistor is operated in the linear region with $V_D = 0.5$ V, calculate the gate voltage to obtain a drain current of 2 mA. Assume $L = 10$ μm, $\mu_n = 500$ cm^2/V-sec, and $Z = 100L$. Calculate the g_m of the device and the maximum frequency of operation f_T. (b) What changes will occur in the characteristics of the MOSFET when the gate metal is replaced by p-polysilicon doped with $N_a > 10^{19}$ cm^{-3} operated at the same current in the linear region with $V_D = 0.5$ V?

16.13 An n-channel MOSFET with $Z/L = 15$, a SiO$_2$ thickness of 800 Å, and $\mu_n = 600$ cm^2/V-sec is to be used as a controlled resistor. (a) What charge density is required for the device to present a dc resistance of 3.5 KΩ between the source and the drain in the saturation region? (b) What value of $(V_G - V_{th})$ is required to obtain the desired resistance?

16.14 Calculate the drain current, g_D, and g_m of an n-channel MOSFET with $Z/L = 10$, $V_{th} = 0.5$ V, $\mu_n = 500$ cm^2/V-sec, and oxide thickness of 0.12 μm with (a) $V_D = 0.2$ V, (b) $V_D = 2$ V, and (c) in the saturation region. Assume $V_G = 4$ V in all the calculations.

16.15 Consider the MOSFET of Prob. 16.14 and a bipolar transistor to be evaluated for use in a linear amplifier with a quiescent current of 2 mA. Calculate the ratio of the transconductances of the two devices and answer which of the two will be more suitable for use in the amplifier.

16.16 A p-channel MOSFET has a 0.1-μm thick SiO$_2$ which contains 8×10^{10} positive charges cm^{-2}. The substrate is n-silicon with $N_d = 1.5 \times 10^{16}$ cm^{-3}, and the gate is heavily doped with boron. The source and substrate are grounded. (a) Calculate the flat-band and the threshold voltages. (b) For $Z = 2L$, $V_G = -4$ V, and $V_D = -0.4$ V, the drain current $I_D = 20$ μA. Calculate the value of Z required to obtain a drain current of 2.5 mA with $V_G = -5$ V and $V_D = -0.5$ V. Assume the gate length to be 10 μm and p$^+$-source and drain electrodes extend 1.25 μm sideways under the gate oxide. (c) Calculate the substrate bias that will be needed to shift the threshold voltage of the above device to -2.7 V.

16.17 Consider an n-channel MOSFET operating in the saturation region. Show that the electron transit time τ_t through the channel can be approximated by Eq. (16.65).

16.18 Combining Eqs. (16.74) and (16.76) show that the surface potential ϕ_s is given by Eq. (16.77).

16.19 A CCD is fabricated on a p-silicon substrate with $N_a = 3 \times 10^{14}$ cm^{-3} and SiO$_2$ thickness of 1500 Å. The electrode area is 10 μm × 20 μm. (a) Calculate the surface potential and the depletion region width for two electrodes biased at 10 V and 20 V, respectively. Assume $V_{FB} = 0$ and $Q_{sig} = 0$.

(b) Repeat (a) after 10^6 electrons are introduced into the cell. (c) Calculate the fringing field at the electrode boundary in (a) if the interelectrode spacing is 3 μm.

16.20 A 1200 Å thick oxide layer is grown on a 5 Ω-cm resistivity p-silicon. If a gate voltage of 10 V is applied and a signal charge of 1.6×10^{-8} C cm^{-2} exists in the well, determine (a) the surface potential and (b) the electric field at the Si-SiO$_2$ interface.

17
CIRCUIT MODELS FOR TRANSISTORS

INTRODUCTION

When a transistor is used in a circuit, it becomes necessary to characterize the device in terms of its electrical behavior. Description of the electrical behavior of a device as viewed at its terminals or at connecting points to a circuit constitutes a circuit model for the device. We use circuit models for semiconductor devices because they are simpler to use than the equations that describe the behavior of the device under consideration.

Circuit models may be classified according to the type of excitation used and the range of operation. Thus, we may divide circuit models into two broad categories: large-signal and small-signal. These models can be developed using one of two approaches. The first approach is to start either from the graphical characteristics or from the equations that express the terminal behavior of the device in terms of its physical construction and operating principles. In the second approach, we start from the physical laws that govern the internal behavior of the device along with a description of its structure. The second approach is more general and requires an exact solution for the set of five equations, discussed in Sec. 5.6, and thus requires the use of a computer. In recent years, great progress has been made in understanding the physical phenomena that occur in a device by applying numerical methods to the modeling of the behavior of transistors. However, this approach is beyond the scope of this book and will not be discussed here.

This chapter provides a brief account of circuit models for bipolar and field-effect transistors. In this discussion, our main emphasis will be on small-signal incremental models that are used in linear circuits where the transistor is operated as an active device. The chapter starts with a two-port network description of a bipolar transistor. Section 17.2 is devoted to circuit models for bipolar transistors where the Ebers-Moll, the charge control, and the Gummel-Poon models have been discussed. Models for JFETs and MESFETs form the subject matter of Sec. 17.3. The chapter ends with a brief reference to circuit models for MOS transistors in Sec. 17.4.

17.1 TWO-PORT NETWORK DESCRIPTION OF A BIPOLAR TRANSISTOR

Transistors are often used in linear circuits as small-signal amplifiers. In such circuits, the transistor is always operated in its active region. This involves establishing a suitable point of operation on the output I-V characteristics around which the current and the voltage will swing when an ac signal is applied to the input of the device. Such a point on the output characteristics is called a *quiescent* or a *Q point*, and it is specified by the dc output current and the output voltage. The process of establishing the Q point is known as *biasing*. Transistor biasing is beyond the scope of this book, and the reader is referred to J. Millman and A. Grabel for details [1].

Once the operating point has been selected, the transistor can be represented by an incremental model whose equivalent circuit can be drawn and analyzed with the help of Kirchhoff's laws. If the input ac signal voltage v_i is small compared to kT/q, the output signal voltage v_o will be linearly related to v_i. When this is the case, the circuit is said to be incrementally linear and the behavior of the transistor can be characterized by the linear two-port network shown in Fig. 17.1. Here, the biased transistor is represented by a black box, and the network is characterized by four variables: the input and output voltages v_i and v_o and the two currents i_i and i_o. In this description, it is no longer necessary to know about the internal physics of the device.

Network theory offers several equivalent possibilities to describe the two-port properties of a network. Since there are six possible choices of independent variables, six different sets of two-port parameters are possible. Out of these, only the h parameters have found wide acceptance in the analysis of small-signal bipolar transistor circuits [1].

The two-port parameter models provide a general characterization of any incrementally linear, three-terminal device. However, these models have several disadvantages. First, each two-port parameter is a complex number that varies with frequency, temperature, and the operating point in a complicated manner. Second, the parameters are derived from measurements and have no simple relation with the internal physics of the device. For these reasons, the two-port parameter models have found only limited use. More accurate models can be obtained directly or indirectly from the considerations of the basic phsyics of the device and are discussed below.

FIGURE 17.1 Two-port network representation of a biased transistor with assigned variables.

17.2 MODELS FOR BIPOLAR TRANSISTORS

Modeling of a transistor begins with solving the basic device equations for carrier transport in semiconductors (see Sec. 5.6). For an accurate solution, the equations can be solved numerically using a computer with donor and acceptor profiles and appropriate boundary conditions as the input data. The numerical solutions not only predict the terminal behavior, but also describe the electron-hole distribution in the device, and thus provide a detailed study of the device behavior. These solutions, though accurate, require a large amount of computational time. For this reason, analytical expressions for the solutions of the device equations have been derived using different approximations. In this section, we will discuss only analytical models for the bipolar transistor. These will include the Ebers-Moll model, the charge control model [2], and the integral charge control model proposed by Gummel and Poon [3].

17.2.1 The Ebers-Moll Model

The Ebers-Moll model is an exact circuit representation of the Ebers-Moll equations (Sec. 13.3.3). That is,

$$I_E = -I_{ES}\left[\exp\left(\frac{V_{EB}}{V_T}\right) - 1\right] + \alpha_I I_{CS}\left[\exp\left(\frac{V_{CB}}{V_T}\right) - 1\right]$$
$$= I_N - \alpha_I I_{INV} \tag{17.1a}$$

and

$$I_C = \alpha_N I_{ES}\left[\exp\left(\frac{V_{EB}}{V_T}\right) - 1\right] - I_{CS}\left[\exp\left(\frac{V_{CB}}{V_T}\right) - 1\right]$$
$$= -\alpha_N I_N + I_{INV} \tag{17.1b}$$

Here I_N and I_{INV} represent the normal and the inverse components of the terminal currents, respectively. The model based on these equations is shown in Fig. 17.2 for p-n-p and n-p-n transistors. Note that the transistor is represented by two diodes each in parallel with a current source that supplies a current that is lin-

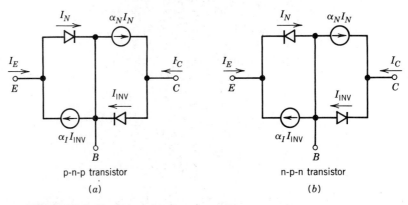

FIGURE 17.2 The Ebers-Moll circuit model for p-n-p and n-p-n transistors.

MODELS FOR BIPOLAR TRANSISTORS

early related to the diode current of the opposite junction. Thus, the model has four parameters, namely, the diode current constants I_{ES} and I_{CS} and the normal and inverse current gains α_N and α_I. Since $\alpha_N I_{ES} = \alpha_I I_{CS}$, only three of the four parameters have to be known.

The Ebers-Moll model is a large-signal model and is valid for all possible combinations of the emitter-base and collector-base junction voltages. However, as already mentioned in Chapter 13, the model is valid only for the intrinsic transistor. This means that the diodes in Fig. 17.2 are ideal diodes, and carrier generation and recombination in the junction depletion regions and high-level injection effects have been ignored. Furthermore, static conditions are assumed to prevail, which implies that all time delays and capacitive charging effects are absent. Thus, the model cannot be used to describe real transistors under dynamic operating conditions. It can, however, be extended to real transistors by including additional components to account for dynamic operation and second-order effects.

(a) SMALL-SIGNAL EQUIVALENT CIRCUITS FROM THE EBERS-MOLL MODEL

A simple small-signal equivalent circuit of a bipolar transistor operating in the normal active region of operation can be obtained from the circuit model of Fig. 17.2. Let us consider a p-n-p transistor biased in the normal active region with a quiescent emitter current I_E and an emitter-base junction voltage V_{EB}. Let a small-signal voltage v_{eb} be applied to the emitter-base junction. The instantaneous values of the emitter current and the emitter-base junction voltage can then be expressed as

$$i_E = I_E + i_e, \qquad v_{EB} = V_{EB} + v_{eb} \tag{17.2}$$

and similarly for the collector-base junction

$$i_C = I_C + i_c \quad \text{and} \quad v_{CB} = V_{CB} + v_{cb} \tag{17.3}$$

Here the instantaneous total values of the terminal currents and voltages are denoted by lower case variables with upper case subscripts; the incremental values are denoted by lower case variables with lower case subscripts; and the dc values are given by upper case variables with upper case subscripts. In the normal active region of operation, the collector-base junction has a large reverse bias, and replacing V_{EB} in Eq. (17.1a) by $(V_{EB} + v_{eb})$ gives

$$i_E = I_E + i_e = -I_{ES}\left[\exp\left(\frac{V_{EB} + v_{eb}}{V_T}\right) - 1\right] - \alpha_I I_{CS}$$

For all practical values of V_{EB}, the $\alpha_I I_{CS}$ term is small compared to the first term and can be ignored. Furthermore, assuming that v_{eb} is small compared to V_T, we can write

$$I_E + i_e \simeq -I_{ES}\left[\exp\left(\frac{V_{EB}}{V_T}\right) - 1\right] - \frac{v_{eb} I_{ES}}{V_T}\left[\exp\left(\frac{V_{EB}}{V_T}\right)\right]$$

This relation yields

$$i_e = \frac{v_{eb}}{r_e} \tag{17.4a}$$

where

$$r_e = \frac{V_T}{|I_E - I_{ES}|} \simeq \frac{V_T}{|I_E|} \tag{17.4b}$$

represents the emitter-base junction resistance for incremental operation. Eliminating the $\exp(v_{EB}/V_T)$ term from the two Ebers-Moll equations, i_C can be written as

$$i_C = -\alpha_N i_E + I_{CO}$$

Neglecting the second term and differentiating with respect to i_E leads to

$$\frac{di_C}{di_E} = -\alpha_N - i_E \frac{\partial \alpha_N}{\partial i_E} = -\tilde{\alpha}$$

where $\tilde{\alpha}$ represents the transistor alpha for the small-signal operations. Identifying the increments di_C and di_E by their respective small-signal values (i_e and i_c), we obtain the relation

$$i_c = -\tilde{\alpha} i_e \tag{17.5}$$

Figure 17.3a shows the transistor small-signal equivalent circuit obtained from Eqs. (17.4a) and (17.5). Note that in this circuit, the forward-biased emitter-base diode is replaced by its equivalent resistance r_e, and its controlled source is replaced by $\tilde{\alpha} i_e$. But the reverse-biased collector-base diode and its controlled source are open-circuited.

The equivalent circuit of Fig. 17.3a can be further improved to include the effect of base-width modulation and base resistance as shown in Fig. 17.3b. Base resistance effects are approximated with the addition of the small-signal resistance r_b. As we have already seen in Sec. 13.4.2, base-width modulation produces a finite output conductance $1/r_c$, and its effect at the input is approximated by the controlled-voltage source μv_{cb}.

The equivalent circuit in Fig. 17.3b is still a low-frequency model. Frequency effects may be incorporated by including the collector-base junction capacitance C_c and the emitter-base capacitance $C_e = C_{je} + C_{De}$, where C_{je} is the emitter-base junction depletion region capacitance and C_{De} is the diffusion capacitance. These modifications result in the T-equivalent circuit shown in Fig. 17.3c.

17.2.2 The Charge Control Model

This model is based on the charge control equations considered in Sec. 13.6. In contrast to the Ebers-Moll model which is valid only for static conditions, the charge control model is a dynamic model and, hence, is more suitable for studying the time dependence of various terminal currents.

The charge control equations for a p-n-p transistor are reproduced below:

$$i_E = \frac{Q_N}{\tau_N} + \frac{Q_N}{\tau_{BN}} + \frac{dQ_N}{dt} + \frac{dQ_{VE}}{dt} - \frac{Q_I}{\tau_I}$$

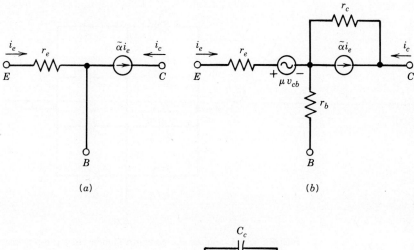

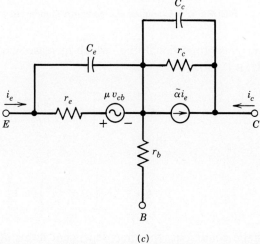

FIGURE 17.3 Small-signal equivalent circuits of a transistor in common base configuration. (a) A simple circuit obtained from the Ebers-Moll model, (b) inclusion of the base resistance and base-width modulation effects, and (c) the high-frequency T equivalent circuit.

$$i_C = \frac{Q_I}{\tau_I} + \frac{Q_I}{\tau_{BI}} + \frac{dQ_I}{dt} + \frac{dQ_{VC}}{dt} - \frac{Q_N}{\tau_N}$$

$$i_B = -\left[\frac{Q_N}{\tau_{BN}} + \frac{dQ_N}{dt} + \frac{Q_I}{\tau_{BI}} + \frac{dQ_I}{dt} + \frac{dQ_{VE}}{dt} + \frac{dQ_{VC}}{dt}\right] \quad (17.6)$$

Figure 17.4 shows a network representation of the charge control equations. The time-dependent elements representing the charges Q_N, Q_I, Q_{VE}, and Q_{VC} are shown as capacitors with a line across them to indicate that these elements store charge but are voltage dependent. When these charge-storing elements are omitted, the circuit of Fig. 17.4 is equivalent to the Ebers-Moll model of Fig. 17.2. The equivalence between these two models can be highlighted by comparing the charge control equations under dc steady state conditions with the Ebers-Moll equations. From Eq. (13.62), we see that Q_N is directly proportional to the excess concentration p_E which rises exponentially with the emitter-base voltage V_{EB}.

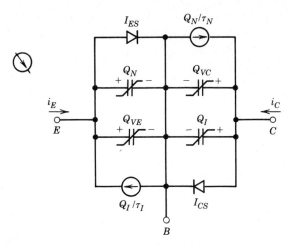

FIGURE 17.4 Complete charge control model for a p-n-p transistor.

Similarly, Q_I is proportional to p_C which increases exponentially with V_{CB}. Consequently, it is possible to write

$$Q_N = Q_{NO}\left[\exp\left(\frac{V_{EB}}{V_T}\right) - 1\right]$$

and

$$Q_I = Q_{IO}\left[\exp\left(\frac{V_{CB}}{V_T}\right) - 1\right] \tag{17.7}$$

Under static conditions, the time-dependent terms dQ_N/dt and dQ_{VE}/dt in the expression for i_E in Eq. (17.6) are zero, and comparing the resulting expression with its corresponding term in Eq. (17.1a), it is easy to see that

$$I_{ES} = -Q_{NO}\left(\frac{1}{\tau_N} + \frac{1}{\tau_{BN}}\right)$$

and

$$\alpha_I I_{CS} = -\frac{Q_{IO}}{\tau_I} \tag{17.8}$$

Similar relations can be obtained for I_{CS} and $\alpha_N I_{ES}$, and the parameters α_N and α_I can be expressed in terms of the charge control parameters τ_N, τ_{BN}, τ_I, and τ_{BI} (see Prob. 17.2).

We would like to emphasize once again that the charge control model does not include the details of the charge distributions in the various regions of the device. It is a quasi-static model and provides a fairly accurate picture of the real situation over a time scale that is large compared to the base transit time τ_N. However, the model will not give correct results when the time scale of observation is comparable to τ_N.

MODELS FOR BIPOLAR TRANSISTORS

Let us use the charge control model to obtain a transistor equivalent circuit for small-signal dynamic operation in the normal active region. Since the collector-base junction is reverse biased, $Q_I = -Q_{IO}$ which is constant and independent of the bias voltage. Hence $dQ_I/dt = 0$, and if we ignore Q_{IO}, the base and collector currents from Eq. (17.6) become

$$i_B = -\frac{Q_N}{\tau_{BN}} - \frac{dQ_N}{dt} - \frac{d}{dt}(Q_{VE} + Q_{VC})$$

and

$$i_C = -\frac{Q_N}{\tau_N} + \frac{dQ_{VC}}{dt} \quad (17.9)$$

We now consider the consequences of incremental changes in the emitter-base junction voltage. Let $v_{EB} = V_{EB} + v_{eb}$ where v_{eb} is a time-varying incremental voltage superimposed on the dc bias. The charge $Q_N(t)$ can now be written as

$$Q_N(t) = Q_{NO}\left[\exp\left(\frac{V_{EB}}{V_T}\right)\exp\left(\frac{v_{eb}}{V_T}\right) - 1\right]$$

When v_{eb} is small compared to V_T, $\exp(v_{eb}/V_T) \simeq 1 + v_{eb}/V_T$, and we obtain

$$Q_N(t) = \bar{Q}_N + q_N(t) \quad (17.10)$$

where

$$\bar{Q}_N = Q_{NO}[\exp(V_{EB}/V_T) - 1] \quad (17.11a)$$

and

$$q_N(t) = \frac{v_{eb}}{V_T}[\bar{Q}_N + Q_{NO}] \simeq \frac{\bar{Q}_N}{V_T}v_{eb} \quad (17.11b)$$

Here $\bar{Q}_N$ is the charge stored at the bias point, and $q_N(t)$ is the charge associated with the incremental voltage v_{eb}.

The instantaneous base and collector currents are

$$i_B = I_B + i_b$$

and

$$i_C = I_C + i_c \quad (17.12)$$

Since the time varying components of charges are zero at the bias point, from Eqs. (17.9) and (17.12), we can write

$$I_B = -\frac{\bar{Q}_N}{\tau_{BN}} \quad (17.13a)$$

and

$$i_b = -\frac{q_N}{\tau_{BN}} - \frac{dq_N}{dt} - C_{je}\frac{dv_{eb}}{dt} - C_c\frac{dv_{cb}}{dt} \qquad (17.13b)$$

where C_{je} and C_c represent the emitter-base and the collector-base junction capacitances, respectively, at the bias point. Similarly, for the collector current we have

$$I_C = -\frac{Q_N}{\tau_N} \qquad (17.14a)$$

and

$$i_c = -\frac{q_N}{\tau_N} + C_c\frac{dv_{cb}}{dt} \qquad (17.14b)$$

From Eqs. (17.11b) and (17.13a), we obtain

$$q_N(t) = \frac{I_B}{V_T}\tau_{BN}v_{be} \qquad (17.15)$$

where $v_{be} = -v_{eb}$ is the base to emitter voltage. Substituting for q_N from Eq. (17.15) into Eqs. (17.13b) and (17.14b) leads to the following relations:

$$i_b = -\frac{I_B}{V_T}v_{be} - \frac{I_B}{V_T}\tau_{BN}\frac{dv_{be}}{dt} + C_{je}\frac{dv_{be}}{dt} + C_c\frac{dv_{bc}}{dt} \qquad (17.16a)$$

and

$$i_c = -\frac{I_C}{V_T}v_{be} - C_c\frac{dv_{bc}}{dt} \qquad (17.16b)$$

where we have used the fact that $I_C/I_B = \tau_{BN}/\tau_N$. We can also write

$$-\frac{I_B}{V_T} = \frac{I_E}{(1+h_{FE})V_T} = \frac{1}{(1+h_{FE})r_e} = \frac{1}{r_\pi} = g_\pi$$

and

$$\frac{|I_C|}{V_T} = g_m \qquad (17.17)$$

After substituting these values into Eqs. (17.16a) and (17.16b), the following equations for i_b and i_c are obtained:

$$i_b = \frac{v_{be}}{r_\pi} + (C_{De} + C_{je})\frac{dv_{be}}{dt} + C_c\frac{dv_{bc}}{dt} \qquad (17.18a)$$

$$i_c = g_m v_{be} + C_c\frac{dv_{cb}}{dt} \qquad (17.18b)$$

MODELS FOR BIPOLAR TRANSISTORS

where

$$C_{De} = \frac{I_B}{V_T}\tau_{BN} = \frac{I_C \tau_N}{V_T} = g_m \tau_N \qquad (17.18c)$$

is the emitter-base junction diffusion capacitance.

The equivalent circuit resulting from Eqs. (17.18a) and (17.18b) is the hybrid-π equivalent circuit shown in Fig. 17.5. Here the signs of the incremental currents i_c and i_b are ignored because the sign is of no consequence to us. This circuit can be further improved by including the effects of the base resistance and the base-width modulation. From our analysis in Sec. 13.4, we have seen that an increase in the reverse bias at the collector causes an increase in the collector current following a reduction in the base width W_B. Furthermore, a decrease in W_B also causes a decrease in the stored charge Q_N, and hence, the base current also decreases. Let us now investigate the effect of these changes on the circuit of Fig. 17.5.

In terms of the small-signal parameters, the variation in i_C with the reverse bias v_{CB} can be written as

$$\left|\frac{\partial i_C}{\partial v_{CB}}\right| = \left(\frac{\partial i_C}{\partial v_{EB}}\right)\left|\frac{\partial v_{EB}}{\partial v_{CB}}\right| = \mu_1 g_m = g_o \qquad (17.19)$$

where $\mu_1 = |\partial v_{EB}/\partial v_{CB}|$ is a dimensionless parameter that may be calculated for a transistor with a known impurity profile. The variation in Q_N with v_{CB} can be estimated from Eq. (17.9) after neglecting the dQ_{VC}/dt term:

$$\left|\frac{\partial Q_N}{\partial v_{CB}}\right| = \left|\frac{\partial}{\partial v_{CB}}(i_C \tau_N)\right| = \mu_1 \tau_N g_m = \mu_1 C_{De} \qquad (17.20)$$

Similarly, using Eqs. (17.13a), (17.14a) and (17.19), we get

$$\left|\frac{\partial i_B}{\partial v_{CB}}\right| = \left|\frac{\partial}{\partial v_{CB}}\left(\frac{Q_N}{\tau_{BN}}\right)\right| = \left|\frac{\partial i_C}{\partial v_{CB}}\left(\frac{\tau_N}{\tau_{BN}}\right)\right| = \mu_1 g_m \delta = \delta g_o \qquad (17.21)$$

where $\delta = \tau_N/\tau_{BN}$. The variations calculated in Eq. (17.19) through Eq. (17.21) can be easily incorporated into the incremental equivalent circuit of Fig. 17.5.

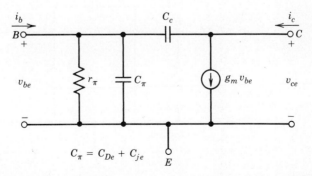

FIGURE 17.5 The basic hybrid-π equivalent circuit for a bipolar transistor obtained from the charge control model.

Following Eq. (17.19), the variations in the collector current i_C can be represented by the relation $\Delta i_C = g_o \Delta v_{CB}$, which in terms of small-signal increments becomes

$$i_c = g_o v_{CB} \tag{17.22}$$

This current flows between the emitter and collector in response to a change in the collector-base junction voltage and can be modeled by a current generator activated by v_{cb} placed between the collector and emitter terminals. Since in the normal active mode, $v_{cb} \simeq v_{ce}$, the same objective can be achieved by connecting a resistance $1/g_o$ between the collector and emitter terminals as shown in Fig. 17.6. The change in the stored charge Q_N given by Eq. (17.20) is modeled as a capacitance in parallel with C_c. Finally, the variation in the base current calculated in Eq. (17.21) can be modeled as a current generator $\delta g_o v_{cb}$ between the emitter and the base terminals. This is equivalent to connecting a resistance $(\delta g_o)^{-1}$ between the base and the collector terminals [4].

To account for the effect of the base resistance into the small-signal equivalent circuit, the dependence of the static base resistance r_B on I_C must be determined. Once this dependence is known, the effect of base spreading resistance is then taken into account by connecting a resistance r_b in series with r_π between the points B and B' in Fig. 17.6. The value of r_b is given by the relation [4]

$$r_b = r_B + I_C \frac{dr_B}{di_C} \tag{17.23}$$

Note that the terminal B' in Fig. 17.6 is the base terminal of an intrinsic transistor and is not accessible for direct measurements. It is also evident from the circuit that the current generator in the output is no longer activated by the applied base-to-emitter voltage v_{be}, but is controlled by the internal node pair voltage $v'_{be} = (v_{be} - i_b r_b)$. At very high frequencies, the base resistance and the junction capacitances behave like a distributed transmission line, and the effect of the base spreading resistance can be modeled by using an equivalent R-C network in place of r_b.

Although the hybrid-π equivalent circuit is the most frequently used model, the complete circuit shown in Fig. 17.6 is seldom used in hand calculations. In many cases, several elements in the circuit have negligible effect and can be left

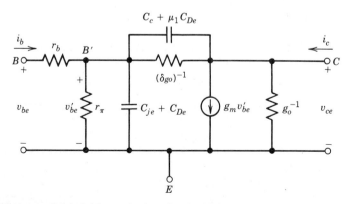

FIGURE 17.6 A complete hybrid-π equivalent circuit which includes the effects of base resistance and the base-width modulation.

out. When all the elements have to be included, the calculations are accomplished using a computer.

17.2.3 The Gummel-Poon Model [3]

The Gummel-Poon model is based on an integral charge control relation that relates the terminal characteristics of the device to the total charge of majority carriers in the base. It is a large-signal model and takes into account many physical effects like base-width modulation and base pushout (see Sec. 14.5). Therefore, this model is suitable for a more detailed study of bipolar transistors than the other two models, and it is widely used in the computer-aided design program known as SPICE [5] (*s*imulation *p*rogram with *i*ntegrated *c*ircuit *e*mphasis).

We have seen in Chapter 14 that by neglecting recombination in the base region, the emitter-to-collector current of a p-n-p transistor is given by Eq. (14.13), which can be rewritten as

$$I_{pC} = \frac{(qAn_i)^2 D_B}{Q_B}\left[\exp\left(\frac{V_{EB}}{V_T}\right) - \exp\left(\frac{V_{CB}}{V_T}\right)\right] \tag{17.24}$$

Here

$$Q_B = qA \int_0^{W_B} n(x)\,dx \tag{17.25}$$

is the total majority carrier charge in the base region. It is possible to split Eq. (17.24) into its normal and inverse components I_{pN} and I_{pINV}. Thus

$$I_{pC} = I_{pN} - I_{pINV} \tag{17.26}$$

where

$$\begin{aligned} I_{pN} &= \frac{Q_{BO}}{Q_B}\frac{(qAn_i)^2}{Q_{BO}} D_B\left[\exp\left(\frac{V_{EB}}{V_T}\right) - 1\right] \\ &= I_s \frac{Q_{BO}}{Q_B}\left[\exp\left(\frac{V_{EB}}{V_T}\right) - 1\right] \end{aligned} \tag{17.27a}$$

and

$$I_{pINV} = \frac{I_s Q_{BO}}{Q_B}\left[\exp\left(\frac{V_{CB}}{V_T}\right) - 1\right] \tag{17.27b}$$

In the Gummel-Poon model, the base charge Q_B is bias dependent and is written in the following form:

$$Q_B = Q_{BO} + Q_{VE} + Q_{VC} + B\tau_N I_{pN} + \tau_I I_{pINV} \tag{17.28}$$

Here Q_{BO} is the fixed base charge given by the dopant impurity distribution; Q_{VE} and Q_{VC} are the charges associated with the emitter-base and collector-base junction depletion regions; and $B\tau_N I_{pN}$ and $\tau_I I_{pINV}$ are the minority carrier charges stored because of the normal and the inverse components of the current, respectively. The parameter B is the base-pushout factor that remains unity at low values of I_{pC}, but becomes larger than unity at high-injection levels because of the Kirk effect.

The base current i_B is modeled as

$$i_B = -\frac{dQ_B}{dt} - I_{EB} - I_{CB} \tag{17.29}$$

where I_{EB} is the current flowing from the emitter into the base and I_{CB} is the collector-to-base current. These currents can be expressed as

$$I_{EB} = I_1\left[\exp\left(\frac{V_{EB}}{V_T}\right) - 1\right] + I_2\left[\exp\left(\frac{V_{EB}}{\eta_1 V_T}\right) - 1\right] \tag{17.30a}$$

and

$$I_{CB} = I_3\left[\exp\left(\frac{V_{CB}}{\eta_2 V_T}\right) - 1\right] \tag{17.30b}$$

In these equations, η_1 and η_2 are the emitter-to-base and collector-to-base junction ideality factors, respectively. The first term in Eq. (17.30a) represents the current caused by minority carrier recombination in the base region. The second term represents the sum of two components. The first component is the current due to minority carriers injected from the base into the emitter, and the second component is the current caused by the electron–hole pair recombination in the emitter-base junction depletion region. For the first component, $\eta_1 = 1$ and for the second $\eta_1 = 2$. Similarly, the current I_{CB} in Eq. (17.30b) has two components: one with $\eta_2 = 1$ and the other with $\eta_2 = 2$.

The total emitter current i_E is given by the relation

$$i_E = I_{pC} + I_{EB} + \frac{dQ_{VE}}{dt} + \tau_N \frac{dI_{pN}}{dt} \tag{17.31}$$

and the collector current i_C is obtained from the condition $i_C + i_E + i_B = 0$. Figure 17.7 shows the circuit representation of the Gummel-Poon model for a p-n-p transistor. The first term in Eq. (17.31) represents the diode with the current I_{pN} and the controlled source I_{pINV}. The second term represents the two diodes D_1 and D_2, while the third and fourth terms account for the effects of the emitter-to-base junction capacitance C_{je} and the diffusion capacitance C_{De}, respectively. The various elements on the collector side can be similarly understood. The model shown in Fig. 17.7 accounts for many physical phenomena inside the transistor. Because the base charge Q_B is voltage dependent, the effects of high-level injection in the base and quasi-saturation of the collector are included.

17.3 CIRCUIT MODELS FOR JFETs and MESFETs

17.3.1 Dynamic Models for Field-Effect Transistors

The current-voltage characteristics of the JFET in Sec. 15.2 were derived only for the static condition, which means that the voltages applied to the transistor

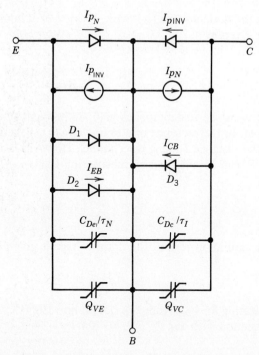

FIGURE 17.7 Circuit representation of the Gummel-Poon model for a p-n-p transistor. After H. K. Gummel and H. C. Poon, reference 3.

are assumed to change slowly with time. For voltages that change rapidly with time, the static analysis is inadequate and must be modified to include the following two effects:

1. When viewed from the gate and source terminals, the transistor structure acts as a parallel plate capacitor, with the gate and channel forming two plates whose separation is the width of the gate-to-channel junction depletion region. Thus, when the gate-to-source voltage is changed, a capacitive gate current will flow.
2. The majority carriers require a finite transit time to traverse the channel from the source to the drain. If the gate voltage V_G changes significantly during this time, the static expression for the drain current may no longer be valid.

We will now include these two effects in the analysis. Since most applications of JFETs and MESFETs involve operation in the saturation region of the drain current, our discussion of dynamic effects will be mainly confined to this region of operation. In order to evaluate the dynamic gate current, we compute the total charge under the gate. Considering the single-gate n-channel JFET of Fig. 15.4b, the gate charge Q_G is

$$Q_G = qN_d Z \int_0^L W(x)\,dx \tag{17.32}$$

When the drain voltage v_D is zero, $W(x) = W$ is constant throughout the channel and is given by

$$W = \sqrt{\frac{2\varepsilon_s}{qN_d}(V_i - v_G)}$$

where v_G is the instantaneous gate voltage and V_i is the built-in voltage of the junction. In the presence of a drain voltage v_D, the depletion region width varies with the distance x along the channel. However, a rough estimate of its average value $\overline{W}$ can be obtained by substituting $V(x) = v_D/2$ in Eq. (15.12) giving

$$\overline{W} = \left[\frac{2\varepsilon_s}{qN_d}\left(V_i - v_G + \frac{v_D}{2}\right)\right]^{1/2} \tag{17.33}$$

In the saturation region, the drain voltage is given by

$$v_D = V_{Dsat} = V_P - (V_i - v_G) \tag{17.34}$$

Substituting for V_{Dsat} from Eq. (17.34) into Eq. (17.33), and combining the resulting equation with Eq. (17.32), we obtain

$$Q_G = ZL[q\varepsilon_s N_d(V_i - v_G + V_P)]^{1/2} \tag{17.35}$$

The instantaneous gate current i_G is now given by the relation

$$i_G = \frac{dQ_G}{dt} \tag{17.36a}$$

Also from Eq. (15.17), the instantaneous saturated drain current can be written as

$$i_D = i_{Dsat} = G_o\left\{\frac{V_P}{3} + \left[\frac{2}{3}\left(\frac{V_i - v_G}{V_P}\right)^{1/2} - 1\right](V_i - v_G)\right\}$$

$$= f(v_G) \tag{17.36b}$$

The circuit model in Fig. 17.8a describes the dynamic behavior of the JFET given by the above two relations. Note that the element storing the charge Q_G is voltage dependent and is shown as a capacitor with a line across it. Also note that a nonlinear voltage-dependent current source has been used to model the saturated drain current.

After the channel is pinched off near the drain, a depletion layer separates the drain from the channel. When the voltage v_D is raised above V_{Dsat}, the additional voltage $(v_D - V_{Dsat})$ appears across this depletion layer. Since some of the field lines originating on the drain terminate at the gate, there is an electrostatic coupling between the gate and the drain. The dynamic consequence of this coupling is taken into account by adding a voltage-dependent capacitance between the gate and the drain as shown in Fig. 17.8b. This capacitance is usually small compared to the gate-to-source capacitance.

The effect of a finite transit time through the channel will cause a decrease in the current relative to its static value. If the gate voltage varies so rapidly that the gate charge Q_G undergoes a significant change during the majority carrier transit

CIRCUIT MODELS FOR JFETs and MESFETs

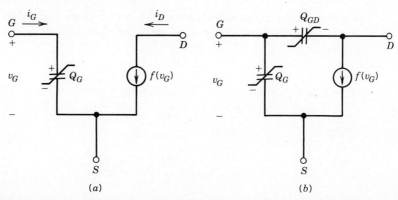

FIGURE 17.8 Dynamic JFET models in the region of saturated drain current. (a) Basic model and (b) a model that accounts for capacitive coupling between the drain and the gate.

time through the channel, the gate current i_G may become comparable to the drain current i_D, and the current gain of the transistor is reduced. Therefore, the device performs the useful function of amplification only for gate voltages that vary slowly enough so that dQ_G/dt changes only by a small amount during one transit time.

17.3.2 Small-Signal Equivalent Circuits

A small-signal equivalent circuit for the JFET can be obtained from the large-signal dynamic model of Fig. 17.8a. Let the transistor be biased by a dc gate voltage V_G, where the drain current is I_D, and let v_g and i_d represent the small-signal components of the gate voltage and the drain current, respectively. Thus, the total gate-to-source voltage can be expressed as the sum of the dc and small-signal components

$$v_G = V_G + v_g \tag{17.37}$$

The resulting drain and gate currents can be expressed as

$$i_D = I_D + i_d = G_o \left\{ \frac{V_P}{3} + \left[\frac{2}{3} \left(\frac{V_i - V_G - v_g}{V_P} \right)^{1/2} - 1 \right] (V_i - V_G - v_g) \right\} \tag{17.38a}$$

and

$$i_G = I_G + i_g = \frac{dQ_G}{dt} = ZL \frac{d}{dt} [q\varepsilon_s N_d (V_i - V_G - v_g + V_P)]^{1/2} \tag{17.38b}$$

Assuming that v_g is small compared to $(V_G - V_i)$, the following expression for i_d is obtained (see Prob. 17.7):

$$i_d \simeq G_o \left[1 - \left(\frac{V_i - V_G}{V_P} \right)^{1/2} \right] v_g = g_m v_g \tag{17.39}$$

where g_m is the transconductance of the device in the saturation region. Since in Eq. (17.38b) the bias voltage V_G is constant, $I_G \propto dV_G/dt = 0$, and if v_g is small compared to $(V_G - V_i)$, we obtain

$$i_g = C_{GS}\left|\frac{dv_g}{dt}\right| \qquad (17.40a)$$

where

$$C_{GS} = \frac{ZL}{2}\left(\frac{q\varepsilon_s N_d}{V_i - V_G + V_P}\right)^{1/2} \qquad (17.40b)$$

is the incremental gate to source capacitance. Equations (17.39) and (17.40a) can be interpreted in terms of the equivalent circuit shown in Fig. 17.9a.

In the above development, we had assumed that after the onset of pinch-off the drain current is independent of the drain voltage. Furthermore, we had ignored the electrostatic coupling between the drain and the gate. Let us now examine the consequences of removing these limitations. In Sec. 15.3.2 we saw that, even in the saturation region, the drain current is a slowly varying function of the drain voltage v_D. Therefore, we write

$$i_D = f(v_G, v_D)$$
$$\Delta i_D = \frac{\partial i_D}{\partial v_G}\Delta v_G + \frac{\partial i_D}{\partial v_D}\Delta v_D \qquad (17.41a)$$

and replacing the incremental values with the small-signal variations gives

$$i_d = g_m v_g + g_D v_d \qquad (17.41d)$$

The effect of electrostatic coupling between the drain and the gate can be represented by an incremental gate-to-drain capacitance C_{GD}. The equivalent circuit with finite drain conductance and the capacitance C_{GD} is shown in Fig. 17.9b.

The circuit of Fig. 17.9b is sufficiently accurate for relatively long gate length JFETs operating at low frequencies. However, for submicron gate length MES-

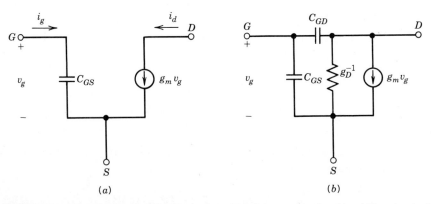

FIGURE 17.9 Small-signal equivalent circuits for a JFET. (a) A basic circuit and (b) a circuit that includes the effects of drain-to-gate capacitance and the finite drain conductance.

FETs, several other effects have to be included. An ac equivalent circuit for a MESFET valid at microwave frequencies is described by C. A. Liechti [6], who has also discussed the physical origins of various circuit elements. A modified version of this circuit is shown in Fig. 17.10 after omitting the parasitic drain and gate series resistances. In this circuit $(C_{GD} + C_{GS})$ represents the total gate-to-channel capacitance, and C_{DC} physically results from drain-to-channel feedback [7]. The resistances R_1 and $1/g_D$ under the gate show the effect of channel resistance. Note that the current source is controlled by the total voltage drop across C_{GS} and R_1, rather than by the voltage drop across C_{GS} alone. The transconductance is complex $[g_m \exp(-j\omega\tau_t)$ where τ_t is the transit time]. The resistance R_s and the capacitance C_{DS} are the parasitic elements. This equivalent circuit may be used to determine the current cutoff frequency f_T and the unity gain frequency $f_{\max}$.

17.4 MODELS FOR MOS TRANSISTORS

17.4.1 Large-Signal Dynamic Model

From our discussion in Chapters 15 and 16 it is clear that the JFETs and MOSFETs show similar electrical behavior. Thus, the large-signal models for MOS transistors can be developed in the same manner as for JFETs. For example, a large-signal static model for the MOSFET can be directly obtained from the relations that express the drain current as a function of gate, drain, and substrate voltages. The static model can then be extended to include dynamic effects by adding components that store charge whenever the gate and the drain voltages are changed. Since a MOS structure is more complex than a JFET structure, a larger number of capacitances are present in the former compared to the latter.

Figure 17.11a shows a schematic n-channel MOS transistor indicating the locations of various circuit elements. The corresponding circuit model for the device is shown in Fig. 17.11b. Here the symbols S and B are used for the source and substrate, respectively. The capacitance C_{GB}, C_{GS}, and C_{GD} are related to the intrinsic device structure, as is the current source $f(v_G)$. The source-to-substrate n$^+$-p junction diode is modeled as an ideal diode in parallel with a capacitor C_{SB}

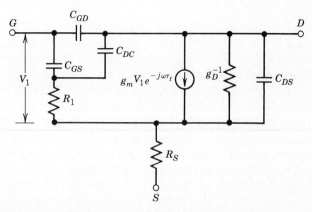

FIGURE 17.10 An ac equivalent circuit of a MESFET. (From W. R. Curtice, reference 7, Fig. 2, p. 1694. Copyright © 1983 IEEE. Reprinted by permission.)

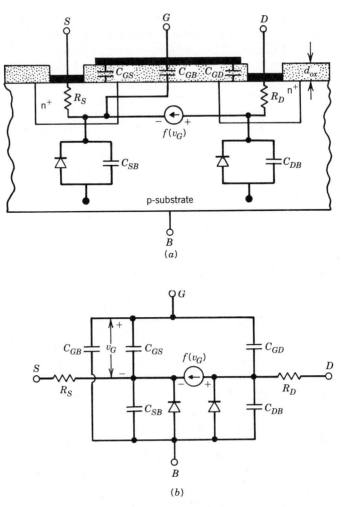

FIGURE 17.11 Circuit model of a MOS transistor. (*a*) Physical origin of the various circuit elements and (*b*) a large-signal model for an n-channel MOSFET. (From D.V. McCaughan and J. C. White, MOS transistors and memories, Chapter 3A in *Handbook on Semiconductors,* vol. 4, T. S. Moss and C. Hilsum (eds.), North-Holland, Amsterdam, p. 305. British Crown Copyright 1981. Reprinted by permission of the Controller of Her Britannic Majesty's Stationery Office.)

which represents the depletion region capacitance of the n$^+$-p junction. Similarly, the drain-to-substrate diode is represented by an ideal diode in parallel with a capacitor C_{DB}. The resistances R_S and R_D are the series resistances associated with unmodulated paths near the source and the drain, respectively.

The intrinsic capacitance C_{GB} represents the series capacitance of the oxide layer and the substrate depletion region. In all the physically based MOS models, C_{GB} is obtained by differentiating the gate charge Q_G with respect to the gate-to-substrate voltage v_{GB}. Similarly, the gate-to-source capacitance C_{GS} is obtained by differentiating Q_G with respect to the gate voltage v_G, and C_{GD} is obtained in the same way. For a MOS transistor operating in the linear region of the I-V characteristics, C_{GS} and C_{GD} are nearly the same and each one is equal to half the oxide capacitance $ZL\varepsilon_{ox}/d_{ox}$. However, in the saturation region, C_{GS} is approximately two-thirds of the oxide capacitance, and C_{GD} becomes negligibly small because of weak coupling between the gate and the drain.

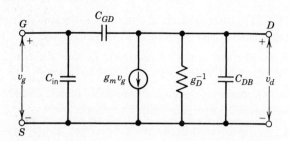

FIGURE 17.12 A small-signal equivalent circuit for MOSFET in common source configuration.

17.4.2 Small-Signal Equivalent Circuit

A small-signal equivalent circuit for the MOS transistor can be derived from the large-signal dynamic model of Fig. 17.11b. The drain current i_D may be expressed as a function of the voltages v_G and v_D, as was done for the JFET, giving

$$i_d = g_m v_g + g_D v_d$$

In many operating conditions, the substrate is connected to the source so that the source-to-substrate diode and the capacitance C_{SB} are short-circuited. The reverse biased drain-to-substrate diode may be assumed to be open-circuited, and the parallel capacitances C_{GS} and C_{GB} may be combined to obtain a single-input capacitor C_{in}. The simplified equivalent circuit model obtained in this way is shown in Fig. 17.12. In this circuit, the parasitic fringing capacitances and the series resistances R_S and R_D have not been included. These can be included when a more accurate model is desired.

MOS integrated circuits containing a large number of transistors are complex circuits. In the design of such circuits, computer studies must be made to predict the steady state and dynamic characteristics of a structure comprising a number of devices. In order to be able to predict the overall performance of the circuit, the MOS circuit model must include as complete a model as possible. This means that the model should be sufficiently sophisticated to include the second-order effects such as the electric field dependence of carrier mobility, subthreshold leakage currents, short-channel effects, and the effects of interface states.

A number of computer programs are available for practical use [5, 8]. These simulation programs are applied to complete circuits by division of the circuit into small units which are suitable for detailed analysis. One of the widely used computer program is SPICE. In this program, the MOSFET is represented by the model shown in Fig. 17.11b. The SPICE model is useful for predicting the transient response of logic circuits and the phase and amplitude response of analog MOS circuits.

REFERENCES

1. J. MILLMAN and A. GRABEL, *Microelectronics,* McGraw-Hill, New York, 1987, Chapter 10 and Appendix C.

2. R. BEAUFOY and J. J. SPARKES, "The junction transistor as a charge-controlled device," *A.T.E. Journal* 13, 310 (1957).

3. H. K. GUMMEL and H. C. POON, "An integral charge control model of bipolar transistors," *Bell Syst. Tech. J.* 49 827 (1970).

4. R. S. MULLER and T. I. KAMINS, *Device Electronics for Integrated Circuits*, 2d ed., John Wiley, New York, 1986, Chapter 7.

5. P. ANTOGNETTI and G. MASSOBRIO (eds.), *Semiconductor Device Modeling with SPICE*, McGraw-Hill, New York, 1988.

6. C. A. LIECHTI, "Microwave field-effect transistors—1976," *IEEE Trans. Microwave Theory Tech.* MIT-24, 279 (1976).

7. W. R. CURTICE, "The performance of submicrometer gate length GaAs MESFET's," *IEEE Trans. Electron Devices* ED-30, 1693 (1983).

8. D. G. ONG, *Modern MOS Technology: Processes, Devices, and Design*, McGraw-Hill, International Edition, 1986, Chapter 12.

ADDITIONAL READING LIST

1. P. E. GRAY and C. L. SEARLE, *Electronic Principles: Physics, Models, and Circuits*, John Wiley, New York, 1969.

2. F. VAN DE WIELE, W. L. ENGL, and P. G. JESPERS (eds.), *Process and Device Modeling for Integrated Circuit Design*, Noordhoff, Leyden, 1977.

3. W. L. ENGL (ed.), *Process and Device Modeling*, North-Holland, Amsterdam, 1986, Chapters 8, 12, and 13.

PROBLEMS

17.1 From the small-signal equivalent circuit shown in Fig. 17.3a, draw a small-signal model driven by the base current i_b and use this model to obtain the simple hybrid π-circuit containing two elements r_π and $g_m v_{be}$.

17.2 Using relations (17.7) for Q_N and Q_I and assuming dc conditions, obtain the results of Eq. (17.8). Obtain similar relations for I_{CS} and $\alpha_N I_{ES}$, and show that

$$\alpha_N = \frac{\tau_{BN}}{\tau_{BN} + \tau_N} \quad \text{and} \quad \alpha_I = \frac{\tau_{BI}}{\tau_{BI} + \tau_I}$$

17.3 Show that the voltage V_{CE} of a saturated transistor is given by the relation

$$V_{CE} = \frac{kT}{q} \ln \left[\frac{Q_{IO}(Q_N + Q_{NO})}{Q_{NO}(Q_I + Q_{IO})} \right]$$

17.4 For an n-p-n transistor, the following parameters are measured at $I_C = 5$ mA and $T = 25°C$; $h_{ie} = 600\ \Omega$, $h_{fe} = 100$, and $h_{re} = 2 \times 10^{-5}$. Calculate the values of g_m, r_π, r_b, and g_o and show that $h_{fe} = g_m r_\pi$.

17.5 Consider a silicon p-n-p transistor with uniformly doped emitter, base, and collector regions, and a base-width of 1 μm. The minority carrier lifetime in the base is 0.1 μsec and $D_B = 12.5$ cm^2/sec. The emitter-base junction is forward biased, $C_{je} = 5.3$ pF, and the collector-to-base junction capacitance is given by $C_c = 5.1 \times 10^{-12}(1 - V_{CB})^{-1/2}$ farad. Assume an emitter efficiency of 0.99, $V_{CB} = -8$ V, $I_C = -2$ mA, and $T = 27°C$. Calculate all the hybrid-π parameters in the circuit of Fig. 17.5.

17.6 For the transistor of Prob. 17.5 take $\mu_1 = 10^{-5}$ and assume that $r_B = (50 - 6\sqrt{I_C})\Omega$ where I_C is expressed in mA. Determine r_b and all the remaining parameters in the hybrid-π equivalent circuit shown in Fig. 17.6.

17.7 Assuming that v_g in Eq. (17.38a) is small compared with $(V_G - V_i)$, obtain Eq. (17.39). Also obtain Eq. (17.40) from Eq. (17.38b).

17.8 Modify the JFET equivalent circuit shown in Fig. 17.9b by including the source and the drain series resistances R_s and R_D, respectively. Assuming that $(g_D)^{-1} \gg (R_S + R_D)$, and neglecting all the capacitances, show that the effective transconductance g'_m is given by the relation $g'_m = g_m(1 + g_m R_S)^{-1}$.

17.9 The drain current for an n-channel MOSFET operating below pinch-off is given by

$$i_D = \frac{Z\mu_n}{L}C_{ox}\left[v_G - V_{th} - \frac{v_D}{2}\right]v_D$$

where $v_D = V_D + v_d$ and $v_G = V_G + v_g$. Show that the incremental drain current i_d for this situation can be written as

$$i_d = g_m v_g + g_D v_d$$

Obtain the values of g_m and g_D.

17.10 For a silicon n-channel MOSFET operated with saturated drain current, determine the total channel charge $Q_c = Z\int_0^L Q_n(x)\,dx$ as a function of gate bias, and show that the gate-to-source capacitance is given by $C_{GS} = 2ZLC_{ox}/3$.

17.11 Define the electron transit time through the channel of an n-channel MOSFET by the relation $\tau_t = -Q_c/I_D$ and show that $\tau_t = 2C_{GS}/g_m$. Assume that the device is operated with the saturated drain current.

18
POWER RECTIFIERS AND THYRISTORS

INTRODUCTION

Power devices are used in high-power rectification and control. The important point about these devices is that heat dissipation is the limiting factor in their operation. Applications of semiconductor power devices became popular with the introduction of silicon-controlled rectifier in 1957. Since then, their power-handling capabilities have improved considerably, and now single devices capable of handling more than 10 MW are commercially available.

The spectrum of power devices spans a wide range of devices, including bipolar power transistors and power JFETs and MOSFETs. These devices have already been considered in earlier chapters. This chapter presents a brief account of the structure and operation of high-power devices. Section 18.1 deals with power rectifiers. The electrical behavior of these devices is similar to that of microwave p-i-n diodes (Sec. 11.2), but the design considerations are different. Section 18.2 is devoted to silicon-controlled rectifiers and thyristors where the basic physics of p-n-p-n structure is emphasized. Some important derivatives of the basic p-n-p-n structures are considered in Sec. 18.3, and the next section discusses the bidirectional thyristors. The chapter ends with a brief reference to the field-controlled thyristor in Sec. 18.5.

18.1 POWER RECTIFIERS

18.1.1 Basic Considerations

An ideal rectifier should only pass current in the forward-bias state without any voltage drop and should provide an open circuit when the device is reverse biased. Semiconductor diodes deviate considerably from this ideal behavior. For both p-n junction and Schottky barrier diodes, a voltage drop occurs when the diode is forward biased, and a finite current flows in the reverse-biased state. Furthermore, the reverse breakdown voltage limits the blocking capability of the diode. At high forward currents, the voltage drop across the series resistance of the diode causes considerable power loss and must be minimized. Although it is true that, for a given forward current the voltage drop across a Schottky barrier junction is lower than that across a p-n junction, the field of power control is dominated by p-n junction devices. This is because in order to obtain a high blocking capability in a power rectifier, the structure must have a thick high-resistivity zone to support the reverse voltage. Such a highly resistive region will cause a large forward drop in the forward on-state, which is undesirable. In a p-n junction diode, minority carrier injection in the high-resistivity zone modulates the conductivity of this region and reduces the forward voltage drop. Conductivity modulation is not possible in a Schottky barrier diode because it is a majority carrier device.

All present-day power rectifiers are silicon p^+-n-n^+ or p^+-p-n^+ diodes. Figure 18.1 shows a typical impurity profile of a p^+-n-n^+ diode. The device is fabricated by deep diffusion of p-and n-type dopants on the opposite sides of the low doped n-Si substrate. This structure behaves like a p-i-n diode.

18.1.2 Reverse Breakdown Voltage

Avalanche breakdown imposes an ultimate limit on the reverse voltage of a junction rectifier. However, the blocking capability is often lowered below its ideal value by imperfections in the bulk and surface breakdown. At breakdown the central n-region of the diode in Fig. 18.1 is punched through, and we can calculate the breakdown voltage if the doping in this region is known (see Sec. 8.5). The calculated value of the breakdown voltage cannot be achieved unless the dopant distribution in the starting material is uniform over the entire area of the device. The conventional crystal growth methods can cause dopant variations as large as ±20 percent across the wafer. This reduces the breakdown voltage consid-

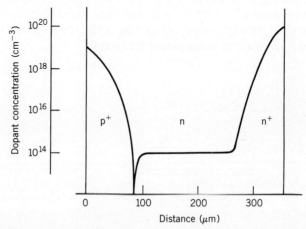

FIGURE 18.1 Typical impurity profile for a p^+-n-n^+ rectifier diode. (From P. Leturcq, Power junction devices, Chapter 7A in *Handbook on Semiconductors,* vol. 4, T. S. Moss and C. Hilsum (eds.), North-Holland, Amsterdam, p. 833. Copyright © 1981. Reprinted by permission from Elsevier Science Publishers, Physical Science and Engineering Div., and the author.)

erably below its anticipated value. Neutron transmutation doped Si is practically free from the resistivity variations (see Sec. 19.1), and it is used as a starting material in the fabrication of high-power devices.

A serious difficulty in achieving a high blocking capability also arises from the electric field concentration at the surface which leads to premature surface breakdown. A commonly used method of reducing the tangential electric field at the surface is edge contouring and beveling of the surface which can be accomplished using mechanical lapping or chemical etching. Figure 18.2 shows a sketch of a reverse-biased p^+-n junction with two different types of bevel angles. In Fig. 18.2a, the diode cross-sectional area is decreasing from the heavily doped region to the lightly doped region, and the bevel angle is defined to be positive. The case of a relatively large negative bevel angle is shown in Fig. 18.2b. Here the cross-sectional area increases in going from the heavily doped side to the lightly doped side, and the peak electric field at the surface is increased. However, for very small negative bevel angles, the depletion layer near the surface on the heavily doped side widens considerably, and the tangential component of the surface field is reduced below the peak field in the bulk. Figure 18.3 shows the fraction of the ideal breakdown voltage plotted as a function of the effective bevel angle θ_{eff} given by the relation [1]

$$\theta_{eff} = 0.04\theta \, (x_n/x_p)^2 \tag{18.1}$$

where θ is the negative bevel angle and x_p and x_n are the plane junction depletion region widths in p- and n-regions, respectively. Since negative beveling requires very small angles, it is not frequently used. For a p^+-n-n^+ device with a positively beveled p^+-n junction and a positively beveled n^+-region, it is possible to completely suppress the surface breakdown, and the breakdown voltage can be optimized to the full bulk breakdown voltage [2].

Mobile ion contamination on the surface can degrade the electrical characteristics of power devices by causing high leakage current and unstable breakdown. In order to minimize these effects, the surface is passivated by coating it with insulating or semiconducting materials. The development of glasses for surface passivation of power devices has been an important step in producing stable devices. An alternative to beveling is the use of guard ring structures [3] which may be obtained using planar technology.

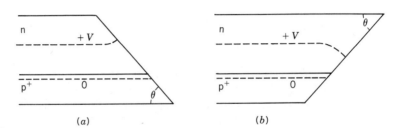

FIGURE 18.2 Sketches showing the position of depletion layer edge in a reverse-biased p^+-n junction with (a) positive bevel angle and (b) negative bevel angle. (From P. Leturcq, in *Handbook on Semiconductors*, vol. 4, North-Holland, Amsterdam, p. 838. Copyright © 1981. Reprinted by permission of Elsevier Science Publishers, Physical Science and Engineering Div., and the author.)

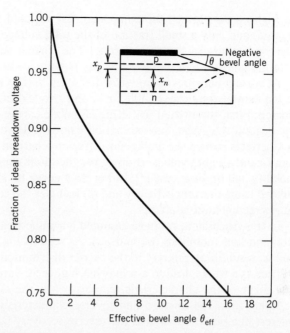

FIGURE 18.3 Normalized breakdown voltage plotted as a function of bevel parameter θ_{eff} for negative beveled structures. (From M. S. Adler and V. A. K. Temple, reference 1, p. 957. Copyright © 1976 IEEE. Reprinted by permission.)

18.1.3 Forward Characteristics

In the forward direction, power rectifiers behave similar to a p-i-n diode in the on-state. This is because high-level injection is reached in the central zone at fairly low currents, although low-level injection prevails in the p^+-and n^+-end regions for current densities on the order of 10^3 A cm^{-2}. Assuming a symmetrical p-i-n structure with equal mobilities for electrons and holes, we can write the diode current I as

$$I = I_M + 2I_L \tag{18.2}$$

where I_M is the current caused by electron and hole injection from the end regions into the central zone, and I_L is the current resulting from carrier injection from the central zone into each of the end regions. The voltage V_a across the diode is given by the relation

$$V_a = V_j + V_M \tag{18.3}$$

where V_j is the voltage drop across the junctions and V_M is the nearly ohmic voltage drop across the central zone. At low values of I, the mean conductivity of the central zone is proportional to the excess carrier concentration in this region, and the diode I-V characteristic has the form

$$I = I_s \exp\left[\frac{q(V_a - V_M)}{2kT}\right] \tag{18.4}$$

At high current levels, the component $2I_L$ injected into the end regions is increased, and only a small fraction of the total voltage is available for modulating the conductivity of the central zone. The current then rises much more slowly with the voltage than predicted by Eq. (18.4).

At current densities in excess of 10^3 A cm^{-2}, the electron–hole concentrations in the central zone exceed 10^{17} cm^{-3}, and two additional effects have to be considered. First, the mutual scattering of mobile carriers reduces the carrier mobility. Second the Auger recombination causes a decrease in the carrier lifetime. Both effects reduce the ambipolar diffusion length in the central zone, causing a significantly higher voltage drop across the diode than expected for the constant mobility and lifetime case [4]. In Fig. 18.4 curve (a) shows the characteristic calculated using the simple theory and (b) includes the carrier-carrier scattering and Auger recombination effects.

In this discussion, we have assumed a symmetrical structure with equal electron and hole mobilities (μ_n and μ_p). As seen in Fig. 11.7, unequal values of μ_n and μ_p result in asymmetry in the carrier distribution in the central zone. Asymmetries as a whole, lead to a somewhat higher forward voltage drop, but the general shape of the characteristic remains nearly the same.

Apart from the forward and reverse characteristics, the other parameter of interest in a power rectifier is the reverse recovery time which limits the operating frequency. As discussed in Sec. 9.4, the electric field in the midzone of the p^+-n-n^+ diode speeds up the removal of electron–hole pairs when the diode is switched from a forward to a reverse voltage. The recovery time can be further reduced by making the central zone thin and reducing the carrier lifetime in this region using either gold doping or electron bombardment. Note that a central thin zone reduces the diode breakdown voltage, and a low lifetime in this region causes an increase in the forward voltage drop. Thus, high blocking capabilities,

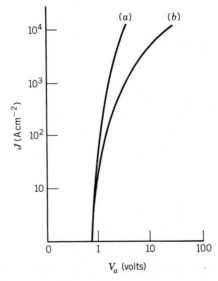

FIGURE 18.4 Theoretical I-V characteristics of a power rectifier calculated according to (a) simple theory and (b) theory which takes into account carrier-carrier scattering and Auger recombination effects. (From J. Burtscher et al, reference 4, Fig. 9, p. 44. Copyright © 1974. Reprinted by permission of Pergamon Press, Oxford.)

18.2 THYRISTORS

18.2.1 Basic Structure and Terminal Characteristics

Thyristor is the name given to a general family of three terminal devices that display intrinsic regenerative action in their operation. The basic thyristor is a four-layer p-n-p-n structure, but many derivatives of this basic structure have been developed. These devices are widely used in static power converters such as switches, choppers, inverters, and cycloconverters which provide static means of dc and ac power control. Commercially available thyristors have current ratings in excess of 5 KA, and the voltage ratings extend beyond 10 KV. This section provides a brief account of these devices. A comprehensive discussion of various thyristor structures and their fabrication technology is given by S. K. Ghandhi [3].

Figure 18.5 shows the cross-section and a typical impurity profile for a diffused p-n-p-n device. The device consists of four regions p_1, n_1, p_2, n_2 and three junctions J_1, J_2, and J_3. The p_1-region is referred to as the anode, the n_1 and the p_2-regions are called bases, and the n_2-region as the cathode. Fabrication of the device starts with a high-resistivity *n*-type Si wafer that forms the n_1-base. Simultaneous *p*-type diffusions are performed into the two sides of the wafer, resulting in junctions J_1 and J_2 which are nearly symmetrical. The heavily doped n_2-region is then produced on one side either by diffusion or by alloying. Finally, ohmic contacts are made to the anode, cathode, and p_2-region which serves as the gate electrode. The device in Fig. 18.5 can be operated either as a diode with voltage applied between the anode and the cathode, or as a three-terminal device with an appropriate bias on the gate. Note that the lightly doped n_1-base is wide and supports most of the applied voltage when the device is biased.

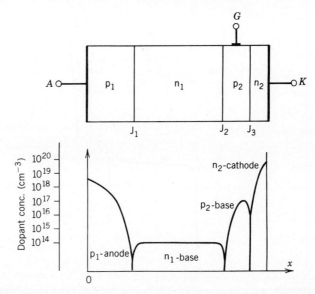

FIGURE 18.5 Schematic cross-section and typical impurity profile of a p-n-p-n thyristor.

The circuit symbol and current-voltage characteristics of a p-n-p-n diode are shown in Fig. 18.6. When the anode is made negative with respect to the cathode, J_1 and J_3 are reverse biased, and J_2 is forward biased. Since the two sides of J_3 are heavily doped, the breakdown voltage of this junction is small (typically about 15 to 30 V). Consequently, as the bias is increased, practically all of the applied voltage appears across J_1 leading to a small reverse current as shown by the line OD. The device is then in its reverse blocking mode. At a voltage V_{BR}, breakdown occurs and the anode current rises steeply with voltage as shown by the line DE.

Under the forward-biased condition, the anode is made positive with respect to the cathode. The junctions J_1 and J_3 are now forward biased, and J_2 is reverse biased. Thus, as the voltage V_{AK} is raised gradually, J_2 gets increasingly reverse biased and a small leakage current flows through the device as shown by OA. The device is now in its forward blocking mode. At a voltage V_{BF}, the device switches over to a low-voltage state, and subsequently, the forward current increases, leading to the forward conducting mode indicated by BC. Figure 18.6 also shows a switching current I_S and a holding current I_h. Here I_S is the minimum current that can switch the device from the forward blocking state to the conducting state. In order to keep the device in the conducting mode, the forward current must be maintained above I_h.

From the above discussion, it is clear that the p-n-p-n structure is a bistable device that can be maintained in an off-state when the current remains below I_S, and can be operated in the forward on-state when the current is made to exceed I_h. The device can be switched over from the off- to on-state or vice versa. The forward breakover voltage V_{BF} at which switching occurs is a function of the gate current and can be varied by varying the gate drive.

18.2.2 Reverse and Forward Breakover Voltages

The reverse blocking capability of a p-n-p-n diode is primarily determined by the width and the dopant concentration of the lightly doped region. However, the

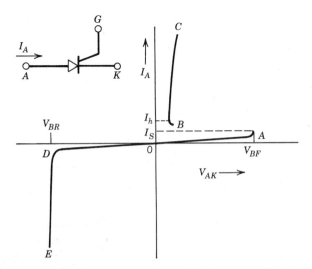

FIGURE 18.6 Circuit symbol and I-V characteristics of a basic p-n-p-n device with gate terminal kept open.

(a) THE REVERSE BREAKOVER VOLTAGE

In the reverse-biased state, J_2 is forward biased and J_1 supports practically all the applied voltage. This is because in many designs, the p_2-base is shorted to the cathode so that J_3 has no ability to block. The blocking ability of J_3 is also limited by its low breakdown voltage. Thus, for a sufficiently large reverse bias, the device behaves as a p_2-n_1-p_1 transistor with an open base and the p_2-emitter shorted to the cathode as shown in Fig. 18.7. The p_1-region acts as the collector. Since the depletion region of J_1 extends mainly into the lightly doped n_1-region, the punch-through voltage V_{PT} of this region places a limit on the blocking capability of the device. For a one-sided abrupt junction with a uniformly doped n_1-region of donor concentration N_{d1} and width W_{n1}, we get

$$V_{PT} = \frac{qN_{d1}}{2\varepsilon_s} W_{n1}^2 \tag{18.5}$$

The breakdown may also occur by an avalanche mechanism at J_1, well before the punch-through voltage is reached. For a one-sided p_1-n_1 junction in Si, the avalanche breakdown voltage can be obtained from Eq. (8.30) giving

$$V_B = 5.3 \times 10^{13} N_{d1}^{-3/4} \text{ volts} \tag{18.6}$$

where N_{d1} is expressed in cm^{-3}. The actual blocking capability of the thyristor is always lower than that given by Eq. (18.6). In Chapter 13 we saw that breakover in a common emitter transistor with an open base occurs when the collector-base junction multiplication factor M has reached a value, such that $M\alpha_T\gamma = 1$. Referring to the p_2-n_1-p_1 transistor in Fig. 18.7, and assuming γ to be unity, we observe that the breakover condition becomes

$$M\alpha_T = 1 \tag{18.7}$$

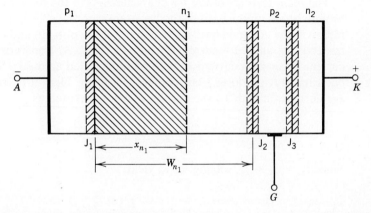

FIGURE 18.7 A p-n-p-n thyristor in reverse blocking mode showing the junction depletion regions.

where the transport factor α_T is given by

$$\alpha_T = \left[\cosh\left(\frac{W_{n1} - x_{n1}}{L_p}\right)\right]^{-1} \tag{18.8}$$

and M is obtained from the relation

$$M = \frac{1}{[1 - (V_j/V_B)^m]} \tag{18.9}$$

Here x_{n1} is the depletion region width, L_p is the hole diffusion length in the n_1-region, and m lies between 4 and 6. The reverse breakover voltage V_{BR} is obtained by substituting V_{BR} for V_j in Eq. (18.9) and by combining the resulting equation with Eq. (18.7). Thus

$$V_{BR} = V_B(1 - \alpha_T)^{1/m} \tag{18.10}$$

Note that α_T varies with the reverse bias, and as long as the base width $(W_{n1} - x_{n1})$ required to satisfy Eq. (18.7) remains large compared to L_p, the factor α_T is very small and V_{BR} is close to the avalanche breakdown voltage V_B.

The structural and chemical uniformity of the starting material, along with surface passivation problems, affect the reverse blocking capability of a thyristor in the same way as in the case of a power rectifier. The starting material is always doped by neutron transmutation doping, and the problems associated with surface breakdown are minimized by beveling the junctions J_1 and J_2.

(b) THE FORWARD BREAKOVER VOLTAGE

In the forward blocking state, J_2 is reverse biased and J_1 and J_3 are forward biased. Thus, at low currents, the n_2-p_2-n_1 structure behaves as an n-p-n transistor with n_2 acting as the emitter. Similarly, the p_1-n_1-p_2 section acts as a p-n-p transistor with p_1 acting as the emitter. This configuration is shown in Fig. 18.8. Note that each transistor is supplied with a base current by the collector of the other device, and J_2 serves as the common collector junction between the two transistors.

The main features of the I-V characteristics in the forward blocking state can be understood on the basis of the two-transistor model (Fig. 18.8b). Let α_1 and α_2 represent the common base-current gains of the p_1-n_1-p_2 and n_2-p_2-n_1 transistors, respectively, and let a current I_g enter the gate. By transistor action, the anode current I_A causes a current $\alpha_1 I_A$ to be transported across J_2. Similarly, the cathode current I_K produces a current $\alpha_2 I_K$ across J_2. Thus, if I_{CO} represents the leakage current through J_2, the current I_A can be written as

$$I_A = \alpha_1 I_A + \alpha_2 I_K + I_{CO} \tag{18.11}$$

Since $I_K = I_A + I_g$, solving for I_A yields

$$I_A = \frac{I_{CO} + \alpha_2 I_g}{1 - (\alpha_1 + \alpha_2)} \tag{18.12a}$$

THYRISTORS

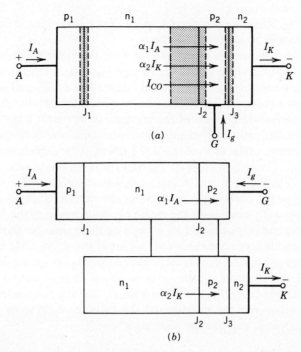

FIGURE 18.8 (a) A forward-biased thyristor and (b) its two-transistor analogue.

When the device is used as a diode, $I_g = 0$ and

$$I_A = \frac{I_{CO}}{1 - (\alpha_1 + \alpha_2)} \tag{18.12b}$$

It is evident from the above relations that I_A remains low when the sum $(\alpha_1 + \alpha_2)$ is small compared to unity. This situation corresponds to the forward blocking state where the voltage V_{AK} is supported by the junction J_2. As mentioned in Sec. 13.4, the α of a transistor is a strong function of the emitter current. Thus, as the current through the device is increased, either by raising the anode-to-cathode voltage or by increasing I_g, the alphas of the two transistors improve until a point is reached where $(\alpha_1 + \alpha_2) = 1$. When this happens, I_A becomes very high and can no longer be controlled by I_g. This new situation corresponds to the on-state of the thyristor and is considered in the next section.

The condition $(\alpha_1 + \alpha_2) = 1$ determines the forward breakover voltage. Since $\alpha_1 = M\gamma_1\alpha_{T1}$ and $\alpha_2 = M\gamma_2\alpha_{T2}$, the breakover condition becomes

$$M(\gamma_1\alpha_{T1} + \gamma_2\alpha_{T2}) = 1 \tag{18.13}$$

and substituting for M from Eq. (18.9) after writing $V_j = V_{BF}$, we obtain

$$V_{BF} = V_B[1 - (\gamma_1\alpha_{T1} + \gamma_2\alpha_{T2})]^{1/m} \tag{18.14}$$

Here the emitter efficiency γ_1 of the p_1-n_1-p_2 transistor may be assumed to be unity, but γ_2 of the n_2-p_2-n_1 transistor is less than unity because both sides of J_3 are heavily doped. Moreover, V_B in Eq. (8.14) is the avalanche breakdown voltage of J_2. Since the junctions J_1 and J_2 are nearly symmetrical, their breakdown voltage will be almost the same.

A comparison of Eq. (18.14) with Eq. (18.10) shows that V_{BF} is always lower than the reverse breakover voltage V_{BR}. The temperature dependence of V_{BR} and V_{BF} is caused by the temperature dependence of V_B and $\gamma \alpha_T$. The avalanche breakdown voltage V_B increases slowly with temperature, and α_T also increases with temperature because of the increase in diffusion length. However, in Si transistors, the emitter efficiency γ increases rapidly with temperature because of a rapid increase in the ratio of thermal current to the depletion region recombination current. Consequently, both V_{BR} and V_{BF} decrease rapidly with temperature [3].

The forward and reverse breakover voltages of a thyristor can be made to approach each other by keeping the current gain α_2 of the n_2-p_2-n_1 transistor extremely small over the entire blocking range of the forward bias. This is most conveniently achieved by using a cathode short as shown in Fig. 18.9. Here the cathode contact extends over a part of the p_2-base. At low forward currents, J_1 is forward biased, J_2 is reverse biased, and J_3 has only a small forward bias because of the presence of the cathode short. Hence, only a small fraction I'_K of the cathode current I_K flows through J_3, and the remainder is shunted by the metal overlap over the p_2-base and the cathode. Analogous to Eq. (18.11), the anode current I_A can now be written as

$$I_A = \alpha_1 I_A + \alpha_2 I'_K + I_{CO} \tag{18.15}$$

With no gate current into the device, $I_A = I_K$, and we obtain

$$I_A = \frac{I_{CO}}{1 - \alpha_1 - \alpha_2(I'_K/I_A)} \tag{18.16}$$

When there is no cathode short, $I'_K = I_A$, and Eq. (18.16) reduces to Eq. (18.12b). However, in the presence of the cathode short, $I_A \gg I'_K$ and

$$I_A \simeq \frac{I_{CO}}{(1 - \alpha_1)} \tag{18.17}$$

The forward breakover voltage V_{BF} is now obtained from the condition $\alpha_1 = M\gamma_1 \alpha_T = 1$, and approaches the reverse breakover voltage V_{BR} in Eq. (18.10) if γ_1 is assumed to be unity.

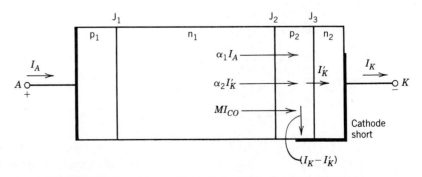

FIGURE 18.9 Schematic cross-section of a shorted cathode thyristor.

18.2.3 The Forward Conducting State

The behavior of the thyristor during the forward blocking and conducting states can be understood with the help of Fig. 18.8b. In the forward blocking state, J_1 and J_3 are forward biased, and J_2 is reverse biased. Holes are then injected from the p_1-emitter into the n_1-base. They diffuse through this region and, after collection over J_2, go to p_2 where they provide excess majority carriers in the base of the n_2-p_2-n_1 transistor. This excess hole charge causes an injection of electrons from n_2 into the p_2-region over the forward-biased junction J_3. These electrons, after collection over J_2, provide excess majority carriers in the n_1-base of the p_1-n_1-p_2 transistor. This electron charge causes a further injection of holes from p_1 to n_1. Thus, the situation is one of positive feedback, and the current builds up until the device is turned on. From Eqs. (18.12a and b) we see that I_A becomes infinite when $(\alpha_1 + \alpha_2) = 1$. However, in actual situations, switching occurs well before this condition is reached. To obtain the criterion for switching, let us consider that the gate current is increased by a small amount ΔI_g. Differentiating Eq. (18.12a) with respect to I_g, we obtain

$$\frac{\Delta I_A}{\Delta I_g} = \frac{i_A}{i_g} \simeq \frac{\tilde{\alpha}_2}{1 - (\tilde{\alpha}_1 + \tilde{\alpha}_2)} \tag{18.18}$$

where $\tilde{\alpha}_1$ and $\tilde{\alpha}_2$ are the small-signal current gains of the p_1-n_1-p_2 and n_2-p_2-n_1 transistors, respectively. In any transistor with a dc current gain α_1, the $\tilde{\alpha}_1$ is given by

$$\tilde{\alpha}_1 = \frac{\partial I_C}{\partial I_E} = \alpha_1 + I_E \frac{\partial \alpha_1}{\partial I_E} \tag{18.19}$$

Thus, $\tilde{\alpha}_1$ is always larger than α_1. It is apparent from Eq. (18.18) that starting from the forward blocking state with a small current gain, and increasing the values of $\tilde{\alpha}_1$ and $\tilde{\alpha}_2$ either through an increase in the gate current or through an increase in the forward voltage, the device becomes unstable when $(\tilde{\alpha}_1 + \tilde{\alpha}_2)$ approaches unity.

The condition $(\tilde{\alpha}_1 + \tilde{\alpha}_2) = 1$ is initially achieved by making the avalanche multiplication factor M larger than unity, and this requires that J_2 be reverse biased. However, as the sum $(\tilde{\alpha}_1 + \tilde{\alpha}_2)$ reaches unity, I_A tends to infinity. Since I_A is limited to a finite value by the external circuit, M must decrease in order to keep $(\tilde{\alpha}_1 + \tilde{\alpha}_2)$ below unity. This means that the reverse bias across J_2 will start decreasing and will reach zero to make $M = 1$ as required by Eq. (18.9). Eventually, J_2 will get slightly forward biased to limit the collection of minority carriers from the n_1- and p_2-bases.

With a forward bias on J_2, the total voltage across the device is the voltage drop across one p-n junction plus the voltage drop of a saturated transistor. From this point onward, any further increase in the bias causes an increase in hole injection from the p_1-emitter across J_1, and electron injection from the n_2-emitter across J_3. These electrons and holes flood the n_1- and p_2-regions and the device behaves, like a p-i-n diode with the p_2-base almost shorted to the n_2-emitter and the n_1-base acting as an intrinsic zone. Thus, the I-V characteristics of thyristors in the on-state are similar to those of power rectifiers.

18.2.4 The Turn-On Process

Switching a thyristor from its off- to on-state requires that the forward current be raised to a level where $(\tilde{\alpha}_1 + \tilde{\alpha}_2)$ becomes unity. This can be achieved by a number of methods such as raising the forward voltage slowly until it exceeds V_{BF}, applying the anode voltage rapidly, or holding the device below V_{BF} and raising its temperature. These methods of triggering are widely used for a two-terminal p-n-p-n diode. In the case of a three-terminal thyristor, the most commonly used method to trigger the device into the on-state is to apply a forward current to the gate by biasing it positive with respect to the cathode. Figure 18.10 shows the I-V characteristics of a thyristor for different values of the gate current. It is seen that the switching voltage decreases as I_g is increased.

Let us now consider the thyristor in Fig. 18.11 biased in the forward blocking state through a resistive load R_L. Let a positive current step I_g be applied to the gate (Fig. 18.12a). The resulting anode current and the voltage V_{AK} are shown in Fig. 18.12b. It is seen that a turn-on delay t_d is required before the anode current rises to a significant value. This interval is followed by a rise time t_r, during which conduction is established, and by a spreading time t_s in which steady state is achieved.

Initially, when the step current I_g is applied to the gate, holes injected into the p_2-base produce electron injection from n_2 over the junction J_3 and the n_2-p_2-n_1 transistor begins to conduct. These electrons flow to n_1 accompanied by an immediate injection of holes from the p_1-emitter of the p_1-n_1-p_2 transistor which are transported to the p_2-base by the transistor action. Thus, electrons and holes begin to cross the depletion region of the blocking junction J_2. Eventually, both the transistors start conducting, and the end of the delay phase is reached when $(\tilde{\alpha}_1 + \tilde{\alpha}_2)$ approaches unity. At this point, the regenerative process starts, and the depletion region of J_2 begins to collapse. This causes a rapid drop in the anode-to-cathode voltage V_{AK} which continues to fall during the time interval t_r. The rise time t_r can be approximated by [3]

$$t_r = \sqrt{t_{n1} t_{p2}} \tag{18.20}$$

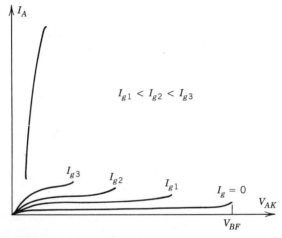

FIGURE 18.10 Effect of gate current on the forward I-V characteristics of a thyristor. (From Gentry, Gutzwiller, Holonyak, and Zastrow, *Semiconductor Controlled Rectifiers: Principles and Applications of p-n-p-n Devices*, copyright © 1964, p. 153. Reprinted by permission of Prentice Hall, Inc., Englewood Cliffs, New Jersey.)

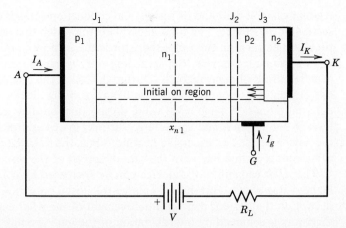

FIGURE 18.11 Cross-sectional view of a thyristor during turn-on through a gate current step I_g.

where t_{n1} represents the hole transit time through the n_1-base, and t_{p2} is the electron transit time through the p_2-base. Note that it is necessary to prolong the gate action up to the end of the period t_r; otherwise the device may fail to turn on.

At the end of the period t_r, only a small part of the device area close to gate contact (Fig. 18.11) is turned on. This is because the gate current causes the injected carriers to crowd at the inner emitter edge of the gate, and the depletion region of the blocking junction quickly collapses in this region as the excess carriers flood the n_1- and p_2-base regions. Once this area begins to conduct, the gate loses its control. Subsequently, carriers from this on-region flow to the adjacent

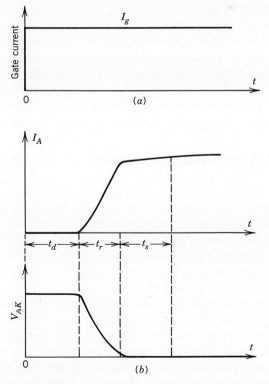

FIGURE 18.12 Turn-on characteristics of a thyristor showing (a) the gate current step and (b) the anode current and the anode voltage V_{AK}.

region by lateral drift and diffusion, and conduction spreads over the entire cross-section of the device. The time interval during which this occurs is the spreading time t_s. The turn-on time of a thyristor is the sum of t_d, t_r, and t_s.

In a large-area device, the spreading time t_s may be on the order of tens or even hundreds of microseconds. During this time, the rate of rise dI/dt of the anode current does not depend on the behavior of the device but is limited by the external circuit. A large dI/dt value during turn-on may lead to a high current level in the on-area and cause large turn-on losses and probably destruction of the device because of excessive heating. Thus, the dI/dt rate must be limited to a value that is compatible with the rate of spread of the on-area.

The dI/dt capability of a device can be increased by overdriving the gate. A higher value of gate current will enlarge the initial on-region and cause a local reduction in the current density. Improvement of the dI/dt capability of large-area thyristors is achieved by interdigitation of the gate geometry to increase the peripheral length of the emitter [3]. This allows a large initial on-area but leads to increased cost. The amplifying gate thyristor (Sec. 18.3) offers another solution to the problem.

18.2.5 The dV/dt Effect

A blocking thyristor can switch to its forward on-state at voltages well below the breakover voltage V_{BF} when the anode-to-cathode voltage is applied abruptly. The anode voltage at which the device will turn on will depend on the rate of rise of this voltage. This phenomenon is called the dV/dt effect. The reason for the dV/dt effect is that, under transient conditions, the reverse-biased junction J_2 can pass a current $C_{j2}(dV/dt)$ where C_{j2} is the depletion region capacitance of J_2. Thus, with a rapidly varying anode voltage, the displacement current across J_2 can cause $(\tilde{\alpha}_1 + \tilde{\alpha}_2)$ to approach unity. The device will then switch into forward conduction. Power thyristors must be designed to have a high dV/dt rating in order to avoid turning the device on prematurely. An effective approach to increase the dV/dt rating is to use a shorted cathode structure like the one shown in Fig. 18.9. The displacement current across J_2 then flows into the short provided to the p_2-base and does not reach the n_2-emitter of the n_2-p_2-n_1 transistor. As a result, the $\tilde{\alpha}_2$ of this transistor is not affected by the displacement current.

18.2.6 Turn-Off and Reverse Recovery

The simplest way to turn the thyristor off is to reduce its current below the holding current I_h. However, this does not remove the stored charge in the base regions, and the device may be turned on by the dV/dt or other effects. In order to fully restore the thyristor to its blocking state, the n_1- and p_2-bases must be completely emptied of the excess carriers. When the turn-off is affected by reducing the current to zero, the excess carriers are lost only by recombination and large values of turn-off time result.

The usual method to turn off a thyristor is to reverse the anode current by applying a negative voltage to the anode. Figure 18.13 shows the current and the voltage waveforms of the recovery process through a resistive load R_L. As soon as the anode voltage is reversed, the slopes of the excess carrier concentrations at J_1 and J_3 change to accommodate the reverse current. Figure 18.14 shows the electron–hole distributions in the two bases during the turn-off transient. Dur-

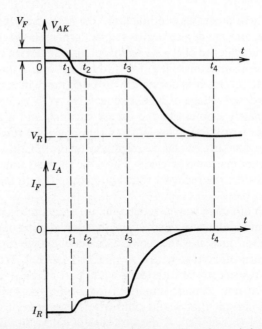

FIGURE 18.13 Turn-off voltage waveforms of a thyristor through a resistive load. (*a*) The anode voltage and (*b*) the anode current. After reference 3.

ing the period $0 - t_1$, the three junctions remain forward biased. Stored charge is removed from the two bases, and a constant current $I_R = -V_R/R_L$ flows (Fig. 18.13). At $t = t_1$, the excess electron concentration at the depletion layer edge of J_3 reaches zero, and the voltage across the device begins to reverse its polarity. During the period $(t_2 - t_1)$, the depletion layer builds up across J_3, and the process is terminated by the avalanche breakdown of this junction. From t_2 to t_3, the reverse current again remains constant at a value $(V_R - V_{B3})/R_L$ where V_{B3} is the avalanche breakdown voltage of J_3. As time elapses, the stored charge from the n_1-base is removed, and at $t = t_3$, the minority carrier concentration at the edge of the depletion region of J_1 falls to zero, and this junction also starts reverse biasing. From this point onward, the device behaves as a floating base p_1-n_1-p_2 transistor with p_2 as the emitter. During the interval $(t_4 - t_3)$, the p_2-base injects holes across the forward-biased junction J_2 into the n_1-base. At the same time, holes from the n_1-base are removed across the depletion region that builds up at

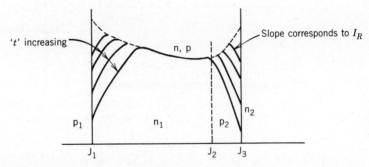

FIGURE 18.14 Excess carrier concentrations in the two base regions of a thyristor at various times after reversing the anode current.

J_1. These processes continue until the reverse current decays to a sufficiently low value, and the device begins to block when biased in the forward direction. This state is achieved at $t = t_4$. At this point, the device is in its reverse blocking mode with J_1 reverse biased, J_2 forward biased, and J_3 in avalanche breakdown.

The reverse bias during the turn-off transient is kept well below the avalanche breakdown voltage of J_1. Consequently, at $t = t_4$, neither of the two bases can be completely emptied of the excess carriers, and a substantial amount of stored charge is still left. This leftover charge can be removed only by recombination. If the forward voltage is reapplied before all this charge has recombined, the thyristor may turn on again. Thus, the turn-off time of a thyristor is considerably larger than the recovery time and may be five to ten times the carrier lifetime in the n_1-base.

An effective way to reduce the turn-off time is to decrease the excess carrier lifetime in the base regions by either gold doping or electron bombardment. This, however, increases the forward on-state voltage drop. An alternative is to assist the turn-off voltage with a reverse gate current. When the gate is reverse biased with respect to the cathode, a small part of the stored charge is removed by the gate current. A more important role of the gate current is that it diverts the forward current that would otherwise flow through the cathode when a forward voltage is reapplied. A limiting case of gate-assisted turn-off is the gate turn-off thyristor described in the next section.

18.3 SOME SPECIAL THYRISTOR STRUCTURES

(a) REVERSE CONDUCTING THYRISTOR

Figure 18.15 shows the cross-sectional view and the I-V characteristics of a reverse conducting thyristor (RCT). The device employs both the cathode and anode shorts. To facilitate ohmic contact to the n_1-base, an n^+ surface layer is diffused before diffusing the p_1-anode. The forward characteristic of the device is similar to that of a normal thyristor. However, in the reverse direction J_1 and J_3 are shorted, and the characteristic corresponds to that of the forward-biased p-n junction J_2.

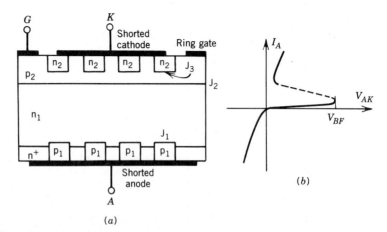

FIGURE 18.15 The reverse conducting thyristor: (a) cross-sectional view and (b) the I-V characteristics. (From S. K. Ghandhi, reference 3, p. 207. Copyright © 1977. Reprinted by permission of John Wiley & Sons, Inc., New York.)

SOME SPECIAL THYRISTOR STRUCTURES

Since both the cathode and anode are shorted to their adjacent bases in the RCT, the forward breakover voltage V_{BF} is almost independent of α_1 and α_2 and approaches the avalanche breakdown voltage V_B of J_2. The avalanche breakdown voltage is only a slowly increasing function of temperature. This allows the device to be operated at higher temperatures handling larger currents in the on-state than a normal thyristor. In addition, the reverse recovery of a RCT is fast because the minority carriers can be swept out of the two bases through paths provided by the shorts. The limitation of the device is that it has no reverse blocking capability.

(b) AMPLIFYING GATE THYRISTOR

The cross-sectional view and equivalent circuit of an amplifying gate thyristor (AGT) are shown in Fig. 18.16. This structure is used to improve the dI/dt capability of the device by increasing the initial turn-on area. When a small trigger current is applied to the central gate G, the auxiliary thyristor n_2'-p_2-n_1-p_1 will turn on rapidly because its cathode is closest to the gate contact. The cathode current of this thyristor serves as the gate current of the main thyristor n_2-p_2-n_1-p_1. Thus, the auxiliary thyristor acts as an amplifying device and supplies a large current to turn on the main thyristor. This action considerably reduces the turn-on time of the device.

(c) GATE TURN-OFF THYRISTOR

A gate turn-off thyristor (GTO) is able to switch directly from the forward on-state to a blocking state by application of a large negative gate current without reversing the cathode-to-anode voltage. In principle, a thyristor can be turned off if the charge is removed from the p_2-base at a faster rate than the rate at which the charge arrives. When this happens, the central junction J_2 can no longer remain in saturation. The p_2-base receives a current $(\alpha_1 I_A)$ from the anode. Part of this current $I_K(1 - \alpha_2)$ flows out to the cathode as the base current of the

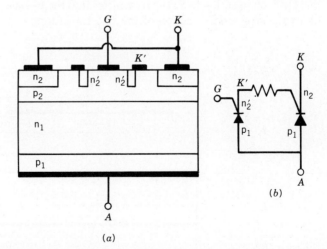

FIGURE 18.16 (a) Cross-sectional view and (b) equivalent circuit of the amplifying gate thyristor. (From S. K. Ghandhi, reference 3, p. 220. Copyright © 1977. Reprinted by permission of John Wiley & Sons, Inc., New York.)

n_2-p_2-n_1 transistor in its on-state. Hence, if I_g is the gate current, then turn-off will occur when

$$I_g + I_K(1 - \alpha_2) \geq \alpha_1 I_A \tag{18.21}$$

Since $I_A = I_K + I_g$, the turn-off gain β_{off} is obtained as

$$\beta_{\text{off}} = \frac{I_A}{I_g} \leq \frac{\alpha_2}{(\alpha_1 + \alpha_2 - 1)} \tag{18.22}$$

Here I_g is flowing out of the gate, but it has been assumed to be positive. It is evident from Eq. (18.22) that, in order to have a large value of the turn-off gain, α_2 should be as close to unity as possible. At the same time, α_1 must be kept small by making the n_1-base thick and by reducing the carrier lifetime in this region. Because of these features, GTOs have fast recovery characteristics, but a relatively large forward voltage drop in the on-state compared to conventional thyristors.

As an example, let us consider a thyristor with $\alpha_1 = 0.3$, $\alpha_2 = 0.85$, and $I_A = 1$ A. From Eq. (18.22) we have

$$I_g \geq (\alpha_1 + \alpha_2 - 1)I_A/\alpha_2 = \frac{(1.15 - 1)}{0.85} = 0.176 \text{ A}$$

Thus, I_g must exceed 0.176 A and β_{off} must be lower than 5.67.

(d) LIGHT-ACTIVATED THYRISTOR

In a light-activated thyristor or light-activated switch (LAS), the gate current is provided by the optical generation of electron–hole pairs in the device. This is achieved by irradiating the central n_2-emitter and the underlying base region with photons of energy higher than the band gap of the semiconductor as shown in Fig. 18.17. The light generates electron–hole pairs that are separated by the reverse-biased junction J_2. The electrons flow to the n_1-base while the holes move to the p_2-zone. This flow causes hole injection from the p_1-emitter into the n_1-base and electron injection from the n_2-emitter into the p_2-base. Thus, the device turns on by the regenerative action of the two transistors.

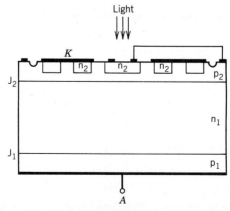

FIGURE 18.17 Schematic cross-section of a light-activated thyristor. (From S. K. Ghandhi, reference 3, p. 226. Copyright © 1977. Reprinted by permission of John Wiley & Sons, Inc., New York.)

18.4 BIDIRECTIONAL THYRISTORS

Thus far, we have been concerned only with unidirectional devices. A number of applications require control of energy that is derived from the ac power lines. Several bidirectional structures have been developed for this purpose, and two of these are described below.

(a) DIAC (DIODE AC) SWITCH

The DIAC is a bidirectional diode switch that has on- and off-states for positive and negative voltages, respectively. A cross-sectional view and the I-V characteristic of the device are shown in Fig. 18.18. The device may be considered as two symmetrical four-layer diodes p_1-n_1-p_2-n_2 and p_2-n_1-p_1-n_4 connected in antiparallel with the anode of each diode shorted to the cathode of the other. When A_1 is positive with respect to A_2, the left-hand side diode p_1-n_1-p_2-n_2 is forward biased, while the right-hand side diode is reverse biased. As the applied bias reaches the forward breakover voltage V_{BF}, the left-hand diode is turned on, but the right-hand one remains in the blocking mode. When A_1 is made negative with respect to A_2, and the bias is increased to V_{BF}, the right-hand diode turns on while the other remains in the off-state. Figure 18.18b shows the resulting I-V characteristics.

(b) TRIAC (TRIODE AC) SWITCH

The TRIAC is a three-terminal device that can conduct load current in both directions. The device structure is similar to that of a DIAC, except that an additional n_3-region is incorporated which is shorted to the p_2-base and acts as

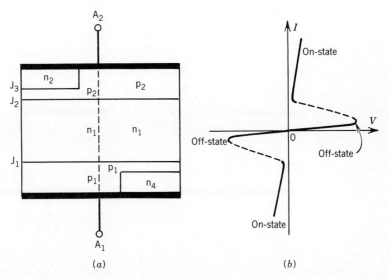

FIGURE 18.18 The DIAC switch: (a) cross-sectional view and (b) the I-V characteristic.

the gate (Fig. 18.19a). The device can switch the current in either direction by applying a low-voltage, low-current pulse of either negative or positive polarity between the gate and one of the terminals A_1 and A_2. Two of its four triggering modes are shown in Fig. 18.19b and c.

In Fig. 18.19b, A_1 has a positive bias with respect to A_2, and G is also biased positive with respect to A_2. Here the device behavior is identical to that of a conventional p_1-n_1-p_2-n_2 thyristor, with the p_2-base acting as gate because the junction J_5 is reverse biased and is inactive. Therefore, the right-hand side thyristor p_2-n_1-p_1-n_4 remains in the off-state. When A_1 is positive with respect to A_2, but the gate is negative (Fig. 18.19c), the junction J_4 between n_3 and p_2 is forward biased, and electron injection occurs from n_3 into the p_2-base. These electrons are collected by the reverse-biased junction J_2 and flow into the n_1-base as majority carriers. This electron current acts as the base current for the p_1-n_1-p_2 transistor section, and as a consequence, the p_1-emitter injects holes into the n_1-base. These holes flow into the p_2-region and act as the base current of the n_2-p_2-n_1 transistor. Because of this regenerative action, the left-hand thyristor p_1-n_1-p_2-n_2 turns on. The other two modes are obtained for positive and negative gate biases when A_1 is made negative with respect to A_2. In both modes, the right-hand side p_2-n_1-p_1-n_4 section is turned on. The explanation for the device's operation under these modes is left as an exercise [3].

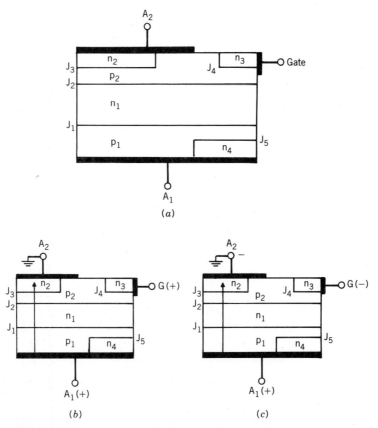

FIGURE 18.19 The TRIAC swithch: (a) schematic cross-section, (b) and (c) the current direction in two of the four triggering modes.

18.5 FIELD-CONTROLLED THYRISTOR [5]

The operating principle of a field-controlled thyristor (FCT) is different from that of a p-n-p-n thyristor. Figure 18.20a shows a cross-sectional view [6] of a typical FCT. It is a p^+-ν-n^+ rectifier with a p-grid forming a collar around the ν-region (Fig. 18.20a). When the anode is made positive with respect to the cathode, the p^+-ν junction is forward biased. If the grid is kept open, holes and electrons are injected into the ν-zone and modulate the conductivity of this region, causing a very low voltage drop across the device. This is the on-state of the thyristor. If the grid G is reverse biased with respect to the cathode, the current that was flowing to the cathode is diverted to the grid (Fig. 18.20b). As the reverse bias on the grid is increased, the channel between the grid is pinched off by the depletion region of the p-ν junction, and a potential barrier is established between the anode and the cathode as shown in Fig. 18.20c. The potential barrier

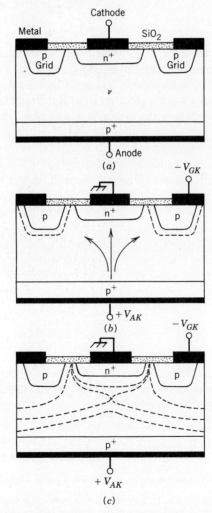

FIGURE 18.20 The field-controlled thyristor: (a) cross-sectional view, (b) cathode current flowing to the reverse-biased grid, and (c) equipotentials in the depletion region under forward blocking state. (From D. E. Houston et al, reference 6, pp. 905, 906 and 907. Copyright © 1976 IEEE. Reprinted by permission.)

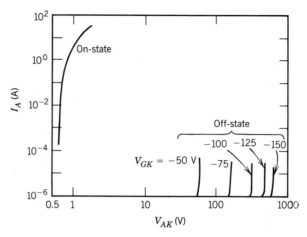

FIGURE 18.21 Forward characteristics of a FCT. (From S. M. Sze, *Semiconductor Devices: Physics and Technology,* p. 156. Copyright © 1985. Reprinted by permission of John Wiley & Sons, Inc., New York.)

prevents the injection of electrons from the cathode to the anode and maintains the device in the forward blocking state. For a given value of the grid-to-cathode voltage V_{GK}, as the anode voltage is increased, the height of the potential barrier is lowered, allowing electron injection from the cathode and hole injection from the anode. When the anode is made negative with respect to the cathode, the device is reverse biased and its characteristic is similar to the reverse characteristic of the conventional thyristor.

The forward I-V characteristics of a FCT are shown in Fig. 18.21. The voltage drop in the forward conducting state is the voltage across the forward-biased p^+-ν junction and is lower than the on-state voltage of a p-n-p-n thyristor. Furthermore, the anode-to-cathode forward blocking voltage increases with increasing reverse bias (V_{GK}) and is not affected significantly by temperature. Thus, FCTs are capable of operating with a junction temperature up to 200°C. These devices have much higher switching speeds, and turn-off times less than 0.5 μ sec have been obtained. Because of these features, FCTs hold considerable promise for the medium-power range in solid-state power-conditioning systems.

REFERENCES

1. M. S. ADLER and V. A. K. TEMPLE, "A general method for predicting the avalanche breakdown voltage of negative bevelled devices," *IEEE Trans. Electron Devices* ED-23, 956 (1976).

2. K. P. BRIEGER, W. GERLACH, and J. PELKA, "The influence of surface charge and bevel angle on the blocking behavior of a high-voltage p^+-n-n^+ device," *IEEE Trans. Electron Devices* ED-31, 733 (1984).

3. S. K. GHANDHI, *Semiconductor Power Devices,* John Wiley, New York, 1977.

4. J. BURTSCHER, F. DANNHAUSER, and J. KRAUSSE, "Recombination in thyristors and silicon rectifiers," *Solid-State Electron.* 18, 35 (1975).

5. S. M. SZE, *Physics of Semiconductor Devices,* John Wiley, New York, 1981.

6. D. E. HOUSTON, S. KRISHNA, D. E. PICCONE, R. J. FINKE, and Y. S. SUN, "A field terminated diode," *IEEE Trans. Electron Devices,* ED-23, 905 (1976).

ADDITIONAL READING LIST

1. A. BLICHER, *Thyristor Physics,* Springer-Verlag, New York, 1976.
2. P. LETURCQ, "Power junction devices," in *Handbook on Semiconductors,* T. S. Moss (ed.), Vol. 4, North-Holland, Amsterdam, 1981, Chapter 7A.
3. M. KUBAT, *Power Semiconductors,* Springer-Verlag, Berlin, 1984.
4. M. S. ADLER, K.W. OWYANG, B. J. BALIGA, and R. A. KOKOSA, "The evolution of power device technology," *IEEE Trans. Electron Devices* ED-31, Centennial special issue, 1570–1591 (1984).
5. B. J. BALIGA, *Modern Power Devices,* John Wiley, New York, 1987.

PROBLEMS

18.1 The p^+-n junction of a Si p^+-n-n^+ rectifier can be approximated by an abrupt junction, and the dopant concentration in the p^+-region is two orders of magnitude higher than that in the n-region. Determine the negative bevel angle θ that will make the surface breakdown voltage 80 percent of the ideal breakdown voltage.

18.2 Examine if the plots of Fig. 18.4 can be used to determine the values of J_s, V_j, and V_M. If possible, determine the values of these parameters at a current density of 10^2 A cm^{-2} from curves *a* and *b*. Explain why the deviation of *b* from *a* becomes more pronounced with increasing current.

18.3 The n_1-region of a Si four-layer thyristor has a width of 100 μm and a uniform donor concentration of 10^{14} cm^{-3}. Determine the avalanche breakdown and the punch-through voltages. Calculate the reverse breakover voltage of the device assuming $L_p = 100$ μm, $\gamma = 1$, and $m = 4$.

18.4 For the thyristor of Prob. 18.3, calculate the forward breakover voltage assuming the low-level injection current gain α_2 of the n_2-p_2-n_1 transistor to be 0.13.

18.5 The mid n_1-region of a Si four-layer diode has a resistivity of 30 Ω-cm, while the other regions are heavily doped compared to this zone. All regions are sufficiently wide to avoid punch-through. The low-level injection current gains are $\alpha_1 = 0.45$ and $\alpha_2 = 0.4$. Assuming $m = 5$, calculate the values of V_{BR} and V_{BF}.

18.6 In Prob. 18.5, calculate the value of the forward breakover voltage when the structure is cathode shorted, and 90 percent of the anode current is shunted by the cathode short.

18.7 A certain p-n-p-n thyristor switches to the forward conducting state when $\alpha_1 + \alpha_2 = 0.8$. The device current at this point is 1 mA. Calculate the $(\partial \alpha / \partial I)$ term assuming that it has the same value for the p_1-n_1-p_2 and n_2-p_2-n_1 transistors.

18.8 In a Si four-layer diode, the n_1-region has a width of 100 μm and the p_2-zone is 50 μm wide. The average doping concentrations in the two regions are 10^{14} cm^{-3} and 2×10^{15} cm^{-3}, respectively. Estimate the rise time t_r of the thyristor.

18.9 Assume that $\alpha_1 = 0.9$ in the thyristor of Prob. 18.8. The device is turned off from an initial current of 100 mA to a holding current of 1 mA by reversing the cathode-to-anode voltage. The current I decays as $I = I_F \exp(-t/\tau_p)$ where $I_F = 100$ mA is the initial current, and τ_p is the hole lifetime in the n_1-region. Calculate the turn-off time of the thyristor.

18.10 Figure P.18.10 shows the cross-sectional view and I-V characteristics of an asymmetrical p^+-n-ν-p-n^+ thyristor. Explain why V_{BR} is lower than V_{BF} in this structure. Mention some advantages of this device over the normal p-n-p-n thyristor.

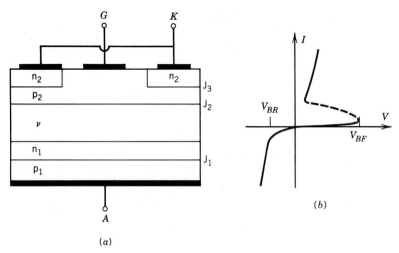

FIGURE P.18.10 (*a*) Schematic cross-section and (*b*) I-V characteristics of an asymmetrical thyristor.

PART 5
SEMICONDUCTOR TECHNOLOGY AND MEASUREMENTS

19
TECHNOLOGY OF SEMICONDUCTOR DEVICES AND INTEGRATED CIRCUITS

INTRODUCTION

A large number of processes are involved in the fabrication of semiconductor devices. As a starting point, the semiconductor material to be used is prepared in single-crystal form and then sliced into wafers that are further processed to obtain defect-free and highly polished surfaces. Subsequently, desired impurities are introduced into selected regions of a wafer, sometimes in conjunction with oxidation and masking, and sometimes without these steps to form appropriate junctions. This step is then followed by a metallization process to realize ohmic contacts, separation of devices into individual dies, lead attachment, and device encapsulation (packaging).

This chapter presents a brief outline of the technology associated with semiconductor processing. Since more than 98 percent of all present-day semiconductor devices are silicon based, this material has been taken as an example, although the technology of other semiconductors is mentioned where necessary. Section 19.1 describes crystal growth and wafer preparation, and Sec. 19.2 presents the methods of junction formation. Growth and deposition of SiO_2 and other dielectric layers are considered in Sec. 19.3, and the planar process for devices is described in Sec. 19.4. A brief account of masking and lithography is given in Sec. 19.5, while the next two sections are devoted to the patterning and metallization processes. The last three sections are concerned with bipolar and MOS integrated devices. Our discussion in this chapter is rather brief; more details of the topics can be found in reference 1 and the other books suggested in the reading list.

19.1 CRYSTAL GROWTH AND WAFER PREPARATION

19.1.1 Crystal Growth Methods

The semiconductor material used to fabricate devices is a single crystal with very high purity and is in the form of flat circular slices. The preparation of these slices prior to device fabrication involves a fairly complex procedure and is discussed here for single-crystal silicon.

Silicon is an abundantly available material on the earth's surface. Sand available in many localities is silica (SiO_2) containing less than 1 percent impurities. This silica is treated chemically to obtain highly purified polycrystalline Si which is then used as the source material for single-crystal growth. Two methods are presently used to grow single-crystal silicon for transistors and integrated circuits: the crucible grown process, often referred to as the Czochralski process; and the float zone process. Figure 19.1 shows the basic arrangement of the Czochralski (CZ) process. Here the polysilicon material is kept in either a quartz or a graphite crucible and is melted by heating it to its melting point using rf or

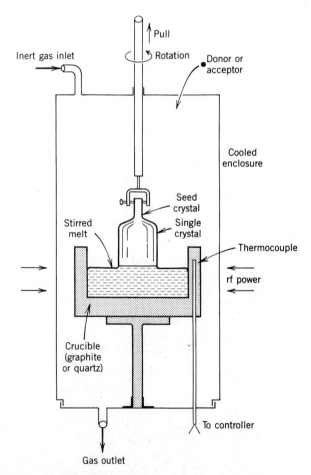

FIGURE 19.1 Schematic diagram of the Czochralski crystal puller. (From S. K. Ghandhi, *Semiconductor Power Devices*, p. 266. Copyright © 1977. Reprinted by permission of John Wiley & Sons, Inc., New York.)

resistive heating. The crystal growth is started by dipping a seed of single-crystal silicon into the melt. The seed is suspended over the crucible in a holder attached to a rotating arm and is inserted into the melt by lowering the arm. As the bottom of the seed begins to melt, the downward motion of the arm is reversed and the crystal is pulled from the melt by slowly rotating and lifting the seed as the crystal grows at its lower end. The arm continues its upward movement forming a larger crystal, and the crystal growth terminates when the melt in the crucible is depleted. During the pulling operation, provision is also made to rotate the crucible. This provides stirring of the melt and avoids asymmetries in the heating. Material with the desired impurity concentration is obtained by adding appropriate impurities to the melt prior to crystal growth. The atmosphere around the crystal puller is made inert by enclosing the entire assembly in a chamber that is flushed with an inert gas such as argon.

The oxygen content of crystals grown using the CZ process may range from 5×10^{17} to 2×10^{18} cm^{-3} depending on growth conditions. The high oxygen content of these crystals results from the reaction of the silicon melt with the quartz crucible. The oxygen may become electrically active during subsequent processing. For this reason, CZ-grown Si is generally not used in devices requiring material whose resistivity is more than 10 Ω-cm.

The CZ technique can also be used to grow single crystals of Ge, GaAS, and other compound semiconductors. In pulling crystal of a semiconductor such as GaAs from the melt, it is necessary to prevent volatile As atoms from vaporizing. One effective approach available is the so-called liquid-encapsulated CZ growth. In this method, a molten dense layer of B_2O_3 is used which floats over the molten GaAs and prevents As evaporation.

The float zone (FZ) process uses a rod of ultrapure polycrystalline Si. As shown in Fig. 19.2, the rod is maintained in a vertical position by means of

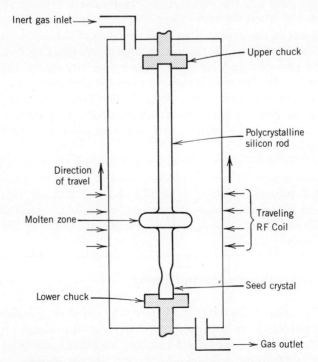

FIGURE 19.2 Schematic of a typical float zone apparatus.

two chucks and is enclosed in a chamber in which an inert ambient is maintained by a flow of argon. An rf heater coil is placed around the chamber. A single-crystal seed is clamped at the lower end of the rod, which is rotated around its axis during the growth process. The coil melts a small length of the rod starting with the seed crystal. The molten zone is then slowly moved upward along the length of the rod by moving the rf coil. As the coil is moved upward, recrystallization of the molten zone at the bottom occurs while the new material begins to melt at the top. The recrystallized region assumes the crystal structure of the seed. The molten zone is held together by surface tension of the liquid, and the diameter of the growing crystal is controlled by the motion of the heater coil. The desired impurity level is obtained by starting with an appropriately doped polycrystalline material.

The absence of a crucible in the FZ process offers many advantages over the CZ process. Not only is impurity contamination from the walls of the crucible eliminated, but also many volatile impurities can be removed by crystal growth in reduced atmospheric pressure. Oxygen levels in FZ-grown crystals are only on the order to 10^{15} cm^{-3}, and resistivities in excess of 30 K Ω-cm can be achieved. However, it is difficult to obtain large-diameter crystals using this method.

A moving molten zone can also be used to purify the starting material of undesired impurities. At the solidifying interface between the melt and the solid, the impurity concentration in the two phases will be different. An important quantity that identifies this property is the distribution coefficient k_o defined by

$$k_o = \frac{C_s}{C_l} \qquad (19.1)$$

where C_s and C_l represent the equilibrium concentrations of impurity atoms in the solid and the liquid phases, respectively. Almost all the important impurities in Si and Ge have values of k_o less than unity. Thus, the impurity content is lower in the crystal than in the melt, and this allows us to remove impurities from the crystal. As the molten zone moves along the bar, impurities are rejected by the solid and are driven along with the molten material until $C_s = k_o C_l$. When the first zone is passed, a considerable part of the bar will have the original concentration of the impurity. However, after many repeated passes, most of the impurity atoms move to the upper end of the bar which can be cut away leaving a highly purified material. This process of purifying a crystal is called *zone refining*.

The distribution coefficient that controls the zone refining process is important not only in the FZ process, but also during the growth from the melt in the CZ process. However, it should be noted that the actual crystal pulling rates are considerably higher than those required to establish thermodynamic equilibrium. Consequently, more impurities are frozen in the growing solid than are given by Eq. (19.1).

19.1.2 Neutron Transmutation Doping

The semiconductor crystals grown by either of the above two processes show considerable radial microresistivity variations. These variations result from the fluctuations in the growth rate that are caused mainly by temperature variations adjacent to the solid-liquid interface. Microresistivity variations are less important in CZ-grown crystals because of the higher background impurity level of this material. However, high-resistivity Si crystals required for power devices like thyristors are always grown using the FZ process. For n-type material, phosphorus is the commonly used dopant, and it has a k_o value of about 0.3. A low value of k_o leads to much larger resistivity variations at both microscale and macroscale (i.e., from center to periphery) levels. These large resistivity fluctuations reduce the blocking capability of power rectifiers and thyristors below the theoretical maximum.

The above-mentioned resistivity variations can be virtually eliminated by using neutron transmutation doping (NTD). In this method, a high-resistivity Si rod (or a wafer) is exposed to the flux of thermal neutrons of a nuclear reactor. Native Si contains about 3 percent of silicon isotope ^{30}Si. Neutron bombardment causes this material to change to ^{31}Si which then undergoes a transition to a stable phosphorus isotopes ^{31}P with the liberation of β rays

$$^{30}\text{Si}_{14}(n,\gamma) \rightarrow {}^{31}\text{Si}_{14} \rightarrow {}^{31}\text{P}_{15} + \beta \qquad (19.2)$$

The reaction has a half-life of 2.62 hours. There is also a secondary process that results in further conversion of a small portion of ^{31}P which then transmutes to ^{32}S. This reaction has a half-life of about 14.3 days and remains insignificant unless a large neutron flux is used to heavily dope the material.

The accuracy of doping in the NTD method depends mainly on maintaining the accuracy of a precalculated neutral irradiation dose. The penetration depth of thermal neutrons in Si is about 90 to 100 cm so that doping is very uniform through the material. It is possible to obtain uniformly doped slices with a radial resistivity variation of about ±1 percent. The lattice defects caused by the irradiation can be annealed out using a suitable thermal treatment.

19.1.3 Wafer Preparation

The semiconductor ingot obtained from the crystal puller has a cylindrical shape. After removing the seed and the tang ends, the first operation is surface grinding. This process is used to precisely define the diameter of the material and is accomplished with a rotating cutting tool that makes multiple passes down the rotating ingot until the desired diameter has been obtained. A flat is then ground along the length of the crystal, and the surface orientation is determined by cutting several slices and measuring the orientation using the X-ray diffraction method. The cutting saw is then reset until the correct orientation is achieved.

After this operation, the crystal is cut into slices called *wafers*. The slicing is accomplished using the inside diameter of a ring-shaped saw blade made of stainless steel with diamond bonded on the inner rim. The cutting process is water cooled, and because of the width of the blade, more than one-third of the crystal is lost as sawdust during the cutting process. The sawing operation also leaves a damaged layer (about 20 to 50 μm thick) on the wafer which is later removed by lapping and etching. A mechanical two-sided lapping operation, performed under pressure using a mixture of Al_2O_3 and glycerine, produces a wafer with flatness uniform to within 2 μm. Considerable damage is left on the surface after lapping which is removed by chemical etching of the wafer. For Si, etching solutions containing mixtures of HF, HNO_3, and acetic acid have been used.

The final operation is polishing. The etched wafers are mounted on large circular stainless steel polishing plates, and either wax or vacuum is used to hold them. These plates are then mounted on a polisher, and the wafers are pressed against a tough polishing pad made of artificial fabrics. A polishing solution may be used that chemically etches and mechanically polishes the wafer simultaneously, and one side of the wafer is polished to a mirror-like finish. After a thorough cleaning, the wafers are ready to be used for device fabrication.

Silicon wafers used in device fabrication have either a {111} or {100} crystal orientation. Besides the doping striations mentioned above, the material may have a number of other defects like dislocations and point defects described in Sec. 2.3.

The availability of large-diameter wafers and the ability to fabricate devices with increasingly smaller geometries have significantly increased the number of devices that can be fabricated on a single wafer. Since the processing cost of a large wafer is only slightly higher than that of a small-size wafer, the cost per function of circuits has been reduced considerably. However, the use of larger wafers requires that the wafer thickness also be increased in order to avoid their breakage during processing. As for the material cost, it must be realized that about two-thirds of the grown crystal is lost during the sawing, lapping, and etching operations. Significant saving in the cost of the starting material would result if the single crystal could be grown in the form of slices. Use of a technology known as *edge-defined film-fed growth* (EFG) represents a step in this direction. This technique allows the growth of continuous ribbons of Si of a preselected width and thickness which can then be cut into substrate wafers of the desired size. Silicon ribbon crystals grown by the EFG process have been tried as substrates in solar cells and photodiodes. However, in terms of crystal perfection, this material has not matched the Si obtained by the CZ and FZ processes.

19.2 METHODS OF P-N JUNCTION FORMATION

19.2.1 Melt Grown Junctions

One of the simplest ways to make a p-n junction is the direct doping of the melt during the growth of the crystal. Let us assume that initially n-type material is being grown from the melt. At some point during the growth, we add a larger amount of acceptor impurity than the donor impurity present in the melt. The crystal grown after the addition of the acceptor impurity will be p-type, and a p-n junction will be formed. The growing crystal may be reconverted to n-type with a still higher amount of donor impurity. Through this technique, both n-p-n and p-n-p transistors can be grown, but the method suffers from poor reproducibility.

Another way of making a p-n junction during growth from the melt is the rate growing method. In this method, both the donor and acceptor impurities are simultaneously present in the melt, and one of these will be predominantly incorporated in the growing crystal as determined by the rate of growth. For example, the distribution coefficient of B in Ge decreases with increasing growth rate, while that of Sb increases. Consequently, by growing crystal at a faster rate, more Sb will be incorporated in the growing crystal than B and the crystal will be n-type. However, at lower growth rates more B atoms will be incorporated, and the crystal will be p-type. Thus, a p-n junction can be made simply by changing the crystal growth rate from the melt. The method of growing p-n junction from the melt is not suitable for large-scale production and is no longer in use.

19.2.2 Alloying

Alloyed junctions are made by heating a semiconductor slice in contact with an impurity that becomes liquid at the temperature used for heating and dissolves some of the semiconductor. As the liquid is cooled, the semiconductor recrystallizes with impurity atoms substituted in the semiconductor lattice. As a typical example, we describe the fabrication process of a p-n junction formed by alloying Ge with In. Figure 19.3a shows the schematic of the setup. A small pellet of In is placed on the surface of a slab of n-type Ge. The semiconductor slab is then placed on a carbon strip heater that is covered with a glass bell jar. An inert ambient is maintained in the space covered by the bell jar by introducing an inert gas like argon or nitrogen. When the temperature of the strip heater is raised to about 500°C, the In pellet on the surface of the Ge slab melts and dissolves some of the Ge and forms a small puddle of a molten In-Ge mixture (Fig. 19.3c). When the temperature is lowered, the molten mass begins to solidify. The initial portion of the recrystallized material will be a single crystal of p-type Ge doped with In. As the solidification proceeds, the remaining mass becomes increasingly rich in In. Finally, when all the Ge is consumed, the material frozen at the outer surface of the recrystallized mass is pure In, which serves as the ohmic contact to the p-type Ge in Fig. 19.3d.

In the case of Si, p-n junctions can be made in a similar way by alloying Al to n-type Si. Alternatively, Au with about 0.1 percent Sb can be prepared in the form of a thin disk which may then be alloyed to p-type Si to produce the n-region of the p-n junction.

The alloy process is a simple and efficient method of making p-n junctions. However, it does not permit tight control on the area and the depth of the junction and thus has found only limited use.

19.2.3 Solid-State Diffusion

Solid-state impurity diffusion is the most widely used technique for making p-n junctions, especially in silicon. The diffusion of dopant impurity atoms into a semiconductor is governed by the same basic laws as the diffusion of free carriers

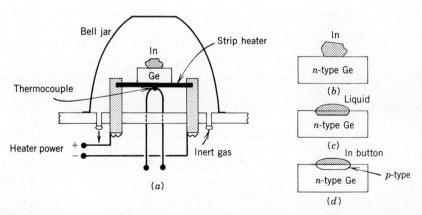

FIGURE 19.3 Schematic diagram showing p-n junction formation by alloying. (a) Experimental setup, (b) In-pellet on Ge before melting, (c) melting of In to form a small puddle of In-Ge mixture, and (d) the p-n junction formed after recrystallization.

described in Sec. 4.7. However, the diffusion coefficient of impurity atoms is several orders of magnitude smaller compared to the diffusion coefficient of free carriers. As a result, the impurity atoms are essentially immobile in the semiconductor at normal temperatures, and temperatures on the order of 1000 °C are required to supply the energy needed for appreciable diffusion of common dopant atoms.

When a semiconductor (e.g., Si) is heated to a high temperature and a flux of dopant impurity atoms is incident at its surface, the dopant atoms migrate into the crystal because of their concentration gradient. At a high temperature, many Si atoms move out of their regular lattice sites creating a high density of vacancies. At the same time, some Si atoms move into the voids between the atoms and create self-interstitials. The impurity atoms diffusing into the semiconductor may occupy either the vacant lattice sites, or the voids between the atoms. The impurities that occupy the vacant sites replace the regular atoms in the lattice and are known as *substitutional impurities,* and those that occupy the voids are called *interstitial impurities.* When the crystal is cooled to room temperature, interstitial atoms may return to substitutional positions and become electrically active. In Si, group V and Group III, as well as many other impurities, occupy predominantly substitutional sites. An important exception is Au which largely diffuses via an interstitial mechanism and, even after cooling, remains predominantly in the interstitial position.

The diffusion process in a semiconductor is described by a second-order partial differential equation known as Fick's law:

$$\frac{\partial N}{\partial t} = \nabla \cdot (D \nabla N) \tag{19.3}$$

where N and D represent the concentration and the diffusion coefficient of the diffusant, respectively. The above equation is usually simplified by making the assumption (generally incorrect) that D is constant. With this assumption and considering one-dimensional diffusion, we obtain

$$\frac{\partial N}{\partial t} = D \frac{\partial^2 N}{\partial x^2} \tag{19.4}$$

The solution of this equation with appropriate boundary and initial conditions yields the impurity distribution in the semiconductor. The equation can be solved analytically for two important cases. In the first case, the surface concentration during diffusion is kept constant at N_o, and in the second case, the total amount Q of the diffusant is kept constant. These cases are now considered.

(a) DIFFUSION WITH CONSTANT SURFACE CONCENTRATION

In this case, $N = N_o$ at $x = 0$ and $N(x,t) = 0$ at $t = 0$. Since the wafer is approximately 500 μm thick whereas the diffusion depth is typically on the order of 1-5 μm, the wafer can be assumed to be infinitely thick. This provides us with an additional boundary condition $N = N_B$ at $x = \infty$. Under these conditions, the solution of Eq. (19.4) is given by

$$N(x,t) = N_o \operatorname{erfc}\left(\frac{x}{2\sqrt{Dt}}\right) + N_B \tag{19.5}$$

METHODS OF P-N JUNCTION FORMATION

where N_B is the background concentration, erfc represents the complementary error function, and $2\sqrt{Dt}$ is the diffusion length. The total number of impurity atoms per unit area of the semiconductor surface is

$$Q(t) = \int_0^\infty [N(x,t) - N_B]\,dx = \frac{2N_o}{\sqrt{\pi}}\sqrt{Dt} \qquad (19.6)$$

(b) DIFFUSION FROM A CONSTANT SOURCE

In this case, a thin layer of dopant with a total amount of dopant Q per unit area is deposited on the surface of the semiconductor. Thus, Q can be represented as a delta function at $t = 0$. The impurity distribution resulting from this type of diffusion is given by

$$N(x,t) = \frac{Q}{\sqrt{\pi Dt}}\exp\left(-\frac{x^2}{4Dt}\right) + N_B \qquad (19.7)$$

which is called the *Gaussian distribution*.

As an example, let us consider phosphorus diffusion performed at 1150°C for two hours in a uniformly doped p-Si sample with $N_B = 2 \times 10^{16}$ cm^{-3}. Let us calculate the junction depth (x_j) by assuming that the surface concentration is maintained at a constant level of $N_o = 4 \times 10^{19}$ cm^{-3}.

From the plots of Fig. 19.4 we obtain $D = 1 \times 10^{-12}$ cm^2 sec^{-1} at 1150°C. Substituting the values of N_o, N_B, D, and t in Eq. (19.5), we obtain

$$4 \times 10^{19}\,\mathrm{erfc}\left(\frac{x_j}{2\sqrt{(1 \times 10^{-12} \times 7200)}}\right) = 2 \times 10^{16}$$

giving $x_j \simeq 4.1$ μm.

(c) DIFFUSION TECHNIQUES

Diffusion may be carried out either in an open-tube furnace or in an evacuated sealed-tube system. The open-tube diffusion process is the preferred method, and in order to improve the controllability of the impurity profile, it is carried out in two steps: predeposition and drive-in.

During the predeposition step, a carefully controlled amount of the desired impurity is introduced into the semiconductor. To achieve this, the semiconductor samples are placed in a carefully controlled high-temperature quartz-tube furnace, and a gas mixture carrying a dopant is made to flow over the samples. The number of dopant atoms that diffuse into the semiconductor is related to the partial pressure of the dopant impurity in the gas mixture. For a given impurity, there is a maximum concentration that can enter the crystal at a given temperature. This maximum concentration is called the *solid solubility*. To obtain controlled conditions during a predeposition, the partial pressure of the impurity in the gaseous ambient is kept high enough to guarantee that the dopant concentration at the surface is determined by the solid solubility of the impurity.

The compound used as a dopant source may be a solid, liquid, or gas. Liquid sources are commonly used for diffusion in Si and Ge, and a schematic diagram of a typical diffusion system is shown in Fig. 19.5. The quartz tube has an inlet for the gases at one end and is connected to a vent at the other end. A liquid

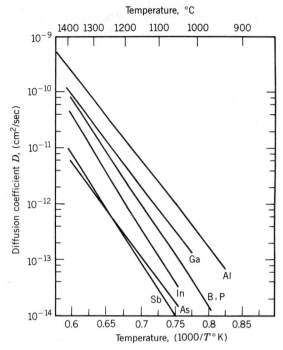

FIGURE 19.4 Diffusion coefficients of some common impurities in silicon. (From S. K. Ghandhi, *Semiconductor Power Devices*, p. 281. Copyright © 1977. Reprinted by permission of John Wiley & Sons, Inc., New York.)

compound containing the dopant is placed in a bubbler held at a constant temperature, and its vapors are transported to the mouth of the furnace by maintaining a weak but controlled flow of N_2 through the bubbler. N_2 is also used as a carrier gas and is mixed with the source vapor at the entrance of the diffusion tube. The chemical reaction at the semiconductor surface usually occurs under oxidizing conditions such that, in addition to N_2, some O_2 is also introduced.

In the case of phosphorus diffusion in Si, the preferred source is $POCl_3$, and the following reaction occurs in the hot zone of the furnace:

$$4POCl_3 + 3O_2 \rightarrow P_2O_5 + 6Cl_2 \uparrow$$

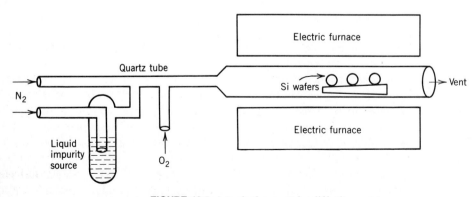

FIGURE 19.5 A typical open-tube diffusion system.

The P_2O_5 forms a glass on the Si wafer and is then reduced to phosphorus by Si

$$5Si + 2P_2O_5 \rightleftarrows 5SiO_2 + 4P$$

The phosphorus that is released diffuses into the Si, and Cl_2 leaves through the vented port.

The liquid source BBr_3 is commonly used for p-type diffusion in Si. In this case, B_2O_3 is formed by oxidation and is then reduced to boron by reacting with Si

$$4BBr_3 + 3O_2 \rightarrow 2B_2O_3 + 6Br_2 \uparrow$$

and

$$3Si + 2B_2O_3 \rightleftarrows 3SiO_2 + 4B$$

The B_2O_3 deposits as a glassy layer on Si, which then serves as a source of boron. The Br_2 resulting from the reaction leaves by the exhaust vent.

Besides the liquid sources, gaseous dopants such as PH_3, AsH_3, and B_2H_6 can also be used. These gases can be directly mixed with the carrier gas. Solid dopants may be used in powder form. They are heated near the entrance of the furnace, and a carrier gas is then used to transport them to the surface of the semiconductor. A solid source can also be used in the form of wafers made from a compound of the desired dopant. An example is the BN source used for boron diffusion in Si.

The drive-in diffusion is performed after predeposition to reduce the surface concentration and to push the impurity atoms further away from the surface into the bulk of the sample. The dopant supply from the source is then shut off, and the semiconductor sample is heated in an oxidizing ambient in the presence of the carrier gas. A protective layer of oxide on the surface of the wafer prevents the escape of impurities through the surface and keeps the total impurity content constant during the drive-in process.

Impurity profiles: During the predeposition step, the surface concentration is maintained constant near its solid solubility limit. Thus, all the conditions for diffusion from a constant source are satisfied; the impurity distribution is described by Eq. (19.5). Predeposition is carried out for short periods at lower temperatures. The drive-in diffusion generally produces an impurity distribution much deeper than that resulting from the predeposition step. As a result, it is a common practice to approximate the distribution at the beginning of the drive-in process as a sheet of dopant at the surface of the semiconductor. The total dopant content per unit area Q is obtained from Eq. (19.6) at a specific value of predeposition time $(t = t_o)$. We then assume that the drive-in diffusion simply redistributes the impurity Q, and the resulting impurity profile is described by Eq. (19.7). In the case of p-diffusion into an n-type material or vice versa, the junction depth $(x = x_j)$ from the surface is obtained by setting $N(x,t) = 0$ in Eq. (19.7).

Sealed-tube diffusion: Sealed-tube diffusion is used for impurities like As and Sb which cannot be diffused by open-tube techniques. In this method, a number of wafers are loaded into a boat and placed in a quartz tube together with the dopant source. The tube is then evacuated, sealed off, and placed in a diffusion furnace at a temperature at which the impurity develops sufficient vapor pressure to cause appreciable diffusion.

The diffusion coefficients of some common impurities in Si are shown in Fig. 19.4 as a function of temperature. With the knowledge of these coefficients, and having similar information for other semiconductors, diffusion profiles can be calculated and junction depths determined using either Eq. (19.5) or (19.7). These results, however, give only the first-order estimate, and several deviations from the ideal behavior occur. The reasons for these deviations and their effect on the impurity profiles are discussed by J. C. C. Tsai [2].

19.2.4 Ion Implantation

Ion implantation is frequently used as an alternative to predeposition for introducing dopant atoms into the desired region of a semiconductor. The implanted impurities are then diffused further using a drive-in step. The major advantage of ion implantation is that it offers the ability to precisely control the amount of dopant and its depth below the surface. Moreover, it is a low-temperature process that eliminates deformation of wafer caused at high temperatures.

In the ion implantation technique, a beam of ionized atoms is accelerated through a desired potential (10 to 500 KV) and is made incident on a semiconductor target. Figure 19.6 shows the setup of a typical ion implantation system. A gas containing atoms of the desired impurity is introduced into a chamber where the atoms are ionized by collisions with high-energy electrons. The ion beam emerging from the chamber contains not only the desired dopant ions, but also ions of unwanted species. The dopant ions are separated from the unwanted ions by passing the beam through a strong magnetic field that bends the desired ions through a 90° turn. The selected ions are then accelerated using an electric field and made to strike the semiconductor target which is kept at ground potential. The beam is deflected horizontally and vertically so that it sweeps across the target in a raster pattern to ensure homogeneous doping. Both the beam and the target are maintained in a high vacuum.

When the accelerated ions enter the semiconductor, they lose their kinetic energy through collisions with the electron cloud of the semiconductor atoms and the positively charged nuclei of the atoms. An ion comes to rest when its kinetic energy is reduced to zero. The total path length of the ion in the semiconductor is called its *range*, and the penetration depth of the ion in the target is called its *projected range R_p*. For implantation made in an amorphous material, the concentration of the implanted atoms as a function of distance from the surface is given by the Gaussian distribution

$$N'(x) = \frac{Q_o}{\Delta R_p \sqrt{2\pi}} \exp\left[-\frac{(x - R_p)^2}{2(\Delta R_p)^2}\right] \tag{19.8}$$

where ΔR_p is the standard deviation of R_p and $Q_o = Jt/q$ is total implanted dose. Here J is the ion current density and t is time. For a crystalline target, the above relation is not correct, and the impurity distribution shows a strong orientation dependence with respect to the ion beam direction. In the main crystallographic directions, the ions find more open and regular paths where interactions occur mainly with the electron cloud, and hence, the penetration is much deeper. This channeling effect is substantially reduced by tilting the target crystal with respect to the direction along which channeling would occur. Even with this adjustment, some ions still channel because the incident beam is scattered in the material.

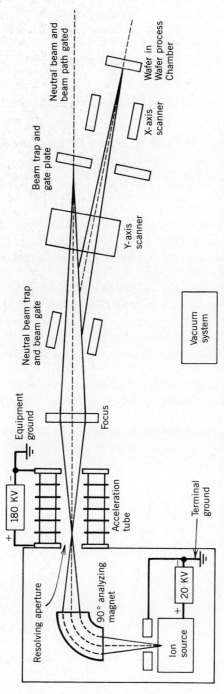

FIGURE 19.6 Schematic of typical ion-implantation equipment. (From Peter Gise/Richard Blanchard, *Modern Semiconductor Fabrication Technology*, copyright © 1986, p. 95. Reprinted by permission of Prentice Hall, Inc., Englewood Cliffs, New Jersey.)

Consequently, tails are frequently found in implanted profiles. Figure 19.7 shows the measured boron distribution implanted using a dose of 10^{15} ions cm^{-2}. Departure from the Gaussian distribution is evident.

To get an idea of the orders of magnitude of the quantities involved, we consider B ions implanted in Si for 10 minutes with a peak concentration of 10^{20} ions cm^{-3}. Let us calculate Q_o and the ion current density assuming a Gaussian distribution with $R_p = 0.5$ μm and $\Delta R_p = 0.07$ μm.

From Eq. (19.8), we get the peak concentration

$$N' = 10^{20} = \frac{Q_o}{0.07 \times 10^{-4}\sqrt{2\pi}}$$

This gives $Q_o = 1.75 \times 10^{15}$ cm^{-2} and an ion current density

$$J = \frac{1.6 \times 10^{-19} \times 1.75 \times 10^{15}}{10 \times 60} = 0.46 \ \mu\text{A cm}^{-2}$$

Ion implantation produces damage in the target crystal by displacing atoms from their regular lattice sites. Thus, a highly disordered layer is created near the surface of the semiconductor. This radiation damage is annealed out by a heat treatment in the 400°C to 800°C range.

Methods of controlling the area of implantation on a wafer rely on masking those portions of the surface in which doping is not desired (see Sec. 19.4). Masks against ion implantation can be made using either insulators (e.g., Si_3N_4 and Al_2O_3) or metals (e.g., Mo and Au). Since very high temperatures do not occur during the implantation process, ordinary photoresist may also be used for masking.

Ion implantation is essentially a shallow doping process and cannot be used in devices where deep diffusions are required. It is widely used in MOS devices for

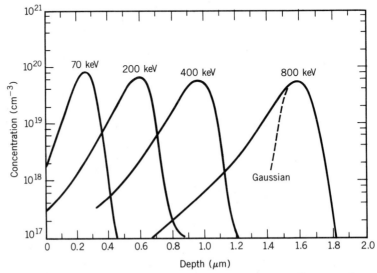

FIGURE 19.7 Measured boron distribution at an implanted dose of 10^{15} ions cm^{-2}. (From W. K. Hofker, Implantation of boron in silicon, *Philips Research Reports*, supplement no. 8, 1975, Fig. 4.1(b), p. 46. Copyright © 1975, Philips International B.V., Eindhoven, The Netherlands, Reproduced by permission.)

threshold voltage control, in CMOS devices for n-and p-well doping, and in the base and emitter dopings of bipolar transistors.

19.2.5 Epitaxy

The term *epitaxy* is applied to processes used to grow a crystalline layer of a semiconductor on a crystalline substrate in such a way that the layer grown has the same lattice structure as the substrate. Epitaxial growth provides an alternative to the diffusion process for obtaining appropriately doped semiconductor regions. An added advantage of this process is that the impurity concentration in the grown layer can be adjusted quite independently of that in the substrate.

The main processes of epitaxial growth are vapor phase epitaxy (VPE), liquid phase epitaxy (LPE), and molecular beam epitaxy (MBE). VPE has been used to grow Si and III-V compounds, whereas LPE has found use mainly with the III-V compounds. MBE is more versatile and allows a very precise control of doping profile.

(a) VPE

Silicon epitaxy occurs almost exclusively using the chemical vapor deposition (CVD) technique which has been used to deposit Si from $SiCl_4$, silane (SiH_4), SiH_2Cl_2, SiH_3Cl, and some other compounds. Figure 19.8 shows a schematic of the growth system using $SiCl_4$. Liquid $SiCl_4$ is kept in a bubbler whose temperature is carefully controlled near 0°C. H_2 gas is passed through the bubbler. The temperature of the bubbler and the flow rate of H_2 through it determine the concentration of the $SiCl_4$ in the mixture. The $SiCl_4$ + H_2 vapor is mixed with more H_2 near the mouth of the furnace and is passed through the quartz tube. The Si wafers are placed on a graphite susceptor kept in the quartz tube which is heated to a temperature in excess of 1100°C by an rf induction coil. The following reaction occurs on the wafer surface:

$$SiCl_4 + 2H_2 \rightleftharpoons Si + 4HCl \qquad (19.9)$$

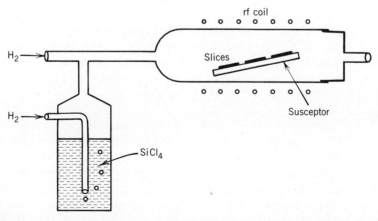

FIGURE 19.8 An epitaxial growth system using $SiCl_4$. (From P. A. H. Hart, reference 5, p. 163. Copyright © 1981. Reprinted by permission of Elsevier Science Publishers, Physical Science and Engineering Div., and the author.)

The Si is deposited on the silicon substrate, and the HCl gas leaves through a suitable vent. The rf coil heats the graphite but not the quartz tube, so there is no deposition on the walls of the tube. Since the reaction occurs in a chamber, the arrangement of Fig. 19.8 is called a *reactor chamber* or simply a *reactor*.

The reaction represented by Eq. (19.9) is reversible. The reverse reaction etches Si and is used to clean the substrate prior to the start of deposition. This etching is performed by passing HCl through the reactor. The growth process is started by stopping the HCl flow and introducing the $SiCl_4 + H_2$ mixture into the reactor. The availability of nucleation sites on the substrate surface is a necessary condition for the deposition to take place. It is, therefore, common to use a surface slightly off the {111} orientation. This situation encourages the nucleation process to occur at the step corners formed by the (111) plane surfaces.

In order to grow doped Si with a desired resistivity and conductivity type, dopant atoms are introduced into the gas system. Gases like PH_3, AsH_3, and SbH_3 are used for *n*-type doping, whereas diborane (B_2H_6) is used for *p*-type doping.

One serious problem with the $SiCl_4$ process is the etch-back effect that arises from the reversibility of the reaction. Some atoms are removed from the layer during the growth because of the reverse reaction. These atoms mix with the dopant gas and modify its composition. This process causes the dopant concentration in the grown Si layer to vary until an equilibrium situation is reached for sufficiently thick layers. Another problem is caused by the fact that the deposition temperatures are in the 1150–1300°C range. These high temperatures may cause significant impurity diffusion across the boundary, such that the interface between the substrate and the grown layer does not remain abrupt.

Deposition of epitaxial silicon using SiH_4 is performed by the pyrolysis of this compound in the temperature range 1000–1100°C. A mixture of SiH_4 and H_2 is fed directly into the reactor where the following reaction takes place:

$$SiH_4 \rightarrow Si + 2H_2 \uparrow$$

Since this reaction occurs at a lower temperature, there is less migration of impurities from the substrate into the growing layer.

The VPE of III-V compounds and their ternary alloys may be performed either in a horizontal or a vertical reactor. Figure 19.9 shows a vertical reactor for the growth of GaAs, GaP, and their ternary alloy GaAsP. GaAs (or GaP) substrates are held on a rotating wafer holder at a temperature of about 800°C. Anhydrous HCl mixed with H_2 is made to react with molten Ga within the reactor to form $GaCl_3$. This gas is then mixed with H_2, PH_3, AsH_3, and the dopant gas and is passed over the substrates. Variation in the As and P ratio of GaAsP can be obtained by changing the proportion of AsH_3 and PH_3 in the mixture.

(b) LPE

LPE has been used extensively to fabricate multilayer structures of ternary and quarternary compounds for use in LEDs and lasers. This technique is based on the fact that a mixture of a semiconductor with a second element may melt at a lower temperature than the semiconductor itself and hence may be used to grow a crystal from solution at the temperature of the mixture. We will illustrate this method by considering the growth of a GaAs layer on a GaAs substrate as an example. The melting point of GaAs is 1238°C. However, when GaAs is mixed with Ga metal, depending on the relative proportions of the two substances, the

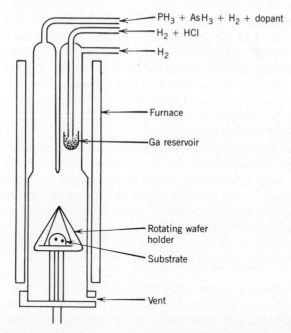

FIGURE 19.9 A vapor-phase vertical epitaxial reactor used for growing GaAs and GaAsP. (After B. G. Streetman, reference 3, p. 22.)

mixture can have a considerably lower melting point. Thus, to grow GaAs, a Ga solution containing GaAs and the desired dopants is placed onto a GaAs single-crystal substrate which is kept in a graphite boat. The boat is kept at a temperature between 800 and 900°C in a furnace in H_2 ambient. If the furnace is now cooled slowly, the GaAs crystallizes from the solution onto the substrate. After the desired growth, the remnants of the solution adhering to the substrate are removed by wiping and dissolution in a suitable solvent.

The same technique can be used to grow other materials at appropriate temperatures. For example, InP is grown from a solution containing In, InP, and appropriate p- or n-type dopants. Similarly, InGaAsP is grown from an In solution containing InP with As and Ga in the desired proportion.

An LPE growth system that uses a multi-bin boat is shown [4] in Fig. 19.10. This system is used for the sequential deposition of several semiconductor layers during one growth cycle. The graphite boat has a movable slide whose upper surface serves as the bottom of the three reservoirs that contain different solutions.

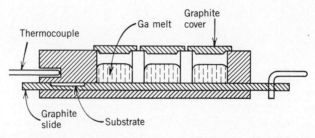

FIGURE 19.10 A multiple-bin graphite boat used for sequential deposition of several layers of semiconductors in one growth cycle. (From H. Kressel and J. K. Butler, reference 4, p. 290. Copyright © 1977. Reprinted by permission of Academic Press, Orlando, Florida.)

The substrate is placed in a recessed area in the slide outside the reservoir. A rod fitted to the graphite slide is used to move the substrate sequentially into the bottom of each reservoir. During the growth process, the temperature of the graphite crucible is raised to the desired value, and the rod is pulled to bring the substrate in contact with the first solution. After the specified growth period, the substrate is moved to the next solution, and any excess solute left over from the first solution is removed. In this way, different types of layers can be deposited sequentially.

In the epitaxial growth using either VPE or LPE, it is extremely important to use substrates that are free from structural imperfections. Any defects left at the wafer surface will enlarge and propagate during the growth process. Consequently, crystal perfection and cleanliness of the reactor are essential requirements for successful growth of epitaxial layers. Close lattice match between the substrate and the grown layer is also necessary.

(c) MOLECULAR BEAM EPITAXY (MBE)

MBE uses an evaporation method as a means of obtaining the epitaxial layer of desired constitution. The substrate is held in an ultra-high vacuum, and the molecular or atomic beams of the constituents are made to impinge on its surface. Figure 19.11 depicts the arrangement for growth of AlGaAs epitaxial layers on a GaAs substrate [3]. The components Al, Ga, and As of the compound and the dopants Sn and Be are heated in separate cells. Each of the source vessels has a controllable shutter port. Collimated beams of these substances are directed onto the substrate which is held at a temperature between 520 and 600°C. The rate at which the atomic beams of the constituent atoms strike the substrate surface can be precisely controlled, and high-quality films can be grown using this method. Because ultra-high vacuum and close controls are involved, MBE remains a rather sophisticated setup. However, its main advantages are the low substrate temperature that minimizes out diffusion from the substrate and a very precise control of the dopant's profile.

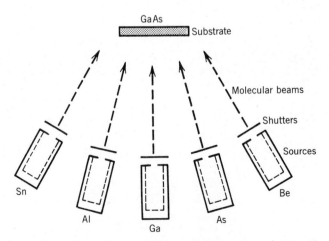

FIGURE 19.11 Schematic diagram illustrating the growth of AlGaAs using MBE. (From B. G. Streetman, *Solid State Electronic Devices*, 3rd ed., copyright © 1990, p. 23. Reprinted by permission of Prentice Hall, Inc., Englewood Cliffs, New Jersey.)

19.3 GROWTH AND DEPOSITION OF DIELECTRIC LAYERS

Dielectric layers play an important role in semiconductor technology. They are used for masking against diffusion, for electrical insulation, for junction passivation, and sometimes for mechanical protection of the structure. The most frequently used dielectric is SiO_2. Other important dielectric materials are Si_3N_4 and polysilicon. This section presents a brief account of how to prepare layers of these materials.

19.3.1 Silicon Dioxide

The most commonly used method of producing SiO_2 layers on Si is thermal oxidation. Other methods that may be used to deposit SiO_2 layers on semiconductors other than Si include chemical vapor deposition (CVD), plasma CVD, and physical techniques such as evaporation and sputtering.

(a) THERMAL OXIDATION OF SILICON

Thermal oxidation of Si is performed in an open-ended quartz tube kept in a resistance-heated furnace in the temperature range of 900 to 1200°C. The oxidizing ambient can be either dry oxygen, wet oxygen, or steam. Figure 19.12 shows the schematic of a wet oxidation system. Si wafers are cleaned using a sequence designed to remove all contaminants. The cleaned wafers are dried and loaded into a quartz boat which is then placed inside the furnace heated to the desired temperature. High purity oxygen enters the inlet side of a bubbler that contains high purity water maintained at a constant temperature below its boiling point. The O_2 becomes saturated with water vapor and exits through the outlet of the bubbler into the furnace. This gas mixture oxidizes the Si surface with the following chemical reactions:

$$Si + O_2 \rightarrow SiO_2$$

$$Si + 2H_2O \rightarrow SiO_2 + 2H_2 \uparrow$$

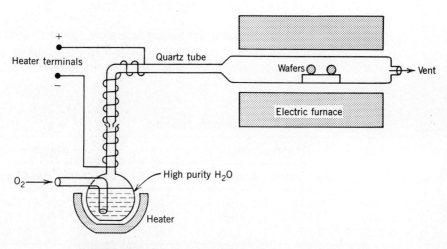

FIGURE 19.12 Cross-sectional view of a wet oxidation system.

In the case of dry oxidation, the bubbler is omitted, and O_2 from the source is passed directly through the quartz tube. For steam oxidation, the O_2 supply at the inlet of the bubbler is blocked, and the water in the bubbler is heated to its boiling point. The resulting steam flows through the furnace and oxidizes the Si wafers.

The wet oxidation rate is much higher than the dry oxidation rate, making the wet oxidation method more suitable for thick SiO_2 layers (exceeding 1 μm). However, oxides grown in the dry ambient are denser and tend to have fewer structural defects and a lower density of interface states. A standard practice is to grow a thin layer of dry oxide, followed by wet oxidation which is again followed by dry oxidation.

The initial SiO_2 layer on Si is formed by the reaction of the surface Si atoms with O_2. Once a SiO_2 layer is formed, further oxidation occurs at the Si-SiO_2 interface such that the oxidizing species diffuse through the oxide to react with Si atoms at the interface. For thin SiO_2 layers, the oxidizing species can easily diffuse through the oxide, and the oxide growth is limited by the reaction rate at the Si-SiO_2 interface. In this situation the oxide thickness varies linearly with time. In the case of thicker oxides, the oxidation reaction is limited by the diffusion of the oxidizing species through the previously formed oxide.

The temperature and time dependence of the oxide thickness d_{ox} can be expressed by the relation [5]

$$d_{ox} = \frac{A}{2}\left[\left(1 + \frac{t + t^*}{A^2/4B}\right)^{1/2} - 1\right] \quad (19.10)$$

where A and B are coefficients that depend on the oxidation temperature, and t^* is an initial value-fit parameter. For short oxidation times, the surface reaction limits the oxide growth and Eq. (19.10) reduces to the linear relation

$$d_{ox} = \frac{B}{A}(t + t^*) \quad (19.11)$$

For long oxidation times, such that $(t + t^*) \gg A^2/4B$, the oxide thickness is given by

$$d_{ox} = \sqrt{B(t + t^*)} \simeq \sqrt{Bt} \quad (19.12)$$

which corresponds to diffusion-limited oxide growth. Table 19.1 lists the values of B, B/A, and t^* for {111} Si in the case of dry and wet oxidation processes at three different temperatures [5]. A large difference between the oxidation rates

TABLE 19.1
Thermal Oxidation Parameters for {111} Si*

T(°C)	Dry Oxidation			Wet Oxidation		
	$B(\mu m^2/h)$	$B/A(\mu m/h)$	$t^*(h)$	$B(\mu m^2/h)$	$B/A(\mu m/h)$	$t^*(h)$
800	0.0011	0.003	9	—	—	—
1000	0.012	0.071	0.37	0.29	1.27	—
1200	0.045	1.120	0.027	0.72	14.40	—

*From B. E. Deal and A. S. Grove, J. Appl. Phys. 36, 3770 (1965). Reprinted by permission of American Institute of Physics.

GROWTH AND DEPOSITION OF DIELECTRIC LAYERS

in the two cases is evident. The oxide thickness is somewhat less in the case of {100} Si.

The addition of traces of Cl_2 (from HCl) during oxidation serves several useful purposes. Sodium and several other impurities form chlorides that are volatile at the oxidation temperature and evaporate leaving a much cleaner oxide layer than otherwise. Such oxides have a lower density of oxide charges and interface states.

Let us consider that a 500 Å thick SiO_2 layer is grown in dry oxygen at 1000°C followed by an additional 0.2 μm thick SiO_2 layer grown in wet oxygen at the same temperature. We need to know the total time for oxidation.

Solving Eq. (19.10) for t, we obtain

$$t = \frac{d_{ox}^2}{B} + \frac{d_{ox}}{B/A} - t^*$$

Using the parameters for dry oxidation from Table 19.1, and writing $d_{ox} = 0.05$ μm, we get

$$t_1 = \left[\frac{(0.05)^2}{0.012} + \frac{0.05}{0.071} - 0.37\right] \text{hour} = 32.5 \text{ min}$$

Next, we determine the time required to grow the same value of d_{ox} in wet oxidation at 1000°C. Using the parameters for wet oxidation, we get

$$t_2 = \left[\frac{(0.05)^2}{0.29} + \frac{(0.05)}{1.27}\right] \times 60 = 2.88 \text{ min}$$

The time required to grow a $(0.05 + 0.2) = 0.25$-μm thick SiO_2 layer in wet oxidation will be

$$t_3 = 60\left[\frac{(0.25)^2}{0.29} + \frac{(0.25)}{1.27}\right] = 24.7 \text{ min}$$

Therefore, the total time $t = t_1 + (t_3 - t_2) = 54.32$ min

(b) CHEMICAL VAPOR DEPOSITION (CVD)

In some situations, thermal growth of SiO_2 is not possible. A typical example is the case where a substrate material other than Si is used. When thermal growth is not possible, a CVD technique offers an important alternative. For example, SiO_2 layers can be prepared by the reaction of SiH_4 and O_2 at temperatures between 250 and 500°C

$$SiH_4 + 2O_2 \rightarrow SiO_2 + 2H_2O \uparrow$$

The reaction can take place in a reactor similar to that shown in Fig. 19.8. A gas mixture containing either N_2 or Ar with O_2 and about 1 percent SiH_4 is used. The SiO_2 growth rate depends on the substrate temperature and the rate of gas flow. This method also allows the deposition of SiO_2 layers doped with dopants such as As, P, and B. The doped oxide can then be used as a diffusion source.

Layers of SiO_2 prepared at low temperatures are not very dense but can be densified by a heat treatment at 900°C for about half an hour. Denser oxide layers can be prepared with a reaction of SiH_4 and CO or CO_2 at temperatures between 900 and 1100°C. The gas passed through the reactor is a mixture of H_2 or N_2 with CO_2 (or CO) + 1 percent SiH_4. The relevant chemical reaction in the case of CO_2 is

$$SiH_4 + 4CO_2 \rightarrow SiO_2 + 2H_2O \uparrow + 4CO \uparrow$$

(c) PLASMA CVD

Plasma aided depositions produce SiO_2 layers at temperatures below 400°C. The deposition process is performed in a reactor shown in Fig. 19.13. SiH_4 and N_2O are mixed in Ar, and the mixture is passed through the reactor chamber. A vacuum pump reduces the pressure in the chamber, and the rf voltage applied between the cathode and the anode produces an Ar-plasma. In this plasma, SiO_2 is formed by the following reaction:

$$SiH_4 + 4N_2O \rightarrow SiO_2 + 4N_2 \uparrow + 2H_2O \uparrow$$

(d) SPUTTERING

In this process, ions of a gas like Ar, produced in a glow discharge, are accelerated through a potential gradient and are then bombarded on a target. Atoms near the surface of the target become volatile and are transported as a vapor to the substrate where they condense and form a film of the material. A schematic diagram of a sputtering system is shown in Fig. 19.14. The material to be sputtered is made the cathode, while anode holds the substrate. Sputtering can be performed using either a dc or an rf power source. For SiO_2 deposition, a SiO_2 block is made the cathode, and the substrates are placed on the anode which is kept at the ground potential (Fig. 19.14). An rf voltage at about 10 MHz is applied between the cathode and the anode. The electrons in the plasma have a much higher mobility than the Ar-ions in the rf field. Consequently, electrons flow to the SiO_2 cathode, which then becomes negatively charged. Ar-ions bombard the cathode target, and SiO_2 is deposited on the substrates. These layers are very dense and can be used as insulating layers in various applications. A problem common to all deposited SiO_2 layers is that the interface between the SiO_2

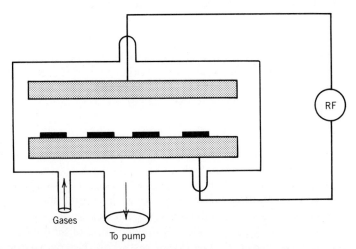

FIGURE 19.13 Horizontal plate plasma reactor for deposition of SiO_2. (From P. A. H. Hart, reference 5, Fig. 55, p. 182. Copyright © 1981. Reprinted by permission of Elsevier Science Publishers, Physical Science and Engineering Div., and the author.)

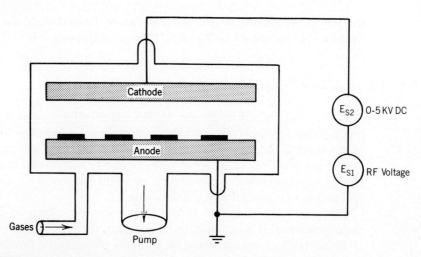

FIGURE 19.14 Schematic diagram of an rf diode sputtering setup. (From P. A. H. Hart, reference 5, p. 192. Copyright © 1981. Reprinted by permission of Elsevier Science Publishers, Physical Science and Engineering Div., and the author.)

and the semiconductor is not as perfect as the Si-SiO$_2$ interface for thermally grown oxide.

19.3.2 Silicon Nitride

Silicon nitride (Si$_3$N$_4$) is used to complement SiO$_2$. Because of its less open structure, Si$_3$N$_4$ layers are impervious to water, sodium, and many other impurities that have a high mobility in SiO$_2$. This material, therefore, has been used as a mask against selective oxidation and diffusion. It is also used as a final protective layer against moisture. A disadvantage of Si$_3$N$_4$ is that Si-Si$_3$N$_4$ interfaces have a higher density of interface states than Si-SiO$_2$ interfaces.

Si$_3$N$_4$ layers are deposited using a CVD technique in a reactor containing SiH$_4$ and NH$_3$ with either H$_2$, N$_2$, or Ar as the carrier gas. The chemical reaction that occurs between 700 and 1100°C is the following:

$$3SiH_4 + 4NH_3 \rightarrow Si_3N_4 + 12H_2 \uparrow$$

The epitaxial reactor of Fig. 19.8 can also be used for this deposition process.

19.3.3 Polysilicon Layers

Polysilicon is used as the gate electrode in MOS devices, as the diffusion source to form shallow junctions, and as the material for large-valued resistors and interconnections in bipolar technology. Layers of this material can be deposited by the pyrolysis of SiH$_4$ at atmospheric pressure in an epitaxial reactor. The chemical reaction and the conditions are the same as in the case of epitaxy, except that now a noncrystalline material like an oxide or a nitride is used as a substrate in place of the single-crystal silicon.

Thin layers of semi-insulating polycrystalline oxygen doped silicon (SIPOS) are often used for junction passivation. A layer of SIPOS covered by an SiO$_2$ layer when applied across a p-n junction can shield against potential changes on the top of the sandwich. SIPOS films can be formed using the CVD method through the reaction of N_2O and SiH_4 at approximately 700°C.

19.4 THE PLANAR TECHNOLOGY

A large majority of silicon devices are fabricated using the planar process. In this process, dopant atoms are introduced into selected areas of Si from one surface of the wafer in order to form n- and p-type regions in the starting material. The technology is called *planar* because the device fabrication process is carried out from one surface plane. As an illustration of the process, we present the fabrication details of a p-n junction planar diode.

The basic steps involved in the fabrication are shown in Fig. 19.15. The starting material is an n^+-substrate ($\approx$150 μm thick). A thin layer of n-type Si (2-10 μm thick) is grown on the substrate by the epitaxial process to obtain the sample shown in Fig 19.15a. The wafer is now cleaned, and a SiO_2 layer is grown

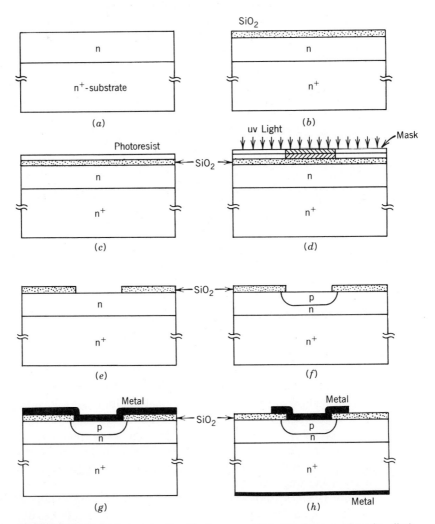

FIGURE 19.15 Various steps involved in the processing of a planar p-n junction diode.

by thermal oxidation (Fig. 19.15b). The top surface of the wafer is then coated with a photosensitive material known as *photoresist*. This material is applied to the wafer in liquid form and then baked in an oven leaving a thin coating of resist on the SiO_2 (Fig. 19.15c). The area of the diode to be fabricated is defined by a photomask. The mask, in its simplest form, is a glass plate that has transparent and opaque regions in a definite sequence. The mask is placed on the photoresist, and the entire structure is exposed to UV light as shown in Fig. 19.15d. The photoresist is polymerized in the regions where the light has fallen but remains unpolymerized on the surface beneath the opaque region. When the exposed photoresist is developed, the unpolymerized area is washed away leaving a solid film of resist in areas that were polymerized. The wafer is now etched in an etching solution containing HF. The oxide layer is etched away from the areas not covered by the photoresist but remains unaffected beneath the photoresist. The polymerized photoresist is then removed by dissolving it in an organic solvent leaving behind the oxide layer with windows in the desired areas (Fig. 19.15e). The wafer is now subjected to a boron diffusion to produce the p-region. The diffusion coefficient of boron in SiO_2 is very small compared to that in Si. Consequently, the SiO_2 layer prevents the diffusion of impurities into the semiconductor that is covered by the SiO_2, and the diffusion occurs only in the windows created by removal of the oxide. The p-n junction formed after the diffusion step is shown in Fig. 19.15f. Note that the diffusion also occurs laterally under the oxide. Next, a thin film of Al-metal is deposited on the top surface as shown in Fig. 19.15g. The metallized area is now coated with photoresist, and another mask is used to define areas where the metal is to be preserved. By keeping these areas covered with the photoresist, the wafer is etched in an acid solution to remove the undesired metal. The photoresist is then dissolved in an organic solvent. A contact metal is also deposited on the back surface of the wafer, and ohmic contacts to both sides of the junction are made by appropriate heat treatment. The final device structure is shown in Fig. 19.15h. Using the basic steps described here, we can fabricate the planar n-p-n transistor shown in Fig. 19.16, as well as a complex integrated circuit involving several components.

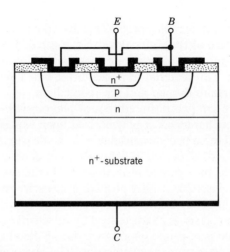

FIGURE 19.16 Cross-sectional view of a planar n-p-n bipolar transistor.

19.5 MASKING AND LITHOGRAPHY

The photomasking process can be divided into two distinct areas: the generation of a mask whose image is to be transferred to the Si wafer; and the transfer of an image from the mask to the surface of the wafer. Both processes are briefly described in this section.

19.5.1 Production of a Mask

In the early days of semiconductor fabrication technology, the circuits were rather simple, and the original artwork for the mask could be made by a draftsperson. The drawing was usually prepared from rubylith, which is a double-layered sheet (red as the top layer and white as the bottom layer). The red layer could be cut to the desired pattern with a razor knife. This pattern might be 100 to 1000 times the actual mask size, and it was reduced to the proper size using a photographic camera. For most present-day integrated circuits, mask making is automated by using computer-controlled drawing boards and other equipment. In either case, copies of the circuit patterns are photographically reduced until they are ten times the ultimate size. The final mask is made from the 10× plate, using a step and repeat camera that has a reduction factor of 10. The step and repeat process results in rows and columns of identical images being transferred to a glass plate called a *master*. The primary mask is used to make contact copies again using photosensitized glass plates. These copies are then used for the actual image transfer to a semiconductor wafer.

Another method utilizes computer-controlled light flashes to generate the desired pattern on a photographic film by writing with a light pencil. The resulting pattern is then reduced and handled in a step and repeat system to create the production mask. Yet another method involves the use of an electron beam exposure system that can directly write the pattern in its final size onto an electron-sensitive photoresist in a hard surface mask.

19.5.2 Lithography

(a) PHOTOLITHOGRAPHY

The photolithographic process involves the transfer of an image from the mask to the surface of the wafer through the use of UV light and a photoresist. Photoresists are chemical compositions containing a light-sensitive material in suspension. This material responds to the blue light of a mercury vapor lamp but is insensitive to the red and the yellow light.

There are two distinct types of photoresists: negative and positive. Negative resists get hardened or polymerized upon exposure to the UV light. Hence, the exposed resist becomes insoluble in the developer, whereas the unexposed resist is dissolved. Exposure to light softens or depolarizes a positive photoresist. Consequently, the exposed resist dissolves in the developer, and the unexposed resist remains insoluble. Either positive or negative resist can be used in most applications, but in some special cases, one type is preferred to the other.

A photoresist is characterized by its sensitivity, adhesion, and etch resistance. Adhesion is of prime importance during etching of the resist-protected material in order to minimize undercutting. The photosensitivity denotes the response of the resists to different light intensities. The sensitivity of positive resist is less than that of a negative resist.

A commonly used technique for coating the wafer with photoresist involves the use of a spinner. The wafer is placed on a flat holder on the top of a rotating shaft. A vacuum is used to hold the wafer to the holder. An appropriate amount of resist is placed at the center of the wafer with the help of a dropper. The shaft is then rotated, and as the wafer spins, the resist spreads uniformly over the wafer's surface. After coating with the resist, the wafer is removed and baked in an oven to remove the excess solvent. After cooling to room temperature, the wafer is exposed to the UV light through an appropriate mask and then developed in a developer. Next, it is again baked in an oven to increase the adherence of the resist to the surface of the wafer. The wafer is now ready for the etch step.

The simplest method of photoresist exposure through a mask is contact printing in which the mask remains in contact with the photoresist. If the wafer has a few particles protruding from its surface, these will locally damage the mask and cause a failure at the same spot in the next wafer. This problem can be avoided, at least partially, by proximity printing in which the mask is separated from the wafer by a small distance of about 20 μm. Projection printing is an even better solution for the mask damage problem. In projection printing the mask is separated from the wafer, and its image is projected on the surface of the wafer using an optical system.

When the incident radiation passes through a photomask close to the edge of an opaque feature, fringes are observed near the edge of the geometric shadow and some light enters the shadow region because of diffraction phenomena. Diffraction effects limit the linewidth that can be obtained using the UV light. The effect of light penetration into the shadow region of an open window in the resist is different for the positive and the negative resists. Overexposure of positive resist causes an increase in the size of the window while the window size is decreased in the case of overexposure of negative resist.

The practical limit of both contact printing and projection printing with UV light is reached at a linewidth of about 1 to 2 μm. Smaller features can be imaged only by reducing the diffraction effects using shorter wavelength radiation like an electron beam and X-rays.

(b) ELECTRON-BEAM LITHOGRAPHY

Electron-beam lithography can generate resist geometries smaller than 1 μm, and the technique can be highly automated. Two types of E-beam systems are in use: a scanning E-beam pattern generator, and an E-beam projection system. In a scanning E-beam pattern generator, the electron beam is produced in an electron-optical column and is focused on a fine spot on the target. The target may be either a mask which is to be made or a Si wafer on which the image is directly written. The beam can be deflected by means of a magnetic field, and its position can be accurately controlled by a computer.

E-beam and X-ray lithographies can use similar resists. In the case of an E-beam, the energy required to effect the desired chemical change after exposure is delivered by the electrons. However, in the case of X-rays, it is believed that photoelectrons produced by X-ray photons cause the chemical reaction. A commonly used resist in E-beam (and also in X-ray) lithography is the positive resist polymethyl methacrylate (PMMA). This resist has a very good contrast and can be used to obtain submicron feature devices. Negative resists for E-beam exposure show less contrast than PMMA and are less suitable for obtaining small dimensions.

Achievable resolution in E-beam exposure is believed to be limited by the scattering of high-energy electrons entering the resist. These electrons cause a blurring of the image in the resist. In a 1:1 E-beam projection system, UV light is made incident on a photocathode through the mask. This light causes the emission of electrons from the photocathode. These electrons are accelerated by an electric field between the mask and the semiconductor wafer and are focused on the resist. High image resolution and parallel processing of several devices are some of the advantages of this system.

(c) X-RAY LITHOGRAPHY

X-ray lithography is similar to the optical lithography except that short wavelength X-rays are used instead of the UV light. Furthermore, the exposure of the resist on the wafer is performed by proximity printing, keeping a gap of 25 to 40 μm between the wafer and the mask. The short wavelength of X-rays reduces diffraction effects leading to high image quality with sharp edges. An X-ray lithography mask is made from a thin membrane of mylor or silicon which is transparent to X-rays. Gold is used for masking on this layer. Gold is evaporated uniformly on the membrane, and an electron beam exposure system is used to define the mask area.

19.6 PATTERN DEFINITION

After the lithographic step, windows in the masking oxide are provided by selective removal of the material from the unmasked regions. Photolithography is also used for patterning the metal and other layers applied to the semiconductor (see Fig. 19.15g). Etching techniques used for the selective removal of undesired dielectric and metallic layers include: wet chemical etching, electrochemical etching, sputter etching, plasma etching, and reactive ion etching. A brief account of these techniques is provided in this section.

(a) WET ETCHING

Wet chemical etching is the most widely used technique for the selective removal of regions of semiconductor material, metal, SiO_2, and Si_3N_4. Numerous solutions exist for the selective etching of different materials. For example, SiO_2 is etched in a buffered solution of $HF + NH_4F$, Si_3N_4 is etched in hot H_3PO_4, while Al is etched in either H_3PO_4, HNO_3, or acetic acid. A common method of etching is to immerse the wafer in the etching solution at a predetermined temperature. The etch time is determined by the etching rate of the solution.

Problems commonly encountered during the selective wet etching include: undercutting, photoresist liftoff, and enhanced etching along the grain boundaries. Undercutting is rather difficult to control and limits the use of wet chemical etching to devices with dimensions larger than 2.5 μm.

(b) ELECTROCHEMICAL ETCHING

In electrochemical etching a voltage is applied between the etchant and the material to be etched, and etching is done at a controlled rate. The associated problems are similar to those that occur in the case of wet chemical etching.

(c) SPUTTER ETCHING

Sputter etching involves the removal of material by ion bombardment. The material to be etched is made the cathode, and Ar-ions are used. The setup is similar to that shown in Fig. 19.14. Since ions impinge the surface nearly vertically, no undercutting occurs. This makes sputter etching more suitable for etching patterns with small widths and separations. Problems associated with sputter etching are the possibility of etching of the photoresist and redeposition of sputtered material along the edges of the photoresist. Radiation damage to SiO_2 from the ions is the main disadvantage for semiconductors because this damage cannot always be removed by annealing.

(d) PLASMA ETCHING

Plasma etching is a dry-chemical etching technique in which reactive gases like CF_4 and CCl_4, which contain halogen atoms, are used for etching Si, SiO_2, Si_3N_4, and certain other materials. After filling the reactor with the reactive gas (at a pressure of 0.1-3 torr), the plasma is produced by an rf field (usually at 13.5 MHz). The fragments of the reactive gas produced in the plasma react with the material to be etched and form volatile compounds that are exhausted from the reactor. The wafers are placed in a Faraday cage and are shielded from the charged species.

Plasma etching has several advantages over wet chemical etching. Attack on photoresist is slow, there is less undercutting, and the definition of the pattern is better than in the case of wet chemical etching. However, good temperature control is necessary for obtaining close tolerances.

(e) REACTIVE ION ETCHING

Reactive ion etching is a combination of sputter etching and plasma etching techniques. Here a plasma is produced between two parallel plates in the reactor, and the wafers to be etched are kept in the electric field between the plates. Consequently, etching takes place partly by chemical reaction of the material with fragments of the reactive gas and partly by removal of the material by ion bombardment. The gas pressure in the reactor and the sample geometry influence the etching rate. Undercutting can be made very small as in the case of sputter etching, but radiation damage is introduced in the SiO_2.

19.7 METAL DEPOSITION TECHNIQUES

After a device has been fabricated on a semiconductor substrate, ohmic contacts to the various regions of the device must be made by metal deposition. In the case of an integrated circuit, all the devices forming the circuit must be suitably connected together with metallization to perform the intended circuit functions. This section briefly describes the various methods of metal deposition along with their important features.

The metal that is used to provide ohmic contacts for devices and interconnections for integrated circuits has to meet several requirements. For example, the chosen metal must have high electrical conductivity and should be able to make good ohmic contacts to the semiconductor. Furthermore, it should have good adherence to the underlying surface and must not corrode under normal operating conditions. No single metal is known to perfectly meet all these requirements for a given semiconductor. However, in the case of silicon, Al meets most of the requirements to a sufficient degree and is widely used for device interconnections, and in many cases, for making ohmic contacts. Pure Al has a tendency to react with Si, and, to prevent this reaction some Si is added to Al-metallization.

In some instances Al does not meet all the requirements of metallization, and multilayered structures are often used. A typical example of such a structure that is used for Si is the Au-based system PtSi-Ti-Pt-Au. Here Au is the main conductor, and each of the other layers meets some of the requirements mentioned above.

Among the numerous methods used for metal deposition are vacuum deposition using resistive heating or electron beam, sputtering, and plating. Plating is used only occasionally in Si technology and may involve either electroplating or electroless plating. Electroless Ni-plating is sometimes used to obtain ohmic contacts on relatively heavily doped Si.

Metal deposition using resistive heating is performed in a vacuum deposition system. The source and substrate are placed in a chamber that is evacuated using an oil diffusion pump. Evaporation of the metal takes place either from a filament or from a boat that is heated by resistance heating. The method is simple, and a number of materials can be deposited using this technique. Al is normally deposited using a tungsten filament. One of the disadvantages of this method is that the contamination level of the deposited material may sometimes become so high as to influence the functioning of the device. Contamination may come from the filament or from the ambient in the vacuum chamber. If layers of different materials are to be deposited, a separate filament is needed for each deposition.

A typical electron-beam evaporation system is shown in Fig. 19.17. Here the material to be evaporated is placed in a crucible that is kept in a vacuum chamber. A high-intensity beam of electrons is generated in an electron gun and is focused on the material to be evaporated. The beam melts the material and causes its evaporation. Since the electron beam comes in contact only with the metal and not with the crucible, contamination is considerably reduced. However, electron-gun evaporation often results in radiation damage to the substrate, and this must be annealed out using a suitable heat treatment cycle.

Sputtering is performed in a low-pressure gas discharge (usually in Ar). The setup is the same as that in Fig. 19.14. The material to be sputtered is made the cathode, while the anode holds the substrate and is usually grounded. Either a dc or rf voltage can be employed, although rf sputtering is more common. The advantage of sputtering is that it can be used to deposit alloys without compositional changes. However, a low deposition rate, damage to the substrate, and substrate temperature rise during sputtering are the disadvantages of this process.

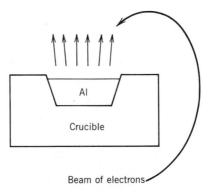

FIGURE 19.17 Cross-sectional view of an electron beam evaporation system. (From Peter Gise/Richard Blanchard, *Modern Semiconductor Fabrication Technology,* copyright © 1986, p. 144. Reprinted by permission of Prentice Hall, Inc., Englewood Cliffs, New Jersey.)

Subsequent to metal deposition, the semiconductor wafer is placed in a furnace to alloy the metal to the semiconductor to ensure good electrical contact. It is then diced into separate pieces, each containing one or more semiconductor devices. Each chip is then soldered to a package, and wires from the package leads are connected to the metal on the semiconductor. Finally, the package is sealed with a ceramic or a metal cover.

19.8 GENERAL REMARKS ON INTEGRATED DEVICES

In an integrated circuit, a number of active and passive components are combined to form one package that performs the desired circuit function. From the perspective of their fabrication technology, integrated circuits can be divided into two categories: monolithic and hybrid. In a monolithic circuit, all the active and passive components are fabricated on a single-crystal wafer of the semiconductor, usually Si. A hybrid circuit, on the other hand, may contain one or more monolithic circuits or individual transistors bonded to an insulating substrate with passive components, like resistors and capacitors, connected externally with appropriate interconnections. A monolithic circuit has the advantage that since all its components are contained on a single chip, many such units can be processed together. For example, a chip may contain thousands of transistors and other components all interconnected properly. Such circuits are batch fabricated on a 3- or 4-inch diameter wafer, and each wafer may contain hundreds of monolithic circuits. These circuits are then separated by sawing or by scribing, and each of them is mounted on a suitable header, contacted, and packaged. Our discussion here and in the following two sections will be limited to monolithic circuits.

The integrated circuit (IC) era has revolutionized electronics and has made many instruments and applications possible that were inconceivable with discrete devices. Miniaturization of components and the batch fabrication of hundreds of circuits has brought advantages such as decreased system size, reduced assembly cost, and lower power requirements per circuit. Besides these factors, reducing the size of each unit in the batch has also led to improved reliability and improvements in high-frequency performance and switching speed. Reliability has been increased mainly because all devices and interconnections are made on a single chip, and this greatly minimizes failures due to the soldered interconnections of discrete components. Operating speed has improved because compactness reduces the capacitance of interconnections and decreases the signal transfer time that results in smaller transmission delays.

Following their applications, monolithic ICs can be divided into two main classes: analog and digital. An analog IC performs linear operations such as signal amplification. Typical examples are simple amplifiers, operational amplifiers, and analog communication circuits. A digital IC involves logic circuits that perform binary arithmetic and include memory circuits that store information. By far, the IC's greatest impact has been in the digital field, and digital ICs have found applications in computers, calculators, microprocessors, and so on.

Both digital and analog ICs can be produced using either bipolar technology or MOS technology. However, MOS technology is almost exclusively used for digital circuits, while bipolar technology is prevalent in the case of analog circuits. The most essential features of bipolar integration are presented in the next section, and several aspects of MOS integration in the subsequent section. For more information the reader should consult A. G. Milnes [6].

19.9 BIPOLAR INTEGRATION

19.9.1 Methods of Isolation

The starting material in bipolar IC fabrication is an *n*-type Si layer that is epitaxially grown on a *p*-type substrate with a 10 Ω-cm resistivity. The thickness and resistivity of n-layer depend on the application and may range from 0.5 to 12 µm and 0.2 to 5 Ω-cm, respectively. An important step in the fabrication process is the provision for electrical isolation between the components. In bipolar ICs, isolation is provided either by a reverse-biased p-n junction or by an isolating dielectric material.

(a) P-N JUNCTION ISOLATION

The p-n junction isolation technique involves diffusing a pattern of *p*-type moats into the n-epitaxial layer to obtain areas of *n*-type islands that are isolated from each other. Figure 19.18a shows the starting wafer. This wafer is first oxidized, and windows are etched through the oxide on the *n*-type side using photolithography and oxide etching. Next, boron diffusion is performed through the windows so that the boron atoms diffuse through the epitaxial layer and join with the p-substrate. The resulting *n*-type island shown in Fig. 19.18b is now ready for components to be fabricated in it. All such islands are electrically isolated from each other when the p-n junction is reverse biased by the application of a negative potential to the p-substrate.

A bipolar n-p-n transistor can be fabricated in the *n*-type island of Fig. 19.18b by first performing a *p*-type diffusion through a window in the oxide and then by performing an emitter diffusion. A major disadvantage of this transistor is that it has a high collector series resistance because the *n*-type epitaxial layer is lightly doped. The collector series resistance is considerably reduced if a selective n$^+$-diffusion is performed in the p-substrate before growing the epitaxial layer. The diffused n$^+$-layer is known as a buried layer (Fig. 19.18c). This buried layer pro-

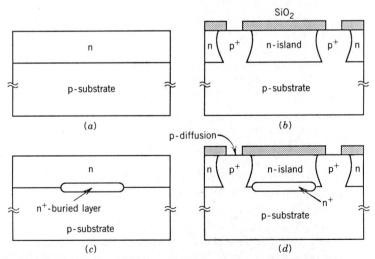

FIGURE 19.18 Illustration of p-n junction isolation: (*a*) p-substrate with n-epitaxial layer, (*b*) *p*-type isolation diffusion, (*c*) substrate with n$^+$-buried layer, and (*d*) *p*-type isolation diffusion including buried n$^+$-layer.

vides a low-resistance path for electrons flowing from the active region of the transistor to the collector ohmic contact. Figure 19.18d shows an isolated region with an n^+-buried layer.

In the above process of n-p-n transistor fabrication, three diffusion steps are needed: the isolation diffusion, and one each for the base and the emitter. The process complexity can be reduced by using the collector diffusion isolation (CDI) structure shown in Fig. 19.19. Here a *p*-type layer rather than an *n*-type layer is grown over the surface of the p-substrate which also contains an n^+-buried layer. Thus, the base is directly formed by the p-epitaxial layer, and no base diffusion is needed. The p-layer growth process is followed by an overall p-diffusion over the wafer. This p-diffusion produces a drift field in the otherwise uniformly doped base, thus enhancing the high-frequency performance of the transistor. Subsequent to the p-diffusion, a deep n^+-diffusion is performed selectively through the *p*-type epilayer which serves as the collector. An n^+-emitter diffusion is now selectively performed as shown in Fig. 19.19.

A disadvantage of the p-n junction isolation method is the capacitance inherent at the isolating p-n junction. Moreover, the substrate has to be maintained at a negative potential to keep the junction under reverse bias. These shortcomings are eliminated in the dielectric isolation.

(b) DIELECTRIC ISOLATION

The process sequence for oxide isolation is shown in Fig. 19.20. Here the starting material is an *n*-type Si-substrate. The process begins with an n^+-diffusion over the entire surface of the substrate. The wafer is then oxidized, and windows are etched in the oxide using photolithography and an etching step. The etching of the SiO_2 is followed by the etching of the Si in the windows to obtain etched channels as shown in Fig. 19.20a. A SiO_2 layer is now thermally grown to cover the wafer, and polysilicon is deposited over the oxide layer (Fig. 19.20b). The structure is now turned upside down, and the *n*-type Si is lapped off to obtain the isolated regions of the *n*-type Si shown in Fig. 19.20c. As required, additional diffusions are now performed in the islands to fabricate the desired components. This type of isolation process eliminates the substrate bias and results in low parasitic capacitance. However, it requires extra processing steps and thus becomes rather expensive.

Another method of isolation is the so-called V-groove process shown in Fig. 19.21. This process is a combination of dielectric and p-n junction isolations.

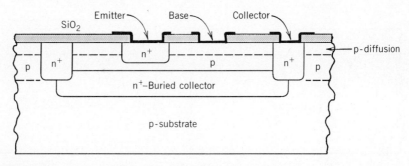

FIGURE 19.19 Collector diffusion isolation. (From B.T. Murphy, V. J. Glinski, P. A. Gray, and R. A. Pederson, Collector diffusion isolated integrated circuits, *Proc. IEEE,* vol. 57, p. 1523. Copyright © 1969 IEEE. Reprinted by permission.)

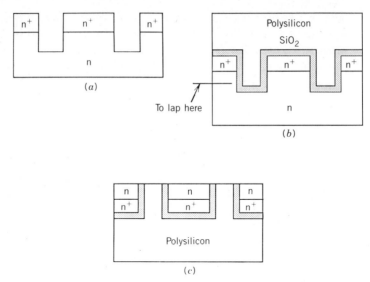

FIGURE 19.20 Oxide isolation: (*a*) etching of channels, (*b*) deposition of polysilicon, and (*c*) lapping-off of the substrate to form islands.

When Si is etched chemically, the etching rate along the (111) plane is considerably higher than that along the (100) plane. A Si wafer with {100} surface orientation is etched using an anisotropic etchant. The etched surface develops a groove with (111) side walls as shown in Fig. 19.21a. The depth of the groove is determined by the width of the oxide window. The wafer is now thermally oxidized to cover it with a layer of SiO_2, and polysilicon is deposited to fill the groove. The resulting isolated pockets are shown in Fig. 19.21b. These pockets can now be used to fabricate the desired components.

The isoplanar process developed by Fairchild Semiconductor is yet another method that combines p-n junction and oxide isolations. The process begins with the Si wafer shown in Fig. 19.18c. A Si_3N_4 layer is first deposited on the surface of the *n*-type epitaxial layer. Using photoresist, a window is etched through the Si_3N_4 layer, and the *n*-type epitaxial layer beneath the window region is etched through halfway using a silicon etchant. The wafer is then subjected to thermal oxidation. Since no oxide growth occurs in the areas covered by Si_3N_4, the oxide

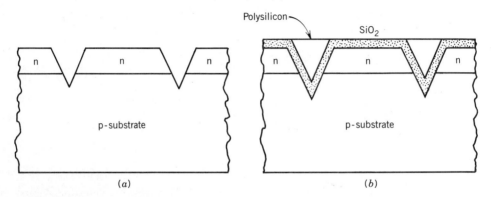

FIGURE 19.21 The V-groove isolation process: (*a*) preferential V-groove etching of (100) substrate and (*b*) thermal oxidation.

BIPOLAR INTEGRATION

fills only the etched areas as shown in Fig. 19.22. The process has the disadvantage that it can be carried out only for thin epitaxial layers, and long oxidation times are required.

19.9.2 Circuit Components

The most important component of a bipolar IC is the n-p-n transistor. Besides this device, the other components are: resistors, capacitors, p-n junction diodes, and lateral p-n-p transistors. These components will be described in this section.

(a) N-P-N TRANSISTORS

The top view and cross-section of an integrated circuit n-p-n transistor are shown in Fig. 19.23. The fabrication process of this device involves the following steps in sequence: Diffusion of an n^+-buried layer into the p-substrate, growth of the n-type epitaxial layer, isolation diffusion, base diffusion, emitter diffusion, contact window opening, and etching of the contact metallization pattern. Note that in the integrated circuit transistor, all three terminals (emitter, base, and collector) are made available on the top surface of the chip. This feature is in contrast to the discrete n-p-n device shown in Fig. 19.16 where the collector contact is taken at the bottom of n^+-substrate. The requirement for making all the contacts on the top surface of the IC transistor forces the collector current to traverse the n-type region laterally. If the n^+-buried layer is not used, then this current will flow through the lightly doped n-epitaxial layer, resulting in a high collector series resistance. Incorporation of the buried layer causes a considerable decrease in the collector series resistance, as has already been discussed in the last section.

When the p-substrate is included, the IC transistor has a four-layer p-n-p-n structure. This structure acts as a thyristor and may turn on if the emitter-base and the collector-substrate junction are simultaneously forward biased. This possibility is avoided by connecting the substrate to the most negative potential in the circuit so that the collector-substrate junction is always reverse biased.

Another problem with the transistor of Fig. 19.23 is that it has a large parasitic capacitance resulting from the substrate-collector junction. The capacitance can be decreased by using either V-groove or isoplanar isolation and is largely eliminated in the case of oxide isolation.

Complex integrated circuits can be fabricated using n-p-n transistors with proper interconnections. However, it is also possible to fabricate multiple transistor structures in the same isolated island. For example, Fig. 19.24 shows a mul-

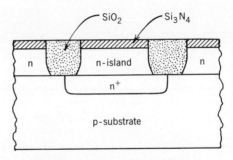

FIGURE 19.22 Isoplanar isolation.

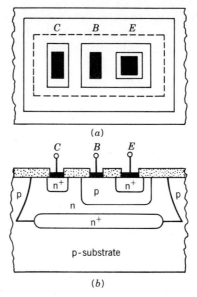

FIGURE 19.23 An integrated circuit n-p-n transistor: (a) top view and (b) cross-section.

tiple emitter transistor in which the collector and base regions are shared by a number of transistors.

(b) DIODES

It is a simple matter to build p-n junction diodes in a monolithic IC. However, no separate diffusion is performed to fabricate the diode elements, and it is a common practice to use the existing transistor structure as a diode. There are five different ways to connect a transistor as a diode, namely: (1) an emitter-base junction diode with the collector shorted to the base, (2) an emitter-base junction diode with an open collector, (3) a collector-base junction diode with the emitter shorted to the base, (4) a collector-base junction diode with an open emitter, and (5) a collector-base junction diode with the emitter shorted to the collector. Of these five possibilities, the most commonly used connection is the emitter-base diode with the collector shorted to the base; a cross-sectional view of the structure is shown in Fig. 19.25. This configuration is essentially a short-base diode that has a high switching speed and requires a low junction voltage for a given forward current (see Prob. 19.10).

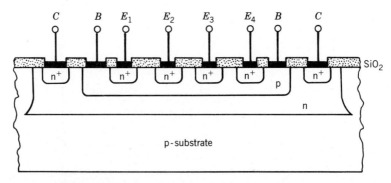

FIGURE 19.24 Cross-sectional view of multi-emitter transistor.

BIPOLAR INTEGRATION

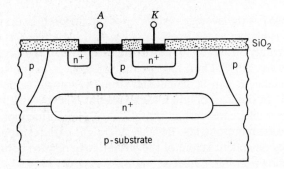

FIGURE 19.25 Cross-sectional view of a emitter-base diode with collector shorted to base.

(c) RESISTORS

In monolithic ICs resistors are usually composed of a layer diffused into a semiconductor of opposite polarity. No additional diffusion step is required to fabricate resistors, and they are obtained by using shallow diffusions for transistor base and emitter regions. Figure 19.26a shows a resistor fabricated by a p-diffusion in one of the n-type islands. The diffusion is performed at the same time when the base diffusion for the n-p-n transistor is made, and the resistor is obtained by contacting the two ends of the diffused region of length L and width W. The resistance R of this structure can be expressed as

$$R = \bar{\rho}\frac{L}{x_j W} = \rho_s \frac{L}{W} \tag{19.13}$$

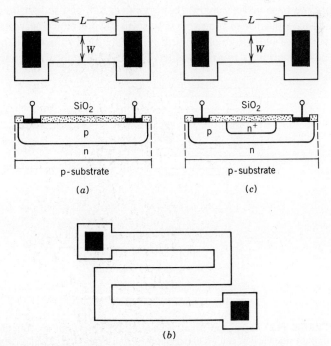

FIGURE 19.26 Integrated resistors: (a) base diffusion resistor with top and side views, (b) meander pattern, and (c) pinched-base buried resistor.

where $\bar{\rho}$ is the average resistivity of the diffused layer, x_j is its thickness, and $\rho_s = \bar{\rho}/x_j$ is known as the sheet resistance of the layer. ρ_s is expressed in $\Omega/\square$ and can be readily measured for a thin diffused layer using the four-point probe technique (see Sec. 20.2). Thus, for a given diffusion, ρ_s is known fairly accurately, and the resistance R can be calculated from the known value of ρ_s and the aspect ratio L/W. In order to obtain a high value of R, the width W should be made as small as possible, considering limitations of photolithographic resolution and heat dissipation requirements. Because a straight-path resistor can frequently give rise to an inconvenient layout of the circuit, long resistor bodies are folded in a meander as shown in Fig. 19.26b.

Values of ρ_s for base diffusion are generally in the range of 100 to 500 $\Omega/\square$. Consequently, to reduce the amount of space used for resistors and to obtain high-resistance values, it is often necessary to obtain surface layers with higher sheet resistance than is available during the standard base diffusion. One way to obtain higher ρ_s values is to use the pinched base resistor design shown in Fig. 19.26c which is fabricated by diffusing an n^+-emitter region over the center of the base resistor. Typical values of sheet resistance that can be obtained with this design are $2 - 10$ K $\Omega/\square$. An alternative method for obtaining high resistances is to use ion implantation to form shallow regions of very high sheet resistance. Resistor values below 100 Ω are made during the emitter diffusion for which values of ρ_s lie in the 4 to 10 $\Omega/\square$ range.

(d) CAPACITORS

Two types of capacitors are commonly used in a monolithic circuit: the junction capacitor and the dielectric (i.e., MOS) capacitor. A junction capacitor utilizes the capacitance of a reverse-biased p-n junction and is formed during one of the diffusion steps for the n-p-n transistor. The collector-base junction between the p-type base diffusion and the n-epilayer can be used without an emitter diffusion if the capacitor is required to sustain a high blocking voltage. However, for low values of blocking voltage and relatively high values of capacitance, the emitter-base junction is more suitable. Figure 19.27a shows top and side views of a ca-

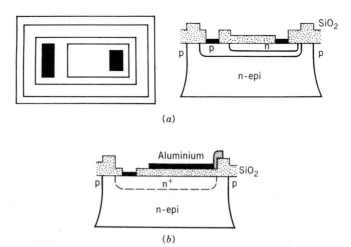

FIGURE 19.27 Integrated circuit capacitors: (a) a junction capacitor showing top and side views and (b) side view of the MOS capacitor. (From Peter Gise/Richard Blanchard, *Modern Semiconductor Fabrication Technology*, copyright © 1986, Fig. 10.18(b), p. 167 and Fig. 10.19, p. 168. Reprinted by permission of Prentice Hall, Inc., Englewood Cliffs, New Jersey.)

pacitor made using the emitter-base junction of the n-p-n transistor. The main disadvantages of junction capacitors are their bias dependence and finite leakage current through the junction that increases the loss factor of the capacitor. Voltage-independent capacitors with a low loss factor can be obtained using MOS structures. Figure 19.27b shows the side view of a MOS capacitor with an SiO_2 dielectric and the n^+-emitter diffusion as one plate of the capacitor.

(e) P-N-P TRANSISTORS

Sometimes it becomes necessary to have p-n-p and n-p-n transistors on the same monolithic IC chip. The most common method of obtaining a p-n-p device in the n-island is the structure shown in Fig. 19.28a. This structure is called a *lateral transistor* because here the current flows laterally from the emitter to the collector. The main advantage of this structure is that its fabrication does not require any additional processing steps except those required for the n-p-n transistor. The p-emitter and collector regions are simultaneously diffused during the base diffusion, and the n^+-base contact region is diffused during the emitter diffusion of the n-p-n device.

The performance of a p-n-p lateral transistor is considerably inferior to that of a vertical transistor. A simple one-dimensional analysis of the device of Fig. 19.28a is possible by splitting the emitter current into its lateral component I_{pL} and the vertical component I_{pV}. Assuming that the device is in its normal active mode, and that the base width W_B is small compared to the hole diffusion

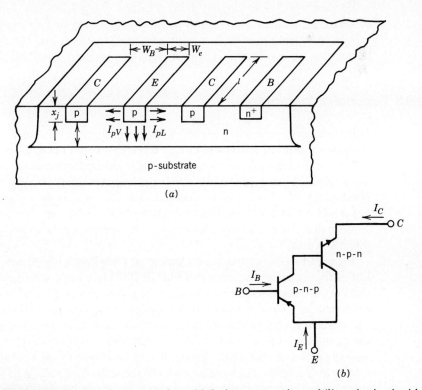

FIGURE 19.28 The lateral p-n-p transistor: (a) device cross-section and (b) use in circuit with an n-p-n transistor.

length L_p in the n-type base, the components I_{pV} and I_{pL} of the emitter current can be written as

$$I_{pL} \simeq 2qD_p l x_j \frac{p_E}{W_B} \qquad (19.14)$$

$$I_{pV} \simeq qD_p l W_e \frac{p_E}{L_p} \qquad (19.15)$$

Here D_p is the hole diffusion coefficient in the n-type base; l and W_e are the length and width of the emitter region, respectively; and p_E is the excess hole density at the base side of the emitter-base junction depletion region. In writing Eq. (19.14) it has been assumed that l is very large compared to W_e, while a long base region is assumed in Eq. (19.15). The common emitter current gain of the transistor is

$$h_{FE} = \frac{I_C}{I_B} \approx \frac{I_{pL}}{I_{pV}} = \frac{2L_p x_j}{W_B W_e} \qquad (19.16)$$

The current gain of the device is controlled by the vertical injection and rarely exceeds 10. Besides lowering the current gain, the vertical component I_{pV} also degrades the frequency response of the device. The vertical injection can be significantly reduced by incorporating a buried n^+-layer below the emitter and the collector regions. This feature results in a considerable improvement in the current gain and frequency response over the device shown in Fig. 19.28a.

Because of its low current gain, a lateral p-n-p transistor is often used in conjunction with a vertical n-p-n transistor as shown in Fig. 19.28b. The n-p-n transistor boosts the overall current gain of the device, and the composite unit behaves as a p-n-p transistor with a large current gain (see Prob. 19.14).

19.9.3 Functional Integration

Functional integration is a higher order integration in the sense that in a functional structure, the individual circuit elements cannot be readily identified. This is because the components have electronic interactions that may result from sharing a common junction. This type of approach not only opens up new electronic possibilities, but it also causes a considerable increase in the packing density. An important example of functional integration is the integrated injection logic (I^2L). This device is particularly important in low-power applications such as digital watch circuits.

Figure 19.29 shows the basic element of I^2L along with its equivalent circuit. Let us first consider the circuit of Fig. 19.29b. Here the collector of a p-n-p transistor is connected to the base of another n-p-n transistor. Thus, the p-n-p transistor acts as a current source that supplies the operating power for the circuit. The current source can be short-circuited by grounding the input point B. When the input at B is open-circuited, the current source injects a large current into the base of the n-p-n transistor, driving it into saturation. Consequently, the output voltage at C reaches its low value equal to the saturation voltage of the n-p-n transistor. With the input B shorted to ground, the n-p-n transistor remains at cutoff, and the voltage at C has its high value of nearly V_{CC} (the supply voltage). Thus, it

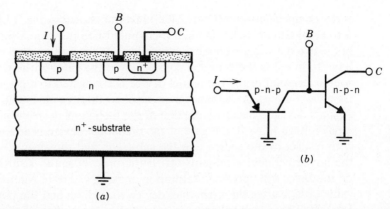

FIGURE 19.29 The integrated injection logic circuit: (a) cross-section of the device and (b) its equivalent circuit.

is clear that this circuit acts as a positive logic circuit whose output is low when the input is high and vice versa.

Figure 19.29a illustrates how the circuit might be fabricated in a simple way. The starting material is an n-type epitaxial layer grown on an n^+-substrate. Two p-type diffusions are performed, followed by an n^+-diffusion in one of the pockets. The p-n-p current source is the lateral device, and the n-p-n transistor is operated upside down with the n^+-substrate serving as the emitter and the diffused n^+-layer as the collector. Note that the emitter of the n-p-n transistor is common with the base of the p-n-p transistor, and the base of the n-p-n transistor is the collector of the p-n-p transistor. Thus, the two transistors are integrated (or merged) into a single unit—hence, the name *integrated injection logic* or *merged transistor logic* (MTL). Large logic arrays can be made from this type of element. Examples are given by Milnes [6].

19.10 MOS INTEGRATION

MOS transistors are most widely used in digital integrated logic gates and memory arrays. These devices have a number of advantages over bipolar transistors:

1. MOS fabrication processes involve fewer manual handling steps, fewer masks, and fewer high-temperature steps.
2. The packing density of MOS devices is high, and the cost per circuit function is lower than that in bipolar devices.
3. The impedance levels in MOS circuits tend to be high, and this results in lower power dissipation compared to bipolar logic circuits.
4. Both n- and p-channel devices can be easily fabricated on the same substrate so that complementary symmetry circuits are possible with low-power dissipation.

19.10.1 Complementary MOS Inverter

Important features of MOS logic can be understood by considering an inverter circuit. There are several versions of this circuit, but a particularly useful device is the complementary MOS (CMOS) inverter shown in Fig. 19.30a. It consists of a combination of n-channel and p-channel MOS transistors such that the drains of the two devices are connected together and form the output of the circuit. The input terminal is the common connection to the gates of the transistors, and the source terminal of the p-channel device is connected to a positive voltage. Both transistors are of the enhancement type, so that the threshold voltage of the p-channel device is negative and that of the n-channel device is positive. When the input voltage $V_{in} = 0$, the gate of the p-channel device is at a negative potential with respect to the source and the substrate, and this device is turned on while the n-channel device remains off. If the bias voltage V is sufficiently large, the on-transistor will operate in saturation, making $V_{out} \simeq V$. Alternatively, a positive value of V_{in} causes the n-channel device to turn on and the p-channel device to turn off. The output voltage measured across the n-channel device is then essentially zero. Thus, it is clear that the circuit operates as an inverter making the output low when the input is high and making the output high when the input is low. Since the two transistors are in series, and one of them is turned off for either condition, little current is drawn in the steady state. However, charging current will flow during the switching from one state to the other, and this causes some power dissipation in the circuit. Because of its small power dissipation, the CMOS circuit is particularly useful in low-power consumption circuits, such as electronic watch circuits.

The device technology of the CMOS circuit requires the fabrication of both p- and n-channel enhancement-type devices with nearly the same magnitude of the threshold voltage. A simple IC implementation of the circuit is shown in

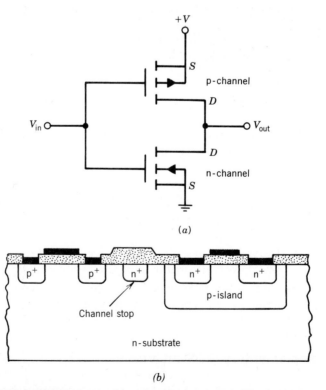

FIGURE 19.30 The CMOS inverter: (a) basic circuit and (b) structure with p-channel and n-channel devices formed together.

Fig. 19.30b. The starting material is an n-type Si substrate. A p-diffusion is performed in the substrate to isolate a p-type island in which the n-channel device is fabricated by subsequent diffusion of n^+-source and drain regions. The acceptor concentration in the p-type island is a critical parameter that is adjusted using ion implantation to obtain the desired value of the threshold voltage. The p-channel device is formed by performing a p^+-source and drain diffusion in the n-substrate. Generally, an n^+-channel stop is also diffused during the source and drain diffusion of the n-channel device. This channel stop prevents the formation of an inversion layer between the n- and p-channel devices under interconnection metallization.

Another type of MOS inverter circuit is shown in Fig. 19.31. Here two n-channel devices are used; one is an enhancement-type device, and the other is a depletion-type device. When the input voltage V_{in} is zero, the depletion mode device is on and the enhancement mode device is off, making $V_{out} \approx V$. However, when V_{in} has a sufficiently large positive value, the enhancement mode device is turned on and the gate voltage of the depletion mode device becomes nearly zero. Consequently, most of the voltage V appears across the depletion mode device, and the output voltage approaches zero. A similar inverter circuit can be made using two p-channel MOS transistors. The D-MOS and V-MOS transistors described in Sec. 16.8 have been used to fabricate MOS inverter circuits.

19.10.2 Silicon on Sapphire Devices

Silicon on sapphire (SOS) devices form a useful extension of the Si MOS process. In this technology, both n- and p-channel MOS transistors can be fabricated on Si films epitaxially deposited on insulating substrates such as sapphire (Al_2O_3) or spinal ($MgO + Al_2O_3$). Figure 19.32 shows a cross-sectional view of n- and p-channel MOS transistors made on a sapphire substrate. A thin film (≈ 1 μm thick) of Si is epitaxially grown on the sapphire substrate by the CVD process using the

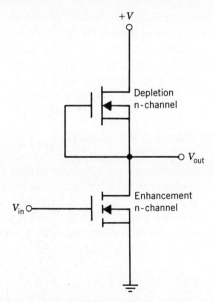

FIGURE 19.31 A MOS inverter with n-channel depletion and enhancement-type devices.

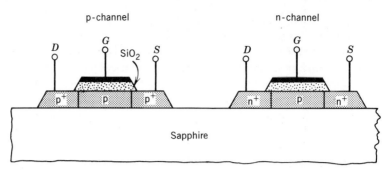

FIGURE 19.32 Side view of p- and n-channel enhancement mode silicon on sapphire MOS transistors. (From B. G. Streetman, *Solid State Electronic Devices*, 2nd ed., copyright © 1980, p. 348. Reprinted by permission of Prentice Hall, Inc., Englewood Cliffs, New Jersey.)

pyrolysis of SiH_4. The film is then etched into desired islands using standard photolithographic techniques. It is then covered with a layer of SiO_2, and windows are etched through the oxide for performing the n^+- and p^+-diffusions to obtain areas for the source and drain regions. Diffusion is continued until the n^+- and p^+-regions extend entirely through the film to the substrate as shown in Fig. 19.32. Since the Si film is very thin and the junctions are vertical, the junction capacitance is reduced to a very low value. Moreover, the presence of the insulating substrate minimizes the stray capacitances. Because of this overall reduction in capacitances, the SOS devices have a high switching speed.

In Fig. 19.32, the n-channel device operates like a normal enhancement-type transistor, and a positive gate voltage in excess of the threshold voltage V_{th} produces an inversion layer at the surface. However, the mode of operation of the p-channel device is different. Here at zero gate bias, the metal semiconductor work function difference and the oxide charge are able to fully deplete the *p*-type epitaxial region under the gate. This condition is called *deep depletion*. A negative gate voltage removes the depletion region and causes an accumulation of holes at the Si surface beneath the gate. Thus, a conducting channel is formed between the source and the drain. The I-V characteristics of this device are similar to those of the conventional p-channel device. An alternative to the deep depletion approach is to deposit islands of lightly doped *n*- and *p*-type Si films in selected areas and to fabricate conventional enhancement-type devices on these islands.

The SOS technology is attractive for CMOS fabrication. However, these devices show inferior performance compared to devices made on single-crystal Si substrates because of high interface states density at the Si-sapphire interface.

19.10.3 Nonvolatile Memory Devices [7]

Many semiconductor memory systems utilize devices in which the stored information is lost when the power supply is switched off. It is often desirable to have nonvolatile memory wherein the information, once written, remains valid even when the power is removed from the device. Two groups of nonvolatile memory devices, namely, the dual dielectric MIOS devices and floating-gate avalanche-injection MOS (FAMOS) devices, are described below.

(a) DUAL DIELECTIC MIOS DEVICES

A MIOS (metal-insulator-oxide-semiconductor) device uses a double dielectric layer on an Si substrate, and the charge is stored at the interface between the two dielectrics. The first layer is a thin SiO_2 layer, and the second layer is a different dielectric that may be either Si_3N_4 or Al_2O_3. Devices with an Si_3N_4 layer are superior to those employing Al_2O_3 as the second dielectric and are the most popular.

Figure 19.33 shows the cross-section of a p-channel MNOS device. The device is a MOSFET in which the SiO_2 layer is replaced by a double dielectric consisting of a thin (≈ 30 Å) layer of SiO_2 and a relatively thick (500 to 1000 Å) layer of Si_3N_4. The writing operation is performed by applying a sufficiently high-voltage (≈ 30 V) pulse to the gate. Following the application of the pulse, a high electric field is created in the oxide, and electrons from the conduction band of Si tunnel into the traps located at the SiO_2-Si_3N_4 interface. The negative charge trapped at the interface causes a change in the threshold voltage of the transistor. Upon removal of the pulse, the trapped electrons cannot flow back to the semiconductor, and the information (i.e., the changed threshold voltage) can be retained for a long period of more than a year.

To switch the device back to the low-threshold voltage state, a negative voltage pulse is applied to the gate. This reversal of polarity causes electron tunneling from the interface traps into the semiconductor and hole tunneling in the opposite direction. With a sufficiently large negative voltage, the device can be brought back to its original state or, if desired, a complete reversal of the threshold voltage to a negative value is possible.

Typical switching speeds during write and erase operations are on the order of 10 μsec and increase as the SiO_2 thickness is reduced and the trap density at the SiO_2-Si_3N_4 interface is increased. To increase the trap density, a few atomic layers of metal such as tungsten (or gold) are processed into the dual-dielectric interface.

(b) FLOATING-GATE DEVICES

The cross-section of a floating-gate avalanche MOS (FAMOS) device is shown in Fig. 19.34. It is a p-channel Si-gate MOS transistor in which a heavily doped polysilicon layer is inserted between two SiO_2 layers and is completely isolated from the Si-substrate. The SiO_2 thickness between the Si substrate and the polysilicon is on the order of 0.1 μm, while the SiO_2 covering the floating gate is much thicker. To charge the floating gate of the device, a voltage sufficiently large to cause avalanche breakdown of the drain-substrate junction is applied to the drain. Following the application of this voltage, a large number of electron–

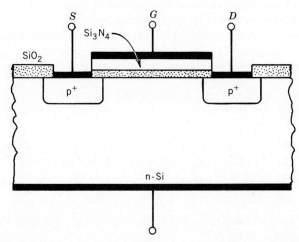

FIGURE 19.33 Cross-section of a metal-nitride-oxide-semiconductor (MNOS) memory device.

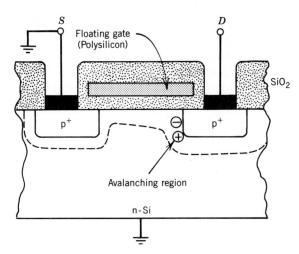

FIGURE 19.34 Side view of a floating gate avalanche MOS (FAMOS) device. (From D. Frohman-Bentchkowsky, FAMOS—A new semiconductor charge storage device, *Solid-State Electronics,* vol. 17, p. 518. Copyright © 1974. Reprinted by permission of Pergamon Press, Oxford.)

hole pairs are produced in the junction depletion region by avalanche multiplication. Some of the electrons generated in the avalanching depletion region acquire sufficient energy to surmount the Si-SiO$_2$ energy barrier and are swept by the electric field in the SiO$_2$ toward the polysilicon gate. This injection process is called *avalanche injection*.

Because of avalanche injection, the floating gate accumulates negative charge which is a function of the amplitude and duration of the drain voltage. After removal of the drain voltage, charge injection is stopped, but the negative charge on the floating gate is retained for several years since no discharge path is available for the accumulated electrons. The negative charge on the floating gate creates strong inversion in the channel. Thus, while the uncharged gates are normally off, the charged gates are on, and the presence or absence of the charge on the floating gate can be sensed by measuring the conductance between the source and the drain regions. Erasure of FAMOS cannot be performed electrically because the gate is not electrically accessible. However, the initial equilibrium condition of no charge on the floating gate can be restored by exposing the device to UV or X-ray radiation for several minutes.

Charge removal by electrical means is possible if a second polysilicon gate is used on the top of the SiO$_2$ layer. By application of a large positive voltage to the external gate, the stored charge can be removed and the device can be brought back to its initial state of no charge at the floating gate.

REFERENCES

1. P. E. GISE and R. BLANCHARD, *Modern Semiconductor Fabrication Technology,* Prentice-Hall, Englewood Cliffs, N.J., 1986.

2. J. C. C. TSAI, "Diffusion," Chapter 7 in *VLSI Technology,* 2d, ed., S. M. Sze (ed.), McGraw-Hill, New York, 1988.

3. B. G. STREETMAN, *Solid State Electronic Devices,* Prentice-Hall, Englewood Cliffs, N.J., 1979.

4. H. KRESSEL and J. K. BUTLER, *Semiconductor Lasers and Heterojunction LEDs,* Academic Press, New York, 1977.

5. P. A. H. HART, "Bipolar transistors and integrated circuits," Chapter 2 in *Handbook on Semiconductors,* T. S. Moss and C. Hilsum (eds.), Vol. 4, North Holland, Amsterdam, 1981.

6. A. G. MILNES, *Semiconductor Devices and Integrated Electronics,* Van Nostrand, New York, 1980.

7. Y. NISHI and H. IIZUKA, "Nonvolatile memories," in *Applied Solid State Science,* Suppl. 2A, D. Kahng (ed.), Academic Press, New York, 1981.

ADDITIONAL READING LIST

1. S. M. SZE (ed.), *VLSI Technology,* 2d, ed., McGraw-Hill, New York, 1988.
2. S. K. GHANDHI, *VLSI Fabrication Principles,* John Wiley, New York, 1983.
3. S. M. SZE, *Semiconductor Devices: Physics and Technology,* John Wiley, New York, 1985.
4. M. J. B. BOLT and J. G. SIMMONS, "The conduction properties of SIPOS," *Solid-State Electron.* 30, 533–542 (1987).

PROBLEMS

19.1 A Si crystal is to be pulled from the melt and is doped with phosphorus ($k_o = 0.35$). How many grams of phosphorus should be added to 1 kg of Si to obtain a donor concentration of 10^{16} cm^{-3} during the initial growth?

19.2 A 20 gm Si crystal containing 1 part in 10^4 of arsenic ($k_o = 0.27$) is melted in a crucible and is repulled on a small seed. The resulting crystal is to have an average arsenic content of 5 parts in 10^5. What fraction of the melt must be discarded to achieve the desired result?

19.3 A uniformly doped *n*-type Si sample of 1 Ω-cm resistivity is subjected to boron diffusion at a temperature of 1150°C with a constant surface concentration of 2×10^{19} cm^{-3}. How long should the diffusion be carried out to obtain a junction depth of 4 µm?

19.4 After a predeposition step, it is found that 7.5×10^{15} phosphorus atoms cm^{-2} are introduced in a *p*-type silicon sample doped with 2×10^{16} acceptor atoms cm^{-3}. Calculate the junction depth when the drive-in diffusion is performed at 1200°C for two hours.

19.5 A 0.5 Ω-cm resistivity, 6-µm thick epitaxial layer is grown on a p-Si substrate of 10 Ω-cm resistivity. Compute the drive-in time for a 1250°C isolation diffusion using boron with a constant surface concentration of 10^{19} atoms cm^{-3}.

19.6 Boron is implanted into *n*-type Si at 300 KeV with an ion dose of 4.5×10^{13} cm^{-2}. Assume a Gaussian distribution with $R_p = 0.735$ µm and $\Delta R_p = 0.115$ µm. Calculate (a) the peak concentration, (b) the average resistivity, and (c) the sheet resistance of the layer. Assume $\mu_p = 200$ cm^2 V^{-1} sec^{-1}.

19.7 A Si sample has a surface area of 1 cm². Calculate the weight of 0.2 μm thick SiO_2 layer on this sample and the weight of the Si consumed during the growth of the SiO_2. Density of SiO_2 is 2.15 gm cm^{-3}.

19.8 Determine the SiO_2 thickness obtained after 50 min wet oxidation of Si at 1000°C followed by 2 h dry oxidation at 1200°C.

19.9 A Si sample is covered with a 0.2-μm thick SiO_2 layer. What will be the time required to grow additional 0.2-μm thick SiO_2 in dry O_2 at 1200°C?

19.10 The five different ways in which a bipolar transistor may be connected to form a monolithic diode are described in Sec. 19.9.2. Consider a n-p-n transistor with $I_{EO} = 1 \times 10^{-10}$ A, $I_{CO} = 3I_{EO}$, and $\alpha_N = 0.98$. If a forward current of 2 mA is required to flow through the diode, determine which of the five diodes will have the lowest forward voltage drop and calculate its value at 300 K.

19.11 The sheet resistance of a boron-diffused Si layer is 150 Ω/□, and the junction depth is 4 μm. Determine the total number of dopant atoms per unit area of the sample and the average resistivity of the diffused layer assuming $\mu_p = 250$ cm² V^{-1} sec^{-1}. A 3 KΩ resistor is to be fabricated using this diffusion with $W = 25$ μm. Determine the aspect ratio and design the resistor so as to maintain a nearly square shape.

19.12 The emitter-base junction of an Si n-p-n transistor is approximated as linearly graded with a grading factor of 10^{22} cm^{-4} and a built-in voltage of 0.8 V. Determine the area required to obtain a capacitance of 0.05 μF.

19.13 An Si p-n-p lateral transistor fabricated on a 30-μm thick epitaxial layer grown on a p-substrate has the following parameters: $x_j = 3$ μm, $W_B = 2.5$ μm, $W_e = 1.5$ μm, and $L_p = 12$ μm. Calculate h_{FE} of the transistor. Also calculate the value of h_{FE} when the epitaxial layer thickness is reduced to 5 μm.

19.14 Consider the composite p-n-p transistor circuit of Fig. 19.28b. If α_1 and α_2 denote the common base-current grains of the p-n-p and n-p-n units, respectively, determine α and h_{FE} for the composite unit and show that $\alpha \simeq \alpha_2$.

19.15 The insulating layer above the floating gate of an MNOS memory device has a dielectric constant of 25 and a thickness of 1200 Å. The same for the lower insulator are 3.5 and 100 Å, respectively. For both insulators the current density J is given by $J = \sigma \mathscr{E}$ where $\sigma = 10^{-7}$ Ω^{-1} cm^{-1}. Estimate the threshold voltage shift when a 15 V pulse is applied to the gate for 1.2 μsec.

20
SEMICONDUCTOR MEASUREMENTS

INTRODUCTION

Device fabrication presumes the availability of an appropriately doped semiconductor. Before the device processing begins, the starting material has to be characterized in terms of its basic electrical parameters. The parameters of main interest are the conductivity type (n or p) and the resistivity, mobility, carrier concentration, and excess carrier lifetime. In this chapter, laboratory techniques for measuring these parameters are presented. The methods described refer to measurements on bulk semiconductor samples. Measurements involving fabricated devices are not included. A comprehensive account of the various semiconductor measurements is given by W. R. Runyan [1].

20.1 CONDUCTIVITY TYPE

A simple method, which is widely used to determine the conductivity type of a semiconductor specimen, is the hot probe method. Figure 20.1 shows the schematic of the experimental arrangement. Two fine metal probes, preferably made of stainless steel or nickel, are placed on the semiconductor sample and a zero-centered galvanometer is connected between them. One of the probes is kept at room temperature, and the other is heated to about 80°C. It is convenient to use a miniature soldering iron tip as the hot probe. The hot probe heats the semiconductor immediately beneath it so that the kinetic energy of free carriers in this region is increased. Therefore, the carriers diffuse out of the hot region at a faster rate than they diffuse into this region from the adjacent low-temperature regions. If the semiconductor is n-type, electrons will move away from the hot probe leaving a positive charge region of donors, and the hot probe becomes positive with respect to the cold probe. The current then will flow from the hot probe to the

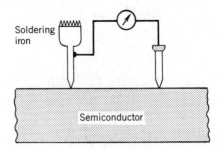

FIGURE 20.1 Experimental setup for the hot probe test.

cold probe. In a *p*-type semiconductor, the direction of the current flow is reversed. Thus, the polarity of the hot probe indicates whether the semiconductor is *n*- or *p*-type.

The hot probe method is not applicable to semiconductors in which the electron and hole concentrations are nearly equal. In such a case, any material that has a higher electron mobility will always be identified as *n*-type.

20.2 RESISTIVITY

20.2.1 Four-Probe Methods

The most commonly used technique in the semiconductor industry for measuring resistivity is the four-point probe method. This method is nondestructive and can be used to measure the resistivity of ingots, as well as wafers. The arrangement is shown in Fig. 20.2. Four collinear metal probes with sharpened tips are placed on the semiconductor. A constant current I is passed through the two outer probes, and the potential difference V developed across the inner two probes is measured using a high-input impedance voltmeter. In this way the metal–semiconductor contact resistance does not affect the results. For collinear probes with separation s_0 placed on a semi-infinite medium, the resistivity ρ is given by

$$\rho = 2\pi s_0 \frac{V}{I} \tag{20.1}$$

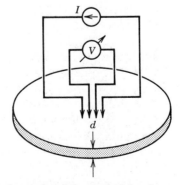

FIGURE 20.2 Schematic of a four-point probe method for the measurement of resistivity. (Reproduced with permission from "Semiconductors and Electronic Devices" by Adir Bar-Lev, Prentice Hall, 1979.)

This relation is also valid for a sample of finite dimensions, provided that the specimen thickness d is larger than $5s_0$, and all edges of the sample are away from the probes by a distance larger than $10s_0$.

If the sample is in the form of a wafer and its thickness is less than $5s_0$, a multiplicative correction factor F_1 is necessary. For the case when the sample is placed on a nonconducting surface, the factor F_1 is shown in Fig. 20.3 as a function of the ratio d/s_0. If the back of the wafer is in contact with a conducting surface, F_1 will be different. Another correction factor F_2 is needed when the edge of the sample is less than $10s_0$ away from the probes. For probes centered in a circular wafer of finite diameter $2r$, this correction factor is shown in Fig. 20.4. With the known values of F_1 and F_2, the true resistivity ρ is obtained from the relation

$$\rho = 2\pi s_0 \frac{V}{I} F_1 F_2 \tag{20.2}$$

It is evident from Fig. 20.3 that material deeper than $3s_0$ beneath the surface does not affect the measurements appreciably. Thus, when the probes are closely spaced, only a thin layer of the material is measured.

For the case when the wafer thickness d is small compared to s_0, and it is semi-infinite in the other two dimensions, Eq. (20.1) becomes

$$\rho = \frac{\pi}{\ln 2} \frac{V}{I} d = 4.53 \frac{V}{I} d \tag{20.3}$$

This relation is used to measure the resistivity of thin diffused layers in semiconductors that are isolated from the substrate by a p-n junction. Typical examples are the diffused layers obtained in the base and emitter diffusions of bipolar transistors. In these layers only the average resistivity is measured. Instead of average resistivity $\bar{\rho}$, one simply measures the layer's sheet resistance ρ_s defined by

$$\rho_s = \frac{\bar{\rho}}{d} = 4.53 \frac{V}{I} \tag{20.4}$$

Note that ρ_s represents the resistance of an arbitrary-size square of that layer and is expressed in ohm per square ($\Omega/\square$).

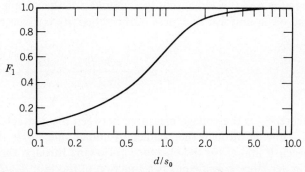

FIGURE 20.3 Correction factor F_1 for a thin wafer placed on a nonconducting surface as a function of d/s_0. (From G. Knight Jr., Measurement of semiconductor parameters, in *Handbook of Semiconductor Electronics*, L. P. Hunter (ed.), p. 20.4. Copyright © 1962. Reprinted by permission of McGraw-Hill, Inc., New York.)

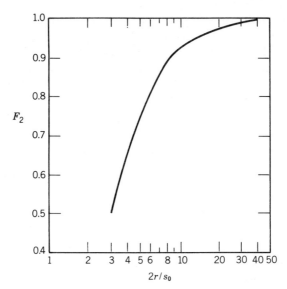

FIGURE 20.4 Correction factor F_2 for probes centered on a circular wafer of finite diameter $2r$ as a function of the ratio $2r/s_0$. (From F. M. Smits, Measurement of sheet resistivities with four-point probe, *Bell Syst. Tech. J.* vol. 37, part I, May 1958. Table III, p. 716. Reprinted with permission. Copyright © 1958 AT&T.)

The electrical circuitry for a four-point probe measurement scheme is simple and requires only a probe head, ammeter, voltmeter, and a constant current source. The measurements may be made using either a dc or an ac current. Contact resistance may become significant in the case of high-resistivity samples, and to lower its value, it is sometimes desirable to apply a forward bias to all the contacts [2]. A number of metals with high Young's modulus are suitable for making the probes, but silicon carbide (SiC) and tungsten are favored because of their hardness.

Another method of measuring resistivity using four contacts is the van der Pauw method which can be used to measure ρ for small and odd-shaped semiconductor samples. Figure 20.5 shows a semiconductor slice of arbitrary shape and thickness d. Four small contacts A, B, C, and D are made on the periphery of the sample. Two of these contacts are used to pass a current, and the voltage is measured across the remaining two contacts. If the sample is of uniform resistivity and has a uniform thickness d, the resistivity is given by [3]

$$\rho = \frac{\pi d}{2 \ln 2} \left(\frac{V_{DC}}{I_{AB}} + \frac{V_{BC}}{I_{AD}} \right) f \qquad (20.5)$$

where V_{DC} is the voltage measured between the contacts D and C for the current I_{AB} flowing through the contacts A and B, and V_{BC} is the voltage measured between B and C for the current I_{AD} flowing through the contacts A and D. The van der Pauw parameter f is a function of the ratio $V_{DC}I_{AD}/V_{BC}I_{AB}$ and is plotted in Fig. 20.6. The sheet resistance of the sample is obtained from Eq. (20.5) by writing $\rho_s = \rho/d$. The contacts in the van der Pauw method must be on the periphery, and their area should be very small.

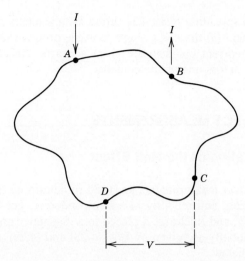

FIGURE 20.5 Top view of a sample of arbitrary shape with four contacts, A, B, C, and D made on the periphery.

20.2.2 Spreading Resistance

The resistivity of a semiconductor sample can also be obtained from the spreading resistance of a point-contact probe placed on the semiconductor. For a flat circular contact of radius a_0 uniformly affixed to the surface of a semi-infinite medium of resistivity ρ, the spreading resistance R_{sp} is given by

$$R_{sp} = \frac{\rho}{2a_0} \tag{20.6}$$

Instead of attempting to calculate R_{sp} from Eq. (20.6), one usually proceeds to construct a tip, then measures R_{sp} for several known resistivities of homogeneous samples, and prepares a plot of R_{sp} as a function of ρ. This plot is then used to determine the resistivity of an unknown material from the measured value of R_{sp}.

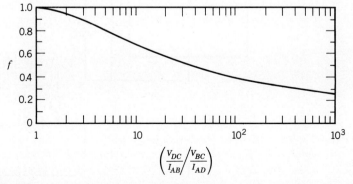

FIGURE 20.6 The correction factor f for the van der Pauw arrangement of Fig. 20.5. (From L. J. van der Pauw, reference 3, Fig. 7, p. 6. Copyright © Philips International B. V., Eindhoven, The Netherlands, 1958. Reprinted with permission.)

Two spreading resistance probe arrangements are shown in Fig. 20.7. In arrangement (a) there is a single movable probe, whereas in (b) both the input and output current contacts are point contacts. An additional probe measures the voltage across one of the contacts.

20.3 HALL EFFECT MEASUREMENTS

20.3.1 Interpretation of the Hall Effect

Hall effect measurements are used to determine the majority carrier type, concentration, and mobility in a semiconductor. For a rectangular bar of length l, width W, and thickness d placed in a magnetic flux B, the Hall constant and the Hall mobility are given by Eqs. (4.28) and (4.29), respectively, which are reproduced below

$$R_H = \frac{V_H d}{I_x B} \tag{20.7a}$$

$$\mu_H = \frac{l}{W} \frac{V_H}{V_x B} \tag{20.7b}$$

where V_H is the measured Hall voltage, V_x is the applied voltage along the length l, and I_x is the current. The sign convention for the rectangular geometry is shown in Fig. 20.8.

For a semiconductor with comparable concentrations of electrons and holes, R_H is given by Eq. (4.31)

$$R_H = \frac{(\mu_p^2 p_o - \mu_n^2 n_o)}{q(\mu_p p_o + \mu_n n_o)^2} \tag{20.8}$$

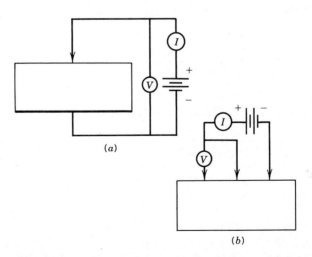

FIGURE 20.7 Two different spreading-resistance probe arrangements. (From W. R. Runyan, reference 1, p. 83. Copyright © 1975. Reprinted by permission of Texas Instruments.)

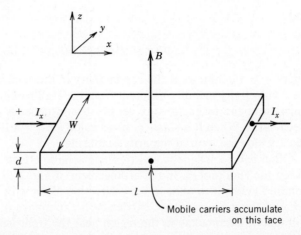

FIGURE 20.8 Sign convention and terminology for a rectangular Hall sample.

Also from Eqs. (4.29) and (4.30), it can be seen that

$$\mu_H = \left| \frac{\mu_p^2 p_o - \mu_n^2 n_o}{\mu_p p_o + \mu_n n_o} \right| \tag{20.9}$$

where n_o and p_o are the thermal equilibrium electron and hole concentrations, respectively. For a nearly intrinsic semiconductor, R_H obtained in Eq. (20.7a) is expressed by Eq. (20.8) and the measured Hall mobility in Eq. (20.7b) is represented by Eq. (20.9). However, for a strongly n-type semiconductor, we obtain

$$R_H = -\frac{1}{qn_o} \quad \text{and} \quad \mu_H = \mu_n \tag{20.10a}$$

and for a strongly p-type semiconductor

$$R_H = \frac{1}{qp_o} \quad \text{and} \quad \mu_H = \mu_p \tag{20.10b}$$

It is clear from these relations that for a material with comparable electron and hole concentrations the interpretation of R_H and μ_H is difficult. But if the material is strongly extrinsic, a measurement of R_H and μ_H determines the majority carrier concentration and the mobility, respectively.

In the case of a metal or a degenerate semiconductor the mobile carriers responsible for conduction are restricted to a narrow range of energy near the Fermi level. Therefore, virtually all of them may be assumed to have the same velocity, and Eqs. (20.10a) and (20.10b) correctly give the value of R_H. However, in a nondegenerate semiconductor, the mobile carriers have energy spanning the whole range from zero to several kT. Because of the energy dependence of carrier scattering, the drift velocity will be different for carriers with different energy ranges. As a consequence, the Hall effect is a function of an average over the velocities of the carriers, and R_H and μ_H are given by

$$R_H = \frac{-r_H}{qn_o}, \quad \text{or} \quad R_H = \frac{r_H}{qp_o} \tag{20.11a}$$

and

$$\mu_H = |R_H \sigma| = r_H \mu_C \qquad (20.11b)$$

where $\sigma = 1/\rho$ represents the conductivity of the semiconductor and μ_C denotes the conductivity mobility different from the Hall mobility. The parameter r_H is a function of the magnetic flux density B and depends on the carrier scattering mechanism. When B is small, r_H has a value $3\pi/8 = 1.18$ for the lattice scattering and 1.93 for the ionized impurity scattering.

Other galvanomagnetic and thermomagnetic effects can produce voltages between the Hall probes. In the arrangement of Fig. 20.8, we would expect a temperature gradient to exist in the y direction along with the Hall field $\mathcal{E}_y$. This is because the more energetic carriers will experience a greater $\mathbf{v} \times \mathbf{B}$ force and will tend to accumulate in the region near the front face of the bar and raise its temperature with respect to the region closer to the back surface. The resulting temperature gradient gives rise to a thermoelectric field in the y direction and results in an additional voltage along this direction that is distinct from the Hall voltage. This effect is called the *Ettingshausen effect.* The Ettingshausen voltage adds with the Hall voltage and, in some cases, may seriously interfere with the Hall voltage measurements.

In the case where there is a thermal gradient along the length of the sample, it will result in a net transport of more energetic carriers from the hot end to the cold end along the x direction, even when the bias current I_x is zero. The $\mathbf{v} \times \mathbf{B}$ field acting on these carriers will produce an electric field in much the same way as the Hall field is produced when the current I_x flows. This phenomenon is known as the *Nernst effect.* Just like the Hall field, the Nernst field is accompanied by a temperature gradient along the y direction and gives rise to an additional thermoelectric voltage analogous to the Ettingshausen voltage. The generation of a thermoelectric voltage due to the Nernst effect is called the *Righi Leduc effect.*

20.3.2 Measurement Procedure

The simplest arrangement for measuring the Hall voltage is the rectangular geometry shown in Fig. 20.8. As mentioned above, a number of spurious voltages are included in this measurement. Another voltage may be present when the probes are not exactly electrically opposite to each other. All these spurious voltages, except the Ettingshausen voltage, are eliminated if four readings are taken by reversing the direction of the bias current I_x and the magnetic flux B. The true Hall voltage is then obtained by taking the average of the four readings.

The contacts used for measuring the Hall voltage in Fig. 20.8 should be infinitesimally small so that they do not distort the current flow. The bridge-shaped sample shown in Fig. 20.9 is often used to reduce the distortion of the current. The ears on the sample allow a large area to be used for contacting the sample without severly distorting the lines of current flow through the specimen. The Hall voltage may be measured between contacts 1 and 2, and then between 3 and 4, with an average taken.

In some cases, it may not be convenient to cut the specimen in the form of a rectangular bar. A flat sample of arbitrary shape similar to the one shown in Fig. 20.5 can be used. The sample is placed in a magnetic field with the direction of the field normal to its surface. A current I_{AC} is passed between the two oppo-

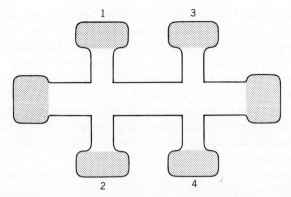

FIGURE 20.9 A bridge-shaped Hall sample with two pairs of ear contacts.

site contacts (A and C), and voltage is measured between contacts B and D as shown in Fig. 20.10a. The voltage V_{BD} is first measured without the magnetic field and then with the field, while the current I_{AC} is held constant for both the measurements. The magnitude and the sign of the change in voltage ΔV_{BD} caused by the magnetic field is then obtained, and the Hall constant is given by

$$R_H = \frac{\Delta V_{BD}}{I_{AC} B} d \qquad (20.12)$$

If the resistivity is also measured on the same sample using the procedure described in the last section, the Hall mobility is obtained from the relation

$$\mu_H = \frac{R_H}{\rho} \qquad (20.13)$$

When the contacts are of finite size and are not at the circumference of the sample, an error will arise and a correction factor will be needed. The influence of

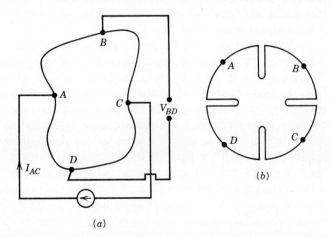

FIGURE 20.10 (a) Van der Pauw arrangement for measuring the Hall constant of an arbitrary shaped sample, and (b) a clover-shaped sample that eliminates the effect of contacts. (From van der Pauw, *Philips Res. Reports*, reference 3, Fig. 9, p. 8. Copyright © Philips International B.V., Eindhoven, The Netherlands, 1958. Reprinted by permission.)

the contact can be further eliminated by using the clover-shaped sample shown in Fig. 20.10b. This sample has several advantages compared to the bridge-type sample of Fig. 20.9. First, it gives a higher Hall voltage for the same amount of power dissipation, and second, it has greater mechanical strength. The proper sample shape is usually obtained by cutting the material with an ultrasonic machine or by chemical etching.

Magnetic fields used in Hall effect measurements in the laboratory environment are in the few kilogauss range (10^4 gauss = 1 weber/m^2). The sample current is usually measured with an ammeter, while the Hall voltage can be measured using a simple potentiometer and galvanometer. This simple equipment may fail because of pronounced noise, and the use of a differential amplifier is more desirable is such cases [4].

20.4 DRIFT MOBILITY

20.4.1 The Haynes-Shockley Experiment

The conductivity mobility obtained from the Hall voltage measurement is the mobility of majority carriers. The mobility obtained by measuring the rate of drift of carriers in an electric field is called the *drift mobility*. The classic Haynes-Shockley experiment directly measures the drift mobility of minority carriers in a semiconductor.

A schematic diagram of the experimental arrangement is shown in Fig. 20.11. It consists of a bar of an n-type semiconductor with ohmic contacts at the two ends. Two rectifying contacts (E and C) are also placed on the bar as indicated. Each of these contacts may be made by converting a small spot on the n-type semiconductor to the p-type. The contact E is used to inject minority carriers into the semiconductor, and C is used to collect them. The variable dc voltage V_1 provides the electric field for the drift of minority carriers. A pulsed voltage source is connected between E and the ground, and the contact C is reverse biased through a resistance R_1. The oscilloscope connected across R_1 is synchronized with the pulse generator.

A sharp positive pulse is applied to the contact E. As soon as the pulse is delivered, a positive pulse appears on the oscilloscope because of the voltage-dividing action of the bar. Thus, the initiation time of the pulse can be noted from the oscilloscope. The positive pulse at E causes injection of holes into the semiconductor, and to maintain the space-charge neutrality an equal number of electrons is drawn through the ohmic contact at the end of the bar. The electron–hole cloud initiated around E moves toward the collector contact C. As these carriers drift down the bar, the total number of carriers in the pulse decreases because of electron–hole pair recombination. In addition, holes diffuse to both sides of the original pulse, and the rectangular pulse spreads into a bell-shaped curve as shown in the lower two sketches in Fig. 20.11b. When the pulse passes under the contact C, holes are collected by the contact causing a current in the resistor R_1. The voltage drop across R_1 as a function of time is displayed on the oscilloscope as a bell-shaped curve.

The arrival time of the center of the pulse determines the mobility, and the pulse spread can be used to calculate the diffusion coefficient of the minority carriers. The continuity equation of injected holes into the n-type bar can be

DRIFT MOBILITY

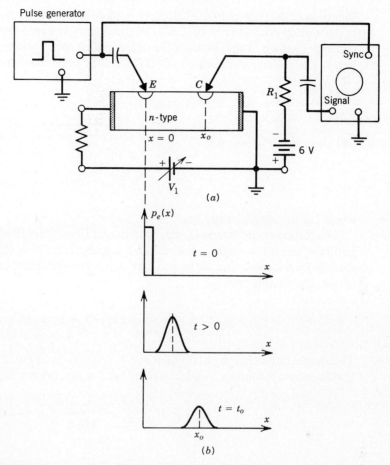

FIGURE 20.11 The Haynes-Shockley experiment: (a) schematic diagram and (b) development of minority carrier pulse during its drift toward the collector point C.

written as (see Eq. 5.61)

$$\frac{\partial p_e}{\partial t} = -\frac{p_e}{\tau_p} - \mu_p \mathscr{E} \frac{\partial p_e}{\partial x} + D_p \frac{\partial^2 p_e}{\partial x^2} \qquad (20.14)$$

Neglecting recombination and putting $\mathscr{E} = 0$, we obtain the solution of Eq. (20.14):

$$p_e(x,t) = \frac{P_o}{2\sqrt{\pi D_p t}} \exp\left(-\frac{x^2}{4D_p t}\right)$$

where P_o represents the number of injected holes per unit area of the surface and t is time. When an electric field $\mathscr{E}$ is applied, the pulse travels along the positive x axis with a velocity $v_d = \mu_p \mathscr{E}$, and the corresponding solution is obtained by writing $(x - v_d t)$ for x. Moreover, because of recombination, the carrier concentration decreases with time as $\exp(-t/\tau_p)$. It is thus clear that an appropriate solution of Eq. (20.14) is

$$p_e(x,t) = \frac{P_o}{2\sqrt{\pi D_p t}} \exp\left[-\frac{(x - v_d t)^2}{4D_p t} - \frac{t}{\tau_p}\right] \qquad (20.15)$$

Referring to the plots of Fig. 20.11b, let $t = 0$ be the time when the pulse is applied at the emitter contact ($x = 0$). Let x_o be the spacing between the emitter and the collector contacts, and t_o be the time of the arrival of the pulse center at the collector. The average drift velocity of holes is then obtained from Eq. (20.15) as

$$v_d = \mu_p \mathscr{E} = \frac{x_o}{t_o}$$

and since $\mathscr{E} = V_1/l$, we get

$$\mu_p = \frac{l x_o}{V_1 t_o} \tag{20.16}$$

where l is the length of the bar.

The diffusion coefficient D_p can be obtained from the half-width of the distribution at which the height is half its peak amplitude at $t = t_o$ (Fig. 20.12a). From Eq. (20.15) it is clear that at the collector point, the maximum occurs at the time $t_o = x_o/v_d$. At this time

$$p_e(x_o, t_o) = \frac{P_o}{2\sqrt{\pi D_p t_o}} \exp(-t_o/\tau_p) \equiv H_o$$

The shape of the charge distribution at $t = t_o$ is shown in Fig. 20.12a. The curve is symmetrical around x_o, and its height $H(x)$ at any point x is given by

$$H(x) = H_o \exp\left[-\frac{(x - x_o)^2}{4 D_p t_o}\right] \tag{20.17}$$

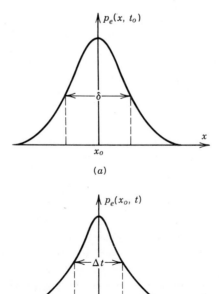

(a)

(b)

FIGURE 20.12 The excess holes pulse (a) as a function of distance x at $t = t_o$ and (b) as a function of time at $x = x_o$.

The ratio H/H_o becomes one-half at a width $\delta/2$ such that

$$\frac{\delta^2}{16 D_p t_o} = \ln 2 = 0.69 \tag{20.18}$$

The display on the oscilloscope is not a plot of the concentration versus distance centered around x_o, but rather a plot of concentration as a function of time as the charge distribution passes beneath the collector. When the drift field is sufficiently large, the pulse shape displayed by the oscilloscope will be nearly symmetrical about $t = t_o$ as shown in Fig. 20.12b. The width δ is then obtained from the relation $\delta = v_d \Delta t$ where Δt is the time interval between the two points in Fig. 20.12b at which the height is half its peak value. Thus, substituting for δ in Eq. (20.18) and writing $v_d = x_o/t_o$, we obtain

$$D_p \simeq \frac{(x_o \Delta t)^2}{11 t_o^3} \tag{20.19}$$

Equations (20.16) and (20.19) enable us to verify the Einstein relation for holes. The minority carrier lifetime τ_p can be obtained from the decrease in the pulse area with time. The decrease in the pulse amplitude with time is due primarily to carrier recombination. Thus, the amplitude of the pulse also decreases exponentially with time and can be directly used to determine τ_p.

Considerable error in the determination of drift time can arise if the proper conditions for the measurement are not established. The drift mobility obtained from Eq. (20.16) is correct only as long as the diffusion spread of the pulse remains reasonably small during the drift of holes from the contact E to the collector at C. This condition requires the use of relatively large drift fields that may lead to serious heating effects. A solution to this problem is to apply the drift voltage in the form of pulses, with pulse width substantially larger than the drift time of minority carriers.

The initial Haynes-Shockley experiment was performed on low doped Ge samples that had excess carrier lifetime in the millisecond range. Thus, the emitter-to-collector spacing could be made on the order of 1 cm. Semiconductors with low minority carrier lifetime require smaller emitter-to-collector spacings, and preparation of the sample becomes difficult in such cases. A structure for performing the Haynes-Shockley experiment on Si, fabricated using planar technology, is described by S. Middelhoek and M. J. Geerts [5]. Figure 20.13 shows a cross-sectional view of the structure. It consists of an n-silicon substrate to which two n^+-ohmic contacts are made. Between these contacts, one elongated p-type emitter and three elongated p-type collectors are placed by standard diffusion process. Since the minority carrier lifetime in Si may be shorter than 1 μsec, the distance between the emitter and the collector is only 100 μm. This small distance requires narrow emitter and collector contacts.

Although the Haynes-Shockley experiment leads to an accurate measurement of minority carrier mobility, the same cannot be said about the diffusion coefficient. Because of the carrier recombination and trapping at the surface and because diffusion occurs in three dimensions, the measurement of diffusion coefficient in this experiment is not very reliable.

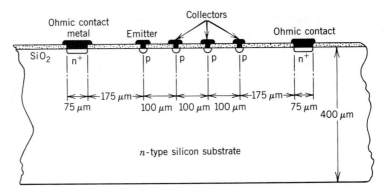

FIGURE 20.13 Cross-section of the planar silicon structure for the Haynes-Shockley experiment. (From S. Middelhoek and M. J. Geerts, reference 5, p. 32. Copyright © 1978 IEEE. Reprinted by permission.)

20.4.2 The Time of Flight Method

In this method, the carrier drift velocity is determined by measuring the transit time of the carriers through a region of uniform electric field that extends across a sample of known width. Figure 20.14a shows a typical structure that uses an n-type semiconductor. The device is fabricated by making a Schottky barrier contact at one end and producing an n^+-region at the other end of the semiconductor to which an ohmic contact is made. A similar structure with one rectifying and the other p-p$^+$ contact can also be made on a p-type semiconductor.

The n-region in Fig. 20.14a is completely depleted by applying a large reverse bias to the left-hand side contact. When the bias voltage is large compared to the punch-through voltage, the field in the whole region becomes nearly constant. A very short pulse of high energy (1-10 KeV) electrons from an electron gun now bombards the sample through the reverse-biased contact. The electron bunch dis-

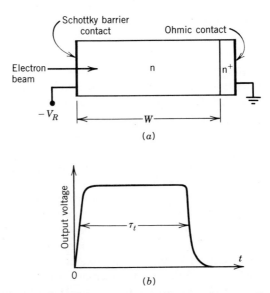

FIGURE 20.14 The time-of-flight method for measuring the drift velocity of carriers: (a) schematic cross-section and (b) the output voltage waveform.

sipates its energy by creating secondary electron–hole pairs in the n-region close to the rectifying contact. The generated holes are collected at the metal contact while the electrons drift toward the n$^+$-region. The drift of electrons into the n-region induces a current in the external circuit that continues to flow until the electrons are collected at the n$^+$-contact. Consequently, the duration of the current pulse is essentially equal to the transit time of the electrons through the n-region. The specimen is mounted in a coaxial sample mount, and voltage proportional to the induced current is observed directly on a sampling oscilloscope from which the transit time is determined [6]. A typical output waveform may look like the one shown in Fig. 20.14b. The field-dependent transit time $\tau_t(\mathscr{E})$ is obtained as the half-amplitude width of the curve shown in the figure, and the electron drift velocity is given by

$$v_d(\mathscr{E}) = \frac{W}{\tau_t(\mathscr{E})} \tag{20.20}$$

where W is the width of the drift region. The same structure can be used to determine the hole drift velocity if the n$^+$-face is exposed to the pulsed beam. In that case, the secondary electrons are collected at the n$^+$-contact, but the holes drift to the other contact and induce a current in the external circuit. This technique has been widely used to measure the electron and hole drift velocities as a function of electric field in Si and GaAs.

20.5 MINORITY CARRIER LIFETIME

Approaches that have been reported in the literature for measuring the excess carrier lifetime can be classified into two categories. In one approach, a suitable test sample of n- or p-type conductivity is used to observe the decay rate of excess carriers. The other approach is to employ a finished device like a p-n junction diode or a MOS capacitor, and to correlate some device parameters with the lifetime. In this chapter we are concerned mainly with lifetime measurements in bulk material. Among all the methods, the photoconductive decay method is the simplest one and is widely used.

20.5.1 Photoconductive Decay

In this method, the excess electron–hole pairs are produced by irradiating the specimen with light of suitable wavelength. The conductance of the sample is directly proportional to the number of carriers and is increased by absorption of the light. When the light is removed, the electron–hole pairs decay because of recombination, and the excess conductance $\Delta G(t)$ decays exponentially with time

$$\Delta G(t) = \Delta G(0) \exp(-t/\tau) \tag{20.21a}$$

or

$$G(t) - G_o = [G(0) - G_o] \exp(-t/\tau) \tag{20.21b}$$

where $G(0)$ is the initial conductance of the sample at $t = 0$, $G(t)$ is the conductance at any time t, G_o is the sample conductance in the absence of light, and τ is the excess carrier lifetime. A constant current is passed through the sample which is equipped with ohmic contacts at the two opposite faces, and the decay of excess conductance given by Eq. (20.21b) is measured by monitoring the voltage drop across the sample. For a specimen operated under constant current, the voltage $V(t)$ is inversely proportional to the conductance and can be written as

$$\frac{1}{V(t)} - \frac{1}{V_o} = \left[\frac{1}{V(0)} - \frac{1}{V_o}\right] \exp(-t/\tau)$$

Here $V(0)$ is the voltage at $t = 0$ and V_o represents the voltage drop across the nonilluminated sample. When we write $\Delta V(0) = V_o - V(0)$, the above relation becomes (see Prob. 20.10)

$$V(t) = V_o - \frac{\Delta V(0) \exp(-t/\tau)}{1 - \frac{\Delta V(0)}{V_o}[1 - \exp(-t/\tau)]} \quad (20.22\text{a})$$

or

$$\Delta V(t) = \frac{\Delta V(0) \exp(-t/\tau)}{1 - \frac{\Delta V(0)}{V_o}[1 - \exp(-t/\tau)]} \quad (20.22\text{b})$$

where $\Delta V(t) = V_o - V(t)$. The decay rate of the voltage $\Delta V(t)$ defines a time constant τ_v such that $d\Delta V/dt = -(\Delta V(t)/\tau_v)$, and it can be shown (see Prob. 20.11) that if $\Delta V(0)$ is small compared to V_o, the lifetime τ is related to τ_v by the relation

$$\tau = \tau_v\left[1 - \frac{\Delta V(0)}{V_o}\right] \quad (20.23)$$

Note that when $\Delta V(0)/V_o$ is negligible compared to unity, the voltage $\Delta V(t)$ in Eq. (20.22b) decays exponentially with time and $\tau = \tau_v$.

Figure 20.15 shows the schematic circuit arrangement for measuring photoconductance decay [7]. The light source should be of appropriate wavelength to produce electron–hole pairs in the semiconductor. A filter made of the same semiconductor as the specimen is often interposed between the light source and the sample. The filter eliminates the short wavelength photons that are too strongly

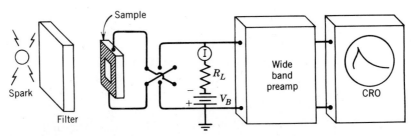

FIGURE 20.15 Schematic circuit arrangement for photoconductance decay measurement. (From IRE standards on measurement of minority-carrier lifetime, reference 7, p. 1296. Copyright © 1961 IEEE. Reprinted by permission.)

absorbed by the semiconductor and create electron–hole pairs near the front surface, causing enhanced recombination at this surface. The region surrounding the end contacts is shielded from the light, and masking strips are placed across the sample as shown in Fig. 20.15. Use of these strips minimizes the transport of carriers to the contacts before normal recombination can occur. The constant current source may be either a battery or a low-noise power supply in series with a resistance R_L whose value should be at least 20 times higher than the sample resistance in the dark. The photoconductive decay pattern should be observed for both polarities by reversing the direction of the sample current. The oscilloscope may be directly triggered by the light source.

The time constant of the voltage decay is obtained from the oscilloscopic display. An exponential curve may be drawn on the oscilloscope screen, and the variable sweep can be used to match the voltage decay curve to this curve. Alternatively, the decay rate can be obtained by generating a second curve from a photocell and a variable but known RC time constant, and by matching this curve to the photo decay curve using a dual-trace oscilloscope.

Figure 20.16a shows a typical voltage decay curve. The initial faster decay is caused by higher order modes associated with surface recombination and must be ignored. Consequently, no measurement of the decay should be made until the voltage ΔV has decayed to less than 60 percent of its peak value.

When trapping is present, excess electrons and holes will contribute to conductance in a manner that will change as trapping proceeds. While one type of excess carriers is trapped, the other type remains in circulation and contributes to a

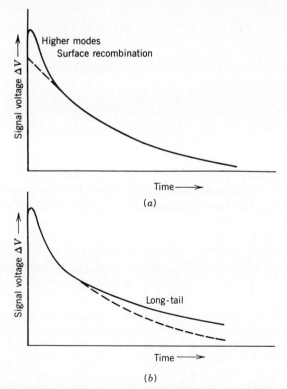

FIGURE 20.16 Oscilloscope display of photoconductance decay. (a) Exponential decay and (b) long-tail decay caused by trapping of minority carriers.

long tail in photoconductance as shown in Fig. 20.16b. This tail declines slowly as the traps are finally emptied. Trapping will disappear in the presence of a steady background light which will empty the traps.

The shortest lifetime that can be measured using photoconductive decay is limited by the fall time of the light source. However, the maximum measurable lifetime is determined by the size of the sample. For a sample of finite dimensions, the excess carriers will diffuse to the surface and will recombine there. The surface recombination can seriously affect the interpretation of the photoconductive decay curves. Therefore, it is recommended to calculate the surface contribution or use a large enough sample to make surface recombination negligible in comparison with the bulk recombination.

In order to calculate the effect of surface recombination, let us first consider a semiconductor sample of large surface area, but of a small thickness d such that the surface recombination is important only on the surfaces normal to d. Let the sample be uniformly illuminated, and let the light be switched off at $t = 0$. Assuming the semiconductor to be n-type, we can write the continuity equation for excess holes as

$$\frac{\partial p_e}{\partial t} = -\frac{p_e}{\tau} + D_c \frac{\partial^2 p_e}{\partial x^2} \qquad (20.24)$$

where x represents the distance along the sample thickness and D_c is the carrier diffusion coefficient. To solve Eq. (20.24), we use the method of separation of variables and write

$$p_e(x, t) = X(x)T(t) \qquad (20.25)$$

Substituting Eq. (20.25) into Eq. (20.24) and separating the variables lead to the following equation:

$$\frac{1}{X}\frac{d^2 X}{dx^2} = \frac{1}{D_c T}\left(\frac{dT}{dt}\right) + \frac{1}{D_c \tau} \equiv -m^2 \qquad (20.26)$$

The left-hand side of Eq. (20.26) is a function of x alone, while the right-hand side is a function of t alone. This is possible only when each side is constant equal to $-m^2$. The two equations can be solved for X and T, and it can be shown (see Prob. 20.13) that

$$p_e(x, t) = [A_1 \cos mx + B_1 \sin mx] \exp\left[-\left(\frac{1}{\tau} + m^2 D_c\right)t\right]$$

where A_1 and B_1 are constants of integration. By the symmetry of the problem, it is clear that the spatial dependence of $p_e(x)$ must be an even function of x. Consequently, $B_1 = 0$, and we obtain

$$p_e(x, t) = A_1(\cos mx) \exp\left[-\left(\frac{1}{\tau} + m^2 D_c\right)t\right] \qquad (20.27)$$

MINORITY CARRIER LIFETIME

If we take the origin at the center of the thickness d, then Eq. (20.27) is required to satisfy the following boundary condition at the surface $x = d/2$:

$$J_p(d/2) = -qD_c\left(\frac{\partial p_e}{\partial x}\right)\bigg|_{x=d/2} = qSp_e(d/2)$$

or

$$\frac{\partial p_e}{\partial x}\bigg|_{x=d/2} = -\frac{S}{D_c}p_e\left(\frac{d}{2}\right) \tag{20.28}$$

where S represents the surface recombination velocity. Substituting for p_e from Eq. (20.27) into Eq. (20.28) we obtain

$$md\left(\tan\frac{md}{2}\right) = \frac{S}{D_c}d \tag{20.29}$$

This equation has an infinite number of solutions for the various values of S. In the limit when S becomes infinitely large, we can write $md/2 = (2n+1)\pi/2$ or

$$m = (2n+1)\pi/d \tag{20.30}$$

where n has values 0, 1, 2, and so on. From Eqs. (20.27) and (20.30) an effective time constant τ_{eff} can be defined so that

$$\frac{1}{\tau_{\text{eff}}} = \frac{1}{\tau} + m^2 D_c = \frac{1}{\tau} + \frac{(2n+1)^2\pi^2 D_c}{d^2}$$

The higher order modes corresponding to $n > 0$ will decay rapidly leaving a single exponential given by

$$\frac{1}{\tau_{\text{eff}}} = \frac{1}{\tau} + \frac{\pi^2 D_c}{d^2} \tag{20.31}$$

For a rectangular sample of sides b, c, and d, the effect of surface recombination on the remaining surfaces can be calculated in a similar manner, and the lifetime τ is given by the relation

$$\frac{1}{\tau} = \frac{1}{\tau_{\text{eff}}} - \pi^2 D_c\left(\frac{1}{b^2} + \frac{1}{c^2} + \frac{1}{d^2}\right) \tag{20.32}$$

Here the diffusion coefficient D_c is the ambipolar diffusion coefficient given by Eq. (5.62), and reduces to the minority carrier diffusion coefficient D_p (or D_n) for strongly extrinsic material. Note that Eq. (20.32) is based on the assumption of an infinite surface recombination velocity. Consequently, the surface of the sample should be sandblasted to obtain the highest possible value of S. Similar corrections have also been calculated for samples of circular cross-sections [1].

20.5.2 Other Methods

(a) PHOTOCONDUCTIVITY MODULATION

In this method, sinusoidal light excitation is used instead of the light pulses, and the phase angle θ between the photoconductivity of the sample and the incident light is interpreted in terms of an effective lifetime τ_{eff} obtained from the relation

$$\tan \theta = \omega \tau_{eff} \tag{20.33}$$

The angular frequency ω of the excitation is chosen such that $\omega \tau_{eff} < 0.3$. To measure the phase angle θ, the intensity-modulated light falls on both the sample and a detector that introduces negligible phase shift between the incident light and the electrical signal. When the frequency is relatively low, the signal from a photocell can be processed through a parallel RC network and compared with the signal from the sample. The amplitude of the signal and the RC time constant are now so adjusted that the two signals are in phase, and a null occurs in the detector for $RC = \tau_{eff}$.

(b) LUMINESCENCE DECAY AFTER EXCITATION FROM A LASER PULSE

In this method the specimen is excited with a laser pulse which creates electron–hole pairs in the semiconductor. After the pulse is terminated, the excess carriers decay by nonradiative recombination as well as by radiative recombination. The intensity of the emitted radiation is proportional to the concentration of excess carriers. Thus, by monitoring the luminescence as a function of time, one determines the decay rate of excess carriers from which the lifetime is obtained. The method requires no electrical connection and can be used to measure lifetime over a wide range starting from milliseconds to a fraction of a nanosecond [8].

20.6 DIFFUSION LENGTH

The minority carrier lifetime can also be obtained from the measured diffusion length. Diffusion length is generally determined from steady-state measurements in which some parameter related to steady-state concentration of excess carriers is measured [9]. A commonly used measurement employs geometries in which the distance between the excitation source and the detector can be varied at will. In recent years, the electron beam of a scanning electron microscope (SEM) has been used to generate excess electron–hole pairs in the semiconductor. The electron beam has the advantage of providing a very small spot size and allows the measurement of very short diffusion length. For studying recombination in semiconductors, the most widely used mode of operation of the SEM is the electron beam induced current (EBIC) mode. In this mode of operation, the electron beam produces excess electron–hole pairs in the semiconductor near the surface. The generated minority carriers then diffuse to a potential barrier of a p-n junction or of a Schottky barrier contact and, after collection, produce a current that reflects the number of carriers collected by the potential barrier.

Figure 20.17 shows two configurations that can be used for diffusion-length measurements in semiconductors [8]. The simple configuration in Fig. 20.17a assumes a semi-infinite sample and a semi-infinite Schottky barrier contact. The electron beam is incident at a distance r' from the contact. It has been shown that

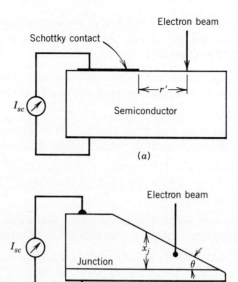

FIGURE 20.17 Two different configurations for the measurement of minority carrier diffusion length using SEM in EBIC mode. (After M. S. Tyagi and R. Van Overstraeten, reference 8, p. 585.)

for $r' \gg L_c$, the short-circuit current I_{sc} varies as $(1/r'^{3/2}) \exp(-r'/L_c)$, provided that the penetration depth of the beam is small compared to r' and the surface recombination velocity is infinite. Thus, by measuring I_{sc} as a function of r', it is possible to calculate the diffusion length L_c.

A method that increases the accuracy of measurements is the use of the angle-lapped structure shown in Fig. 20.17b. A shallow p-n junction is produced in the semiconductor by diffusion of an appropriate dopant. The structure is then lapped through an angle θ, and ohmic contacts are attached on the two sides of the junction. An electron beam normal to the junction plane excites electron–hole pairs in the semiconductor and produces a short-circuit current I_{sc}. Let us confine our attention to the case where the semiconductor is uniformly doped, so that L_c becomes independent of position and the surface recombination velocity is uniform over the surface. Moreover, we keep $\theta < 10°$ and $x_j > 3L_c$. When these conditions are satisfied, the current I_{sc} decreases exponentially with x and the diffusion length is obtained from the slope of $\ln I_{sc}$ versus x plot giving

$$L_c = \frac{dx}{d(\ln I_{sc})} \sin \theta \qquad (20.34)$$

Since the angle θ is small, and L_c is very sensitive to θ, the angle must be known accurately. Diffusion lengths as short as 0.2 μm have been measured using this method.

REFERENCES

1. W. R. RUNYAN, *Semiconductor Measurements and Instrumentation*, McGraw-Hill, New York, 1975.

2. C. C. ALLEN and W. R. RUNYAN, "An ac silicon resistivity meter," *Rev. Sci. Instr.* 32, 824 (1961).

3. L. J. VAN DER PAUW, "A method of measuring specific resistivity and Hall effect of discs of arbitrary shape," *Philips Res. Reports* 13, 1 (1958). Also *Philips Tech. Rev.* 20, 220 (1959).

4. G. H. HEILMEIER, G. WARFIELD, and S. E. HARRISON, "Measurement of the Hall effect in metal-free phthalocyanine crystals," *Phys. Rev. Lett.* 8, 309 (1962).

5. S. MIDDELHOEK and M. J. GEERTS, "The Haynes-Shockley experiment with silicon planar structures," *IEEE Trans. Education* E-21, 31 (1978).

6. C. B. NORRIS and J. F. GIBBONS, "Measurement of high-field carrier drift velocities in silicon by a time-of-flight technique," *IEEE Trans. Electron Devices* ED-14, 38 (1967).

7. Institution of Radio Engineers, "IRE Standards on measurement of minority-carrier lifetime," *Proc. of the IRE.* 49, 1292 (1961).

8. M. S. TYAGI and R. VAN OVERSTRAETEN, "Minority carrier recombination in heavily-doped silicon," *Solid-State Electron.* 26, 577 (1983).

9. T. WILSON and P. D. PESTER, "An analysis of the photoinduced current from a finely focused light beam in planar p-n junctions and Schottky barrier diodes," *IEEE Trans. Electron Devices* ED-34, 1564 (1987).

ADDITIONAL READING LIST

1. P. F. KANE and G. B. LARRABEE, *Characterization of Semiconductor Materials,* McGraw-Hill, New York, 1970.

2. A. G. R. EVANS and P. N. ROBSON, "Drift mobility measurements in thin epitaxial semiconductor layers using time-of-flight techniques," *Solid-State Electron.* 17, 805–812 (1974).

3. T. S. MOSS (ed.), *Handbook on Semiconductors,* Vol. 3, *Materials, Properties and Preparation,* North-Holland, Amsterdam, 1981.

PROBLEMS

20.1 The four-point probe readings for a large and very thick n-type Si sample are $V = 10$ mV and $I = 1$ mA. The spacing between the collinear probes is 1 mm. Determine the resistivity and dopant concentration of the sample.

20.2 A 0.2-mm thick semiconductor wafer has a diameter of 2.5 cm and is placed on an insulating plate. A four-point probe measurement made in the center of the wafer has $V/I = 80\ \Omega$. The spacing between the probes is 0.5 mm. Determine the resistivity and the sheet resistance of the wafer. What will be the diameter of the sample for which the measurement made at the center will give $V/I = 115\ \Omega$?

20.3 Consider an arbitrary shaped p-type Si sample of 0.15 mm thickness with four contacts A, B, C, and D made on the periphery as shown in Fig. 20.5. If $V_{DC}/I_{AB} = 20\ \Omega$ and $V_{BC}/I_{AD} = 100\ \Omega$, determine the resistivity and the sheet resistance of the sample.

PROBLEMS

20.4 A noninjecting metal probe of 1 mm diameter is placed on a plane surface of a semiconductor of 25 Ω-cm resistivity. The outer surface of the semiconductor has an ohmic contact. A 2 V battery is connected between the probe and the ohmic contact. Neglecting the voltage drop across the metal and the ohmic contact, determine the current flowing through the circuit.

20.5 The hole mobility in the sample of Prob. 20.3 is found to be 400 cm^2/V-sec. The sample is placed in a magnetic field of 5 K Gauss normal to the sample surface, and a current $I_{AC} = 5$ mA is passed. Determine the magnitude of the Hall voltage developed across the terminals B and D.

20.6 Hall measurement is made on a bridge-type sample similar to the one shown in Fig. 20.9. The following data are obtained: the length of the sample is 2.5 cm, the distance between the two Hall probes is 0.5 cm, $W = 0.2$ cm, $d = 0.1$ cm, and $I_x = 5$ mA. The voltage between probes 1 and 3 is $V_{13} = 0.125$ V and $B = 1$ Wb-m^{-2}. The average value of Hall voltage $V_{12} = 2.3$ mV and $V_{34} = 2.4$ mV. Determine (a) the type and concentration of majority carriers, (b) the Hall mobility, and (c) the conductivity mobility assuming $r_H = 1.18$.

20.7 In the arrangement of the Haynes-Shockley experiment, the mobility is measured on an n-type Ge rod with a resistivity of 10 Ω-cm and $\tau_p = 5$ μsec. The emitter and collector contacts are 1 cm apart, and the drift field is 5 V/cm. Determine the pertinent parameters and find out if this will be a good choice of the experimental setup.

20.8 A sample of n-type Ge is used in the Haynes-Shockley experiment with the following data: $l = 1$ cm, $x_o = 0.8$ cm, $V_1 = 2$ V, $t_o = 2.2 \times 10^{-4}$ sec, and $\Delta t = 95$ μsec. Calculate μ_p, D_p, and the sample temperature assuming the validity of the Einstein relation.

20.9 The following data are obtained from the Haynes-Shockley experiment at 300 K: $l = 2$ cm, $x_o = 1.2$ cm, $V_1 = 10$ V, $t_o = 150$ μsec, and $\Delta t = 33$ μsec. Calculate the mobility and the diffusion constant and determine if these data satisfy the Einstein relation.

20.10 Starting from Eq. (20.21b) and assuming a constant current I through the sample, obtain Eq. (20.22a).

20.11 Determine the time constant τ_v for the voltage decay $\Delta V(t)$ in Eq. (20.22b) and show that when $\Delta V(0)$ is small compared with V_o the lifetime τ is related to τ_v by Eq. (20.23).

20.12 In a typical lifetime measurement setup, the dark resistance of the sample is 200 Ω and the resistance of the illuminated sample just before removing the light at $t = 0$ is 180 Ω. Assume a constant sample current of 2 mA, and the time constant for the decay of the voltage $\Delta V(t)$ is $\tau_v = 0.5$ μsec. Calculate the bulk lifetime and the sample voltage 0.3 μsec after removal of the light. Neglect surface recombination.

20.13 Solve Eq. (20.26) for X and T, and using the symmetry of the problem, obtain Eq. (20.27).

20.14 A semiconductor wafer of thickness d has a large surface area and a bulk lifetime of 1 μsec. Determine the value of d for which the measured lifetime is 10 percent lower than the bulk lifetime when (a) $S = 10^6$ cm/sec and (b) $S = 20$ cm/sec. Assume a diffusion coefficient of 25 cm^2 sec^{-1} where necessary.

20.15 Consider a rectangular n-type Si sample of length 0.2 cm, width 0.1 cm, and thickness 0.05 cm. The measured lifetime of the sample is 5 μsec. Calculate the bulk lifetime if $D_p = 12$ cm^2 sec^{-1} and $S = 10^6$ cm sec^{-1}.

APPENDIX A
LIST OF SYMBOLS

Symbol	Description
A	Area (cm^2)
a	Lattice constant (Å)
a	Half-channel thickness (cm)
a_0	Radius (cm)
B	Magnetic flux density (Wb/m^2)
B	Base pushout factor
b	Lattice translation vector (Å)
C	Capacitance (F)
C_j	Junction capacitance (F)
C_D	Diffusion capacitance (F)
c	Velocity of light (cm/sec)
C_c	Collector capacitance (F)
C_n, C_p	Auger coefficients for electrons and holes (cm^6 sec^{-1})
D	Electric flux density (C/cm^2)
D	Diffusion coefficient (cm^2/sec)
d	Thickness (cm)
D_B, D_E, D_C	Minority carrier diffusion coefficients in base, emitter, and collector, respectively (cm^2/sec)
D_a	Ambipolar diffusion coefficient (cm^2/sec)
D_s	Density of surface states (cm^{-2} eV^{-1})
E	Energy (eV)
E_C	Lower edge of conduction band (eV)
E_d, E_a	Donor, acceptor level energies (eV)
E_F	Fermi level (eV)
E_g	Energy band gap (eV)
E_i	Intrinsic Fermi level (eV)
E_i^*	Threshold energy for secondary pair generation (eV)
E_I	Ionization energy of dopant atoms (eV)
E_r	Optical phonon energy at zero momentum (eV)
E_t	Activation energy of recombination center (eV)
E_V	Upper edge of valence band (eV)

APPENDIX A LIST OF SYMBOLS

Symbol	Description
$\mathscr{E}$	Electric field (V/cm)
$\mathscr{E}_b$	Electric field at junction breakdown (V/cm)
$\mathscr{E}_c$	Critical field (V/cm)
$\mathscr{E}_d$	Electric field in drift region (V/cm)
$\mathscr{E}_p$	Peak electric field (V/cm)
$\mathscr{E}_T$	Threshold electric field for intervalley electron transfer (V/cm)
$\mathbf{F}$	Force (kg-m/sec^2)
$F_e(E), F_h(E)$	Electron and hole distribution functions
f	Frequency (Hz)
f_c	Cutoff frequency (Hz)
f_{max}	Maximum frequency of oscillation (Hz)
f_T	Current gain-bandwidth product frequency (Hz)
g, G	Conductance (A/V)
G_A	Auger generation rate (cm^{-3} sec^{-1})
G_B, G_E	Base and emitter Gummel numbers (cm^{-4} sec)
g_D	Drain conductance (A/V)
g_m	Transconductance (A/V)
g_o	Output conductance (A/V)
G_{th}, G_L	Thermal and photogeneration rates (cm^{-3} sec^{-1})
h	Planck constant (J-sec)
h_{FE}	Dc common emitter current gain
h_{fe}	Small-signal common emitter current gain
I	Current (A)
i	Instantaneous current (A)
I_A, I_K	Anode and cathode currents (A)
I_E, I_B, I_C	Emitter, base, and collector currents (A)
I_{CO}, I_{EO}	Collector and emitter reverse saturation currents (A)
I_D	Drain current (A)
I_g	Gate current (A)
I_L	Light-generated current (A)
I_s	Reverse saturation current (A)
I_S	Switching current (A)
J	Current density (A/cm^2)
J_{nD}, J_{pD}	Electron and hole diffusion current densities (A/cm^2)
J_{nf}, J_{pf}	Electron and hole drift current densities (A/cm^2)
$\mathbf{k}, k$	Wave vector (cm^{-1}), wave number
k	Boltzmann constant (J/K)
K	Grading constant (cm^{-4})
k_o	Distribution coefficient
K_s	Thermal conductivity (W/cm-K)
L, l	Length (cm)
L_a	Ambipolar diffusion length (cm)

APPENDIX A LIST OF SYMBOLS

Symbol	Description
L_B, L_C, L_E	Minority carrier diffusion lengths in base, collector, and emitter (cm)
L_{Di}, L_D	Debye length (Å)
l_r	Mean free path for optical phonon scattering (Å)
L_n, L_p	Electron and hole diffusion lengths (cm)
L, L_s	Inductance (H)
M	Avalanche multiplication factor
m	Mass (kg)
m	Exponent of power law distribution
m_e, m_h	Electron, hole conductivity effective masses (kg)
m_e^*, m_h^*	Electron, hole density of states effective masses (kg)
m_o	Free electron mass (kg)
N	Total impurity concentration (cm^{-3})
N_a, N_d	Acceptor and donor concentrations (cm^{-3})
N_B	Background dopant concentration (cm^{-3})
N_C	Effective density of states at the conduction band edge (cm^{-3})
N_V	Effective density of states at the valence band edge (cm^{-3})
N_t	Concentration of recombination centers (cm^{-3})
n	Electron concentration (cm^{-3})
n_e	Excess electron concentration (cm^{-3})
n_i	Intrinsic electron concentration (cm^{-3})
n_{ie}	Effective intrinsic electron concentration (cm^{-3})
P	Crystal momentum (kg-cm/sec)
P	Power (W)
P_t	Tunneling probability
p	Hole concentration (cm^{-3})
p_e	Excess hole concentration (cm^{-3})
Q_d	Depletion region charge per unit area (C/cm^2)
Q_I	Excess charge stored in the base of a bipolar transistor in inverse operation (C)
Q_N	Excess charge stored in the base of a bipolar transistor in normal operation (C)
Q_B	Total base charge per unit area in a bipolar transistor (C/cm^2)
Q_{BO}	Charge density in the surface depletion region of a MOS structure at the onset of strong inversion (C/cm^2)
Q_n	Charge density of electrons in the inversion layer (C/cm^2)
Q_{ox}	Oxide charge density (C/cm^2)
Q_s	Surface charge density in the semiconductor (C/cm^2)
q	Electronic charge (C)
R	Carrier recombination rate (cm^{-3} sec^{-1})
R	Resistance (Ω)
R^*	Richardson constant (A/cm^2-K^2)

Symbol	Description
R_D, R_S	Drain and source resistances (Ω)
R_s	Diode series resistance (Ω)
R_{th}	Thermal resistance (K/W)
r, r'	Position vectors (cm)
r, r_o	Radius (cm)
r_B	Base-spreading resistance (Ω)
r_b	Small-signal base resistance (Ω)
r_c	Collector series resistance (Ω)
r_e	Small-signal emitter junction resistance (Ω)
r_j	Small-signal junction resistance (Ω)
r_J	Radius of curvature (cm)
S	Surface recombination velocity (cm/sec)
S	Ionicity index
S_e	Emitter stripe width (cm)
s	C-V index
s_0	Probe spacing (cm)
T	Temperature (K)
T_e	Electron (or hole) temperature (K)
T	Time period of a wave (sec)
t	Time (sec)
t_d	Delay time (sec)
U	Excess carriers recombination rate (cm^{-3} sec^{-1})
u	Wave velocity (cm/sec)
V	Potential energy (eV), voltage (V)
V_a	Applied voltage (V)
V_A, V_K	Anode and cathode voltages (V)
V_B	Breakdown voltage
V_{CB}, V_{EB}	Collector-base and emitter-base junction voltages (V)
V_{Br}	Breakdown voltage of collector-base junction (V)
V_D	Drain voltage (V)
V_F, V_R	Forward and reverse voltages (V)
V_{FB}	Flat-band voltage (V)
V_G	Gate voltage (V)
V_g	Voltage corresponding to band gap E_g (V)
V_i	Built-in voltage (V)
V_j	Junction voltage (V)
V_P	Pinch-off voltage (V)
V_{PT}	Punch-through voltage (V)
$V_T = kT/q$	Thermal voltage (V)
V_{th}	Threshold voltage (V)
v	Velocity (cm/sec)
v	Small-signal voltage (V)

APPENDIX A LIST OF SYMBOLS

Symbol	Description
$\mathbf{v}_d$	Drift velocity (cm/sec)
v_s	Saturation velocity (cm/sec)
v_{th}, v_{rms}	Thermal velocity (cm/sec)
W	Depletion region width (cm)
W^*	Width constant (cm/V$^{1/2}$)
W_B	Base-width (cm)
W_b	Junction depletion region width at breakdown (cm)
W_C, W_E	Collector and emitter widths (cm)
W_D	Drain depletion region width (cm)
W_m	Maximum depletion region width (cm)
X_C, X_E	Depletion region widths of collector-base and emitter-base junctions (cm)
x	Position (cm)
x_j	Junction depth (cm)
x_n	Depletion region width on the n-side of the junction (cm)
x_p	Depletion region width on the p-side of the junction (cm)
y	Distance (cm)
y	Junction admittance (A/V)
Z	Channel-width (cm)
Z_{in}	Input impedance (Ω)
α	Absorption coefficient (cm^{-1})
α	Common base dc current gain
$\tilde{\alpha}$	Common base small-signal current gain
α_i	Effective ionization coefficients (cm^{-1})
α_n, α_p	Electron, hole ionization coefficients (cm^{-1})
α_N, α_I	Normal, inverse common base-current gains
α_T	Base transport factor
$\boldsymbol{\beta}$	Wave vector for lattice wave (cm^{-1})
β	Common emitter current gain
γ	Emitter efficiency
γ_p	Force constant (kg/sec^2)
γ_T	Temperature coefficient of breakdown voltage (K^{-1})
δ	Thickness (cm)
θ	Angle (rad)
Θ	Body factor (V$^{1/2}$)
λ	Wavelength (Å)
λ	Characteristic length for dopant diffusion (cm)
ω	Angular frequency (rad/sec)
ω_I, ω_N	Inverse and normal angular cutoff frequencies (rad/sec)
Ψ	Wave function
ψ	Time-independent wave function
η	Efficiency, ideality factor, grading constant

Symbol	Description
ε_o	Permittivity in vacuum (F/cm)
ε_i	Insulator permittivity (F/cm)
ε_s	Semiconductor permittivity (F/cm)
μ_n, μ_p	Electron and hole mobilities (cm^2/V-sec)
μ_B	Bulk mobility (cm^2/V-sec)
μ_C	Conductivity mobility (cm^2/V-sec)
μ_s	Surface scattering mobility (cm^2/V-sec)
$q\chi$	Electron affinity (eV)
$q\chi_m$	Metal electronegativity (eV)
v	Frequency (Hz)
σ	Conductivity $(\Omega\text{-cm})^{-1}$
σ_i	Intrinsic conductivity $(\Omega\text{-cm})^{-1}$
σ_{cn}, σ_{cp}	Electron, hole capture cross-sections (cm^2)
ρ	Resistivity (Ω-cm)
ρ_s	Sheet resistance ($\Omega/\square$)
$\phi(t)$	Time-dependent part of wave function
ϕ	Electrostatic potential (V)
ϕ_B	Schottky barrier potential on n-type semiconductor (V)
ϕ_B'	Schottky barrier potential on p-type semiconductor (V)
ϕ_F	Fermi potential (V)
ϕ_m, ϕ_{sc}	Metal, semiconductor work functions (V)
ϕ_s	Surface potential (V)
τ	Minority carrier lifetime (sec)
$\bar{\tau}$	Carrier mean freetime (sec)
τ_a	Ambipolar lifetime (sec)
τ_A	Auger lifetime (sec)
τ_B	Minority carrier lifetime in the base of a bipolar transistor (sec)
τ_c	Relaxation time (sec)
τ_d	Electron transit time through drift zone (sec)
τ_F	Effective lifetime in forward direction (sec)
τ_f	Fall time (sec)
τ_I, τ_N	Inverse and normal injection charge control parameters (sec)
τ_n, τ_p	Electron, hole bulk lifetimes (sec)
τ_o	Carrier generation-recombination lifetime in the depletion region (sec)
τ_r, τ_{nr}	Radiative, nonradiative lifetimes (sec)
τ_R	Effective reverse lifetime (sec)
τ_t	Transit time (sec)

APPENDIX B
PHYSICAL CONSTANTS

Quantity	Symbol	Value
Angstrom unit	Å	$1 \text{ Å} = 10^{-8}$ cm $= 10^{-10}$ m
Avogadro number	N	6.023×10^{23}/mol
Boltzmann constant	k	8.620×10^{-5} eV/K $= 1.381 \times 10^{-23}$ J/K
Electronic charge	q	1.602×10^{-19} C
Electron rest mass	m_o	9.109×10^{-31} kg
Electron volt	eV	$1 \text{ eV} = 1.602 \times 10^{-19}$ J
Gas constant	R	1.987 cal/mole-K
Permeability of free space	μ_o	1.257×10^{-6} H/m
Permittivity of free space	ε_o	8.850×10^{-12} F/m
Planck constant	h	6.626×10^{-34} J-s
Proton rest mass	m_p	1.673×10^{-27} kg
$h/2\pi$	$\hbar$	1.054×10^{-34} J-s
Thermal voltage at 300 K	V_T	0.02586 V
Velocity of light in vacuum	c	2.998×10^{10} cm/s
Wavelength of 1-eV quantum	λ	1.24 μm

Appendix C
INTERNATIONAL SYSTEM OF UNITS

Quantity	Unit	Symbol	Dimension
Length	meter	m	
Mass	kilogram	kg	
Time	second	s	
Temperature	kelvin	K	
Current	ampere	A	
Voltage	volt	V	
Force	newton	N	kg-m/s^2
Pressure	pascal	Pa	N/m^2
Energy	joule	J	N-m
Power	watt	W	J/s
Electric charge	coulomb	C	A-s
Frequency	hertz	Hz	s^{-1}
Resistance	ohm	Ω	V/A
Conductance	siemens	S	A/V
Capacitance	farad	F	C/V
Magnetic flux	weber	Wb	V-s
Magnetic induction	tesla	T	Wb/m^2
Inductance	henry	H	V-s/A

APPENDIX D
UNIT PREFIXES

Multiple	Prefix	Symbol
10^{12}	tera	T
10^{9}	giga	G
10^{6}	mega	M
10^{3}	kilo	k
10^{2}	hecto	h
10	deka	da
10^{-1}	deci	d
10^{-2}	centi	c
10^{-3}	milli	m
10^{-6}	micro	μ
10^{-9}	nano	n
10^{-12}	pico	p

APPENDIX E
PROPERTIES OF SOME IMPORTANT SEMICONDUCTORS AT 300 K

Group	Semiconductor	T_m (°C)	d (g/cm³)	a (Å)	$\varepsilon_s/\varepsilon_o$	E_g (eV)	m_e/m_o	m_h/m_o	Mobility at 300 K (cm²/V-sec) Elec.	Holes
III-V	GaP	1467	4.13	5.45	10.2	2.26	0.13	0.60	300	150
	InAs	942	5.66	6.06	12.5	0.36	0.028	0.33	22600	200
	InSb	525	5.77	6.47	17.7	0.18	0.013	0.18	100000	1700
	InP	1070	4.78	5.86	12.1	1.35	0.07	0.40	4000	600
II-VI	CdS	1750	4.84	5.83	8.9	2.42	0.20	0.70	250	—
	CdSe	1350	5.75	6.05	10.6	1.73	0.13	0.40	650	—
	CdTe	1098	5.86	6.48	10.9	1.50	0.11	0.35	1050	100
	ZnSe	1515	5.26	5.67	8.1	2.58	0.17	0.60	600	—
	ZnTe	1238	5.70	6.1	9.7	2.25	0.15	0.60	530	900
IV-VI	PbS	1077	7.5	5.93	17.0	0.37	0.10	0.10	500	600
	PbSe	1062	8.10	6.15	23.6	0.26	0.07	0.06	1800	930
	PbTe	904	8.16	6.46	30.0	0.29	0.24	0.30	1400	1100

Explanation of symbols:
T_m = melting point, d = density, a = lattice constant, E_g = band gap, $\varepsilon_s/\varepsilon_o$ = relative permittivity, m_e = electron effective mass, m_h = hole effective mass, m_o = free electron mass.

APPENDIX F
IMPORTANT PROPERTIES OF SiO₂ AND Si₃N₄ at 300 K

Properties	SiO_2	Si_3N_4
Structure	Amorphous	Amorphous
Melting point (°C)	~1600	—
Density (g/cm³)	2.2	3.1
Dielectric constant	3.9	7.5
Dielectric strength (V/cm)	10^7	10^7
Energy gap (eV)	9	5
Thermal conductivity (W/cm-K)	0.014	—
dc resistivity at 25°C (Ω-cm)	10^{14}–10^{16}	~10^{14}

APPENDIX G
ELECTRICAL NOISE

The term *noise* refers to random fluctuations of current flowing through a device and has its origin in the corpuscular nature of current flow. If the fluctuating current or voltage, generated in the device, is amplified by a low-frequency amplifier and the amplified signal is fed to a loudspeaker, the loudspeaker produces a hissing sound—hence the name "noise." It is a common practice to call spontaneous fluctuations noise, regardless of whether an audible sound is produced.

Noise is an important problem in science and engineering because the spontaneous fluctuations set a lower limit to the signal to be amplified or to the quantities to be measured. For example, no signal with an amplitude lower than the noise amplitude can be amplified by an electronic device.

The discrete nature of current carriers is responsible for noise in electronic devices. In p-n junction devices there are three fundamental sources of noise: thermal or Johnson noise, shot noise, and generation-recombination noise. Thermal noise results from the random motion of charge carriers in neutral regions of the semiconductor. This type of noise exists in any conductor that is in thermal equilibrium at a temperature higher than $T = 0°K$. The noise power is independent of frequency, at least up to optical frequencies. For this reason, it is termed a *white noise*. The random movements of charge carriers appear as a fluctuating emf whose mean square value $\overline{e_n^2}$ for a conductor of resistance R in a frequency interval Δf is given by the relation $\overline{e_n^2} = 4kTR\Delta f$.

Shot noise results from the random movement of charge carriers over the potential barrier of a p-n junction. The current I flowing through a biased p-n junction can be viewed as the sum total of many short-duration current pulses, each caused by the passage of a single electron (or hole) through the junction depletion region. This noise is also practically white and for a band width Δf, it can be represented by a current source whose mean square value $\overline{i_n^2}$ is given as $\overline{i_n^2} = 2qI\Delta f$.

In a semiconductor, electrons and holes appear and disappear randomly by generation and recombination processes. As a consequence, the resistance R of the semiconductor sample shows fluctuation $\delta R(t)$. If a dc current I is passed through the sample, a fluctuating voltage $\delta V(t) = I\delta R(t)$ develops across it. This noise is called *generation-recombination noise*.

ANSWERS TO SELECTED PROBLEMS

Answers are provided for most of the problems which have numerical solutions. A solutions manual is available to teachers on request from the publisher.

CHAPTER 1

1.1 $E = 3.976 \times 10^{-19}$ joule = 2.485 eV and $E = 3.976 \times 10^{-16}$ joule = 2485 eV.
1.2 1223 Å, 1032 Å, 979 Å, 956 Å.
1.3 2.18×10^8 cm/sec^{-1}. No, there is no excited state where the relativistic treatment becomes necessary.
1.4 (a) $\lambda = 1.88 \times 10^{-12}$ Å, (b) $\lambda = 1.45$ Å.
1.6 $E = 4.14 \times 10^{-8}$ eV, $n = 1.51 \times 10^{20}$.
1.7 $\Delta x = 0.53$ Å.
1.12 (a) $E_1 = 4.19$ eV, $E_2 = 16.76$ eV and $E_3 = 33.51$ eV, 1.2×10^8 cm sec^{-1}, (c) $\langle x \rangle = 1.5$ Å, $\langle p_x \rangle = 0$.
1.13 $E = 10.15$ eV, $\lambda = 1224$ Å.

CHAPTER 2

2.3 Simple cubic = 1, bcc = 2, fcc = 4.
2.4 (a) ($\sqrt{3}/2$)a, (b) a/$\sqrt{2}$ and (c) ($\sqrt{3}/4$)a.
2.6 1.578×10^{15} cm^{-2}, 1.82×10^{15} cm^{-2}.
2.11 Acoustical phonon kT, optical phonon 6.04×10^{-3} eV.
2.12 1.33×10^{-24} Kg m sec^{-1}, and 6.03 eV.

CHAPTER 3

3.3 (a) 4.99×10^{22} cm^{-3}, (b) 4.42×10^{22} cm^{-3}, (c) 2.21×10^{22} cm^{-3}.
3.5 5.45 Å. It will decrease on cooling.
3.8 0.93 eV for electron and 0.12 eV for hole.
3.9 $\lambda_m = 0.867$ μm, $\bar{\lambda} = 0.82$ μm.
3.11 (a) 1.45×10^{19} states cm^{-3}, $E_F = 4.37$ eV above the bottom of the band, (b) zero.
3.13 $n_o = 1 \times 10^{15}$ cm^{-3}, $p_o = 2.25 \times 10^{5}$ cm^{-3}, $(E_F - E_i) = 0.287$ eV, and $n = 3.31 \times 10^{12}$ cm^{-3}.
3.16 99.99 percent, $(E_C - E_F) = 0.298$ eV.
3.17 (a) $p_o = 5 \times 10^{16}$ cm^{-3}, $n_o = 8 \times 10^{9}$ cm^{-3}, (b) 120.6 °C.
3.18 1.2 eV and 0.8 eV.
3.19 $(E_i - E_F) = 0.351$ eV.
3.20 (a) At 150 K, $p_o \simeq 6 \times 10^{15}$ cm^{-3} and n_o is negligibly small. (b) 268 °C, (c) $p_o = 2 \times 10^{15}$ cm^{-3}, $n_o = 1.125 \times 10^{5}$ cm^{-3}.
3.21 (a) -0.0073 eV, (b) 0.041 eV, (c) 0.055 eV, (d) 0.064 eV.
3.24 $r_o = 13.46$ Å, and $\Delta r = 43$ Å.

CHAPTER 4

4.3 $p_o = 8.25 \times 10^{6}$ cm^{-3}, $n_o = 3.88 \times 10^{5}$ cm^{-3}, $(N_a - N_d) = 7.86 \times 10^{6}$ cm^{-3}.
4.4 (a) 2.56×10^{-4} sec, (b) 10 eV.
4.5 1.79×10^{-4} A cm^{-2}.
4.7 One electron per atom.
4.8 8.47×10^{-5} gram.
4.9 2.59×10^{5} Ω-cm for Si and 9.47×10^{8} Ω-cm for GaAs.
4.10 -0.074 K^{-1}.
4.11 54.5 Ω.
4.12 $N_a = 2.32 \times 10^{16}$ cm^{-3}, $\rho_o = 0.18$ Ω-cm.
4.13 1.462 eV.
4.14 (a) $n_o = 4.63 \times 10^{14}$ cm^{-3}, $p_o = 4.86 \times 10^{5}$ cm^{-3}, (b) $N_a = 4.66 \times 10^{14}$ cm^{-3}, (c) 459 K.
4.15 (a) $N_a = 8.8 \times 10^{12}$ cm^{-3}, (b) 3.69×10^{10} cm^{-3}, (c) 3.81×10^{7} cm^{-3}.
4.16 53.4 mV.
4.18 1.89×10^{-6} cm, 1.89×10^{7} cm sec^{-1}, 1.84×10^{-6} eV.

CHAPTER 5

5.1 (a) 10.376 mW, (b) 3.31 mW, (c) 3.57×10^{16} and 3.09×10^{16} photons/sec.
5.2 (a) 8.59×10^{13} cm^{-3}, (b) 0.115 (Ω-cm)$^{-1}$, low level injection.
5.5 1.078×10^{8} Ω-cm, 2.3×10^{-6} sec.
5.6 1.15×10^{14} cm^{-3}, 1.15×10^{20} cm^{-3} sec^{-1}.

ANSWERS TO SELECTED PROBLEMS 651

5.7 3.57 Ω-cm, 2.8×10^4 percent.
5.10 $\tau_p = 2.3 \times 10^{-7}$ sec and $\tau_p = 1.6 \times 10^{-5}$ sec.
5.12 (a) 6.58×10^{-7} sec, (b) 1.02×10^{-6} sec.
5.13 (a) 2×10^{19} pairs cm^{-3} sec^{-1}, (b) $\sigma = 0.218$ $(\Omega\text{-cm})^{-1}$, $(E_{Fn} - E_i) = 0.288$ eV and $(E_i - E_{Fp}) = 0.186$ eV.

CHAPTER 6

6.3 (a) 1.32×10^{30} cm^{-6}, (b) 3.58×10^{11} cm^{-6}.
6.4 -1.76×10^{-3} V K^{-1}, 0.74 V.
6.5 0.308 V, 298.2 K.
6.7 (a) 0.79 V, (b) 1.02×10^{-4} cm, (c) -3.4×10^4 V cm^{-1}.
6.8 $|x_p| = 3.65 \times 10^{-4}$ cm, $|x_n| = 1.82 \times 10^{-5}$ cm, $|\mathscr{E}_m| = 5.56 \times 10^4$ V cm^{-1}.
6.9 0.5 V.
6.10 (a) 0.26 V, (b) No, the junction cannot be used as a rectifier.
6.11 (a) $V_i = 0.6$ V, $W = 1.23 \times 10^{-5}$ cm, (b) $N_a = 4.54 \times 10^{15}$ cm^{-3}, $N_d = 5.9 \times 10^{14}$ cm^{-3}.
6.12 0.8 V, 6.02×10^{15} cm^{-3}.
6.13 It is an abrupt p-n junction.
6.15 (a) 1.06×10^5 V cm^{-1}, 3.82×10^5 V cm^{-1}, 99.5 pF, 52.4 pF, (b) 3.57×10^5 Hz, 4.92×10^5 Hz.
6.16 (a) 6.06×10^{17} cm^{-3}, 3.67×10^{17} cm^{-3}, (b) $K = 1.84 \times 10^{21}$ cm^{-4}.

CHAPTER 7

7.2 19.86 μA, 3.49×10^{-9} A.
7.6 1.84×10^{-5} cm.
7.7 (a) $P_e(x_n) = 1.92 \times 10^{14}$ cm^{-3}, $n_e(-x_p) = 4.94 \times 10^{14}$ cm^{-3}, (b) In the n-region, 8.26×10^{-4} cm from x_n.
7.9 0.348 V, 39.66 A cm^{-2}.
7.10 312.16 K, 0.586 eV.
7.11 (a) 12.36 mV, (b) 0.517 V.
7.12 79.86 A, 411 mV and 589 mV.
7.13 15.38 mA, 231 mV.

CHAPTER 8

8.1 (a) 6.7×10^{-7} cm, (b) 1.12×10^{-6} cm, (c) 1.43×10^{-6} cm.
8.2 0.594 eV.
8.3 $A' = 8.78 \times 10^8$ A cm^3 V$^{-5/2}$.
8.4 $\beta^* E_g^{3/2} = 5.26 \times 10^6$ V/cm, $m^* = 0.033\ m_o$.
8.5 1.5×10^{-3} V K^{-1}.
8.6 (a) 8.03×10^7 cm sec^{-1}, (b) 5.45×10^7 cm sec^{-1}, (c) 9.27×10^7 cm sec^{-1}.

8.8 552 V, 276.4 V.
8.10 5.1×10^{15} cm^{-3}.
8.12 2.13×10^{14} cm^{-3}, 76.6 μm.
8.13 107.4 V.
8.14 198 V.
8.16 9.89×10^{13} cm^{-3}.

CHAPTER 9

9.1 (a) $(12.93 - j0.57)$ Ω, (b) $9.8\sqrt{1 - j0.754}\,\Omega$, (c) $(0.617 - j0.617)$ Ω.
9.2 (a) $(0.054 - j1.18)$ Ω, (b) 0.5 V.
9.3 8.27×10^{-7} sec, 1.65×10^{-9} C, 3.2×10^{-8} F.
9.6 $\tau_p = 1.5 \times 10^{-6}$ sec, $\tau_F = 7.41 \times 10^{-8}$ sec.
9.7 8.35×10^{-8} F.
9.8 4.54×10^{-9} C, 0.218 V.
9.10 (a) 0.317 V at $t = 1$ μsec, (b) 2×10^{-8} C and 4.3×10^{-9} C.
9.11 2.96×10^{-6} sec.
9.12 1.62×10^{-7} sec.

CHAPTER 10

10.1 $\phi_n = 0.145$ V, $\phi_p = 0.11$ V; $V_i = 0.712$ V, $V_v = 0.254$ V.
10.3 (a) 0.425 V, (b) 28 Ω, (c) 0.515 V.
10.4 $f_{r0} = 1.428 \times 10^9$ Hz, $f_{x0} = 4.277 \times 10^8$ Hz.
10.6 $V_i = 0.59$ V, $\phi_B = 0.85$ V, $W_o = 7.2 \times 10^{-5}$ cm.
10.8 $V_i = 0.55$ V, $\phi_B = 0.776$ V, $W_o = 4.01 \times 10^{-5}$ cm.
10.9 $\phi_B = 0.93$ V, $V_i = 0.704$ V, $W = 1.04 \times 10^{-4}$ cm.
10.10 $D_s = 6.52 \times 10^{12}$ states cm^{-2} eV^{-1}, $q\phi_0 = 0.5$ eV above E_V.
10.11 $V_i = 0.846$ V, $\phi_B = 0.963$ V, $N_d = 5.08 \times 10^{15}$ cm^{-3}, $q\phi_m = 5.03$ eV.
10.13 $V_i = 0.605$ V, $\phi_m = 4.85$ V, $I_p/I = 7.93 \times 10^{-5}$.
10.14 $I_s = 2.7 \times 10^{-11}$ A, $\eta = 1.103$, R* = 93.36 A cm^{-2} K^{-2}.
10.15 $V_i = 1.148$ V.
10.16 $V_i = 0.02$ V.

CHAPTER 11

11.1 $f_c = 1.658 \times 10^{11}$ Hz, $f = 1.658 \times 10^9$ Hz.
11.2 $m = -3/2$, $C_j = 102$ pF.
11.3 (a) 26.3 GHz, (b) 64.49 GHz, (c) 317.19 GHz.
11.4 (a) 36.6 GHz, (b) 91.1 GHz, (c) 421.6 GHz.
11.5 $C_j(0) = 1.26 \times 10^{-9}$ F cm^{-2}, $V = -9.6$ V, RC = 3.25×10^{-4} sec and 8.03×10^{-3} sec.
11.6 10.6 μm.
11.7 $V_M = 18.2$ mV and from Eq. (11.20) $V_a = 0.727$ V.

ANSWERS TO SELECTED PROBLEMS 653

11.9 (a) $L_A = 8.43 \times 10^{-11}$ H, $C_A = 18.25$ pF, (b) 4.058 GHz.
11.10 11.33 watt.
11.11 $N_d = 2.31 \times 10^{15}$ cm^{-3}, $V_{FB} = 85.98$ V.
11.12 $f = 10.7$ GHz, $G = -0.006$ Ω^{-1}.
11.13 $W = 9.6 \times 10^{-4}$ cm, $P_{dc} = 0.215$ Watt, $V_1 = 13.64$ V.
11.15 Transit time mode at 6 GHz and LSA mode at 12 GHz.
11.16 2.415 psec.

CHAPTER 12

12.3 $V_{oc} = 0.654$ V, $\Delta E_g = 24$ meV.
12.4 $P_m = 17.14 \times 10^{-3}$ watt, $R_{opt} = 13.84$ Ω, FF = 0.767, $\eta = 15.3$ percent.
12.5 (a) 1.1 V and 0.8 A, (b) 0.514 V, 1.8 A.
12.6 656 mV.
12.7 $I_{sc} = 61.3$ mA, $V_{oc} = 325$ mV, 8.08 percent.
12.9 0.435 mA.
12.10 47.46 mA.
12.12 $\eta_i = 39.8$ percent, $\eta_r = 82.7$ percent.
12.13 0.61 percent, 1.09×10^4 cm^{-1}.
12.14 The GaP diode will appear brighter.
12.16 $S = 1.5 \times 10^4$ cm sec^{-1}, $\Delta a/a = 7.5 \times 10^{-4}$.
12.17 4.93 Ω.

CHAPTER 13

13.3 $\alpha = 0.989$, $h_{FE} = 89.9$.
13.5 (a) 536 mV, 4.5×10^{12} cm^{-3}, (b) $\gamma = 0.93$, $\alpha_T = 0.999$, $h_{FE} = 13.1$.
13.8 27.17 μsec.
13.9 6.17 μm, $\alpha = 0.999$.
13.10 32.91 μm, $L_B = 98.6$ μm, $\tau_B = 8.35 \times 10^{-6}$ sec.
13.11 14.82 V.
13.12 -47.3 mV.
13.13 $h_{FE} = 98$, $W_B = 2.857$ μm.
13.14 $I = 23.92$ μA, $V_{EB} = 64.38$ mV, $V_{BC} = 4.91$ V.
13.16 66.28 V.
13.17 20.5 V, $V_{CEO} = 26.19$ V.
13.18 $Q_N = 8.1 \times 10^{-11}$ C, $Q_I = 1.557 \times 10^{-10}$ C; both junction voltages are positive.

CHAPTER 14

14.1 36.96 MHz, $0.927 \angle -26.35$.
14.2 $G_B = 2.93 \times 10^{13}$ cm^{-4} sec, $G_E = 4.5 \times 10^{13}$ cm^{-4} sec, $h_{FE} = 1.54$.
14.6 $m = 0.646$, $\tilde{\alpha}/\alpha_0 = 0.832 \angle -56.99$.

14.7 1.08×10^{16} cm^{-3}.
14.8 $f_\alpha = 101.5$ MHz, $f_\beta = 1$ MHz, $f_T = 100.5$ MHz, $W_B = 1.94$ μm.
14.10 32.77 and 20.88.
14.11 3.02 A.
14.12 60.12 μA.
14.13 2×10^{-8} sec and 4×10^{-7} sec.
14.14 2.96 °C/watt.
14.15 (a) 4.41 Ω, (b) $\tau_d = 24$ psec, $\tau_s = 445.5$ nsec.

CHAPTER 15

15.2 (a) 3.8 V, (b) 3.82 mA, (c) 0.717 mA.
15.3 (a) 0.781 V, 3.04 V, 1.58×10^{-4} A/V, (b) 7.51×10^{-5} A/V and 4.28×10^{-5} A/V, (c) 3.23×10^{-5} A/V.
15.4 (a) 4.56 V, (b) 321.5 MHz.
15.5 (a) 2.7 V, (b) 301.6 Ω, (c) 7.515 mA/V.
15.6 26 V.
15.7 $V_P = 4.05$ V, $I_D = 16.57$ μA, $g_D = 2.61 \times 10^{-5}$ A/V, $g_m = 1.48 \times 10^{-5}$ A/V and $g_m(\text{sat}) = 4.09 \times 10^{-5}$ A/V.
15.9 57.4 μA.
15.11 (a) $\mu_1 + \mu_2$, (b) $g_m = g_{m1} + g_{m2}$.
15.12 $I_D = 10.61$ mA, $g_m = 3.23$ mA/V, $f_T = 6.25$ GHz.

CHAPTER 16

16.1 (a) -0.329 V, (b) -0.658 V.
16.2 0.612 V.
16.3 (a) -2.076 V, (b) -3.11 V, $W_m = 7.23 \times 10^{-5}$ cm.
16.5 (a) -1.089 V, (b) -2.147 V.
16.6 -1.67 V and -0.611 V.
16.7 1096 Å, 7.07×10^{11} charges cm^{-2}.
16.8 11.12 nF, 17.04 nF, 3.97 nF and 17.26 nF.
16.9 (a) -5.01 V, -5.77 V, (b) -9.645 V, -10.4 V, (c) -0.375 V, -1.135 V.
16.12 (a) $V_{th} = 0.76$ V, $V_G = 3.074$ V, $g_m = 8.63 \times 10^{-4}$ A/V, $f_T = 184.4$ MHz, (b) $V_{th} = 2.01$ V.
16.13 (a) -3.17×10^{-8} C cm^{-2}, (b) 1.47 V.
16.14 (a) 100.7 μA, 5.03×10^{-4} A/V, 2.88×10^{-5} A/V, (b) 0.719 mA, 2.157×10^{-4} A/V, 2.876×10^{-4} A/V, (c) 5.03×10^{-4} A/V.
16.15 7.58×10^{-4} A/V for MOSFET, 7.73×10^{-2} A/V for BJT.
16.16 (a) 0.68 V and -1.775 V, (b) 0.104 cm, (c) 0.96 V.
16.19 (a) 8.71 V, 8.92×10^{-4} cm, (b) 5.5 V, 4.86×10^{-4} cm, (c) 3.33×10^4 V/cm.
16.20 (a) 6.93 V, 8.4×10^4 V/cm.

ANSWERS TO SELECTED PROBLEMS

CHAPTER 17

17.4 $g_m = 1.946 \times 10^{-1}$ A/V, $r_\pi = 513.75$ Ω, $r_b = 86.25$ Ω, $g_o = 3.89 \times 10^{-6}$ A/V.

17.5 $r_\pi = 910.65$ Ω, $g_m = 7.73 \times 10^{-2}$ A/V, $C_\pi = 36.24$ pF, $C_c = 1.7$ pF.

17.6 $r_b = 37.27$ Ω, $g_o = 7.73 \times 10^{-7}$ A/V, $(\delta g_o)^{-1} = 9.11 \times 10^7$ Ω.

CHAPTER 18

18.1 0.0295°.

18.3 $V_B = 1496.8$ V, $V_{PT} = 759.63$ V, $V_{BR} = 548$ V.

18.4 160 V.

18.5 1097 V and 846 V.

18.6 1080.7 V.

18.7 100/ampere.

18.8 1.248 μsec.

18.9 167 μsec.

CHAPTER 19

19.1 0.581 mg.

19.3 1.63 hr.

19.4 7.28 μm.

19.5 55.4 min.

19.6 (a) 1.56×10^{18} cm^{-3}, (b) 0.102 Ω-cm, (c) 694 Ω/□.

19.7 4.3×10^{-2} mg, 2×10^{-2} mg.

19.8 0.508 μm.

19.9 2.846 hrs.

19.10 396 mV for $V_{EB} = 0$.

19.11 1.667×10^{14}, 0.06 Ω-cm, $L/W = 20$.

19.12 0.878 cm^2.

19.13 18.78, 3.17.

19.14 4.25 V.

CHAPTER 20

20.1 6.28 Ω-cm and 7×10^{14} cm^{-3}.

20.2 7.037 Ω-cm, 351.86 Ω/□, 0.23 cm.

20.3 3.26 Ω-cm, 217.35 Ω/□.

20.4 8 mA.

20.5 21.75 mV.

20.6 (a) $p_o = 1.33 \times 10^{16}$ cm^{-3}, (b) 470 cm^2/V-sec, (c) 398.3 cm^2/V-sec.

20.7 $t_o = 1.05 \times 10^{-4}$ sec, spread of the pulse = 7.19×10^{-2} cm.

20.8 1818 cm^2/V-sec, 49.3 cm^2/sec, T = 41.5 °C.
20.9 1600 cm^2/V-sec, 42.24 cm^2/sec.
20.12 4.5 × 10^{-7} sec, 0.378 V.
20.14 (a) 4.71 × 10^{-2} cm, (b) 1.897 × 10^{-2} cm.
20.15 7.25 × 10^{-6} sec.

INDEX

Abrupt approximation, 166
Abrupt junction, 166
  depletion:
    region capacitance, 171–174
    region width, 170, 171
  one-sided, 170, 227
  space charge, 169
Absorption:
  coefficient, 130, 338
  optical, 130, 338
ac admittance (diode), 243, 244
Acceptor:
  atom, 72
  doping, 71, 72
  impurity, 71, 77
  ionization energy, 77
  level, 76
Accumulation:
  of electrons, 460
  of holes, 471, 477
Accumulation region, 460
ac equivalent circuit:
  bipolar transistor, 520, 521, 525, 526
  JFET and MESFET, 462–463, 531, 532, 533
  MOSFET, 497, 498, 535
  p-n junction diode, 243–244
Acetic acid, 592
ac impedance, 314
Acoustic modes, 42, 43
  longitudinal, 43
  phonon, 43
  transverse, 43
Activation energy, 141
Active region (bipolar transistor):
  inverse, 392, 440
  normal, 380, 392, 397, 405
Air mass one, 341, 376
Air mass two, 342
Air mass zero, 342, 375
AlAs, 64, 349

AlGaAs, 65, 357, 369, 370, 371
  growth, 357, 582
  liquid phase epitaxy, 372
  molecular beam epitaxy, 582
Alkali ions, 483
Alloyed junction:
  diode, 166, 269, 571
  transistors, 383
Alloying, 571
Alloy process, 571
Alpha of a transistor:
  calculation, 388, 389
  defect, 389
  definition, 382
  dependence:
    on current, 394
    on voltage, 389, 396
  drift transistor, 417
  frequency dependence, 413, 420, 421
AlSb, 64
Aluminum, 288, 470
  acceptor in Si, 71, 72
  contacts, 593
  metallization, 593, 594
AMO (air mass zero) spectrum, 342, 375
AM1 spectrum, 341, 347, 376
AM2 spectrum, 342
Ambipolar:
  diffusion, 151
  diffusion coefficient, 151, 309, 310, 631
  mobility, 151, 309
  semiconductor, 73
  transport, 151, 310
Amorphous:
  semiconductor, 62
  silicon, 350, 351
  solid, 29, 30
Amplification:
  current, 382, 389

  voltage, 381, 412
Amplifier, 332, 333
  transistor, 517
  tunnel diode, 269
Amplifying gate, 555
Angular momentum, 4, 15
Anisotropic:
  etchant, 505, 598
  etching, 598
Anisotype heterojunction, 291–294
Anode, 314, 331, 543, 544
Anode–cathode voltage, 550, 551
Annealing, 145, 496, 503, 578
Antimony, 63, 70, 74, 77, 575
Antireflection coating, 345, 346, 350, 353
Argon bombardment, 586, 593
Aspect ratio, 602, 612
Asymmetrical junction, 170, 227
Atomic number, 17, 64
Atomic structure, 3–4, 16–19
Atomic theory, 3–5, 16–19
Atomic weight, 18, 64
Au, 77, 144, 288, 297. *See also* Gold
Auger lifetime, 142–143, 419
Auger process, 134
Auger recombination, 141–144, 419
  band-to-band, 141–143, 344
  trap-aided, 143–144
Avalanche breakdown, 213, 214–215, 225, 226, 312
  bipolar transistor, 396, 399, 403–405
  p-n junction, 225–230
Avalanche breakdown voltage:
  calculation, 227–228
  doping dependence, 228–229
  temperature dependence, 229–230
Avalanche excitation, 358
Avalanche gain, 356

INDEX

Avalanche injection, 502, 509, 610
Avalanche multiplication, 224–225
  factor, 225–226, 229, 232, 389, 396, 404, 545–547
Avalanche noise, 356, 357
Avalanche phase delay, 313, 314
Avalanche photodiode, 355–357
Avalanche region, 313, 316, 317–319, 357
Avalanche shock front, 321
Avalanche transit time diode, *see* IMPATT diode
Averages (statistical), 19
Average thermal velocity, 21, 224, 284
Avogadro number, 643

Back surface field, 345, 347
Backward diode, 270
Balance of drift and diffusion, 164–165
Balmer series, 5
Band bending, 292, 471–472, 474
Band diagram, *see also* Energy band
  one dimension, 44–53
  three dimensions, 54
Band gap, 48
  of common semiconductors, 126, 646
  definition, 47–48
  determination, 101, 130–131
  electrical, 96
  optical, 95, 131
  temperature dependence of, 83, 98, 224
Band gap narrowing, 94–95, 344
  electrical, 96, 344, 419
  optical, 95
Band-to-band transitions, 74, 357
BARITT diode, 301, 323–327
Barrier:
  to injection, 162, 165, 327
  to tunneling, 216–217, 219–220, 264–265
Barrier height, 272, 274–275, 278, 287
  effective, 287
  measured, 288
  metal silicide, 277
Barrier layer, 162, 272
Barrier lowering, 280–281, 286, 287
Barrier penetration, 216
Base, 379
  impurity distribution, 384, 414–415
Base (BJT):
  contact, 383, 396
  current, 380–381, 383, 400, 405, 408, 422
  drift in, 384, 412, 414
Base region, 379, 384, 385
Base resistance, 396–397, 520, 525, 526
Base transit time, 418–419, 422, 423, 522
Base transport factor, 382, 388–389, 420

Base width, 383, 384, 394, 426
Base-width modulation, 395, 398, 520, 525, 526
Basic equations, 148–150, 518
Basis vectors, 30
Batch fabrication, 595
Beta (β) of a BJT:
  calculation, 388–389, 417–418
  of a composite transistor, 604, 612
  definition, 389, 403
  dependence on:
    biasing, 389, 396
    injection, 394
    temperature, 420
  relation to f_T, 422
Beveled structures, 540
Beveling, 540, 546
Biasing, 162–163
Bidirection thyristors, 557–558
Binding energy, 5, 27
  of an atom, 28
  of donors and acceptors, 70, 71, 77
  of an electron, 5, 28
Bipolar junction transistor (BJT):
  amplification, 381, 382, 389
  configurations, 397–398
  currents, 380–381
  equivalent circuits, 518–529
  fabrication, 383–384, 589, 599
  high frequency, 425–427
  parameters, 381–382, 388–389, 417–418
  power, 427–434
  switching, 435–440
  switching times, 438–440
  transient effects, 437–440
Bloch functions, 50, 51
Body centered cubic (bcc) crystal, 33, 34, 56
Body factor, 489, 490–491
Bohr model (of an atom), 4–5, 70, 100
Boltzmann approximation, 23
Boltzmann constant, 20, 638, 643
Boltzmann distribution, 19–20
Boltzmann relation, 167
Boltzmann statistics, 19–21, 61
  conditions for validity, 61, 81, 83
Bombardment flux, 569, 593
Bond:
  chemical, 27
  covalent, 28
  dangling, 275
  ionic, 28
  metallic, 29
  molecular, 29
Bonding of atoms, 27–29
Boron (B):
  acceptor in Si and Ge, 71
  diffusion coefficient in Si, 574
  diffusion in Si, 575, 589, 596
  energy level in Si, 77
  ion implantation, 503, 578
  source material for diffusion, 575
Boron compounds, 63

Boron nitride (BN), 63, 64, 574
Boron phosphide (BP), 63, 64
Bose-Einstein statistics, 23–24, 43, 57
Bragg reflection, 41, 50, 52, 53
Bravais lattice, 30
Breakdown:
  avalanche, 215–216, 225–230
  edge, 540
  surface, 234, 539–540
  Zener, 214, 221–223
Breakdown condition, 226–227, 308
Breakdown diode, 234–235
Breakdown field, 312–313
Breakdown voltage:
  common emitter, 405
  graded junction, 228
  junction curvature effect, 231–232
  orientation dependence, 225, 228
  p-i-n diode, 308
  punched-through diode, 230–231
  silicon p-n junction, Eq. (8.30), 545
  temperature dependence, 223–224, 230
  universal expression, 228
Breakover voltage:
  bipolar transistor, 404–405
  thyristor:
    forward, 546–547
    reverse, 545–546
    temperature dependence, 548
Built-in field, 162, 384, 414
Built-in voltage, 162, 166–167, 272, 292
Bulk-effect devices, 327, 329–333
Bulk negative differential resistance, 327
  devices, 332–333
  materials, 328, 333
  mechanism, 328–330
Buried channel CCD, 510, 511
Buried layer, 596, 599, 600

Cadmium selenide (CdSe), 65, 73
  properties, 646
Cadmium sulphide (CdS), 65, 73
  barrier heights, 288
  properties, 646
  unit cell, 65
Cadmium telluride (CdTe), 65, 73, 328
  properties, 646
Capacitance:
  abrupt p-n junction, 171–173
  arbitrary junction, 186
  charge storage, 171–172, 181
  collector junction, 423, 426, 438, 524, 526
  depletion layer, 173, 182, 281
  diffusion, 240, 243–245, 528
  emitter-base junction, 423, 438, 520, 524, 526
  linearly graded junction, 181–182

INDEX

MOS, 470–471, 474, 475–482, 505, 603
  packaging, 244, 302, 425
  parasitic, 503, 535, 599–600
Capacitance-voltage measurement, 287
Capacitor, integrated, 602–603
Capture cross sections, 138, 140, 141
Capture of electrons and holes, 137–138, 197
Carbon (C), 47, 63
  properties, 63
Carrier concentration, 78–86
  constant product, 71, 82
  equilibrium, 85–86
  excess, see Excess carriers
  gradient, 122–124
  intrinsic, 69, 82–83
  relation to doping, 85–86
  temperature dependence, 90–92
Carrier confinement, 370–371
Carrier extraction, 131, 133, 197
Carrier injection, 130–132
Carrier lifetime, 133–134
Carriers, majority and minority, 71
Carrier transport, 103–122
Cathode:
  p-n-p-n diode, 543
  thyristors, 544
Cathode contact, 330, 548
Cathode short, 548
CCD, buried channel, 511
CCD structure, 507, 510–511
Centers, recombination and trapping, 136–137, 140
Channel (FET), 444, 484, 486
  conductance, 450, 492
  diffused, 487
  doping, 452, 496
  induced, 485
  length, 446, 486
  pinch-off, 445–446, 486–487
  resistance, 446, 486–487
  thickness, 446
  width, 446, 486
Channel stops, 606–607
Charge carriers, see Electron; Hole
Charge conservation principle, 247
Charge control equations:
  bipolar transistor, 405–408
  junction diode, 246–248
Charge control model (BJT), 516, 520–526
Charge coupled devices, 505–511. See also CCD
Charge neutrality, 85, 87
Charge packet, 505, 510
Charge storage, 507
  diode, 259–260
  in a p-n junction, 172, 181, 246–248
  in a transistor (BJT), 406–408, 438–440, 521–523
Charge transfer, 507–508
  efficiency, 509, 510
  mechanisms, 509–510
Chemical vapor deposition (CVD), 579, 583, 585–586, 587, 607
Chlorine, 28, 585
Circuit configuration (BJT), 397–398
Circuit models:
  BJT, 518–528
  JFET, 462–463, 528–533
  MOSFET, 497–498, 533–535
Circuit symbols:
  BJT, 381
  diode, 379
  JFET, 463
  MOSFET, 498, 606–607
  thyristor, 544
Clean contacts, 279
Cleaving, 279
CMOS, 579, 606, 607–608
Coatings, protective, 540, 587
Coherent light, 364
Cohesive energy, 27
Collector (BJT):
  breakdown, 404–405
  characteristics, 398–400, 401–403
  charging time, 423
  contact, 383–384, 589, 599
  current, 380–381, 387–388
  depletion region delay time, 422, 423
  ideality factor, 528
  injection, see Inverse active mode of BJT
  multiplication factor, 382, 389, 396, 404
  region, 379
  resistance, 423, 599
Collisions (scattering), 53, 103, 104, 129
Color:
  bands, 362
  of light emission, 360–362
Common-collector configuration, 398
Common-emitter:
  breakdown voltage, 405
  configuration, 398, 400–403
  current gain, 389, 403
Compensation, 72, 73, 570. See also Self compensation
Complementary error function (erfc), 257, 573
Complementary MOS, see CMOS
Complex conjugate, 8
Composite transistor, 603–604, 612
Compound semiconductors:
  II-VI, 65–66, 73
  III-V, 63–65, 72, 73
  IV-VI, 66, 73
  oxide, 66
Computer, 378, 590
Computer methods, 516
Conductance:
  channel, 44
  diode, 243–245
Conduction band, 48, 55, 62, 74
  effective density of states, 81
  effective mass, 82
  electron density, 81
Conduction band edge, 74
  in heavily doped silicon, 94–95
Conduction process, see also Diffusion; Drift
  impurity band, 114–115
Conductivity:
  extrinsic, 112–113
  high field, 118–121
  intrinsic, 112
  low field, 110–114
  thermal, 320
Conductivity modulation, 306, 310, 539
Conductivity type, 613–614
Conductors, see Metallization
Confinement (in lasers), 369–371
  barrier, 370
  carrier, 369–370
  optical, 370–371
Constant current source, 248–249
Constant source diffusion, 573
Contact, metal-semiconductor, 270–290
Contact potential, see also Built in voltage
  heterojunction, 292
  Schottky barriers, 272
Contact printing, 591
Contacts to semiconductors:
  ohmic, 273, 289–291
  rectifying, 271–279
  solar cells, 345
Continuity equation, 149, 189, 246, 339
Continuous wave (CW) laser, 373, 374
Controlled rectifiers, see Thyristor
Controlled source, 520, 526, 533
Conversion efficiency, 316–317, 323, 326, 333, 343, 346, 347, 349, 350
Coordination number, 33, 35
Copper:
  heat sink, 320, 333, 432
  impurity in germanium, 144
Covalent bond, 28
Covalently bonded semiconductors, 63–64, 275, 280
Critical angle, 360
Critical field:
  for breakdown, 231, 312–313
  for velocity saturation, 121
Cross-section (capture), 138, 140, 141
Crystal defects, 36–39, 232–233
Crystal growth:
  Czochralski technique, 566–567
  epitaxial technique, 579–582
  float zone process, 567–568
Crystalline solids, 29
Crystal momentum, 51, 54
Crystal planes and directions, 35–36
Crystal structure, 30–35
Cubic crystal:
  body centered, 30, 33–34

Cubic crystal (*Continued*)
  close packed, 32, 33
  face-centered, 31, 33
  simple, 30, 31
Current:
  diffusion, 122, 123–124, 147, 165
  drift, 108, 110, 123–124, 147, 165
Current density, 108, 110, 123–124
  equations, 149
Current triggering (thyristor), 550
Current-voltage characteristics:
  BJT, 385–388, 390, 397–403
  ideal diode, 191–193
  JFET, 447, 449–450, 453
  MOSFET, 487, 489–491, 593
  p-n junction diode, 198, 199–201
  p-n-p-n diode, 544
  Schottky diode, 284, 286
  solar cell, 339, 341–342
  thyristor, 550, 554, 560, 562
  tunnel diode, 266–268
Current-voltage measurement, 287
Cutoff frequency:
  BJT, 422–423
  JFET, 462
  MOSFET, 498
  varactor diode, 302
Cutoff region (BJT), 392–393, 435–436
Cutoff wavelength, 352
C-V measurement, 287
Cylindrical junction, 231–232
Czochralski technique, 556–567

Dangling bond, 275
de Broglie hypothesis, 6
de Broglie wavelength, 6, 19, 25
Debye length:
  extrinsic, 477
  intrinsic, 178
Decay time, *see* Fall time
Deep depletion:
  in CCDs, 505
  in SOS, 608
Deep level impurities, 136, 144, 259
  incorporation of, 144, 259, 542, 554
Degeneracy of level, 84
Degenerate semiconductor, 96, 221, 265, 366
Density, 646, 647
Density of states:
  conduction band, 80–82
  effective mass, 82
  function, 78, 80
  valance band, 80, 82
Depletion (surface), 471
Depletion approximation, 168, 171, 176, 281
  of holes, 474
Depletion layer:
  capacitance, 171–172, 181–182, 281–282
  charge, 171, 181, 281, 480
  curvature, 231–232
  width, 170, 180, 281, 477–478
Depletion mode FET, 456, 485

Depletion region, *see* Depletion layer
Detection (of optical radiation), 351
Detector (photo), 351–357
Detailed balance principle, 125, 138
Device:
  analysis, 148
  equations, 148–150
  modeling, 518–535
DH laser, 369–371
Diac, 557
Diamond:
  band gap, 47, 63
  lattice, 34, 56
  structure, 34–35, 62
dI/dt effect, 552
  capability, 552
Diborane (B_2H_6), 580
Dichlorosilane (SiH_2Cl_2), 579
Dielectric constant, 126, 281
  image force, 281
  semiconductors, 70, 126, 646
Dielectric isolation, 597–598
Dielectric relaxation time, 281, 330
Differential capacitance, 172
Differential negative resistance, 266, 327
Diffraction, 591
Diffused-epitaxial transistor, 384–385, 589
Diffused junction, 182–183
Diffused structures:
  BJT, 384–385, 432, 589
  guard ring, 289, 356–357, 540
  JFET, 451–452
  junction laser, 366–367
  MOSFET, 484, 503, 504, 606
  p-n-p-n diode, 543
  resistor, 601–602
  solar cell, 341, 345–346
Diffusion:
  ambipolar, 150–151, 631
  capacitance, 240, 243–245, 520, 525
  carrier, 121–122, 623–625
  coefficient, constant, 122, 124, 624, 625
  conductance, 243–244
Diffusion, solid state:
  coefficients in Si, 574
  coefficient in SiO_2, 589
  constant source, 573
  constant surface concentration, 572
  drive-in, 573
  furnace, 574
  predeposition, 573
  profiles, 575
  sealed-tube, 575
  source materials, 573–575
  techniques, 573–575
Diffusion current, 122, 152, 164–165
Diffusion equation, 572
Diffusion length, 154, 632
Diffusion process, 121, 623–625
Digital IC, 595

Diode, *see also* p-n junction
  ac equivalent circuit, 243–244
  backward, 270
  BARITT, 323–327
  breakdown, 234–235
  forward characteristics, 199–201
  Gunn, 327–333
  ideal, 191, 193
  IMPATT, 311–320
  laser, 366–374
  light emitting, 357–363
  long-base, 191
  metal-semiconductor, 271–289
  Misawa, 318
  p-i-n, 306–311
  p-n-p-n, 543–554
  Read, 312–317
  rectifier, 538–542
  reverse characteristics, 199
  Schottky barrier, 271–289
  short-base, 191–192
  TRAPATT, 320–323
  tunnel, 263–269
  varactor, 301–306
  Zener, 234–235
Diode equation, 191, 200–201
Dipole layer (TED), 330
Direct band gap, 73–74, 134, 347, 357, 363, 372
Direct band-to-band transitions, 74, 134, 357
Direct recombination, 134–136
Dislocations, 37–38
  edge type, 38
  screw type, 38, 39
Dispersion, 41–43
  relations, 42, 43
Displacement current, 242, 243, 552
Distributed feedback laser, 371, 372
Distribution coefficient, 568, 611
Distribution function, 19
  Bose-Einstein, 23–24
  Fermi-Dirac, 21–23
  Maxwell-Boltzmann, 19–21
Domain (Gunn diode), 330, 331–333
Donor:
  action in semiconductors, 70–71
  atoms, 70, 72–73, 77
  binding energy, 70, 77
  isovalent, 73, 357–385
  level, 76, 77
  in silicon, 70, 77
  in III-V compounds, 72–73
  in II-VI compounds, 73
Donor-electron interaction, 93
Dopant:
  isovalent, 72–73, 357–358
  n-type, 70–71, 72–73
  p-type, 71–72, 73
Doping:
  compound semiconductors, 72–73
  definition, 69
  distribution, 570
  neutron transmutation, 568–569
  profile, 575

INDEX

Double acceptor, 73
Double diffused epitaxial transistor, 384, 589
Double heterostructure laser, 369–371
DOVETT diode, 327
Drain, 444, 484
  conductance, 450–451, 492
  series resistance, 454, 492
Drain voltage limitations, 455, 497
Drift:
  carrier, 102–108, 116–121, 150–151
  component in p-n junction, 162–165, 193–196, 207–209
  current, 108, 110–111
  mobility, 102–104, 622–627
  region in IMPATT diode, 313–319
  transistor, 384, 414–420
  velocity, 104–105, 118–121, 315, 327, 423, 624, 627
Drive-in diffusion, 573
Dry etching, 592–593
Dual-gate MESFET, 465–466
dV/dt effect, 552
Dynamic behavior:
  bipolar transistors, 420–424, 437–440
  p-n junction diodes, 240–244, 248–258
  Schottky barrier diodes, 288

Early effect, 395, 399
E-beam evaporation, 594
Ebers-Moll equations, 388, 390, 392
Ebers-Moll model, 516, 518–520
Edge breakdown, 234, 539–540
Edge crowding, 397, 427, 428, 432
  consequences of, 429
Edge-defined film-fed growth, 570
Edge dislocation, 38
Edge leakage current, 199, 286
Effective bevel angle, 540
Effective density of states:
  conduction band, 81, 126
  valence band, 82, 126
Effective dopant concentration, 227, 228
Effective lifetime, 146, 631, 632
Effective mass, 51–53, 70, 78, 646
  conductivity, 103–105, 126
  density of states, 78–82, 126
  direction dependence, 54
Efficiency:
  Gunn diode, 333
  IMPATT diodes, 316–317
  injection, 359
  light emitting diodes, 360
  quantum, 338, 352, 353
  solar cell, 343, 345–347, 348
Einstein relation, 124, 167, 184, 625, 635
Electric field:
  built-in, 162, 384, 414
  effect on drift velocity, 104–105, 118–121, 329
  effect on energy bands, 217–218

Hall, 116–117
  in junction region, 162, 168–170
  maximum in junction, 168, 170, 221, 222, 227
  relation to electrostatic potential, 123–124, 147
Electroluminescence, 337, 358
Electroluminescent devices, 66, 357–374
Electron:
  beam evaporation, 594
  beam lithography, 591–592
  capture at a center, 136–141, 143–146, 196–197
  carrier, 67–73
  concentration, 81–87
  effective mass, 51–53, 70, 78, 103–105, 126, 646
  energy levels, 4–5, 45–50, 364–365
  gun, 594
  mean free path, 103–104, 224–225, 285
  potential energy, 77
  quasi-Fermi level, 147, 164–165, 204, 366–367
  scattering, 103–108
  trap, 136–140, 143, 629–630
  velocity, 103–105, 116, 118–121, 224, 329, 626–627
Electron affinity, 126, 271, 275, 291–292, 369, 471
Electronegativity, 19, 276
Electron-electron, interaction, 93–94
Electron-hole pair, 68–69
Electron irradiation, 145
Electron temperature, 119–120, 224
Electrostatic potential, 75, 123, 149, 164, 168, 474
Elemental semiconductors, 62–63
Elements, periodic table of the, 17–19
Emission:
  probabilities, 138
  spectra, 5, 367–368
  spontaneous, 364–365, 368
  stimulated, 364–365, 368
  from a trap, 136–138, 197, 629–630
Emitter (BJT):
  contact, 383, 384–385
  injection efficiency, 382–383, 388, 393–394, 428–429
  multiple, 599–600
  polysilicon, 417
  region, 379
  stripe geometry, 425–427
Energy:
  Fermi, 22–23, 87–88
  kinetic, 4, 9, 48
  potential, 4, 9, 78
Energy band:
  definition, 44–47
  in diamond, 47
  in sodium, 46–47
  two dimensional representation, 74–75

Energy band diagram:
  BARITT diode, 323–324
  bipolar transistor, 385–386
  CCD, 505
  heterojunction, 292–293
  laser diode, 367, 370–371
  metal-semiconductor contact, 273–274, 276, 278, 291
  MIS diode, 278, 349
  MOS diode, 471, 472, 475, 505
  photodiode, 354
  p-n junction, 164
  solar cell, 347, 348
  tunnel diode, 64–65
Energy gap, see Band gap
Energy levels:
  carbon, 47
  hydrogen, 5
  sodium atom, 46
  sodium crystal, 46, 47
Energy momentum plot:
  direct-gap semiconductor, 74
  indirect-gap semiconductor, 74
  one dimensional lattice, 42, 43
Enhancement MOSFET, 485
Epitaxial growth (epitaxy):
  liquid-phase (LPE), 373, 580–582
  molecular beam (MBE), 582
  vapor phase (VPE), 362–363, 579–580
Epitaxial layer:
  definition, 579
  use in:
    BJT's, 384–385, 423, 589
    IC's, 599–601
    lasers, 369–373
    varactor, 303
Equilibrium conditions, see also specific devices
  carrier concentrations, 82, 85–86
  definition, 19–20
  p-n junction, 161–162, 165–167
Equivalent circuit:
  IMPATT diode, 315
  MESFET, 463, 533
  MOSFET, 498, 534–535
  solar cell, 344
  thyristor, 547
  tunnel diode, 269
  varactor diode, 302
Esaki diode, 263. See also Tunnel diode
Etching:
  to achieve isolation, 597–598
  anisotropic, 346, 505
  chemical, 569, 592
  dry, 593
  electrochemical, 592
  of SiO_2 layer, 373, 384, 589, 592
Etch techniques, 592–593
Ettingshausen effect, 620
Eutectic temperature, 277, 289
Evaporation, 594
Excess carriers, 129, 131–143, 147
Expectation value, 9
Extraction of carriers, 131, 133, 197
Extrinsic Debye length, 477

Extrinsic semiconductor, 69–73, 86, 140

f_T cut off frequency:
   BJT, 422
   JFET, 462
   MOSFET, 498

Fabrication methods:
   BJT, 383–384, 425, 589, 599, 603
   Gunn diode, 333
   IC, 596–603, 606–608
   junction laser, 366, 373–374
   MOSFET, 484, 503–505, 606–608
   p-n junction diodes, 570–582
   Schottky barrier diode, 288–289
   varactor diode, 306

Fabry-Perot cavity, 366, 368, 374
Face centered cubic (fcc) lattice, 31, 33
Fall time, 255, 258, 438
FAMOS, 609–610
Fast recovery diode, 259–260
Fat zero, 510
Fermi-Dirac distribution, 21–23, 61, 96
Fermi level, 22–23
   extrinsic semiconductor, 88
   intrinsic semiconductor, 87–88
   pinning, 275, 279
Fermi potential, 473
FET, *see* Field effect transistor
Fibers (optical), 357
Fick's equation, 572
Field-controlled thyristor, 559–560
Field-dependent mobility model, 457–458
Field doping, *see* Channel stops
Field-effect transistor:
   JFET, 444–455
   MESFET, 455–464
   MOSFET, 484–504
Field emission, 214, 216–220, 282, 285. *See also* Tunneling
Filament, 594
Fill factor, 342, 343, 375
Fine line lithography, *see* Electron, beam lithography; X-ray lithography
Fixed oxide charge, 483
Flat-band capacitance, 477
Flat-band condition, 476–477, 480
Flat-band voltage, 324, 480, 481
Floating gate device, 609–610
Float zone crystal, 568
Float zone process, 566, 567–568
Flux, 121, 122, 135, 152, 569
Focused ion beam, 577
Forbidden gap, 47, 50, 55, 62
Forward bias, 162, 164–165, 273. *See also specific devices*
Forward-blocking state (thyristor), 544, 546
Forward breakover voltage, 546–548
Forward conducting state, 549
Four-layer diode, *see* p-n-p-n devices, diode
Four point probe, 602, 614–616

Free particle, 11, 48–49
Frenkel defect, 37
Frequency limitations (BJT), 413–414, 420–423
Fresnel loss, 360
Furnaces:
   crystal growth, 566–567
   diffusion, 574
   epitaxy, 581
   oxidation, 583

GaAlAs, *see* AlGaAs
GaAsP, 65, 333, 360, 361–362
GaInAs, 355, 372
GaInAsP, 333, 357, 372–374
Gain-bandwidth product, 421, 422
Gallium antimonide, 63, 269
Gallium arsenide (GaAs), 63, 66
   absorption coefficient, 338
   band gap, 126, 362
   drift velocity, 329
   Fermi level (intrinsic), 101
   ionization coefficients, 225
   ionization energy of impurities, 77
   properties, 126
   resistivity *vs.* impurity, 113
   velocity field characteristic, 329
Gallium arsenide devices:
   avalanche breakdown voltage, 228
   built-in voltage, 179
   Gunn diode, 327–333
   laser, 366–368
   metal–semiconductor contact, 288
   p-n junction diode, 185, 201
   Schottky barrier diode, 288, 297
   solar cell, 347–348
   tunnel diode, 269
Gallium nitride (GaN), 63, 361, 362
   band gap, 362
Gallium phosphide (GaP), 63, 357, 360, 361, 362, 363
   avalanche breakdown voltage, 228
   lattice constant, 99, 646
   metal–semiconductor barrier height, 288
   properties, 646
Galvanomagnetic effects, 115, 620
Gate:
   JFET, 444, 452
   MESFET, 455–456
   MOSFET, 484, 503–504
   thyristor, 543, 555
Gate-assisted turn-off, 554
Gate oxide breakdown, 494
Gate turn-off thyristor, 554, 555–556
Gaussian diffusion, 573
Gaussian distribution, 182, 573, 576, 578, 623–624
Gaussian function, 573, 623
Gauss's law, 148, 475
Generation:
   current, 191, 197

   optical, 130–131, 338–340, 351–357
   rate, 71, 132
   thermal, 67–69, 132–134, 136–140, 196–197, 479
Generation recombination:
   current, 198, 199
   noise, 648
   process, 196–198
Germanium (Ge), 28, 63, 66, 67, 146, 199, 357
   absorption coefficient, 338
   band gap, 63, 126
   drift velocity, 120
   effective mass, 126
   intrinsic carrier concentration, 98, 126
   ionization coefficients, 225
   lattice constant, 126
   mobility, 111
   properties, 126
   resistivity *vs.* dopant concentration, 113
Germanium devices:
   avalanche breakdown voltage, 228
   built-in voltage, 179
   metal–semiconductor barrier height, 288
   phonon energy, 224, 225
   p-n junction diode, 184, 199–200, 571
   tunnel diode, 269
Gold (Au):
   alloying metal, 571, 594
   doping in Si, 77, 259, 542, 554
   energy levels in Si, 77, 144
Graded base region, 384, 414
Graded base transistor, 384, 414–420
Graded channel JFET, 452
Graded junction, 177–182, 185, 259, 303–305
Graded semiconductor, 123–124
Gradient (carrier), 121–122
Gradient voltage, 180, 182
Grading constant, 177
Gradual channel approximation, 448–449, 450, 488, 490
Grain boundary, 29
Grain size, 350
Ground state, 5, 15, 70, 364, 365
   radius, 5, 70
Group velocity, 41
Grown junction, 570
Growth of semiconductor materials, 566–568
Guard ring, 289, 356–357, 540
Guard ring structure, 289, 540
Gummel number, 416–418
   effect of heavy doping, 419, 420
Gummel-Poon model, 516, 518, 527–528
Gunn effect, *see* Transferred electron devices
Gunn oscillations, 327

Half life, 569

INDEX

Half width, 624, 627
Hall constant, 117–118, 618, 619
Hall effect, 115–118
Hall field, 117–118
Hall measurements, 620–622
Hall mobility, 117, 618–620
Hall probes, 620, 621
Hall scattering factor, 619, 620
Hall voltage, 116, 618, 620
Hamiltonian, 9
Harmonic oscillator, 12–13, 40, 43
 classical, 12
 quantum mechanical, 13
Haynes-Shockley experiment, 622–626
 germanium sample, 622
 silicon sample, 625–626
Heat of reaction, 279, 280
Heat treatment, 145, 503, 578
 of Scottky barrier, 289
Heavily doped polysilicon, 417, 480–481, 587
Heavily doped semiconductors, 92–98
Heavy doping effects:
 in bipolar transistor, 419–420
 in solar cell, 344
Heine tail states, 279
Heisenberg uncertainty principle, 7, 25, 94
Heterojunction, 291–295, 327
 isotype, 291, 295
Heterojunction devices:
 avalanche photodiode, 357
 lasers, 369–371
 photodiode, 355
 solar cell, 349–350
Hexagonal crystal, 33
 close packed (hcp), 32, 33, 65
HF (etch of SiO_2), 592
HgSe, 65
HgTe, 65, 66
High energy particles, 144, 345
High-field domain, 330
High-field property, *see* Nonlinear conductivity
High-frequency capacitance curve, 476, 479
High-frequency operation:
 bipolar transistor, 420–424
 p-n junction diode, 244
High level injection, 131–132
 bipolar transistor, 427–431
 definition, 131
 p-n junction diode, 203–205, 210
High-low junction, 290, 318
High mobility semiconductors, 285, 455
High mobility valley, 328
Holding current, 544
Hole:
 capture at a center, 137–138, 197–198
 concentration, 81–82, 86
 effective mass, 126, 646
 mobility, *see* Mobility
 as a particle, 53, 68–69, 76

quasi-Fermi level, 147, 165, 204, 366–367
 trap, 136–140
Hole affinity, 292, 369
Hot electron effects, 120–121, 224–225, 327–329
Hot electron injection, 502, 610
Hot electrons, 120
Hot spot, 434
Hybrid integrated circuit, 595
Hybrid-π model, 525–526, 536
Hydrofluoric acid, *see* HF
Hydrogen atom:
 Bohr model, 4–5
 energy levels, 5
 quantum mechanical model, 14–16
Hydrogen gas, 579–580, 581
Hyperabrupt junction, 303, 304
Hyperabrupt varactor diode, 303–305

IC, *see* Integrated circuit
Ideal diode:
 assumptions, 188–189
 characteristic, 193
 equation, 191
Ideality factor, 201, 286, 287, 528
Idealized JFET, 448–451
Idealized MOSFET, 487–491
Ideal MOS capacitor, 471–472
Ideal MOS curve, 476–479
IGFET, *see* MOSFET
I^2L, 604–605
Illuminated junction, 338–340
Image force, 280
 barrier lowering, 280–281
 permittivity, 281
Impact ionization, 311
IMPATT diode, 311–320
 equivalent circuit, 314–315
 operation, 311–314
 performance, 319–320
 Schottky barrier, 319–320
Implantation, 496, 576–578
 boron, 503, 578
Impurity:
 concentration, 69–73, 567–568
 distribution coefficient, 568
 levels in GaAs and Si, 77
 scattering, 107. *See also* Acceptor; Donor
Impurity, incompletely ionized, 84, 90–92
Impurity band conduction, 114–115
Impurity ionization energy, 70, 93, 100, 101
InAs, 63, 65, 328, 646
Indirect gap semiconductors, 74, 134, 363
Indirect recombination, 136–141
Indirect transitions, 74, 357–358
Indium (In), 63, 71, 77
 alloy in Ge, 571
Indium antimonide (InSb), 63, 64, 65
 lattice constant, 646
 properties, 646

Indium arsenide (InAs), 63, 65, 328
 lattice constant, 646
 properties, 646
Indium phosphide (InP), 63, 64, 328, 333, 349, 355, 372–374
 band gap, 355, 646
 drift velocity, 329
 lattice constant, 646
 properties, 646
Induced channel MOSFET, 484
Inductance, 45, 268, 302, 314–315
Infrared:
 detector, 66, 351, 352
 Laser, 371–372
 LED, 357
 region, 353, 362, 371
Injection:
 carrier, 130–132
 efficiency of a junction, 359, 360, 382–383, 393–394
 electroluminescence, 358–359
Injection laser, 366–374
Input capacitance, 498, 535
Input current, 517
Input impedance, 268
Instability at breakdown, 232, 234
Instantaneous current and voltage, 519, 523, 530
Insulated gate FET, 469. *See also* MOSFET
Insulator, 54–55, 61, 62, 469
Integrated circuit:
 bipolar, 596–605
 devices, 599–605
 economics, 595
 hybrid, 595
 monolithic, 595
 MOS, 605–610
Integrated injection logic (I^2L), 604–605
Interdigitated structure, 427, 465, 552
Interface charge, 480
Interface states, 279, 280, 293, 295
Interface trapped charge, 481
Interface traps, 293, 493
Interfacial layer, 278, 279, 280, 292, 293
Intermetallic (III–V) compounds, 63–65, 72–73
Intermetallic (II–VI) compounds, 65–66, 73
International system of units, 644
Interstitial atom, 37
Interstitial site, 37
Intrinsic concentration, 69, 82–83
Intrinsic excitation, 352
Intrinsic Fermi level, 82, 87–88
Intrinsic germanium, 127
Intrinsic semiconductor, 67–69, 86, 92, 127
Intrinsic silicon, 127
Intrinsic temperature, 86
Intrinsic transistor, 380, 390, 392, 397, 519
Inverse active mode of BJT, 392, 440

Inverse alpha, 390, 393
Inversion, 199, 472, 474
    p-n junction, 366–367, 370
    population, 364–365, 366
Inversion layer, 472
Inversion layer charge, 472
Inversion regime, 478
Inverted alpha cutoff frequency, 440
Ion channeling, 576
Ionically bonded semiconductors, 65–66, 275, 280
Ionic bond, 28
Ionicity, 279
Ionic semiconductor, 62
Ion implantation, 496, 576–578
    system, 577
Ion implanted devices, 503, 579
Ionization, threshold energy, 224
Ionization coefficient, 224, 225, 230, 356
Ionization integral, 225
Ionized impurities, 69–73, 83–85, 90–91, 162
Irradiation, high energy particles, 145
    electrons, 145
    neutrons, 144
Isoelectronic semiconductors, 64
Isolation methods in IC's:
    dielectric, 597–598
    p-n junction, 596–597
Isoplanar process, 599
Isotype heterojunction, 291, 295, 298
Isovalent dopants, 72–73, 357, 362
Isovalent doping, 361
I-V characteristics, *see* Current voltage characteristics

JFET, 443–455
    field dependent velocity model, 457–458
    idealized model, 448–451
    saturated velocity model, 459
    two region model, 458–459
Joule heating, 53, 119, 497
Junction, *see* p-n junction
Junction breakdown, 213
Junction capacitance, 171–173.
    See also Capacitance
Junction conductance, 243–245
Junction curvature, 231
    effect on breakdown voltage, 232
Junction field effect transistor, *see* JFET
Junction isolation, 596–597
Junction resistance, 239

Kinetic energy, 4, 9, 48, 78
Kinetic theory of gases, 284
Kirchhoff's circuit laws, 517
Kirk effect, 430, 528

Large-signal model, 516, 518
Large-signal operation (IMPATT), 316–317

LAS, 556
Laser:
    characteristics, 368
    conditions, 364–366
    fabrication, 372–373
    heterojunction, 369–371
    light output, 368
    p-n junction, 366–368
    structures, 373–374
Lasers (semiconductor), 337, 363–374
Lateral diffusion, 231, 589
Lateral transistor, 599, 603–604, 612
Lattice, crystal, 30
Lattice constant of semiconductors, 126, 646
Lattice matching, 291, 349, 355, 373
Lattice scattering, 106
Lattice translation vector, 30
Lattice vibrations, 39–44
Lead selenide (PbSe), 66, 73, 646
Lead sulphide (PbS), 66, 73
    laser, 372
    lattice constant, 646
    properties, 646
    unit cell, 66
Lead telluride (PbTe), 66, 73
    laser, 372
    lattice constant, 646
    properties, 646
LED, *see* Light-emitting diodes
Lifetime, excess carriers:
    Auger, 141–143
    definition, 133
    effective, 146, 631, 632
    high level, 136, 141, 143
    low level, 135, 140, 142
    measurement, 627–631
    radiative, 135–136, 359–360
    reduction, 144–145, 259, 542, 554
    space charge generation, 197
Light activated thyristor, 556–557
Light-emitting diodes (LED), 357–363
    configurations, 363–364
    materials, 360–362
    performance, 363
Limited space charge accumulation (LSA) mode, 332, 333
Linear IC, 595
Linearly graded junction, 177–182
    breakdown voltage, 228
    built-in voltage, 179
    capacitance, 181–182
    condition for rectification, 177–178
    cylindrical, 231–232
Linear region of FET characteristics, 448, 450, 489–490
Line defect, 36
Liquid-encapsulated Czochralski, (LEC) growth, 567
Liquid-phase epitaxy (LPE), 373, 580–582
List of symbols, 637–642

Lithography:
    electron beam, 591–592
    optical, 590–591
    X-ray, 592
Load line, 435, 436
Long-channel JFET, 448–455
Long-channel MOSFET:
    basic operation, 486–487
    criteria, 500
    output characteristics, 492–493
    subthreshold behavior, 493
Longitudinal modes, 41–43, 106
Longitudinal optical phonon, 44
Long-wavelength cutoff, 352
Lorentz force, 116
Low-high-low diode structure, 318
    doping profile, 318
Low-level injection, 132, 189
Low-mobility satellite valley, 328
LSA mode, 332, 333
Luminescence (electro), 337, 358
Luminescent devices, 357–374
Lyman series, 5

Majority carrier:
    concentration, 85–87, 90–91
    current, 193–196
    definition, 71
Mask:
    generation, 590
    photolithographic, 590–591
    X-ray, 592
Masking, 384, 503, 589–592
Mass, *see* Effective mass
Materials for visible LED, 360–362
Matrix mechanics, 7
Maximum current, 320, 426
Maximum depletion region width, 478
Maximum electric field in a junction, 168–170, 180, 185, 281, 294
Maximum frequency of oscillation, 423, 462
Maximum obtainable power, 343
Maxwell-Boltzmann statistics, 19–21, 61
Maxwell equations, 148
Maxwellian distribution (of velocities), 21, 94, 284
Mean free path, 103–104, 224–225, 285
Mean free time, 103, 106, 127
Measurement of barrier height, 287
Mechanical damage, 213
Melting point of semiconductors, 126, 646
Merged transistor logic (MTL), 605
MESFET, 301, 455–466, 468
Metal:
    bonding, 29
    energy band, 45–47
    silicides, 277
    work function, 54, 271–280, 470, 480–481
Metallization, 589, 593–595

INDEX

Metallurgical junction, 168
Metallurgical total base width (BJT), 395
Metal-overlap structure, 289
Metal-oxide-semiconductor (MOS) structure, 470–512, 606–610
Metal-semiconductor avalanche photodiode, 356–357
Metal-semiconductor contact, 270–291
Metal-semiconductor IMPATT diode, 319
Metal-semiconductor ohmic contact, 289–291
Metal-semiconductor photodiode, 354–355
Microplasmas, 232–233, 356
Microresistivity variations, 568
Microwave device:
  Gunn diode, 327–333
  IMPATT, 311–320
  p-i-n diode, 306–311
  transistor, 425–427
  varactor, 301–306
Microwave frequency range, 301
Mid-region voltage, 310, 541
Miller indices, 35–36
Minimum channel length, 500
Minority carrier:
  concentration, 81–82, 85–86, 133–134, 147
  current in a junction, 189–191, 193–196, 207–209, 246–247
  definition, 71
  distribution:
    in a BJT, 385–388, 392–393, 405
    in a junction, 188, 193–194, 207
  extraction, 131, 133, 197
  injection, 130–132
  lifetime, 133–134
  pulse, 622–623
  storage time, 255–257
MIOS, 608–609
Misawa diode, 318
MIS diode (MOS diode), 470–483
MIS solar cell, 348–349
MIS transistor, 469
MNOS device, 409
Mobile carriers, see Electron; Hole
Mobile ionic charge, 383
Mobility (carrier):
  ambipolar, 151, 309
  conductivity, 108, 110–111, 620
  definition, 105
  drift, 622–627
  Hall, 117–118, 618–621
  inversion layer, 494–495
  ionized impurity, 107
  measurement, 622–627
  relation to diffusion coefficient, 124
  surface, 494–495
  typical values for semiconductors, 126, 646

Mobility variation:
  doping, 109–110
  electric field, 118–120
  temperature, 109–110
Models, see Circuit models
Modes (laser), 368
Modes of operation of Gunn diodes, 332–333
Modified Read diode, 318
Molecular beam epitaxy (MBE), 579, 582
Molecular bond, 29
Momentum:
  angular, 4, 15
  carrier, 9, 48, 74, 78
  crystal, 51–54
  operator, 8
  relation to wave vector, 51
Monolithic circuit:
  capacitor, 602–603
  definition, 595
  diode, 600
  resistor, 601–602
  transistor, 599–600, 603–604
MOS Capacitor, 470–482, 603. See also MIS diode
MOSFET, 301, 469, 484–505, 533–535
  electrical symbol, 498
  equivalent circuit, 498, 535
  normally-off, 484–485
  normally-on, 484–485
  output characteristics, 483, 485, 487
  structure, 484, 503–504
  types, 384–385
MOST, see MOSFET
Mott theory, 271
Multilayer metallization, 594
Multi-level impurities, 77
Multiplication (avalanche), 224–225
Multiplication factor, 225–226, 229, 232, 389, 396, 404, 545–547

n-channel CCD, 507–508
Nearest neighbor distance, 32, 34–35, 56, 67
Negative beveling, 540
  breakdown voltage, 541
Negative differential mobility medium, 329–331
Negative photoresist, 590
Negative resistance, 266, 268, 327
Nernst effect, 620
Net dopant concentration, 86, 87
Neutrality, 85, 87, 90
  quasi, 151–152
Neutral region, 162, 165, 167–168
Neutron transmutation doping, 568–569
Newton's second law, 51
Nitric acid (HNO_3), 592
Nitrogen center in GaP, 72–73, 357–358
$n_o L$ product (Gunn diode), 331, 332, 333
n^+-n junction, 185, 291

Noise:
  avalanche, 356
  generation recombination, 444, 648
  shot, 648
  thermal, 648
Nondegenerate electron gas, 83
Nondegenerate semiconductor, 82–83, 84, 87, 89
Nonequilibrium condition, 129, 275
Nonlinear conductivity, 118–121
Nonradiative recombination, 134, 136, 358
Nonvolatile memory devices, 608–610
Normal active region (BJT), 380, 392, 397–405
Normal current gain, 390–391
Normalized capacitance, 476, 478–480
Normally-off transistor:
  MESFET, 455–456
  MOSFET, 484–485
Normally-on transistor:
  MESFET, 455–456
  MOSFET, 484–485
Normal modes of vibration, 39–43
n^+-n-p^+ junction, 173, 230, 259, 539
n^+-p junction, 170, 262
n-p-n transistor, 384–385
n^+-p-n^+ transistor, 384, 431–432
n-type semiconductor, 70–71, 76
Nucleus, 4, 14
Number (coordination), 33, 35

Off-state (BJT):
  current, 435
  resistance, 436
Ohmic contact, 187, 190, 195, 289–291, 571
Ohm's law, 118, 119
One sided abrupt junction, 170–171
On state (BJT):
  resistance, 436
  voltage, 436
Open-circuit voltage (solar cell), 342, 347, 375
Open-tube diffusion, 572–573
Operators, 8–9
Optical absorption coefficient, 130, 338
Optical confinement, 370–371
Optical fiber communication, 357
Optical generation of carriers, 130–132, 338–340, 351–357
Optical lithography, 590–591
Optical modes, 42–43
Optical phonon energy, 224–225
Optical phonon scattering, 106, 120, 224
Optical turn on (thyristor), 556
Optoelectronic devices:
  lasers, 363–374
  LEDs, 357–363

Optoelectronic devices
(*Continued*)
  photodetectors, 351–357
  solar cells, 337–350
Orbit (Bohr), 4–5, 15–16
Overlapping energy bands, 54
Overlay transistor, 427
Oxidation rate, 584
Oxidation of Si, 583–585, 597, 598
Oxide charge, 480–481, 483
Oxide etching, 373, 384, 589, 592
Oxide isolation, 597–598
Oxide masking, 384, 503, 589–592
Oxide thickness, 584
Oxygen, 583
  introduction during crystal growth, 567–568

P, *see* Phosphorus
Package capacitance, 244, 302
Packaging, 595
Packing in crystals, 31–33
Packing efficiency, 56
Parabolic energy barrier, 219–220
Parasitic elements, 425, 503, 535
Parasitic transistor action, 499, 502
Paschen series, 5
Pattern generation, 590
Pattern transfer, 590–592
Pauli exclusion principle, 16, 21, 94
PbS, *see* Lead sulphide
PbSe laser, 372
p-channel MOS, 484–485, 606, 608
Peak current, 266, 268
Peak luminosity, 361–362
Periodic boundary condition, 79
Periodic table of elements, 17–19
Phase velocity, 41
Phonon, 43–44
Phonon assisted Auger recombination, 143
Phonon mean free path, 224–225
Phonon scattering, *see* Lattice scattering
Phosphine, 575, 580
Phosphorus, 100
  donor, 70
  diffusion in Si, 574
  distribution coefficient in Si, 611
  energy levels in Si, 77
  source material for diffusion, 574–575
Phosphorus doped oxide, 585
Photoconductive decay measurement, 627–631
Photoconductivity, 130, 351–352, 627–628
Photoconductor, 351–352
Photocurrent, 338–340, 351–355
Photodetectors, 351–357
Photodiode, 353–355
  avalanche, 355–357
Photoelectric measurement, 287
Photogeneration, 130
Photolithography, 590–591
Photon, 6, 23, 44

Photoresist, 590–591
  negative, 590
  positive, 590
Photovoltaic effect, 338–340
Physical constants, 643
Physical models, 516–517
Pinch-off (channel), 446–447, 487
  condition, 448, 490
  point, 447, 486, 487
  voltage, 448
p-i-n diode:
  breakdown voltage, 308
  detector, 353–354
  microwave diode, 306–311
  photodiode, 353–354
  power rectifier, 538–542
Pinning of Fermi level, 275, 279
Planar process, 588–589
Planck's constant, 4, 643
Planes, crystal, 35–36
Plasma-assisted chemical vapor deposition (PCVD), 583, 586
Plasma assisted etching, 593
Plated heat sink, 333
Platinum, 144
PMMA, 591
PMOS, *see* p-channel MOS
p-n junction:
  abrupt, 166–176
  alloyed, 166, 571
  breakdown, *see* Breakdown
  capacitance, 171–174, 181–182
  current–voltage characteristics, 198, 199–201
  diffused, 182–183
  dynamic behavior, 239–245, 248–258
  forward bias, 162–163, 199–201
  graded, 177–182
  grown, 570
  reverse bias, 163, 199
  surface leakage current, 199, 202
  transient effects, 248–258
  transition region, 162, 168–171, 180–182
p-n junction diode, *see* Diodes
p-n junction solar cells:
  GaAs, 347–348
  silicon, 345–347
p^+-n-n^+ diode, 173, 230, 259, 539
p^+-π-n^+ diode, 307
p-n-p transistor, *see* Bipolar transistor
p-n-p-n devices:
  bidirectional, 557–558
  diode, 543–549
  thyristor, 550–557
pn-product:
  in a biased junction, 204
  effect of heavy doping, 95–97
  equilibrium, 71, 82
  in presence of excess carriers, 147
Point contact diode, 271
Point defects, 37
Poisson equation, 148–149, 168, 329, 449, 477
Polycrystalline silicon, *see*

Polysilicon
Polycrystalline solids, 29–30
Polysilicon, 470, 480, 587, 597, 598
  emitter, 417
p^+-p junction, 290
Population inversion, 364–365, 366
Positive bevel, 540
Potential, electrostatic, 75, 123, 149, 164, 168, 474
  and energy bands, 75
Potential barrier, 216–220
Potential distribution:
  abrupt junction, 168–170
  graded junction, 180
Potential energy, 4, 9, 78
Potential well, 13–14, 48, 507–508
Power-frequency limitation, 320, 426–427
Power rectifier, 538–543
Power transistor, 427–435
Precipitates, 38, 233
Predeposition, 573
Primitive unit cell, 30, 32
Principal quantum number, 15
Principle of detailed balance, 125, 138
Probability, 9, 13, 19, 20
  density, 9
Projected range, 576
Projection printing, 591
Proton bombardment, 374
Proximity printing, 591
Pt-Si diode, 327
p-type semiconductor, 71–72, 76
Puller, crystal, 566–567
Punched-through diode, 230–231, 320–321
Punch-through condition, 501
Punch-through effect, 192, 216, 259
Punch-through factor, 231
Punch-through voltage, 323, 396, 545

Quality factor, 302
Quantization, 4, 13
Quantum efficiency, 338, 352, 353, 376
Quantum mechanical operators, 8
Quantum mechanics, 6–9
Quantum numbers, 13, 15–17
Quarternany compound, 333, 372
Quartz:
  crucible, 566, 567
  furnace tube, 573, 579, 583
Quasi-Fermi levels, 147, 165, 204
Quasi-neutrality, 151, 152, 153, 203
Quasi-neutral region, 195, 198
Quasi-saturation, 431

Radial resistivity variations, 568–569
Radiation damage, 144–145, 503, 578
Radiative lifetime, 135, 359–360, 376
Radiative recombination, 134–136, 359–360
Radiative transitions, 74, 357–358

INDEX

Radio frequency (rf) heating, 567, 568, 579
Radius of curvature, 231
Random thermal motion, 103, 121
RC time constant, 461, 462, 626, 632
Reach-through diode, ,536
Reactor, epitaxial, 580–581
Read diode, 312–317, 318
Recombination, carrier:
 Auger band-to-band, 141–143
 direct band-to-band, 134–136
 indirect via trap levels, 136–141
 surface, 146, 630–631
 trap assisted Auger, 143
Recombination-generation process, 196–198
Recombination lifetime, 133–134
Recombination-process, 132–133
Recovery:
 diode, 255–258
 gate turn on devices, 556
 thyristor, 552–554
Rectifier, 538–542
 design considerations, 542
Reduced zone representation, 50
Reference diode, 235
Reflection coefficient, 338
Refractive index, 360–371
Registration tolerance, 503
Relative eye sensitivity, 361–362
Relaxation time, 102–105
Resistance:
 base, 396–397, 520, 525, 526
 collector, 423, 599
 negative, 266, 268, 327
 series, 198, 204, 210, 244, 268, 302, 344–345
 sheet, 602, 615, 616
 thermal, 434, 435, 442, 455
Resistivity, 112, 113, 613
 measurement, 614–618
Resistor:
 diffused, 601–602
 ion implanted, 602
 pinched-base, 602
Resonant cavity (optical), 366
Resonant circuit, 332
Resonant frequency, 315
Response speed, 353
Reverse blocking state (p-n-p-n diode), 544–545
Reverse breakdown:
 avalanche, 215–216, 225–230, 312
 surface, 234, 539–540
 Zener, 214, 221–223
Reverse breakover voltage, 545–546
Reverse conducting thyristor, 554–555
Reverse recovery transient (p-n junction), 252, 258
Reverse saturation current, 191, 199, 284, 286
Richardson constant, 284, 287, 295
Righi-Leduc effect, 620
Rise time:
 BJT, 438–439
 thyristor, 550
Runaway, thermal, 455

Safe operation area, 435
Sapphire, 607
Satellite valley, 328
Saturated drain current, 450, 453, 490–491
Saturated velocity model, 459
Saturation current, reverse:
 BJT, 390–391
 p-n junction, 191, 199
 Schottky barrier, 284, 286
Saturation region (BJT), 392, 435, 437
Saturation velocity, 120–121, 315–316, 423, 457, 461
Saturation voltage (FET), 477–448, 490
Sb, see Antimony
Scaled-down MOSFET, 502
Scattering mechanisms, 105–108
Schottky barrier:
 chemically prepared surfaces, 278
 cleaved surfaces, 279
 reactive interfaces, 277
Schottky barrier devices:
 photodiode, 354–355
 solar cell, 348
 transistor, see MESFET
Schottky barrier diode:
 current transport, 282–285
 current-voltage characteristics, 286
Schottky barrier height:
 Bardeen approximation, 275
 effect of surface states, 275–276
 measured values, 288
 measurement, 287
 reactive interface, 277
 Schottky approximation, 272
Schottky barrier lowering, 280
Schrödinger wave equation, 9, 25
Second breakdown (BJT), 433–434
Seed crystal, 567, 568
Selective diffusion, 588–589
Self-aligned MOS transistors, 503
Self compensation, 73
Semiconductor controlled rectifier, see Thyristor
Semiconductor crystal:
 extrinsic, 69–73, 76–77
 heavily doped, 92–98
 intrinsic, 67–69, 74–75
 n-type, 70–71
 p-type, 71–72
Semiconductor materials:
 doping, 69–73, 570
 growth, 566–568
 for lasers, 371–372
 for LEDs, 360–362
Semiconductor properties, 126, 646
Semiconductors:
 classification, 62
 compound, 63–66
 electronic, 62
 elemental, 62–63
 ionic, 62
 oxide, 66
Semiconductor surfaces, 145–146, 470–474
Semi-insulating GaAs, 455
Series inductance, 268, 302
Series resistance:
 p-n junction diode, 198, 204, 210, 244
 solar cell, 344–345
Shallow levels, 77
Sheet resistance, 602, 615, 616
Shell, atomic, 16, 17
Shockley-Read-Hall (SRH) recombination, 136
Short-channel effects, 456, 498–502
Short-circuit current (solar cell), 342
Shorted cathode structure, 548
Shot noise, 648
Signal charge, 506
Silane (SiH_4), 579–580, 585–586
Silica, 566. See also Quartz; Silicon dioxide
Silicide, 277
 transition metal, 277
Silicon:
 Czochralski, 566–567
 float zone, 567–568
 neutron doped, 568–569
 polycrystalline, 587
 semi-insulating, 587–588
Silicon (Si), 28, 63, 66, 286, 306, 345, 357, 384
 absorption coefficient, 338
 amorphous, 349
 band gap, 63, 126, 362
 bond picture, 67
 crystal growth, 566–568
 crystal structure, 34, 67
 drift velocity, 120
 impurity levels, 77
 intrinsic, 67–68
 intrinsic carrier concentration, 83, 98
 ionization coefficients, 225, 229
 mobility, 111
 properties, 126
 resistivity, 113
 technology (planar), 588–589
Silicon carbide (SiC), 362, 616
Silicon controlled rectifiers, see Thyristor
Silicon diode:
 avalanche breakdown voltage, 228, 229
 built-in voltage, 179
 forward characteristics, 201
 gradient voltage, 180
 reverse characteristics, 200
Silicon dioxide (SiO_2):
 dielectric isolation, 597–598
 growth and deposition, 583–586
 masking, see Oxide masking
 passivation, 67
 properties, 647

Silicon gate technology, 503
Silicon nitride (Si_3N_4), 587, 609, 647
Silicon on sapphire (SOS) devices, 607–608
Silicon tetrachloride ($SiCl_4$), 579, 580
Simple cubic lattice, 30, 31
SIPOS, 587
Si-SiO_2 diode, 470–472, 480–482
Si-SiO_2 system, 483
Slip plane, 38
Small-signal alpha, 382, 413, 420–421, 520, 549
Small-signal diode admittance, 243–244
Small-signal equivalent circuit:
  bipolar transistor, 521, 525, 526
  IMPATT diode, 315
  JFET, 463
  MOSFET, 498, 535
  varacter diode, 302
Sodium, 45–46
  electron distribution, 46
  energy band diagram, 46, 47
Sodium chloride (NaCl), 28, 33, 66
Sodium ion, 28, 33, 483
Solar cell, 337–350
  equivalent circuit, 344
  ideal efficiency, 345–346
  I-V characteristics, 339, 342
  materials, 345
Solar cells:
  GaAs, 347
  heterojunction, 349–350
  MIS, 348–349
  p-n junction, 340–344
  Schottky barrier, 348
  silicon, 345–347
  thin film, 350
Solid solubility, 573
Solid state diffusion, 571–576
SOS, *see* Silicon on sapphire (SOS) devices
Source series resistance, 454, 492, 533, 537
Space charge build-up (Gunn diode), 330–331
Space charge generation-recombination current, 197–198
Space charge limited current, 192, 501
Space charge region:
  capacitance, 171–173, 181–182, 281
  definition, 161–162
Space charge resistance, 317, 335
Spacing of crystal planes, 36
Specific contact resistance, 291
Spectral lines, 5
Spectral width, 366, 368
Spherical coordinates, 15
Spherical junction, 231
SPICE, 527, 535
Spin, electron, 16
Spontaneous emission, 364, 367
  rate, 365

Spreading resistance, 617
  probe, 618
Spreading time, 552
Sputter-etching, 592–593
Sputtering, 586, 594
Square-law FET characteristic, 491
Stacking fault, 38
Statistical mechanics, 19–24
Step and repeat camera, 590
Step junction, *see* Abrupt junction
Step recovery diode, 260
Stimulated emission, 364–365, 367
Storage time:
  bipolar transistor, 438–440
  p-n junction diode, 255–257
Stored charge approximation, 250
Stripe geometry:
  bipolar transistor, 425–427
  laser, 373
Strong inversion, 474, 477
Subcritically doped diode, 333
Submicron devices, 591
Substitutional impurity, 70
Substitutional site, 37, 71
Substrate bias (MOSFET), 491, 496
Substrate current, 501
Subthreshold current, 493, 499–500
Subthreshold region, 493
Surface accumulation, 471–472
Surface channel CCD, 510
Surface contouring, *see* Beveling
Surface FET, 469
Surface mobility, 494
Surface potential, 474, 486, 489, 505, 506
Surface recombination velocity, 145–146, 190, 211, 631
Surface scattering, 494
Surface space charge region, 470–474
Surface state density, 278
Surface states, 145–146, 275–276, 278–279
Susceptor (graphite), 579
Switching:
  bipolar transistor, 435–440
  p-n junction diode, 258–260
  Schottky diode, 288
  thyristors, 552–555
Switching transients in junction diodes, 248–258
Symbols:
  bipolar transistor, 381
  JFET, 463
  junction diode, 379
  list of, 637
  MESFET, 463
  MOSFET, 498
  thyristors, 544, 555
  tunnel diode, 268

Ta_2O_5, 345
TED, *see* Transferred electron devices
Temperature coefficient:
  breakdown voltage, 223, 229
  diode forward voltage, 202–203, 288

Temperature dependence:
  carrier concentration, 90–92
  carrier mobility, 108–110
  intrinsic carrier concentration, 83
  p-n junction diode characteristics, 201–203
  Schottky diode characteristics, 288
Tensor, 54
Ternany compound, 66, 369, 372
Tetrahedral configuration, 67
Thermal conductivity, 320, 497
Thermal current, 265–266
Thermal generation of carriers, 67–69, 132–134, 136–140, 196–197, 479
Thermal noise, 648
Thermal oxidation of Si, 583–585, 589, 597, 598
Thermal relaxation time, 506, 507
Thermal resistance, 434, 435, 442, 455
Thermal runaway, 455
Thermal scattering, *see* Lattice scattering
Thermal velocity, 21, 103–104, 119, 137–138
Thermionic emission:
  current, 284
  process, 282–283, 287, 295
  theory, 283–284
Thermionic emission-diffusion theory, 284
Thermionic field emission, 282, 285
Thin film solar cell, 350
Three-phase CCD, 507–508
Threshold:
  current density (laser), 368
  energy for secondary pair generation, 224–225
  field for Gunn devices, 329
Threshold voltage:
  control, 495–496
  JFET, 450
  MESFET, 456
  MOSFET, 479, 481–482, 485
  temperature dependence, 497
Threshold voltage shift, 491
Thyristor:
  amplifying gate, 555
  bidirectional, 557–558
  conducting, 549
  field controlled, 559–560
  forward blocking mode, 544
  gate assisted turn-off, 554
  gate turn-off, 555–556
  light-activated, 556
  reverse blocking mode, 544
  reverse conducting, 554–555
Tight binding approximation, 44–48
Time:
  delay, 437, 438
  fall, rise (BJT), 438–439
  mean free, 103, 106, 127
  photoconductive decay, 628
  relaxation, 102–105

storage (delay), 255–257, 438–440
transit, *see* Transit time
Time constant (RC), 461, 462, 626, 632
Tin-α, 63, 64
Total internal reflection, 360
Transconductance:
   BJT, 524–525
   JFET, 450–451, 531–532
   MOSFET, 492, 535
Transfer efficiency (CCD), 609–610
Transferred-electron devices, 327–333
   domain formation, 330
   fabrication, 333
   materials, 333
   mechanism, 328–329
   modes of operation, 332–333
   performance, 333
   velocity field characteristics, 329
Transferred-electron effect, 327
Transient behavior:
   BJT, 437–440
   p-n junction diode, 248–258
   Schottky diode, 288
Transistor:
   bipolar, 378–442
   drift, 414–419
   graded channel, 452
   inverted vertical, 605
   JFET, 443–455
   lateral, 603–604
   MESFET, 455–466
   microwave, 425–427
   monolithic (IC), 599, 603–607
   MOSFET, 484–505
   power, 427–435
   switching, 435–440
   unipolar, 443
Transistor action, 378–379
   parasitic, 502
   in p-n-p-n device, 546–547
Transition metal oxides, 66
Transition metal silicides, 277
Transition region, *see also* Space charge region
   population inversion in, 366–367
Transitions:
   direct, 74, 134, 357
   indirect, 74, 357–358
Transit time:
   base (BJT), 406, 418, 419
   FET, 461, 497
   IMPATT, 314, 326
Transit time mode (domain), 332
Transmutation doping, 568–569
Transport factor (BJT), 382, 388–389, 420
Transverse electric field, 494
Trap, 136
TRAPATT diode, 301, 320–323

Trapping, 136–140, 510, 629–630
Triac, 557–558
Triangular barrier, 219–220
Triggering methods, 550
Truncated cone structure, 364
Tungsten, 594, 616
Tunnel current, 221, 266–267
Tunnel diode:
   characteristic, 266–268
   equivalent circuit, 268–269
   materials, 269
   operating principle, 264–265
   performance, 269
Tunneling, 214, 216–220, 264–268, 285
Tunneling probability, 216–219
Tunneling process, 214, 216, 264–266
Turn-off time (thyristor), 554
Turn-off voltage, 450
Turn-on delay time, 550
Turn-on time, 550–552
Turn-on and turn-off transients:
   bipolar transistor, 437–440
   junction diode, 248–258
Twinning, 38
Two-dimensional models, 459–461
Two-region model, 458–459
Two-sided abrupt junction, 168–171, 227
Two-transistor analogy (p-n-p-n diode), 546–547
Types of MOSFETs, 484–485

Ultraviolet light, 589, 590–591
Ultraviolet radiation, 610
Ultraviolet region, 362
Uncertainty principle, 7, 25, 94
Unipolar transistor, *see* Field effect transistor
Unit cell, 30–31, 32, 33, 34, 67
Unit prefixes, 645
Unit quantum efficiency, 338, 376
Units, international system of, 644

Vacancy, 37, 572
Vacancy interstitial pair, 37
Vacant sites, 572
Vacuum level, 54, 271, 272, 292, 471
Valence band, 48, 55, 62
   effective density of states, 81–82
   hole concentration in, 81–82
Valence bond model, 66–73
Valence electrons, 17, 27
Valency, 17
Valley, conduction band, 338
Valley current, 266, 269
Valley voltage, 266, 268, 269
Vapor-phase epitaxy (VPE), 362–363, 579–580
Vapor pressure, 65, 575

Varactor diode, 301–306
   equivalent circuit, 302
   hyperabrupt, 303–305
   linearly graded, 303, 304
Velocity:
   drift, 104–105, 118–121, 315, 327, 423, 457, 497, 624, 627
   saturation, 120–121, 315–316, 423, 457, 461
   surface recombination, 145–146, 190, 211, 631
   thermal (rms), 21, 103–104, 119, 137–138
Velocity field characteristic, 329
Velocity field relationship, 120
Visible LED, 361
Visible region, 357, 362
Visible spectrum, 354
V-MOS, 504–505, 607
Voltage controlled negative resistance, 268, 327
Voltage regulator, 234–235

Wafer, 565, 569
   preparation, 569–570
   shaping, 569
Water, 583
Wave equation, 9, 25
Wave functions, 9, 11
   stationary state, 11
Wave mechanics, 7, 9
Wave number, 48
Wave vector, 54, 73–74, 78–80, 328
Wet chemical etching, 569, 592
Wide band gap semiconductor, 294, 349, 370
Width of depletion region, 170, 180, 281, 477–478
WKB method, 219
Work function, 54, 271–280, 290, 292–293, 470
   difference, 480–481, 485, 487
Wurtzite lattice, 65

X-ray lithography, 592
X-ray mask, 592
X-ray photoresist, 591

Zener breakdown, 214, 221–223
Zener diode, 234–235, 263
Zener tunneling, 216–220, 266
Zinc blende lattice, 65
Zinc oxide (ZnO), 65
Zinc selenide (ZnSe), 65, 73, 288, 328, 333, 349, 362, 646
Zinc sulphide (ZnS), 65, 66, 288, 362
Zinc telluride (ZnTe), 65, 73, 362, 646
Zone melting, 568
Zone refining, 568